11th Aceh International Workshop and Expo on Sustainable Tsunami Disaster Recovery (AIWEST-DR 2018)

Advancing Science and Technology Innovation for Sustainable Disaster Resilience and Risk Reduction

IOP Conference Series: Earth and Environmental Science
Volume 273

Banda Aceh, Indonesia
10 - 12 October 2018

Editors:

Ella Meilianda
Syamsidik
Sebastien Penmellen Boret
Yunita Idris

ISBN: 978-1-5108-9326-9
ISSN: 1755-1307

Printed from e-media with permission by:

Curran Associates, Inc.
57 Morehouse Lane
Red Hook, NY 12571

Some format issues inherent in the e-media version may also appear in this print version.

Institute of Physics
Dirac House, Temple Back
Bristol BS1 6BE UK

Phone: 44 1 17 929 7481
Fax: 44 1 17 920 0979

techtracking@iop.org

Additional copies of this publication are available from:

Curran Associates, Inc.
57 Morehouse Lane
Red Hook, NY 12571 USA
Phone: 845-758-0400
Fax: 845-758-2633
Email: curran@proceedings.com
Web: www.proceedings.com

TABLE OF CONTENTS

COMMUNITY PREPAREDNESS, DISASTER EDUCATION, LOCAL AND INDIGENOUS KNOWLEDGE

GIS AND APPLIED TECHNOLOGY FOR DISASTER

RECOVERY AND HUMAN SECURITY

IOP Conf. Series: Earth and Environmental Science **273** (2019) 011001 doi:10.1088/1755-1315/273/1/011001

Preface

The 11th Aceh International Workshop and Expo on Sustainable Tsunami Disaster Recovery, AIWEST-DR 2018, was organised from 10 to 12 October 2018. The bi-annual AIWEST-DR was conducted this time with the theme "*Advancing Science and Technology Innovation for Sustainable Disaster Resilience and Risk Reduction*". Tsunami and Disaster Mitigation Research Center (TDMRC) of Syiah Kuala University was profoundly enjoying the collaboration with the International Research Institute of Disaster Science (IRIDeS) of Tohoku University in Japan in a joint-organizing committee.

AIWEST-DR conference ensures opportunities for academics, professionals, researchers and students who are working on disaster-related research to communicate the science they produce not only among themselves, but also for the practitioners, governmental officers, and communities in disaster-prone areas. We also want to ensure the Sendai Framework for Disaster Risk Reduction (SFDRR) be resonant in the mind of all stakeholders.

AIWEST-DR 2018 proceeding contains papers written by academics, professionals, researchers and students from at least six countries including Indonesia, Japan, Germany, Thailand, Malaysia and Saudi Arabia. Various research works related to the conference theme were presented during the conference. From a total of 62 full papers accepted upon review, they are categorized into the following specific topics:

A. Tsunami, climate change and others coastal hazard
B. Geohazard and hazard characteristics and modelling
C. Community preparedness, disaster education, local and indigenous knowledge
D. GIS and applied technology for disaster
E. Recovery and human security

Topic A (Tsunami, climate change and others coastal hazard) consists of 13 papers which topics are related to the sources or forcing factors that create natural hazards, ranging from flash flood, land and coastal water-borne disasters, as well as extreme events derived from climate change.

Topic B (Geohazard and hazard characteristics and modelling) encapsulates 14 full papers discussing about the results of investigations ranging from seismic activities, coastal nearshore processes, landslides and volcanic activities, using various advanced methods and mathematical modelling.

Topic C (Community preparedness, disaster education, local and indigenous knowledge) deals with 18 full papers on social aspects of disasters, such as disaster education, local and indigenous knowledge, risk communication, media engagement, community preparedness and religious-based mitigation concept.

Topic D (GIS and applied technology for disaster) discusses about various approaches of data analysis spatial tools and applied technology for mapping, smartphone application for disaster, rapid assessment and geotechnical modelling for escape hill, which are presented in 8 full papers.

Topic E (Recovery and human security) specifically dedicated to topics on human security, such as influence of natural hazard event to the community at risk and insurance program for farmers, and the aspect of economic recovery and long-term recovery.

IOP Conf. Series: Earth and Environmental Science **273** (2019) 011001 doi:10.1088/1755-1315/273/1/011001

The organization of AIWEST-DR 2018 would not have been possible without the support of many institutes and individuals. First and foremost, we are very grateful to the financial sponsors of AIWEST-DR 2018: *the Ministry of Research, Technology and Higher Education of Republic of Indonesia* through the Center of Excellence Program granted for TDMRC; the internal funding for conference by Syiah Kuala University, and from *IRIDeS of Tohoku University*.

The conference was the 2nd time being organized by good collaboration between the Tsunami and Disaster Mitigation Research Center (TDMRC) of Syiah Kuala University and IRIDeS of Tohoku University. We also thank to our colleagues and extended institutional networks who were kindly supporting and guiding us through a steering committee. Most importantly, we were humbly thankful for all the keynote and invited speakers, paper reviewers, participants, colleagues, staff and students in the organizing committee and beyond who had been tirelessly preparing for the success of the conference and to keep working afterwards on preparing this proceeding.

Last but not least, we thank all the authors for their excellent contributions to these proceeding and all the participants for their contributions to this successful conference, either in the form of oral presentation, posters, expo material, or in the form of their attention to the presentations and contributions to the discussions.

We hope that everyone enjoys reading the proceeding and that the papers inspire you in your research or other academic and professional activities.

Dr. Ella Meilianda
Dr. Syamsidik
Dr. Sébastien Penmellen Boret
Dr. Yunita Idris

Editors

IOP Conf. Series: Earth and Environmental Science **273** (2019) 011002 doi:10.1088/1755-1315/273/1/011002

Peer review statement

All papers published in this volume of *IOP Conference Series: Earth and Environmental Science* have been peer reviewed through processes administered by the proceedings Editors. Reviews were conducted by expert referees to the professional and scientific standards expected of a proceedings journal published by IOP Publishing.

IOP Conf. Series: Earth and Environmental Science **273** (2019) 012001 doi:10.1088/1755-1315/273/1/012001

Is flash flood cycle? A preliminary climate study on Teunom fluvial system

M Irham[1,*], Y Ilhamsyah[1,2], Sugianto[3], A Deli[3], S Syahreza[4,5]

[1]Dept. of Marine Sciences, Faculty of Marine and Fisheries, Syiah Kuala University, Banda Aceh - Indonesia 23111
[2]Applied Climatology, Graduate School of Bogor Agricultural University, Bogor - Indonesia 16680
[3]Faculty of Agriculture, Syiah Kuala University, Banda Aceh Indonesia 23111
[4]Physics Department of Mathematics and Science Faculty of Syiah Kuala University, Banda Aceh - Indonesia 23111
[5]Tsunami and Disaster Mitigation Research Center (TDMRC) of Syiah Kuala University, Banda Aceh - Indonesia

[*]Corresponding author: irham@unsyiah.ac.id

Abstract. The atmospheric dynamics in the Teunom region have caused flashflood several times in last decade, especially in the Krueng Tenom river basin. This problem has been speculated for many scientists who predicted that such flashflood were caused by either regional climate change or the uncontrolled land use in this area. However, it should also be understood the possibility that flash floods may be caused by regional atmospheric cycle. Therefore, this study aims to understand how the climate approach on temperature and rainfall as meteorology process as a way to comprehend whether the flashflood cycle. The result of the research shows that flashflood is actually cyclical, however, due to micro climate change that happened and it is predicted that land use change or land conversion and illegal mining accelerates flashflood cycle so that early warning system is difficult to be implemented.

1. Introduction

The Krueng Teunom watershed is the main fluvial system in the Aceh Jaya area with an area of ~268,279 ha located at $4^0$26'00.94" up to $4^0$44'09.60" North and $95^0$48'17.31" up to $95^0$59'06.50" East and its upstream is in the border area of Pidie district (Figure 1). Forest degradation occurring in the upper reaches and around the Krueng Teunom watershed has caused flooding and abrasion of river banks during the rainy season [1] and the decrease in water volume during the dry season [2]. The area around the Krueng Teunom watershed in early January 2017 has already experienced two floods and in 2016 there were four big floods, one of which was flash flood. Floods in 2016 are more than twice the floods that occurred in 2012 to 2015. Meanwhile, between 1999 and 2011 only occurred three flash flood and on average once floods a year. Economic data showed a loss of Rp. ~ 8 billion due to floods that occurred in 2016 and some public facilities did not fully function to support the economic activities which impact the people in the region [3].

According to BPBD Aceh [4], the main cause of repeated floods in this area is the result of a narrowed river flow, sedimentation [5], and land conversion [6]. On the other hand, the data from

IOP Conf. Series: Earth and Environmental Science 273 (2019) 012001 doi:10.1088/1755-1315/273/1/012001

Meteorology, Climatology and Geophysics Agency [7] show that it has been a trend of increasing rainfall 0.3% per year especially the last five years. Furthermore, extreme weather due to warming of sea surface temperatures and Autralia and Asia monsoon [8], [9] combined with land use change cause excess rainfall in this area so that the capacity of the river is saturated [10], therefore the water overflows and causes flooding [11].

Figure 1. Spatial landscape of Kreung Teunom, Aceh Jaya (not scalable) (Google Earth Map)

In Teunom, the flood is the most common occurred along the river after heavy rainfall. The drainage system overflow in this area leads a major problem especially along a heavily populated areas where the major livestock and crops grow [12], [13]. The flooding pattern has also different from time to time and may last for weeks [14]. Severe thunderstorms and heavy rainfall mostly trigger flash flood that poses the greatest of live and significant property loss [15]. All of the problems believed to happen because of the changes on local climate trend as well as changes in land use and land conversion [16].

In the policy document of the Department of Public Works and the Provincial Water Resources Office of Aceh, Krueng Teunom watershed is a category of degraded basin and one of a top priority to be treated and it needs a high priority to improve its condition. Therefore, this purpose of this paper is to understand how the local climate approach on temperature and rainfall as meteorology process as a way to comprehend whether the flashflood cycle.

2. Research Method

This study will be based primarily on secondary data collected in the Teunom River study area. The data collected was intended to have a better understanding of the processes that contributed to the development of environmental features that may also improve the ability to explain phenomena observed in nature.

To evaluate the weather, either temperature or rainfall, the data from BMKG and NASA climate stations was used. The 20 years temperature and rainfall data will be evaluated and analysed using linear regression. By applying this method, the temperature and rainfall related to climate change can be characterized and construed. The development of method suitable for the collection of data that contribute to understanding these processes is a still-evolving science because temperature and weather are complex processes. Therefore the rainfall and temperature data are fundamental requirement for proper management of climate for early warning system.To analyse the temperature and rainfall trend in the Krueg Teunom study area, the 12 stations for 20 years of records were collected. The data is readily available at BMKG and NASA in the area of study. From these data, the temperature and rainfall trend can be investigated and analysed.

3. Result and Discussion

The results show that there is an increase in temperature trend over the last 20 years as shown in Figure 2. The anomaly temperature data also shows that the cycle of temperature change over the last 10 years is sharper (0.6%) than the previous 2 decade changes (0.2%) showing trend sinusiodal. This means that there is a temperature rise of about 0.013^0C every year starting from 2009. The rising in temperature lead to changing in weather pattern.

Weather patterns that determain the amount of rain, however, the amount and time over wich precipitation occurs for any given area is not constant. Even though the water cycle is a balnce system, sometime the quantity flowing into one area is greater than the capacity of the system to hold it with in natural confines. This can be seen in the Teunom area (Figure 1) and rainfall data (Figure 3), where combination factors along with extraordinary precipitation lead to flooding.

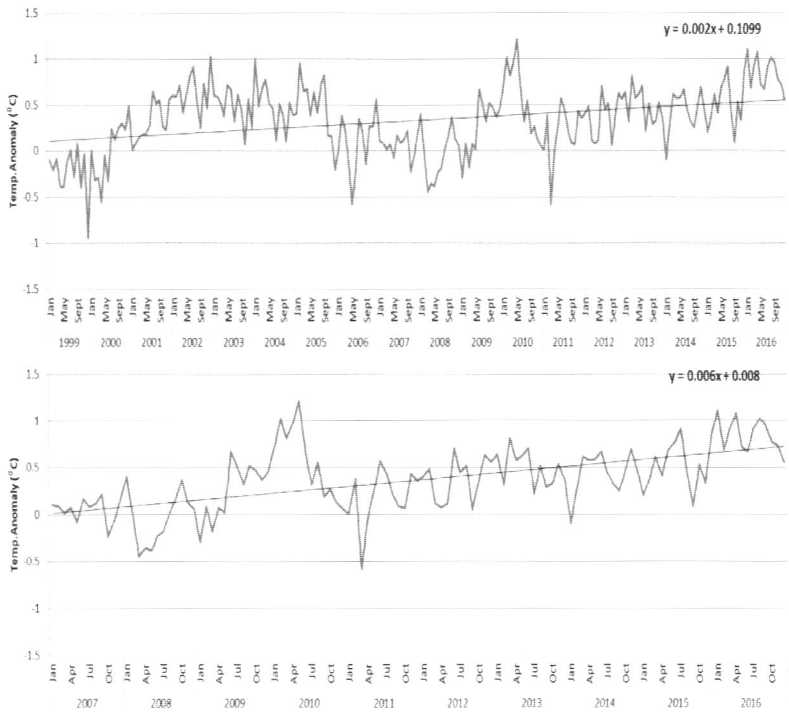

Figure 2. The anomaly temperatures change in Teunom over the last 20 and 10 years

Rainfall data that occurred in Teunom for the last 15 years shows that the rainfall cycles appearances normal; however, there is little change when combined with El-nino signal (Figure 3). It can be seen that the rainfall cycle is anomalous in the last ten years which is marked by the presence of extreme rain (rain above normal) accompanied by extreme drought. It is estimated that this situation causes the occurrence of extreme and the flash floods is not well cycled. This situation is exacerbated by the use and conversion of land in the unrolled Teunom area. Therefore, this situation has created local microclimate change in Teunom and its surrounding areas.

The trend in Figure 3 shows that rainfall fluctuations were very uneven over the past seven years (starting in 2009), causing floods to occur at any time of the year. This non-periodic flood is indicated by a red circles (Figure 3) compared to the 2009 flood event (yellow circle in Figure 3) [3]. When viewed from the global Elnino (ENSO) trend, it is difficult to correlate the number of floods in the Teunom area due to the small area coverage. The results of the field survey and data from the Aceh

Jaya government [3] show that there is a large utilization of forest land into plantations which are generally oil palm. Forests in the Teunom region, which for the last ten years have been around 82% of the area of Teunom, now account for about 23% [3]. Therefore, this land use is suspected to be the cause of non-cyclical floods as happened ten years earlier. Land use (land conversion) is also estimated to have caused local climate change in the Teunom area as shown in Figure 2.

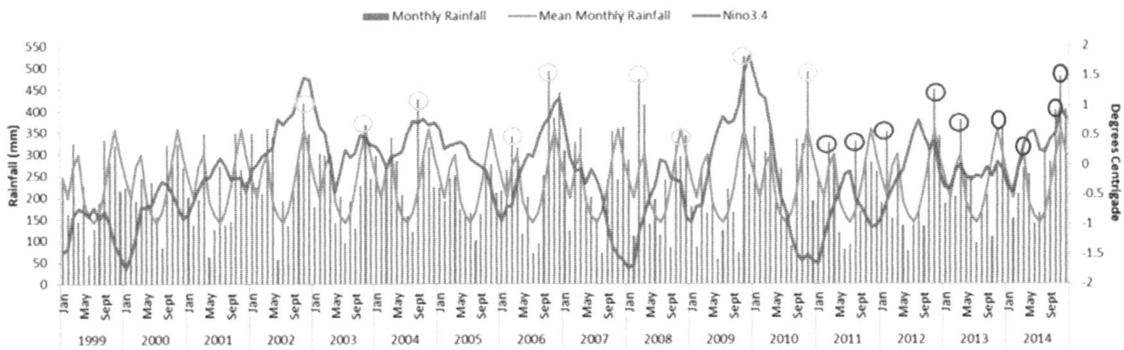

Figure 3. The graph of monthly and mean monthly rainfall of Teunom combined with El-Nino trend

Most flash flood are caused by repeatedly slow moving thunderstorms over the same area. In the last 15 years, the repeated of thunderstorms with high intensity in Teunom area triggered cycle of flash flood as it shown in Figure 3. Even though the flash flood is a natural phenomenon, the cause of this run off may come from the changes in local micro climate trigger by human intervention. The micro climate change that occurs in Teunom area is a result of the many changes in land use and land cover that used to be functioned as a buffer and now it has been converted into plantation land. Hence, the change in land use and land cover is actually the result of economic interests that are not based on local wisdom. Therefore, changes in the rainfall cycle and irregular temperature increases will have implications for the difficulty of implementing early prevention of flashflood disasters that occur.

4. Conclusion

It is concluded that flash flood is actually cyclical, however, due to micro climate change that happened and it is predicted that land use change and land conversion accelerates flash flood cycle so that early warning system is difficult to be implemented.

5. References

[1] Irham, M., Setiawan, I., Febriani, I., and Sugianto, S. (2017). The Spatial Distribution of Bed Sediments at Krueng Cut Estuary: The Sieve Analysis Approach. Jurnal Rekayasa Kimia dan Lingkungan, 12(2), 69 – 76.
[2] Irham, I., Fadla, Y., and Setiawan, I. (2018). The spatial distribution of suspended sediment analysis along Krueng Cut River, Banda Aceh. IOP Conference Series: Earth and Environmental Science 106 (1), 1 – 6.
[3] PEMDA Aceh Jaya (2016). Rencana Strategis PEMDA Aceh Jaya.
[4] BPDB Aceh (2016). Laporan penanggualangan banjir provinsi Aceh. PEMDA Provinsi Aceh.
[5] Irham, M. (2016). The spatial distribution of bed sediment on fluvial system: A mini review of the aceh meandering river. Aceh Journal of Science and Technology, 5 (2), pp.82-87.
[6] Irham, M., Giardino, J. R., Guneralp, I. (2013). Two dimensional morphological dynamic modeling of a complex fluvial system: The spatial distribution of the bed sediment of Brazos River, TX. GSA Regional Meeting, Austin, TX.

[7] BMKG (2016). Laporan tahunan kondisi curah hujan Indonesia. http://www.bmkg.go.id.

[8] Setiawan, I., Rizal, S., Haditiar, Y., Purnawan, S., Yuni, SM., and Irham, M. (2018). Study of current circulation in the Northern Waters of Aceh. IOP Conference Series: Earth and Environmental Science 176 (1), 1 – 6.

[9] Irham, M., Miswar, E., Ilhamsyah, Y., and Setiawan, I. (2018). The nortern tidal dynamic of Aceh waters: A 3D numerical model. IOP Conference Series: Material Science and Engineering (MSE) 352 (1), 1 – 6.

[10] Coppola, E., Verdecchia, M., Giorgi, F., Colaiuda, V., Tomassetti, B., and Lombardi, A. (2014). Changing hydrological conditions in the Po basin under global warming. Science of the Total Environment, 43, pp, 1183-1196.

[11] Irham, M. and Setiawan, I. (2018). The numerical model of the sediment distribution patern at Lampulo National Fisheries Port. IOP Conference Series: Earth and Environmental Science 106 (1), 1 – 6.

[12] Davidson, S.K., Hartley, A.J., Weissmann, G.S., Nichols, G.J. and Scuderi, L.A. (2013). Geomorphic elements on modern distributive fluvial systems. Geomorphology, 180-181: 82-95.

[13] Dumanski, S., Pomeroy, J.W., and Westbrook, C.J. (2015). Hydrological regime changes in a Canadian prairie basin. Hydrological Processes, 26 (18): 3893 - 3904.

[14] Nelson, J.M., Bennet, J.P. and Wielle, S.M. (2003). Tool in fluvial geomorphology: Flow and sediment transport modeling. Chapter 18, pp. 539 - 576. John Wiley and Son Ltd, England.

[15] Kron, W. (2011). Increasing weather losses in Europe: What they cost the insurance industry? CESifo Forum 2, 73 - 87.

[16] Pedersen, M.L., Kristensen, K.K. and Friberg, N. (2014). Re-Meandering of Lowland Streams: Will Disobeying the Laws of Geomorphology Have Ecological Consequences?. PLoS ONE 10(3): e0118939. doi:10.1371/journal.pone.0118939.

Acknowledgment

Authors express gratitude to the Research and Community Service Institutions of Syiah Kuala University for financial assistance under 'Unsyiah Feature Research (*Penelitian Unggulan Unsyiah*)'. We also thank to Syiah Kuala University through the facility support at Ocean Modelling Laboratory and Geology Research Laboratory during the research.

Greater Aceh, Indonesia Enters Climate Change: Climate on Extreme ENSO 2015-2016

Y Ilhamsyah[1,2], A Farhan[3], M Irham[1], I Setiawan[1], Y Haditiar[1] and Irwandi[4,5]

[1] Dept. of Marine Sciences, Faculty of Fisheries and Marine Sciences, Syiah Kuala University, Banda Aceh 23111, Indonesia
[2] Applied Climatology, Graduate School of Bogor Agricultural University, Bogor 16680, Indonesia

[3] Dept. of Physics Education, Faculty of Education and Teacher Training, Syiah Kuala University, Banda Aceh Indonesia 23111

[4] Dept. of Physics, Faculty of Math. and Natural Sciences, Syiah Kuala University, Banda Aceh Indonesia 23111
[5] Tsunami and Disaster Mitigation Research Center, Syiah Kuala University, Banda Aceh Indonesia 23111

E-mail: yopi.ilhamsyah@unsyiah.ac.id

Abstract. The objective of the study is to analyse the rainfall variability and temperature over Greater Aceh and its correlation to ENSO event. Long-term rainfall observation data spanning from 1982-2018 was retrieved from meteorological station of Blang Bintang, Greater Aceh. The Pearson correlation was employed to find the relationship between rainfall and El Niño and Southern Oscillation (ENSO) occurrence. It is shown from frequency analysis that rainfall decreased in the 90s. After the tsunami as climate change become unequivocal the rainfall trend increase as a result of land use change and much worst as severe ENSO 2015 arrived over the region which statistically significant was found between those two corresponding events. During El Niño, rainfall is less while La Niña more rainfall than normal drop over the region. To cope with climate change issues where ENSO is very likely to occur in future time, an appropriate mitigation is required to conduct for instance: managing forest resources and land use by reducing deforestation and land conversion, the control of fossil fuel burning, and the most important thing is to establish a renewable energy as an effort to implement sustainable green energy in the future as well as climate risk reduction.

1. Introduction

Aceh Besar District (Greater Aceh-red) is one of the districts in Aceh Province, Indonesia. Greater Aceh is located at 05.37°N and 95.53°E in far northwest of Sumatera Island and geographically border the Indian Ocean and the Malacca Strait in the west and east, respectively. A unique topographic condition is found over Greater Aceh since it is surrounded by three highlands (*lhee sagoe*), i.e., hilly areas in the east, Mount *Seulawah* in the southeast, and *Bukit Barisan* Mountain in the west with lowland stretched in the center in the valley of these three highlands. Therefore, it contribute to the formation of specific meteorological characteristics over the region, especially rainfall variability. Since it is located at the edge of vast sea areas, Greater Aceh have a humid and warm tropical climate. According to Köppen climate classification, Greater Aceh similar to many regions in Indonesia is categorized into wet tropical climate (Am) where rainfall variability is the greatest [1, 2].

The weather and climate in the western and eastern parts of Greater Aceh are strongly influenced by local interactions between ocean and topographic condition of *Bukit Barisan* Mountain. Since it lies in windward region, climate conditions in the western of Greater Aceh is wetter than the eastern region. Sea breeze bringing air mass from the ocean become quickly saturated due to the orographic lift and contribute to moderate until heavy rainfall in the afternoon in the west. Different characteristics are found in the eastern part of Greater Aceh where located in the leeside of *Bukit Barisan* Mountain, air that have converted to adiabatic cooling in the western region descends and turn into adiabatic warming in the eastern region establishing a rain shadow area which create climate conditions along eastern region of Greater Aceh warmer and drier [3, 4].

Nevertheless, rain is still possible to occur since there exists a gap in the north that are connected to the Indian Ocean and Malacca Strait which allow the movement of water vapour carried by the sea breeze to the mountainous areas in the southwest. For instance: during daytime, the air that descends from *Bukit Barisan* is warmer and drier leading Greater Aceh to experience a hot and humid day while the sea breeze formed in the north in the afternoon carry water vapour towards hilly areas in the south and southwest. The air become saturated and turn into adiabatic cooling which cause clouds to form and rain is produced making the air temperature in the eastern region of Greater Aceh is cooler in the afternoon.

The worst ever recorded El Niño 2015 caused world's weather to change than usual. In the following year during La Niña, the world once again went into a period of a changing climate. Following strong El Niño and La Niña (refer to ENSO-El Niño and Southern Oscillation), it is clear that climate change existed and threatening life on Earth. In Indonesia, ENSO episodes which are associated with cooling and warming of surface Pacific Ocean often linked to the drought and floods, respectively [5, 6, 7, 8]. Since Indonesia is dominated by three rainfall zones, ENSO occurrence over the region is very likely on typical monsoonal and local climatic zones which are mostly located in the central and eastern part of Indonesia [9, 10, 11]. In the west where Semi-Monsoonal type climatic zone is dominant, the influences of ENSO is about as likely as not [12, 13]. As ENSO signal is strengthening as a result of climate change, however, its impact is very likely to occur as what Greater Aceh suffer from strong ENSO 2015-2016 events. The purpose of the study is to analyze the characteristics of rainfall variability and temperature and its correlation to ENSO signal in Greater Aceh.

2. Materials and Methods

Temperature and rainfall data from 1982 to 2018 were extracted from National Weather, Climate, and Geophysics Services (BMKG) under Data Online Website. The data originated from first class weather station of Blang Bintang, Greater Aceh. The ENSO signal was retrieved from KNMI Climate Explorer under Monthly Climate Indices representing by NINO3.4 where normal condition range between -0.5 and 0.5°C, El Niño and La Niña are recorded if they are above and below the normal ranges, respectively. The signal came from average anomalies of monthly HadISST sea-surface temperature. The Pearson correlation is employed to find the relationship between rainfall and ENSO.

Table 1. ENSO and Rainfall Variability

Climate índices (°C)	*ENSO event*	*Rainfall variability*
-0.5 < Nino3.4 <0.5	Normal	Normal
0.5 < Nino3.4 < 1.0	Weak El Niño	Normal
1.0 < Nino3.4 < 2.0	Moderate El Niño	Below normal
2.0 < Nino3.4 < 3.0	Strong El Niño	Far below normal
-1.0 < Nino3.4 < -0.5	Weak La Niña	Normal
-2.0 < Nino3.4 < -1.0	Moderate La Niña	Above Normal
-3.0 < Nino3.4 < -2.0	Strong La Niña	Far above normal

BMKG classify the climate condition into several terms, i.e., below normal, normal, and above normal. Climate is considered normal if it ranges between the averages. Normal climate ranges between 85% and 115% from the averages. Climate is considered below and above normal if they are lower and higher than 85 % and 115 % from the averages, respectively (BMKG, unpublished data).

3. Results and Discussion

The mean annual air temperature in Greater Aceh is 27.07°C. It is shown that no significant difference is found in the mean monthly air temperature. The mean maximum air temperature of 27.86 °C reach its peak on June and mean minimum air temperature of 25.97 °C is recorded on December (Figure 1). Because Greater Aceh is situated in the intertropical convergence zone, this area experiences intense solar radiation in nearly vertical direction, thus, little variability of temperature between the warmest and coldest months often describe changes in cloud cover instead of sun position. For instance, it is shown that the highest temperature is recorded in the month where rainfall is less in this case on June (Figure 1 and 2). Because the variability in mean air temperature is small, although June is the highest air temperature, it is, however, not the month of the lowest rainfall. The lowest rainfall is actually recorded at the peak of dry season on August with an average of 36 mm (Figure 2).

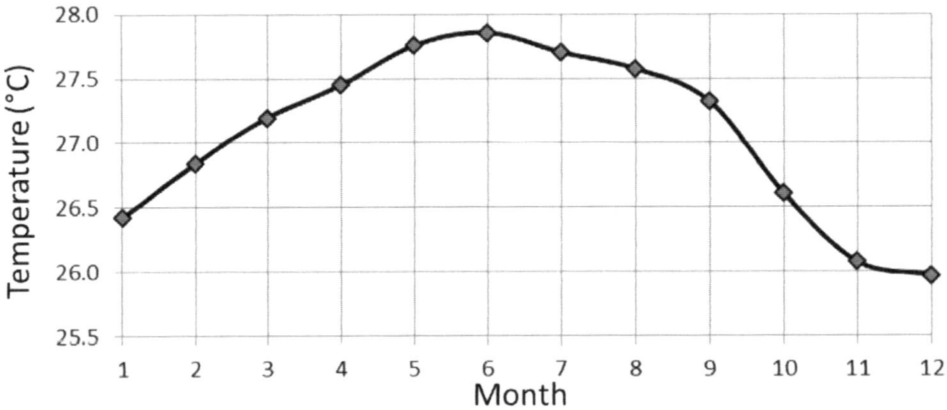

Figure 1. Mean Monthly temperature (°C) over Greater Aceh.

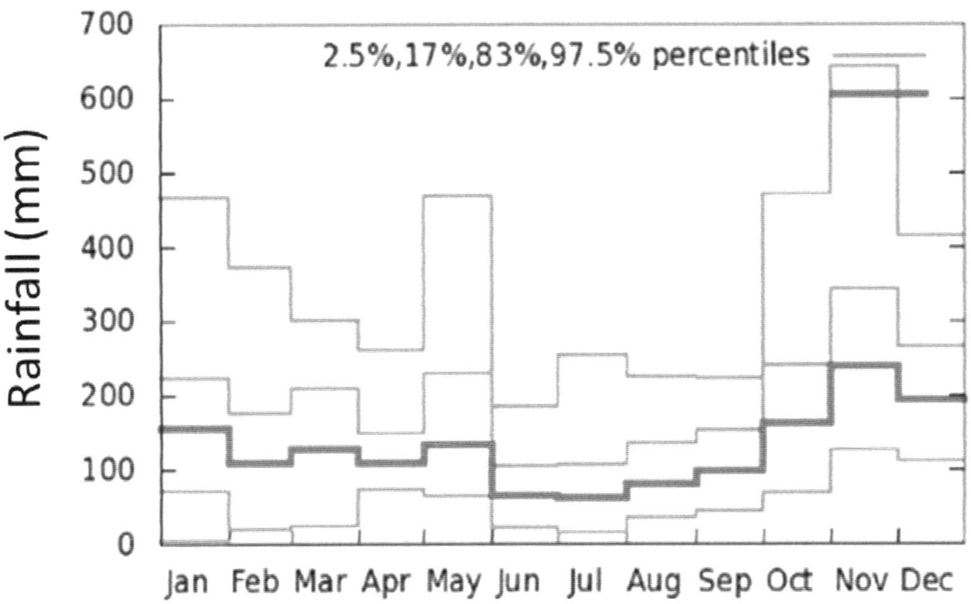

Figure 2. Mean Monthly Rainfall (mm) over Greater Aceh.

IOP Conf. Series: Earth and Environmental Science **273** (2019) 012002 doi:10.1088/1755-1315/273/1/012002

Greater Aceh has two main seasons, the dry season from June to August and the rainy season lasting from December to February, interrupted by two transition periods (March - May and September - November). However, it is noted that two peaks of rainy season fall over Greater Aceh, those are on May and November. A typical semi-monsoonal climatic pattern is observed over Greater Aceh where the region does not have a very distinct wet and dry season. Wet characteristics of climate occurred almost throughout year. The peak of the rainy season fall on November with an average monthly rainfall of 250 mm while in the dry season, a monthly average of 60 mm is observed on July. By understanding a dominant type climatic zone and its pattern accompanied by total amount of monthly rainfall and temperature over Greater Aceh based on their climatological variability (at least 30 years of data), a normal condition of climate is well-identified which might possibly be set up as baseline to determine the anomaly condition of the climate over Greater Aceh.

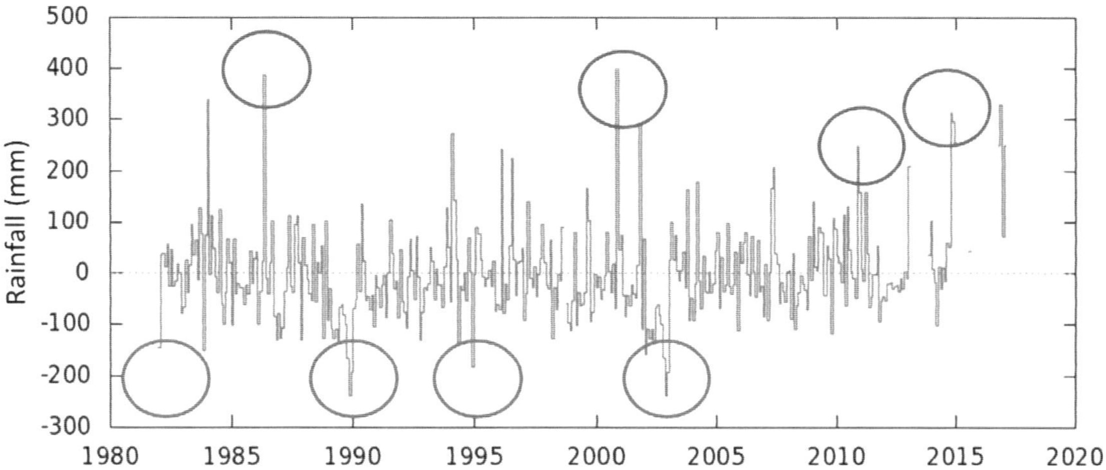

Figure 3. Rainfall anomaly (mm) during 1982-2018 over Greater Aceh. Blue and brown circle indicate flood and drought events due to La Niña and El Niño, respectively.

Time series of rainfall anomaly for three decades showed the increase of rainfall in the last 15 years. It is shown that the signal of flood event appeared frequently. Rainfall above normal was known to have been accompanied by La Niña occurrence particularly during moderate La Niña 2010-2011 and 2016. The increase of rainfall over tropics was reported in the IPCC AR5 where based on their ensemble model results, the tropical region is projected to receive a lot of rainfall in the future [14]. This condition is triggered by high sea-surface temperature as well as strong surface convection. Such factors then develop cumulonimbus clouds that may reach the stratosphere. The deep dark clouds might cover a whole tropics leading such hydrometeorological hazards as floods/inundation, wind and hailstorm, lightning to be very likely and intense in frequency to happen both nowadays and future time. In the 80s and 90s, the El Niño signal was repeatedly detected as rainfall decrease below normal over the periods (Figure 3).

The trends of rainfall variability were shown to decrease in the 80s and 90s and raise in middle 2000s up to present time which is confirmed by regular emergences of El Niño and La Niña events, respectively. It is also shown that ENSO play a major role in influencing the rainfall variability in Greater Aceh despite time-series correlation is weak. The statistical performance showed that a p-value of 0.6470 which is less than 0.95 revealed that the significant correlation exists between Nino3.4 and the onset that fall on October-December. Since ENSO is a global climate anomaly, its teleconnection might alter throughout the globe which was confirmed by field statistical test among Nino3.4, Greater Aceh rainfall, and global air temperature where a statistically significant connection (p-value < 5.0%) was found in the map of the present study (Figure 4).

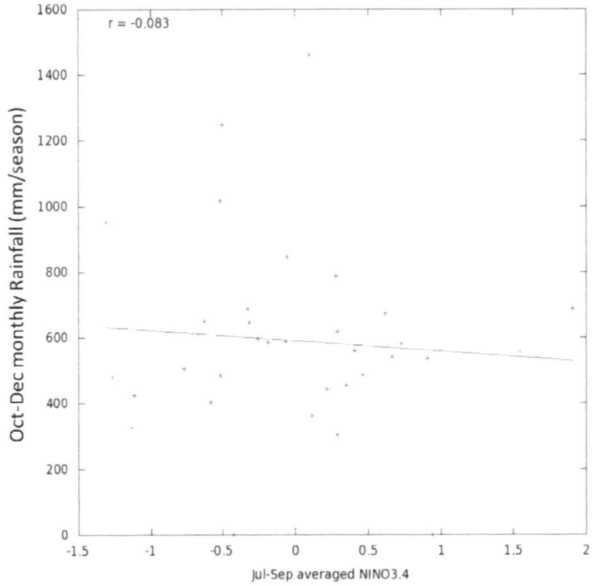

Figure 4. Time series correlation between rainfall variability and ENSO Signal (Nino3.4) over Greater Aceh

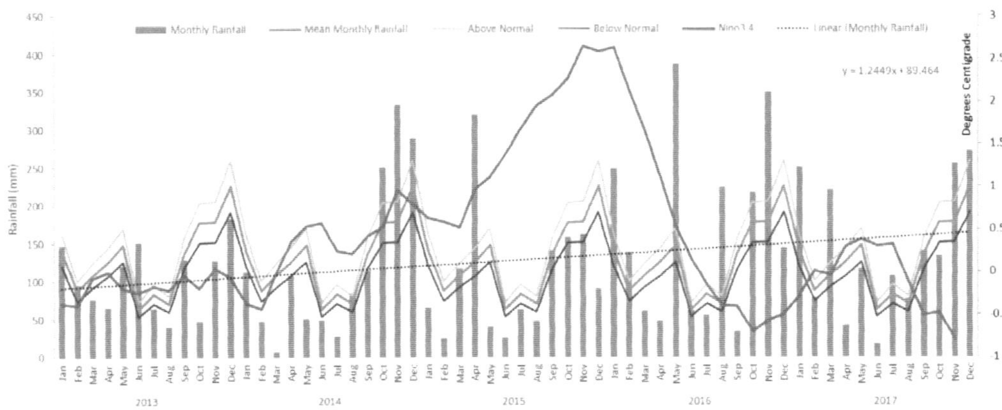

Figure 5. Characteristics of monthly rainfall (mm) and ENSO signal (°C) in Banda Aceh during 2013-2017.

In 2015, the El Niño signal started appearing above normal on April, raise severely in the following months, and establish on November-January. The intensification was accompanied by the decrease of the monthly rainfall (Figure 5). At the peak of rainy season on December, the total amount of rainfall was far below normal which resulted in the retreat of the onset in Banda Aceh. During dry season, El Niño affected to the drought due to the lack of water resources as rain drop below normal. It is also noted that the emergence of El Niño signal lead the first peak of the rainy season in Banda Aceh that occurred on May to fall far below normal. In addition, a lot of rain was received on April prior to the growth of intense El Niño. It is interesting to notice that more and less rainfall at the opening of El Niño period could be an indicator of typical severe El Niño years in Banda Aceh. In 2016, the development of La Niña on October lead the onset of the rainy season to advance. Following extreme

La Niña years, the worst condition due to high rainfall terrify the Greater Aceh. When annual rainfall arise, flooding are haunting as the climate change arrive over the region.

The disaster risk reduction by minimizing the climate risk by means of performing proper adaptation and mitigation is urgently implemented. Anthropogenic effect derived from the excessive utilization of fossil fuel, exploiting forest values, land cover change, forest fire have polluted the environment. The release of greenhouse gas emission have been uncontrolled since industrial revolution which have reached an alarming level leading global climate processes to disturb. In consequence, climate anomaly of ENSO become more frequent which further threating the terrestrial life. The monitoring of weather and climate condition, climate field school, dissemination of the research output as well as building public awareness associated with extreme weather and its climate change impact in many sectors, e.g., health, agriculture, tourism, fisheries etc are sort of adaptation action that could be executed to address the climate change issues. The awareness as well as the preservation of the environment and the development of green energy foremost are such mitigation that could be introduced to cope with an inconvenient truth of global warming.

4. Concluding remarks

Climate variability is very likely to occur in Banda Aceh. In addition to retreat the onset, strong El Niño cause annual rainfall to drop leading to drought. On the contrary, La Niña period bring surplus water that is potential to water reserves to cope with water scarcity during dry season in the region. It is however, La Niña is well-known for the years of flooding due to the excessive rainfall, urban built-up, lack of catchment areas, and mismanagement foremost.

5. References

[1] Ilhamsyah Y 2012 *Aceh International Journal of Science and Technology* **1(1)** 60-66
[2] Adi RF, Sugiarto Y, Arifah A, Arini EY and Atmaja T 2017 *IOP Conf. Ser. Earth Environ. Sci.* **54(1)** 012031
[3] Sandi 1987 *Iklim Regional Indonesia* (Jakarta: FMIPA UI) p 50
[4] Whiteman CD 2000 *Mountain Meteorology: Fundamentals and Applications* (UK: OxfordUniversity Press) p 376
[5] Vernimmen RRE, Hooijer A, Aldrian E and Van Dijk A 2012 *Hydrol. Earth Syst. Sc.* **16 (1)** 133-146
[6] Aldrian E, Karmini M and Budiman 2011 *Adaptasi dan Mitigasi Perubahan Iklim di Indonesia* (Jakarta: Pusat Perubahan Iklim dan Kualitas Udara BMKG) p 60
[7] Ilhamsyah Y 2012 *DEPIK Jurnal Ilmu-Ilmu Perairan, Pesisir dan Perikanan* **1(3)** 165-174
[8] Setiawan AM, Lee WS and Rhee J 2017 *Int. J. Climatol.* **37(13)** 4700-4719
[9] Aldrian E and Susanto RD 2003 *Int. J. Climatol.* **23** 1435–1452
[10] Aldrian E, Gates LD and Widodo FH 2007 *Theor. Appl. Climatol.* **87(1-4)** 41-59
[11] Aldrian E, Dümenil-Gates L, Jacob D, Podzun R and Gunawan D 2004 *Clim. Dyn.* **22 (8)** 795-814
[12] Ilhamsyah Y, Fadli N, Setiawan I and Haridhi HA 2014 AACL Bioflux **7** 508-515
[13] Ilhamsyah Y 2013 *DEPIK Jurnal Ilmu-Ilmu Perairan, Pesisir dan Perikanan* **2(3)** 200-209
[14] Intergovernmental Panel on Climate Change 2013 *Working Group I Contribution to the IPCC Fifth Assessment Report Climate Change 2013: The Physical Science Basis.* (Geneva: IPCC)

Acknowledgments

Authors express gratitude to Ministry of Research, Technology, and Higher Education Indonesia under Domestic Postgraduate Scholarship Program 2014 (BPPDN 2014) for financial support of the research.

IOP Conf. Series: Earth and Environmental Science **273** (2019) 012003 doi:10.1088/1755-1315/273/1/012003

Addressing Disaster Risk Reduction by Analyzing Sea Level Rise in the Aceh Waters

Ichsan Setiawan[1,2], Yopi Ilhamsyah[1,3], Edy Miswar[1], Muhammad Irham[1*]

[1]Dept. of Marine Sciences, Faculty of Marine and Fisheries, Syiah Kuala University, Banda Aceh Indonesia 23111
[2]Center for Marine and Fisheries, Syiah Kuala University, Banda Aceh Indonesia 23111

[3]Applied Climatology, Graduate School of Bogor Agricultural University, Bogor Indonesia 16680

[*]Corresponding author: irham@unsyiah.ac.id

Abstract. Eustatic changes produce changing on global sea levels, such as changes in water volume in the oceans world. This change indirectly affects the balance of water and land and it is very harmful for the archipelago countries which one of them is Indonesia. The two most important effects are the temperature and mass of water stored on land and sea as fresh water in rivers, lakes, glaciers, polar ice caps, and sea ice. Therefore the purpose of this study is to see how far the sea level changes in Aceh waters in relation to disaster risk reduction. To achieve this goal, the method used is to analyze the record of sea level data in the waters of Aceh during the last 25 years. The analysis of sea surface change data in Aceh shows that there has been a linear increase in sea level since 2008 in parallel with the rise in temperature so that it is expected to lose ground in some areas of the Aceh coast.

1. Introduction

Sea Level Rise is a phenomenon of rising sea levels as a result of climate change which is an important issue today. Moreover for Indonesia, the archipelagic country has +18.110 islands and coastline along the 108,000 km. The results of the Intergovernmental Panel on Climate Change (IPCC) investigation estimated that within 100 years starting from 2000 the sea level would rise as high as 15-90 cm with a 48 cm height increase certainty. In Indonesia alone, a projected sea level rise of that magnitude would result in loss of land due to sea level rise of 30120 km^2 in 2050 and 90260 km^2 in 2100 [1-3].

Coastal region which is the boundary between land and sea make it as a sufficiently potential area of disaster where most of the people of Indonesia live along this area. This area is socially inhabited by more than 110 million people or about 60% of the population of Indonesia. Temperature changes cause the physical condition of the atmosphere to become more unstable and cause anomalies of long-lasting weather parameters that cause climate change potentially to have sea level rise.

The previous research in Aceh water related to this research such as tidal dynamic, wave, climate change, and coastal morphology surveyed [4-12] have shown the tendency of sea level rise. None of the topics, however, address the disaster risk reduction associated with the sea water dynamic in Aceh water. Therefore in this study will be discussed how to overcome disaster risk reduction by analyzing sea level rise in Aceh Waters. This is done with the intention of ensuring that early evaluation of the sea level rise impact on coastal areas will be used as a reference in adaptation strategy to reduce disaster risk [1-3]. The purpose of this study is to see how far the sea level changes in Aceh waters in relation to disaster risk reduction.

2. The Methods of the Research

The sea level anomaly data was extracted from NASA's MEaSUREs program under NASA Sea Level Portal that is available at https://sealevel.nasa.gov/data-analysis-tool. The data originate from satellite altimetry of TOPEX/Poseidon, Jason-1, Jason-2, and Jason-3. It is monthly gridded data with the spatial resolution of 1/6 degree. The data range between January 1993 and June 2017. In the present study, we analyze the cumulative changing of the sea level in the western, northern, and southern Aceh Waters.

3. Results and Discussion

Figure 1 shows the changes in monthly average sea level at three stations of Aceh water. These three stations represent the northern coastal areas (Bireuen), the west coast (Lhoknga) and the south Coast (Calang). The result of sea level anomaly trend shows that for the last 25 years there is a tendency of sea level rise up, where significant increases began in 2008. It appears that over the past 25 years sea level in Aceh waters has risen by approximately 4 cm which indicates that there is an increase of 0.16 cm per year. The increase in water level is in line with the estimated world sea level rise trend as reported by NOAA (National Oceanic and Atmospheric Administration) and CSIRO (Commonwealth Scientific and Industrial Research Organization) (Figure 2).

On this occasion, various factors affect the volume and mass of the oceans that result in long-term changes in the sea level rise. This condition can be related to two influencing factors, namely sea temperature rise (because the volume of water depends on temperature) and sea mass change where water stored on land and sea as fresh water in rivers, lakes, glaciers, polar ice caps, and sea ice. This condition occurs due to the dominant influence of thermal expansion so that the volume of seawater increases. On a long time scale, changes in the shape of the ocean and the distribution of terrestrial will affect the sea level (Figure 3). The sea-surface temperature anomaly Bireuen Station is larger than Calang Station. In general, this is in line with the Sea level anomaly which is also larger than Bireuen Station than Calang Station.

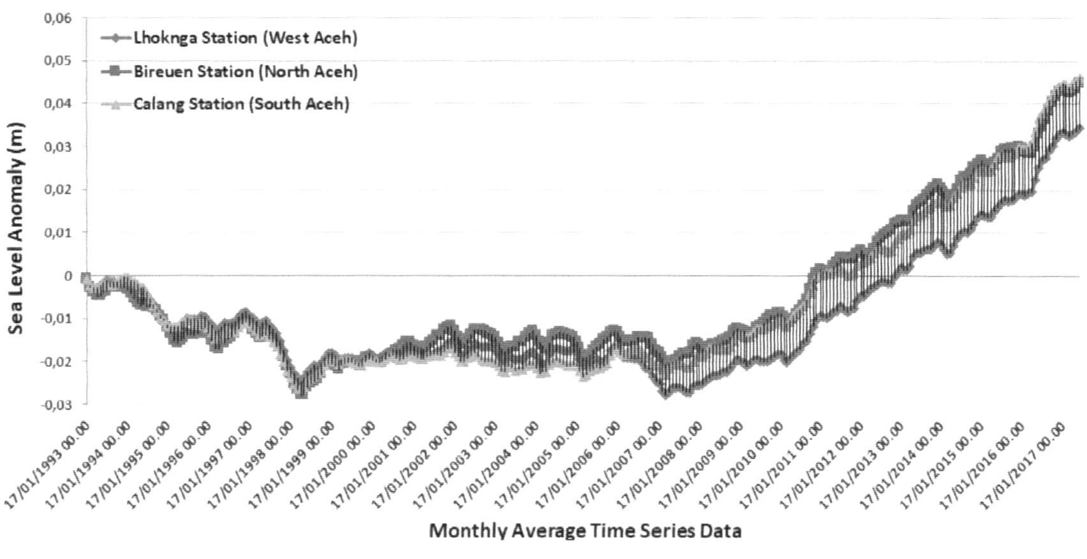

Figure 1. Sea level anomaly of the Aceh Waters (in meters).

The rise of sea level of 1.6 mm per year in the waters of Aceh requires us to rethink to anticipate the possibility of an unwanted disaster. Therefore, adaptation strategy management is a strategy that must be applied to anticipate the bad possibility that will happen. Consequently, the implementation of disaster risk management to anticipate the possibility of adverse consequences of sea level rise from the analysis results obtained need to be immediately implemented.

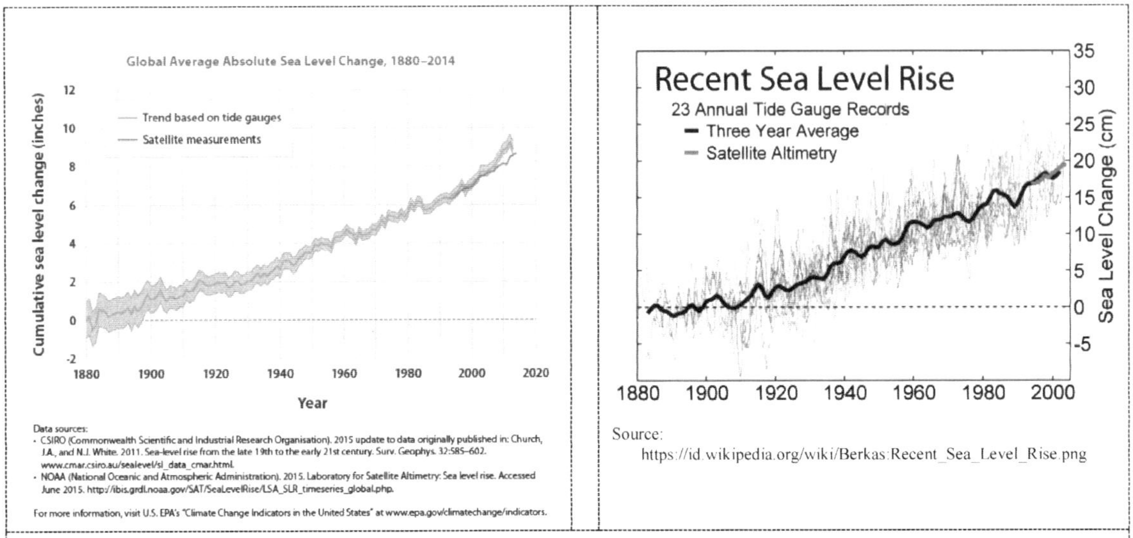

Figure 2. The trend of rising of the global sea level anomaly for the last 130 years.

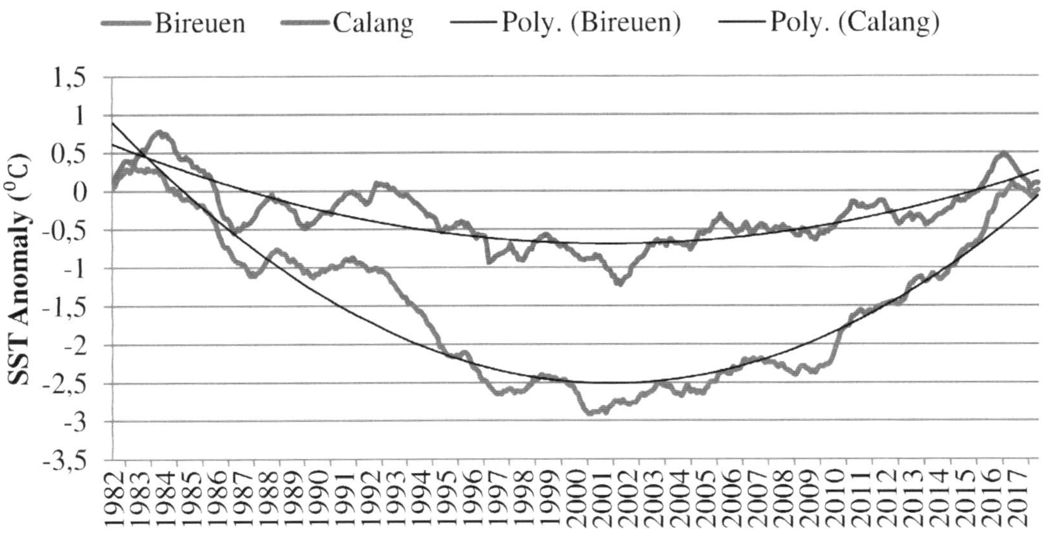

Figure 3. Sea-surface temperature anomaly (in centigrade).

IOP Conf. Series: Earth and Environmental Science **273** (2019) 012003 doi:10.1088/1755-1315/273/1/012003

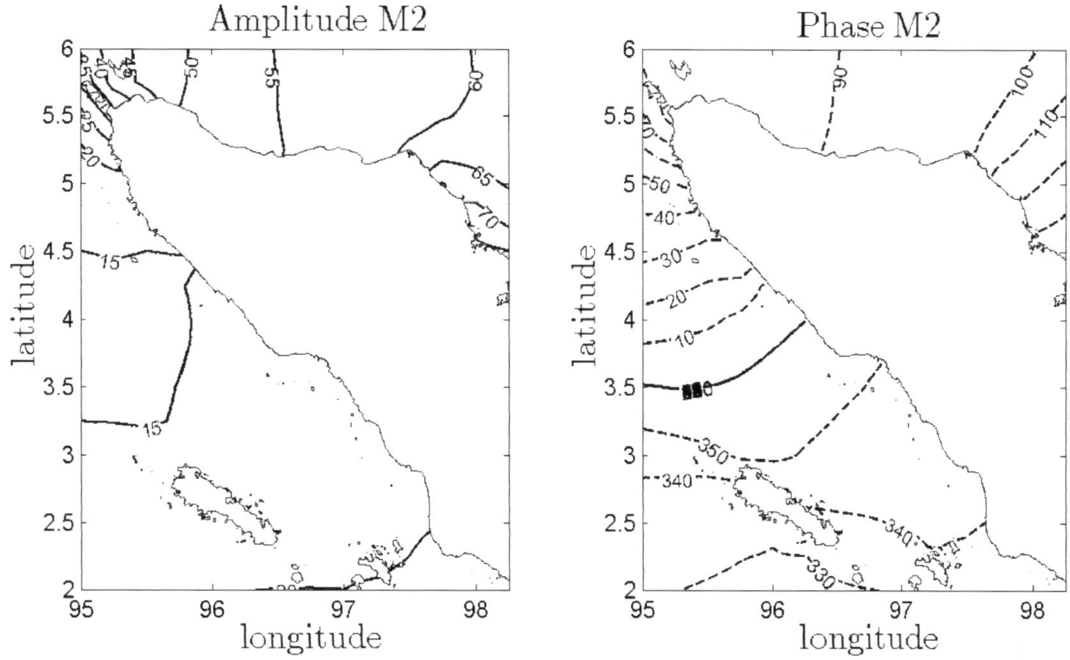

Figure 4. M2 Amplitude in centimeter and M2 Phase in degree.

Figure 4 shows the results of M2 Amplitude and M2 Phase [13]. Based on the amplitude result, the waters of North Aceh is greater than the West and South water. Figure 4 shows a close relation of amplitude and phase of Aceh water to the sea level rise data (Figure 1) which indicates that the sea level rise of North Aceh is larger compared to the West and South water. While the phase differences can be used to predict the mass flow patterns of water. The sea water movement is influenced by the difference between high to low tide.

Table 1. Tides harmonic component of Aceh waters

Location	M2		S2		K1		O1		Tide Type
	A (cm)	g (°)	A (cm)	g (°)	A (cm)	g (°)	A (cm)	g (°)	
Lhoknga	0.3023	81.36	0.1348	105.48	0.0944	208.33	0.0421	178.63	Mixed tide prevailing semidiurnal
Bireuen	0.4637	87.21	0.2075	119.54	0.1035	213.70	0.0437	181.73	Semidiurnal tide
Calang	0.1515	29.70	0.0748	51.64	0.0847	197.74	0.0432	170.39	Mixed tide prevailing semidiurnal

Table 1 shows the constant harmonic components of Lhoknga, Bireuen and Calang water. Based on these constants, the harmonic of semidiurnal tide and mixed tidal prevailing semidiurnal are found in Lhoknga and Calang water that represents west and south Aceh water. However, the Bireuen waters have the harmonic semidiurnal tide, which is located in the Malacca Strait. This location has a higher tidal elevation when compared to Lhoknga and Calang water which is directly opposite to the Indian

Ocean. The location and tidal height are basically very influential on sea level rise. Therefore, based on the sea level rise data in Figure 1, there is a greater increase in Bireuen water (northern waters) than Lhoknga and Calang waters (wester southern water).

The consequecy of the condition of Aceh sea level for the previous 25 years, based on data and water dynamic of Aceh water, we can assume that the northern part of Aceh coastal area will likely to lose ground faster than western and southern part of Aceh. Therefore, the management strategy for disaster risk reduction will have to be the focus on the northern part of Aceh coast rather than soutern one. In a long run, Aceh coastal area will still losing ground due to sea level rise either northern part or western and southern part of Aceh and both of them should be evaluated yearly for disaster management strategy.

4. Conclusions

For the last 25 years, the rising sea level in Aceh has reached about 4 cm or an increase of 0.16 cm per year. This linear increase will plunge the coastal areas of Aceh. Therefore, a preliminary evaluation of sea level rise can be a management strategy for disaster risk reduction to be implemented in Aceh's coastal areas.

References

[1] Mimura N, Harasawa H (Eds) 2000 Data book of Sea-Level Rise 2000 *Centre for Global Environment Research* National Institute for Environmental Studies, Environmental Agency of Japan.
[2] Susandi A, Herlianti I, Tamamadin M 2008 *Impact of climate change on sea levels in the Banjarmasin* Region ITB [in Indonesian].
[3] Soenarno 2003 *Review of aspects of spatial planning in the management of sea and coastal areas* Delivered at the 43rd ITS Dies Natalies General Seminar [in Indonesian].
[4] Ilhamsyah Y, Fadli N, Setiawan I, Haridhi HA 2014 Notes: coral reef bleaching in Weh Island, Indonesia, a natural climate variability or global climate change impact? *AACL Bioflux* **7(6)** 508-515.
[5] Irham M, Fadhla Y, Setiawan I 2018 The spatial distribution of suspended sediment analysis along Krueng Cut River, Banda Aceh *IOP Conf. Ser.: Earth Environ. Sci.* **106** 012066 doi: 10.1088/1755-1315/106/1/012066.
[6] Irham M, Miswar E, Ilhamsyah Y, Setiawan I 2018 The northern tidal dynamic of Aceh waters: A 3D numerical model *IOP Conf. Ser.: Mater. Sci. Eng.* **352** 012043.
[7] Irham M, Setiawan I 2017 The study of flow resulting from wave on Lhonga beach, Aceh Besar *OmniAkuatika* **13(1)** 5-12.
[8] Irham M, Setiawan I 2018 The numerical model of the sediment distribution pattern at Lampulo National fisheries port *IOP Conf. Ser.: Earth Environ. Sci.* **106** 012067 doi: 10.1088/1755-1315/106/1/012067.
[9] Purnawan S, Mailala NA, Karina S, Muhammad, Setiawan I, Ilhamsyah Y 2018 The beach slopes and grain size distribution at Anoi Itam and Pasir Putih Beaches, Sabang City *IOP Conf. Ser.: Earth Environ. Sci.* **176** 012013 doi: https://doi.org/10.1088/1755-1315/176/1/012013.
[10] Purnawan S, Setiawan I, Haridhi HA, Irham M 2018 Granulometric analysis at Lampulo Fishing Port (LFP) substrate, Banda Aceh, Indonesia *IOP Conf. Ser.: Earth Environ. Sci.* **106** 012070 doi: 10.1088/1755-1315/106/1/012070.
[11] Setiawan I, Irham M 2018 Wave trajectory study on the coast of Lhoknga, Aceh Besar, Indonesia: a numerical model approach *Civil Engineering Dimension* **20(1)** 30-34.
[12] Setiawan I, Rizal S, Haditiar Y, Ilhamsyah Y, Purnawan S, Irham M, Yuni SM 2018 Study of Current Circulation in the Northern Waters of Aceh *IOP Conf.* Ser.: *Earth Environ. Sci.* **176** 012016 doi: 10.1088/1755-1315/176/1/012016.
[13] Egbert GD, Erofeeva SY 2002 Efficient inverse modeling of barotropic ocean tides *Journal of Atmospheric and Oceanic Technology* **19(2)** 183-204.

Acknowledgments
Authors express gratitude to the Research and Community Service Institutions of Syiah Kuala University for financial assistance under 'Associate Professor Research Grants', with contract number: 288/UN11/SP/PNBP/2018. We also thank to Syiah Kuala University through the facility support at Ocean Modelling Laboratory during the research.

IOP Conf. Series: Earth and Environmental Science **273** (2019) 012004 doi:10.1088/1755-1315/273/1/012004

Identification of Paths and Places of Tsunami Evacuation Based on P646 Fema for Tourism Beaches in the Regency of Gunungkidul, Yogyakarta

Sarwidi[1], Limpat Aji[2], and Iman Satyarno[3]

[1]Master Program of Civil Engineering, Universitas Islam Indonesia, Jl. Kaliurang Km 14,5 Yogykarata, 55584, Indonesia
[2]Department of Civil Engineering, Universitas Gunung Kidul, Jl. KH. Agus Salim 170 Kepek, Wonosari, 55813, Indonesia

[3]Department of Civil Engineering, Universitas Gadjah Mada, Jl. Grafika No. 2 Kampus UGM, Yogyakarta, 55281, Indonesia

E-mail: sarwidi@uii.ac.id

Abstract. The maximum number of monthly visitors to the tourism coasts in Gunungkidul Regency between 2012 and 2015 is 25,160 people. Those Indian Ocean beaches that are in the southern sides of the Island of Java are prone to tsunami disaster. However, there were no systematic and scientific study to make analysis of the tsunami disaster mitigation to those tourism beaches. Therefore, a series of policies with comprehensive rescue strategies and efforts to minimize the risk of the tsunami disaster is required. The purpose of this study is to identify the paths and places of tsunami evacuation for the tourism beaches in Gunungkidul Regency based on P646 of FEMA (Federal Emergency Management Agency) and compare them with the result of the identification of paths and places of tsunami evacuation conducted by BPBD (Local Disaster Management Authority) of Gunungkidul Regency, Yogyakarta. This purpose also includes the want of the authors to implement the method of this study to be the policy of tsunami mitigation for similar tourism beaches in the Special Territory of Yogyakarta as well as similar beaches in Indonesia. This study uses quantitative method with descriptive analysis technique that is analyzing the paths and places of tsunami evacuation based on P646 of FEMA. Given location elevation is the basic ingredient of the analysis, and therefore, elevation data, tsunami arrival time, projected number of visitors to support the analysis need to be gathered. The analysis of the evacuation areas is required to determine the evacuation places that are safe points in the evacuation process. An elevation of a safe point elevation can be a point that is beyond the reach of tsunami waves or safe areas within the tsunami pool areas. The results of this study are the height of the gathering point / safe point 25 meters above sea level to be the basis for saving themselves, if connected with the speed of walking (weak condition), then the arrival time / arrival time of the tsunami is not more than 0.5 hours. This is still within limits in accordance with the Federal Emergency Management Agency (FEMA P-646, 2008), the location of the gathering point / tsunami evacuation safe point can be used as the first rescue of tsunami victims by considering the arrival time of the tsunami. On the other hand, BPBD accommodate tsunami victims using public facilities even though they are far away and heed the arrival time of the tsunami.

1. Introduction

Gunungkidul Regency has a coastal area with a length of ± 72 km, which covers 6 sub-districts (Purwosari, Panggang, Saptosari, Tanjungsari, Tepus and Girisubo) and consists of 19 villages bordering the sea. There are 5 beaches which are areas of human activity that are quite high both from the tourism sector, fisheries and trade. The coastal areas include Baron, Kukup, Krakal, Sundak, and Indrayanti [5]. The location of all coastal tourism objects in this study can be seen in Figure 1.

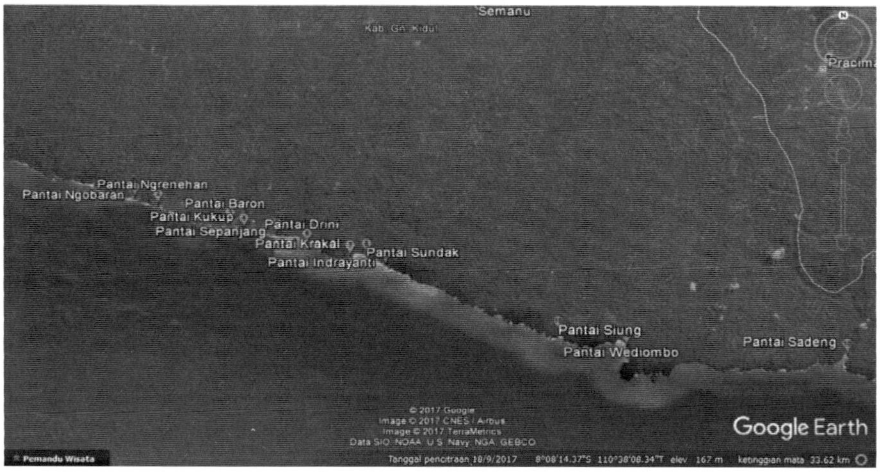

Figure 1. Map of the location of coastal tourism objects in this study (http://Google Earth)

The number of visitors to each beach in Gunung kidul Regency at the peak holiday conditions in this study [5], can be seen in Table 1.

Table 1. Number of Visitors on peak holidays

No	Beach Name	Number of Peak Vacation Visitors (per day) (people)
1	Baron	167.112/8 = 20.889
2	Kukup	
3	Krakal	34.173/8 = 4.271
4	Sundak	
5	Indrayanti	
		201.285/8 = 25.160

Approximately 28% (150 of 530) of cities / districts [4] in Indonesia have a high risk of tsunamis, including the Region of Yogyakarta Special Region. Gunungkidul District has a high risk of tsunami; the policy of the disaster risk reduction with a comprehensive rescue strategy is needed in order to minimize the risk of casualties during a tsunami. In addition to the need for Tsunami Early Warning Systems, tsunami disaster risk reduction efforts are also needed, namely in the form of identification of tsunami evacuation routes and paths in tsunami prone areas. Beach tourism objects in Gunungkidul require evacuation sites and shelter when the tsunami disaster occurs. Identification of lanes and evacuation sites that is used as a refuge can be in the form of fields, hills, both natural hills and artificial hills that are designated as lanes and evacuation sites, existing buildings, and new buildings specifically made for the purpose of lanes and evacuation sites.

The parameters in determining the tsunami paths and evacuation locations base on relevant codes and standards, namely FEMA P646. This FEMA P646 guideline provides an explanation of the analysis of time, speed and maximum distance that must be determined by considering the number and place of evacuation, as well as the visitor population as an important parameter so that the existence of the evacuation site is optimum in accommodating the surrounding population to the maximum distance.

The purpose of the study was to identify tsunami evacuation pathways and places on coastal tourism objects in Gunungkidul Regency based on FEMA P646, and compare the results of identification of tsunami evacuation routes and places conducted by Gunungkidul District BPBD in accordance with FEMA P646 guidelines.

2. Method

A tourism object is a place or state of nature that has tourism resources that are built and developed so that it has an attraction and is sought as a place visited by tourists [5]. The object of research for planning tsunami pathways and evacuation sites based on FEMA P646 is coastal tourism objects in Gunungkidul, namely Baron, Kukup, Krakal, Sundak, and Indrayanti [4].

In analysing the paths and evacuation sites (TE) for identification of tsunami evacuation routes and paths, the road network is the basic material of the analysis carried out. Since the evacuation site (TE) is needed in the evacuation process, an analysis of the evacuation area needs to be carried out. The analysis process is carried out by first determining the safe area. Safe areas can be areas that are outside the reach of the tsunami wave or areas that are inside the tsunami inundation area [7]. For evacuations carried out by directing refugees to areas that are beyond the reach of the tsunami, horizontal evacuation is called. While evacuation carried out by directing refugees to a safe area within the tsunami coverage area is called vertical evacuation.

In this study, the data that has been collected, both through field research and library research is analyzed quantitatively combined with qualitative as a supplement to the explanation of the conclusions obtained. Quantitatively Google Earth is used, while qualitatively using the FEMA P646 Guidelines. The stages of data analysis in the study through the following 5 steps.

1) Analysis of population distribution
   Estimates of the number and distribution of populations in tourist areas are very important to know for evacuation planning, meaning that tsunami evacuation sites can be placed correctly. Service areas related to evacuation capacity can also be known, so that it can be known the needs of the number and capacity of evacuation sites needed, population distribution data, among others, are as follows.
   a. The population in each tourism object includes men, women, persons with disabilities, and others.
   b. The maximum number of visitors to each tourist attraction.
   c. Number and type of buildings such as houses, schools, offices, places of worship, etc. in each tourist attraction.
   d. Height (elevation) of the land surface, distance from the coastline using the Altimeter and Google Earth Applications
2) Analysis of time, speed and distance of evacuation
   Within the estimated time of arrival of the tsunami or ETA (estimated time of arrival), not all can be used as an evacuation time, but there is time to detect a tsunami, preparation time and time to climb to a safe position.
   Empirical calculations for analysis of time, speed and distance of evacuation can be seen at the maximum distance of evacuation sites based on warning times [6].
3) Analysis of direction and behavior of evacuation
   The potential of the road lane in the tourism object is used for tsunami evacuation routes by looking at the map of the road network in the area of the tourism object. In tsunami evacuation, residents walk from population distribution points to the nearest evacuation route, then to the tsunami vertical evacuation site [7].
4) Analysis of the capacity of tsunami vertical evacuation sites and existing tsunami vertical evacuation sites
   If a safe tsunami evacuation site is identified, an analysis of the capacity of the tsunami vertical evacuation site is carried out, which includes the need for evacuation space and the effectiveness of

evacuation spaces. If a tsunami evacuation site is identified, the criteria for tsunami vertical evacuation are determined based on FEMA P646 [7].

5) Digitize the map
This process is carried out to get a thematic map that will be derived data using Google Earth, then the results are used as a reference in making evacuation routes.

3. Results and Discussion

The Regional Disaster Management Agency (BPBD) of Gunungkidul Regency in 2012 created a tsunami evacuation map for the Beaches of Baron, Kukup, Krakal, Sundak, and Indrayanti, then with the Google Earth map application the length of the evacuation route from the coast to the point of gathering and the capacity of the place Tsunami evacuation of each tourism object can be searched, more details can be seen in Table 2.

Table 2. Details of Gunungkidul District BPBD Evacuation Map

No	Beach Name	Maximum number of visitors	l (m)	H (m)	L (m^2)	V_1 (weak) (km/hr)	ETA (minute)	Space Need (1 m^2 = 2 orang)
1.	Baron	20.889	2.773	75	1.729	3.22	51.67	3.458
2.	Kukup		2.507			3.22	46.71	
3.	Krakal	4.271	3.276	75	± 500	3.22	61.04	± 1000
4.	Sundak		2.605			3.22	48.54	
5.	Indrayanti		3.277	75	± 500	3.22	61.06	± 1000

1) Determination of Location of Evacuation / Safe Points with Altimeter
Altimeter is a tool to measure the height of a point from the surface of the sea. In this study to determine the height of the evacuation gathering point as the first safe point due to the tsunami using an altimeter application. Collecting points are determined at a minimum of 25 meters above sea level [4] as the first safe point for tsunami evacuation by considering the potential of the area for evacuation and distance not too far from the coast location. Details of height and location of evacuation points can be seen in Table 3 and Figure 2.

Table 3. Elevation of Evacuation / Gathering Points

No	Beach Name	Elevation of Gathering Point AMSL (m)
1	Baron dan Kukup	37
2	Kukup	37
3	Krakal	34
4	Sundak	34
5	Indrayanti	34

IOP Conf. Series: Earth and Environmental Science **273** (2019) 012004 doi:10.1088/1755-1315/273/1/012004

Figure 2. Height and location of Baron Beach and Kukup Beach gathering points

Figure 3. Height and location of Krakal, Sundak and Indrayanti Beach gathering points

2) *Data and Analysis of Projected Number of Visitors in 2025*

The number of tourists visiting coastal tourism objects in Gunungkidul Regency makes the potential for tsunami victims very large. The current tsunami evacuation site provided by the local government, if we calculate the comparison with the number of visitors to the coastal tourism object in 2015, the capacity / need for an evacuation room is still insufficient. As shown in Table 2. Based on data obtained from Statistic Gunungkidul Regency, it is known that the growth rate of the number of visitors to coastal tourism reaches 30% [5]. The growth rate is used as a reference to calculate the projected number of visitors in the next 10 years, assuming the growth rate is the same every year. Then it can be calculated the projection of the number of visitors to the beach tourism object in the next 10 years from 2015. As shown in Table 4.

Table 4. Projected Number of Visitors in 2025

No.	Retribution Post	Year 2015	Year 2025
1	Baron, Kukup	167.112	2.303.781
2	Tepus (Krakal, Sundak, Indrayanti)	34.173	471.104
		201.285	2.774.885

3) *Determination of Path, Distance and extent of tsunami evacuation gathering points*

In determining the evacuation route, distance and extent of gathering points using Google Earth, it will get direct path, distance and extent of the gathering points of tsunami evacuation. As shown in Table 5.

Table 5. Height and distance of tsunami evacuation gathering points

No	Beach Name	Elevation of Gathering Point AMSL (m)	Distance (m)
1	Baron	37	1.044
2	Kukup	37	778
3	Krakal	34	1.429
4	Sundak	34	758
5	Indrayanti	34	1.430

Figure 4. Map of the location of evacuation points of Baron Beach and Kukup Beach

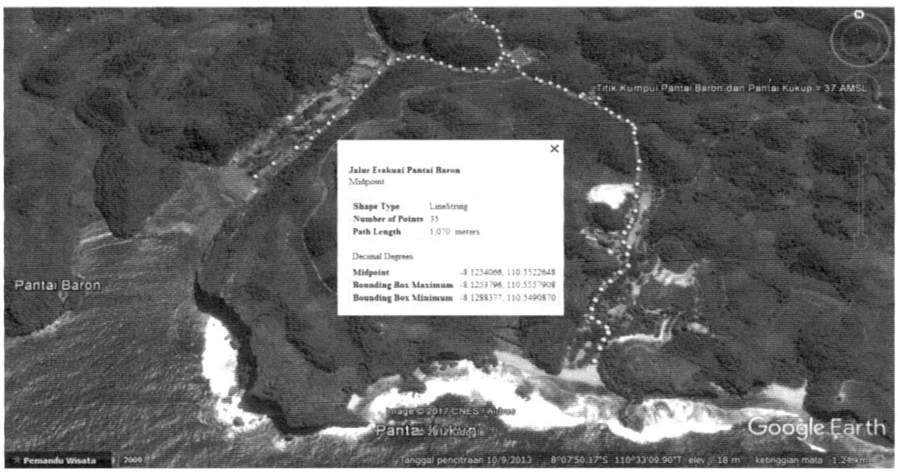

Figure 5. Map of the Baron Beach evacuation route

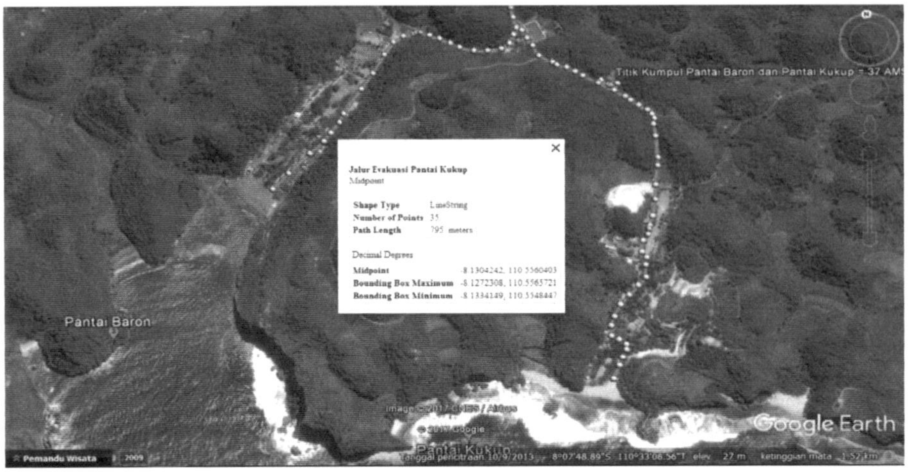

Figure 6. Map of Kukup Beach evacuation routes

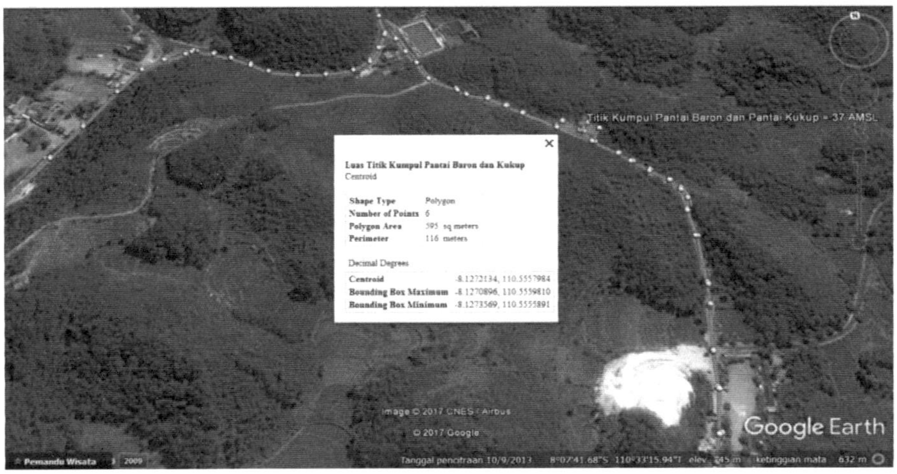

Figure 7. Wide map of evacuation locations for Baron Beach and Kukup Beach

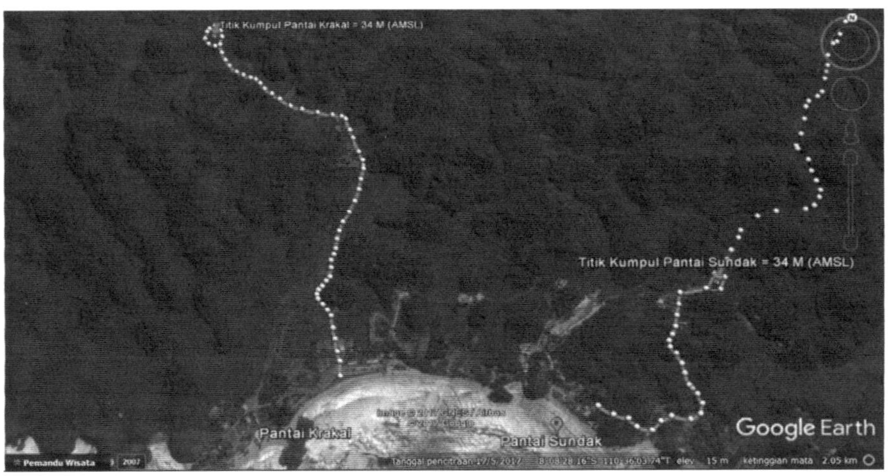

Figure 8 Map of Krakal Beach evacuation points and Sundak Beach

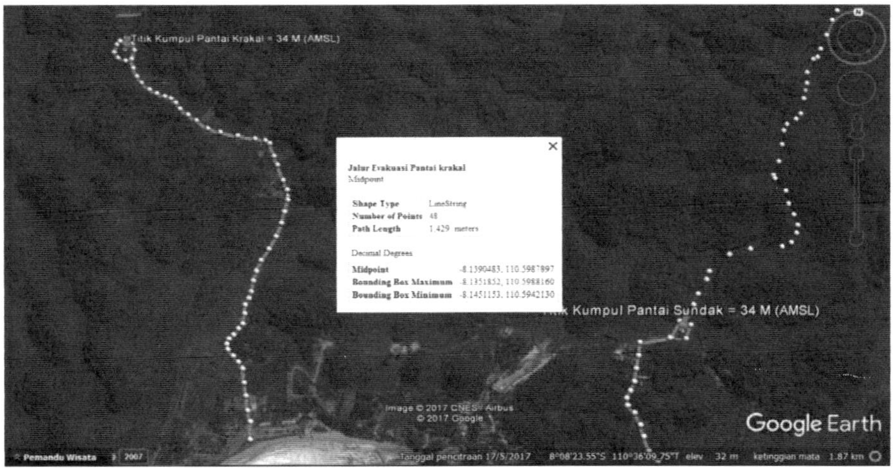

Figure 9. Map of Krakal Beach evacuation routes

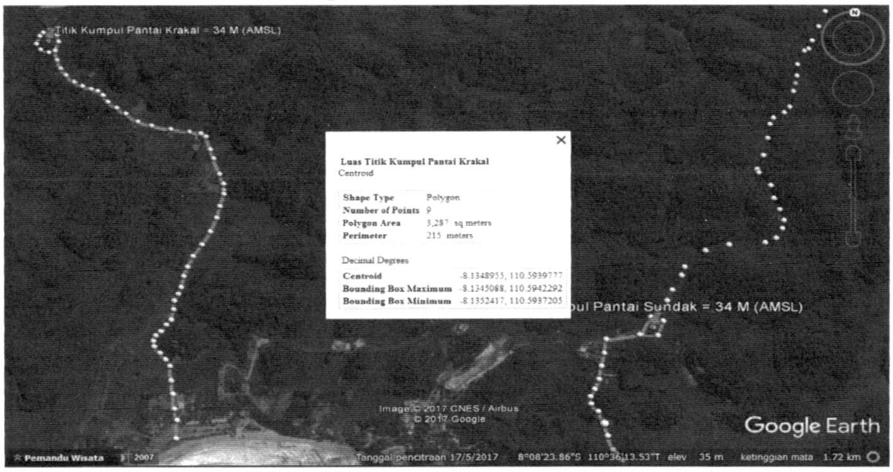

Figure 10. Broad map of Krakal Beach evacuation points

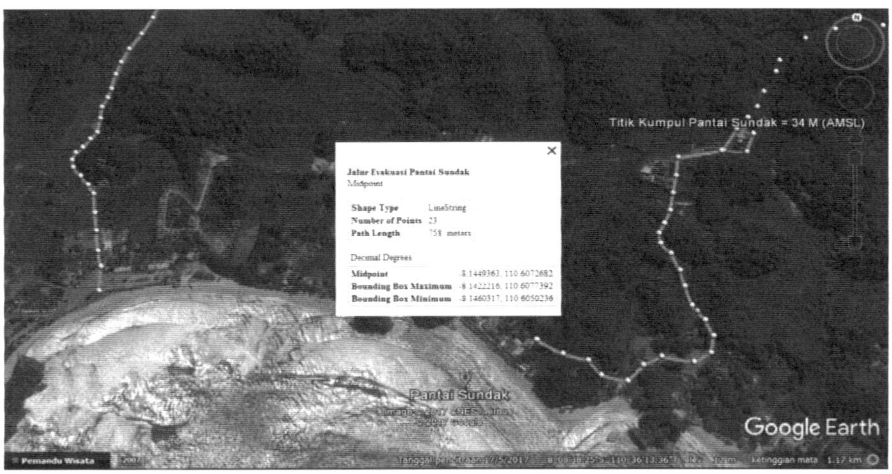

Figure 11. Map of Sundak Beach evacuation routes

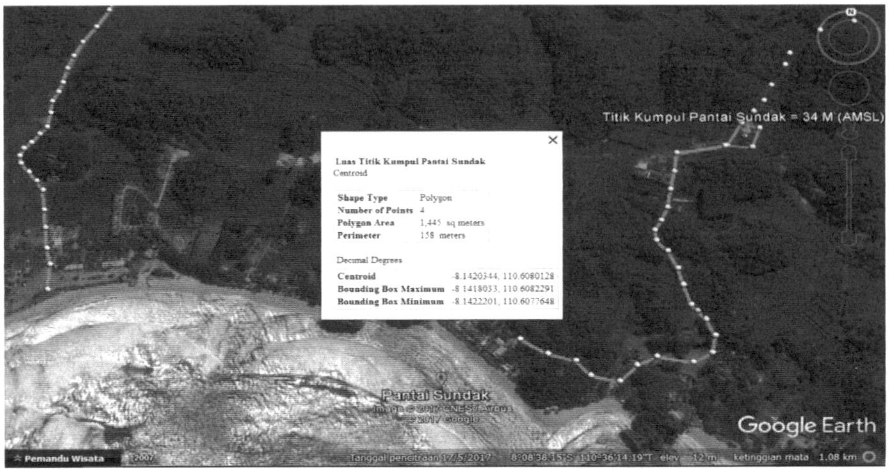

Figure 12. Wide map of Sundak Beach evacuation points

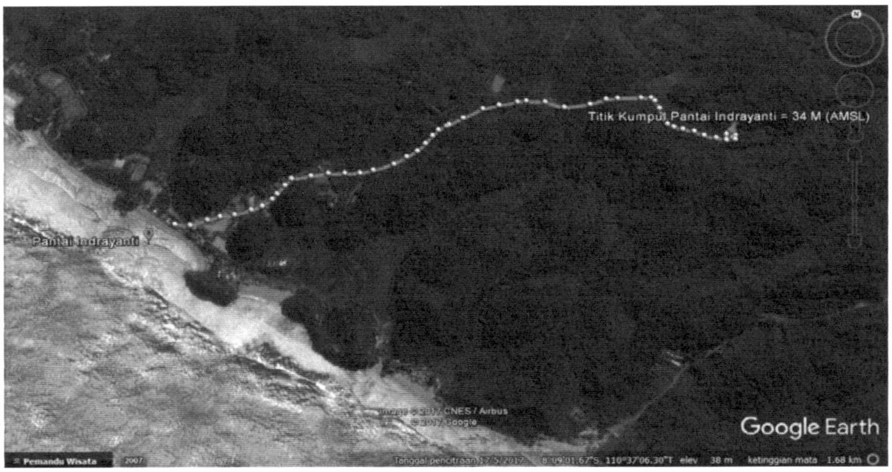

Figure 13. Wide map of Sundak Beach evacuation points

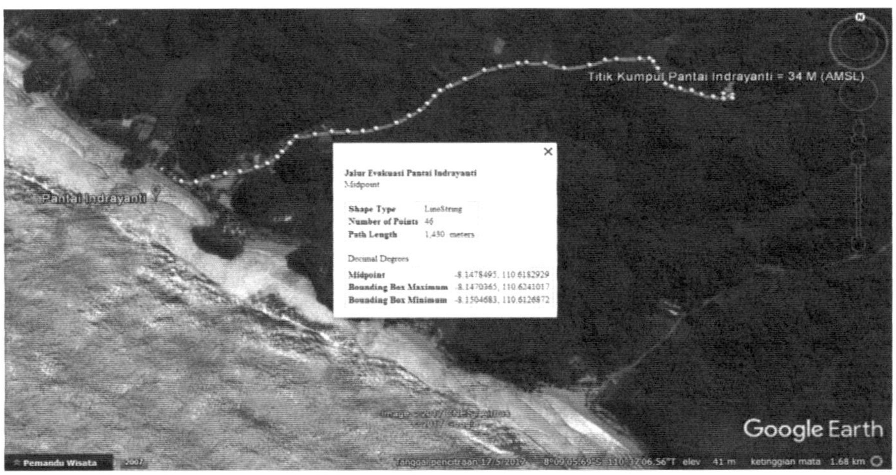

Figure 14. Map of Indrayanti Beach evacuation routes

IOP Conf. Series: Earth and Environmental Science **273** (2019) 012004 doi:10.1088/1755-1315/273/1/012004

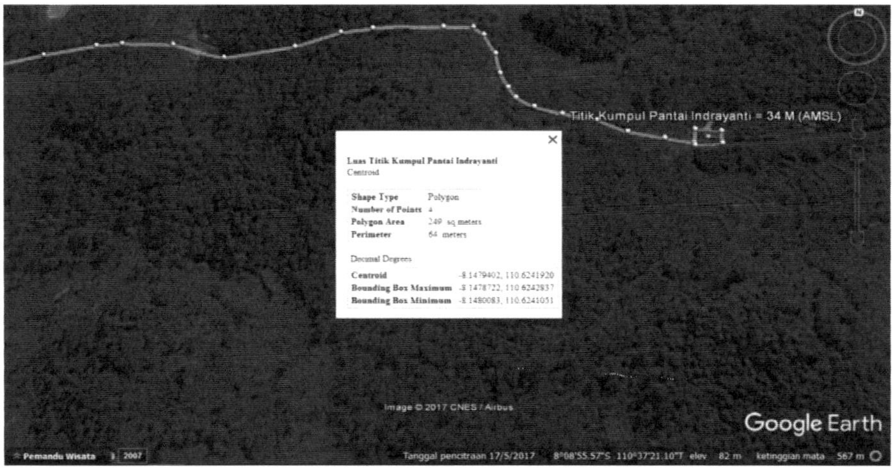

Figure 15. Map of Indrayanti Beach evacuation points

4) Determination of the arrival time / arrival time of the tsunami

According to the Federal Emergency Management Agency [4], the average capacity of healthy people can run at a speed of 6.44 km per hour and the speed of people walking with physical limitations is 3.22 km per hour, so the arrival time / arrival time tsunamis (ETA) can be determined, as shown in Table 6.

Table 6. Warning time based on safe point distance

No	Beach Name	Distance (m)	Speed of people walking (weak) (km/hour)	Tsunami arrival time (minutes)
1	Baron	1044	3.22	19.45
2	Kukup	778	3.22	14.50
3	Krakal	1429	3.22	26.63
4	Sundak	758	3.22	14.12
5	Indrayanti	1430	3.22	26.64

Gunungkidul BPBD tsunami evacuation map shows that the evacuation distance taken takes a long time and the need for evacuation is still insufficient for the number of victims in the event of a tsunami. For this reason, this research base on [4] that the height of a tsunami reaches a height of 25 meters, which is then used as a reference to determine the first safe point in the rescue process in the event of a tsunami as shown in Table 7.

Table 7. Perbandingan Rincian Peta BPBD dan Titik Kumpul Penelitian

		Peta BPBD			Penelitian		
No	Beach Name	Panjang (m)	Ketinggian (m)	ETA (minute)	Panjang (m)	Ketinggian (m)	ETA (minute)
1.	Baron	2.773	75	51.67	1044	37	19.45
2.	Kukup	2.507		46.71	778		14.50
3.	Krakal	3.276	75	61.04	1429	34	26.63
4.	Sundak	2.605		48.54	758		14.12
5.	Indrayanti	3.277	75	61.06	1430	34	26.64

Determination of the maximum distance based on the warning time, illustration of the calculation, if the speed of the person walking with physical limitations is 3.22 km per hour and if the golden time after the tsunami warning is 30 minutes, the Evacuation Place (TE) can be placed at a distance maximum 1.61 km from the departure point. This will result in an average maximum distance of 3.22 km between two Evacuation Places (TE). Likewise, the 15 minutes warning time assumption, Evacuation Place (TE) will be located at a maximum distance of 0.8045 km from the departure point, and 1.61 km distance between the two Evacuation Points (TE). Longer warning times will require a longer distance.

Determination of vertical evacuation locations considering evacuation behavior and natural high places, when determining the structure of natural vertical evacuation, refugee behavior must be considered. Most of their population in coastal communities has been educated to go to high places. The natural tendency for refugees to migrate away from the coast. Therefore, vertical evacuation structures must be located on the inland side of the evacuation zone and must take advantage of natural topography that will tend to attract refugees towards them. In this study, the results of height and distance obtained have taken into account these provisions. Height, distance and extent of tsunami evacuation gathering points. Side effects of hazards around each building are considered in determining the location of vertical evacuation structures. potential hazards of buildings including waves, debris carried through waves. If possible, the vertical evacuation structure must be far from potential hazards that could result in additional damage to structures and safety for occupants. Perhaps due to the limited availability of buildings, and restrictions on travel and population mobility in the community, some vertical evacuation structures need to be in places that would be considered less than ideal.

Evacuation routes in this study use existing road lines, considering that the accesses are easily seen and passed. These routes aim to stay away from the wave direction and must be clearly shown through signs, such as away from the coastline, avoid crossing bridges, utilizing existing lines, towards roads with greater widths so as not to occur bottle necks, and avoid obstacles or obstacles. In addition, the mass movement of each block is directed so that it is not mixed with other blocks to avoid congestion. In the consideration of accessibility, vehicles are prohibited from parking on the road, so that congestion or congestion does not occur on the main road. Thus, this study has considered both its accessibility and away from the coastline to determine the tsunami evacuation site.

For horizontal evacuation, the evacuation point is the intersection between the road and the tsunami immersion limit. But in the study there is no tsunami wave inundation map, so what is done is to determine the point with the altimeter randomly, but consider the distance, the height that is safe from the tsunami wave, then from these points a service area of how many minutes is made, meaning the location whichever can reach the evacuation point within 60 minutes. 60 minutes is the time available to evacuate.

An outline comparison is that in this study the safe location / point is not the same as the location for determining the safe point on the tsunami evacuation map contained in BPBD, because it is based on [4]. If in this study the tsunami evacuation location / safe point was used as the first rescue site for tsunami victims by considering the arrival time of the tsunami, and could be used as a temporary tsunami victim shelter. In BPBD for tsunami victims' shelter, they use public facilities even though they are far away and heed the arrival time of the tsunami.

4. Conclusion

The height value of the 25-meter gathering point from the sea level becomes the basis for saving yourself. If connected with the speed of walking (weak condition), then the arrival time / arrival time of the tsunami is obtained not more than 0.5 hours. This is still within limits in accordance with the Federal Emergency Management Agency guidelines [6]. If the extent of the tsunami evacuation gathering point is associated with the maximum number of visitors between 2012-2015, the evacuation point as a whole cannot accommodate the number of visitors / refugees if a tsunami occurs.

In addition, the location / safety point of the evacuation in the study is not the same as the location for determining the safe point on the tsunami evacuation map contained in BPBD. In this study the tsunami evacuation safe location / point was used as the first rescue place for tsunami victims by considering the arrival time of the tsunami, and could be used as a temporary tsunami victim shelter. In BPBD for tsunami victims' shelter, they use public facilities even though they are far away and heed the arrival time of the tsunami.

5. References

[1] H. Amano, N. Sawaki, I. Akasaki, Y. Toyoda, Appl. Phys. Lett. **48** (1986) 353.
[2] S. Nakamura, Jpn. J. Appl. Phys. **30** (1991) L1705.
[3] B. Kuhn, F. Scholz, phys. stat. sol. (a) **188** (2001) 629.
[4] BAKORNAS 2007 *Introduction of Disaster Characteristics and Mitigation Measures in Indonesia. Second Edition*. Directorate of Mitigation, Lakhar Bakornas PB. Jakarta
[5] DISBUDPAR 2016 *Potential of Culture and Tourism of Gunungkidul Regency*, Gunungkidul, Yogyakarta (2016)
[6] P646 of FEMA 2008, *Guidelines for Design of Structures for Vertical Evacuation from Tsunamis*. California, United States: Federal Emergency Management Agency (FEMA)
[7] Satyarno I 2013 *Utilization of Existing Buildings as Vertical Evacuation Facility in Tsunami Potential Areas in Indonesia* in F. Wang et al. (eds.), 2013. Progress of Geo-Disaster Mitigation Technology in Asia, Environmental Science and Engineering, DOI: 10.1007 / 978-3-642-29107-4_14, © Springer-Verlag Berlin Heidelberg
[8] Satyarno I 2013 *Assessment of the 2004/2011 Earthquake and Tsunami in Indonesia and Japan: Future Lessons and Challenges in the Proceeding of the 1st International Conference on Sustainable Civil Engineering Structures and Construction Materials* (SCESCM), Gadjah Mada University, Yogyakarta

Acknowledgments
The authors would like to thank the Head of the Domestic Postgraduate Education Scholarship Bureau (BPP-DN), the Directorate General of Higher Education, the Ministry of Education and Culture, the Republic of Indonesia for scholarships given for 2 (years) study period of the second author in the field study of Earthquake Engineering Management in the Master Program of Civil Engineering of Universitas Islam Indonesia.

IOP Conf. Series: Earth and Environmental Science **273** (2019) 012005 doi:10.1088/1755-1315/273/1/012005

Adaptive Urban Design Principles for Land Subsidence and Sea Level Rise in Coastal Area of Tambak Lorok, Semarang

I Akbar[1,*], H W Poerbo[2] and W K Soedarsono[2]

[1]Magister Student of Urban Design, Department of Architecture, School of Architecture, Planning and Policy Development, Institute of Technology Bandung Jl. Ganesha No. 10 Bandung 132, West Java, Indonesia
[2]Lecturer, Department of Architecture, School of Architecture, Planning and Policy Development, Institute of Technology Bandung Jl. Ganesha No. 10 Bandung 132, West Java, Indonesia

[*]Corresponding author: ivanakbar1991@gmail.com

Abstract. In recent years, the upward trend of sea level rise caused by climate change continues to increase. This condition threatens urbanization that occurs in urban areas, especially in coastal area. Coastal is a strategic location for various activities such as ports, recreation, fisheries and agriculture, but it is vulnerable to changes caused by community and natural activities. Semarang is one of the coastal city located in the north of Central Java. Dense and slum settlements are dominating the coastal area of Semarang, one of them is fisherman's settlement in Tambak Lorok. This area has been undergoing land subsidence for a long time and is now threatened by sea level rise due to climate change. The purpose of this study is to formulate adaptive urban design principles on land subsidence and sea level rise in Tambak Lorok, Semarang. This study is carried out with qualitative method using the approach of research and development in building a formula that applies to the conditions formed by land subsidence and sea level rise. The design principles formulated from this study are general principles that serve as guidelines for designing coastal areas that adaptive to land subsidence and sea level rise. We believe, the adaptation of urban space can reduce the impact of the disaster and create a resilient coastal area to reduce the risk of natural-induced disasters.

1. Introduction

Sea level rise (SLR) poses a threat to people living in low elevation coastal zones (land up to 10 m height above mean sea level) [1], as well as community and natural systems that support this area. In the Fifth Assessment Report (AR5), the Intergovernmental Panel on Climate Change (IPCC) projected, the upper end of the range of possible global average SLR, to reach 98 cm by the end of this century in accordance with changes of MSL in 1986-2005 [2]. Extreme seawater conditions arise as a combination of average sea level, astronomical waves, storm surges (driven by tropical storms), and dynamic wave components (especially in open coastlines) that lead to run-ups on the coast and overtopping built and natural coastal defenses, such as embankments, sea walls, or sand dunes.

Coastal is a strategic area for various activities such as port, recreation, fisheries, agriculture, industry, settlements, and etc. Behind all the advantages, coastal area are vulnerable to changes caused by coastal activities. Residents living in coastal areas are also vulnerable because they rely on natural resources in coastal areas. Changes in sea level will adversely affect coastal communities by increasing the flood rick and/or coastal and cliff erosion, and that will also have ecological and

IOP Conf. Series: Earth and Environmental Science **273** (2019) 012005 doi:10.1088/1755-1315/273/1/012005

economic impacts on valuable marine ecosystems (such as productive estuaries, coastal wetlands and coral reefs). One of the cities located in coastal area is Semarang. Semarang is the capital of Central Java which is located on the North coast of Java Island. The topography of Semarang consists of two plains, the lowlands and coastal areas in the North; and hilly areas in the South. The North, which consists of city center, port, airport and train station, is a relatively flat area. This section has a relatively higher population and more industrial and business areas compared to the South.

The main problems found in Semarang are land subsidence and tidal flooding. Tidal flooding is defined as a flood that hit areas with low elevations in coastal areas, including estuaries and deltas, which are inundated by brackish water or sea water. Inundation due to rob is one of the environmental problems faced by cities located in coastal areas such as Semarang. Tidal flood inundation is also influenced by various modifications of landforms by human activities. Modifications to the landforms of coastal plains such as reclamation greatly affect the possibility of coastal disasters. Tidal flooding inundation area in Semarang has the potential to expand with the phenomenon of land subsidence. Not to mention the threat SLR as one of the impacts of climate change. If these conditions continue without intervention or prevention, it can be ascertained that there will be a decline in environmental quality which directly reduces the quality of life of the coastal communities of Semarang. There needs to be an adaptation effort to reduce negative impact from the upcoming disaster. The purpose of this study is to formulate adaptive urban design principles on land subsidence and sea level rise in Tambak Lorok, Semarang. These adaptive principles can be considered by the government in redesigning Tambak Lorok area in accordance with its physical and socio-economic conditions while also anticipating the future disasters.

Adaptation to disasters and climate change is mentioned in the New Urban Agenda (NUA) article on disaster mitigation, which is to strengthen the city's resilience to reduce the risks and impacts of disasters. Take actions to address climate change by reducing greenhouse gas emissions [3]. Not only in NUA article, on Sustainable Development Goals (SDGs) No. 13 is also mentioned about the urgency to take action to combat climate change and its impacts by regulating emissions and promoting the development of renewable energy. The Government has sought to integrate SDGs indicators into the RPJMN, as well as climate change. One of the general policy directions of the RPJMN 2015-2019 is about improving environmental quality, mitigating natural-induced disasters and handling climate change. Plans related to disaster adaptation and mitigation have been integrated in various cross-cutting programs with the aim of reducing GHG emissions by 26% and improving climate resilience in the regions [4].

2. Study Area

The study area is located in Tambak Lorok which located in Tanjung Mas Subdistrict, North Semarang and has long been known as a fisherman village. Semarang coastal cultivation is very productive because of abundant sea resources. This situation attracted fishermen from around Central Java to settle around Tanjung Mas. Since the 1980s, local community have turned open spaces in the form of fish pond and swamps into settlements. City stakeholders also built infrastructure such as transportation (ports, stations and airports), industrial, and commercial zones in coastal areas to support the growth of Semarang. The existence of fishing villages for decades shows culture and behavior that are familiar with sea phenomena such as high tide and coastal flooding. The existence of fishing settlements is also closely related to fishing locations, distribution and coastal areas, where these areas must be easily accessed by the public with a good transportation system and road network, enriched with various economic, social and cultural activities without having to damage environmental conditions.

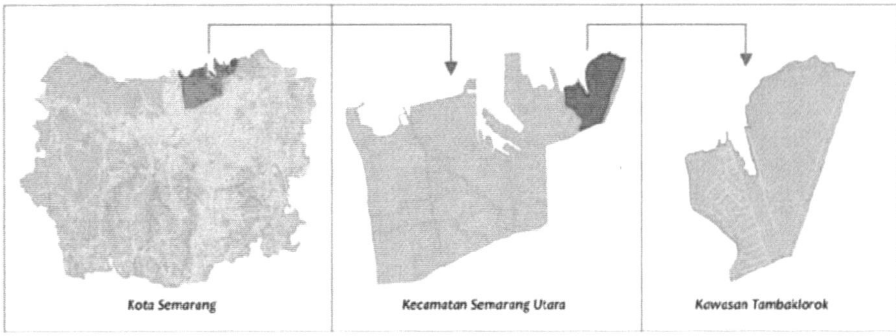

Figure 1. Study area in Tambak Lorok, North Semarang District (Courtesy of RTPLP)

3. Methodology

The methodology used in this paper is qualitative using triangulation method. Triangulation refers to the use of multiple method or data sources in qualitative research to develop a comprehensive understanding of phenomena [5]. Data sources used in this study are collected from literature studies, government policies through planning document and regulations, and existing conditions from the study area. Literature study focuses on the land subsidence, SLR, and adaptive measures in respond to disaster. Literature study is carried out by describing the cause and effects of land subsidence and SLR caused by climate change, both large scale and site spesific. The resulting findings are then further elaborated into the criteria for designing adaptive urban areas by evaluating government policy and existing conditions.

4. Land Subsidence

Land subsidence is a gradual and/or sudden change in land shape to a lower level due to the movement of soil constituent material. Some causes of land subsidence include consolidation of soil layers, underground mining, groundwater extraction and building loads [6]. Land subsidence due to excessive use of ground water is also experienced by several major cities in Asia such as Tokyo, Osaka, Shanghai, Bangkok, and etc [7]. According to that paper, land subsidence occurred in Bangkok was due to excessive use of groundwater and the structure of Bangkok's soil layer which contains a thick layer of soft clay. The impact of this land subsidence includes increased risk of flooding, settlement problems on soil and building structures, coastline regression, and sea water intrusion into the aquifer layer. This problem is very important because the city is located in a low-lying topography. At present, some of the eastern region of the city has sunk more than 1.0 m below the mean sea level. During rainfall, water needs to be drained by pumping through drainage channels. Recent mitigation measures adopted include groundwater management systems, expansion of tap water supply, and strict groundwater law enforcement, which has resulted in a sharp decline in groundwater use. However, land subsidence will continue for a long time because the soil layer consolidation process still continues.

Land subsidence in North Semarang is believed to be caused by a combination of natural consolidation of young alluvium soil, extraction of groundwater and excessive building and structural loads [8]. In coastal areas, around Tanjung Mas Harbor, the coastline development is quite fast with a rate of around 4 to 16 m / year in the period 1695 to 1991. Since 1695 until now, the development of the coastline has reached around 2-3 km in certain locations [9]. According to Bemmelen [10], muddy sedimentation in the coastal areas of Semarang occurred at least 500 years ago. Therefore, it can be estimated that the consolidation of natural coastal young alluvium soils will have a significant contribution to the relatively large decline in the coastal areas of Semarang. In addition to the relatively young natural alluvium consolidation, the land subsidence in Semarang is also partly due to the extraction of excessive groundwater. Groundwater extraction in the city of Semarang has risen sharply since the early 1990s, especially around industrial estates. According to Marsudi [11], the number of wells registered in 1900 was 16, increasing to 600 wells in 1990 and 1050 wells in 2000.

Due to the excessive extraction of groundwater the groundwater level in Semarang during the 1980s and 1996 periods experienced a significant up to 1.2 to 2.2 m / year. Due to the excessive extraction of groundwater the groundwater level in Semarang during the 1980s and 1996 periods experienced a significant up to 1.2 to 2.2 m / year. Suhelmi [12] concludes that land subsidence played an important role in increasing the inundated area in Semarang from initially 2,162 ha to 3,896 ha in just 10 years.

5. Sea Level Rise

Rising sea level is one of the consequences of climate change, both for the community, the built area, and the environment. Sea level also changes when there is a mass exchange between reservoir, ice or terrestrial and oceanic atmosphere. In the ice age, water is stored in the form of a large ice sheet in the highlands. Variations of the surface of the earth's crust by water and ice change the shape of the earth as a result of the elastic response of the earth's lithosphere and mantle, thereby changing the level of soil and relative sea level. This change in mass distribution changes the gravitational field of the earth, thereby changing the surface of the sea. Relative sea level can also be affected by local tectonic activity and also by the release of soil when groundwater is extracted or sedimentation increases. Sea water density is strongly influenced by temperature. As a result, sea level will change if the ocean temperature changes (due to thermal expansion) without changes in mass.

Some of the initial impacts of SLR that can be identified are erosion, increased inundation due to coastal flooding and saltwater intrusion. Bird [13] illustrates that changes that occur on the coastline are not necessarily the result of rising sea levels. Erosion can only occur in stable water conditions due to soil structure that has not been consolidated and the deficit of shoreline-forming sediments. However, rising sea levels can accelerate and worsen erosion of the coast in areas that experience SLR. Nicholls [14] states that the direct effects of SLR are immersion and increased frequency of flooding in coastal areas, as well as saltwater intrusion. Long-term effects also occur because the coastline adjusts to new conditions, including increased erosion and saltwater intrusion into groundwater. Sea level rises relatively faster than the trend caused by climate change in regions that experience land subsidence. In many areas, human activity exacerbates land subsidence in vulnerable coastal areas, including most river deltas. Therefore, greater attention must also focus on non-climatic components of SLR, especially for beaches that are more vulnerable to land subsidence. The impact of human action tends to be greater where there is a high concentration of population and economic activity, and therefore, has higher potential impact.

Wdowinski [15] in his research on SLR in Miami Beach, Florida, expresses the same thing. The rate of floods increased dramatically due to SLR after 2006, where rain-induced flood increased by 33% and tidal-incuded flood increased by 400%. The increase in flood due to SLR was affected by the gravity-dependent drainage system, until finally the Miami Government replaced the drainage system in 2014. The study also found the rate of SLR in Miami before 2006 was similar to the global rate of SLR, but the rate of SLR increased dramatically beyond the global rate after 2006. From various studies conducted in assessing the impact of SLR, all of them stated an increase in the intensity of floods, shoreline changes and sea water intrusion. Adaptive measures to reduce impacts of SLR should rely on local conditions, because not all locations experience the same rate of SLR and land subsidence.

6. Adaptive Measures

Adaptation and mitigation are two complementary strategies in responding to climate change and its impacts. Mitigation is the process of reducing emissions or limiting the release of greenhouse gases (GHG) into the atmosphere, so it can limit climate change in the future. Adaptation is the adjustment of the natural or community system to a changing or new environment. Adaptation to climate change refers to the adjustment of natural or human systems in response to climate stimuli, actual or expected stimuli, which are moderately harmful or exploit opportunities that might be beneficial (Figure 2). Various types of adaptation can be distinguished, including anticipatory and reactive adaptations, personal and community adaptation; and autonomous and planned adaptation [16]. Both adaptation and mitigation can reduce and manage the risk of climate change impacts on the built environment. However, behind all the benefits resulting from these two actions, adaptation and mitigation can also pose risks. Strategic responses to climate change require consideration of climate-related risks related to the risks and benefits provided by adaptation and mitigation actions [17].

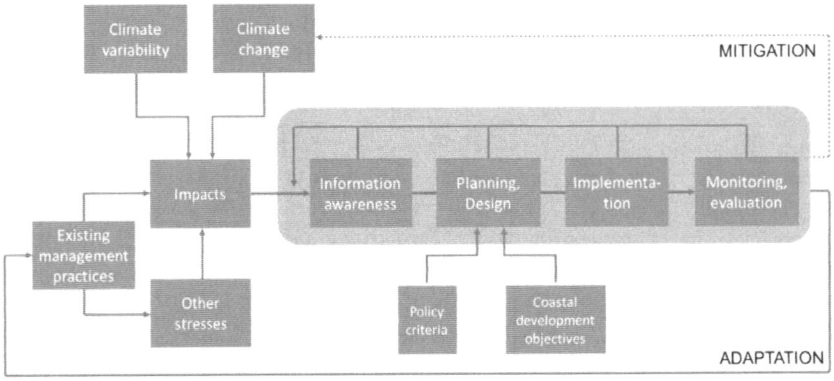

Figure 2. Adaptation and Mitigation Process (Adapted from Klein et al [18])

The basic principles of adaptation to the impact of land subsidence and SLR, such as permanent inundation, are adjustments to physical and non-physical aspects. Physical aspect focuses more on urban geometry, while non-physical focuses on activity, community awareness and education [19]. This study focuses on physical approaches on urban spaces and activities within them. The use of elevated buildings [20][21], green infrastructure - in the form of protection of natural elements, green roof, rain garden, naturalization of rivers, floodplains and blue infrastructure - in the form of rainwater harvesting and ponds retention are phisical adaptation that can reduce the impact of flood and climate change disasters [19][21][23]. Some strategies that can be carried out in the land use are mixed land uses where the function of the building on the ground floor has a lower level of disaster vulnerability than the above. An example is putting residential above commercial functions such as shops, retail and café/restaurants. The ground floor can be used as a commercial area, but the function can change into boat storing when the sea level rises and there is a permanent inundation [22].

Planning for building mass and intensity in coastal areas has a different approach compared to other regions. The narrow shape of the parcel and perpendicular circulation path to the coastline can reduce the damage to the structure by giving way to incoming water. Building relocation located in the coastline, or commonly called coastal borders, is needed to preserve coastal ecosystem and provide a safe space from the threat of tidal waves which can also cause erosion. This coastal border can serve as a buffer area of the waves which can be reduced further with coastal vegetation, such as mangrove forests [23]. In the context of the Water Sensitive Urban Design (WSUD), open space in an adaptive area is designed to be able to accommodate multiple functions. One of the main functions that needs to be accommodated in the open space includes water catchment areas, evacuation area, and as a temporary water retention area [24]. Vegetation in the open spaces has a major role to increase the absorption capacity of the soil by using native plants, because native plants have higher compatibility with the local soil, and deep-rooted vegetations [23]. The main consideration in designing circulation and parking in flood-prone areas is safety and security. Several design principles of circulation and parking include parallel road orientation to the topographic form or water body, circulation to the evacuation point must be easily defined, the addition of alternative pathways at a certain height as an evacuation attempt, [21][23][25] and public parking areas are located in area that have a high level of flood vulnerability [25].

Table 1. Summary of the urban design principles taken from literature.

Element	Principle	Source
Land Use	Mixed land use, with lower level function having low vulnerability to disaster	RIBA, 2009
Intensity and Building Mass	Narrow parcel and perpendicular to shoreline	Watson, 2011
	Use of elevated building mass	Liao, 2016
Open Space	Multi-function open space Use of native vegetation Use of deep-rooted vegetation	Morgan, 2013 Watson, 2011
Circulation and parking	Parallel to topography and shoreline	Mendez, 2014
	Provide clear and easily defined evacuation route on different heights	Watson, 2011 Communities and Local Government, 2009
	Communal parking space located in more vulnerable zone	Communities and Local Government, 2009

7. Design Criteria

Evaluation of the existing conditions using the principles formulated through literature review is done in order to formulate the design criteria for Tambak Lorok, Semarang. During the analysis of secondary data using aerial photograph, a change on the Eastern side of the study area, that was previously fishpond, was found. High tide and SLR caused fishpond to be inundated permanently (Figure 3). In Sayung Subdistrict located adjacent in the East of Semarang, Utami [26] also states that SLR caused changes to the coastline and increased permanent inundation in coastal areas.

Figure 3. Coastal erosion on study area (Courtesy of Google Earth)

Evaluation of the existing condition is carried out by observing the study area using urban design elements by Shirvani [27]. The elements used in this observation are elements related to the principles formulated in the previous chapter such as land use, building mass and intensity, open space, and circulation and parking. As one of many settlement that support economic activities in North Semarang Subdistrict (port and industrial area), land use in Tambak Lorok is dominated by residential. According to the projection of sea level rise carried out by DKP [28], some areas in the Tambak Lorok will experience permanent inundation in 2100. In areas that will experience permanent inundation, land use on the ground floor must be in accordance with the flood zone classification [22]. Residential will be placed on the second floor, so this area will have mixed land use. Restrictions on land use are to ensure that activities in the flood prone zone are not disturbed by inundation due to SLR.

Tambak Lorok is a densely populated settlement with a total population of 8,319 people divided into 5 RWs. Form of settlements that exist in this area are mostly attached houses that stand 1-2 floors high. As an adaptation effort to land subsidence, the inhabitants always increase the height of their houses by 1-1.5 meters every 5-10 years. Green open space is nowhere to be found because of high Building Coverage Ratio (BCR). In addition to the high BCR, there are still many buildings located directly adjacent to the sea (Figure 4). The absence of buffer area causes residents who live in these houses to be displaced during high tide and sometimes it damages their houses. Buildings in the coastal area should be located behind the coastal border which is 100 meters from the highest tide [29]. The building mass in this area must also be oriented towards the shoreline with narrow shape, clear distance between the buildings to ensure optimal sunlight and air circulation, and act as a water passage during high tide; and have open spaces. To get optimum open space, BCR can be further reduced to minimum 80% by increasing the Floor Area Ratio (FAR). Residential buildings located on the ground floor must also be elevated according to the projected inundation and the rate of land subsidence.

Figure 4. Existing houses located on coastline

Based on RTRW Kota Semarang 2011-2031, Tambak Lorok must have at least 20% of green open space. Tambak Lorok, which is part of North Semarang Subdistrict, is also included in the tidal flood and abrasion prone area [30]. The city development document suggests creating blue and green infrastructure in the form of retention ponds and mangrove forest. The provision of retention pond must be at a lower elevation than other regions to utilize the natural properties of water. While open space with evacuation function must be at a higher elevation and easily accessible, for example in the middle of the study area which stands 6 m above sea level. According to Martuti's research [31] on mangrove diversit, in Tugurejo Village, there are several mangrove varieties both planted by the community and those that grow naturally. The most common types of mangroves are *Rhizopora mucronata* and *Avicennia marina*, both types are believed to be suitable and can be planted in the study area which has the same soil structure as Tugurejo Village. On the circulation and parking element, most road network in the region are already perpendicular to the coastline. However, it still needs additional evacuation route that is accessible and safe from potential inundation.

Table 2. Summary of the design criteria in Tambak Lorok, Semarang.

Element	Principle	Design Criteria
Land Use	Mixed land use, with lower level function having low vulnerability to disaster	- Land/building function according to the vulnerability zone classification - No residential function on the ground floor in high-vulnerable zone
Intensity and Building Mass	Narrow parcel and perpendicular to shoreline	- No permanent structures on coastal border zone - Maximum BCR of 80%

Element	Principle	Design Criteria
	Use of elevated building mass	Residential building located on high-vulnerable zone must be elevated according to the potential inundation height
Open Space	Multi-function open space	- Provision of minimum 20% green open space - Floodable open space on lower elevation - Evacuation area located in higher elevation
	Use of native vegetation Use of deep-rooted vegetation	Provision of green belt in the form of mangrove forest consists of *rhizopora mucronata* and *avicennia marina*
Circulation and parking	Parallel to topography and shoreline	Road and streets are mostly perpendicular to the shoreline
	Provide clear and easily defined evacuation route on different heights	Provision of evacuation route to the higher grounds
	Communal parking space located in more vulnerable zone	

8. Conclusion

Land subsidence and SLR are global phenomena that happen due to both natural and human activities. These phenomena will continue to pose challenges in many different coastal environments, one of them is in Semarang. Semarang has long been experiencing land subsidence, especially in North Semarang District, and is now threatened by upcoming SLR. In responding to these conditions, several design criteria formulated in the previous chapter can improve the adaptability of Tambak Lorok settlement. Some design criteria that can be carried out are land use in accordance with the flood zone, moving dwellings to the upper floors, building narrow building plots, to multifunctional open spaces. Without intervention on the urban scale, Semarang is likely to experience economic losses and decrease in environmental and social qualities. Adaptation in urban planning/design can reduce risks and create resilient coastal area while on the same time making mitigation efforts to reduce the occurrence of disasters. This action is also in accordance with SDGs number 13, namely take action to combat climate change and its impacts, which also has been integrated into RPJMN 2015-2019.

Suggestion for further research on adaptive area in coastal area of Tambok Lorok, Semarang, is a research using non-physical adaptation approaches on the local community. Community participation aspects, that consist of public awareness of disaster and climate change, willingness and ability to adapt, can be explored.

References

[1] McGranahan, G.; Balk, D.; Anderson, B. 2007. The rising tide: Assessing the risks of climate change and human settlements in low elevation coastal zones. *Environ. Urban.* **19**, pp 17–37. doi: 10.1177/0956247807076960.

[2] Church, J.A.; Clark, P.U.; Cazenave, A.; Gregory, J.M.; Jevrejeva, S.; Levermann, A.;Merrifield,M.A.;Milne, G.A.; Nerem, R.S.; Nunn, P.D.; et al. Sea level Change. In Climate Change 2013: The Physical Science Basis. Contribution of Working Group I to the Fifth Assessment Report of the Intergovernmental Panel on Climate Change; Stocker, T.F., Qin, D., Plattner, G.-K., Tignor, M., Allen, S.K., Boschung, J., Nauels, A., Xia, Y., Bex, V., Midgley, P.M., Eds.; Cambridge University Press: Cambridge, UK; New York, NY, USA, 2013.

[3] UN-HABITAT. 2017. New Urban Agenda. Accessed 8 September 2018 http://habitat3.org/the-new-urban-agenda/.

[4] Republik Indonesia. 2015. Peraturan Presiden Republik Indonesia No. 2 Tahun 2015 Tentang RPJMN 2015-2019 : Buku 1. Jakarta.

[5] Patton, M.Q. 1999. Enhancing the quality and credibility of qualitative analysis. *Health Sciences Research*, **34**:1189-1208.

[6] Zeitoun, David G. and Eliyahu Wakshal. 2013. Land Subsidence Analysis in Urban Areas. The Bangkok Metropolitan Area Case Study. Springer: London.

[7] Phien-wej, N., Giao, P. H., & Nutalaya, P. (2006). Land subsidence in Bangkok, Thailand. *Engineering Geology*, **82**(4), pp 187–201. doi:10.1016/j.enggeo.2005.10.004.

[8] Andreas, H., Abidin, H. Z., Gumilar, I., Sidiq, T. P., & Yuwono, B. 2017. Adaptation and mitigation of land subsidence in Semarang. *AIP Conference Proceeding*, **1857** (060005). doi:10.1063/1.4987088.

[9] Abidin, H.Z, et al. 2013. Land subsidence in coastal city of Semarang (Indonesia): Characteristics, Impacts and Causes. *Geomatics, Natural Hazards and Risk*, **4**:3, pp 226-240.

[10] Bemmelen, van, R.W., 1949, The Geology of Indonesia, Martinus Nyhoff, The Haque, Netherland.

[11] Marsudi, 2001, Prediksi Laju Amblesan Tanah di dataran Aluvial Semarang, Propinsi Jawa Tengah. Disertasi Doktor, Institut Teknologi Bandung, 176 pp.

[12] Suhelmi, IR. 2012. Kajian Dampak Land Subsidence Terhadap Peningkatan Luas Genangan Rob di Kota Semarang. *Jurnal Ilmiah Geomatika*, **18**:1.

[13] Bird, Eric C.F. 1996. Coastal Erosion and Rising Sea-Level. *Sea-Level Rise and Coastal Subsidence* pp 87-103. Springer: London.

[14] Nicholls, R. J., and Cazenave, A. (2010). Sea-Level Rise and Its Impact on Coastal Zones. *Science*, **328**(5985), 1517–1520. doi:10.1126/science.1185782.

[15] Wdowinski, S., Bray, R., Kirtman, B. P., & Wu, Z. 2016. Increasing flooding hazard in coastal communities due to rising sea level: Case study of Miami Beach, Florida. *Ocean & Coastal Management*, **126**, 1–8. doi:10.1016/j.ocecoaman.2016.03.002

[16] IPCC, 2007: Climate Change 2007: Synthesis Report. Contribution of Working Groups I, II and III to the Fourth Assessment Report of the Intergovernmental Panel on Climate Change [Core Writing Team, Pachauri, R.K and Reisinger, A (eds.)]. IPCC, Geneva, Switzerland, 104 pp.

[17] IPCC, 2014: Climate Change 2014: Mitigation of Climate Change. Contribution of Working Group III to the Fifth Assessment Report of the Intergovernmental Panel on Climate Change [Edenhofer, O., R. Pichs-Madruga, Y. Sokona, E. Farahani, S. Kadner, K. Seyboth, A. Adler, I. Baum, S. Brunner, P. Eickemeier, B. Kriemann, J. Savolainen, S. Schlömer, C. von Stechow, T. Zwickel and J.C. Minx (eds.)]. Cambridge University Press, Cambridge, United Kingdom and New York, NY, USA.

[18] Klein, R.J.T, et al. 2000. Coastal adaptation. Chapter 15 in IPCC Special Report on Methodological and Technological Issues in Technology Transfer. B. Metz, O.R. Davidson, J.W. Martens, S.N.M. Van Rooijen, and L.L. Van Wie McGrory, eds, Cambridge University Press, Cambridge, UK.

[19] Jha, Abhas K, et al. 2011. Cities and Flooding. A Guide to Integrated Urban Flood Risk Management for the 21st Century. The World Bank : Washington DC.

[20] Liao, KH, Anh Tuan Leb and Kien Van Nguyenca. 2016. Urban Design Principles For Flood Resilience: Learning From The Ecological Wisdom of Living With Floods In The Vietnamese Mekong Delta (VMD). *Landscape and Urban Planning Journal* **155** pp 69-78.

[21] Mendez, Naia Landa, 2014. Adaption to Urban Floods by Planning and Design. KTH Royal Institute of Technology. Stockholm.

[22] Royal Institute British Architect. 2009. Climate Change Toolkit:. Design for Flooding. John Wiley & Sons, New Jersey

[23] Watson, Donald and Michele Adams. 2011. Designing for Flood Risk. CABE, London.

[24] Morgan, Celeste, C Bevington, David Levin, Peter Robinson, Paul Davis, Justin Abbot, and Paul Simkins. 2013. Water Sensitive Urban Design in the UK. Ciria. London, UK.
[25] Communities and Local Government. 2009. Planning Policy Statement 25: Development and Flood Risk Practice Guide. London.
[26] Utami, Widya Sari, Petrus Subardjo and Muhammad Helmi. 2017. Studi perubahan garis pantai akibat kenaikan permukaan air laut di Kecamatan Sayung, Kabupaten Demak. *Jurnal Oseanografi* **6**:1, pp 281-287.
[27] Shirvani, Hamid. 1985. Urban Design Process. Van Nostrand Reinhold.
[28] Dinas Kelautan dan Perikanan. 2008. Strategi Adaptasi dan Mitigasi Bencana Pesisir sebagai Akibat Perubahan Iklim terhadap Pesisir dan Pulau-Pulau Kecil. Jakarta.
[29] Republik Indonesia. 2007. UU No. 27 Tahun 2007 Tentang Pengelolaan Wilayah Pesisir dan Pulau-Pulau Kecil. Jakarta.
[30] Pemerintah Daerah Kota Semarang. 2011. Peraturan Daerah No. 14 Tahun 2014 Tentang RTRW Kota Semarang 2011-2031. Semarang.
[31] Martuti, NKT. 2013. Keanekaragaman mangrove di wilayah Tapak, Tugurejo, Semarang. Jurnal MIPA **36**:2, pp 123-130.

IOP Conf. Series: Earth and Environmental Science **273** (2019) 012006 doi:10.1088/1755-1315/273/1/012006

Detecting the Phenomena of Sea Surface Temperature Anomaly by Vector Autoregressive

Miftahuddin[1], Eva Maulia[1], Ichsan Setiawan[2], Asma Gul[3], Fadhli[4], Hidayati[5]

[1]Department of Statistics, Syiah Kuala University, Banda Aceh, Indonesia, 23111
[2]Department of Marine, Syiah Kuala University, Banda Aceh, Indonesia, 23111
[3]Department of Statistics, Shaheed Benazir Bhutto Women University, Peshawar, Pakistan.
[4]Department of Physics, Syiah Kuala University, Banda Aceh, Indonesia, 23111
[5]Aceh Bappeda, Banda Aceh, Indonesia, 23121

miftah@unsyiah.ac.id, evamaulia01@gmail.com, ichsansetiawan@unsyiah.ac.id, asmagul@sbbwu.edu.pk, fadhli@unsyiah.ac.id, hidayati77@ymail.com

Abstract. Climate change is one of the issues and a considerable threat to the environment in the future. These changes tend to fluctuate and vary significantly with time. Mostly, climate change issues are related to global temperature. The Earth's climate system is effected by many parameters. Sea Surface Temperature Anomaly (SSTA) is one of those parameters. The phenomena of SSTA is a necessary indicator for understanding of climate change variability. Aceh province as to the west of Indonesia has location face directly to the Indian Ocean, especially in western region, southwest, and southern. Vector Autoregressive (VAR) approach has shown that SSTA dataset has stationer and non-cointegration properties in the period 2006-2017. Based on this research, it can be concluded that the best model for SSTA with climate parameters (air temperature, rainfall, relative humidity, wind speed and short-wave radiation) is VAR with the 4th optimal lag or VAR(4). The Impulse Response Function (IRF) analysis based on VAR(4) model, which is formed to look at the phenomena of SSTA to climate parameters, shows that wind speed, rainfall and short-wave radiation have a similar pattern of detection on the equilibrium line due to shock from SSTA. It takes around 5 days for the three variables to reach the equilibrium line. Whereas the air temperature and relative humidity variables have no significant effects of shocks that occur in the sea surface temperature anomaly.

1. Introduction

Climate change is one of the big issues of the day and the future. Changes tend to fluctuate and significantly dependent on time-space. Mostly climate change issues are related to global temperatures in climate system. The Earth's climate system is effected by many parameters and sea surface temperature (SST) is one of those parameters where in fact that ocean is larger than continent.

Weather and climate on earth are determined by the amount and distribution of radiation received from the sun [1]. Climate change is the prevailing weather conditions in the region in general (in large spaces) and in the long run (long periods of time). Climate change can be distinguished by space (region), i.e. local and global climate change. Based on time, cyclical changes in climate can be observed such as daily, monthly, seasonal, and yearly. Such changes can be represented as canonical time series and canonical patterns, so there is correlation (canonical) correlation. The feature trend is essentially effected by global climate change such as by increasing atmospheric gas concentration. These climate conditions have high impact on the environment and human activities. This cause melting of polar ice and rising sea levels that leads to impacts on agriculture, economy [2, 3], increased risk of forest fires, and other life spheres such as changes in the surrounding environment, including increasing total temperatures and changing weather conditions in the local environment [4]. Key weather and climate features include air temperature, humidity, rainfall, air pressure, wind, and sunlight duration. Factors affecting climate features that distinguish the climate in a place with climate elsewhere are called climate control. The sun is a climate control that can lead to air movement and ocean currents [5]. Sea Surface Temperature (SST) depends on the amount of heat received from the sun. The anomaly influence of the SST is seen in two forms, e.g. an increase or decrease rainfall.

When the SST anomaly is high, a lot of water vapour is released which trigger the formation of rain-producing cloud convection. SSTA conditions in the western Indian Ocean region also affect the climate in the west of Sumatra [6]. Several studies have shown that convection activity is regulated by SST [7, 8, 9]. These studies show the importance of the role of SST in the dynamics of weather and climate of a region. Indonesia has three types of climate patterns, i.e. the monsoonal climate, equatorial, and local system climate [10]. Global warming has related to El-Nino Southern Oscillation (ENSO) phenomena via SSTA investigated [11, 12]. Model prediction of the ENSO by using VAR has been investigated [13]. Whereas in this research aims to look at the phenomena of Sea Surface Temperature Anomaly (SSTA) with respect to several climate parameters, namely air temperature, rainfall, relative humidity, wind speed and short-wave radiation through VAR approach. The specific purpose of this research is to study the phenomena of Sea Surface Temperature Anomaly (SSTA) to air temperature (AIRT), precipitation/rainfall (PREC RAIN), relative humidity (RH), wind speed (WSPD) and short-wave radiation (SWRAD) in the Indian Ocean by using vector autoregressive.

2. Material and Methods

The type of data used in this study is secondary data, obtained from the National Oceanic and Atmospheric Administration (NOAA) website in the period 2006-2017. The data is multivariate time series with six variables, namely SSTA, AIRT, PREC RAIN, RH, WSPD and SWRAD. The research method used is Vector Autoregressive (VAR) analysis. VAR model has a simple model structure with a small number of variables, where all variables are regarded as endogenous variables with the independent variable is lag [14]. The VAR model is designed for stationary variables that do not contain trend [15]. The general model of VAR is as follows.

$$Y_t = \alpha_0 + A_1 Y_{t-1} + A_2 Y_{t-2} + \dots + A_p Y_{t-p} + \varepsilon_t \quad (1)$$

$$Y_t = \alpha_0 + \sum_{n=1}^{p} A_n Y_{t-n} + \varepsilon_t \quad (2)$$

The outline procedure of statistical modelling as follows:

1. Input data into the R software.
2. Stationary test of the dataset for all variables: If the data on these variables are stationary, then carry the analysis process using VAR. In addition, if the data is not stationary, then the next phase of analysis is to use VECM (Vector Error Correction Model) [16].
3. Determine the length of the lag to be used on the VAR model, using the measures, AIC, SIC, FPE, and HQ values [17]. The optimal length of the lag is selected based on smallest value in these measures.
4. Formulate the VAR model.
5. IRF analysis based on the model obtained to determine the effect of each variable if given shock.

3. Results and Discussion

A picture of the SSTA dataset is displaying by using statistics descriptive analysis Figure 1 shows trend of all the variables in the period 2006-2017.

The Figure 1 it can be seen from the figure that the pattern of SSTA, AIRT, PREC RAIN, RH, and WSPD have similarities except for SWRAD pattern more fluctuating.

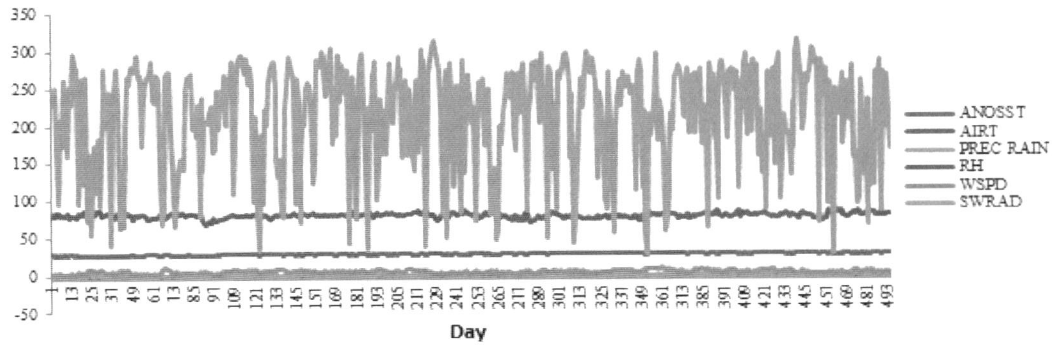

Figure 1. Time series plot of ANOSST (or SSTA), AIRT, PREC RAIN, RH, WSPD and SWRAD for 2006-2017

3.1 Stationarity test

One of the assumptions of VAR analysis is that all the variables must be stationary. There are several ways to measure the data stationary. We have implemented Augmented Dickey Fuller test (ADF) [18] for checking the data stationary. The ADF test statistics is given as:

$$\Delta Y_t = \alpha_0 + \gamma Y_{t-1} + \beta_i \sum_{i=1}^{P} \Delta Y_{t-i+1} + \varepsilon_t \qquad (3)$$

The results for all the variables are reported in table 1.

Table 1. ADF test results

No	VARIABLE	STAT.ADF	p-value
1	SSTA	-3.843	0.02
2	AIRT	-3.739	0.02
3	PREC.RAIN	-6.850	0.01
4	RH	-5.634	0.01
5	WSPD	-4.940	0.01
6	SWRAD	-8.116	0.01

*If p-value < α (0.05) then H_0 is rejected.

Based on table 1, the results show that all the variables are stationary. Therefore, the analysis can be carried out using VAR analysis.

3.2 Determination of Optimal Lag

An optimal lag determination is important in the analysis using the VAR method. A very long or too short lag results can be make the wrong model specification. An optimal lag is selected using Final Prediction Error Correction (FPE), Akaike Information Criterion (AIC), Schwarz Information Criterion (SIC) and Hannan-Quinn Information Criterion (HQ). The values of AIC, SIC and HQ can be found using the following formula.

$$\text{AIC} = -2\left(\frac{1}{T}\right) + 2(n + T) \qquad (4)$$

$$\text{SIC} = -2\left(\frac{1}{T}\right) + n\frac{\log(T)}{T} \qquad (5)$$

$$\text{HQ} = -2\left(\frac{1}{T}\right) + 2n\log\left(\frac{\log(T)}{T}\right) \qquad (6)$$

In VAR method, structural analysis uses Impulse Response Function (IRF). IRF analysis investigate the effect of each variable for agiven shock (suddenly effect). The impact of one variable's shocks on other variables can be traced through IRF. IRF essentially traces the effect of standard deviation shocks on changes in the variable value of the present and future periods [19]. The optimal lag length using the four criteria is displayed in table 2.

Table 2. Optimal lag length determination results

Lag	AIC	HQ	SC	FPE
1	-13.79651	**-13.65391***	**-13.43360***	1.019193e-06
2	-13.81286	-13.54803	-13.13889	1.002728e-06
3	-13.82229	-13.43523	-12.83726	9.934814e-07
4	**-13.85552***	-13.34623	-12.55942	**9.613101e-07***
5	-13.80210	-13.17058	-12.19493	1.014570e-06

*Minimum value.

Table 2 shows that the optimal lag for the VAR, based on the minimum value of AIC and FPE, is the 4th lag or VAR(4) while the optimal lag for the VAR estimate based on the minimum value of HQ and SC is the 1st lag or VAR(1). As there are two optimal lag values thus these are used on the VAR model.

3.3VAR model estimation

VAR model is designed for stationary variables, which is not contain trend [15]. The summary of VAR model for the two lags are reported in table 3.

Table 3. Summary of VAR model results

NO	VAR Model	1st Optimal Lag			4th Optimal lag		
		RSE*	Adjusted R-squared	p-value	RSE	Adjusted R-squared	p-value
1	SSTA	42.68	0.631	2.2e-16	41.74	0.648	2.2e-16
2	AIRT	0.018	0.591	2.1e-16	0.017	0.623	2.2e-16
3	PREC.RAIN	0.465	0.065	1.293e-06	0.458	0.083	1.308e-05
4	RH	0.030	0.455	2.2e-16	0.029	0.494	2.2e-16
5	WSPD	0.451	0.422	2.2e-16	0.435	0.461	2.2e-16
6	SWRAD	0.425	0.135	3.665e-14	0.416	0.176	7.064e-14

*RSE: Residual Standard Error

The results from table 3 reveal that the lag 4 is an optimal value for the VAR model as it gives maximum adjusted R-squared. In addition, it can also be seen that the RSE value of VAR model are smaller with the 4th lag as compared to the RSE with the 1st optimum lag. The comparative results with lag 4 and lag4shows that the best VAR model is produced with the 4th optimum lag. Therefore, IRF analysis are carried out using the VAR model with the 4th optimal lag or VAR(4).

3.4 Impuls Respon Function (IRF)

Impulse Response Function (IRF) analysis is used in the structural analysis in VAR. IRF is implemented to investigate the effect of each variable for a given shock. IRF can trace the impact of variable shock. A graphical structural analysis of IRF in VAR(4) is depicted in the following graph.

The figure 2 shows that a shock on the SSTA variable does not result a high shock to the variables ART and RH. It is obvious as the line is close to the equilibrium line and therefore it does not take long for the variable AIRT and RH to returns stable or on the equilibrium line. For WSPD, PREC RAIN and SWRAD variables, it can be seen that in case of shocks in the SSTA variable it takes about five (5) days for the three variables to reach the equilibrium line. The figure 3 shows that there is a high shock received by the SSTA variable due to the shock of the AIRT variable. The high shock occurs on the 3rd and 5th days, then again approaches the equilibrium line on the 7th day (after one week). For WSPD variables, PREC RAIN and SWRAD have similarity patterns in equilibrium. While in the RH variable shocks are not detected.

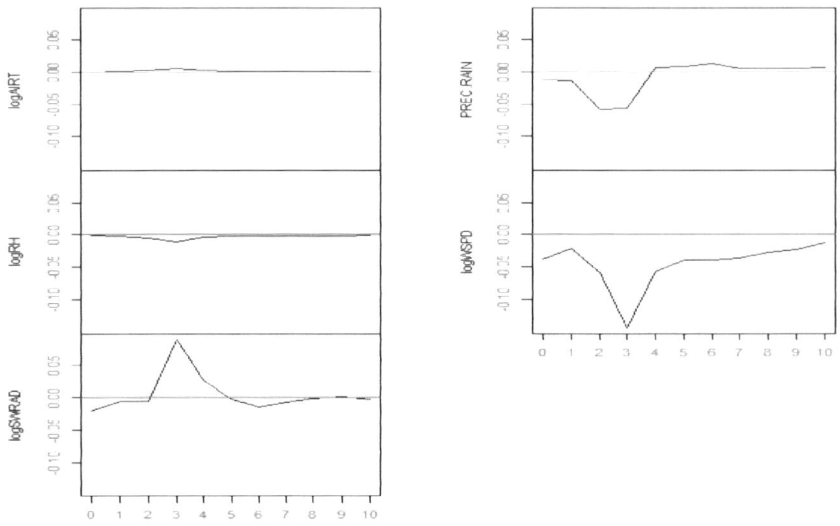

Figure 2. IRF of Sea Surface Temperature Anomaly (SSTA) in VAR(4)

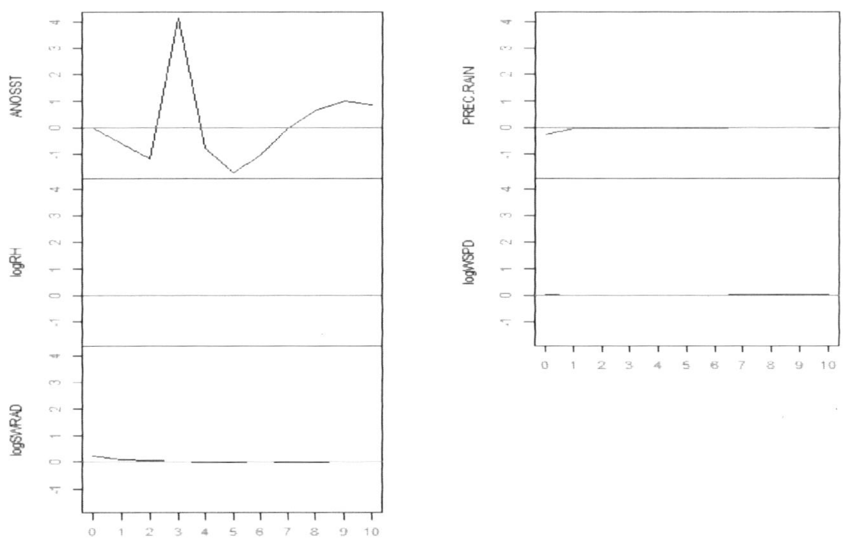

Figure 3. IRF of Air Temperature (AIRT) in VAR(4)

The figure 4 displays that shocks on the PREC RAIN variable provide a high shock response received by the SSTA variable. The high shock occurs on the first day and takes more than 10 days to re-approach the equilibrium line. For WSPD variables, PREC RAIN and SWRAD almost similar patterns are detected in equilibrium. While the RH variable does not respond shock events, when the shock occurs on the PREC RAIN variable.

The figure 5 shows that there is a high shock received by the SSTA variable due to the shock of the RH variable. The high shock occurs on the first day, and then returns to the equilibrium line on the 3rd and 6th days, but again high shock occurs in the following days. For WSPD variables, PREC RAIN and SWRAD, the received shock is not so great. While the AIRT variable does not experience the impact of shock caused by RH variable. In the figure 6 shows that there is a high shock received by the SSTA variable due to the shock of the WSPD variable. The high shock occurs on the 3rd day and takes more than 10 days to re-approach the equilibrium line. For the AIRT variable, PREC RAIN and SWRAD have similarity to the detection pattern on the equilibrium, but have differences at the interval (time interval) in the day of the event. While in the RH variable the shock is undetected.

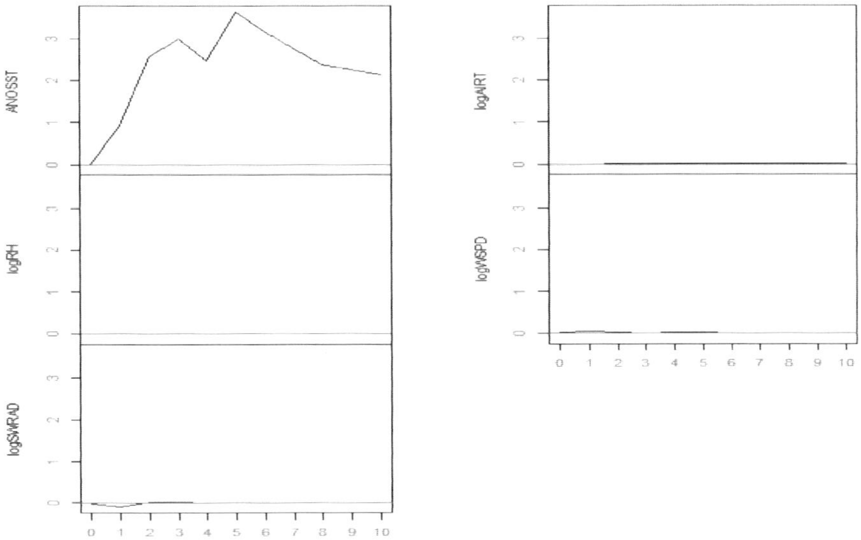

Figure 4. IRF of Precipitation/Rainfall (PREC RAIN) in VAR(4)

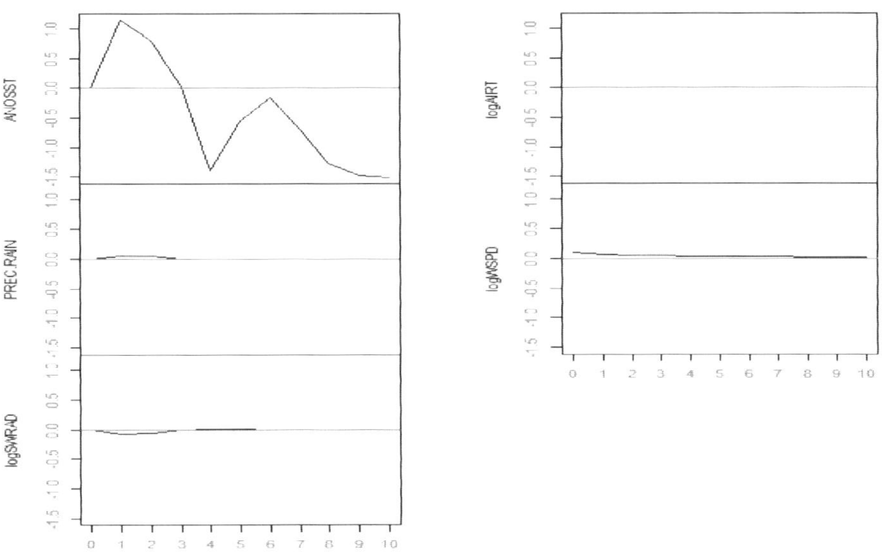

Figure 5. Graph IRF of Relative Humidity (RH) in VAR(4)

The figure 7 depicts that there is a high shock received by the SSTA variable due to the shock of the SWRAD variable. The high shock occurs on the 4th day and takes more than 10 days to re-approach the equilibrium line. For PREC RAIN and WSPD variables have similarity detection patterns on equilibrium. While no shock is detected for the variables AIRT and RH.

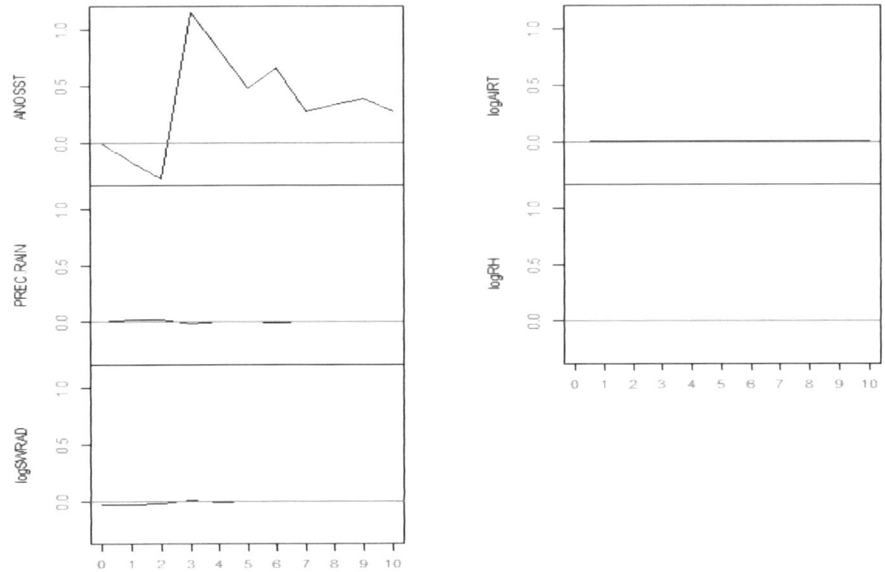

Figure 6. IRF of Wind Speed (WSPD) in VAR(4)

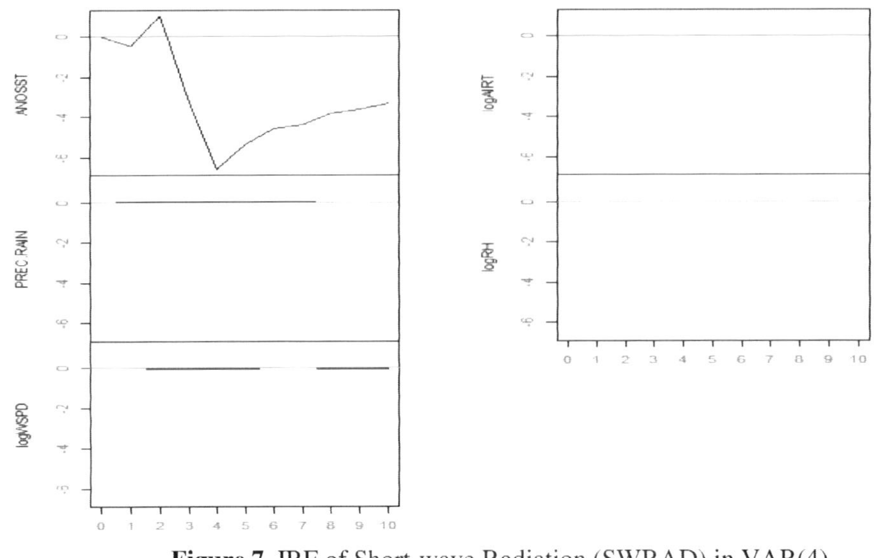

Figure 7. IRF of Short-wave Radiation (SWRAD) in VAR(4)

Overall in SSTA dataset is given different pattern shock with several variables explanatory. For example, the five days are critical point for WSPD, PREC RAIN and SWRAD variables by IRF analysis of Sea Surface Temperature Anomaly (SSTA) in VAR(4).

4. Conclusion

To investigate Sea Surface Temperature Anomaly phenomena using VAR method it can be concluded that the best model for the SSTA and climate parameters (Air Temperature, Precipitation/Rainfall, Relative Humidity, Wind Speed and Short-wave Radiation) is VAR with the 4th optimal lag or VAR(4). The Impuls Respon Fuction (IRF) analysis based on VAR(4) model shows that Wind Speed (WSPD), Precipitation/Rainfall (PREC RAIN) and Short-wave Radiation (SWRAD) have a similar pattern of detection on the equilibrium line due to shock from SST Anomaly (SSTA). Our study reveals that it takes around 5 days (short-time) for the three variables to reach the equilibrium line. However, the variables Air Temperature (AIRT) and Relative Humidity (RH) have no significant effect of shocks that occur in the SSTA. We recommend that the VAR approach is useful to give information related SSTA phenomena in time duration regarding the shock events with several climate parameters given.

5. References

[1] Trenberth, K.E., Fasullo, J.T & Kiehl, J. 2009. Earth's global energy budget. *Bull. Am. Meteorol. Soc.* 90(3): 311–323.
[2] Chen, C.C., McCarl, B.A & Adams R.M. 2000. *Economic Implication of Potential ENSO Frequency and Strength Shifts*. National Assessment of Climate Change, USA: Agriculture Focus Group.
[3] Chen, M., Parton, W. J., Del Grosso, S. J., Hartman, M. D., Day, K. A., Tucker, C. J., & Gao, W. 2017. The signature of sea surface temperature anomalies on the dynamics of semiarid grassland productivity. Ecosphere, 8(12).
[4] He, Z., Wu, R., Wang, W., Wen, Z., & Wang, D. 2017. Contributions of Surface Heat Fluxes and Oceanic Processes to Tropical SST Changes: Seasonal and Regional Dependence. Journal of Climate, 30(11), 4185-4205.
[5] Tjasyono, H.K & Bayong. 1999. *Klimatologi Umum*. Bandung: ITB. (Indonesia)
[6] Pramudia, A., Estiningtyas, W., Susanti, E., dan Suciantini. 2013. Fenomena dan Perubahan Iklim Indonesia serta Pemanfaatan Informasi Iklim untuk Kalender Tanam. Litbang Pertanian (Indonesia).
[7] McBride J.L, Haylock M.R, Nichols N. 2003. Relationships between the maritime continent heat source and the El Niño–Southern oscillation phenomenon. J Clim; 16(2): 905-2,914.
[8] Neale, R.B, and Slingo J.M. 2003. The Maritime Continent and its role in the global climate: A GCM study. J Clim; 16: 834-848.
[9] Miftahuddin. 2016. Fundamental fitting of the SST data using linear regression models. Journal of IEEE: 128-133.
[10] Aldrian, 2011. Aldrian, E., Karmini, M., and Budiman. 2011. Adaptasi dan Mitigasi Perubahan Iklim di Indonesia. BMKG, Jakarta. (Indonesia)
[11]Pan, Y. H., & Oort, A. H. 1983. Global climate variations connected with sea surface temperature anomalies in the eastern equatorial Pacific Ocean for the 1958–1973 period. Monthly Weather Review, 111(6), 1244-1258.
[12]Huang, P., Chen, D., & Ying, J. 2017. Weakening of the Tropical Atmospheric Circulation Response to Local Sea Surface Temperature Anomalies under Global Warming. Journal of Climate, 30(20), 8149-8158.
[13] Chapman D., Cane M. A, Henderson N., Lee D. E., and Chen C. 2015. A Vector Autoregressive ENSO Prediction Model. American Meteorological Society.
[14] Maddala, G. S. 2004. Introduction to Econometrics 3rd ed. Prentice-Hall, Inc. UK
[15] Gujarati, D. 2006. Basic Econometrics 4th ed, Mc. Graw Hill, New York.
[16] Sofyan, H., Maulia, E., & Miftahuddin. 2017. Structure analysis of tax revenue and inflation rate in Banda Aceh using vector error correction model with multiple alpha. In AIP Conference Proceedings (Vol. 1905, No. 1, p. 050029). AIP Publishing.
[17] Maulia, E., & Sofyan, H. 2018. Tax revenue and inflation rate predictions in Banda Aceh using Vector Error Correction Model (VECM). In IOP Conference Series: Materials Science and Engineering (Vol. 352, No. 1, p. 012056). IOP Publishing.
[18] Lutkepohl, H. 2006. New Introduction to Multiple Time Series Analysis, Springer-Verlag, Berlin.
[19] Ariefanto, M. D. 2012. Econometrics of Essence and Application by Using Eviews.

Acknowledges
The authors would like to thank the Department of statistics, Faculty of Mathematics and Sciences, LPPM Syiah Kuala University and Directorate of Research and Community Service (or DPRM) Ristekdikti Jakarta. We also would like to thank the TDMRC and anonymous reviewers for their valuable suggestions.

Coastal Flooding Impacts Induced Sea Level Rise on Banda Aceh Coastal Morphology

Musa Al'ala[1], and Syamsidik[1,2,*]

[1]Tsunami Computation and Visualization Laboratory, Tsunami and Disaster Mitigation
Research Center (TDMRC), Syiah Kuala University, Jl. Prof. Ibrahim Hasan, Gampong Pie, Banda Aceh, 23233, Indonesia
[2]Civil Engineering Department, Syiah Kuala University, Jl. Syeh Abdurrauf, No. 7, Banda Aceh, 23111, Indonesia.

*Corresponding Author : syamsidik@tdmrc.org

Abstract. The research focus is in assessing the impact of sea level rise induced by climate change towards coastal morphology. The process went through numerical modelling combining the flow and swan model generated by Delft3D to mimic morphology process within 100 years in Banda Aceh Coastal Defence. Banda Aceh coastal morphology keeps in recovery process as the coastal line keeps growing after the 2004 devastating morphology destruction until the late 2017. A sea level rise scenario was adopted from IPCC and TOPEX Poseidon with a 7 mm/year increase rate which was recorded between 1992 and 2015. Results recorded Banda Aceh had both sedimented and eroded area. Coastal protection gave the proper protection to Banda Aceh coastal defense while the three of Banda Aceh rivers (Krueng Neng river, Krueng Aceh river, and Floodway) gave sediment supply from the channel. Coastal line retreat between 15 to 100 m was also recorded on the side of Syiah Kuala- Alue Naga beach site, while others had enough sediment supply produced growing shoreline. The predicted regrow coastal area could be utilized as buffer zone as the government had difficulties in relocating the people from the coastal area.

Keywords: Sea level rise; Coastal defense; Morphology; Banda Aceh

1. Introduction

Banda Aceh is situated in a complex ocean system due to the conjunctures of three open seas, i.e., Indian Ocean, Andaman Sea, and Malacca Strait. As a coastal plain area, Banda Aceh is also facing another problem namely coastal flooding. Almost 14 years have passed after the 2004 Indian Ocean tsunami. Several parts of the coastal area have recovered from a severe erosion thanks to their natural ability. However, according to IPCC and recorded data of TOPEX Poseidon, there is a 7 mm/year of sea level rise (SLR) rate taking place between 1992 to 2015. This increase will enhance the vulnerability of the coastal area during future coastal evolution. The rise of climate change induced sea level rise will produced more frequent coastal floods [1]. Today's coastal development in many developing countries still does not properly include SLR impacts on their city development plan. As found in Jakarta, in 2050 several coastal areas will be flooded due to sea level rise and land subsidence [2]. Beside Jakarta, Banda Aceh also has been predicted that the impact of SLR will drawn 2.97% of the coastal area and disturb the city drainage systems in 2067 [3].

This research is aim at investigating the impacts of 100-year SLR on the coastal morphology of Banda Aceh. The impacted area was observed in terms of flooded area, coastal erosion, and sedimentation. We used Delft3D model that adopts Finite Difference Scheme Method with nonlinear shallow water equations. There are also the models which mimic the forecast of morphological change induced by sea level rise in western Mediterranean region, which also combine the mean condition along with extreme conditions [4]. The model domain was built based on topography data provided by BAPPEDA Banda Aceh in cooperation with JICA [5] and field measurement for bathymetry. Data was updated using the scenario of 100-year impacts of SLR with 0.7 m mean sea level increased. We compared both existing morphological change results and impact of the 100-year SLR scenario to investigate the consequences on Banda Aceh coastal morphology. The results are expected to be used as one of the considerations for the government decision-making in future coastal development plan in Banda Aceh. This study is aimed at estimating impacts of the sea level rise induced coastal flooding on coastal (Jetty) structures and coastal area. Numerical simulations of Delft3D were applied to investigate the influence of the gradual sea level rise in 100 years. The result will be presented in a quantity of flooded area that specified into several coastal areas around Banda Aceh.

Figure 1. Coastal morphology condition after tsunami and it's recovery process (Google Earth images)

After the destruction of the 2004 Indian Ocean tsunami, the coastal area kept growing (figure 1). A wide area had been wiped away after tsunami and regrew a few meters back, this indicating that the sediment supplies were available around the area. Banda Aceh consists of three main river channels that end in the Banda Aceh Ocean. The three main rivers are Krueng Neng river, Krueng Aceh river, and Krueng Aceh Floodway and all of them provide supply towards Banda Aceh coastal defense. According to the growing process which has been occuring until 2017, the coastal areas were found to be healing with almost 70 m after a few years (google earth). In detail between 2005 to 2011 Ulee Lheue had 1.91 km^2 area formed naturally [6]. It If the growth of the coastal area was to happen in 100 more years, it would give a hope for coastal development, but if the erosion process was to dominate, then the problem wouldbe completely different.

IOP Conf. Series: Earth and Environmental Science **273** (2019) 012007 doi:10.1088/1755-1315/273/1/012007

2. Study Area

The study area for this research was Banda Aceh where many low-lying area present. The coastal area of Banda Aceh presents mostly a mild slope, and has only less than 2 m of elevation above the sea level. The conditions during high tide (highest water level) were relatively close to the coastal are that are inhabited. The highest tide that happened along with heavy rain could produce both coastal or river floods. Banda Aceh has three river streams which are known to produce sediment during heavy rain in upstream. The human activities around the coastal area were also found to be prone to floods as the distance was relatively close.

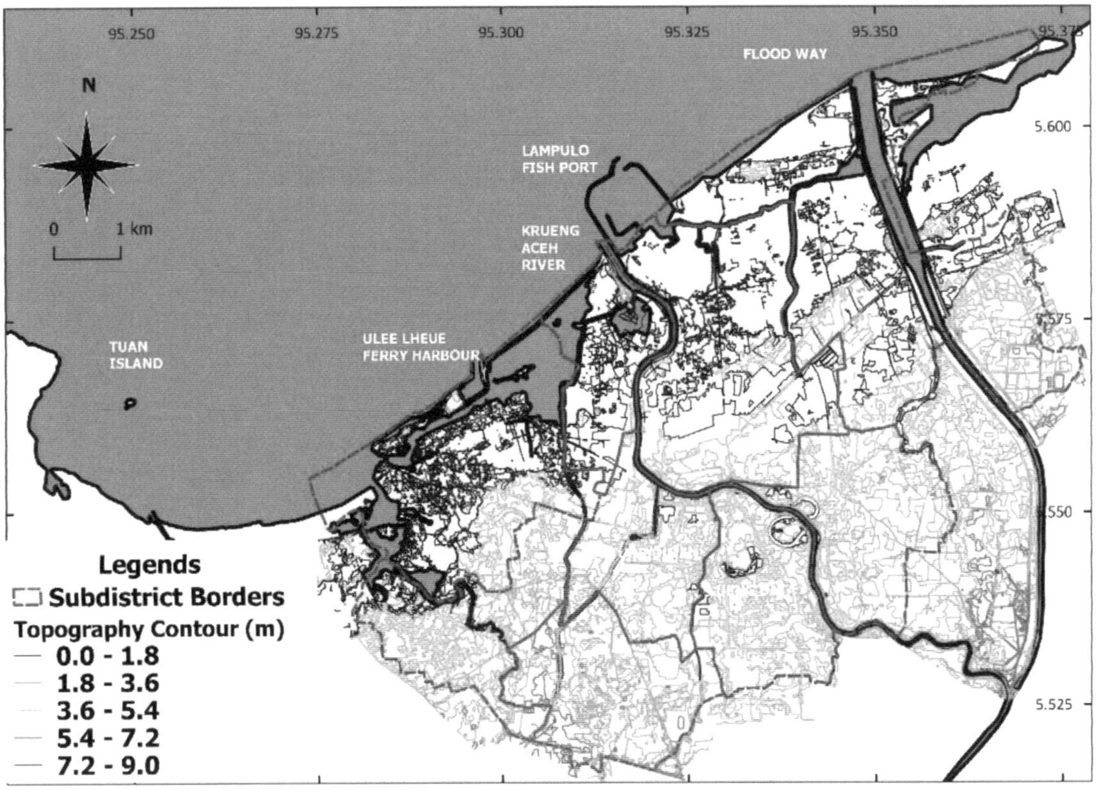

Figure 2. Banda Aceh region and topography condition

(a) (b)

Figure 3. Ulee Lheue coastal area with houses nearby (a) and fisherman port (b).

IOP Conf. Series: Earth and Environmental Science **273** (2019) 012007 doi:10.1088/1755-1315/273/1/012007

Houses and public facilities also keep growing in the coastal area as it provide a cheaper costs compared to inland strategic locations. The coastal habited areas, such as Ulee Lheue are very close to water basins such as the lagoon which is directly connected to the ocean. The facilities were build on a relatively mild slope and low lying area which produce more vulnerability towards hydrometeorological hazards which combined with climate change (such as sea level rise).

3. Methods

Several methodologies were applied in performing this research. Data collection started with field measurements and was supported with available charts (secondary data) for topography and bathymetry data. Numerical simulation is one of the methods that we used to mimic the coastal morphological process which combine flow (tides) and swan (wave) simulation in Delft3D. The simulation domain generated by topography and bathymetry data which represented in 11 x 3 km simulation domain with hydrodynamic components that consist of tide and wave were also generated along with the discharge from the river streams. There are several steps which were performed during the research.

1. Domain data
   Both bathymetry and topography data were collected and digitized from first source (measurement) and secondary data (available nautical chart). The domain was translated into 3 types of grid to speed up the simulation process. first grid contains of overall grid (40x40m) was used to generate the waves and forward it to a "detail grid" (20x20m) and a flow grid (10x10m) for tides and wave simulation for morphological processes.
2. SLR scenario
   The SLR scenario was gained from IPCC (International Panel for Climate Change) from satellite observations of TOPEX/POSEIDON with 4 mm precision between 1992 and 2015 [7]. The hydrodynamic process which also combine with morphology mimics the Banda Aceh morphological change in 100 years projection with 7mm/year sea level rise rate.
3. Hydrodynamic and morphology parameter
   Four harmonics components of tidal were used to simulate the tides. The wave generation were performed by wave generated data of Hs (wave height) and Ts (wave period) which was calculated from the wind data. In the process of producing the morphological process for a century (100 years) we used the morphological factor for time efficiency.

The result of coastal morphology change were analyzed further for coastal line change with spatial analysis and cross profile. Coastal morphological change map and cross profile of coastal change graphic were produced to inform the prediction of coastal condition within the next century.

4. Result and Discussion

During the simulation, sedimentations and erosions processes were projected to happen over the next 100 years. On the eastern part, the shoreline kept growing and created a new shoreline forward to the ocean based on the simulated projection. This happened due to the coastal protection (jetty) which trap the sediment from the eastern part. Mostly the sediments were deposited around the coastal protection which perpendicularly to the shoreline. Areas with this characteristics were found near Floodway estuarine, Lampulo fish port, and Krueng Aceh estuarine. In the long span coastal line such as Syiah Kuala – Alue Naga shoreline, the erosion was occurred instead of sedimentation. The erosion was measured ranging between 5 - 98 m.

Figure 4. Shoreline change prediction in 100 years after morphological process.

Three of the main river streams produce sediment supply through estuarine and disperse in the coastal area. Based on the simulation, the decreasing of th area that was eroded in the coastal area was mainly caused by the coastal protection which spans along the Banda Aceh coastal line. The depositions also appeared around the estuarine, producing shallow depth estuarine. The shallow estuarine made the fishermen boats unavailable to pass (figure 5. b).

Figure 5. Prediction of shoreline change in 2118 compared to 2018 in two sites in Banda Aceh (a. Syiah Kuala – Alue Naga, b. Alue Naga)

On the other site, the erosion occurred in the area of Syiah Kuala which located in Syiah Kuala–Alue Naga beach. The areas that were eroded nearly inhabited by people and also functioning as fish ponds. the erosion happened in the middle of two perpendicular coastal protection (Alue Naga jetty and Lampulo fish port breakwater). While the middle area was eroded, the areas on the eastern and western part were deposited with sediment that were transported by the erosion processes.

IOP Conf. Series: Earth and Environmental Science **273** (2019) 012007 doi:10.1088/1755-1315/273/1/012007

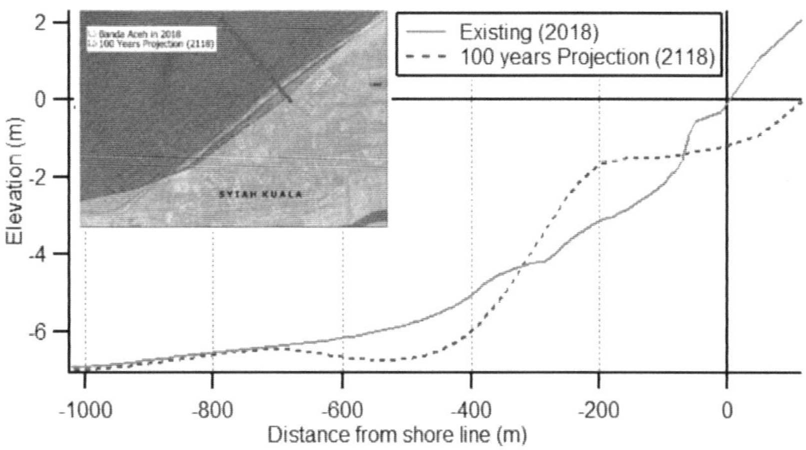

Figure 6. Cross profile Syiah Kuala – Alue Naga beach area.

The erosion went 98 m maximum inland while eroding more than 2 m thick of sandy layer leaving the former beach into water area (figure 6). The new shoreline had form relatively close to the houses even it still had the coastal protection. Erosion prone area were need attention from the government for future development plan since many property developers use the coastal area as a housing area for it's cheaper cost. A century long development in Banda Aceh can produce coastal based houses if there is no regulation regarding the condition between after the tsunami until 2018.

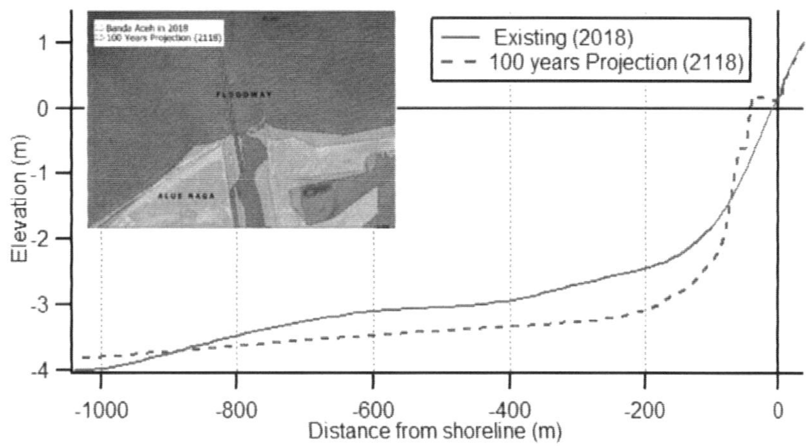

Figure 7. Cross profile Floodway estuarine.

The concerned also goes to the estuarine that functions as a fishermen transport channel, and at the same time plays a role as the floodway, delivering mass discharge when heavy rain happens. The sedimentation appeared to be more than 3 m thick, causing the estuarine to be shallow and defective for the fishermen transportation channel (figure 7). The sediment supply went from the river stream and was also transported from the mild slope bed. Both of the sediment supplies met in the estuarine and deposited there after having some turbulences process.

5. Conclusion

Sea level rise of 0.7 m, in theory, will caused the coastal line retreat automatically, However, in practice, the effect towards the hydrodynamic and morphology process gave more varied result. Most of Banda Aceh coastal area had been protected with coastal protection, i.e., rivertment, breakwater and jetty. More than 3 m thick of sedimentation were produced in the estuarine after 100 years leaving the site very shallow until become not working. The Jetty facility in Alue Naga Estuarine (Floodway) become less functional after affected by the sedimentation. Hence, shoreline retreat also happened in a long span coastal line and mild slope area leaving a 98 m of coastal retreat as the work that need to be done by the government regarding the coastal line area were inhabited by the society. The rise of sea level produced both shoreline retreat and regrow; the coastal development should consider the sea level rise factor. Both erosion and sedimentation could happen, but which one is dominant depends on the sediment supply and the coastal morphology. The gulf bay coastal form will trap more sediment and the coastal with estuarine also has the sediment supply from the river streams.

6. References

[1] Titus J G, Kuo C Y, Gibbs M J, LaRoche M B, Webb M K, and Waddell J O, 1984 *Greenhouse Effect, Sea Level Rise, and Coastal Drainage Systems*, (Journal of Water Resources Planning and Management) Vol. 113, No.2 pp.218-225.

[2] Takagi M, Esteban T, Mikami, and Fujii D 2016 *Projection of coastal floods in 2050 Jakarta*, (Urban Clim,)., vol. 17, pp. 135–145.

[3] Al'ala, M, Syamsidik, Kato, S 2017 "Predicting Impact of SLR on Coastal Flooding in Banda Aceh Coastal Defences" AIP Conference Proceeding **1892**, 100004 (2017); doi: 10.1063/1.5005736.

[4] Enriquez A R, Marcos M, Alvarez-Ellacuria, A, Orfilia A, and Gmis D 2017 Changes in beach shoreline due to sea level rise and waves under climate change scenario application to the Balearic Island (western Mediterranean) (Natural Hazards Earth System Sciences) vol.17, pp 1075-1089.

[5] JICA Report 2005, The study on the urgent rehabilitation and reconstruction support program for Aceh Province and affe cted areas in north Sumatra (urgent rehabilitation and reconstruction plan for Banda Aceh City) in the Republic of Indonesia : final report (1) ; Vol. 2. -Main report.

[6] Syamsidik, Iskandar A and Rasyif, T M 2015. *Progress of Coastal Line Rehabilitation After the Indian Ocean Tsunami Around Banda Aceh Coasts*. R. Shaw (ed.), Recovery from the Indian Ocean Tsunami: A Ten-Year Journey, Disaster Risk Reduction, DOI 10.1007/978-4-431-55117-1_13. pp.175-189.

[7] IPCC 2007, *Summary for Policymakers. Climate change 2007*. Physical Science Basis Contribution of Working Group 1 to Fourth Assessment Report of the Intergovernmental Panel on Climate Change. Cambridge University Press (18pp).

Acknowledgments

We would like to extend our gratitude for Coastal Research Group Syiah Kuala University for the helpful contribution on the research process. This research and publication is supported by PEER Cycle 5 Grant USAID and National Academy of Sciences funding this publication process through PEER Cycle 5, with research project title: "Incorporating Climate Change Induced Sea Level Rise Information into Coastal Cities' Preparedness toward Coastal Hazards" under Federal Award Identification Number (FAIN) : AID-OAA-A-A-11-00012 (Subaward Number 2000007546).

IOP Conf. Series: Earth and Environmental Science **273** (2019) 012008 doi:10.1088/1755-1315/273/1/012008

Reconstruction of the 2004 Tsunami Inundation Map in Banda Aceh Through Numerical Model and Its Validation with Post-Tsunami Survey Data

Tursina[1,3] and Syamsidik[1,2,*]

[1]Tsunami and Disaster Mitigation Research Center (TDMRC), Syiah Kuala University, Jl. Prof. Dr. Ibrahim Hasan, Gampong Pie, Banda Aceh, 23233-Indonesia,
[2]Civil Engineering Department, Syiah Kuala University, Jl. Syeh Abdurrauf, No. 7, Banda Aceh, 23111, Indonesia
[3]PhD Student of Engineering Doctoral Program, Syiah Kuala University, Jl. Syeh Abdurrauf, No. 7, Banda Aceh, 23111, Indonesia

*Corresponding Author: syamsidik@tdmrc.org

Abstract. The National Oceanic and Atmospheric Administration (NOAA) recorded the 2004 tsunami heights in the Indian Ocean using the DART buoy tool and survey field data in the land side of Banda Aceh. Furthermore, 85 tsunami poles around Banda Aceh and Aceh Besar were used to mark the tsunami flow depth. The flow depth of the tsunami data recorded plays an important role in fostering our understanding of the physical tsunami wave generation and inundation using tsunami modeling. This study is aimed at comparing the results of the tsunami flow depth modeling in Banda Aceh with some measurement data from NOAA and Tsunami Poles. The tsunami modeling process was applied using the Cornell Multi-Grid Coupled Tsunami Model (COMCOT) computational program.We used 6 nested grids and the innermost layer covering the Banda Aceh region. The Layer 6 used for the variable Manning roughness coefficient represents the surface land use before the tsunami. We put 139 and 67 observation points on land to observe tsunami flow depth data according to NOAA and Tsunami pole coordinates, respectively. The result shows that NOAA data gave poor agreement with numerical simulations. A better result was obtained when using tsunami poles. We recommend increasing the use of the tsunami poles data as they were produced with a rigorous validation process for the 2004 tsunami.

Keywords: Tsunami Numerical Simulation; Flow Depth; Wave Height; Validation

1. Introduction

A tsunami inundation map was prepared to assist cities and countries in identifying their vulnerabilities to tsunamis. Banda Aceh City was one of the worst impacted areas due to the 2004 Indian Ocean Tsunami. As the most populous city in Aceh Province, city planning based on disaster mitigation is important in this area. Numerical simulations are the most flexible method in generating inundations of the reconstruction process and in estimating future forecasts. This methodology would be valid if the numerical models could be validated using actual data. Numerical models may produce unreliable estimates if initial data used as an inputs in computational tools is unrealistic. This error in fact, would further influence policy makers to commit mistakes in providing decisions in disaster mitigation, and could undoubtedly lead to panic in the community.

The National Oceanic and Atmospheric Administration (NOAA) recorded the 2004 tsunami heights in the Indian Ocean using the DART (Deep-ocean Assessment and Reporting of Tsunami) buoy tool and field survey data measuring tsunami heights in Banda Aceh. Furthermore, 85 tsunami poles around Banda Aceh and Aceh Besar were used to mark the tsunami flow depth [1]. 67 of them were located in Banda Aceh [2]. The post-survey data of the tsunami wave heights plays an important role in fostering our understanding of the physical tsunami wave generation and propagation using tsunami modeling. Research conducted by [3] used data recorded in the DART buoys to verify the tsunami modeling data in the deep ocean, as well as to generate Estimated Times of Arrival (ETA) in several tsunami potential cities in the Aceh Province. Their tsunami modeling results show realistic findings with actual events [3]. When the tsunami is run-up on the mainland, instead the process becomes more complicated. The tsunami inundation phase is a highly nonlinear process [4]. Forecasting tsunami inundation requires high-resolution bathymetry and topographic data, and validation with observed data. In the post-tsunami field survey, the flow depth and wave height was the method applied to measure the tsunami height. The flow depth is the height of the water from the ground to the tsunami water level, and the wave height is the height of water from the mean sea level to the height of the tsunami surface. Tsunami water level on land is indicated by watermarks on the walls and debris on the trees [5].

All of the tsunami processes begins from tsunami generation, propagation and inundation, almost completely simulated by the COMCOT model [6]. However, to evaluate the accuracy of the simulation model, the data validation test is needed. The objective of this study is not to validate the COMCOT model performance, which has been systematically validated with analytical solution in many tsunami cases [7]. This study just aims at evaluating the tsunami height in Banda Aceh. The validation method was adopted from the Aida function [8]. The modeled tsunami height and the observed tsunami height were compared to assess the similarities and the differences between them. In 2014, [7] and [9] validated the modeled tsunami height of the COMCOT model on the 2011 Tohoku Tsunami, and also validated the flow depth of the 2009 Samoan tsunami. They used the Aida function in order to validate the simulation result.

2. Study Area

A few days after the 2004 tsunami, a number of researchers conducted a field survey in Banda Aceh. Tsunami memorial Pole's (tsunami pole) is a tsunami monument that records tsunami flow depth on land, which was measured from the ground based on eyewitnessing. Moreover, the NOAA research team also conducted a survey that observed the tsunami height with measurements of tsunami wave height. The locations of the survey were scattered over the inundation area until the boundaries of the inundation (figure 1). Based on the satellites from Google Earth, the distance of the tsunami inundation range was computed to be 3.40 km.

Figure 1. The distribution of post-tsunami survey location in Banda Aceh

IOP Conf. Series: Earth and Environmental Science **273** (2019) 012008 doi:10.1088/1755-1315/273/1/012008

3. Method

3.1 Numerical Simulation COMCOT

The tsunami modeling process was run using the Cornell Multi-Grid Coupled Tsunami Model (COMCOT) computational program. Six layers of grid resolution were used, and the innermost layer (Layer 6) covered the Banda Aceh region. The thrust fault event on the larger earthquake offshore Sumatra on 26 December 2004 was adopted from Koshimura [10]. Bathymetry data from Layer 1 to Layer 4 used GEBCO data with a 1-minute resolution for the outer layer. Details of topographic data at Layers 5 and Layer 6 were obtained from JICA's measurements. Layer 6 was the smallest grid with the grid cell size of 11.5 m x 11.5 m, and was run with a 1.0 second time step. The computational domain Layer 6 show in figure 2.

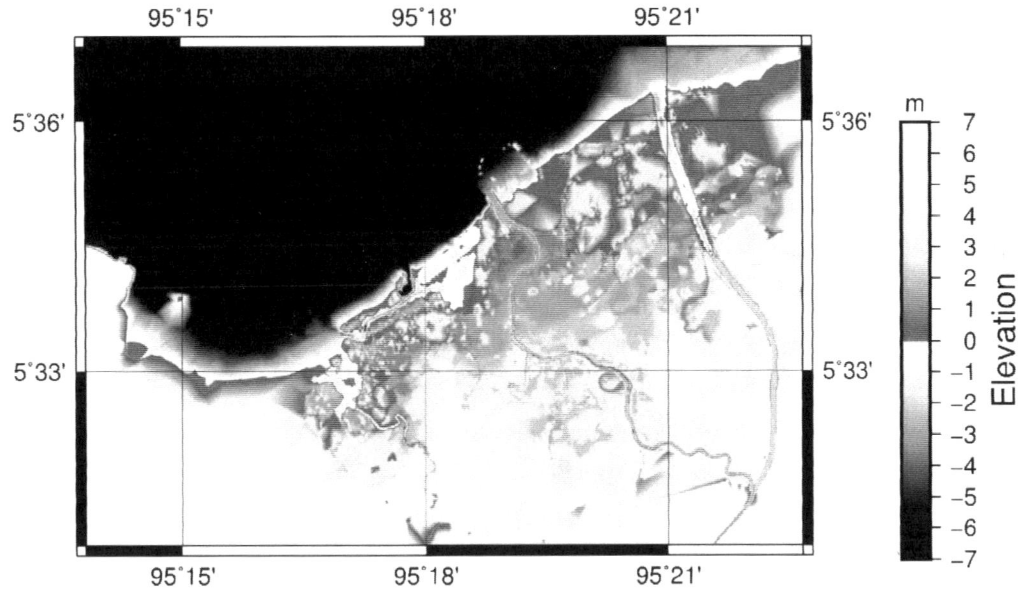

Figure 2. Banda Aceh in computational domain simulation (Layer 6). The highest land elevation is 7.0 m. Around the shore area, the topography elevation less than 3.0 m and cover half of the land side of Banda Aceh

The calculation of the tsunami waves from the open ocean was performed using the shallow water equation (SWE). The SWE's equation was applied from Layer 1 to Layer 5. Meanwhile, for the calculation of the inundation process at Layer 6, the nonlinear shallow water equation including the bottom friction effect was applied. The bottom friction of the Banda Aceh region represents the surface land use before the tsunami with Manning Roughness classification as mentioned in table 1. All of the level Layers were implemented in Spherical Coordinates.

Table 1. Manning Coefficients based on land cover in Banda Aceh before the Tsunami base on [11]

Land Use	Manning's Roughness Coefficient (n)
Coastal Vegetation	0.035
Fish Ponds	0.017
Building	0.04
Sea	0.013
Soil	0.02

3.2 Validation and Verification Flow Depth

The tsunami flow depth of the simulation results was compared with the results of the NOAA and Tsunami Pole surveys to evaluate the performance of the inundation models. There were 139 survey locations where to observe tsunami heights with NOAA data, and 67 points for Tsunami Pole data. Measurements from NOAA were based on wave height and tsunami poles were based on flow depth. The result of the COMCOT simulation is a tsunami flow depth. The illustration in figure 3 clearly delineates the definition of flow depth and wave height.

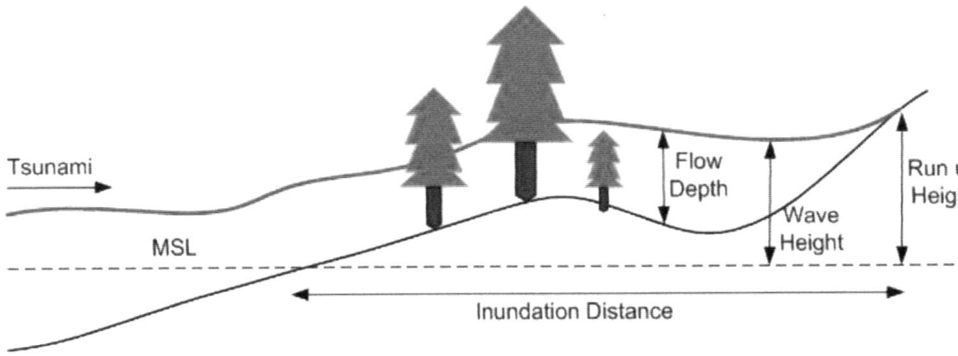

Figure 3. A diagram illustrating some terms used for tsunami measurement modified from [12]

The Aida function [8] was used as a formula to see the agreement between the simulated tsunami flow depth and the observed tsunami flow depth. The accuracy of the simulation results was calculated by using the geometric mean ratio *K* of the observed data *(H-obs*) and simulation data (*H-sim*), as well as the standard deviation of *k*.

At the survey location *i*, the ratio of the tsunami flow depth observed (*H-obs*) to the tsunami flow depth simulated is represented by *Ki*

$$K_i = \frac{H_{obs-i}}{H_{sim-i}}, \tag{1}$$

For *n* being a total number of observation points, geometric average value of *K* for all ratios is defined by the following equation.

$$\log K = \frac{1}{n}\sum_{i=1}^{n} \log K_i \tag{2}$$

If *K*>1, its means that the simulation data is underestimated from the observation. However, if *K*<1, the simulation is overestimated from the observation data. The simulation results match better with the observed data if *K*=1. Takeuchi [13] suggests a range of values that indicates the model is a 'good agreement' if $0.8 \leq K \leq 1.2$ and the value of $k \leq 1.60$. The equation of the standart devation of *k* as follows,

$$\log \kappa = \sqrt{\frac{1}{n}\sum_{i=1}^{n} (\log K_i)^2 - (\log K)^2} \tag{3}$$

4. Results and Discussion

The comparison of simulation models with measurement data is the focus of this study. The 2004 tsunami flow depth simulation with points of survey location can be seen in figure 4. The inundation map results show a pattern that is closer to the post-tsunami survey boundary location with tsunami inundation limits. An average inundation distance of 3.5 km approaches the pattern that was recorded by the Google Earth satellite. This inundation occurs approximately at the ground elevation smaller than 3.0 m or about 70% of the area Banda Aceh. Tsunami inundation from COMCOT simulation results is the maximum flow depth value.

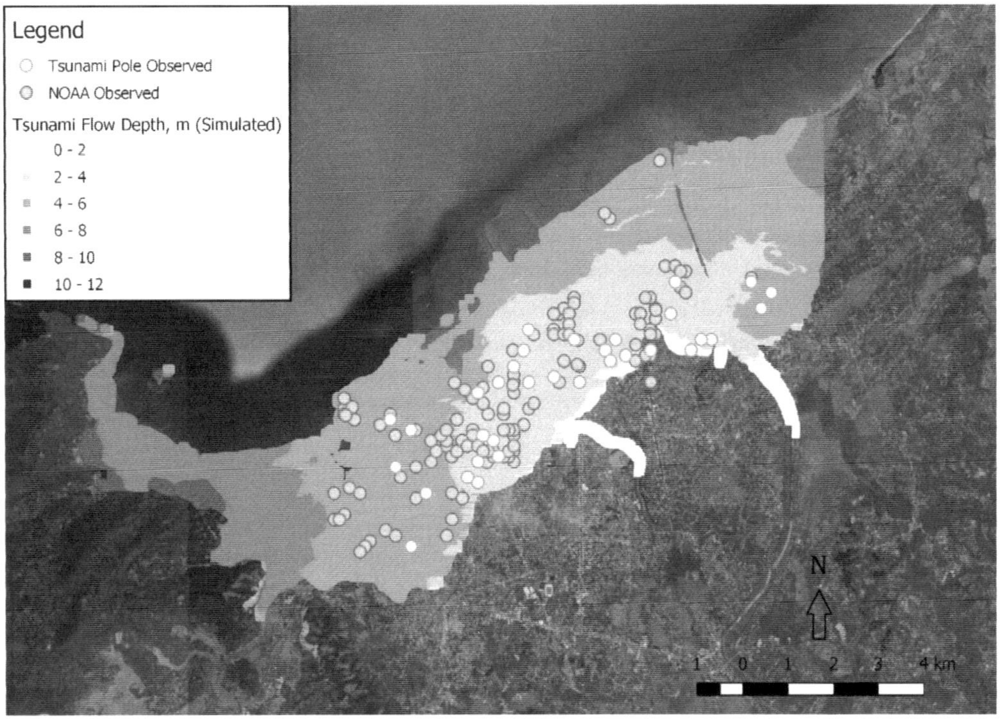

Figure 4. The 2004 tsunami flow depth simulation in Banda Aceh with several survey location by NOAA and Tsunami Pole

The simulation results were then compared with the NOAA and Tsunami Pole measurements. NOAA observed the base on wave height, and tsunami pole recorded the flow depth. NOAA measurement data was converted to flow depth data by reducing the topographic elevation value in the survey location. Table 2 show the Aida validation test for validation flow depth. The results explain that K and k values for NOAA data are outside the range value, indicating *poor agreement*. However, the Tsunami Pole data shows instead *good agreement*, with K and k values closer to the threshold value.

Table 2. K and k value for validation flow depth

	Aida parameters	
Model Results	K	k
NOAA Data (n= *139*)	1.44	1.86
Tsunami Pole data (n= *67*)	0.78	1.56

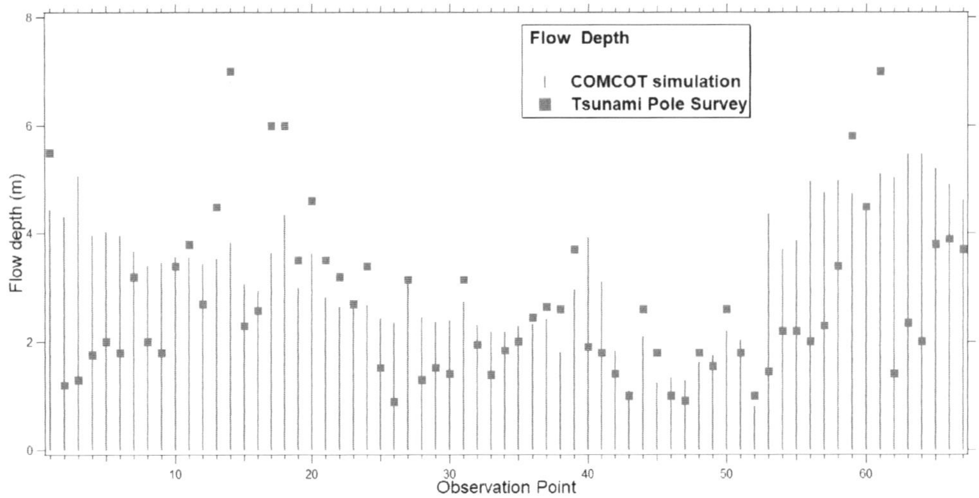

Figure 5. Tsunami flow depth compare by simulation and tsunami pole recorded

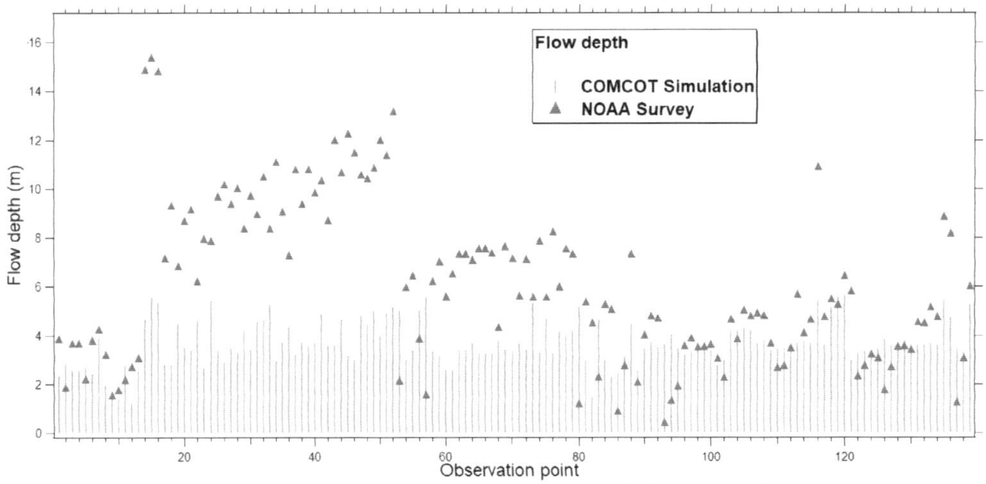

Figure 6. Tsunami flow depth compare by simulation and NOAA survey

Although some point locations were still overestimated and underestimated compared with the observed ones, the tsunami flow depth model comes much closer to the Tsunami Pole one (figure 5). Conversely, with NOAA most of the flow depth simulation values were underestimated compared to the observed ones (figure 6). This is likely due to the fact that the topographic data that interpolated into the 11.5 x 11.5 m grid size was not enough to represent the real elevation when surveyed. The pattern of the flow depth simulation can be observed in the two Figures above, showing quite similar patterns. The value of each point is in the range between 2.0 m and 6.0 m. The data from each point varies greatly compared with the observed one. There is a possibility that the splashing water cannot be modeled properly with the grid size in the model. Even though the determination of the flow depth is measured at the highest limit of water, higher traces caused by splash water can be found in the observed data.

5. Conclusion

High-resolution topographic data and numerical validation results with observed data will produce accurate results in inundation maps. NOAA data gave poor agreement with numerical simulations. A better result was obtained when we using tsunami poles. We recommend to increase the use of the tsunami poles data as they were produced with a rigorous process of eyewitness validation after the 2004 tsunami. Validation model is a continuous process. Even if the model has already been validated, additional testing is also required to constantly bring new knowledge to future studies.

6. References

[1] H. Iemura, M. H. Pradono, A. bin Husen, T. Jauhari, and M. Sugimoto. Information Dissemination For Reality-Based Tsunami Disaster Education The 14th World Conference on Earthquake Engineering October 12-17, 2008, Beijing.
[2] Tursina, Syamsidik, S. Kato, 2017. Projections of tsunami inundation area coupled with impacts of sea level rise in Banda Aceh, Indonesia. AIP Conference Proceedings 1892, 100003 (2017); doi: 10.1063/1.5005735.
[3] Syamsidik, T.M. Rasyif and S. Kato, Development of accurate tsunami estimated times of arrival fortsunami-prone cities in Aceh, Indonesia. Int. J. of Disaster Risk Reduction, 14(4) 403-410 (2015).
[4] L. Tang, Christopher D. Chamberlin, Vasily V. Titov, Developing Tsunami Forecast Inundation Models For Hawaii: Procedures And Testing, 2008, NOAA Technical Memorandum OAR PMEL-141.
[5] USGS (2018). Note from the field: USGS Scientis in Samoa and American Samoa Studying Impacts of Recent Tsunami October-November 2009. Access date October 2018. Link: https://walrus.wr.usgs.gov/news/tsu-terms.html
[6] Liu, P. L.-F., Cho, Y.-S., Briggs, M. J., Synolakis, C. E., & Kanoglu, U. (1995). Run-up of solitary waves on circular island. J. Fluid Mech., 302:259-285, 1995.
[7] W.L. Power, A. Barberopoulou, X. Wang and C. Muller. 2014. Validation of a deaggregation-based approach for tsunami evacuation mapping. GNS Science Report 2014/36. 56 p.
[8] Aida, I., 1978. Reliability of a tsunami source model derived from fault parameters. Journal of Physics of the Earth 26, 57-73.
[9] H. Gokon, S. Koshimura, K. Imai, M. Matsuoka, Y. Namegaya, and Y. Nishimura, 2014. Developing fragility functions for the areas affected by the 2009Samoa earthquake and tsunami. Nat. Hazards Earth Syst. Sci., 14, 3231–3241, 2014.
[10] Koshimura S., Oie T., Yanagisawa H., and Imamura F., 2009. Developing fragility functions for tsunami damage estimation using numerical model and post-tsunami data from Banda Aceh, Indonesia. Coastal Engineering Journal 51(3), 243-273.
[11] Li, L., Qiu, Q., and Huang, Z., 2012. Numerical modeling of the morphological change in Lhok Nga, west Banda Aceh, during the 2004 Indian Ocean tsunami: understanding tsunami deposits using a forward modeling methodNat Hazards 64, 1549-1574.
[12] Hill, E. M., et al. (2012), The 2010 Mw 7.8 Mentawai earthquake: Very shallow source of a rare tsunami earthquake determined from tsunami field survey and near-field GPS data, J. Geophys. Res., 117, B06402, doi:10.1029/2012JB009159.
[13] Takeuchi H., Y. Murashima, F. Imamura, N. Shuto, and K. Yoshida (2005), Verification of tsunami run-up height records of Meiji Sanriku Tsu- nami and Showa Sanriku Tsunami on the coast of Iwate Prefecture using numerical simulation, Hist. Earthq., 20, 155–163.

Acknowledgments

We acknowledge the Partnership Enhanced Engagement in Research (PEER) Cycle 5 funded by USAID andthe National Academy of Sciences (NAS) for the sponsorship in this study. This PEER project under research title: "Incorporating Climate Change Induced Sea Level Rise Information into Coastal Cities' Preparedness towards Coastal Hazards" (#5-395). We would like to thanks Prof. Anawat Suppasri (IRIDeS) for providing the NOAA post tsunami survey data and some discussion on numerical modelling. Also, thanks to Coastal Research Group Syiah Kuala University for the great help. A sincere thank you to Giordano Margaglio for his diligent proofreading of this paper.

IOP Conf. Series: Earth and Environmental Science **273** (2019) 012009 doi:10.1088/1755-1315/273/1/012009

Predicted Rainfall and discharge Using Vector Autoregressive Models in Water Resources Management in the High Hill Takengon

I Ramli[1], S Rusdiana[2], H Basri[3] A A Munawar[4], VAzelia[5]

[1,4,5]Agriculture Engineering Departement, Syiah Kuala University
[2]mathematics and natural science Departement, Syiah Kuala University, Indonesia
[3]Soil Science Departement, Agriculture Faculty, Syiah Kuala University, Indonesia

Abstract. Rainfall is one of climate element that is very important to be studied. The relationship between rainfall and discharge needs to be known to see the size of the discharges so that it depends on the rain and the initial discharge conditions that occur. This study aims to predicted rainfall and discharge by using Vector Autoregressive (VAR), rainfall and discharge data in 2008-2015 is used to predicted rainfall for the next five years 2016-2020. While the data for 2016-2017 is used as a comparison for the predictive data obtained in this study. The results of the actual data accuracy test with predictive data from the VAR (Vector Autoregressive) model, which is done with the NSE (Nash-Sutcliffe efficiency) worth 0.9522. Predicted rainfall value with actual rainfall based on R^2 value (coefficient of determination) is 0.6584 or 66%. While the debit prediction test with actual discharge has a R^2 coefficient of determination of 0.0691 or 6.91%. It is because there are other factors that affect discharge other than rainfall. Rainfall in the Takengon plateau which is predicted to be unstable in the future. This condition is the same as the rainfall in the previous periods which fluctuated at the same point. Rainfall prediction results until 2020, has the highest peak season in December.

1. Introduction

Indonesia is a nation near the equator, where there is a tropical climate or a hot climate so it is vulnerable to drought. At present climate change is happening in the world. Indonesia is one of the countries that are experiencing the impact of climate change. According to the study, the air temperature in Indonesia increased by 0.30 °C since 1900 [1] this increase in temperature occurred throughout the season. Changes in weather and seasons are marked by an increase in rainfall (La Nina) that occurs in an area, while in other regions there is a decrease in rainfall (El Nino) of 2-3% [1].

This climate change has caused floods and droughts that are now happening in Indonesia. A significant drop in rainfall during the dry season, including the El Nino effect, has had a negative impact on food crop production. Drought is a threat that often disrupts agricultural production systems. In recent years drought has not only increased in size and intensity, but also has more and more impacts and distribution of areas affected by drought [2]. Uncertainty of rainfall and rising temperatures quickly emerge as one of the most serious global problems affecting many sectors. It is considered one of the most serious threats to sustainable development with adverse impacts on the environment, human health, food security, economic activity, natural resources and physical infrastructure [3].

According to [4], climate change has begun to occur by looking at one of the climate variables, namely temperature. The average air temperature for twenty years (1992-2011) was 26.77 °C. For the average annual temperature in the period 1992-2001 was 26.41 °C and the average period in 2002-2011 was 27.13 °C. There is an annual average temperature that continues to increase in certain years.

Significantly positive temperature increases from the Kendall analysis test results and will affect water management practices. Increases in temperature and rainfall which result in a little drought, thus posing a serious threat to food security.

The atmospheric conditions in the area with high levels of non-linearity are more difficult to predict than those in high latitudes. This becomes the base point for predicting or predicting the coming rainfall with a certain time series. The statistical approach in time-derived analysis is done by using a statistical model to explain the dynamic behavior of a time slot. A multivariate statistical model is required for a condition in which an event is influenced by more than one variable. The statistical model for rainfall and discharge prediction in this study uses Vector Autoregressive (VAR) in Central Aceh District.

2. Methodology

This research was carried out in Pegasing District, Central Aceh Regency (figure 1) site selection was carried out purposively because the location had rainfall and discharge data for analysis using the VAR model the data used was rainfall and discharge in 2008-2015. While the 2016-2017 data is used as a comparison for predictive data obtained in this study. Monthly rainfall data taken at the Climatology and Geophysics Agency of Aceh Besar Climatology Station from 1992 to 2017 and monthly debit data taken at BPDAS Krueng Aceh in 2008 to 2016.

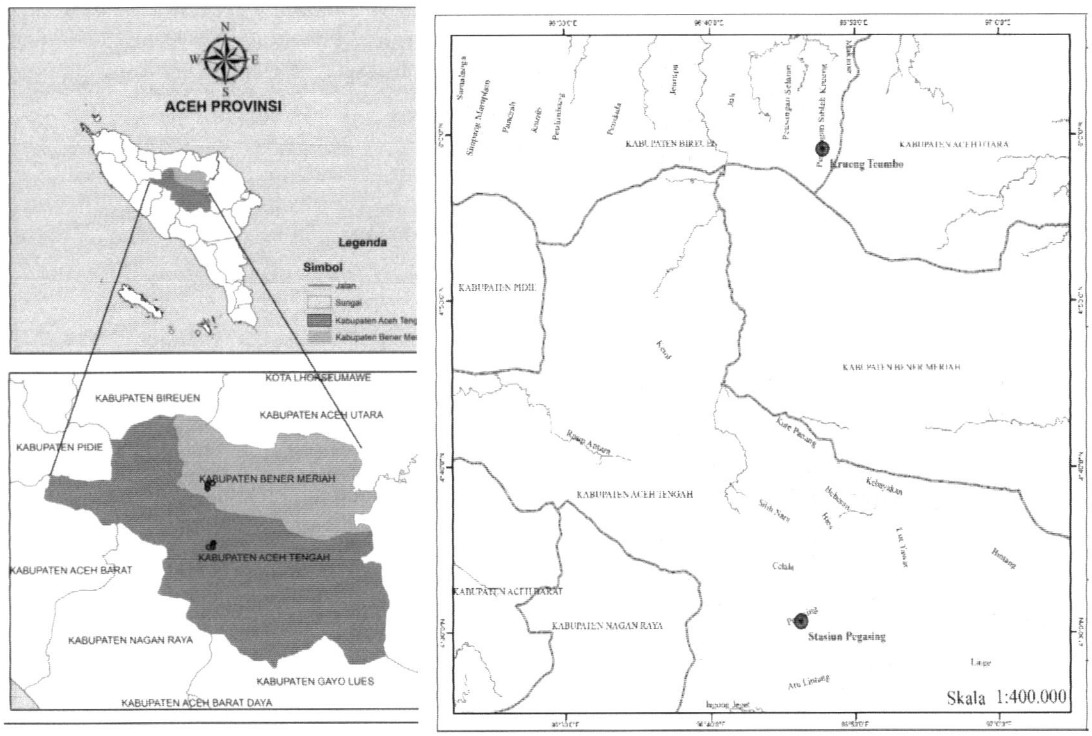

Figure 1. Location of study area research

The research method used in this research is VAR (Vector Autoregressive) analysis. The steps of the research are as follows:

a. Stationary test was conducted on research data. If the data is stationary then the data analysis process uses VAR analysis. If the data is not stationary, a differencing process will be carried out until stationary data is obtained. There are several ways that can be done to measure data stationarity, one of which is to use the Augmented Dickey Fuller (ADF) test [5]. The following ADF test statistics:

$$\Delta Y_t = \alpha_0 + \gamma Y_{t-1} + \beta_i \sum_{i=1}^{P} \Delta Y_{t-i+1} + \varepsilon_t \quad (1)$$

where

ΔY_t = the form of the *first difference* variabel Y
α_0 = Intersep
Y = Variabel tested stationary
p = lag length

The hypothesis for the ADF test is as follows: H0: Data is not stationary H1: Stationary data
If p-value <α then H0 is rejected.

b. Stationary data after differencing, it is necessary to do cointegration test first. If there is no cointegration (long-term relationship) between variables, then data analysis uses VAR (Vector Autoregressive) analysis. However, if there is cointegration between variables, then the next stage of analysis uses a Vector Error Correction Model (VECM) analysis.
c. Next, determining the lag length that will be used in the VAR model. In this study, the lag length is determined by looking at the minimum values of FPE, AIC, SIC and HQ. The criteria that can be used to determine the amount of optimal lag include Final Prediction Error Correction (FPE), Akaike Information Criterion (AIC), Schwarz Information Criterion (SIC) and Hannan-Quinn Information Criterion (HQ), which have a minimum value between various lags submitted. Mathematically, the AIC, SIC and HQ values can be searched using the following formula:

$$\text{AIC} = -2\left(\frac{1}{T}\right) + 2(k + T) \quad (2)$$

$$\text{SIC} = -2\left(\frac{1}{T}\right) + k\frac{\log(T)}{T} \quad (3)$$

$$\text{HQ} = -2\left(\frac{1}{T}\right) + 2k\log\left(\frac{\log(T)}{T}\right) \quad (4)$$

The meaning of the symbols used in equations (2) to (4) are:
1 = The value of the log likelihood function is the same as that $-\frac{T}{2}\left(1 + 2\pi + \log\left(\frac{\varepsilon'\varepsilon''}{T}\right)\right)$; ε\` ε\` \` is *sum of square residual*
T = Number of observations
k = Estimated parameters

d. The general model of VAR is as follows:

$$Y_t = \alpha_0 + A_1 Y_{t-1} + A_2 Y_{t-2} + \ldots + A_p Y_{t-p} + \varepsilon_t \quad (5)$$

or

$$Y_t = \alpha_0 + \sum_{n=1}^{p} A_n Y_{t-n} + \varepsilon_t \quad (6)$$

f. Then structural analysis is carried out as follows:
- Impulse Response Function (IRF), to see how much independent variables are affected by shocks that occur in the dependent variable, some time to come.
- Variance Decompotion (VD), to see how much the independent variable contributes to the dependent variable some time to come.

g. Forecasting based on the VAR model obtained.

Model Accuracy test

The prediction model is then tested for accuracy or validation using a formula correlation coefficient and Nash-Sutcliffe efficiency (NSE) [6]; [7]; [8].

$$R = \frac{\sum_{i=1}^{n}\left(CH_i^{obs} - CH_{obs}^{mean}\right)\left(CH_i^{sim} - CH_{sim}^{mean}\right)}{\sqrt{\sum_{i=1}^{n}\left(CH_i^{obs} - CH_{obs}^{mean}\right)^2\left(CH_i^{sim} - CH_{sim}^{mean}\right)^2}} \tag{7}$$

$$NSE = 1 - \frac{\sum_{i=1}^{n}\left(CH_i^{obs} - CH_i^{sim}\right)^2}{\sum_{i=1}^{n}\left(CH_i^{obs} - CH_i^{mean}\right)^2} \tag{8}$$

Where:

CH_{obs} = actual data
CH_{sim} = Prediction data
$CH_{mean\ obs}$ = actual mean data
$CH_{mean\ sim}$ = Prediction mean data

Then the most accurate and accurate prediction results are selected with the actual data in the field. In general, the simulation model can be seen satisfactory if the value of NSE> 0.50

3. Result and Discussion

The highest dominance in the area of Central Aceh Regency is 1000-2000 masl with a total area of 287,387.21 Ha or 62.65%. So that the region of Central Aceh Regency has diverse agricultural and plantation commodities. While the lowest altitude is at 125-1000 meters above sea level several places are possible to be planted with crops such as coffee, rice, pulses, and horticulture. Central Aceh Regency has a mountainous and hilly topography with an average height varying between 200 - 2,600 meters above sea level.

In general the slope classification of Central Aceh Regency with 14 sub-districts has a land slope of> 40%, 25-40%, 15-25%, 8-15% and 0-8%, divided into very steep, steep, rather steep, sloping and flat. Based on this class the slope or slope of the land in the Central Aceh District is dominated by slope with 15-25% and an area of 208,440.37 Ha or 45.44%. So that with a slope of 15-25% that can only be planted or become land for plantation business. The lowest land slope is with the 0-8% range and 109.434.07 Ha wide area or 23.86% with the flat area. So, the total area in total is 458,707.47 Ha.

Rainfall is an important component in the hydrological process in agriculture, as it is known that plants need water for their primary needs both in quality and quantity. In plant growth, water needs are based on the weather and climate that occur in an area. One of the factors that influence plant growth is rainfall in the area. Rainfall is very important to be studied as an object of research, because rainfall information is very useful in changing weather and climate, one of which is agriculture. Besides that, rainfall also has fluctuating properties on differences in cropping patterns in the agricultural sector which can cause farmers' products to decline.

The large amount of rainfall into the catch area will determine the flow discharge in the watershed, both surface flow, subsurface, base flow and river flow. Rainfall in the Takengon Plateau is very volatile in every year. However there are several months in different years of very high rainfall followed by high flow of water flow. In the month of high rainfall there was a flood in the district of Central Aceh District Linge that occurred in two villages namely Kute Riem and Kute Keramil. Natural disasters are making as many as 57 houses residents submerged in floods. In addition, the disaster damaged hundreds of hectares of rice fields. The incident occurred in the year 2009, the cause is overflowing water of the river Nangka located in the district of Central Aceh with maximum rainfall in that year.

Rainfall characteristics include matters involving annual, seasonal, monthly, or daily distribution, intensity, duration of rain, and frequency of rainy days [9]. In the tropics rainfall is a decisive factor since most agricultural production is produced from dry land farming, where the water required comes from rainfall, so rainfall characteristics are a factor to be considered especially in relation to plants.

Multivariate statistical models will be needed in conditions where an event is affected by more than one variable. One model that is often used is the Vector Autoregressive System (VAR). VAR is one of multivariate time series analysis and useful to see the relation between variables [5]. Stationary test is the first test conducted to determine whether the data is diverse. The results of the rainfall predictions of variable temperature, wind speed, humidity are obtained rainfall = 0.07537952 (rainfall.l1) - 4.84116742 (Temperature.l1) + 6.36430062 (humidity.l1) - 0.11016965 (wind speed.l1) + 0.04866761 (rainfall.l2) + 3.66042192 (Temperature.l2) - 1.28623713 (humidity.l2) - 0.29067619 (wind speed.l2) - 0.08908943 (rainfall.l3) - 4.16258553 (Temperature.l3) - 5.90352425 (rainfall.l3) + 4.00617726 (wind speed.l3) + 300.58327741

If changes in rainfall one month ago increased by 1 mm, will cause changes in rainfall this month increased by 0.07537952 mm. If changes in rainfall two months ago increased by 1 mm, will cause changes in rainfall this month increased by 0.04866761 mm. If changes in rainfall three months ago increased by 1 mm, will cause changes in rainfall this month decreased by 0.08908943 mm.

If the change in temperature one month ago increased by 1°C, it would cause changes in rainfall this month to decrease by 4.84116742 mm. If the change in temperature two months ago increased by 1C, it would cause changes in rainfall this month to increase by 3.66042192 mm. If changes in temperature three months ago increased by 1°C, this would cause changes in rainfall this month to decrease by 4.16258553 mm.

If the change in humidity one month ago increased by 1 RH, it will cause changes in rainfall this month increasing by 6.36430062 mm. If the change in humidity two months ago increased by 1 RH, it will cause changes in rainfall this month decreased by 1.28623713 mm. If the change in humidity three months ago increased by 1 RH, it will cause changes in rainfall this month decreased by 5,90352425 mm.

If changes in wind speed one month ago increased by one knot, this would cause changes in rainfall this month to decrease by 0.11016965 mm. If changes in wind speed two months ago increased by 1 knot, this will cause changes in rainfall this month decreased by 0.29067619 mm. If changes in wind speed three months ago increased by 1 knot, this will cause changes in rainfall this month increasing by 4.00617726 mm.

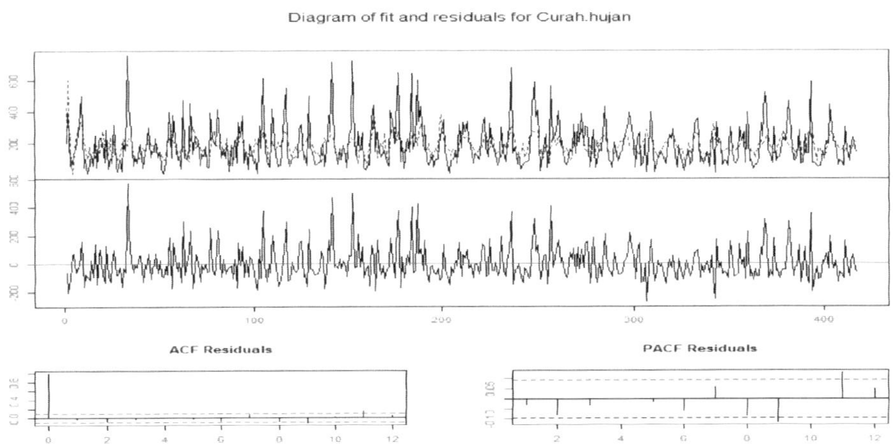

Figure 2. Graph of residual rainfall value from the VAR (3) period model

January 1981 - December 2015.

Plot IRF (Impulse Response Function)

RF analysis aims to see the effect (effect) of each variable if given shock (shock). The impact of the shock of one variable on another can be traced through IRF.

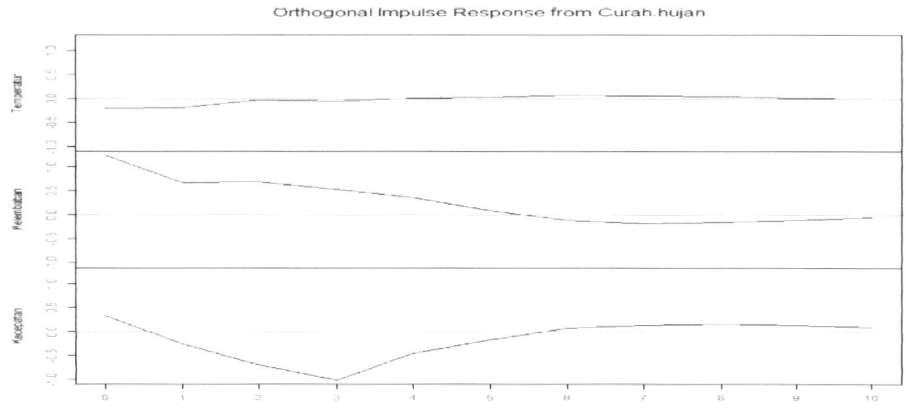

Figure 3. RF rainfall graph

The graph above shows that if there is a shock (shock) in the rainfall variable, it does not cause a shock to the temperature variable, so that the line is right or close to the equilibrium line (Equilibrium), so it does not take long for the variable temperature to stabilize or be in the equilibrium line . As for the rainfall variable to the humidity variable, if there is a shock to the rainfall variable, it takes about 5 months for the humidity variable to return to / close to the equilibrium line. For rainfall variables to speed variables, if there is a shock to the rainfall variable, it will take about 6 months for the wind speed variable to return to / close to the equilibrium line.

Variance decompotion (VD) aims to separate the impact of each of these variables individually, on the response received by a variable. So, variance decompotion can provide information about the relative role of each shock, to the variables in the VAR (Table 1). The variance decompotion analysis shows that in the span of 12 months, the shock to itself results in weaker rainfall fluctuations. While shocks to temperature, humidity and wind speed cause fluctuations in rainfall to increase

Table 1. The variance decompotion analysis at High Hill Takengon

Month	Rainfall	Temperature	Humidity	Wind Speed
1.	1.000	0.000	0.000	0.000
2.	0.962	0.011	0.026	2.633
3.	0.949	0.011	0.038	2.775
4.	0.903	0.013	0.045	3.737
5.	0.873	0.013	0.064	4.827
6.	0.858	0.014	0.075	5.150
7.	0.854	0.015	0.077	5.118
8.	0.854	0.016	0.078	5.150
9.	0.853	0.016	0.077	5.238
10.	0.853	0.016	0.077	5.316
11.	0.851	0.016	0.078	5.366
12.	0.851	0.016	0.078	5.383

Figure 4 is a Climate Data Prediction Plot for the next 5 years (60 Months) using the VAR Model (3) Based on the results of the prediction of rainfall in High Hill Takengon until 2020, the highest peak

(peak season) rainfall occurs in December. The visualization of rainfall at location High Hill Takengon in January 2016 to December 2020 based on the results of the forecasting.

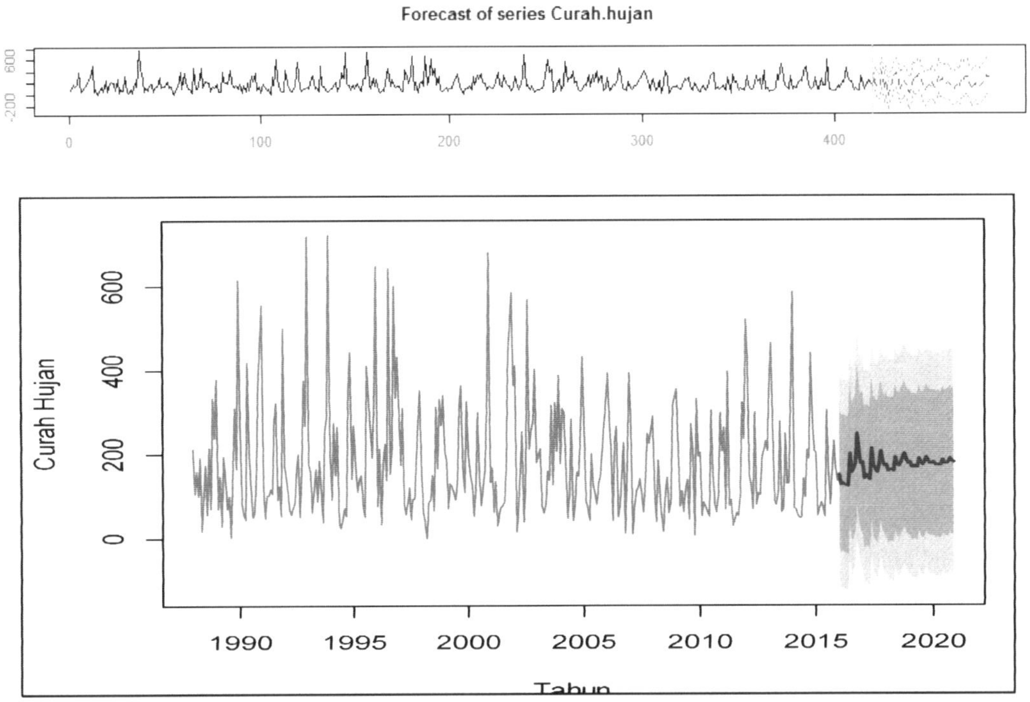

Figure 4. Rainfall Prediction for 2015-2020 at High Hill Takengon

The visualization above shows that rainfall in location High Hill Takengon is predicted to be unstable in the future. This condition is the same as actual rainfall in previous periods which tends to fluctuate at the same point. The blue line is the rainfall prediction data at location High Hill Takengon in January 2016-December 2020. Based on the results of rainfall predictions, the lowest rainfall rate occurred in July 2016, reaching 79.5 (mm). The highest rainfall rate occurred in October 2017 which reached 207.4 mm. The rainfall season based on the results of the prediction occurs in October, November and December. On the part of the lower limit of the confidence interval the value of the minimum predictive value while the upper limit of the interval trusts the maximum predicted value. The picture looks light gray and dark gray. This illustrates dark gray showing 80% of the highest predictive limit or under the opposite. While light gray shows 90% predictions that might occur.

Output of accuracy Models

VAR model on Pegasing for the relationship between rainfall and discharge hypothesis Ho accepted on probability value > 5%. This shows that the discharge is not only affected by rainfall. While the rainfall is influenced by latitude, altitude, distance from the sea, wind direction, mountain rows, differences in terrestrial and ocean temperatures and land area factors. Factors affecting rainfall are the humidity, air, temperature and wind speed [10] [11] suggests climate factors that are strongly associated with the biophysical region are precipitation and temperature.

Vector Autoregressive (VAR) analysis to predict rainfall and discharge that occurs in Central Aceh District, using rainfall data from 2008-2015. From the results, it was found that from the 2008-2015 data, the prediction can only be done in two years, from 2016-2017. This happened because of the limited data available to predict rainfall. Then the accuracy or data validation of the VAR model is tested using the Nash-Sutcliffe efficiency (NSE). The results obtained from NSE are 0.9522. The R^2 which influences the amount of predicted precipitation with actual rainfall is 0.6584 or 66%. The R^2

that most influences from the graph is the 2nd order R^2 graph, where the resulting value is 0.0691 or 6.91%.(Figure 5 and 6). Based on the study, This is because there are other factors that affect discharges other than rainfall. Based on [12] it said that there is a significant correlation between climate, and soil type on river discharge. [13] said that watershed geomorphic characteristic of drainage density, slope variability and colluvium were also shown to have significant impact on stream flow and baseflow. Stream flow quantity and quality are dependent on the physical characteristic of a catchment [14]. The effect of land use at seasonal and monthly runoff depths showed significant variation[15]. Overall, the effects of changes in land use on river flows in the summer are clear. Therefore it was concluded that forest change is a major factor in the variation of river flow in the summer. It shows that the increase in agricultural land can increase the flow in the rainy season and reduce the flow in the dry season but when the forest area is reduced to near zero, the peak flow and average flow will increases. The time series model of regression should consider the relationship between the dependent variable and independent variables. However, in many applications the transfer function and regression model may not be appropriate [16]. Rainfall in the Takengon plateau which is predicted to be unstable in the future. This condition is the same as the rainfall in the previous periods which fluctuated at the same point. Rainfall prediction results until 2020, the peak season rainfall occurs in December.

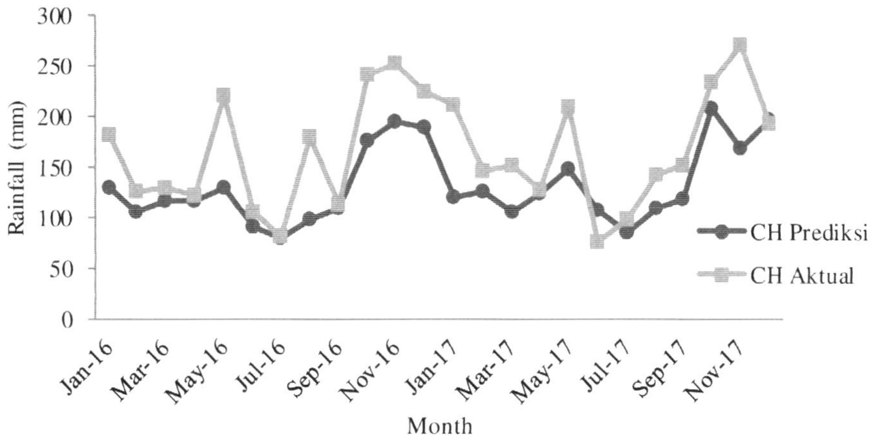

Figure 5. Comparative data on Prediction Rainfall and Actual Rainfall

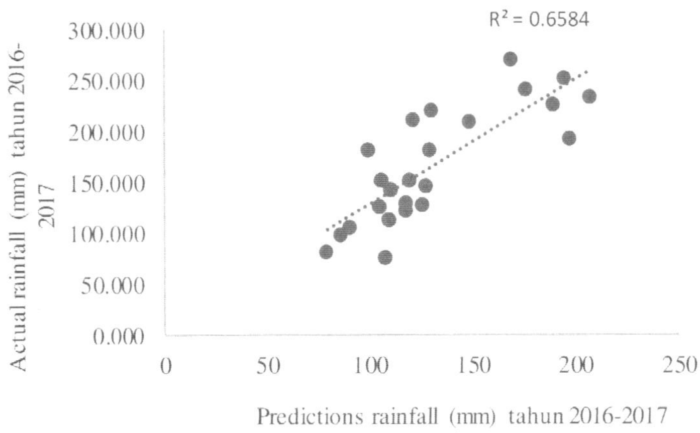

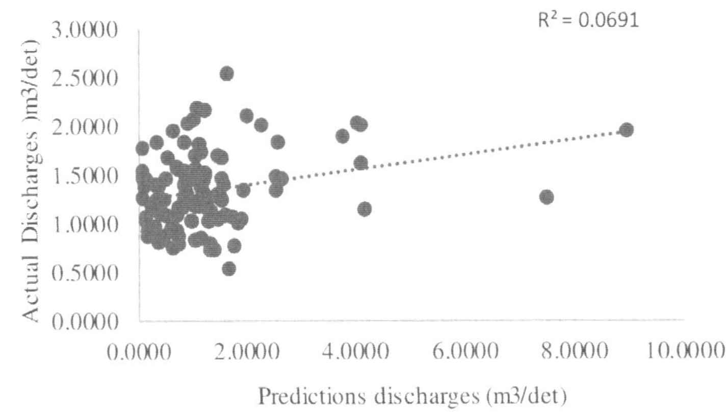

Figure 6. Coefisien determinate (R^2) for Rainfall, discharge from actual and prediction data

4. Result

a. Rainfall prediction in the Takengon plateau with Vector Autoregresive (VAR) model is obtained from the validation results between prediction rainfall and actual rainfall for 2016-2017 with Nash-Sutcliffe efficiency (NSE) test of 0.9522 (good performance). While the coefficient of determination (R^2) that affects the amount of predicted precipitation with actual rainfall is 0.6584.

b. The value of the coefficient of determination for prediction discharges with the actual discharges is 0.0691.

c. Rainfall in the Takengon plateau which is predicted to be unstable in the future. This condition is the same as the rainfall in the previous periods which is fluctuated at the same point. Rainfall prediction results until 2020, the peak season rainfall occurs in December.

References

[1] Hulme, M. and N, Sheard, 1999. *Climate Change Scenarios for Indonesia*. Leaflet CRU and WWF. Climatic Research Unit. UEA, Norwich,UK

[2] Hadiyanto,S. 2007. *Pola Tingkat Kerawanan Kekeringan di Jawa Tengah*. Tesis; Departemen Geografi FMIPA UI

[3] Obert Jiri, Paramu L. Mafongoya1, Chipo Mubaya & Owen Mafongoy. Seasonal Climate Prediction and Adaptation Using Indigenous Knowledge Systems in Agriculture Systems in Southern Africa: *A Review Journal of Agricultural Science*; Vol. 8, No. 5; (2016) ISSN 1916-9752 E-ISSN 1916-9760.

[4] Ichwana, A. Achmad, and S. Chairani. 2015. *Climate trends and dynamic change of land use patterns in Krueng Peusangan Watershed. Aceh-Indonesia.* International Symposium of Geoinformatics (ISYG) 2-3 Desember 2015 di Universitas Brawijaya Malang

[5] Hartini , S. , M, P, Hadi., Sudibyakto dan A, Poniman.. Application of Vector Autoregression Model for rainfall-river discharge analysis. *Jurnal Forum Geografi.* 29: (2015) 1-10

[6] Moriasi, D.N, J.G. Arnold, M.W. Van Liew, R.L Bingner, R.D Harmel, dan T.L Veith 2007. Model Evaluation Guidelines for Systematic Quantification of Accuracy in Watershed Simulations, *Jurnal American Society of Agricultural and Biological Engineers.* ISSN 0001-2351.50 (3): 885-900

[7] Nash JE, dan J.V. Suteliffe 1970. River flow forecasting through conceptual Models 1, a Discussion of Prnciples, *J. Hydrol.* 10 (1):282-290

[8] Nasution, Z., Ichwana, Sumono and Delvian. 2013 Land Use Change and Simulation of Hydrological Response using Integrated NRCS Model and Base flow for Ecohydrological Concept at Krueng Peusangan Watershed, Aceh, Indonesia, *journal of Environmental Science and Water Resources, journal cition report,* ISSN 2277 070 2(7): 210 - 220, August 2013

[9] Nieuwolt,1997 *Tropical Climatology an Introduction to the Climate of the Low Latitude*. John Wiley and Son, New York

[10] Larsen, T.H., Brehm, G., Navarrete, H., Franco, P., Gómez, H., Mena, J.L., Morales, V.,Argollo, J., Blacutt, L., Canhos, V.,2011 *Range shifts and extinctions driven by climate change in the tropical Andes: synthesis and directions*. In: Herzog, S.K.

[11] Rabatel, A., Francou, B., Soruco, A., Gomez, J., Caceres, B., Ceballos, J.L., Basantes, R., Vuille, M., Sicart, J.E., Huggel, C., Scheel, M., Lejeune, Y., Arnaud, Y., Collet, M., 2013 Current state of glaciers in the tropical Andes: a multi-century perspective on glacier evolution and climate change. *J.Cryosphere* 7, 81–102.(2013). http://dx.doi.org/10.5194/tc-7-81-2013

[12] Krakauer, N.Y and M. Temimi. 2011. Stream recession curves and storage variability in small watersheds, *Hydrol. Earth Syst. Sci.,* 15, 2377–2389, 2011 www.hydrol-earth-syst-sci.net/15/2377/2011/ doi:10.5194/hess-15-2377-2011.

[13] Price, K, Jackson CR, dan Parker AJ. 2011. Effects of watershed land use and geomorphology on stream low flows during servere drought conditions in the southern Blue Ridge Mountains, *GA and NC, USA Water Resources Research* 47:W02516.

[14] Abeyou Wale Worqlul, Essayas K. Ayana, Amy S. Collic, Tammo S Steenhuis, 2015, Effect of Physical Catchment Characteristics on River Flow: The Case of Ribb and Gumara Rivers in Upper Blue Nile. International Soil &Water Assesment Tool Conference, Sardinia, Italy, DOI: 10.13140/RQ.2.1.2288.3680

[15] Patts M.A. Odira, Dr. M.O. Nyadawa, B. Okelloh Ndwallah, Ms. Nelly A. Juma, Mr. John and P.Obiero, 2011. Impact of Land Use /Cover dynamics on Streamflow: A Case of Nzoia River Catchment, Kenya, Nile Basin, *Water Science& Engineering Journa*l, Vol.3.

[16] Wei, W.W. 2006 *Time Series Analysis Univariate and Multivariate Methods*. Addsion-Wesley Publishing Co, USA.

Acknowledgement
The authors would like to deliver their gratitude to the Agency for Meteorology, Climatology, Geophysics (BMKG) and BPDAS Krueng Aceh and this research was funded by Syiah Kuala University of 2018 Fiscal Year.

IOP Conf. Series: Earth and Environmental Science **273** (2019) 012010 doi:10.1088/1755-1315/273/1/012010

Numerical Experiments on Tsunami Wave Forces on Open Structures Using Dam-Break Method

Syamsidik[1,2,*], **Mumtaz Luthfi**[1], **Amir Fauzi**[2], and **Eldina Fatimah**[2]

[1]Tsunami and Disaster Mitigation Research Center (TDMRC), Syiah Kuala University, Jl. Prof. Dr. Ibrahim Hasan, Gampong Pie, Banda Aceh, Indonesia, 23233.
[2]Civil Engineering Department, Faculty of Engineering, Syiah Kuala University, Jl. Syeh Abdurrauf, No. 7, Banda Aceh, Indonesia 23111.

[*]Corresponding Author: syamsidik@tdmrc.org, syamsidik@unsyiah.ac.id

Abstract. Open wall structures, such as those found in mosques, sustained large forces generated by a tsunami as in the case of the 2004 Indian Ocean tsunami. Although the surrounding buildings were flattened, there were a number of mosques in the tsunami affected area could still stand in the middle of the area. These motivated this study to look into more details the hydrodynamic forces generated around the open structures. This research was aimed at investigating the tsunami wave forces on open structures and at knowing relation between the tsunami waves velocity and the pressures on the pillars of the open structures. DualSPHysic was used to simulate the tsunami wave forces on the structures. The DualSPHysic is a smoothed particle hydrodynamic model with mesh-less method. We simulated 12 numerical simulations by constructing a-130 m wave flume equipped with dam-break at one end to generate the tsunami-like flows. A sloping bed was placed in front of an open wall structure. Four pillars were set to imitate a simplified model of open wall structure. Numerical observation points were placed at offshore part of the model, in front of the pillars (lee-side) and at the back of the pillars (wake-side). Simulations were performed for about 6 seconds started from releasing the dam gate and before the flow being reflected by the other end of the flume. This study proved that the impacts of the tsunami-like forces are more significant of the front row of the pillars. Meanwhile, tsunami wave forces at the back row of the pillars group are inconclusive to the wave velocity and wave heights.

1. Introduction

Open structures such as mosque's walls were found to remain with minimum damages due to impacts of tsunami waves in several areas. The 2004 Indian Ocean tsunami case revealed many cases where mosques remained standing despite massive destructions of buildings around. This could provide opportunity to utilize the mosque as an alternative place to evacuate. However, there are limited number of studies have investigated the impacts of tsunami waves on mosques that have open wall structures.

In tsunami prone area, tsunami evacuation buildings are not sufficient to accommodate all evacuees in many areas. One of the examples could be found in Aceh. In Banda Aceh, the number of official assigned evacuation buildings are six buildings. In total, all the assigned evacuation buildings could facilitate about 4,500 people at one time. This is far from the number of human exposure should any tsunami hit the city in the future. About one third of the city population could become victims of the tsunami (or about 80,000 people). If only about 2% of the human exposure to be assigned to use the evacuation buildings, the present condition is far from sufficient. Therefore, other alternatives to facilitate safe and sufficient evacuation sites should be considered. Another serious problem is caused by an extremely short time available to perform evacuation and difficulties to access public information/media during the emergency period [9]. The return migration process in Banda Aceh has increased number of coastal population in the city and it means the tsunami risk also has also increased in recent years [10].

One of the most potential alternatives is mosques. A number of mosques in Aceh could be used as tsunami shelters provided their structures being modified to suit the evacuation process and their capacities to sustain from earthquake and tsunami are properly checked. In order to provide a preliminary information on the open structure capacity to sustain tsunami wave forces, a set of numerical experiments were performed.

Thanks to the advances in numerical modelling using Smoothed Particle Hydrodynamic Method, details numerical studies involving complex forms of testing objects now become possible. The development of computer software and hardware in recent years has also facilitated numerous studies in computational fluid dynamics (CFD). This study is one of recent applied researches to support ideas on providing ample tsunami evacuation buildings by using existing public facilities, namely mosques that have open wall structures. This study was aimed at investigating tsunami wave forces on open structures where model of a mosque with open walls was used. We presented the results that show the influence of tsunami wave heights at the offshore, generated by dam-break method, on forces on pillars of open wall structures.

2. Methods

2.1 DualSPHysic

Smoothed Particle Hydrodynamic (SPH) is a relatively new method that enables scientists to precisely simulate wave breaking process and non-linear effects of waves [1]. SPH is a Lagrangian mesh-less method that uses particles movement and their interactions with structures. The Navier-Stokes equations were discretized and integrated to the coordinates of the particles. SPH simulations provide new opportunities to tsunami scientists to look into more detail of the tsunami wave forces on open structures. In recent years, the development of smoothed particle has reached new advancements by the use of GPU (Generic Processing Unit) in the simulation process. DualSPHysic that was developed in early 2010 has been widely acknowledged as one of breakthroughs in the numerical experiments [2]. The GPU system supported the simulations by accelerating the simulations significantly faster than conventional CPU. Mass conservation formulae of SPH was developed to consider the particle movement and acceleration as they were caused by the particle interaction with its neighbouring particle [3]. The momentum equation applied in SPH can be seen in Equation (1) as follows.

$$\frac{dv_a}{dt} = -\frac{1}{\rho}\nabla P + \boldsymbol{g} + \boldsymbol{\Theta} \tag{1}$$

where $\boldsymbol{v_a}$ is vector velocity in x, y, z directions, P is pressure, and ρ is density of the fluid, g is gravitational acceleration in z direction only, and Θ is vector of dissipative terms. Meanwhile, ∇ is a Laplace operator as follows.

$$\nabla = \frac{\partial}{\partial x} + \frac{\partial}{\partial y} + \frac{\partial}{\partial z}, \tag{2}$$

The validity of the DualSPHysic in tsunami-like simulations have been confirmed by Cunningham et al. using wave maker method [4]. In this case, we used a group of cylindrical columns that support a concrete plate to represent the open structure of a mosque. Particles in SPH are treated as circular in two dimensional model and spherical as in three dimensional model. In this study, we used three dimensional model. Furthermore, instead of wave-maker method, we deployed dam-break method to generate long waves, as if in the case of tsunami waves. The GPU enables us to run for millions of particles as it also helps us to conduct a more detail simulation [6].

2.2 Numerical Experiments

We developed a hypothetical wave flume with its dimensions are 130 m in length, 12 m in width and 8 m in height (see Figure. 1). Dam-break gate was imitated at the upstream part of the flume and one frame of open structures that mimic the mosques open wall were made at the downstream part of the flume. The dam-break gate retained water column with varied heights. There were two types of object to be tested, i.e. one was an open-structure wall to imitate a mosque wall and the other one was close-structure wall for comparison. A slope media was made to imitate the coastal slope of 1:50. It was expected the slope to reconstruct non-linear effects of coasts.

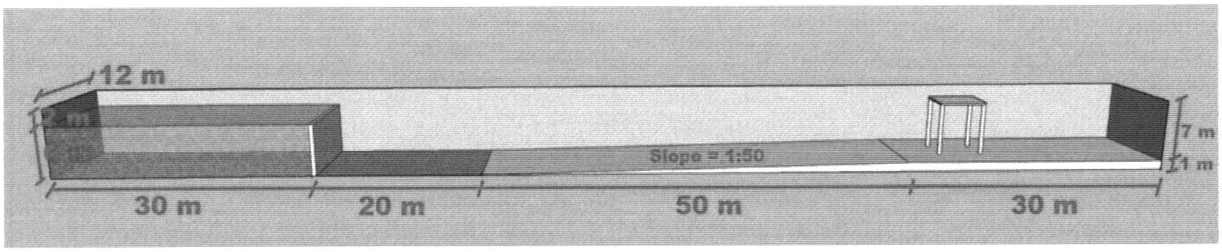

Figure 1. The Wave Flume Dimension for Numerical Simulations.

We employed Wendland Kernell function in the simulation, with particle size of 0.01 m. The simulations were performed using double precision mode in particle interaction, and step algorithm of Verlet was used. Documentation of Verlet scheme can be found at [5]. The simulations were run for 20 seconds with water mass density of 1000 kg/m^3. We set up lattice bound and fluid bound in the simulations. Furthermore, Courant-Friedrich-Levy (CFL) coefficient was 0.2 and γ was 7. Delta-SPH coefficient that controls large wave number was disabled (set to zero). Smoothing particle length was calculated as $\sqrt{(3d_p^2)}$ since this is a 3 dimensional model, where d_p is particle size.

The following table shows the scenarios of the numerical experiment in this study.

Table 1. Scenarios of the numerical experiments using DualSPHysic.

Water depth Behind the Dam Gate (m)	The height of the opening gate (m)	Simulation Number
1.4	1.0	#111
	0.8	#112
	0.6	#113
	0.4	#114
1.2	1.0	#211
	0.8	#212
	0.6	#213
	0.4	#214
1.0	1.0	#311
	0.8	#312
	0.6	#313
	0.4	#314

At the lee-wake sides of the pillars, a series of numerical observation points were placed at every 0.1 m interval of the points in z-direction. Depth average velocities ($\overline{U}$)were calculated as in Equation (3).

$$\overline{U}^t = \frac{1}{h}\int_0^{z=h} U(z)^t \, dz, \tag{3}$$

Here, z is distance from the bed level to the observe points and t is time of the simulation in second, and h is total depth of the water at time t. The $\overline{U}^t$ was calculated started when the wave hit the pillars after 0.1 second to allow the wave depths at significant levels (deeper than 0.25 m). Maximum values of the $\overline{U}^t$ were plotted in graphs to see the tsunami wave velocities influence on forces at the lee-wake sides of the pillars.

3. Results

Based on numerical observation points placed at offshore area and in front of pillars, we observed the influences of tsunami waves generated on lee-side of the pillars. Figure 2 shows the influence of wave heights measured at the onshore part of the flume on forces on pillars. The onshore wave height observation points were placed about 1 cm in front of the pillars. Here, there is no significant different between front row pillars and back-row pillars. The influence of the increasing tsunami wave heights on forces on pillars is linear. The similarity between the front row and back row pillar is most likely due to combination between wave heights and velocities on pillars.

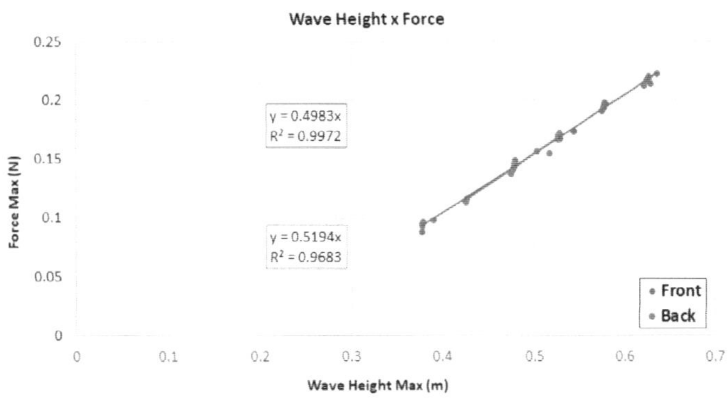

Figure 2. Influence of off-shore tsunami wave heights on forces working on pillars of an open structure.

Figure 3 shows the influence off-shore wave heights (upstream wave height) on forces working on pillars. Here, the front pillars have a steeper increase of forces due to the increase of the generated wave heights released from the gate.

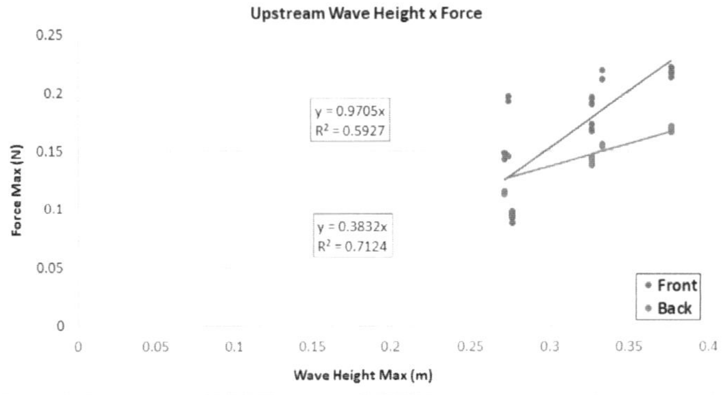

Figure 3. Influence of Off-shore wave heights on forces working on pillars of an open structure.

Figure 4 shows the comparison of the influence of the depth averaged velocity at the lee-side of the both rows of the pillars on maximum forces induced by the tsunami waves. The back row pillars group received smaller forces than the front row pillars. This is due to the eddies generated at the wake side of the front row pillars reduce the forces of the waves at the lee-side of the back-row pillars group. This study used a group of pillars where Morison formulae was not applied to convert the force on cylindrical types pillars. This could underestimate forces generated by long waves, as in the case of tsunami waves [7]. Converting pressures resulted by the DualSPHysic sensors to forces could give smaller values than if the Morison formulate applied [8]. Notwithstanding the use of the Morison formulae, this study could provide important information on pressures exerted by tsunami-like waves.

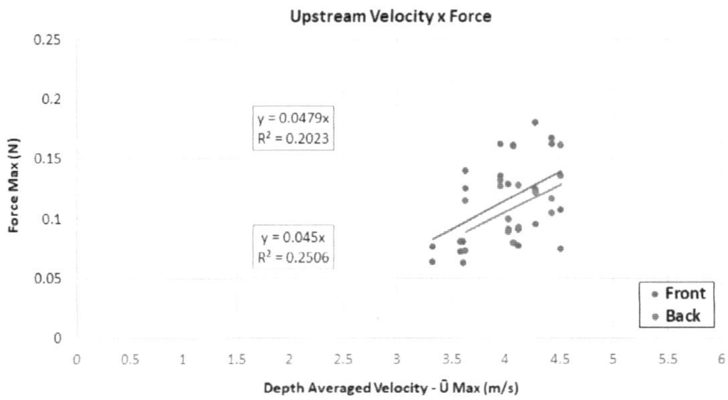

Figure 4. Influence of depth average velocity

4. Conclusions

This study attempted to investigate forces generated by tsunami waves on open wall structures by means of numerical experiments. DualSPHysic was utilized to simulate simplified structure of the open wall that have two rows of pillars, i.e. front-row pillars and back-row pillars. The shape of the pillars is cylindrical. This study concludes that influence of the tsunami wave heights on forces at the front-row pillars is linear with a steeper increase than the back-row pillars. The depth averaged velocity at the front-row pillars generated larger forces at the lee-side of the pillars compared to the back-row pillars group. Further investigations are needed to see the forces generated by tsunami waves with a complete structure of an open wall, as in the case of a mosque.

References

[1] Darlymple R A and Rogers B D 2006 Coastal Engineering **53** (2-3) 141

[2] Crespo A C, Dominguez J M, Rogers B D, Gomez-Gesteira M, Longshaw S, Canelas R, Vacondio R, Barreiro A, Garcia-Feal O 2015 Computer Physics Communications **187** 204

[3] Crespo A C, Dominguez J M, Barreiro A, Gomez-Gesteira M, Rogers B D 2011 Plos ONE **6**(6) e20685

[4] Cunningham L S, Rogers B D, and Pringgana G 2014 ICE Proceeding **167** 126

[5] Verlet L 1967 Phys. Rev. **159** 98

[6] Winkler D, Rezavand M, and Rauch W 2018 Computer Physics Communications **225** 140

[7] Panalaran S, Triatmadja R, and Wignyosukarto B S 216 AIP Conference Proceedings 1755 060004

[8] Bisgard C 2005 Breaking and non-breaking solitary wave impact pressures on a cylinder over a 3-D bathymetry, Master Thesis, Oregon State University, US

[9] Taubenböck H, Goseberg N, Setiadi N, Lämmel G, Moder F, Oczipka M, Klüpfel H, Wahl R, Schlurmann T, Strunz G, Birkmann J, Nagel K, Siegert F, Lehmann F, Dech S, Gress A, and Klein R 2009 Nat. Hazards Earth Syst. Sci. **9** 1509

[10] Syamsidik, Oktari R S, Munadi K, Arief S, and Fajri I Z 2017 Natural Hazards **88**(3) 1503

Acknowledgments

Authors thank to Ministry of Research, Technology and Higher Education (RISTEKDIKTI) of Indonesia through PKLN Research Grant No. SK.60/UN11.2/SP3/2018 Year 2018, titled *Mitigating Impacts Of Tsunami Waves On Coastal Structures And Harbor Facilities*. Authors also thank to Partnership Enhanced Engagement in Research (PEER) Cycle 5 supported by USAID and National Academies of Sciences (NAS) of the United States who have provided two GPUs for the numerical simulations at TDMRC of Syiah Kuala University. The PEER Cycle 5 Grant title is *Incorporating climate change induced sea level rise information into coastal cities' preparedness toward coastal hazards* (#5-395).

Numerical Simulations of Tsunami Wave Properties on Coastal Slopes using One Piston-Wavemaker Method

Mumtaz Luthfi[1], **Syamsidik**[1,2,*], **Amir Fauzi**[2] **and Eldina Fatimah**[2]

[1] Tsunami Computation and Visualization Laboratory, Tsunami and Disaster Mitigation Research Center (TDMRC), Syiah Kuala University, Jl. Prof. Dr. Ibrahim Hasan, Gampong Pie, Banda Aceh, 23233-Indonesia
[2] Lecturer at Civil Engineering Department, Syiah Kuala University, Jl. Syeh Abdurrauf, No. 7, Banda Aceh, 23111, Indonesia

[*]Corresponding author: syamsidik@tdmrc.org

Abstract. Most of the coastal area in the Ulee Lheue Bay of Aceh Besar was affected by a tsunami wave during the 2004 Indian Ocean Tsunami. Such area has an inclined bed floor which varies in slope. Variations of slope can affect tsunami wave properties such as tsunami arrival time, wave height and velocity. With those variations, the degree of the tsunami effects is highly unpredictable. Therefore, it is important to conduct a research to understand the effects of coastal slopes on tsunami waves propagations and wave deformation. Google SketchUp was used for designing the coastal slopes' 3 Dimensional models. For the tsunami waves, solitary waves were generated by using smoothed particle hydrodynamics numerical models through the DualSPHysics software. The computation was based on the meshless method. From the simulations, the waves' properties were obtained and were analyzed how coastal slopes affect the arrival time of tsunami waves, also for their height and velocity. Results show that the steepest coastal slope has the highest velocity, but shortest wave height and longest arrival time. Conversely, the flattest coastal slope has the lowest velocity, but highest wave height and shortest arrival time. Longer tsunami arrival times will benefit people in the evacuation process during the hazard.

Keywords: DualSPHysics, Solitary Waves, Meshless Method, Coastal Slopes

1. Introduction

On December 26th, 2004, a tsunami devastated Aceh, Indonesia. This event generated major loss, especially in terms of human casualties and properties, since it caused over 200,000 fatalities and almost 20 billion USD of economic damage. So far, it has also been the deadliest natural disaster of the 21st century.

Since Aceh is located in very active subduction zones, this area is prone to numerous earthquakes and tsunamis [1]. After 2004, people started to realize how vulnerable Indonesia, especially in coastal cities. This vulnerability highlights the need for developing several strategies to reduce the impact of natural hazards.

Numerical simulations have become one of the main methods which is used in coastal research. Furthermore, numerical modeling is able to provide more economic solutions in terms of funds and preparation time, compared, for instance, to physical modeling. Hence, numerical modelling has become-the best option for running tsunami simulations and other coastal studies.

The purpose of this study is to show the interaction between tsunami waves and coastal slopes with multiple slope variations; as well as to present the effect of the slope on the wave's height, velocity and arrival time. The results could then be used as a reference for designing coastal structures that would be able to mitigate future tsunamis. The height and velocity of the wave could become standards to take into account in the building of future coastal structures. Finally, tsunami arrival times could be used to determine the best evacuation route according to the area's geophysical configuration.

2. Methods

Numerical simulations were conducted using the DualSPHysics software. This software is based on a Smoothed Particle Hydrodynamics named SPH. It is used to study free-surface flows such as tsunami-like waves. DualSPHysics uses the meshless method for hydrodynamic and other types of fluid simulations. SPH can be run by using CPU (Central Processing Unit) or GPU (Graphics Processing Units). GPU is able to process more data and in a faster way compared to CPU [2]. In this simulation, DualSPHysics was executed by using GPU of NVIDIA® GeForce GTX 1080 8GB. The distance between the particles (DP) was set to 0.01 m.

2.1. Model setup

Ten simulation cases were run according to different volumes of water and variation of coastal slopes. The simulation time of each case was 10.0 seconds with a time step of 0.01 seconds. Every case took about 12 hours of running time. All of the cases were using 0.5 m as the sea width with the horizontal length consisted of 2 m of flat area and 10 m of inclined slope. The coastal slopes with multiple variations had ratios of 1:80, 1:100, 1:120, and 1:150. These slopes have variation of water levels regarding on the slope, which are 0.11 m, 0.09 m, 0.08, and 0.06 m. These 16 cases are shown in Table 1.

Table 1. Simulation case variations

Cases		Designed Wave Height (m)
Slopes	Water Level (m)	
1 : 80	0.11	0.06
		0.05
		0.04
		0.03
1 : 100	0.09	0.06
		0.05
		0.04
		0.03
1 : 120	0.08	0.06
		0.05
		0.04
		0.03
1 : 150	0.06	0.06
		0.05
		0.04
		0.03

This research used a 3D DualSPHysics model, where the domain resembles a flume. This flume appeared to be a boundary describing the coastal area of Ulee Lheue Bay, Aceh Besar. This coastal area was designed in 3D models with a 1:10 scale. The design had a 5 meters length of inclined area, and a meter land area, with varied elevation on the even level of the surface in the land/lee side. The

IOP Conf. Series: Earth and Environmental Science **273** (2019) 012011 doi:10.1088/1755-1315/273/1/012011

varied elevation height on the lee side depended on the coastal slopes, varying among 0.125 m, 0.1 m, 0.083 m, and 0.067 m. The slope model was placed 2 meters from the start of the flume.

The paddle was used as a solitary wave generator, and it was placed at the start of the flume with a square-shaped dimension of 0.5 m x 1 m and a thickness of 0.2 m. Sensors as Wave Gauges (WG) were placed in four locations. WG #1 was placed in the Deep Ocean (2 m from the start), WG #2 was placed on the Nearshore (9 m from the start), WG #3 was placed on the Shoreline (12.1 m from the start), and WG #4 was placed on the Land Side/Lee Side (12.6 m from the start). The complete dimension of the design is shown in Figure 1.

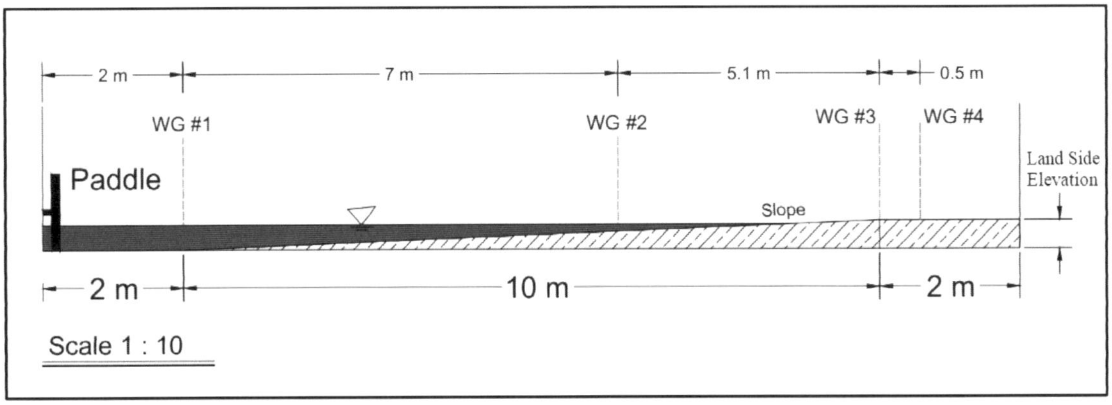

Figure 1. Flume design

For the coastal slopes, Google SketchUp was used for designing these 3D models. The model consisted of coastal slopes with multiple variations of: 1:80, 1:100, 1:120, and 1:150. Furthermore, the models were saved in *.stl format for further use in DualSPHysics as coastal slopes. The SketchUp model is shown in Figure 2.

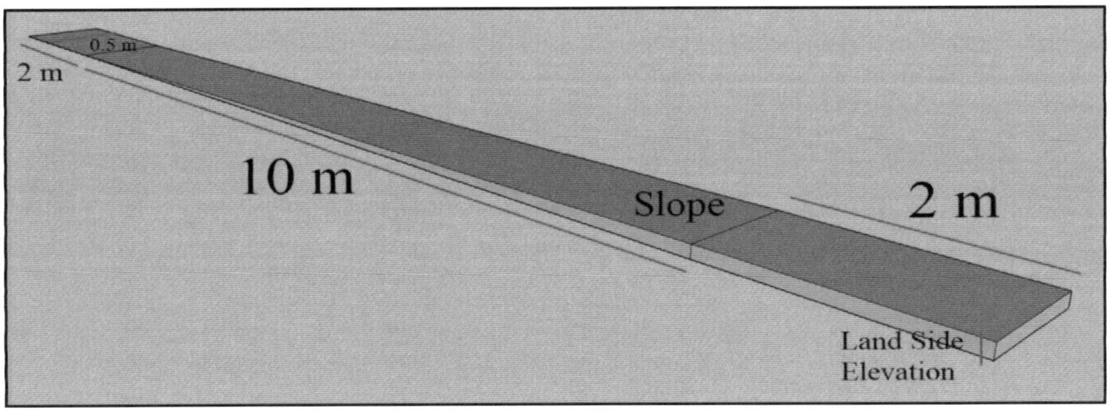

Figure 2. Google SketchUp slope model for the coastal slope

2.2. Solitary Wave

The DualSPHysics cannot generate tsunami wave from an earthquake source, thus the tsunami wave was assumed to be a solitary wave, which is a single wave that travels from the deeper part of the ocean and starts breaking through the shallower area. The best ratio for perfect solitary waves should be 1:10 on the designed wave height (H) over the depth of the sea (h). The wave was generated from a paddle movement start, with a distance (x) equal to 0 m and a time (t) equal to 0 s until time is equal to 5 s. The solitary wave was generated based on a Goring equation which was calculated using Visual

Basic, and converted into Paddle Movement Data in *.txt format. The solitary wave equation was calculated as follows [3]:

$$u_{(xs,t)} = \frac{dx_s}{dt} \quad (1)$$

$$u_{(xs,t)} = \frac{c.\eta_{(xs,t)}}{h+\eta_{(xs,t)}} \quad (2)$$

$$xs_{(t)} = \frac{2H}{k.h} tanh\ [k(c.t - x_s(t))] \quad (3)$$

$$c.\eta_{(xs,t)} = H.sech^2\ [k(c.t - x_s)] \quad (4)$$

Equation (1) defines the movement formula, where x_s is the paddle displacement (in meters) and t is the movement time (in seconds). In Equation (2), the average velocity of the wave was calculated, where c is the wave velocity, h is the water depth, and η is the surface elevation. After that, the paddle displacement over time can be calculated with Equation (3). Finally, the solitary wave can be generated using Equation (4), where H is the designed wave height (in this study equal to 0.03 m) and k is the Ouskirt Coefficient.

After the solitary wave data were calculated, the back and forth wavemaker piston type in DualSPHysics was used as the paddle. The paddle moves for 5 seconds. In t=0 s, the paddle moves until t = 2.5 s for 1.25 m of displacement. Finally, the paddle returns for another 2.5 s to its original position (0 m). The paddle movement graph of 0.06 m designed wave is shown in Figure 3.

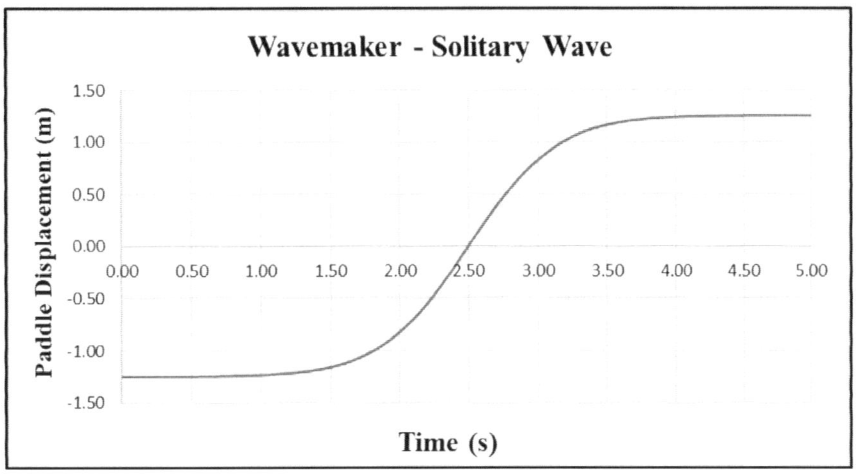

Figure 3. Time series of paddle displacement for solitary wave generation of 0.06 m. Within a 5 s displacement time, the paddle moves horizontally as far as 1.25 m.

2.3. Current Velocity Profile

Wave Gauges (WG) were placed in four places as observers. These wave gauges observed the wave velocity and height at every point. WG #1 was placed in the deep ocean for observing the travel of solitary wave. WG #2 was placed on the nearshore for observing the hydrodynamic process when the wave was going to break. WG #3 and WG #4 were placed in the elevated land area for comparing the differences in wave height between every variation of the case. These sensors recorded wave height and wave velocity data for every 0.1 m of elevation.

The wave travels as a profile of water column. The velocity increases as the water level rise. It starts from zero in the sea bed until it hits the maximum velocity on the surface. This phenomenon is called current velocity profile. To determine the current velocity profile at certain times and places, the depth averaged velocity formula by Soulsby was used as follow [4]:

$$\bar{U} = \frac{1}{h}\int_0^h U(z)\,dz \tag{5}$$

Where $\bar{U}$ is depth averaged velocity (m/s), h is water depth (m), $U(z)$ is the current speed at z-depth (m/s), and z is the observed depth point (m). From Equation (5), trapezoidal rules are used as a mathematical approach to obtain $\bar{U}$. The velocity in the sea bed is assumed to be zero (at z=0) and the maximum value on the surface (at z=h). Thus, if the current speed in the depth of z_1, z_2, z_3,..., z_n (ascending) are U_1, U_2, U_3,..., U_n, then the Equation (5) becomes [4]:

$$\bar{U} = \frac{0.5}{h}\begin{bmatrix} U_1 z_1 + (U_1 + U_2)(z_2 - z_1) + (U_2 + U_3)(z_3 - z_2) \ldots \\ +(U_{n-1} + U_n)(z_n - z_{n-1}) + 2U_n(h - z_n) \end{bmatrix} \tag{6}$$

3. Results

3.1. Paraview animations

The DualSPHysics has *.vtk and *.csv files as its outputs. From the *.vtk files, video animations can be rendered using Paraview. Subsequently, snapshots from animations at certain time steps can be obtained. Those images are presented in Figure 4.

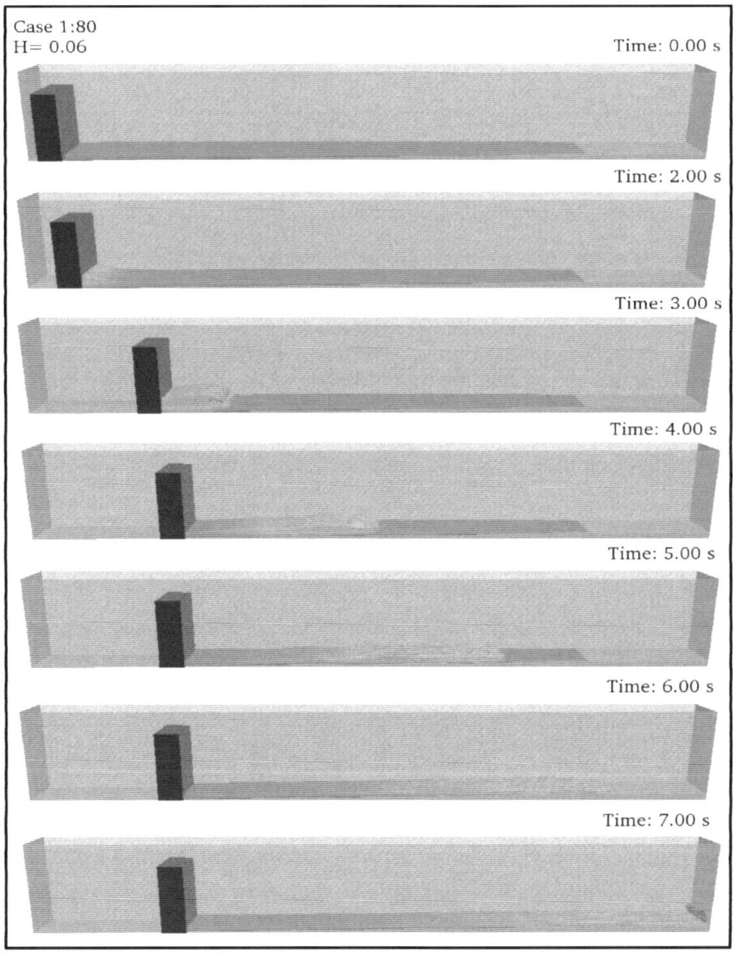

Figure 4. Snapshots of a travelling tsunami wave at certain times, rendered with Paraview

From the snapshots in Figure 4, at t=0 s, the solitary wave doesn't exist yet since the paddle is still in sedentary state and the water volume is still at its initial condition. Afterward, the paddle starts moving based on the solitary wave generation data in Figure 3.

At t=2 s, the wave travels and reaches its maximum peak at t=2.5 s. After that, at t=3 s, the wave still travels through the flume from the deeper part of the sea to the nearshore area at t=4 s. Finally, at t=5 s, the wave hits the shoreline and the lee side of the flume.

3.2. Time series graph of flow depth and wave velocity

From the *.csv files (obtained from wave gauge sensors), wave height and the velocity data can be collected and analyzed. Then time series graphs can be made. Those graphs are shown in Figure 5 and Figure 6.

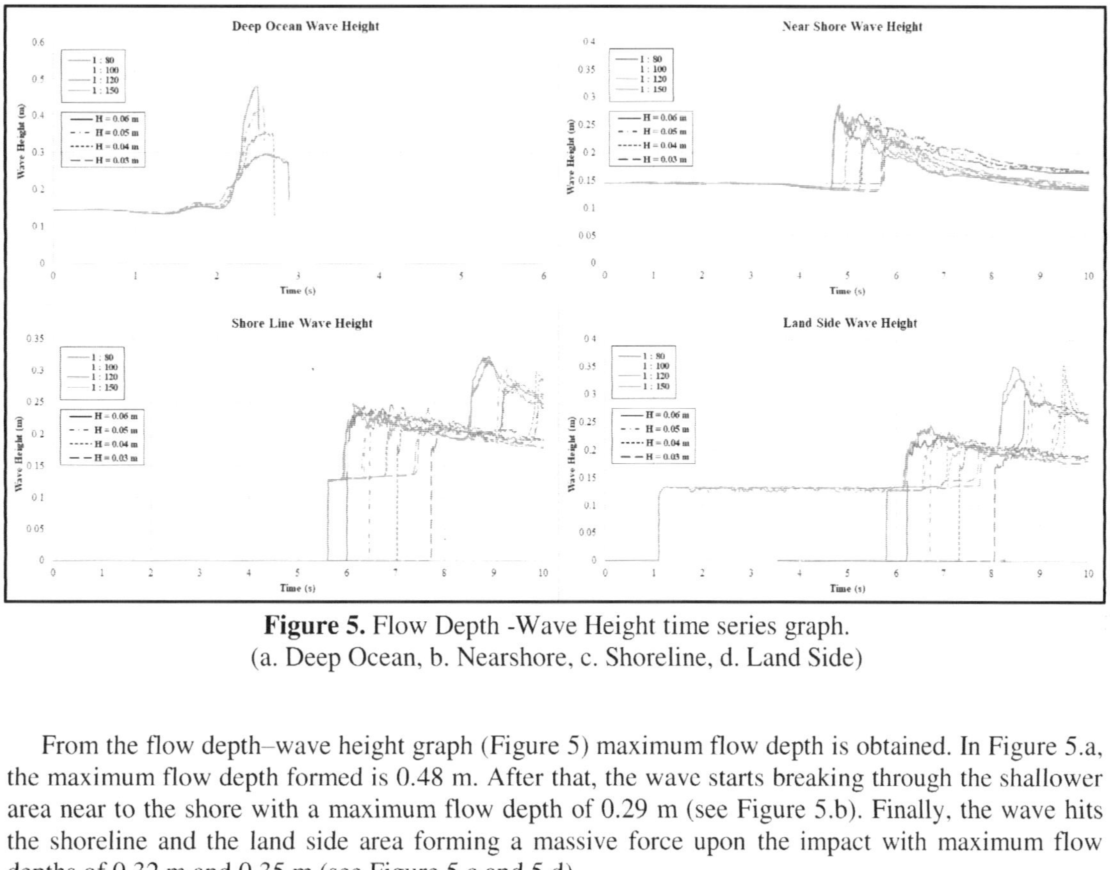

Figure 5. Flow Depth -Wave Height time series graph.
(a. Deep Ocean, b. Nearshore, c. Shoreline, d. Land Side)

From the flow depth–wave height graph (Figure 5) maximum flow depth is obtained. In Figure 5.a, the maximum flow depth formed is 0.48 m. After that, the wave starts breaking through the shallower area near to the shore with a maximum flow depth of 0.29 m (see Figure 5.b). Finally, the wave hits the shoreline and the land side area forming a massive force upon the impact with maximum flow depths of 0.32 m and 0.35 m (see Figure 5.c and 5.d).

Also, based on the velocity graph (see Figure 6), the maximum velocities from each wave gauge are collected. The maximum velocities that occurred in WG #1, WG #2, WG #3, and WG #4 are 1.99 m/s, 3.00 m/s, 2.53 m/s, and 2.35 m/s respectively.

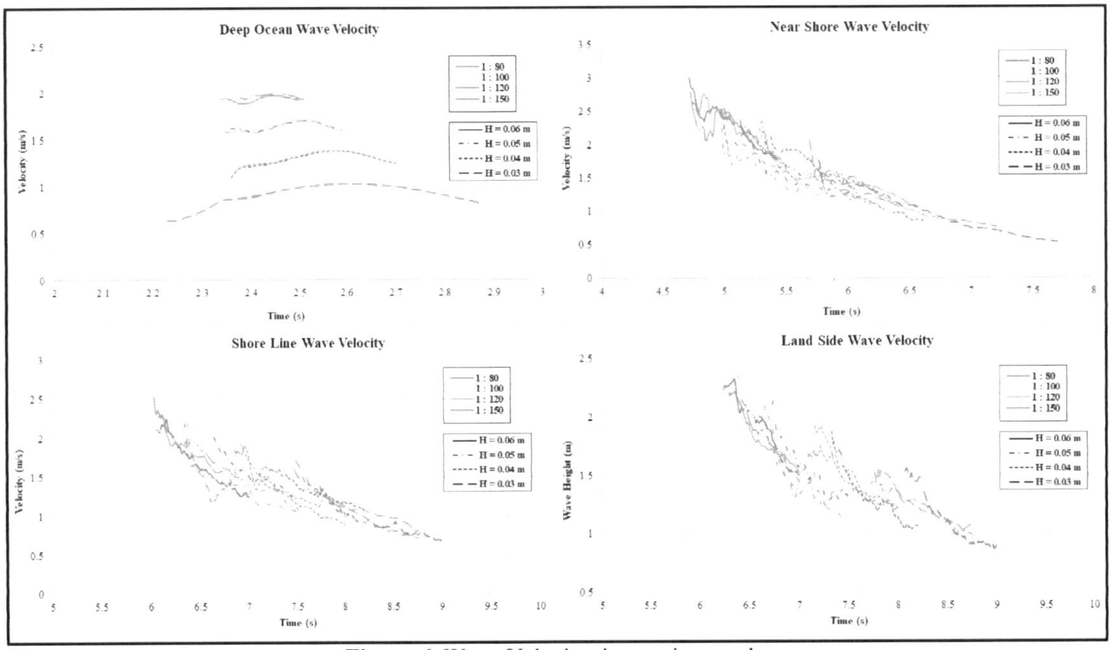

Figure 6. Wave Velocity time series graph.
(a. Deep Ocean, b. Nearshore, c. Shoreline, d. Land Side)

3.3. Depth averaged velocity

Since the wave moves as a water column/profile, the maximum velocity does not represent its significant values. Therefore, the depth averaged velocity ($\bar{U}$) for every time step was calculated by using Equation 6. From these results, the Maximum value of $\bar{U}$ ($\bar{U}_{max}$) can be obtained as the significant velocity of the wave. $\bar{U}_{max}$ in the Land Side of the case is used as a reference for the comparison to $\bar{U}_{max}$ in the Nearshore area in the same time steps. The maximum wave height in the land side is also compared to the wave height in the nearshore area with the same time steps as the $\bar{U}_{max}$. The graphs are presented in Figure 7.

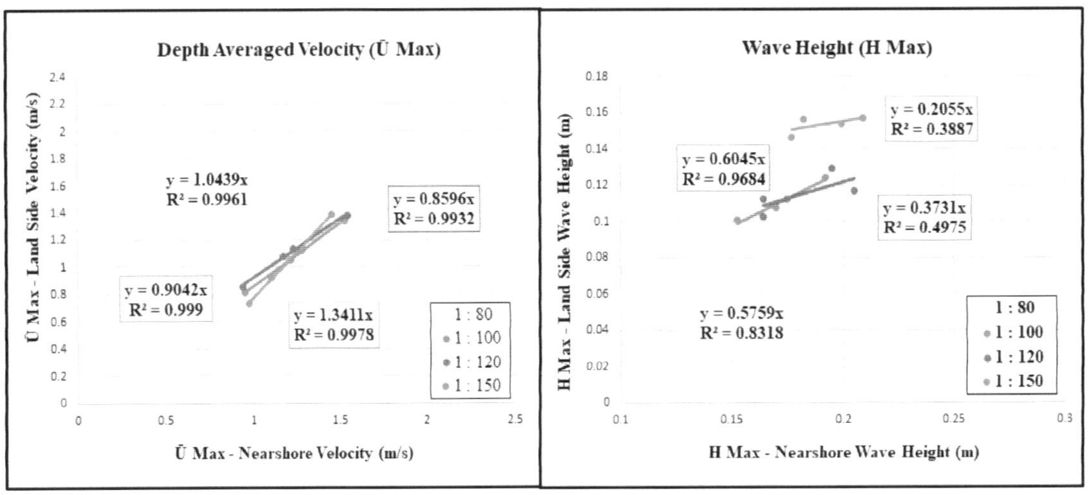

Figure 7. Comparison between two values in the land side and near shore area. (a. $\bar{U}_{max}$ and b. H_{max})

Based on the graph, The highest $\bar{U}_{max}$ value in the land side area is 1.69 m/s with the steepest coastal slope of 1:80. The minimum value is 0.74 m/s with the flattest coastal slope of 1:150 (see Figure 7.a). On the contrary, the lowest wave height that occurs in the land side area came from the steepest slope of 1:80 with the value of 0.07 m, and the highest wave height came from the flattest slope of 1:150 with the value of 0.16 m (see Figure 7.b).

3.4. Graph summary and Estimated Time of Arrival (ETA)

The maximum values of the case parameters are summarized into Table 2. Those parameters are flow depth maximum, depth averaged velocity maximum, and the arrival time of the wave (see Table 2). The maximum values of flow depth and velocity parameters have already been mentioned in the previous section. From Table 2, averaged value of each parameters was calculated based on each slope. So, the averaged parameter values represent every slope it had. The averaged value is shown in Table 3.

The steeper coastal slopes have bigger advantages in diminishing wave velocity compared to flatter slopes. Weaker velocity will benefit most infrastructures from massive damages caused by Tsunami waves. On averages, Banda Aceh and Aceh Besar have coastal slopes of 1:150 [5].

Table 2. Simulation case summaries

Cases		Designed Wave Height (m)	Wave Height Maximum (m)				Depth Averaged Velocity Maximum (m/s)		Arrival Time (s)	
Slopes	Water Level (m)		WG #1	WG #2	WG #3	WG #4	WG #2	WG #4	WG #2	WG #4
1 : 80	0.11	0.06	0.480	0.284	0.274	0.303	1.622	1.691	4.72	6.24
		0.05	0.425	0.268	0.245	0.269	1.346	1.427	4.99	6.73
		0.04	0.357	0.252	0.224	0.213	1.157	1.241	5.30	7.33
		0.03	0.297	0.239	0.209	0.205	0.930	0.963	5.76	8.07
1 : 100	0.09	0.06	0.480	0.280	0.286	0.301	1.528	1.340	4.68	6.27
		0.05	0.424	0.270	0.269	0.284	1.275	1.124	4.98	6.71
		0.04	0.355	0.253	0.229	0.259	1.210	1.051	5.27	7.22
		0.03	0.295	0.231	0.221	0.204	0.945	0.815	5.73	7.91
1 : 120	0.08	0.06	0.480	0.278	0.317	0.329	1.546	1.380	4.70	6.23
		0.05	0.422	0.263	0.304	0.309	1.222	1.130	4.96	6.65
		0.04	0.353	0.251	0.287	0.336	1.166	1.080	5.29	7.16
		0.03	0.294	0.227	0.215	0.203	0.938	0.852	5.73	7.79
1 : 150	0.06	0.06	0.480	0.288	0.323	0.351	1.451	1.392	4.70	6.17
		0.05	0.423	0.263	0.294	0.338	1.268	1.124	4.97	6.59
		0.04	0.353	0.246	0.299	0.354	1.095	0.922	5.28	7.08
		0.03	0.294	0.225	0.220	0.213	0.972	0.739	5.70	7.71

Table 3. Averaged parameter summaries.

Cases	Averaged Wave Height on Land Side (m)	Averaged Ūmax (s)		Averaged Arrival Time (s)	
Slopes	WG #4	WG #2	WG #4	WG #2	WG #4
1 : 80	0.25	1.26	1.33	5.19	7.09
1 : 100	0.26	1.24	1.08	5.17	7.03
1 : 120	0.29	1.22	1.11	5.17	6.96
1 : 150	0.31	1.20	1.04	5.16	6.89

Estimated Time of Arrival (ETA) can be obtained from the *.csv data. The longest times for a wave to get to the nearshore and land side came from the steepest slope of 1:80 which are 5.19 s and 7.09 s. On the other side, the shortest times came from the flattest slope of 1:150 which the averages are 5.16 s and 6.89 s.

4. Conclusion

This study was conducted to investigate numerical simulation results to obtain the role of coastal slope variations to reduce tsunami waves energy; in term of tsunami wave heights, tsunami wave velocity, and wave arrival time. The maximum values recorded were: 0.351 m wave height on land side of a 1:150 slope, 1.691 m/s of velocity on land side of a 1:80 slope, and 7.09 s of arrival time on a 1:80 slope.

The study shows that steeper coastal slopes have longer wave travel times to the shoreline than flatter coastal slope areas. However, wave height in steeper coastal slopes is slightly shorter than in flatter coastal slopes. Steeper slopes also have higher velocities than flatter areas. This would benefit most infrastructures from damage caused by tsunami waves. Longer arrival times would benefit people in evacuation times, and smaller velocities would reduce the impacts of future potential tsunamis in this area. Nevertheless, topography and other related factors are the most critical in influencing tsunami wave impacts.

5. References

[1] Natawidjaja D H and Wahyu T 2007 *The Sumatran Fault Zone-From Source to Hazard* (Journal of Earthquake and Tsunami) **1(01)** pp 21-47.

[2] Crespo A J C, Domíngueza J M, Rogers B D, Gómez M-Gesteira, Longshaw S, Canelas R, Vacondio R, Barreiro A., and García-Feal O 2015 *DualSPHysics: Open-source parallel CFD solver based on Smoothed Particle Hydrodynamics (SPH)* (Computer Physics Communications) **187** pp 204–216.

[3] Altomare C, Crespo A J C, Domínguez J M, Gómez M-Gesteira, Suzuki T, and Verwaest T 2015 *Applicability of Smoothed Particle Hydrodynamics for estimation of sea wave impact on coastal structures* (Coastal Engineering) **96** pp 1–12.

[4] Soulsby R 1997 *Dynamics of Marine Sand* (New York: H.R. Wallingford)

[5] Al'ala M, Syamsidik, and Kato S 2017 *Predicting Impact of SLR on Coastal Flooding in Banda Aceh Coastal Defences* (Banda Aceh: AIP Conference Proceedings 1892) 100004; doi: 10.1063/1.5005736

[6] Tursina and Syamsidik 2018 *Numerical Simulations the Role Elevated Road Construction on Reducing Impacts of Tsunami Waves Using DualSPHysics* (Ikatan Ahli Kebencanaan Indonesia (Iabi))

[7] Rasyif T M, Kato S, Syamsidik, and Okabe T 2017 *Preliminary Study on Performance of a Coupled Hydrodynamic and Sediment Transport Model on Small Domain* (AIP Conference Proceedings) **1892(1)**:100001 DOI:10.1063/1.5005733

[8] Tursina and Syamsidik 2017 *Numerical Simulations of Land Cover Roughness Influence on*

Tsunami Inundation in Ulee Lheue Bay, Aceh-Indonesia (IOP Conference Proceedings) **56** 012009 doi:10.1088/1755-1315/56/1/012009
[9] Altomare C, Lomonaco P, Gonzales J.-Cao, Domíngueza J M, Crespo A J C , and Gómez M-Gesteira 2017 *Generation of trains of tsunami-like solitary waves in DualSPHysics model* (Ourensen: 12th Conference paper of International SPHERIC workshop)
[10] Ren B, Wen H, Dong P, and Wang Y 2014 *Numerical Simulation of Wave Interaction with Porous Structures using an Improved Smoothed Particle Hydrodynamic Method* (Coastal Engineering) **88** pp 88–100

Acknowledgement
Authors would like to thank the Ministry of Research, Technology and Higher Education (RISTEKDIKTI) of Indonesia through the Research Grant No. SK.60/UN11.2/SP3/2018 Year 2018 from PKLN scheme. The title of the PKLN research grant is *Mitigating Impacts of Tsunami Waves on Coastal Structures and Harbor Facilities.*

Also, authors would like to thank the USAID by granting a research grant in the Partnership Enhanced Engagement in Research (PEER) Cycle 5 and National Academy of Sciences (NAS) with the research title: "*Incorporating climate change induced sea level rise information into coastal cities' preparedness toward coastal hazards* (#5-395)." under Federal Award Identification Number (FAIN): AID-OAA-A-11-00012 (sub-award Number 2000007546) which funded this research. Authors are also grateful for the member of Coastal Research Group and TDMRC Syiah Kuala University for support and help.

Chlorophyll-a, and Sea Surface Temperature (SST) as proxies for Climate Changes: Case Study in Batu Ampar waters, Riau Islands

S Antoni[1*], R A Bantan[1], T A Al-Dubai[1], M Z Lubis[2], W Anurogo[2], R D Silaban[2]

[1]Marine Geology Department, Faculty of Marine Sciences, King Abdulaziz University, Jeddah 21589, Saudi Arabia

[2]Geomatics Engineering, Politeknik Negeri Batam, Batam 29461, Indonesia

*Email: suddin@stu.kau.edu.sa

Abstract. The condition of Indonesian oceanography is largely determined by the development of the Monsoon Wind and the Indonesian Cross Flow (ARLINDO). The development of marine remote sensing technology that is so fast, able to make it easier to map the condition of waters in Indonesia in an actual way easily and cheaply. This can make it easier to obtain information about climate change approaches to the physical oceanography conditions in Indonesian waters. The purpose of this research is to study the changes of the physical oceanography parameters resulting from climate changes. Based on the results of the research that for the highest Sea Surface Temperature (SST) exist in 2013 (above 30,75^{o} C). It has increased from the previous year i.e. in 2012 with an increase of SST value of 30 ^{o} C. In 2013 SST is spread high on the Batu Ampar Permit - Harbor Bay ferry port. Analyzing the value of chlorophyll-a, sea surface temperature, and sea level to see global climate change in the waters of Batu Ampar. The analysis shows the highest sea level with an elevation of 3.1 meters, with sea surface temperature in the range of 29,5^{o} C - 29,75 ^{o} C, at the time of recording 00:05-02:55 (August 1, 2017 - September 1, 2017). The results of the analysis obtained from the oceanographic conditions in the waters of Batu Ampar did not experience a fluctuating change in global climate change in the waters. The annual SST interval spacing interval of 0.5-0.75 ^{o}C and the changing phase relationship could have an impact on subsequent climate change and sea level rise to be concrete evidence. Reduced levels of chlorophyll-a each year will have an impact on the heat of the sea level, thus rising sea levels and indicating a climate with hot temperatures (ice at the poles melt).

Keywords: Climate Change, Chlorophyll-a, Sea Surface Temperature (SST), Batu Ampar waters

1. Introduction

One of the impacts of global warming is climate change affecting the condition of the marine ecosystem [1], and also the pattern of wind movement in Indonesia in general with the movement of seasons [2]. Each season has a different direction of wind movement. In marine ecosystems, oceanographic parameters such as Sea Surface Temperature (SST), chlorophyll-a, tidal seawater can be used to determine the relationship of oceanographic conditions to climate change. Chlorophyll-a is a pigment commonly called an important factor contained in phytoplankton in the process of photosynthesis [3-5]. While the photosynthesis process influenced by the temperature either directly or indirectly. Franz et al. (2005) stated that global chlorophyll content has undergone changes in recent years due to the influence of SST. On the other hand, sea surface temperature plays a pivotal role in variations of the tidal conditions [7-9]. However, the water mass movements affect the variability of surface oceanic variables such as SST and chlorophyll-a [10]. Thus, the oceanographic conditions are associated indirectly with the climate changes in the area. Sea surface temperature and chlorophyll-a are important parameters oceanographic sciences that function in the process of improving the aspects of marine and fishery resources. Sea surface temperature can be used as an indicator of the predicted location of upwelling, downwelling, fronts associated with the potential area of tuna [11]. While the values of chlorophyll-a located at sea level is an indicator of fertility and productivity for aquatic [12].

Mapping in Indonesian`s areas especially Batam island is very scarce until recently, however, some maps have already been completed, particularly with regard to sea surface temperature [13]. Therefore, this research aims to study the variations of the oceanographic conditions resulted from climate changes in Batu Ampar waters area, Riau Islands using Sea Surface Temperature (SST) and Chlorophyll-a data obtained from Terra Modis satellite using level 3 data.

2. Materials and Methods

2.1 *Location and research flow diagram*

This research was conducted in August 2017 at Punggur Sea, Batam, Kepulauan Riau-Indonesia, located between latitudes 1 ° 03.2448 and 1 ° 03.3977 N and longitudes 104 ° 08,7102 and 104 ° 08.8133 E (Figure 1), and the research flow diagram can be seen in Figure 2.

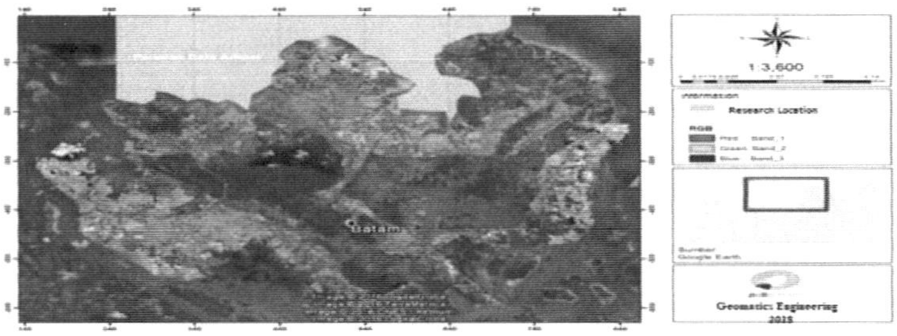

Figure 1. Map of research location

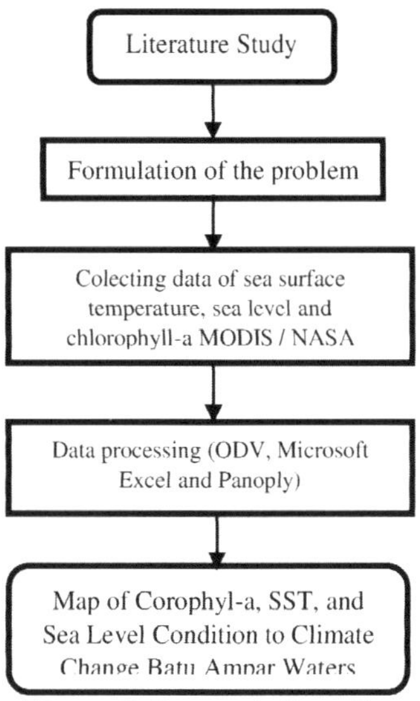

Figure 2. Research flow diagram

2.2 *Chlorophyll –a*

The studied areas at the Batu Ampar waters and Riau Islands (Figure 1), obtained chlorophyll-a data using Terra-MODIS 3 level satellite images with temporal resolution (9 km) from 2012 to August 2016 (during 5 years). The data was obtained from Ocean Color Web website from NASA - USA [14]. Terra-MODIS imagery processed in consideration of 7 cloud-free images collected annually until August 2016. Chlorophyll-a has a concentration using the following equation:

$$C = a^{(a_0+a_1R+a_2R^2+a_3R^3+a_4R^4)}$$

$$a = 0.2830,-2.753,1.457,0.659,-1.403$$

Where,

$R = \log_{10}\left(\frac{R_{rs}443>R_{rs}488}{R_{rs}551}\right)$

R_{rs} is a reflectant sensor

2.3 *Sea Surface Temperature (SST)*

Sea Surface Temperature (SST) data were obtained using Terra-MODIS / NASA 3 level temporal resolution (9 km) images from 2010 to August 2017 (during 7 years). Terra-MODIS imagery processed with the consideration of 8 cloud-free images collected every August each year. Analysis of SST data have done by using a thermal infrared channel consisting of band 10 - band 14, so utilizing each band can be calculated each radian value and brightness temperature. To get the radians value used the equation, that is:

$$rad = (DN-1) * UCC_{band}$$

Where:

rad = Radian value (*Watts*/ m^2sr m)

DN = Digital number on each band

UCC_{band} = Unit conversion coefficients, (watts/ meter^2/ steradian/ micrometer)/ DN

Then convert the radian value to the temperature value, the following equation is used:

$$BT = \frac{C_2}{\lambda c Ln(\frac{C1}{\lambda c^5 rad}+1)}$$

Where:

BT = Brightness temperature (^{0}K)

Λc = Central Wavelength (m)

c1 = Constants (3,741775 x 10^{-22} W-m^3-μ^{-1})

c2 = Constants (0,0143877 m-^{0}K)

rad = Radian Spectral Value (watts/ m^2sr m)

The data obtained is converted into units of celcius where:

$$T\ °Celsius = T\ °Kelvin - 273°$$

$$SST = a + b10*T10 + b11*T11 + b12*T12 + b13*T13 + b14*T1$$

Where:

T10, T11, T12, T13, T14 = *Brightness temperature* of *band*

3. Results and discussion

3.1 *Chlorophyll –a*

Based on the results of data processing obtained, it is showed that the distribution of chlorophyll-a concentration (above 1 mg/m^3) occurred in 2011, while it was (above 2 mg/m^3) in the previous year, ranging from 2 and 2.3 mg/m^3. Whereas the distribution of chlorophyll-a concentration during 2014 ranged between 1.1 - 1.3 mg/mg^3, while in 2013 in the varied from 2.1 and 2.3 mg/mg^3. This result can be expressed as a high-level chlorophyll group [15-16]. Also, it reveals the significant changes in chlorophyll-a levels caused by climate change. Therefore, the occurrence of these changes in chlorophyll-a concentrations can be considered as an indicator of climate change.

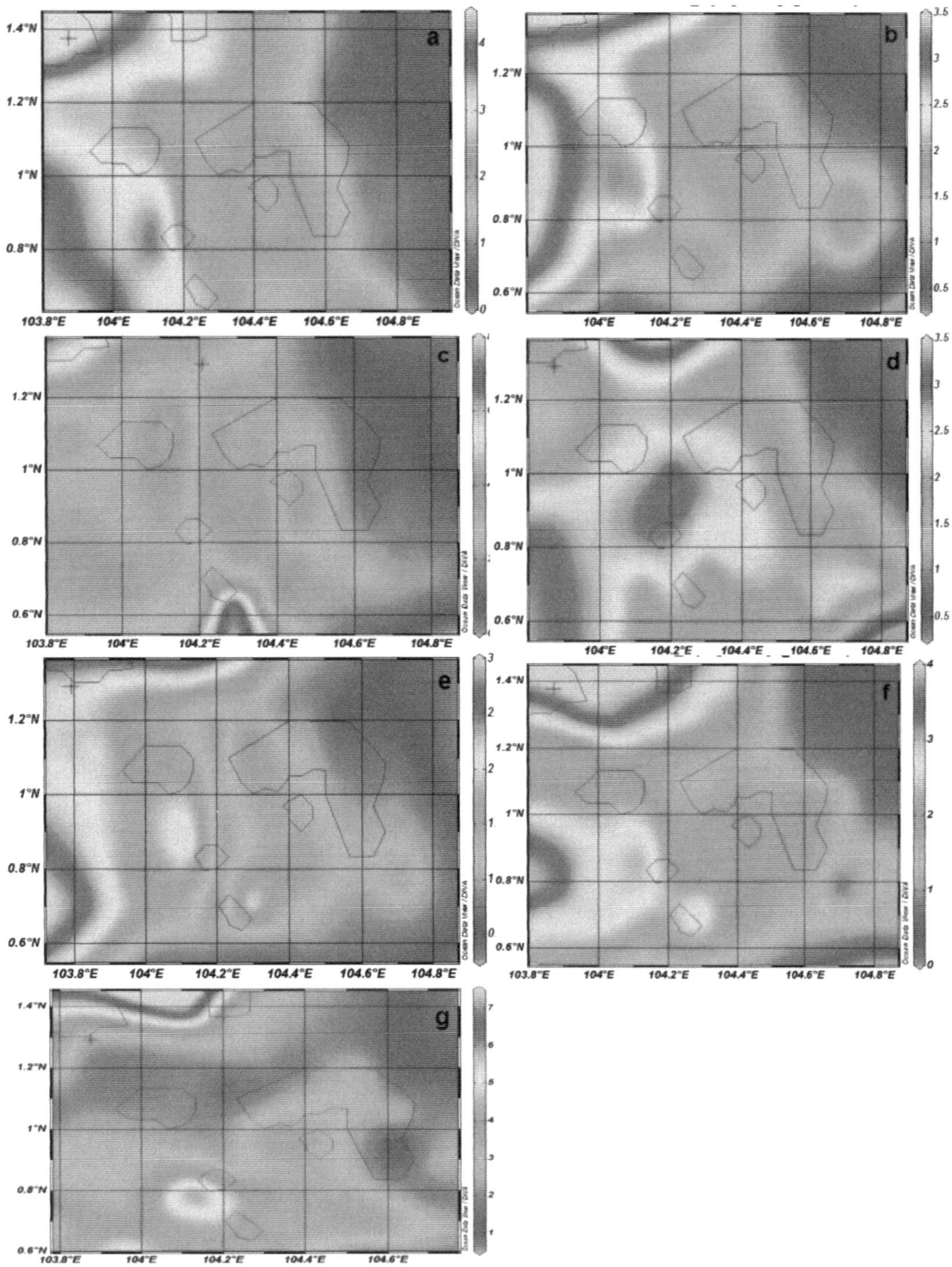

Figure 3. Chlorophyll a (a = 2010, b = 2011, c = Year 2012, d = Year 2013, e = Year 2014, f = Year 2015, g = Year 2016)

In 2015 the Batu Ampar waters (around the harbour bay ferry) show the distribution of chlorophyll-a in the range 1.9-2.1 mg / mg^3, which indicates the chlorophyll-a content of that year is at a moderate level. In 2016, there was no significant change in chlorophyll-a from the previous year, as evidenced by the chlorophyll-a levels in the range of 1.9 mg / mg^3 , but only a different distribution than the previous year in which chlorophyll-a spread western region of Batam city that is in Sekupang waters area.

3.2 *Sea Surface Temperature (SST)*

Based on the results of the research, the highest SST value occurred in 2013 (above 30,75°C). Where increased with an average 3°C from the previous year i.e. in 2012. In 2013, SST was spread highly on the Batu Ampar Permit - Harbor Bay ferry port. The indication of this SST increase occurs unevenly with sea surface temperature (SST) condition with the value of 30.5°C in the northern area of Batam city that is the waters of steel-sekupang.

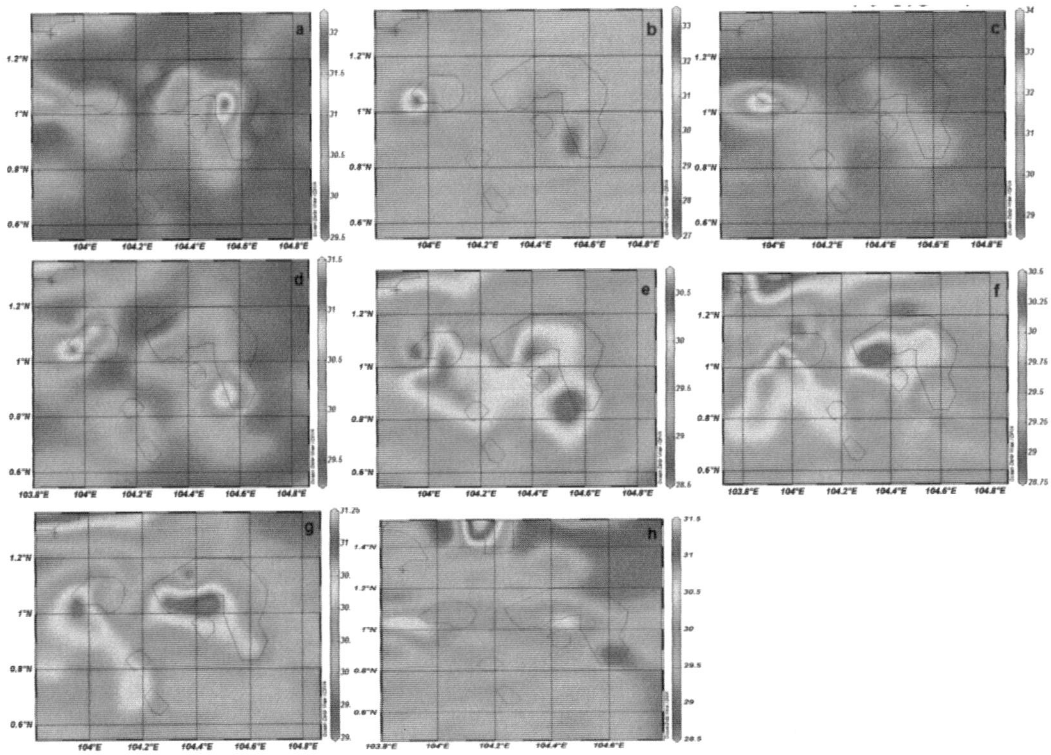

Figure 4. SST (a = 2010, b = 2011, c = Year 2012, d = Year 2013, e = Year 2014, f = Year 2015, g = Year 2016, h = August 2017)

Oceanographic conditions occurred in 2014 where there was an impairment of SST to 30.15°C (previous year was 30,75°C). Hence, this analysis showed the effect of the climatic changes on the variations of SST in the area. This year the SST value dominates in the northern part of Batam (Batu Bay-Harbor Bay Ferry), whereas in the north-western region of Batam (Sekupang) occurs in the range of 29.5°C. For the lowest SST occurred in 2011 (29°C), where there was evenly distributed for the Batu Ampar-Nongsa-Sekupang Waters region (blue tasks). This indication is decreased from the previous year SST value of 29,75°C (purple). This will affect the condition of marine ecosystems located in Sekupang area [17].

In the year 2015, there was an uneven spread of SST scores. For Sekupang and Batam city area, the SST temperature was in the range of 29.2-29.5°C. As for the territorial waters of Batu Ampar and Nongsa was in the range 29.05-29.0°C. The increase of SST occurred in the next year 2016 which happened unevenly for all parts of North of Batam city. For Sekupang-Lubuk steel SST area was 30.6°C. Then at Batu Ampar-Bengkong Waters area, the SST level was 30.4°C. While the Nongsa area, the SST value was 30°C. The increase of SST from 2015-2016, probably can be considered an indicator of changing climate conditions in Batam City every year.

Besides using the data, in this study also used SST Terra Modis / NASA data with resolution 4 km in 2000, 2004, 2007 and 2017 as other comparisons (Figure 5).

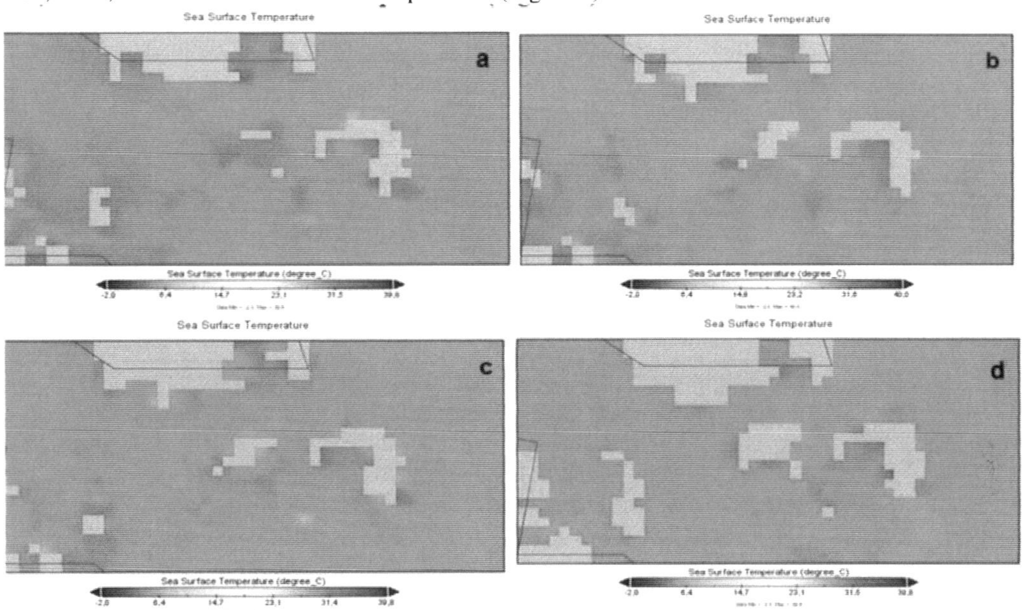

Figure 5. Panoply (a = Year 2000, b = Year 2004, c = Year 2007, d = Year 2017)

From the figure above shows that from 2000, 2004, 2007 and July 2017, the SST values of Batu Ampar water area ranging between 27 °C and 31°C (Figure 5). The results of the analysis using SST anomaly using anomaly data on data recording that is dated 1 August 2017 and 1 September 2017. In that month spread of SST value to Bintan, Region Singapore and Malacca Strait. In September, the SST Spread There was a decline. SST objects in anomaly move towards the southern Riau Islands and more dominantly located in the eastern part of Sumatra include Bangka-Belitung, to Palangkaraya. For the area, the SST Anomaly is in the range of -0.5-0.00°C (Blue light blue) (Figure 6).

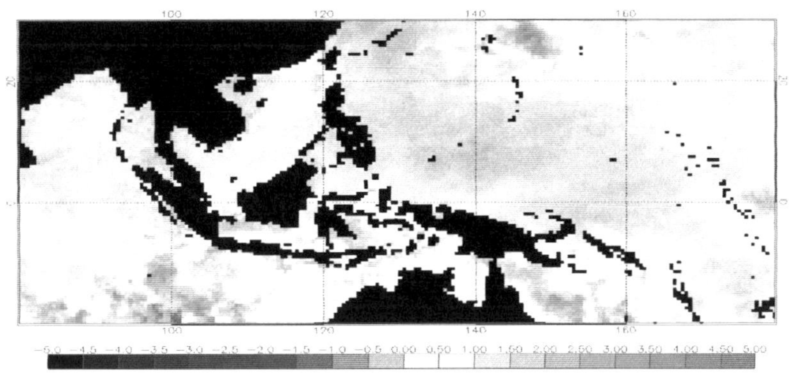

Figure 6. Sea surface temperature anomaly (SST) in August-September 2017.

4 Conclusion

From the results of research conducted using the data of 2000, 2004 and 2007 that the SST concentration decreased not very significant (low), while for the year 2010 -2017 produced that the SST concentration experienced a moderate categorized increase. This is evidenced by the widening of SST coverage in 2010-2017. As for chlorophyll-a in 2010-2016 had a low decrease. The annual SST interval spacing interval of 0.5-0.75 and the changing phase relationship will have an impact on subsequent climate change and sea level rise to be concrete evidence. Reduced levels of chlorophyll-a each year will have an impact on the heat of the sea level, thus rising sea levels and indicating a climate with hot temperatures (ice at the poles melt).

5 References

[1] Henson S A, Beaulieu C, and Lampit R 2016 Observing climate change trends in ocean biogeochemistry: when and where. *Global change biology*, 22(4):1561-1571

[2] Lubis M Z, Anurogo W, Kausarian H, Surya G, and Choanji T 2017 Sea Surface Temperature and Wind Velocity in Batam Waters Its Relation to Indian Ocean Dipole (IOD). *Journal of Geoscience, Engineering, Environment, and Technology*, 2(4):255-263

[3] Feng J F, and Zhu 2012 Changing trends and relationship between global ocean chlorophyll and sea surface temperature. *Procedia Environmental Sciences,* 13:626-631

[4] Surya G, Khoirunnisa H, Lubis M Z, Anurogo W, Hanafi A, Rizky F, Timbang D, Dewi A, Situmorang, Guspriyanto D, Ramadhan W and Mandala G F T 2017 Karakteristik Suhu Permukaan Laut dan Kecepatan Angin di Perairan Batam Hubungannya dengan Indian Ocean Dipole (IOD). *Dinamika Maritim,* 6(1):1-6

[5] Lubis M Z, Daya A P, Suzita S, Silaban R D, Anjani M, Perananda A and Ghazali M. 2017 Karakteristik Kondisi Fisik Oseanografi Menggunakan Citra Landsat 8 di Laut Batam. *Dinamika Maritim*, 6(1):12-17

[6] Franz B A, Werdell P J, Meister G, Bailey S W, Eplee, R E, Feldman G C and Thomas D 2005 The continuity of ocean color measurements from SeaWiFS to MODIS. In Earth Observing Systems X (Vol. 5882, p. 58820W). *International Society for Optics and Photonics*

[7] Yoga B, Bima R, Setyono H, and Harsono G 2014 Dinamika Upwelling dan Downwelling Berdasarkan Variabilitas Suhu Permukaan Laut dan Klorofil-A di Perairan Selatan Jawa. *Jurnal Oseanografi*, 3(1):57-66

[8] Khoirunnisa H, Lubis M Z, and Anurogo W 2017 The Characteristics of Significant Wave Height and Sea Surface Temperature In The Sunda Strait. *Geospatial information*, 1(1).

[9] Lubis M Z and Khoirunnisa H 2016 Dinamika Pantai Praikalogu di Provinsi Nusa Tenggara Barat, Indonesia. *Jurnal Integrasi*, 8(2):125-133

[10] Kunarso K Hadi S Ningsih N S and Baskoro M S 2012 Variabilitas suhu dan klorofil-a di daerah upwelling pada variasi kejadian ENSO dan IOD di perairan selatan Jawa sampai Timor. ILMU KELAUTAN: *Indonesian Journal of Marine Sciences*, 16(3):171-180

[11] Setyono H and Harsono G 2014 Dinamika upwelling dan downwelling berdasarkan variabilitas suhu permukaan laut dan klorofil-a di perairan Selatan Jawa. *Journal of Oceanography,* 3(1): 57-66.

[12] Daya A P and Lubis M Z 2017 Pemetaan Parameter Oseanografi Fisik Menggunakan Citra Landsat 8 di Wilayah Perairan Nongsa Pulau Batam. *Jurnal Integrasi,* 9(1): 09-15

[13] Lubis M Z, Taki H M, Anurogo W, Pamungkas D S, Wicaksono P and Aprilliyanti T 2017 Mapping the Distribution of Potential Land Drought in Batam Island Using the Integration of Remote Sensing and Geographic Information Systems (GIS). In IOP Conference Series: Earth and Environmental Science (Vol. 98, No. 1, p. 012012). IOP Publishing.

[14] Maritorena S, Siegel D A and Peterson A R 2002 Optimization of a semianalytical ocean color model for global-scale applications. *Applied Optics*, 41(15): 2705-2714

[15] Antoni S, Bantan RA, Taki HM, Anurogo W, Lubis MZ., Al Dubai TA, Al-Zubieri AG. 2018 The extent of agricultural land damage in various tsunami wave height scenarios: Disaster management and mitigation. Intl Arch Photogramm Remote Sens Spatial Inf Sci. DOI: 10.5194/isprs-archives-XLII-3-W4-51-2018.

[16] Syahdan M, Atmadipoera A S, Susilo S B and Gaol J L 2014 Variability of surface chlorophyll-a in the Makassar Strait–Java Sea, Indonesia. International J. of Sciences: Basic and Applied Research (IJSBAR), 14(2): 103-116
[17] Lubis M Z Anurogo W, Hanafi A, Kausarian H, Taki H M, Antoni S 2018 Distribution of benthic habitat using Landsat-7 Imagery in shallow waters of Sekupang, Batam Island, Indonesia. Biodiversitas 19:1117-1122.

Acknowledgments
The study and publication was financed by Politeknik Negeri Batam (P2M) For Support All Activities.

IOP Conf. Series: Earth and Environmental Science **273** (2019) 012013 doi:10.1088/1755-1315/273/1/012013

Study of the Effect of River Basin Morphology Change on Threshold Parameters in Cimahi Flood Early Warning System

Ariani Budi Safarina[1], Iin Karnisah[2], Ade Sena Permana[3]

[1]Civil Engineering Department, Universitas Jenderal Achmad Yani, Jalan Terusan Jenderal Sudirman PO BOX 148, Cimahi, 40511, Indonesia
[2]Civil Engineering Department, Politeknik Negeri Bandung, Jalan Kampus Polban Ciwaruga, Bandung, 40559, Indonesia
[3]Electrical Engineering Department, Universitas Jenderal Achmad Yani, Jalan Terusan Jenderal Sudirman PO BOX 148, Cimahi, 40511, Indonesia

Corresponding author's e-mail address: ariani.budi@lecture.unjani.ac.id

Abstract. Flood early warning systems are used for flood disaster preparedness to reduce disaster risks. The problem is damage of water level sensor due to aggradation which also results in river capacity reduced. The purpose of this research is to determine the effect of river basin morphology change on threshold parameters in flood early warning system. Measured hydraulic parameters are flow depth, flow velocity, and riverbed slope. Land cover was conducted using Cimahi river basin land use map to determine runoff coefficient. Rainfall-runoff model used is the Deconvolution Software program to determine unit hydrographs. Results of this study indicate that morphological changes of watershed significantly influence threshold parameters. A 0.5 m rise in river bed can reduce river capacity by 41.8% and narrowing of 16 m-wide river border dropped river capacity from 730 m^3/s to 95.2 m^3/s. An increase of 17 % runoff coefficient, increased direct runoff by 17%. Reducing this river capacity makes threshold parameters in flood early warning system invalid causing flood events cannot be detected earlier by the system.

1. Introduction

Flood early warning systems are used for flood disaster preparedness to reduce disaster risks. The system uses threshold parameters to provide flood information quickly to communities and stakeholders via telemetry networks [1]. The commonly used threshold parameter is the river water level [2]. In Cimahi flood early warning system, there are two parameters of threshold those are river water level and rainfall. The problem that happened was flooding but the warning alarm did not ring. Based on field observations, this occurs because the change in threshold parameters where the water level that causes the flood is lower than the threshold water level that has been set in the system. Similarly, for threshold rainfall parameters, floods occur before the precipitation reaches the threshold limit [3].

Morphology characteristics analysis is essential to any hydrological study, so the determination of stream networks behavior and their interrelation with each other is of great importance in many water resources studies[4]. Parameters of basin morphology such as riverbed slope, flow depth, flow velocity, river border and also land cover affect river capacity or discharge. Based on the study, it is known that there is a logarithmic relationship between land use and the number of abstractions. In the dry season, the amount of abstraction is 16% greater than in the rainy season. For the same watershed slope, abstraction is greater if the amount of forest and gardens are larger. In addition, the abstraction is also influenced by the watershed area with logarithmic relationships, where the abstraction is greater with the increase in the area but in the area above 1000 km^2, the addition is slow or asymptote [5].

River morphology is changing from its natural channel due to human activities, such as sand excavation from the bed, agricultural activity, disposal of municipal waste and construction on the river. The environment of the river should be maintained and would not be extremely disturbed by human activity [6]. River is important part of human being which is continually changing from its evolutions. The river slope, velocity and nature of river are responsible perimeters for changing the river shape and size. Throughout study at Pravara River showing that river stream changes owing to various natural and manmade phenomena [7].

The purpose of this research is to determine the influence of river basin morphology change on threshold parameters in flood early warning system. The river morphology parameters reviewed are bed river elevation, land cover, and river border. Change of bed river elevation affect on river water level measurement while land cover affect the magnitude of runoff coefficient. Changes in river borders occur in urban areas that use river banks for construction and activities. The relationship between changes in river basin morphology parameters to the change of threshold parameters can also be used as a guide in river management and flood control.

2. Methods of The Study

This research was conducted by field data analysis using mathematical model to get correlation between morphology parameters of watershed and threshold parameters of flood early warning system. There are 15 cross-sections of the river that are used scattered in the upstream, midstream and downstream areas of Cimahi river basin, and then one of the cross-section of the river is determined as a threshold. The measured hydraulic parameters are the flow depth, flow velocity and the slope of the river bed. The land cover analysis was conducted using the Cimahi river basin land use map to determine the runoff coefficient. Water level of the river on the threshold cross section and the rainfall threshold are determined using the rainfall-runoff model where the direct runoff discharge is determined by the convolution method which is the function of effective rainfall and unit hydrograph.

Discrete convolution equation for the linear system used is shown in equation 1,

$$Qn = \sum_{m=1}^{n \le m} P_m U_{n-m+1} \qquad (1)$$

Where Q is direct runoff discharge, P is effective rainfall and U is unit hydrograph ordinate.

Landuse of a watershed shows the type of surface or land cover of the watershed. The land surface affects the amount of water infiltrated into the soil, so the type of land surface becomes the determining variable of the flow coefficient or runoff coefficient [8][9]. If $\sum_{m=1}^{M} R_m$ is the total rainfall and r_d is the runoff depth, then the runoff coefficient can be defined as:

$$C = \frac{r_d}{\sum_{m=1} R_m} \qquad (2)$$

In actual case, uniform flow is practically parallel flow. Gradually varied flow may also be regarded as parallel flow, since the change in depth of flow is so mild that the streamlines have neither appreciable curvature nor divergence, that is, the curvature and divergence are so small that the effect of the acceleration components in the cross-sectional plane is negligible [10].

The continuity equation for steady flow with constant ρ is,

$$\iint_{cs} V.dA = 0 \qquad (3)$$

3. Results and Discussion

3.1. Study Area

Cimahi river basin is located in West Bandung Regency in the upstream, Cimahi City in the midstream and Bandung Regency in the downstream. The upstream of Cimahi river is located in a geographical position of 06°46'20'' south latitude and 107°34'42'' east longitude at an elevation of 1,482 m mean sea level and downstream at the geographical position of 06°57'37'' south latitude and 107°32'38'' east longitude at an elevation of 678 m mean sea level. Cimahi river is 30.6 km length and watershed area of 72.2 km^2. Cimahi river basin map is shown in Figure 1 below.

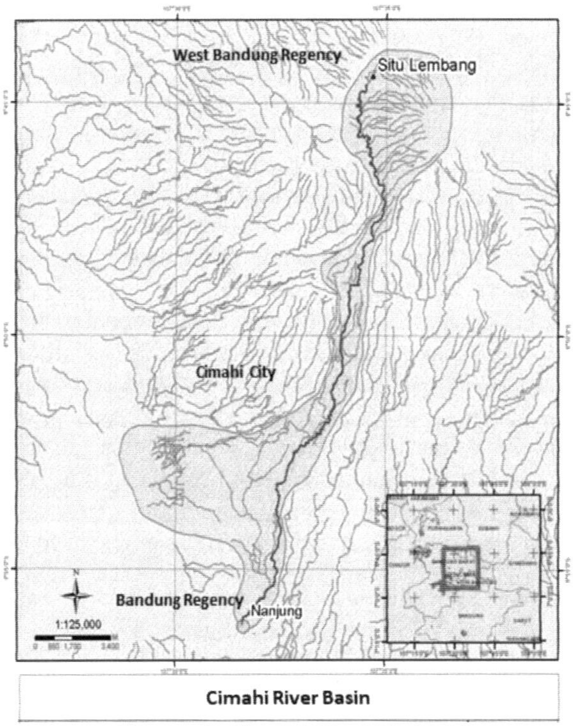

Figure 1. Cimahi River Basin Map

Runoff coefficient of Cimahi city increased from 0.57 in 2011 to 0.63 in 2017 with the highest percentage increase in residential and industrial areas which reduced the area of gardens and agriculture.

3.2. Observed Cross Section of Cimahi River

Five cross sections in the upstream region show the top width is about 48 m with an average channel elevation of 1450 mean sea level with an average slope of 17%. One of those cross sections is shown in figure 2 below,

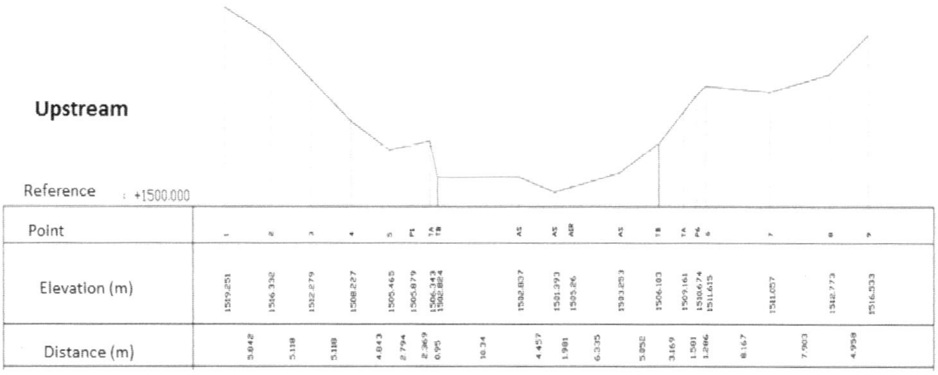

Figure 2. Upstream Cimahi River Cross Section

Five cross sections in the midstream region show the top width is about 12 m with an average channel elevation of 768 mean sea level with an average slope of 10.6%. One of those cross sections is shown in figure 3 below,

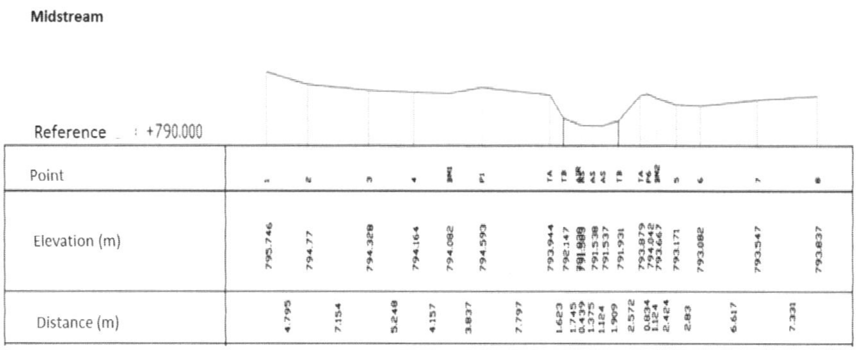

Figure 3. Midstream Cimahi River Cross Section

Five cross sections in the downstream region show the top width is about 19 m with an average channel elevation of 674 mean sea level with an average slope of 2.31%. One of those cross sections is shown in figure 4 below,

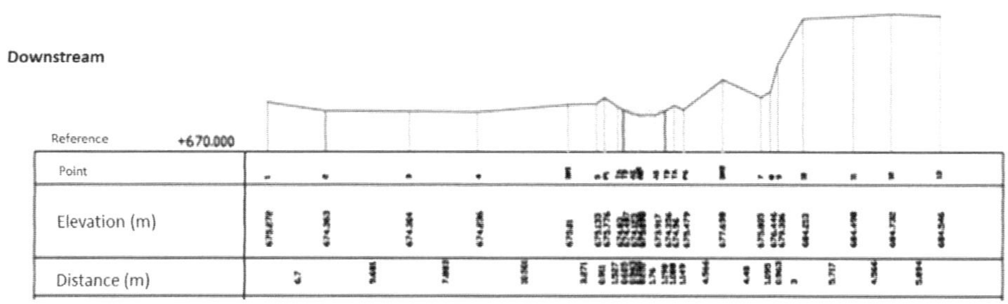

Figure 4. Downstream Cimahi River Cross Section

3.3 The Effect of Changes in River Morphology to Threshold Parameters

3.3.1 Change of River Border. Narrowing of the border on the cross-section of the Cimahi river is very heavy. The top width upstream is 48 m, narrowing in the middle to 12 m and downstream to 19 m. The narrowing of the border makes the cross-sectional capacity of the Cimahi river upstream of 730 m^3 / s decrease to 95.2 m^3 / s in the middle and 59.4 m^3 / s in the downstream. The capacity of the river due to the narrowing of this border resulted in the Cimahi river not being able to accommodate a number of design discharge. With Rational method, using the effective rainfall threshold in the middle which is determined based on the cross-sectional capacity of 29 mm with a duration of 1 hour, design discharge load for the downstream section is 332 m^3 / s, very much greater than its capacity of 59.4 m^3 / s. Change of river border is shown in Table 1.

Table 1. Change of River Border and The Effect to River Capacity

Cross Section	Top Width (m)	Maximum Area (m^2)	Q_{cap} (m^3/s)	Q_{load} (m^3/s); Ieff = 29 mm/h
Upstream	48	890	730	55
Midstream	12	18.5	95.2	184
Downstream	18	18.02	59.4	332

3.3.2 Change of Runoff Coefficient. Change in runoff coefficient is directly proportional to runoff volume. If the runoff coefficient increases, the runoff volume will also increase, resulting in greater discharge load. The rainfall threshold in the downstream area is based on river hydraulic and land use data in 2011 with a runoff coefficient of 0.57 is 29 mm with a discharge capacity of 59.4 m^3 / s, produce design discharge of 332 m^3/s. By using the rational Q = CIA equation, the increase in c by 0.63 will increase the design discharge becomes 367 m^3/s.

3.3.3 Change of River Bed Elevation. Aggradation caused by sedimentation can reduce river capacity. This happens a lot in tropical countries like Indonesia. As shown in Table 2, downstream section of the Cimahi river which is the threshold location of the Cimahi flood early warning system, rise in the elevation of the river bad as high as 10 cm can reduce river capacity by 7.3% and rise in river bed by 50 cm can reduce river capacity by 41.8%. To overcome this aggradation, threshold of the flood early warning system must be periodically updated.

Table 2. Change of River Bed at Threshold Location

Depth (m)	Discharge capacity, Qcap (m^3/s)
0.8	34.6
1.3	**59.4**
1.52	71.12
1.62	76.6
1.72	82.17
2	98.3
2.2	110.3
2.5	129.1
3.5	197.9

the effect of rise in river bed on the river capacity is shown in the regression equation in the following figure 5,

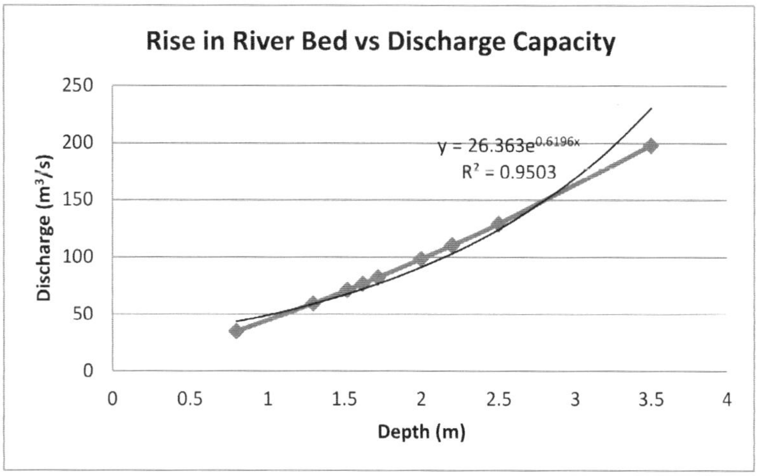

Figure 5. Correlation between Rise in River Bed and Discharge Capacity at Downstream Cross Section

The curve above is exponential curve but its limited by maximum discharge of the river at the cross-section. R^2 of 0.95 is good enough for good correlation because it's closed to one.

4. Conclusion

Changes in morphology and land cover have a very significant effect on the hydraulic parameters of the watershed ie river capacity and depth. Rise in runoff coefficient, increase the load on the river so that the risk of flooding is higher. Narrowing of the river border and the elevation of the river bed, significantly reduce the river capacity. Narrowing the width of the river from 48 m to 18 m, decreasing the discharge capacity from 730 m^3 / s to 95.2 m^3 / s, and increasing the elevation of river bed as high as 50 cm can reduce river capacity by 41.8%. Reducing this river capacity makes threshold parameters in flood early warning system invalid causing flood events cannot be detected early by the system.

References

[1] S.H.M. Fakhruddin Akiyuki Kawasaki and Mukand S.Babel 2015 Community Responses to Flood Early Warning System: Case Study in Kaijuri Union Bangladesh International Journal of Risk Reduction. 14(4): 1-9.

[2] Joel T de Castro, Gabriel M Salistre, Jr.Young-Cheol Byun and Bobby D Gerardo. 2013. Flash Flood Prediction Model Based on Multiple Regression Analysis for Decision Support System Word Congress on Engineering and Computer Science Proceedings Vol II 2013. San Fransisco, USA March 2013. 23-25

[3] Ariani Budi Safarina, Ramli, Muhammad Shiddiq Sayyid Hasharon 2017 Two Alert Flood Early Warning System Method Based on Rainfall-Runoff Model International Journal of Civil Engineering and Technology 8(12): 1032-1044

[4] Abiodun O.Adebola,Adebowale A,Adeseko & Matthew.Ibitoye 2016 Drainage Basin Morphology and Terrain Analysis of the Nigerian Section of Lake Chad River Basin, Nigeria using GIS and Remote Sensing Confluence Journal of Environmental Studies 10: 89-99

[5] Ariani Budi Safarina 2012 Analisa Pengaruh Topografi dan Pola Tataguna Lahan terhadap Abstraksi Daerah Aliran Sungai Berdasarkan Model Rainfall-Runoff Jurnal Riset Geologi dan Pertambangan 22(1):1-11

[6] Sailesh Samanta and Babita Pal 2012 Change Detection of Channel Morphology using GIS Technique International Journal of Advances in Science and Technology 4(2): 53-59

[7] Sainath P>Aher, Shashikant I.Bairagi, Pragati P.Deshmukh, Ravindra D Gaikwad 2012 River Change Detection and Bank Erosion Identification using Topographical and Remote Sensing Data International Journal of Applied Information Systems 2(3): 1-7

[8] L.Alfieri, P.Burek, E.DutraB.Krzeminski, D.MuraroJ.Thielen and F.Pappenberger 2013 GloFAS-Global Ensemble Streamflow Forecasting and Flood Early Warning Journal Hydrology and Earth System Sciences. 17: 1161-1175.

[9] Priya Menon K, Kala L 2017 A Review of Flood Monitoring: Design, Implementation and Computational Modules International Journal of Innovative Research in Computer and Communication Engineering. 5(2) :254-2359

[10] Chow VT, Maidment, Mays Larry W 1988 *Applied Hidrologi*, McGraw-Hill International

Acknowledgments
This research is fully supported by Ristekdikti Competitive Research, SP DIPA-0819/K4/KM/2018, and UNJANI Competitive Research SKEP/174/UNJANI/VIII/2018. The authors fully acknowledged Ministry of Research Technology Higher Education (Ristekdikti) and Jenderal Achmad Yani University for the approved fund which makes this important research viable and effective.

Landslide Susceptibility Mapping (LSM) in Kelud Volcano Using Spatial Multi-Criteria Evaluation

Syamsul Bachri[1]*, Sumarmi[1], Listyo Yudha Irawan[1], Sugeng Utaya[1], Farizki Dwitri Nurdiansyah[1], Alif Erfika Nurjanah[1], Lela Wahyu Ning Tyas[1], Akhmad Amri Adillah[1] and Denny Setia Purnama[1]

[1] Department of Geography, State University of Malang, Jl. Semarang 05, Malang 65145, Indonesia

Corresponding Author: syamsul.bachri.fis@um.ac.id

Abstract. Mount Kelud eruption on February 2014 has a tremendous impact on the surrounding physical environment which ejected more than 200x106m3 of material. Thus, triggered secondary hazard such as landslides in the surrounding area. The purpose of this study is to map landslides susceptibility using Spatial Multi-Criteria Evaluation (SMCE) approach in Kelud mountainous area in particular within KRB 1 and 2. The identification of landslide occurrence was conducted based on remote sensing data and field observation along with four considerations criteria; topography, hydrology, soil, and environmental characteristics. Each factor then reduced into several sub-criteria such as slope, aspect, topographic position index, topographic wetness index, stream power index, rainfall, soil texture, soil structure, COLE index and land use. The SMCE method was also engaged with expert judgment provided by academic university' view and the BPBD agencies. The result showed that the river channel and surrounding areas categorized as landslide high prone area. Furthermore, eruption material found as the sources of landslides occurrences.

Keywords: Kelud Volcano, Landslides, Spatial Multi-Criteria Evaluation, Disaster

1. Introduction

Mount Kelud eruption on February 2014 with VEI 4 has a tremendous impact on the surrounding physical environment and affects the condition of agriculture, livestock, and fisheries. This physical environmental impacts are triggered by eruption material which ejected by Mount Kelud. The material such as ballistic bombs, volcanic ash, and pumice material reached more than 200x106m3 of material [1]. Thus, triggered secondary hazard such as landslides in the surrounding area.

Landslides are described as mass movement of soil or rock that shear displacement along one or may be several slip surfaces [2-3]. Over the last two decades, researcher have investigated landslide hazard and construct maps portraying their spatial distribution [4]. Mapping landslide susceptibility is essential for proper land use planning and disaster management [5]. Different techniques and methods have been developed and applied in landslide susceptibility mapping using both the quantitative or qualitative approach [6] such as probabilistic models [7–12], the logistic regression model [13–17], and multi-criteria analysis [18–21,5]. In this study, the landslide susceptibility assessment employed spatial multi criteria model as a method. Four considerations criteria such as topography, hydrology, soil, and environmental characteristics were used. Each factor then reduced into several sub-criteria such as slope,

IOP Conf. Series: Earth and Environmental Science **273** (2019) 012014 doi:10.1088/1755-1315/273/1/012014

aspect, topographic position index, topographic wetness index, stream power index, rainfall, soil texture, soil structure, COLE index, land use, and distance from river. The SMCE method was also engaged with expert judgment provided by academic university' view and the BPBD agencies.

2. Study Area

Mount Kelud is located at 7^{0}56'00" South Latitude and 112^{0}18'30" East Longitude with elevation level of 1731 . Administratively Mount Kelud distributed within three districts in East Java, namely Blitar, Kediri, and Malang. It is categorized as high intensity of volcanic activity in Indonesia. The eruption characteristic of Kelud is explosive along with phreatic eruption [22]. It is characterized by spilling water that contained of mixed mud and material. In addition, freatomagmatic eruption of Mount Kelud produces ash-lapilli precipitate in the form of fall, then can be followed by flow of fall of pyroclastic material. The eruption period of Mount Kelud is relatively short with the eruption cycle in 20 years. In Fig. 1 shows the location of Mount Kelud.

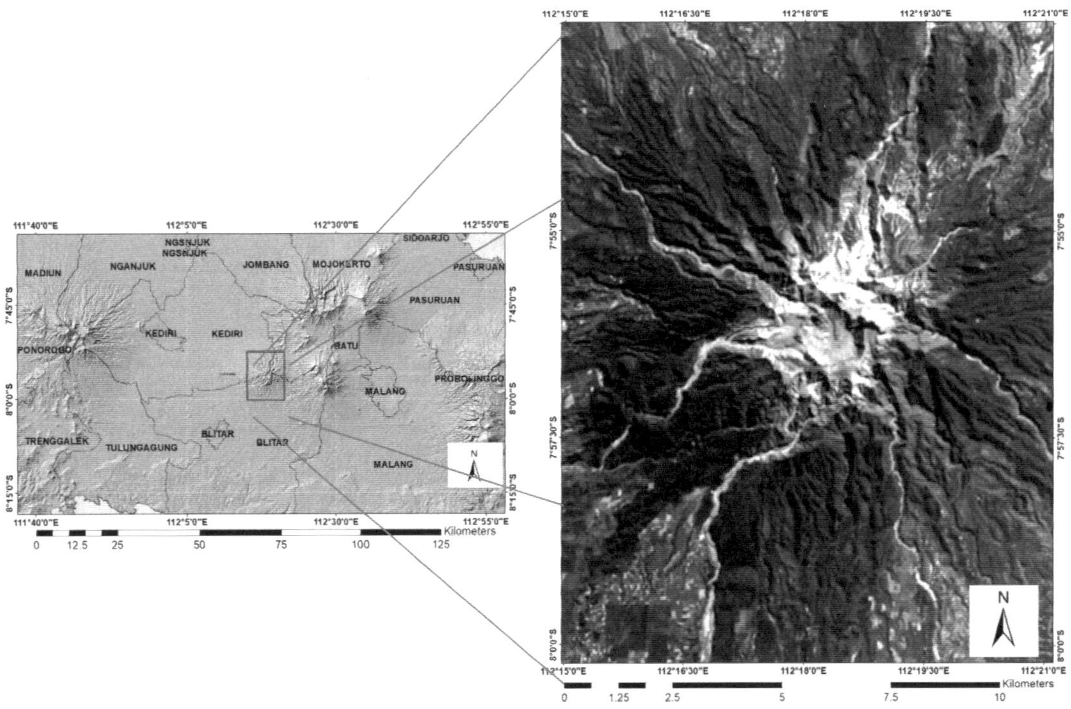

Figure 1. Location of The Study Area Showing Mount Kelud, East Java Province, Indonesia

3. Methodology

3.1. Input Data

This study used nine different GIS layers data to produce Landslide Susceptibility Map. All raster images (30m x 30m) were projected using Datum Universal Transverse Mercator (UTM), with WGS 84 zones. Then, each data layer is classified into several classes using Natural Breaks (Jenks) method.

Natural breaks class based on natural grouping attached to the data. It identifies the points of reference by choosing the class that breaks to the best group of value, same value and maximize the differences between classes [5]. Details of layers data used in this study are described as follows.

3.1.1 Altitude

A digital elevation model (DEM) is a digital model representing an earth's surface. It is one of the important input in modelling dynamic natural phenomena such as landslides, mass-movement, and soil erosion. DEM describes information about morphological relief of earth's surface on a digital format raster that height value on each pixel. To produce this parameter which controlled by several geologic and geomorphological processes [7,21,23,24], ASTER Global-DEM (Advanced Spaceborne Thermal Emission and Reflection Radiometer) data was used. Figure 2 showed the spatial information related to the altitude/ elevation within research area.

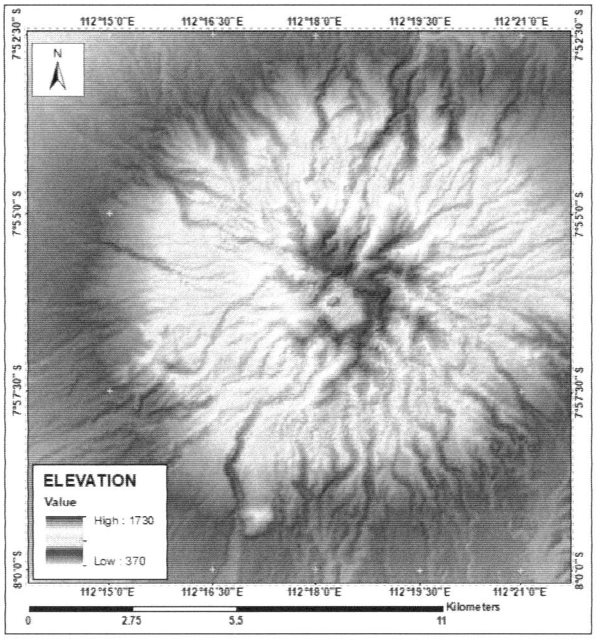

Figure 2. The elevation map.

3.1.2 Slope

The most important parameter in the slope stability analysis is the slope (Lee and Min, 2001). The slope gradient is one primary geomorphometric parameter that figure of geomorphological process of earth's surface. The slope is directly related and affected to the landslides and it is frequently used in preparing susceptibility maps [25–28]. Based on the ASTER GDEM data, the slope map was produced. Below figure showed the spatial distribution of slope gradient.

IOP Conf. Series: Earth and Environmental Science **273** (2019) 012014 doi:10.1088/1755-1315/273/1/012014

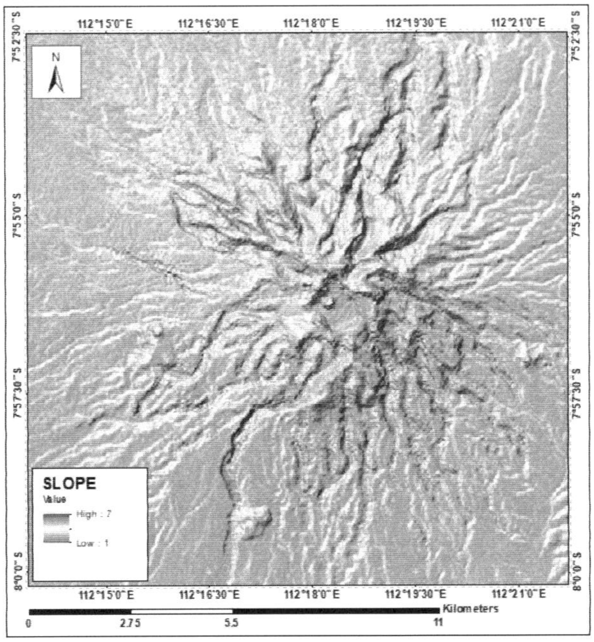

Figure 3. The slope gradient map.

3.1.3 Topographic Position Index (TPI)

The Topographic Position Index (TPI) is calculated as the difference the cell elevation and the mean elevation of neighbouring cells [28]. Applied specific thresholds for TPI values allows for the identification of different topographic landforms, such as ridge, slope, and valley. Since the landslide scarps occur mostly on the ridges, the TPI index may be seen as one of landslide conditioning factor, and used for landslide susceptibility map [29,30]. TPI was calculated using Jennes, et al. (2013) implementation. The result of calculated TPI can be seen in figure 4.

IOP Conf. Series: Earth and Environmental Science **273** (2019) 012014 doi:10.1088/1755-1315/273/1/012014

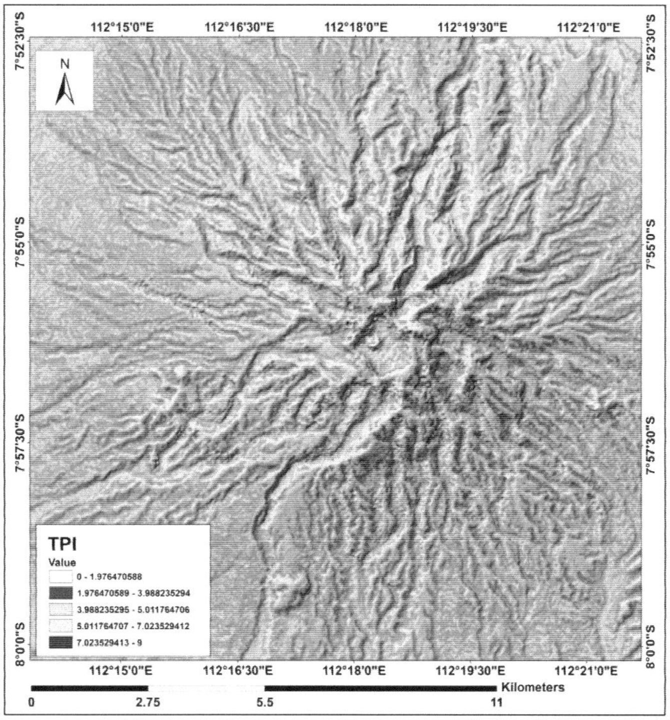

Figure 4. The Topographic Position Index (TPI) Map.

3.1.4 Stream Power Index (SPI)

The massive soil erosion describe the geomorphological process, and it's have direct affect to potential landslide in an area. The Stream Power Index (SPI) used in this study (Fig. 5). This index is used to visualization potential flow erosion and related with landscape processes [31]. The SPI describes potential for flow erosion at the given point of the surface, and controls potential erosive power of water flow [28,31]. The SPI is calculated from following formula:

$$SPI = As * tan\beta \quad (1)$$

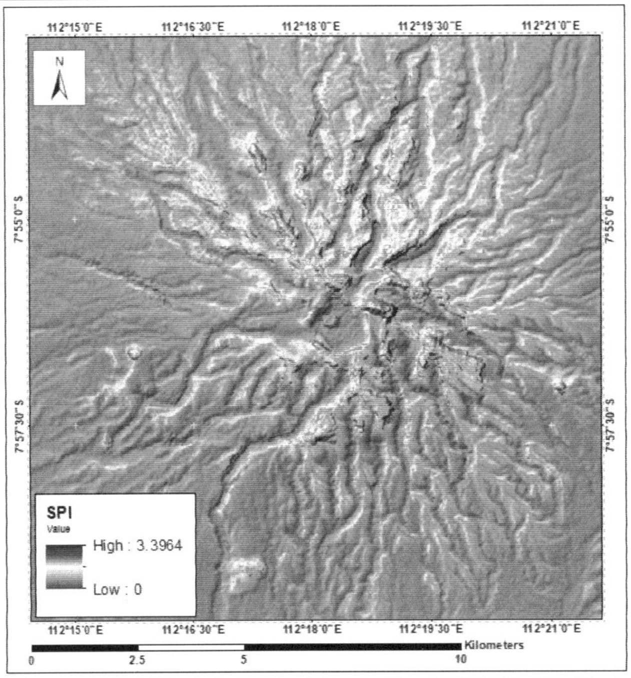

Figure 5. The Stream Power Index (SPI) Map.

3.1.5 Rainfall

We used rainfall data with the duration three years periods from Malang, Kediri, Blitar, Blitar City, Malang City, and Batu City rainfall station in processing landslide susceptibility map. Rainfall data can figure the accumulation of water that can transporting materials and soil, and also triggering the landslides. Spatial distribution of average of three annual rainfall data, we visualized in figure 6.

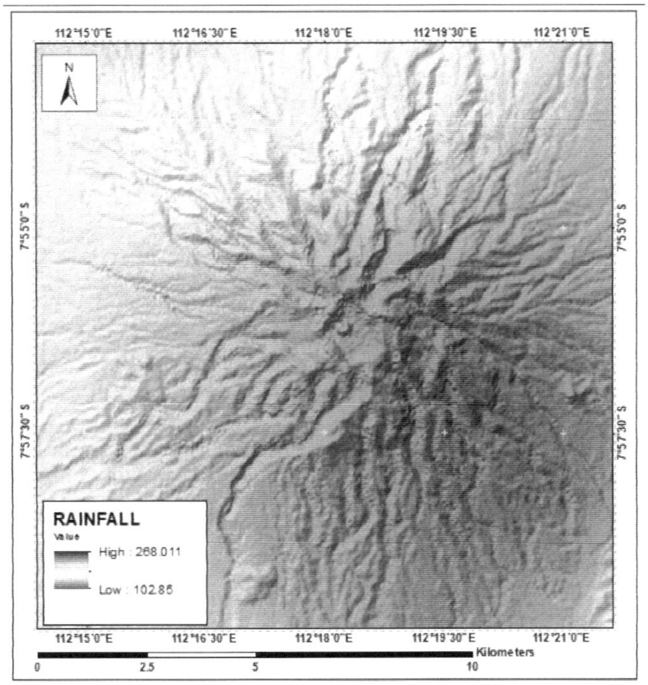

Figure 6. The Rainfall Map

3.1.6 Topographic Wetness Index (TWI)

The Topographic Wetness Index (TWI) model is one of the criteria that indicate hydrological process, related to the accumulation of water flow based on the control of slope factor in an area [28–30]. The slope control of the hydrological process will be highly visible in areas with high relief configurations. Especially at the Mount Kelud area, the high slope configuration encourages control of hydrological processes, which relate to the process of accumulation of water flows. In Figure 7, its show the accumulation of water flows.

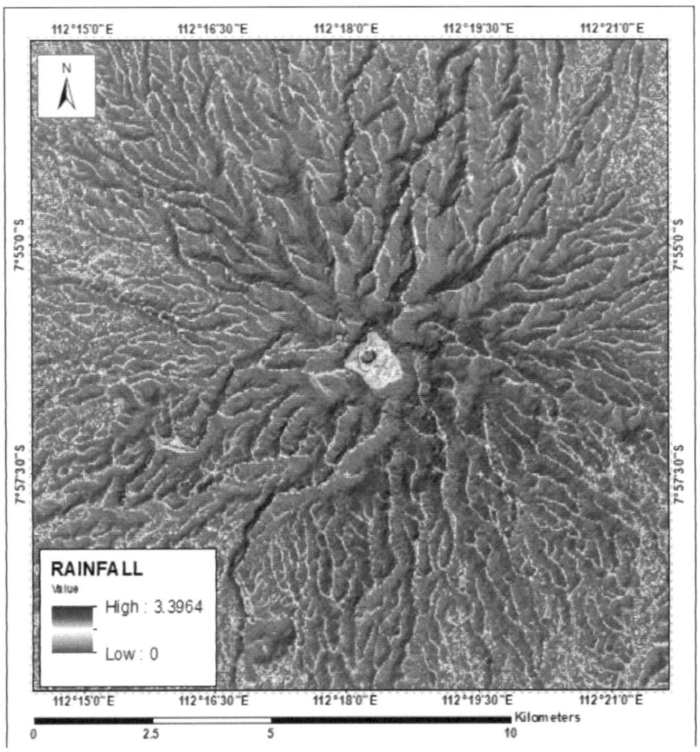

Figure 7. The Topographic Wetness Index (TWI) Map.

3.1.7 Soil Texture

Soil texture describe the materials and physical characteristic in each soil profile. After catastrophic event, in 2014 Kelud eruption ejecting the amount of material, is important to make spatial distribution of soil texture. Soil texture data extracted from the laboratory analyses. The average percentage of soil texture in the study area were 70% sand, 20% silt, and 10% clay. Figure 8 contain the distribution of soil texture in the Kelud area.

IOP Conf. Series: Earth and Environmental Science **273** (2019) 012014 doi:10.1088/1755-1315/273/1/012014

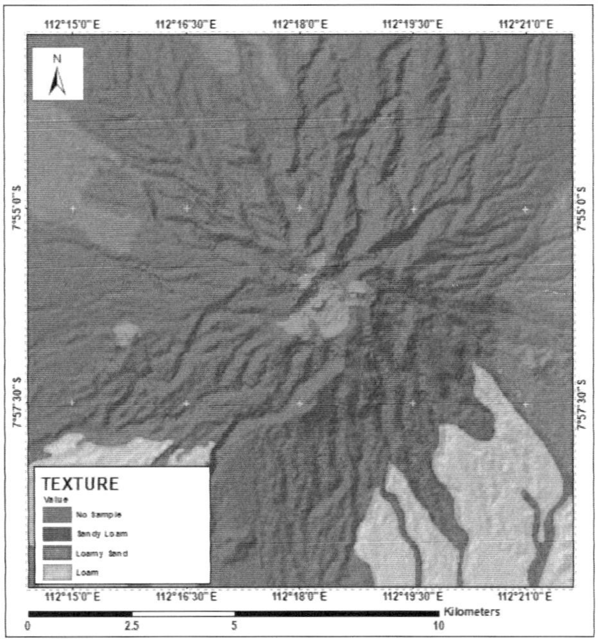

Figure 8. The Soil Texture Map.

3.1.8 COLE Index

The low percentage content of montmorillonite clay fraction causes a low of COLE index in this study area. The low COLE index will have an effect on low soil ability to retain water and soil ability to form aggregates. Figure 9 contain the information of distribution COLE index in research area.

IOP Conf. Series: Earth and Environmental Science **273** (2019) 012014 doi:10.1088/1755-1315/273/1/012014

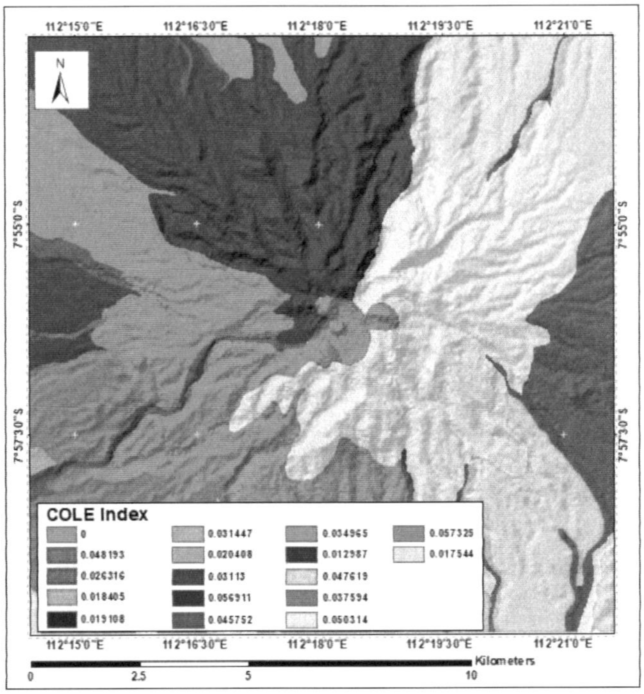

Figure 9. The COLE Index Map.

3.1.9 Land-Use

In this study, land use map was produced from the LANDSAT 8 OLI/TIRS satellite image, and applying a object-based classification scheme. There are eight types of land use are identified in the study area. In the landslides susceptibility modelling, land use map is important to produce information of human induce and environment condition in research area.

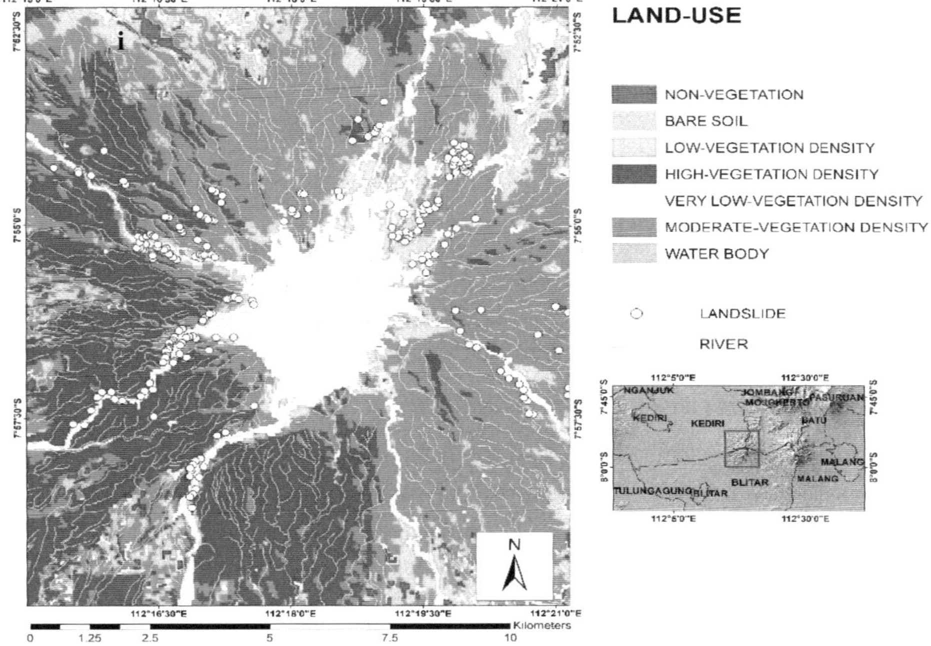

Figure 10. Human-Induced Parameter Land-Use Map.

3.2. Research Proses

3.2.1 Landslide Inventory Map

Creation of landslide inventory maps is an important step in the process of determining and assessing the level of susceptibility. The landslides inventory map (see Fig. 11) can document the occurrence of landslide in an area [32]. The location of the incident, the volume and impact of a landslide can be represented and visualized through landslide inventory map [19]. A landslide inventory map is prepared for several scopes such as (i) documenting the occurrence of landslides in an area, (ii) as a first step towards the assessment of landslide susceptibility, and (iii) to investigate the distribution, type and pattern of landslides that occur [33].

Landslide inventory maps can be done with different techniques. The determination of the techniques used depends on the objectives of the landslide inventory process, the extent of the study area, the time available in the data collection process, the data sources used, and the experience of the researcher [32,34,35]. The selection of manufacturing techniques becomes very important, because it affects the processes and effort undertaken to produce landslide inventory maps that have a high degree of reliability.

The landslide delineation is carryout by visual interpretation using remote sensed data. Applied Google Earth for visual interpretation. In addition, Google Earth provides high-resolution of imagery, the ability to display 3D earth surface has provided the ease and opportunity to detect and mapping landslide distribution in study area. Figure 11 showed the landslide interpretation that we digitized and marked where landslide occurred in this research area. The result of interpretation process, we used this information to validated, captured, and measured the materials of landslides in field observation process.

Figure 11. Interpretation Landslide Based on 3D Visualization in Google Earth.

Field observation process is needed to know the landslide characteristic. Through field observation information will be obtained specifically related to the type and visual characteristics of landslide [34], as well as to determine the validation of landslide inventory maps that have been made. In the process of field observation, we observed the landslide event, and took a picture for every landslide that we found, showed in figure (a). In addition, we took the material of landslide (see figure b and c) , and we classified the type of every landslide events that we found.

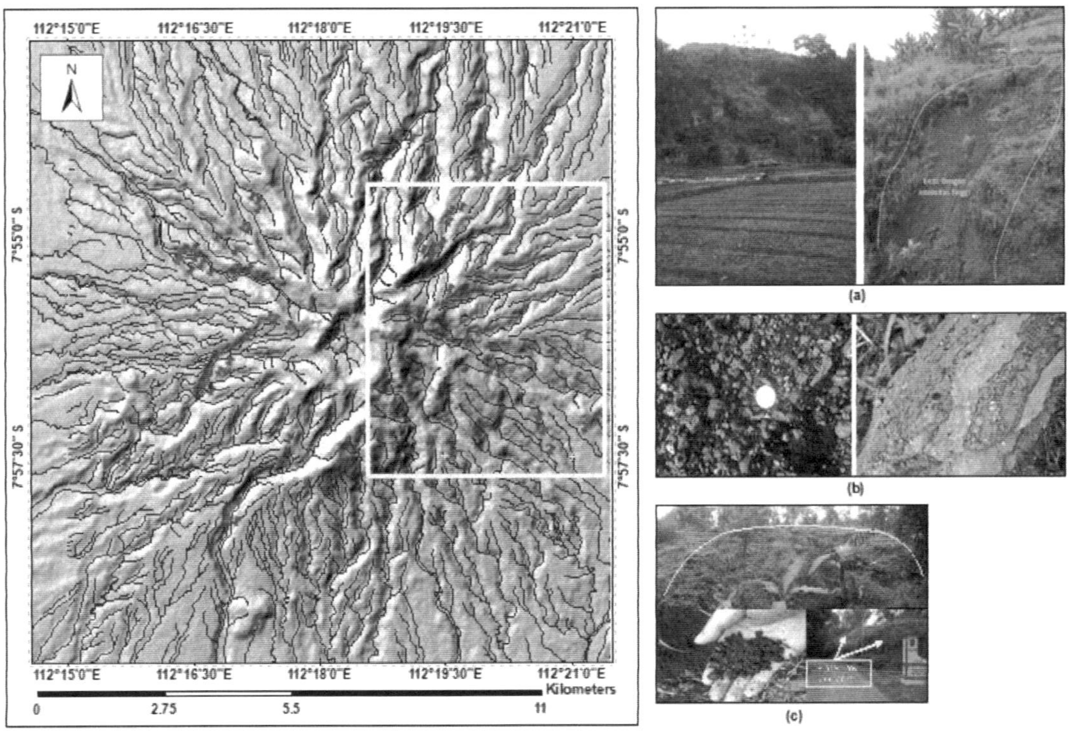

Figure 12. Landslide Inventory Mapping and Field Observatory

3.2.2 LSM Using SMCE

To solve spatial-based problems such as landslide phenomenon, GIS-based Spatial Multi-criteria Evaluation (SMCE) have been used. SMCE is a way of producing policy-relevant information about spatial decision problems for decision makers [21]. The basic steps is to divide the decision problem into small, understandable parts, analyse each of them, and integrate these parts in a logical manner to produce a meaningful solution [36]. SMCE process can be visualized in Figure 12 [21,37].

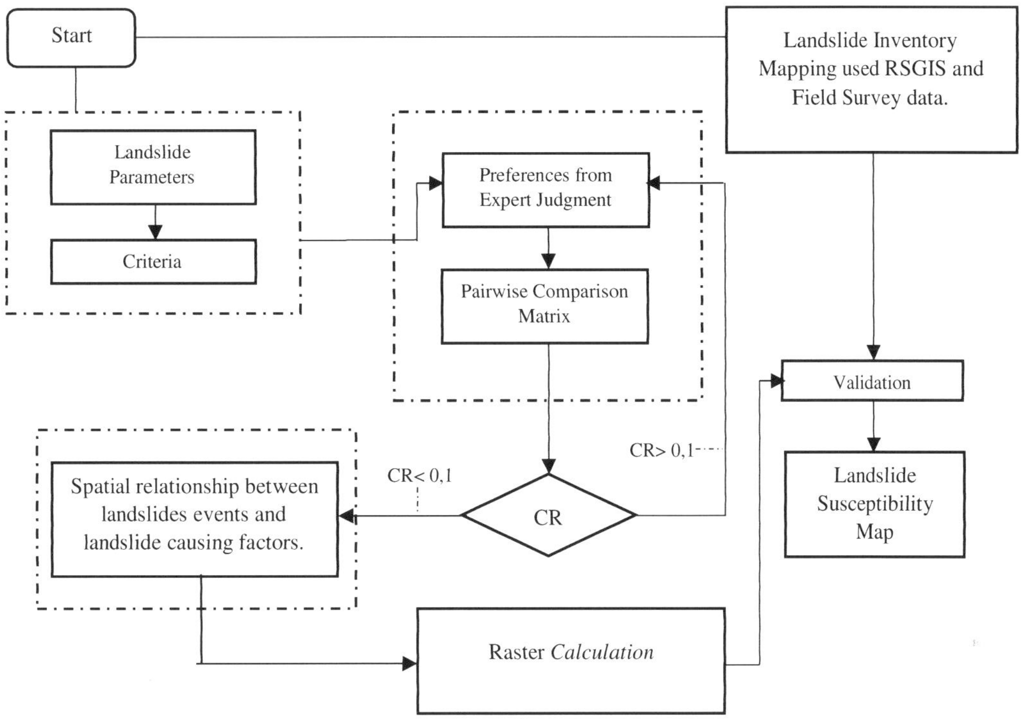

Figure 13. Landslide Susceptibility Process Using Spatial Multi-Criteria Evaluation (SMCE).

SMCE method in this study was also engaged with expert judgment provided by academic university's view and the BPBD agencies. For these method, a pair-wise comparison based weighting was used. Pair-wise comparison method was established by Saaty (1987) in the context of the Analytical Hierarchy Process (AHP), showing in Table 1. According to Pourghasemi, et al., (2012) the AHP consists of three main steps; (1) generating the pari-wise comparison matrix, (2) computing the weigts of the criterion, and (3) estimating the consistency ratio. In making comparison matrix, the AHP method uses a scale with a range of values 0 – 9 to assess relatve preferences for two criteria. The range value can be shown in table as followed:

Table 1. The Range Value in AHP.

Intensity of Importance	Definition	Explanation
1	Equal importance	Two activities contribute equally to objective
3	Weak importance of one over another	Experience and judgment slightly favor one activity over another
5	Essential or strong importance	Experience and judgment strongly favor one activity over another
7	Demonstrated importance	An activity is strongly favored and its dominace demonstrated in practice
9	Absolute importance	The evidence favoring one activity over another is the highest possible order of affirmation
2, 4, 6, 8	Intermediate values between the two adjacent judgments	When compromise is needed

Source: Pourghasemi *et al*, 2014

3.2.3 Landslide Parameters

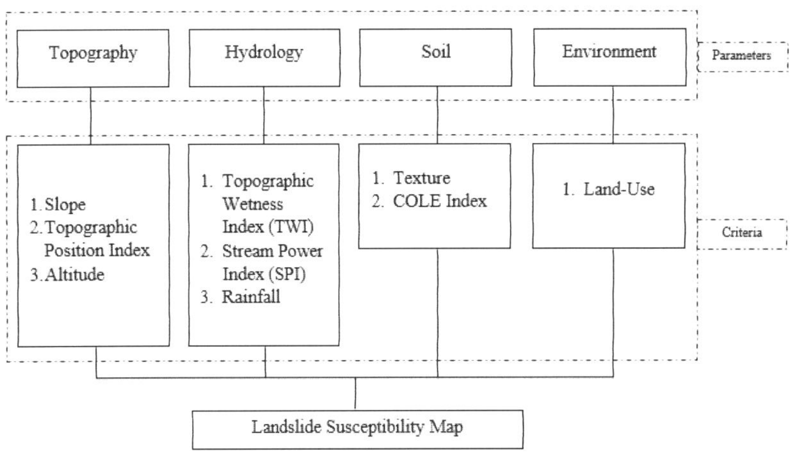

Figure 14. The flowchart of the research

In the diagram above shows the landslide parameters of this research. In the process of assessment of landslide susceptibility have been used several parameters covering topography, hydrology, soil, and environment. Then, each parameters is reduced to several criteria including slope, TPI, altitude, TWI, SPI, rainfall, soil texture, soil COLE index, and Land Use. The layers from each data is then integrated in attempt to build a landslide susceptibility maps.

IOP Conf. Series: Earth and Environmental Science **273** (2019) 012014 doi:10.1088/1755-1315/273/1/012014

3.2.4 Pairwise Comparison Matrix

Table 2. Pairwise Matrix for Landslide Susceptibility Mapping.

LSM Parameters

Parameters	(1)	(2)	(3)	(4)	Eigen Value
Topography (1)	1	1/2	3	3	0.308
Hydrology (2)		1	3	3	0.433
Soil (3)			1	3	0.165
Land-Use (4)				1	0.094

Consistency Ration: 0.08

Topography's Criteria

Elevation

Factors	(1)	(2)	(3)	(4)	(5)	(6)	Eigen Value
0 – 550 (1)	1	½	1/2	1/2	½	½	0.085
551 – 700 (2)		1	1/2	1/2	½	½	0.107
701 – 850 (3)			1	1/2	½	½	0.135
851 – 1000 (4)				1	½	½	0.169
1001 – 1250 (5)					1	1/3	0.204
1251 – 1730 (6)						1	0.300

Consistency Ration: 0.06

Slope

Factors	(1)	(2)	(3)	(4)	(5)	(6)	Eigen Value
0 – 5 (1)	1	½	1/3	1/3	1/3	1/3	0.059
6 – 15 (2)		1	1/3	1/3	1/3	1/3	0.074
16 – 30 (3)			1	1/3	1/3	1/3	0.117
31 – 50 (4)				1	1/3	1/3	0.171
51 – 70 (5)					1	1/2	0.258
> 70 (6)						1	0.322

Consistency Ration: 0.08

Topographic Position Index (TPI)

Factors		(1)	(2)	(3)	(4)	(5)	(6)	Eigen Value
0	(1)	1	2	2	2	2	2	0.275
2	(2)		1	2	2	2	2	0.218
4	(3)			1	2	2	2	0.173
5	(4)				1	2	2	0.138
7	(5)					1	2	0.109
9	(6)						1	0.087

Consistency Ration: 0.04

Hydrology's Criteria

Stream Power Index (SPI)

Factors		(1)	(2)	(3)	(4)	(5)	(6)	Eigen Value
0 – 20	(1)	1	3	3	3	3	3	0.350
20 – 40	(2)		1	3	3	3	3	0.241
40 – 60	(3)			1	3	3	3	0.166
60 – 80	(4)				1	3	2	0.107
80 – 100	(5)					1	2	0.074
>100	(6)						1	0.062

Consistency Ration: 0.09

Rainfall

Factors		(1)	(2)	(3)	(4)	(5)	Eigen Value
175	(1)	1	1/3	1/3	1/3	1/3	0.070
200	(2)		1	1/3	1/3	1/3	0.110
225	(3)			1	1/3	1/3	0.172
245	(4)				1	1/2	0.281
270	(5)					1	0.367

Consistency Ration: 0.09

Topographic Wetness Index (TWI)

Factors		(1)	(2)	(3)	(4)	(5)	Eigen Value
0 – 4	(1)	1	1/3	3	3	3	0.261
0 – 7	(2)		1	3	2	3	0.412

7 – 11 (3)			1	2	2	0.140
11 – 15 (4)				1	2	0.106
15 – 18 (5)					1	0.081

Consistency Ration: 0.06

Soil's Criteria

Texture

Factors	(1)	(2)	(3)	Eigen Value
Loam (1)	1	½	1/3	0.157
Loamy Sand (2)		1	1/3	0.249
Sandy Loam (3)			1	0.594

Consistency Ration: 0.05

Factors	(1)	(2)	(3)	(4)	(5)	(6)	(7)	(8)	(9)	(10)	(11)	Eigen Value
0.01299 (1)	1	1/2	2	2	2	2	2	2	2	2	2	0.128
0.01841 (2)		1	2	2	2	2	2	2	2	22	2	0.143
0.01911 (3)			1	2	2	2	2	2	2	1/2	2	0.108
0.02041 (4)				1	2	2	2	2	2	2	2	0.102
0.02632 (5)					1	2	2	2	2	2	2	0.091
0.03110 (6)						1	2	2	2	2	2	0.082
0.03145 (7)							1	2	2	2	2	0.073
0.37594 (8)								1	2	2	2	0.065
0.04575 (9)									1	2	2	0.058
0.04762 (10)										1	2	0.052
0.04819 (11)											1	0.058

Consistency Ration: 0.06

Land-Use's Criterion

Factors	(1)	(2)	(3)	(4)	(5)	(6)	(7)	Eigen Value
Lahars	1	2	2	2	2	2	2	0.232

Non-Vegetation		1	2	2	2	2	2	0.192
Bare Soil			1	2	2	2	1/2	0.141
Very Low-Veg.				1	2	2	2	0.134
Low-Veg.					1	2	2	0.110
Moderate-Veg.						1	2	0.091
High-Veg.							1	0.100
	Consistency Ration: 0.08							

IOP Conf. Series: Earth and Environmental Science **273** (2019) 012014 doi:10.1088/1755-1315/273/1/012014

3.2.5 Data Standardization

Table 3. Spatial Relationship between Landslides and Landslide Causing Factors.

Factor	Class	No.of pixels in domain	Percentage of domain	No. of landslide	Percentage of landslide	FR		a	b	c	d
Elevation	550	389135	21.88	2	2.35	0.11	0.085	0.0091	0.249	0.308	0.0007
	700	564495	31.74	3	3.53	0.11	0.107	0.0119	0.249	0.308	0.0009
	850	374988	21.09	20	23.53	1.12	0.135	0.1506	0.249	0.308	0.0116
	1000	226558	12.74	32	37.65	2.96	0.169	0.4994	0.249	0.308	0.0383
	1250	153265	8.62	21	24.71	2.87	0.204	0.5848	0.249	0.308	0.0449
	1730	69955	3.93	7	8.97	2.28	0.3	0.6844	0.249	0.308	0.0525
Slope (deg)	0 - 5	504856	28.41	2	2.35	0.08	0.59	0.0489	0.594	0.308	0.0089
	6 - 15	459015	25.83	4	4.71	0.18	0.074	0.0135	0.594	0.308	0.0025
	16 - 30	336505	18.93	11	12.94	0.68	0.117	0.0800	0.594	0.308	0.0146
	31 - 50	204188	11.49	10	11.76	1.02	0.171	0.1750	0.594	0.308	0.0320
	51 - 70	138069	7.77	25	29.41	3.79	0.258	0.9767	0.594	0.308	0.1787
	> 70	134527	7.57	33	38.82	5.13	0.322	1.6514	0.594	0.308	0.3021
TPI	0	179733	10.11	17	20.00	1.98	0.275	0.5442	0.157	0.308	0.0263
	2	162564	9.14	16	18.82	2.06	0.218	0.4490	0.157	0.308	0.0217
	4	429951	24.18	10	11.76	0.49	0.173	0.0842	0.157	0.308	0.0041

	5	540320	30.38	10	11.76	0.39	0.138	0.0534	0.157	0.308	0.00
	7	139962	7.87	8	9.41	1.20	0.109	0.1304	0.157	0.308	0.00
	9	325866	18.32	24	28.24	1.54	0.087	0.1341	0.157	0.308	0.00
SPI	0-20	1129999	63.54	16	18.82	0.30	0.35	0.1037	0.249	0.433	0.01
	20-40	369601	20.78	18	21.18	1.02	0.241	0.2456	0.249	0.433	0.02
	40-60	178372	10.03	27	31.76	3.17	0.166	0.5258	0.249	0.433	0.05
	60-80	75223	4.23	20	23.53	5.56	0.107	0.5952	0.249	0.433	0.06
	80-100	23104	1.30	3	3.53	2.72	0.074	0.2010	0.249	0.433	0.02
	>100	2097	0.12	1	1.18	9.98	0.062	0.6186	0.249	0.433	0.06

Table 4. (*continued*).

Factor	Class	No.of pixels in domain	Percentage of domain	No. of landslide	Percentage of landslide	FR		a	b	c	d
Rainfall	175	132116	11.67	2	2.35	0.20	0.7	0.1412	0.594	0.433	0.03
	200	186247	16.45	10	11.76	0.72	0.11	0.0787	0.594	0.433	0.02
	225	237631	20.98	17	20.00	0.95	0.172	0.1639	0.594	0.433	0.04
	245	243707	21.52	33	38.82	1.80	0.281	0.5070	0.594	0.433	0.13
	270	332822	29.39	23	27.06	0.92	0.367	0.3379	0.594	0.433	0.08
TWI	0-4	757470	43.02	22	25.88	0.60	0.261	0.1570	0.0157	0.433	0.00

IOP Conf. Series: Earth and Environmental Science **273** (2019) 012014 doi:10.1088/1755-1315/273/1/012014

	0-7	564669	32.07	36	42.35	1.32	0.412	0.5441	0.0157	0.433	0.00
	7-11	253185	14.38	16	18.82	1.31	0.14	0.1833	0.0157	0.433	0.00
	11-15	137012	7.78	7	8.24	1.06	0.106	0.1122	0.0157	0.433	0.00
	15-18	48489	2.75	4	4.71	1.71	0.081	0.1384	0.0157	0.433	0.00
Texture	Loam	48415	16.41	5	5.88	0.36	0.157	0.0563	0.75	0.165	0.00
	Loamy Sand	41837	14.18	9	10.59	0.75	0.249	0.1859	0.75	0.165	0.02
	Sandy Loam	204820	69.41	71	83.53	1.20	0.594	0.7148	0.75	0.165	0.08
Cole	0.01299	79561	6.87	16	18.82	2.74	0.128	0.3507	0.25	0.165	0.01
	0.01841	143539	12.39	14	16.47	1.33	0.143	0.1900	0.25	0.165	0.00
	0.01911	65241	5.63	7	8.24	1.46	0.108	0.1579	0.25	0.165	0.00
	0.02041	42961	3.71	6	7.06	1.90	0.102	0.1941	0.25	0.165	0.00
	0.02632	79114	6.83	2	2.35	0.34	0.091	0.0313	0.25	0.165	0.00
	0.03110	37431	3.23	4	4.71	1.46	0.082	0.1194	0.25	0.165	0.00
	0.03145	73728	6.37	3	3.53	0.55	0.073	0.0405	0.25	0.165	0.00
	0.37594	28311	2.44	1	1.18	0.48	0.065	0.0313	0.25	0.165	0.00
	0.04575	112169	9.69	2	2.35	0.24	0.058	0.0141	0.25	0.165	0.00
	0.04762	111689	9.64	3	3.53	0.37	0.052	0.0190	0.25	0.165	0.00
	0.04819	256884	22.18	23	27.06	1.22	0.058	0.0708	0.25	0.165	0.00

Table 5. (*continued*).

Factor	Class	No.of pixels in domain	Percentage of domain	No. of landslide	Percentage of landslide	FR		a	b	c	d
Landuse	Lahar	86320	7.01	11	12.94	1.85	0.232	0.4282	0.094		0.04025
	Non-Vegetation	55602	4.52	6	7.06	1.56	0.192	0.3001	0.094		0.02820
	Bare Soil	7014	0.57	12	14.12	24.78	0.141	3.4936	0.094		0.32839
	Very Low Vegetation	65521	5.32	3	3.53	0.66	0.134	0.0889	0.094		0.00835
	Low Vegetation	192981	15.68	21	24.71	1.58	0.11	0.1734	0.094		0.01630
	Moderate Vegetation	526222	42.75	24	28.24	0.66	0.091	0.0601	0.094		0.00565
	High Vegetation	297322	24.15	8	9.41	0.39	0.01	0.0039	0.094		0.00037

Domain: pixels in study area, domain (%): (domain/total pixels in study area)*100, landslide: number of landslide occurrences, landslide (%): (landslide/ total number of landslide occurrences)* 100, frequency ratio: landslide (%)/ domain (%).

[a] Normalized value.

[b] Parameter's value.

[c] Group's value.

[d] Final weight.

4. Result and Discussion

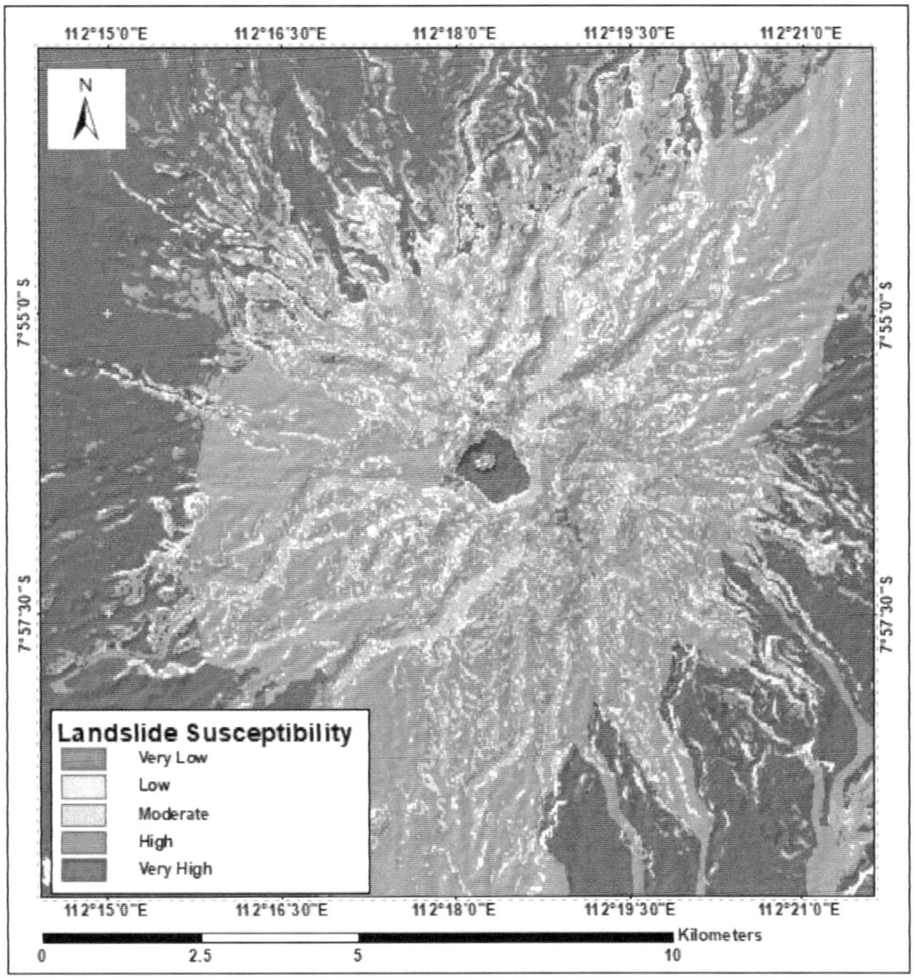

Figure 15. Landslide Susceptibility Map

This study produce landslide susceptibility map which applied AHP method for determine the weight value of each parameters and criterion. The standardization stage of each criterion is based on frequency ratio calculation, where the percentage of landslide points in each class is divided by number of pixels domains from each class within each criterion. We applied Natural Break (Jenks) classification to divide the susceptibility into several class. The class of landslide susceptibility consists of "Very Low', "Low", "Moderate", "High", and "Very High".

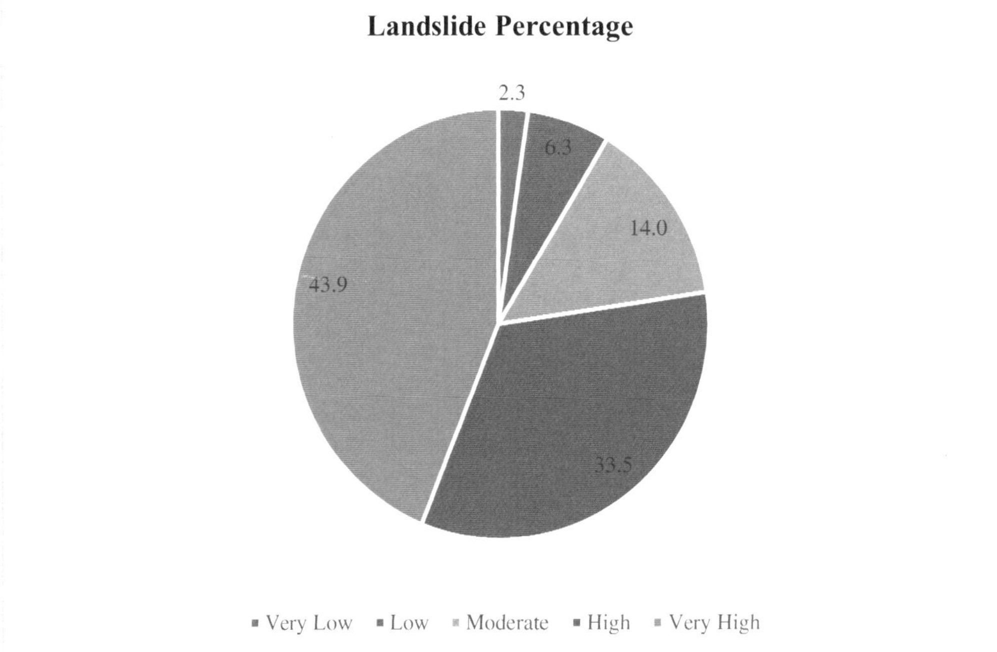

Figure 16. The Pie Diagram of Percentage Landslide Occurrences

Total number of landslide point used in this study is 121, with landslide characteristic has area >100m2. Number of landslide in each class of susceptibility can be seen in the above figure. Based on the graph it is known that the area with very high level of susceptibility has a percentage of landslide incidence of 24, 977%, then high susceptibility class is 27, 957%, moderate susceptibility class is 27, 660%, low susceptibility class is 15,288%, and very low class is 4,116%.

5. Conclusion

The phenomenon of landslide becomes a serious threat post eruption of Mount Kelud in 2014. This is in line with the mapping of landslide susceptibility that has been done. The area around Mount Kelud has a high landslide potential level. Based on research results of 24.977% of the area included into the class of very high susceptibility, and 27.957% including into high susceptibility class. The information generated from this research becomes important to reduce the impact of losses that can be caused by landslides.

References

[1] Blake D 2015 *The 2014 eruption of Kelud Volcano, Indonesia: impacts on infrastructure, utilities, agriculture and health*

[2] Varnes D J 1978 Slope movement types and processes *Spec. Rep.* **176** 11–33

[3] Akgun A and Bulut F 2007 GIS-based landslide susceptibility for Arsin-Yomra (Trabzon, North Turkey) region *Environ. Geol.* **51** 1377–87

[4] Guzzetti F, Carrara A, Cardinali M and Reichenbach P 1999 Landslide hazard evaluation: a review of current techniques and their application in a multi-scale study, {Central} {Italy} *Geomorphology* **31** 181–216

[5] Ahmed B 2015 Landslide susceptibility mapping using multi-criteria evaluation techniques in Chittagong Metropolitan Area, Bangladesh *Landslides* **12** 1077–95

[6] Park S, Choi C, Kim B and Kim J 2013 Landslide susceptibility mapping using frequency ratio, analytic hierarchy process, logistic regression, and artificial neural network methods at the {Inje} area, {Korea} *Environ. Earth Sci.* **68** 1443–64

[7] Dai F C and Lee C F 2001 Terrain-based mapping of landslide susceptibility using a geographical information system: a case study *Can. Geotech. J.* **38** 911–23

[8] Gokceoglu C, Sonmez H, Nefeslioglu H A, Duman T Y and Can T 2005 The 17 March 2005 Kuzulu landslide (Sivas, Turkey) and landslide-susceptibility map of its near vicinity *Eng. Geol.* **81** 65–83

[9] Akgün A and Bulut F 2007 GIS-based landslide susceptibility for Arsin-Yomra (Trabzon, North Turkey) region *Environ. Geol.* **51** 1377–87

[10] Lee S and Pradhan B 2007 Landslide hazard mapping at Selangor, Malaysia using frequency ratio and logistic regression models *Landslides* **4** 33–41

[11] Akgun A, Dag S and Bulut F 2008 Landslide susceptibility mapping for a landslide-prone area (Findikli, NE of Turkey) by likelihood-frequency ratio and weighted linear combination models *Environ. Geol.* **54** 1127–43

[12] Oh H-J, Lee S, Chotikasathien W, Kim C H and Kwon J H 2009 Predictive landslide susceptibility mapping using spatial information in the Pechabun area of Thailand *Environ. Geol.* **57**

[13] Chauhan S, Sharma M, Arora M K and Gupta N K 2010 Landslide Susceptibility Zonation through ratings derived from Artificial Neural Network *Int. J. Appl. Earth Obs. Geoinf.* **12** 340–50

[14] Nefeslioglu H A, Sezer E, Gokceoglu C, Bozkir A S and Duman T Y 2010 Assessment of landslide susceptibility by decision trees in the metropolitan area of {Istanbul}, {Turkey} *Math. Probl. Eng.* **2010**

[15] Pradhan B 2010 Remote sensing and {GIS}-based landslide hazard analysis and cross-validation using multivariate logistic regression model on three test areas in {Malaysia} *Adv. Sp. Res.* **45** 1244–56

[16] Bai S, Lü G, Wang J, Zhou P and Ding L 2011 GIS-based rare events logistic regression for landslide-susceptibility mapping of Lianyungang, China *Environ. Earth Sci.* **62** 139–49

[17] Akgun A 2012 A comparison of landslide susceptibility maps produced by logistic regression, multi-criteria decision, and likelihood ratio methods: a case study at İzmir, {Turkey} *Landslides* **9** 93–106

[18] Abella E A C and Van Westen C J 2007 Generation of a landslide risk index map for {Cuba} using spatial multi-criteria evaluation *Landslides* **4** 311–25

[19] Castellanos Abella E A and Van Westen C J 2008 Qualitative landslide susceptibility assessment by multicriteria analysis: A case study from San Antonio del Sur, Guantánamo, Cuba *Geomorphology* **94** 453–66

[20] Akgun A and Türk N 2010 Landslide susceptibility mapping for Ayvalik (Western Turkey) and its vicinity by multicriteria decision analysis *Environ. Earth Sci.* **61** 595–611

[21] Pourghasemi H R, Pradhan B, Gokceoglu C and Deylami Moezzi K 2012 Landslide Susceptibility Mapping Using a Spatial Multi Criteria Evaluation Model at Haraz Watershed, Iran *Terrigenous Mass Movements* ed B Pradhan and M Buchroithner (Berlin, Heidelberg: Springer Berlin Heidelberg) pp 23–49

[22] Zaennudin A 2009 Prakiraan Bahaya Erupsi Gunung Kelud *Bul. Vulkanol. dan Bencana Geol.*

4 1–17

[23] Lineback Gritzner M, Marcus W A, Aspinall R and Custer S G 2001 Assessing landslide potential using GIS, soil wetness modeling and topographic attributes, Payette River, Idaho *Geomorphology* **37** 149–65

[24] Ayalew L and Yamagishi H 2005 The application of GIS-based logistic regression for landslide susceptibility mapping in the Kakuda-Yahiko Mountains, Central Japan *Geomorphology* **65** 15–31

[25] Akgun A, Dag S and Bulut F 2008 Landslide susceptibility mapping for a landslide-prone area ({Findikli}, {NE} of {Turkey}) by likelihood-frequency ratio and weighted linear combination models *Environ. Geol.* **54** 1127–43

[26] Pourghasemi H R, Pradhan B, Gokceoglu C, Mohammadi M and Moradi H R 2013 Application of weights-of-evidence and certainty factor models and their comparison in landslide susceptibility mapping at Haraz watershed, Iran *Arab. J. Geosci.* **6** 2351–65

[27] Feizizadeh B and Blaschke T 2013 {GIS}-multicriteria decision analysis for landslide susceptibility mapping: comparing three methods for the {Urmia} lake basin, {Iran} *Nat. Hazards* **65** 2105–28

[28] Pawluszek K and Borkowski A 2017 Impact of DEM-derived factors and analytical hierarchy process on landslide susceptibility mapping in the region of Rożnów Lake, Poland *Nat. Hazards* **86** 919–52

[29] Jebur M N, Pradhan B and Tehrany M S 2014 Optimization of landslide conditioning factors using very high-resolution airborne laser scanning ({LiDAR}) data at catchment scale *Remote Sens. Environ.* **152** 150–65

[30] Pourghasemi H R, Moradi H R, Fatemi Aghda S M, Gokceoglu C and Pradhan B 2014 {GIS}-based landslide susceptibility mapping with probabilistic likelihood ratio and spatial multi-criteria evaluation models ({North} of {Tehran}, {Iran}) *Arab. J. Geosci.* **7** 1857–78

[31] Moore I D, Grayson R B and Ladson A R 1991 Digital terrain modelling: {A} review of hydrological, geomorphological, and biological applications *Hydrol. Process.* **5** 3–30

[32] Mondini A C, Viero A, Cavalli M, Marchi L, Herrera G and Guzzetti F 2014 Comparison of event landslide inventories: the {Pogliaschina} catchment test case, {Italy} *Nat. Hazards Earth Syst. Sci.* **14** 1749–59

[33] Guzzetti F, Mondini A C, Cardinali M, Fiorucci F, Santangelo M and Chang K T 2012 Landslide inventory maps: New tools for an old problem *Earth-Science Rev.* **112** 42–66

[34] Guzzetti F and Cardinali M 1990 Landslide inventory map of the {Umbria} region, {Central} {Italy} *Proc. ALPS* **90** 273–84

[35] Guzzetti F, Cardinali M, Reichenbach P and Carrara A 2000 Comparing landslide maps: A case study in the upper Tiber River basin, central Italy *Environ. Manage.* **25** 247–63

[36] Malczewski J 1999 *GIS and multicriteria decision analysis* (John Wiley & Sons)

[37] Sharifi M A 2004 Site selection for waste disposal through spatial multiple criteria decision analysis *J. Telecommun. Inf. Technol.* 28–38

Acknowledgments

The authors are grateful to RISTEKDIKTI (Directorate General of Higher Education, Ministry of National Education of Republic Indonesia), who supported this research activity through a PUPT Research Scheme.

IOP Conf. Series: Earth and Environmental Science **273** (2019) 012015 doi:10.1088/1755-1315/273/1/012015

Probabilistic Seismic Hazard Analysis for Aceh Region

V H Banyunegoro[1,2,*], Z A Alatas[2], A Jihad[2], Eridawati[2] and U Muksin[3]

[1]Department of Physics, Faculty of Natural Sciences, Syiah Kuala University, Jln. Teuku Nyak Arief Darussalam, Banda Aceh, Aceh, Indonesia.
[2]Mata Ie Geophysical Station, Jl. Raya Mata Ie, Banda Aceh, Aceh, Indonesia.
[3]Department of Physics, Syiah Kuala University, Jln. Teuku Nyak Arief Darussalam, Banda Aceh, Aceh, Indonesia.

*corresponding author : vrieslend22@gmail.com

Abstract. A study has been conducted to understand effect caused by earthquakes in Aceh. Using earthquake data from ISC, probabilistic seismic hazard analysis has been done to obtain informations how is the effect caused by earthquake sourced near Aceh. The results are peak ground acceleration, spectral acceleration for period of 0.2 and 1 Hz within return period 500 and 2500 years which equal to probability of exceedance 10% in 50 years and 2% in 50 years. There are several differences compared to results of study that had been conducted by Pusgen earlier in 2017. The results might be used as preliminary information to illustrate level of seismic hazard in Aceh.

1. Introduction

Aceh region is located in Sumatra Island of Indonesia. Every year, there are a lot of earthquake events happened in Aceh. According to ISC catalogs, from 1964 to 2017, there are about 16600 earthquakes happened in Aceh [1]. This can happen because Aceh is surrounded by some earthquake sources including faults, and subduction zone. The 2017 earthquake map and hazards released by PUSGEN [2] mentions that there are at least seven known faults that have recorded events. The latest famous earthquakes are the Pidie Jaya earthquake that struck in December 6th 2016 resulting in about 100 casualties [3] and the Central Aceh earthquake in 2013[4]. This event is really unique because the source is a fault that is relatively passive or has been inactive for a while. Aceh itself has several unknown faults yet has recorded events. These sources are usually named as background source [3]. Considering the amounts of earthquake happened in Aceh, thus it is necessary to know how vulnerable Aceh to effect of earthquakes. Since we know the sources of earthquakes in Aceh, then we can calculate or estimate effect of those earthquakes. We use probabilistic seismic hazard analysis to estimate effect of earthquakes that can struck Aceh region.

2. Data

2.1 Earthquake data

We use earthquake data from ISC catalogs as main source of earthquake data [1]. The area of research is limited by latitude between 2° – 7° N and by longitude between 93° – 100° E. Prior to area of research, the earthquakes used are those that happened within the area. There are about 16.600 events happened in this area between 1964 to 2017. Earthquakes that are considered to be used are not limited by magnitudes, therefore all recorded earthquakes are accounted. ISC earthquake data have various magnitudes. In order to calculate Peak Ground Acceleration, one has to uniform magnitudes into single type of magnitude. The most commonly type of magnitude used is moment magnitude of Mw. Moment magnitude is widely used because it does not saturate at large magnitude [6].

IOP Conf. Series: Earth and Environmental Science **273** (2019) 012015 doi:10.1088/1755-1315/273/1/012015

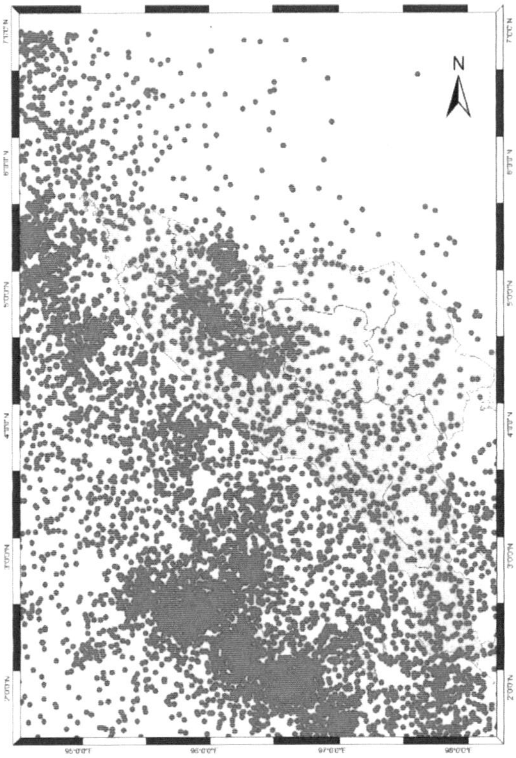

Figure 1. Earthquake data from ISC catalogs [1]

2.2 Source parameters

Probabilistic seismic hazard analysis requires not only earthquake data, but also parameter of possible source of earthquake within particular area. Source parameters used in this research are referred to the 2017 earthquake map and hazards of Indonesia released by PUSGEN [2] as well as the latest fault system in Pidie Jaya[3] and Aceh[4]. Probabilistic seismic hazard analysis accounted all possible source of earthquake.

The first source we all know is subduction zone. Subduction zone is widely known as one of source of earthquake which had been produced many devastating earthquake all over the world. One of them is Aceh earthquake which happened on 26th of December 2004. Next one is faults, which are spread across Aceh. The 2017 earthquake map and hazards released by PUSGEN mentions that there are at least 7 major faults in Aceh [2]. Those faults have experienced several earthquakes larger than magnitude of 6 [7]. The last source of earthquake is background. Background source is earthquake source which is not yet known but has recorded events [8].

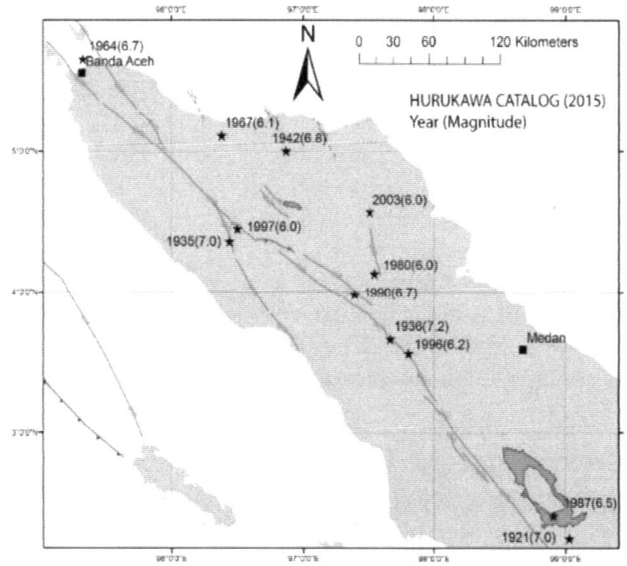

Figure 2. Earthquake sources in Aceh and historical earthquake from Hurukawa catalog [9]

Table 1. Parameter data of earthquake sources in Aceh [2] [10].

No.	Segment	Slip-rate (mm/yr)	Mechanism	Dip	Length (Km)	Top (Km)	Bottom (Km)	Mmax
1	Lhokseumawe	1,0	Strike-slip	90	36	3	20	6,5
2	Seulimeum-North	18,0	Strike-slip	90	143	3	20	7,6
3	Seulimeum-South	7,0	Strike-slip	90	38	3	20	6,9
4	Aceh-North	2,0	Strike-slip	90	65	3	20	7,2
5	Aceh-Central	14,0	Strike-slip	90	140	3	20	7,6
6	Aceh-South	14,0	Strike-slip	90	27	3	20	6,7
7	Lot tawar	0,5	Strike-slip	90	21	3	20	6,6
8	Peusangan	0,5	Strike-slip	90	32	3	20	6,8
9	Tripa 1	16,0	Strike-slip	45	32	3	20	6,8
10	Tripa 2	7,0	Strike-slip	90	103	3	20	7,4
11	Tripa 3	14,0	Strike-slip	90	75	3	20	7,3
12	Tripa 4	7,0	Strike-slip	90	55	3	20	7,1
13	Tripa 5	7,0	Strike-slip	90	21	3	20	6,6
14	Batee A	6,0	Strike-slip	90	44	3	20	7,0
15	Batee B	0,5	Strike-slip	90	50	3	20	7,0
16	Batee C	0,1	Strike-slip	90	37	3	20	6,9
17	Aceh-Andaman Megathrust	40	Thrust		1300			9,0

Faults parameters listed in the 2017 earthquake map and hazards released by PUSGEN [2] do not provide fault width. Fault width is required to do hazard analysis. Then we use Papazachos relations [11] to obtain that.

3. Methods

Probabilistic seismic hazard analysis aims to calculate potential ground motion caused by earthquakes which are sourced nearby certain area. It is really important for designing building codes regarding result of the analysis. There are several steps in order to do probabilistic seismic hazard analysis. Typically it can be written as below [12]

1. Identificating and characterizing all earthquake sources that potentially can produce significant ground motion and make.
2. Finding recurrence relationships, in order to define rate of recurrence of earthquake with certain magnitude at particular source.
3. Calculating the ground motion produced by certain earthquake with some particular magnitude at specific area using attenuation functions.
4. Combining of earthquake location, size and ground motion parameter estimation in order to obtain probability of ground motion parameter will be exceeded in given range of time.

Technically, probabilistic seismic hazard analysis is done by several steps [13]. First one needs to decluster the earthquake data in order to separate which earthquake is happened in which source then we can determine the type of each sources. Declustering the earthquakes is done by using Gardner and Knopoof Scheme [14] which is included in Zmap [15]. In order to avoid overestimate result of hazard analysis, main shock needs to separated. Magnitude of the earthquake needs to be uniformed into moment magnitude, in this research, equation listed in the 2017 earthquake map and hazards released by PUSGEN [2] as well as the Aceh and Pidie Jaya fault system[3][4] are used. Then we find a-b value which is very important in this calculation. The value of a-b is obtained using Gutenberg-Richter equation included in Zmap [15]. Next, one needs to determine the earthquake source parameters in order to do hazard analysis [16]. Then we should choose ground motion equation for each type of source. Probabilistic seismic hazard analysis is done using USGS software developed by Harmsen [17].

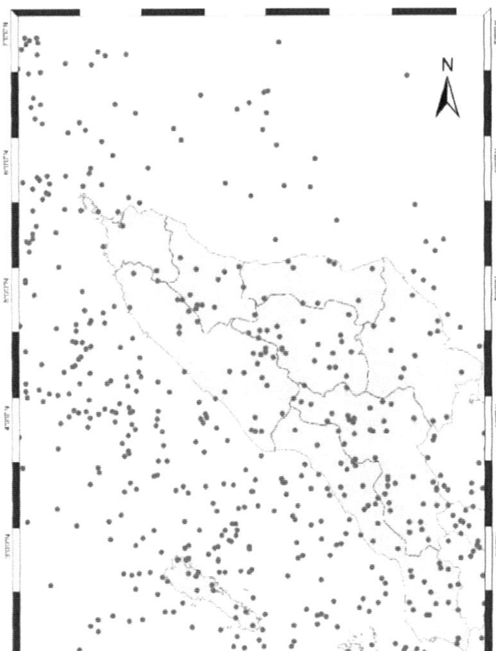

IOP Conf. Series: Earth and Environmental Science **273** (2019) 012015 doi:10.1088/1755-1315/273/1/012015

Figure 3. Earthquake data after being declustered.

4. Results and discussions

Probabilistic Seismic Hazard analysis had been done in order to obtain three parameters in two hazard levels [18] as listed in table 2. Parameters need to obtain are enlisted in Indonesian building codes. Indonesia building codes earlier [19] mentioned that recommended hazard level is 10% probability of exceedance in 50 years which is equal to 500 years earthquake return period. Later version of Indonesia building codes [20] revised the hazard level. It is said that 2% probability of exceedance in 50 years is the recommended hazard level which corresponds to 2500 years earthquake return period. Considering some buildings were built referred to earlier codes, in this research earlier hazard level is also calculated alongside later recommended hazard level.

Table 2. Parameters that will be obtained from PSHA.

Hazard Level	PGA	T=0.2 second	T=1.0 second
10 % PE in 50 years	v	v	v
2% PE in 50 years	v	v	v

The results from probabilistic seismic hazard analysis are distributed spatially, then in order to read the data, the results are plotted into a map. Since the results are in form of grid, then we need to interpolate it using kriging to make contour maps. Three parameters in two hazard levels are individually plotted which resulted in six maps to read.

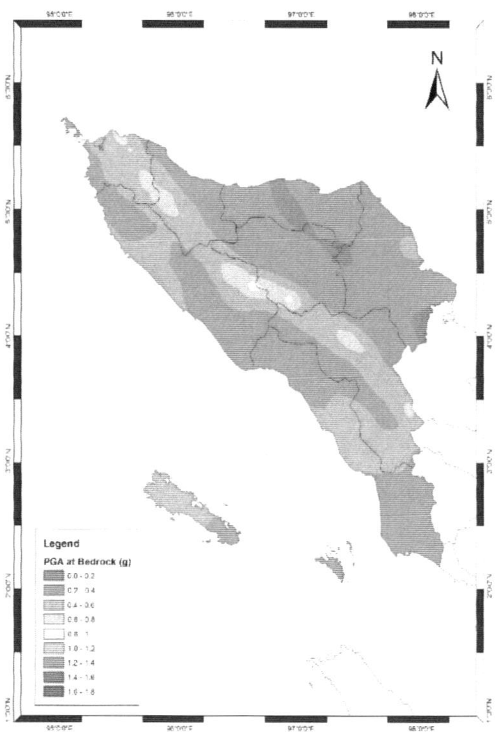

(a)

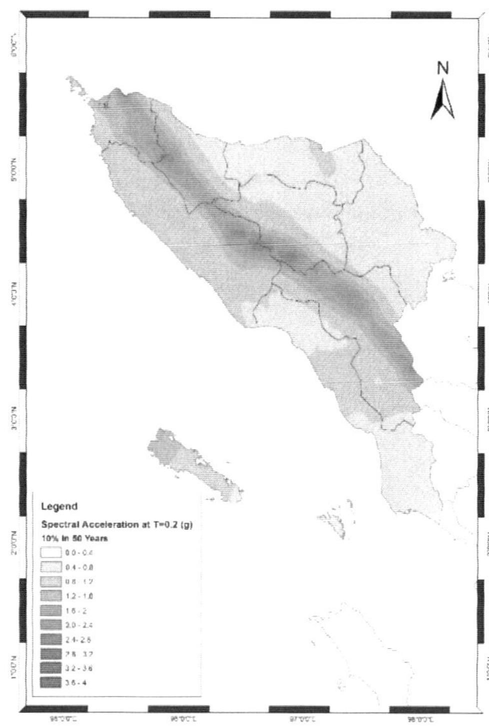

(c)

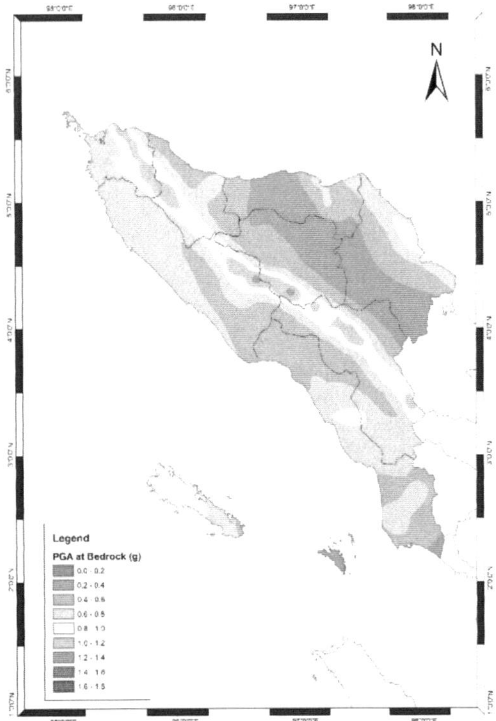
N
Legend
PGA at Bedrock (g)

(b)

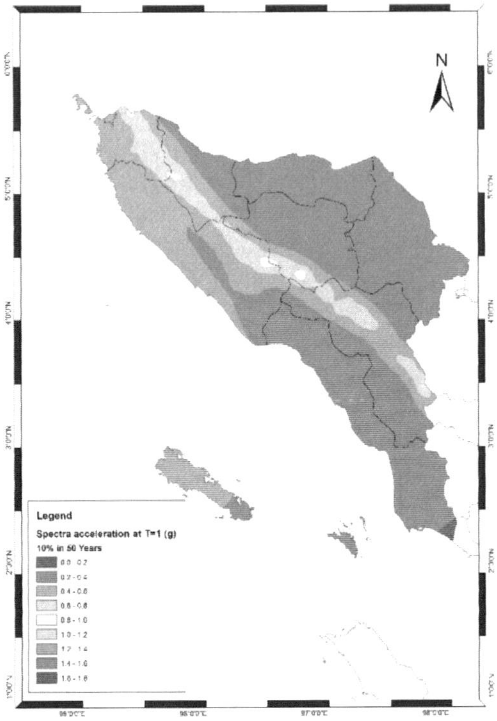
N
Legend
Spectra acceleration at T=1 (g)
10% in 50 Years

(e)

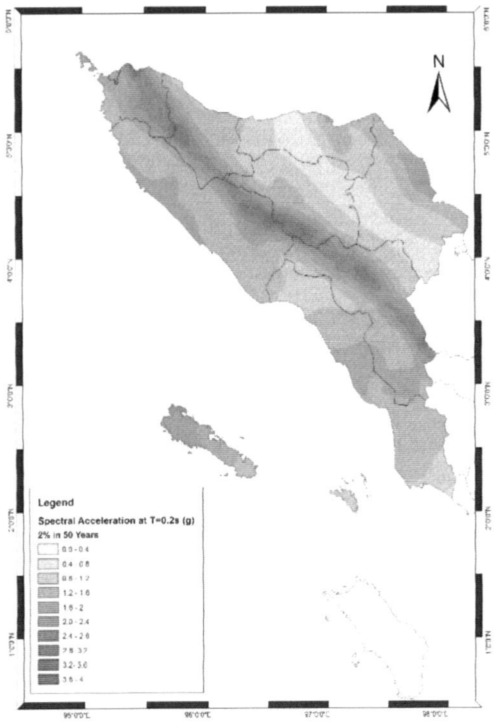
N
Legend
Spectral Acceleration at T=0.2s (g)
2% in 50 Years

(d)

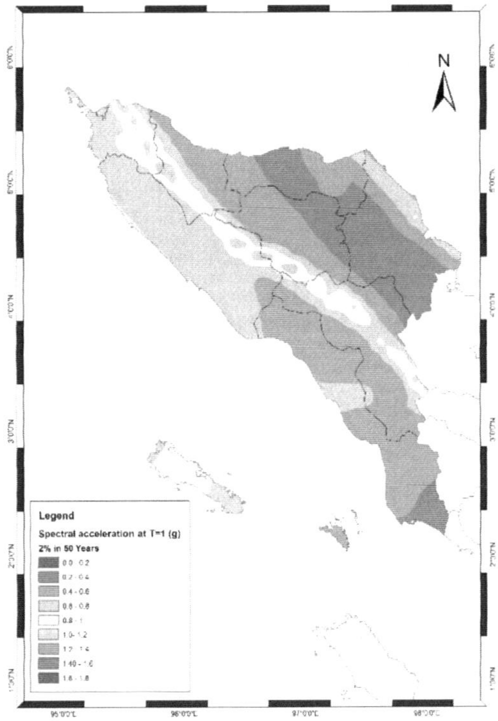
N
Legend
Spectral acceleration at T=1 (g)
2% in 50 Years

(f)

Figure 4. (a) Peak ground acceleration at bedrock with 10% probability of exceedance in 50 years, (b) peak ground acceleration at bedrock with 2% probability of exceedance in 50 years, (c) spectral acceleration map at bedrock for T=0.2 Hz with 10% probability of exceedance in 50 years, (d) spectral acceleration map at bedrock for T=0.2 Hz with 2% probability of exceedance in 50 years, (e) spectral acceleration map at bedrock for T=1 Hz with 10% probability of exceedance in 50 years, (f) spectral acceleration map at bedrock for T=1 Hz with 2% probability of exceedance in 50 years.

Figure 4 (a) shows that peak ground acceleration with 10% probability of exceedance in 50 years in Aceh has high value alongside middle part of Aceh and relatively lower in other part of area. The values vary from 0.1 to 1 g and have similar pattern with figure 4 (b) but (b) has higher value than (a). Figure 4 (c) and (d) share similar features. They have high value line in the middle part of Aceh and slightly lower value in northern part of the high value line. The differences are first, in figure (d) has high value line in northern and southern part of Aceh. Second is the value range, (c) has value ranges between 0.4 and 3.2 g while (d) varies from 0.4 to 3.6 g. Same features as (c) and (d) will be found in figure 4 (e) and (f) but with relatively lower values. Values in figure 4 (e) are ranged from 0.1 to 1 g, while (f) values are between 0.1 and 1.2 g. The island of Simeulue surprisingly has relatively high value in western part of island in all parameters. High value line can be found in middle part of Aceh because there we can find fault system which consists several major faults. While the reason Simeulue island has high value is because it is located near subduction zone. Areas near Langsa and Aceh Singkil also have slightly high value because there we found several earthquakes but the source is yet be defined. In this case, we put earthquakes in this area as earthquake related to background source. The point is, if the return period is longer, the acceleration will be larger.

Previous attempt to calculate peak ground acceleration in Aceh had been conducted by PUSGEN which results are shown below.

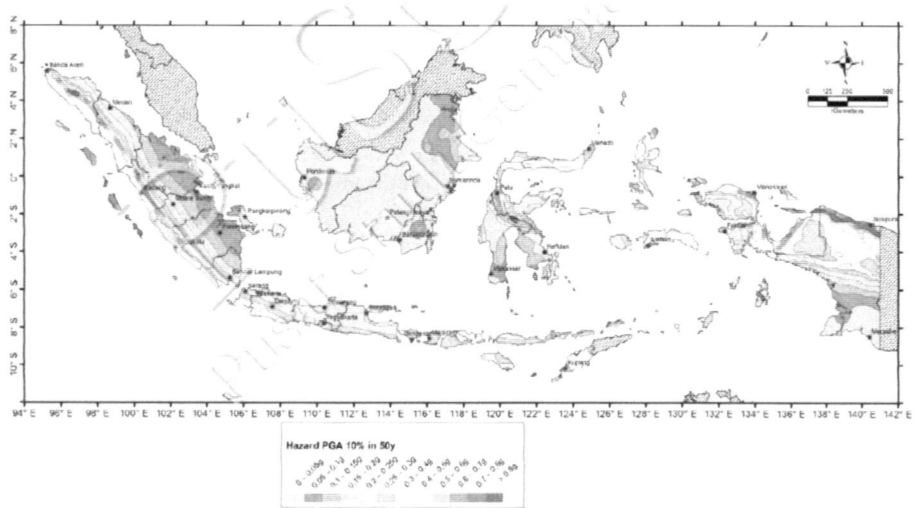

(a)

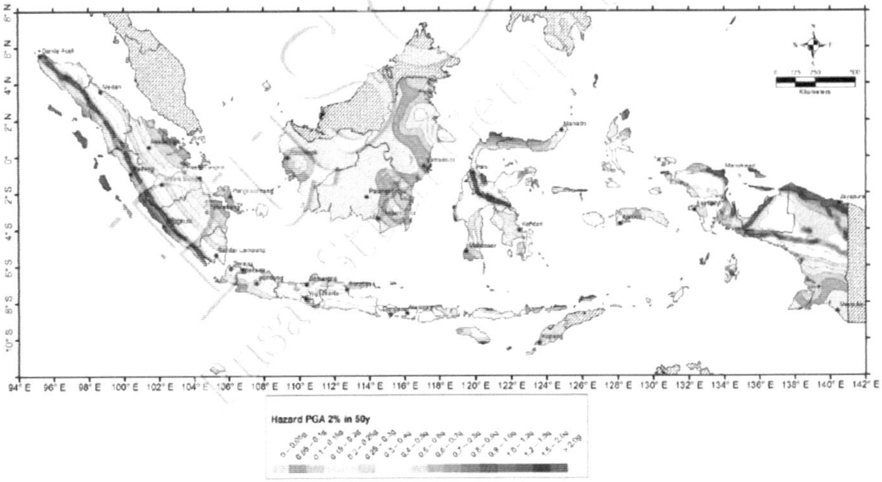

(b)

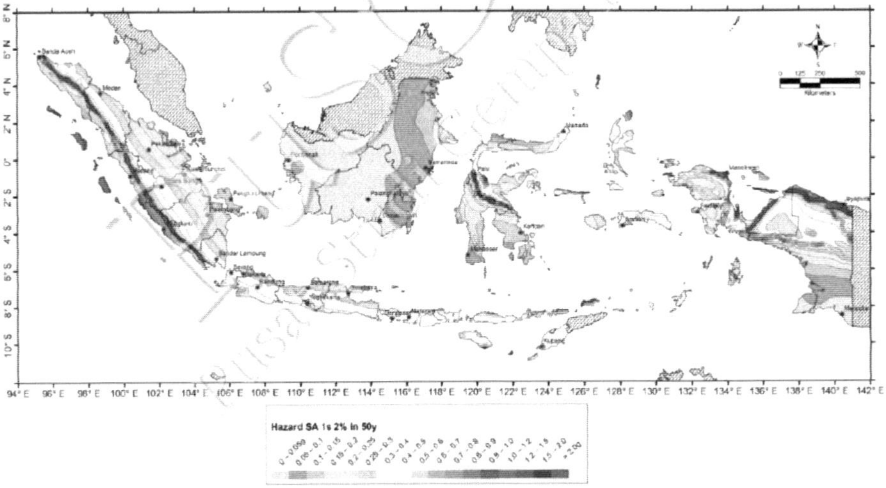

(c)

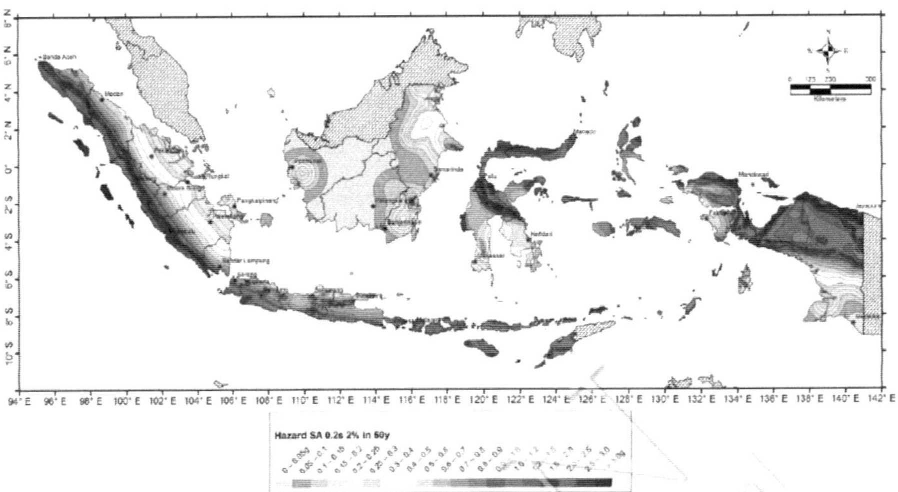

(d)

Figure 5. Results of PUSGEN studies, (a) Peak ground acceleration at bedrock with 10% probability of exceedance in 50 years, (b) peak ground acceleration at bedrock with 2% probability of exceedance in 50 years, (c) spectral acceleration map at bedrock for T=1 Hz with 2% probability of exceedance in 50 years, (d) spectral acceleration map at bedrock for T=0.2 Hz with 2% probability of exceedance in 50 years [2].

PUSGEN results when compared to our results, peak ground acceleration for 10% in 50 years from PUSGEN is slightly lower, while for 2% in 50 years our result values are lower than PUSGEN. These differences are caused by differences earthquake data used. Difference earthquake data will result in different a and b values. In the end it will produce different peak ground acceleration result. Another difference that our study and PUSGEN did is the area that is studied, our study only focused in Aceh while PUSGEN did for wider area.

5. Conclusions

Probabilistic seismic hazard analysis is useful to understand seismic hazard in such area. Our results have slight difference with PUSGEN produced. Difference in results are mainly caused by difference earthquake data used. Technically our results and PUSGEN results have similar features that are consistent in several hazard levels. Overall if certain area is near earthquake source, it will have higher ground motion than area which has larger distance to faults or subduction zone. It is shown by high value line in the middle of Aceh, near Simeulue island and near Langsa and Aceh Singkil. Analysis shows that if return period is longer, the ground motion produced will be larger. Our results which consistent to PUSGEN results, prove that.

6. References

[1] ISC Earthquake Catalogue Internet: https://www.isc.ac.uk/search/catalogue/ accessed June 10th 2018
[2] PUSGEN 2017 *Peta Sumber dan Bahaya Gempa Indonesia Tahun 2017* Bandung Puslitbang PUPR
[3] Muzli, M., Umar, M., Nugraha, A. D., Bradley, K. E., Widiyantoro, S., Erbas, K., Jousset, P., Rohadi, S., Nurdin, I. & Wei, S. 2018 *Seismological Research Letters.* Seismological Research Letters, **89**, 5, 1761-72.
[4] Muksin, U., Bauer, K., Muzli, M., Ryberg, T., Nurdin, I., Masturiyono, M. & Weber, M. 2019. *Journal of Asian Earth Sciences,* **171,** 20-27.
[5] Santoso E, Widiyantoro S and Sukanta I N 2011 *Jurnal Meteorologi dan Geofisika* **12** 129-136
[6] Hanks T C and Kanamori H 1979 *Journal of Geophysical Research* **84** 2348-2350
[7] Hurukawa N, Wulandari B R and Kasahara M 2014 *Bulletin of the Seismological Society of America* **104** 1750-1762
[8] Hutapea M B and Mangape I 2009 *Jurnal Teknik Sipil* **16** 121-132
[9] Natawidjaja D H 2018 *IOP Conf. Series: Earth and Environ. Sci.* 118 012001
[10] McCaffrey R 2009 *Annu. Rev. Earth Planet Sci.*37 345-366
[11] Papazachos B C, Scordilis E M, Panagiotopoulos D G, Papazachos C B and Karakaisis G F 2004 *Bulletin of the Geological Society of Greece* **36** 1482-1489
[12] Reiter L 1990 *Earthquake Hazard Analysis: Issues and Insights* New York Columbia University Press
[13] Taruna R M, Banyunegoro V H and Daniarsyad G 2018 *Int. Conf. In Rehab and Maintenance in Civ. Eng.* 195 03019
[14] Gardner J K and Knopoff L 1974 *Bulletin of the Seismological Society of America* **64** 363-367
[15] Wiemer S 2001 *Seismological Research Letter* **72** 373-382
[16] McGuirre R K 2004 *Earthquake Engineering Research Institute Pub.* MNO-10
[17] Harmsen S 2007 *USGS Software for Probabilistic Seismic Hazard Analysis (PSHA)*
[18] Sengara I W 2008 *The 14th World Conf. on Earthquake Eng.*

IOP Conf. Series: Earth and Environmental Science **273** (2019) 012015 doi:10.1088/1755-1315/273/1/012015

[19] Pusat Penelitian dan Pengembangan Teknologi Permukiman 2002 *SNI-03-1726-2002* Bandung Departemen Permukiman dan Prasarana Wilayah.
[20] Badan Standardisasi Nasional 2012 *SNI 1726:2012* Jakarta.

IOP Conf. Series: Earth and Environmental Science **273** (2019) 012016 doi:10.1088/1755-1315/273/1/012016

SVIM: A Program for Seismic Vulnerability Index Determination and HVSR Data Processing

Aulia Khalqillah[1,2], Umar Muksin[1,2], Musfirah[1], W. A. Ningsih[1,2], and Irwandi[1,2]

[1] Department of Physics, Syiah Kuala University, Aceh, Indonesia
[2] Tsunami and Disaster Mitigation Research Center (TDMRC), Syiah Kuala University, Aceh, Indonesia

Email: aulia.kh@mhs.unsyiah.ac.id or auliakhalqillah.mail@gmail.com

Abstract. The seismic vulnerability of an area is important to be determined in order to reduce the risk of earthquake disaster. Ones need to calculate the values of dominant frequency and amplification from the Horizontal to Vertical Spectral Ratio (HVSR). Currently existing software for HVSR data processing, Geopsy, does not include seismic vulnerability index calculation. Using Geopsy, ones also need additional software for mapping purposes. From that problem, we introduce a Graphical User Interface (GUI) program for seismic vulnerability index determination and Horizontal to Vertical Spectral Ratio (HVSR) data processing and plotting the results into a map of the study area. The program is called SVIM (Seismic Vulnerability Index of Microtremor). The program has been built using Matlab and successfully tested to 21 microtremor data points acquired in Lamteuba, Aceh Besar, Aceh, Indonesia. We compared our results with those resulted from the calculation by using Geopsy and validated using coefficient correlation. The validation shown that the correlation coefficient of seismic vulnerability index was 96,40%, the correlation coefficient of frequency was 84,19%, and the correlation coefficient of amplification was 95,80%. We conclude the SVIM program can be used for seismic vulnerability index determination and HVSR data processing. The mapping part will be included in the SVIM software so that the users can easily obtain the necessary map.

1. Introduction

The seismic vulnerability index (K_g) is important to identify site characteristics of an earthquake prone area. The value of seismic vulnerability index (K_g) explains how vulnerable an area is, when an area influenced by ground motion. To get the K_g value, the parameter of dominant frequency and parameter of amplification factor have to be known. Those parameters are obtained through Horizontal to Vertical Spectral Ratio (HVSR) data processing which was introduced by Nakamura in 1989 [1]. The HVSR data processing make use of ambient noise of microtremor measurements. The ambient noise is recorded by three components seismic sensor consists of two horizontal components (East-West and North-South) and one vertical component (V). This technique can be applied to geophysical prospecting and studies of buildings dynamical behavior [2].

Currently, the existing software for HVSR data processing is Geopsy which does not include the calculation of seismic vulnerability index (K_g). Therefore, we introduce Graphical User Interface

(GUI) for seismic vulnerability index (K_g) determination and HVSR data processing which is called SVIM (Seismic Vulnerability Index of Microtremor). We use 21 measurement microtremor data points acquired in Lamteuba, Aceh Besar, Indonesia to test SVIM program. We compared between HVSR results and Geopsy results and calculated the coefficient correlation to validate the results of SVIM program.

2. HVSR

The Horizontal to Vertical Spectral Ratio (HVSR) method is used to estimate dynamics characteristic of ground surface [3]. This method makes use of ambient noise spectrum ratio between two horizontal components and one vertical component. We use equation (1) to calculate the HVSR:

$$\frac{H}{V} = \sqrt{\frac{NS^2 + EW^2}{2V^2}} \tag{1}$$

where H is horizontal component spectrum, V is vertical component spectrum, NS is North-South component spectrum (horizontal component) and EW is East-West component spectrum (horizontal component) [4]. There are two output parameters that are obtained from HVSR calculation which are dominant frequency (F) and amplification factor (A). The dominant frequency is detected when the amplification factor is maximum. Both parameters can be used to estimate depth of bedrock of sediment [1]. The HVSR method also can use earthquake data for estimation of dynamic characteristics [3].

3. Konno-Ohmachi Smoothing

The smoothing has to be applied to HVSR spectrum to determine amplification factor. The common smoothing spectrum method is called Konno-Ohmachi smoothing which was introduced by Konno and Ohmachi in 1998 [5]. This smoothing can be calculated by using equation (2)

$$W_b(f, f_c) = [\sin(\log_{10}(f/f_c)^b)/\log_{10}(f/f_c)^b]^4 \tag{2}$$

where b is the band-width coefficient, f is the spectrum frequency, and f_c is the cut off frequency. According to Konno and Ohmachi (1998), the band-width value (b) has two conditions. The first condition, the smaller band-width value (b) is good to determine linear relation between peaks ratios and velocity contras (e.g. b value smaller than or equal to 10). The second condition, the higher band-width value (b) is good to reduce the elongation of the peak period (e.g. b value higher than or equal to 20) [5].

4. Seismic Vulnerability Index (Kg)

When an earthquake occurs, the damages of buildings are caused by ground surface response to seismic wave propagation or shakes. The level of damages depends on the intensity, frequency, and duration of earthquakes. The characteristics of ground surface can be identified based on Horizontal to Vertical Spectral Ratio (HVSR) of microtremor which was called as *Quasi-Transfer Spectrum* (QTS). The characteristics value from ground surface can be calculated by using equation (3) which is called seismic vulnerability index (K_g)

$$K_g = A^2/F \tag{3}$$

where the A is amplification factor and F is dominant frequency. The seismic vulnerability index (K_g) explains how vulnerable soil or ground surface is deformed when a ground motion occurs (e.g. earthquake)[6].

5. Methods

The SVIM program has been built using equations (1), (2), and (3). We calculate the dominant frequency and amplification factor by using HVSR method to determine the seismic vulnerability index (K_g). First, we read three components of microtremor data (two horizontal components that are East-West and North-South and one vertical component) in mseed format. Second, three components of microtremor data were filtered by using band pass filter from frequency range 0.1 to 10 Hz. This frequency range is used for ambient noise [7]. Third, the microtremor data are windowed by using Short Term Average/Long Term Average (STA/LTA) anti-trigger for each component. The STA/LTA anti-trigger is the opposite of STA/LTA trigger for earthquake detection [8]. In this case, the STA/LTA anti-trigger is used to ignore a spike of microtremor data based on the range of threshold values. There are some parameters of STA/LTA anti-trigger have to be configured, consist of the value of STA window was adjusted 1 s, the value of LTA window was adjusted 30 s, the minimum threshold value was adjusted 0.2, the maximum threshold value was adjusted 2.0, and the length window was adjusted 25 s for this research. Fourth, the time series of three components of microtremor data were transformed to frequency spectrum by using Fast Fourier Transform (FFT). Fifth, the two horizontal components of spectrum frequency were merged and calculated the average. Sixth, the spectrum of horizontal component and the spectrum of vertical component were calculated by using equation (1) to get the HVSR spectrum. Seventh, the spectrum of HVSR was smoothed by using equation (2) where the band-width value (b) was adjusted equal to 40 in this research. Therefore, dominant frequency and amplification factor were detected automatically. Eighth, the seismic vulnerability index (K_g) was calculated by using equation (3) automatically.

6. Results and Discussion

The Graphical User Interface of SVIM program (figure (1)) has been successfully built. The SVIM program has been tested to determine seismic vulnerability index (K_g) and HVSR data processing by using 21 microtremor data points of Lamteuba, Aceh, Indonesia. The SVIM program has several panels such as filter panel, anti-trigger STA/LTA panel for windowing, smoothing panel, output parameters panel, and option panel for button process. The program also has two plot panels such as waveform plot panel and amplification curve panel. Figure 1 shows the interface when the program is opened and figure 2 shows the interface program after the data processing completed.

The results of seismic vulnerability index determination and HVSR data processing using 21 microtremor data points shown in Table 1. The first example of the results of seismic vulnerability index and HVSR data processing using file c0avf180311031836 (hereafter 1836) is shown in Figure 3. The amplification curve resulted by SVIM program is shown in Figure 3(a) and the amplification resulted by Geopsy is shown in figure 3(b) which show the comparable values between that two software. SVIM program use linear scale and Geopsy use logarithmic scale. When the 1836 file was processed using SVIM program, the dominant frequency was 4.49219 Hz, the amplification factor was 0.39973, and the seismic vulnerability index (K_g) was 0.03557. The 1836 file was also processed using Geopsy, where the dominant frequency was 4.27264 Hz, the amplification factor was 0.470059, and the seismic vulnerability index (K_g) was 0.05171.

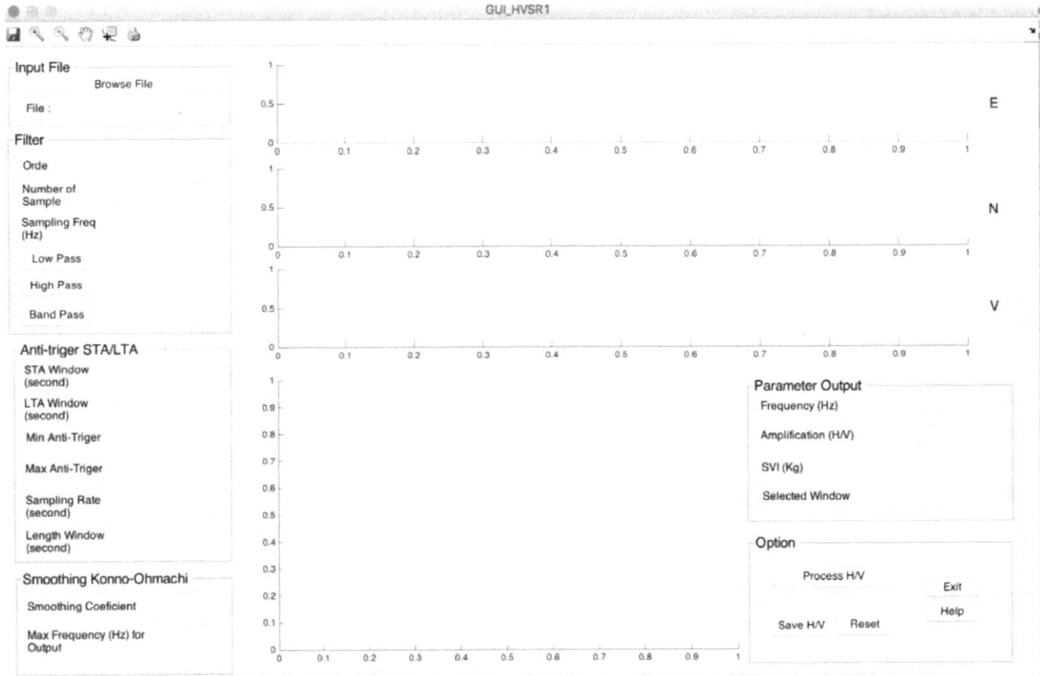

Figure 1. The SVIM program is displayed in Graphical User Interface (GUI)

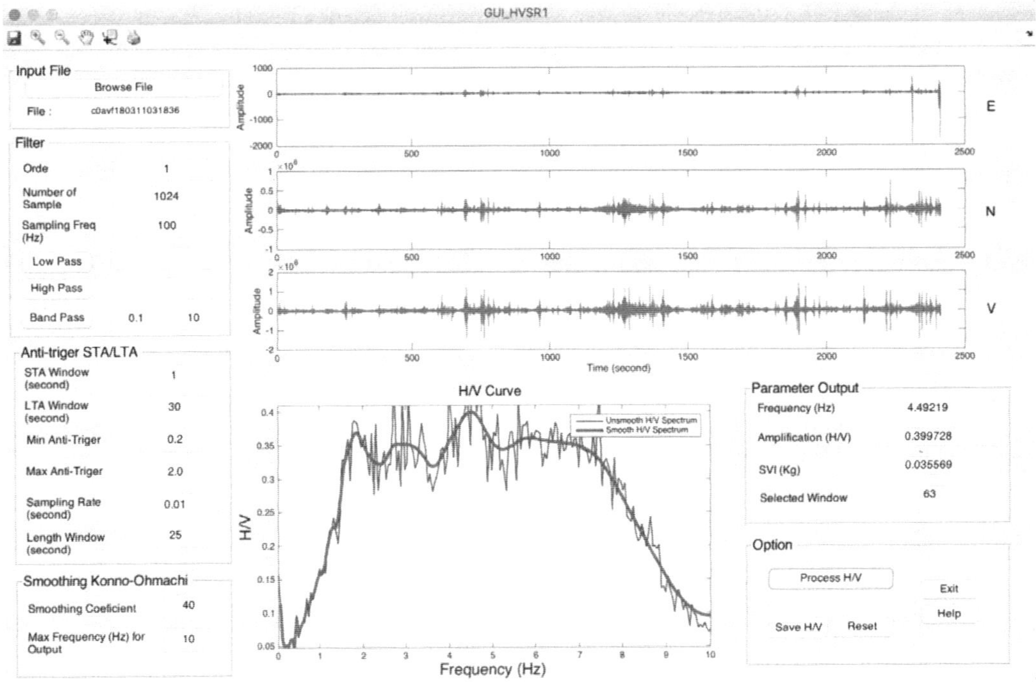

Figure 2. The results from SVIM processing. Seismic vulnerability index was calculated directly by using SVIM program.

IOP Conf. Series: Earth and Environmental Science **273** (2019) 012016 doi:10.1088/1755-1315/273/1/012016

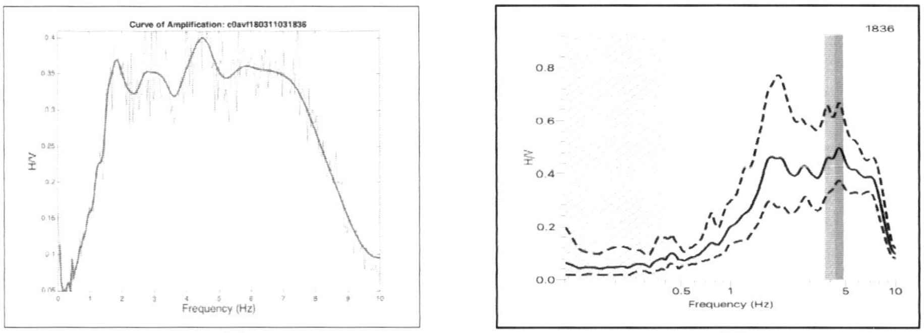

(a) (b)

Figure 3. The first comparison of 21 microtremor data of amplification (H/V) curve between (a) SVIM program and (b) Geopsy results.

Table 1. The comparison result of the dominant frequency, amplification factor and seismic vulnerability index between SVIM program and Geopsy.

File	SVIM			Geopsy		
	Dominant Frequency (Hz)	Amplification Factor	Kg	Dominant Frequency (Hz)	Amplification Factor	Kg
1836	4.49219	0.39973	0.03557	4.27264	0.470059	0.05171
2927	1.85547	0.59313	0.18960	1.85292	0.680417	0.24986
1420	2.68555	0.44583	0.07401	1.99865	0.524459	0.13762
5018	6.10352	1.11742	0.20457	5.65641	1.09648	0.21255
4929	9.7168	0.26959	0.00748	4.34258	0.398673	0.03660
5853	4.39453	0.2846	0.01843	7.06946	0.337485	0.01611
4010	5.22461	0.28362	0.01540	4.40759	0.562528	0.07179
4059	4.39453	0.35189	0.02818	4.2873	0.6375585	0.09481
5058	8.30078	1.667	0.33477	8.44106	1.76995	0.37113
3636	7.61719	1.01597	0.13551	7.38644	1.07741	0.15715
3727	5.66406	0.34078	0.02050	5.52023	0.575192	0.05993
3710	4.49219	0.36554	0.02974	4.02013	0.547666	0.07461
1020	4.49219	0.64772	0.09339	4.44892	0.936314	0.19706
0157	9.42383	0.52956	0.02976	9.51935	0.535112	0.03008
0037	2.19727	0.93717	0.39972	2.3339	0.839148	0.30171
0821	3.17383	0.98102	0.30323	2.92109	1.01594	0.35334
0675	1.80664	1.04548	0.60501	1.32727	1.02244	0.78762
1548	2.19727	1.14103	0.59253	2.04367	1.08544	0.57650
1427	1.85547	0.67811	0.24783	1.90198	0.789858	0.32801
3002	7.12891	0.48998	0.03368	6.23972	0.540737	0.04686
1708	5.46875	0.39985	0.02924	5.43022	0.598081	0.06587

The whole comparison results of SVIM program and Geopsy shown in figure 4, red lines are SVIM results and blue lines are Geopsy results. The results between SVIM and Geopsy are close each other. It can be seen that seismic vulnerability index (figure 4(a)), dominant frequency (figure 4(b)), and amplification factor (figure 4(c)) are almost identical between SVIM and Geopsy results. However, refer to dominant frequency comparison shown in figure 4(b) at the fifth measurement point has different dominant frequency result between SVIM program and Geopsy. The dominant frequency of SVIM result is greater than dominant frequency of Geopsy, whereas the settings of parameters were similar during data processing as explained in the method section.

We also validated the results by calculating correlation coefficient between SVIM and Geopsy results. The correlation coefficient of seismic vulnerability index was obtained 0,964 or around 96,40% (figure 4(d)), frequency was 0,842 or around 84,19% (figure 4(e)), and amplification was 0,9580 or around 95,80% (figure 4(f)). If the correlation coefficient is close to one or 100%, the correlation is strong. Three output parameters (seismic vulnerability index, dominant frequency and amplification factor) have strong correlation. We conclude that the SVIM program can be used for seismic vulnerability index determination and HVSR data processing.

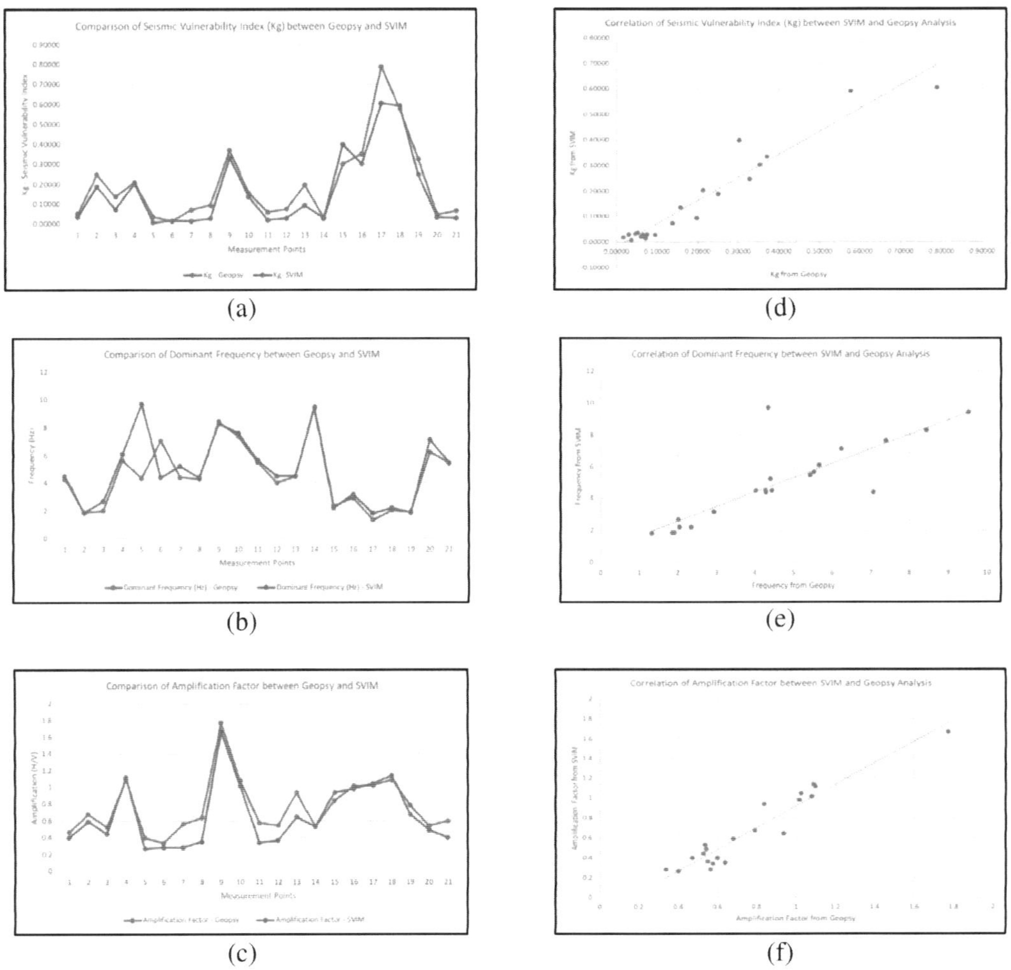

Figure 4. Comparison of parameters and correlation results between SVIM program and Geopsy data processing: (a) seismic vulnerability index, (b) dominant frequency, and (c) amplification factor.

7. Conclusion

The Graphical User Interface (GUI) of SVIM program has been built and successfully tested to determine seismic vulnerability index and HVSR data processing to 21 microtremor data points in Lamteuba, Aceh Besar, Aceh, Indonesia. The program is easy to use for seismic vulnerability index determination and HVSR data processing. The comparison results between SVIM program and Geopsy show that SVIM results almost close to Geopsy results based on comparison the seismic vulnerability index, dominant frequency and amplification factor. SVIM results have been validated by calculating coefficient correlation between SVIM results and Geopsy results. The coefficient correlation of seismic vulnerability index was 96,40%, dominant frequency was 84,19%, and amplification factor was 95,80%. We conclude the SVIM program can be used for seismic vulnerability index determination and HVSR data processing.

8. Acknowledgments

The authors would like to thank to Tsunami and Disaster Mitigation Research Center (TDMRC) Syiah Kuala University for facilitating this project. Also, the authors would like to thank to the Geological Hazard Research Group, TDMRC Syiah Kuala University. Especially, the authors are also grateful to anonymous reviewers for their valuable comments and suggestions which help in improving the manuscript.

References

[1] Nakamura Y. A method for dynamic characteristics estimation of subsurface using microtremor on the ground surface. Railw Tech Res Institute, Q Reports 1989;30.
[2] Mucciarelli M, Gallipoli MR. A critical review of 10 years of Nakamura technique. Boll Geof Teor Appl 2001;42:255–6.
[3] Nakamura Y. On the H/V spectrum. 14th World Conf Earthq Eng 2008:1–10.
[4] Guo Z, Aydin A. A modified HVSR method to evaluate site effect in Northern Mississippi considering ocean wave climate. Eng Geol 2016;200:104–13. doi:10.1016/j.enggeo.2015.12.012.
[5] Konno K, Ohmachi T. Ground-motion characteristics estimated from spectral ratio between horizontal and vertical components of microtremor. Bull Seismol Soc Am 1998;88:228–41.
[6] Nakamura Y. Seismic Vulnerability Indices for Ground and Structures Using Microtremor. World Congr Railw Res 1997:16–9.
[7] Fäh D, Kind F, Giardini D. A theoretical investigation of average H / V ratios Donat Fa. Geophys J Int 2001;145:535–49.
[8] Trnkoczy A. Understanding and parameter setting of STA / LTA trigger algorithm 1 Introduction. 1999.

IOP Conf. Series: Earth and Environmental Science **273** (2019) 012017 doi:10.1088/1755-1315/273/1/012017

Mineral Identification of Sedimentary Deposits of the 2004 Tsunami on the Aceh Besar Coast, Indonesia

Ismail[*] and Nazli Ismail

Department of Physics, Faculty of Mathematics and Natural Science, Universitas Syiah Kuala, Banda Aceh 23111, Indonesia

*E-mail: ismailab@unsyiah.ac.id

Abstract. Understanding tsunami deposit is quite important in order to understand the history of the tsunami and its process deeply. Nonetheless, most tsunami deposit studies only discussed the sediment grain characteristics, not the mineral contents in the deposits. Meanwhile, understanding the minerals in tsunami deposits is essential in order to understand the sediment composition deeply. Thus, we have utilized X-Ray Fluorescence and X-Ray Diffraction to study the sedimentary deposits of the 26 December 2004 Indian Ocean tsunami on the Aceh Besar coast, Indonesia. Our results showed that the sedimentary deposits of tsunami 2004 on the Aceh Besar coast contain 55% of SiO_2, 19% of Fe_2O_3, 12% of Al_2O_3, 9% of CaO, 3% of K_2O, and 2% of TiO_2. The deposits are found to be crystalline with the average crystallite size of 678 Å. There is no heavy mineral found in the tsunami deposit from Aceh Besar coast. Our results revealed that the composition of mineral contained in the sedimentary deposits of tsunami 2004 on the Aceh Besar coast is the same as those from the volcanic ash.

1. Introduction

The mega-earthquake occurred on the 26th of December 2004 was located at 3.3° N and 99.95° E near Banda Aceh city in Indonesia. The focal depth of this earthquake was about 10 km which the seismic moment magnitude was up to 9.1 to 9.3 [1]. According to the previous study, this earthquake initiated slowly at first with a small slip for several seconds. After that, the rupture was expanded with fast speed which extended up to more than 1200 kilometers. The fault slip was up to about 15 meters along the plate boundary offshore of the Sumatra and Nicobar islands [2]. This mega-earthquake had triggered the giant tsunami that caused more than 250 thousand deaths. This disaster has a major impact on the geomorphology and sedimentology of the coast of Sumatra [3, 4].

Identifying tsunami deposit is very important in order to understand the history of tsunamis deeply. The paleoseismic events have been recognized based on their tsunami deposits [5, 6]. A recent study using radiocarbon analyses found an extraordinary 7,400-year stratigraphic sequence of prehistoric tsunami deposits from a coastal cave in Aceh, Indonesia [7]. According to this study, there were at least 11 prehistoric tsunamis struck the Aceh coast between 7,400 and 2,900 years ago and the tsunami occurred about every 450 years [7]. Most tsunami deposit studies only discussed the sediment grain characteristics, not the mineral contents in the deposits [5-8]. Moore et al. studied sedimentary deposits of the 26 December 2004 tsunami on the northwest coast of Aceh, Indonesia [8]. They found some tsunami sediments near Lhoknga beach in Aceh Besar district in Indonesia where the grain size fines both inland and upwards in the sand [8]. Nonetheless, detail information of minerals contained in the tsunami deposit on the Aceh coast is still unknown. On the other hand, tsunami sediments could contain hazard materials such as heavy minerals (As, Pb, Hg, etc.). Thus, it is important to study the minerals contained in the tsunami deposit. The purpose of the study is to identify the minerals contained in the sedimentary deposits of the 26 December 2004 Indian Ocean tsunami on the Aceh Besar coast, Indonesia. Detail results are reported in this paper.

2. Methods

The sedimentary deposits of the 2004 tsunami were obtained from the Lambada Lhok Village (marked by X), Aceh Besar district, Indonesia as shown in figure 1. This village was swept by the 26 December 2004 Indian Ocean tsunami. The location of the sample obtained was about 100 meter from the beach. Four samples of sedimentary deposits were collected that were at 0 – 2 cm depth (sample 01), 10 – 12 cm depth (sample 02), 20 – 22 cm depth (sample 03), and 30 – 32 cm depth (sample 04). X-Ray Fluorescence (XRF) and X-Ray Diffraction (XRD) have been utilized to identify the mineral content in the all samples of sedimentary deposits.

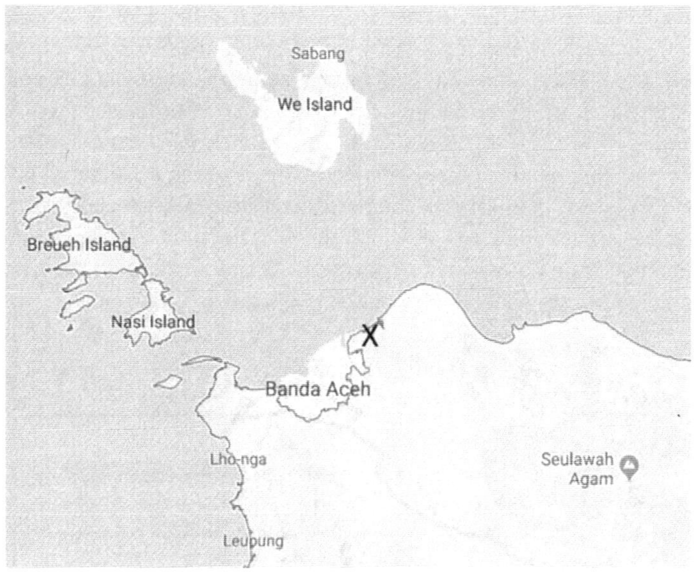

Figure 1. The location of the study area marked by X (Source: Map data @2018 Google).

3. Results and Discussion

The sediment types of all samples were classified as sandy. Our results of XRF measurement were shown in table 1.

Table 1. XRF data from the sedimentary deposits of tsunami 2004 on the Aceh Besar coast.

Kind of Mineral	Sample 01 (0 – 2 cm)	Sample 02 (10 -12 cm)	Sample 03 (20 -22 cm)	Sample 04 (30 -32 cm)
SiO_2 (%)	56.9	57.9	51.4	53.6
Fe_2O_3 (%)	17.2	17.0	20.4	20.5
Al_2O_3 (%)	11.0	12.0	12.0	12.0

CaO (%)	8.5	7.9	9.5	8.3
K_2O (%)	3.0	3.0	2.3	2.5
TiO_2 (%)	1.6	1.5	2.2	1.9
SrO (%)	0.4	0.3	0.4	0.4
MnO (%)	0.3	0.2	0.3	0.3
Eu_2O_3 (%)	0.2	0.2	0.3	0.3

The sedimentary deposit at the ground surface (0 – 2 cm depth) contains 56.9% of SiO_2. The silicon oxide is also found in sedimentary deposits from the 10 – 12 cm depth (57.9% of SiO_2), 20 – 22 cm depth (51.4% of SiO_2), and 30 – 32 cm depth (53.6% of SiO_2). The percentage of silicon oxide contained in each sample is about the same with the average of 55%. The sedimentary deposit also contains Fe_2O_3, Al_2O_3, CaO, K_2O, TiO_2. The percentage of Fe_2O_3 from each sample is about the same. Its average is 19% of Fe_2O_3. It is similar to the compositions of other phases (Al_2O_3, CaO, K_2O, TiO_2). The average percentages of Al_2O_3, CaO, K_2O, TiO_2 from sedimentary deposits are 12%, 9%, 3%, 2%, respectively. Besides the phases mentioned above, there are some other phases contained in the sedimentary deposits (SrO, MnO, Eu_2O_3) but the percentages of those phases are very small.

Our result of XRD measurement for all samples is shown in figure 2. Our XRD data were compared to JCPDS data (Joint Committee on Powder Diffraction Standards) to determine kind of minerals (phases) contained in the sedimentary deposits of tsunami 2004 on the Aceh Besar coast.

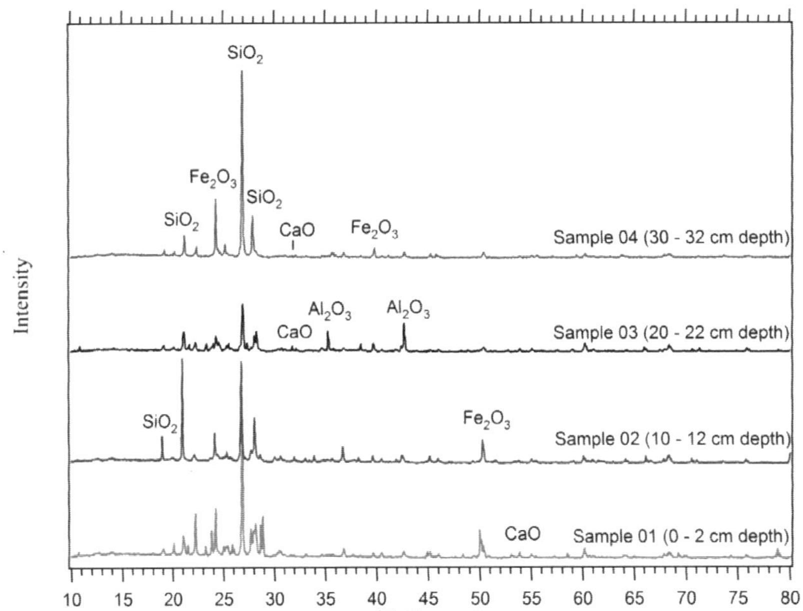

Figure 2. XRD data from the sedimentary deposits of tsunami 2004 on the Aceh Besar coast.

Our XRD data analysis revealed that the sedimentary deposits of tsunami 2004 on the Aceh Besar coast contain SiO_2, Fe_2O_3, Al_2O_3, and CaO. The SiO_2 peak at the Bragg angle (2θ) of 26.88 degrees is the highest peak. This indicates that the SiO_2 has the largest percentage contained in the sedimentary deposits of tsunami 2004. The other sharp peaks were observed at the Bragg angle (2θ) of 24.26 degrees (Fe_2O_3) and 42.73 degrees (Al_2O_3). The K_2O and TiO_2 phases were not observed by XRD because the percentage of these phases are quite small. In general, our XRD data are in good agreement with our XRF data. The crystallite size of the sedimentary deposits of the tsunami was determined by using the following equation [9]:

$$D = \frac{k.\lambda}{\beta.\cos(\theta)} \tag{1}$$

Where D = the crystallite size; k = 0.95 (constant); λ = 0.15406 nm (the wavelength of x-ray used in the experiment); β = the full width at half maximum (FWHM in radian); θ = the Bragg angle (degrees).

The FWHM for each peak was determined by fitting a Lorentzian function to the experimental XRD data as shown in figure 3.

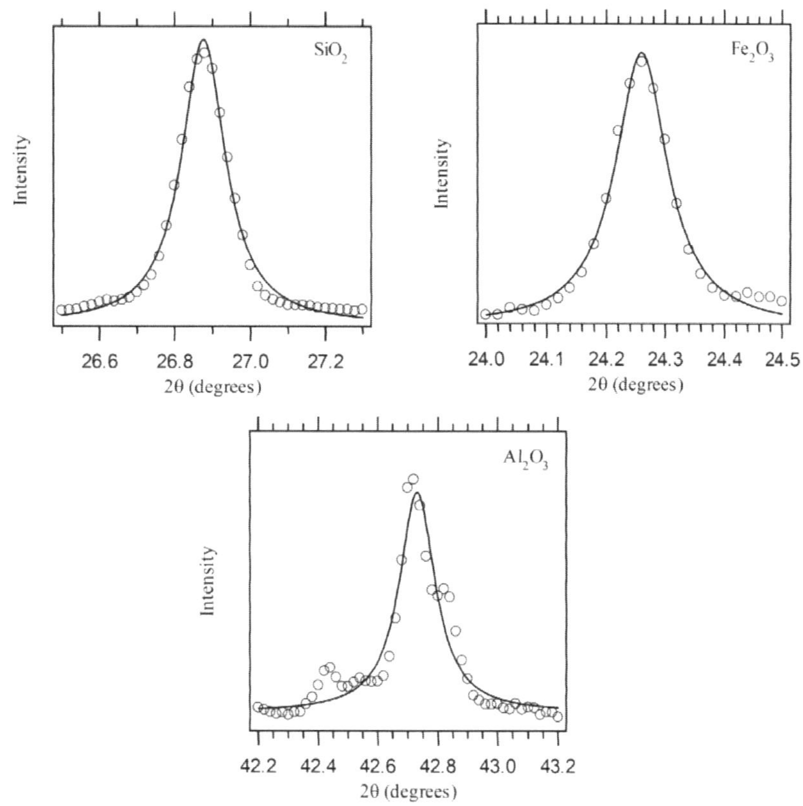

Figure 3. The comparison of the XRD data (circles) from the sedimentary deposits of tsunami 2004 on the Aceh Besar coast and calculated best fit (solid lines).

By using the equation (1) above, we found that the crystallite size of SiO_2 is 61.29 nm (612.9 Å). The crystallite size of Fe_2O_3 was found to be 79.37 nm (793.7 Å) which is a little bit larger than that of SiO_2. The crystallite size of Al_2O_3 was found to be 62.74 nm (627.4 Å) which is almost the same as the crystallite size of SiO_2. The average of crystallite size of the sedimentary deposits of tsunami 2004 on the Aceh Besar coast is 67.80 nm (678 Å). The summary of the crystallite size of all phases is listed in table 2.

Table 2. The crystallite size of the sedimentary deposits of tsunami 2004 on the Aceh Besar coast.

No	Phase	2θ (deg)	FWHM (deg)	D (nm)
1	SiO_2	26.88	0.14	61.29
2	Fe_2O_3	24.26	0.11	79.37
3	Al_2O_3	42.73	0.14	62.74
		Average crystallite size		67.80

Based on our results above, the percentage of mineral contained in each sample is about the same which means that the thickness of sedimentary deposits of tsunami 2004 on the Aceh Besar coast is about 30 cm. The sedimentary deposits of tsunami 2004 on the Aceh Besar coast mainly contain 55% of SiO_2, 19% of Fe_2O_3, 12% of Al_2O_3, 9% of CaO, 3% of K_2O, and 2% of TiO_2. There is no heavy mineral found in sedimentary deposits of tsunami 2004 on the Aceh Besar coast. Veerasingam et al. found that the tsunami deposits from Indian Ocean tsunami 2004 at the southeast coast of India contain about 50% of SiO_2, 12% of CaO, and 1% of TiO_2 [10] which are in good agreement with our study. However, the composition of mineral from sedimentary deposits of tsunami 2004 on the Aceh Besar coast is different than those from tsunami deposits in Sendai Japan [11] as shown in tabel 3. The tsunami deposits in Sendai Japan contains 79% of SiO_2 while it is only 55% of SiO_2 in the sedimentary deposits of tsunami 2004 on the Aceh Besar coast. The compositions of Fe_2O_3 and CaO are also quite different. This suggests that the source type of sedimentary deposits of the 26 December 2004 Indian Ocean tsunami is different than those of tsunami in Sendai Japan.

Table 3. The comparison of mineral contained in the sedimentary deposits of tsunami 2004 on the Aceh Besar coast and tsunami deposit from Sendai Japan.

Kind of Mineral	This Study	Tsunami deposit in Sendai Japan [11]	Volcanic ash from Mount of Merapi [12]
SiO_2 (%)	55	79.39	52.52
Fe_2O_3 (%)	19	0.63	9.17
Al_2O_3 (%)	12	13.61	18.69
CaO (%)	9	1.40	8.96
K_2O (%)	3	2.15	2.10
TiO_2 (%)	2	1.5	1.45

Wahyuni et al. found that the volcanic ash from Merapi Mountain in Yogyakarta contains 52.52% of SiO_2, 18.69% of Al_2O_3, 9.17% of Fe_2O_3, and 8.96% of CaO [12]. The comparison between this study and the mineral contained in the volcanic ash from Merapi Mountain in Yogyakarta is shown in table 3. It is very surprising that the composition of mineral contained in sedimentary deposits of tsunami 2004 on the Aceh Besar coast is about the same as those from volcanic ash from Merapi Mountain eruption in Yogyakarta.

4. Conclusions

The minerals contained in the sedimentary deposits of tsunami 2004 on the Aceh Besar coast are 55% of SiO_2, 19% of Fe_2O_3, 12% of Al_2O_3, 9% of CaO, 3% of K_2O, and 2% of TiO_2. The sedimentary deposits of tsunami 2004 on the Aceh Besar coast are crystalline with the average of crystallite size of 67.80 nm (678 Å). There is no heavy mineral observed in the sedimentary deposits of tsunami. The tsunami deposit from Indian Ocean tsunami 2004 at the Aceh Besar coast is found to be similar to those from the southeast coast of India. We found that the composition of mineral contained in sedimentary deposits of tsunami 2004 on the Aceh Besar coast is the same as those from the volcanic ash from Merapi Mountain eruption in Yogyakarta.

References

[1] Lay *et al.* 2005 *Science* **308** 1127
[2] Ammon *et al.* 2005 *Science* **308** 1133
[3] Meltzner A J, Sieh K, Abrams M, Agnew D C, Hudnut K W, Avouac J P, and Natawidjaja D H 2006 *J. Geophys. Res.* **111** B02407
[4] Chlieh *et al.* 2007 *Bulletin of the Seismological Society of America* **97** S152
[5] Cisternas *et al.* 2005 *Nature* **437** 404
[6] Nanayama F, Satake K, Furukawa R, Shimokawa K, Atwater B F, Shigeno K, and Yamaki S 2003 *Nature* **424** 660
[7] Rubin C M, Horton B P, Sieh K, Pilarczyk J E, Daly P, Ismail N, and Parnell A C 2017 *Nature Commun.* **8** 16019
[8] Moore A, Nishimura Y, Gelfenbaum G, Kamataki T, and Triyono R 2006 *Earth Planets Space* **58** 253
[9] Suryanarayana C and Grant Norton M 1998 *X-Ray Diffraction A Practical Approach* (New York: Plenum Press)
[10] Veerasingam S, Venkatachalapathy R, Basavaiah N, Ramkumar T, Venkatramanan S and Deenadayalan K 2014 *J. Earth Syst. Sci.* **123** 905
[11] Minoura K, Imamura F, Sugawara D, Kono Y, Iwashita T 2001 *Journal of Natural Disaster Science* **23** 83
[12] Wahyuni E T, Triyono S, and Suherman 2012 *J. Manusia dan Lingkungan* **19** 150.

Acknowledgments

This research did not receive any specific grant from funding agencies in the public, commercial, or not-for-profit sector.

Revisiting The 2018 Kalibening Earthquake Sequence In Central Java: Call for the Revision of Earthquake Hazard

Renhard Sipayung[1], Dimas Sianipar[1,2], Setyoajie Prayoedhie[1], Daryono[1], Januar Arifin[1], Andrean V H Simanjuntak[1], Umar Muksin[3], Sesar Prabu[1], Gatut Daniarsyad[1], Haryanta[1], Nugroho Putranto[1], Ali Azimi[1]

[1]Agency for Meteorology Climatology and Geophysics (BMKG), Angkasa I Street No. 2, Jakarta, Indonesia 10720
[2]Meteorological Climatological and Geophysical State College (STMKG), Perhubungan I Street No. 5, Pondok Betung, Banten, Indonesia 15221
[3]Department of Physics, Syiah Kuala University, Banda Aceh

E-mail: Renhard.sipayung@bmkg.go.id

Abstract. As a moderate devastating earthquake that impacted economic loss was about $ 1.68 million US, the 2018 Kalibening earthquake quite shocked where the faulting mechanism still unconfirmed. There was no reference indicating the existence of the faulting zone exactly on the earthquake sequence. We take a benefit from a seismic temporary network with the density ±2 km, which deployed while a macroseismic survey in purposing to evaluate the 2018 Kalibening earthquake sequence. Earthquake relocation using double difference combined with cross-correlation and determining earthquake moment tensor inversion solution using near source seismograms were applied. We suggested the deformation scheme of this earthquake with a thrust faulting with 307.5/ 28.8/ 118.5 (Strike/ Dip/ Rake) as a result from mainshock and supported by aftershock moment tensor solutions. This parameter is consistent to aftershocks relocation results which formed a lineation trending NW-SE appropriate with Strike = 307.5. The cross-section exhibits aftershocks pattern which elongated deeper and formed a slope from SW to NE approximately fit to Dip = 28.8. The results from investigating the background seismicity in Banjarnegara Region using combined catalog (ISC-USGS-BMKG) compared to BJI (single station) showed the sparse and the lacking of InaTEWS seismic network configurations. Finally impacted to losses in earthquake cataloging, leads the low area coverages around Banjarnegara region. Together with our results, we called for the revision of earthquake hazard assessment for Central Java province, especially in Banjarnegara region.

Keywords: earthquake relocation, moment tensor solution, seismic temporary network, Kalibening earthquake, earthquake hazard assessment

1. Introduction

There are seven faults lying in Central Java denoted by the red lines in Fig.1 which confirmed and summarized in Indonesian Seismic Sources and Seismic Hazard Maps 2017 [1]. One of the longest that extending from the eastern section of West Java to East Java is Baribis-Kendeng fault which passes through the northern section of Central Java. This fault is dividing into several segments and named based on the area that passes. Another study ensured there are two major strike-slip faults existing in Central Java, called the Muria-Kebumen Fault (left-lateral, trending southwest-northeast) and the Pamanukan-Cilacap Fault (right-lateral, trending northwest-southeast) denotes by orange straight lines in Fig.1 [2,3]. This two fault separated in the northern area but meeting in the southern area, opposite in slip and trend to each other therewith considered to have indented Central Java's coastline and caused many geologic changes in Central Java. Both of these references are contradictions but prominently they don't announce any faulting mechanism in Banjarnegara Region and the surrounding area.

On 18 April 2018 06:28:35 UTC, an Mw (mB) 4.4) earthquake ruptured onshore at the Kalibening District, Banjarnegara County, Central Java. BMKG reported that Kalibening earthquake epicenter at 7.21°S, 109.65°E with shallow depth as 4 km. The estimated of economic loss was about $ 1.68 million U.S (23.56 billion Rupiah) according to the National Disaster Management Authority of Indonesia (BNPB). That excessive and worst impact became a reason why this disaster is staggering. The stakeholders faced difficulty when determined the generating fault since related to the previous explanations, this moderate earthquake occurred in that unconfirmed faulting zone.

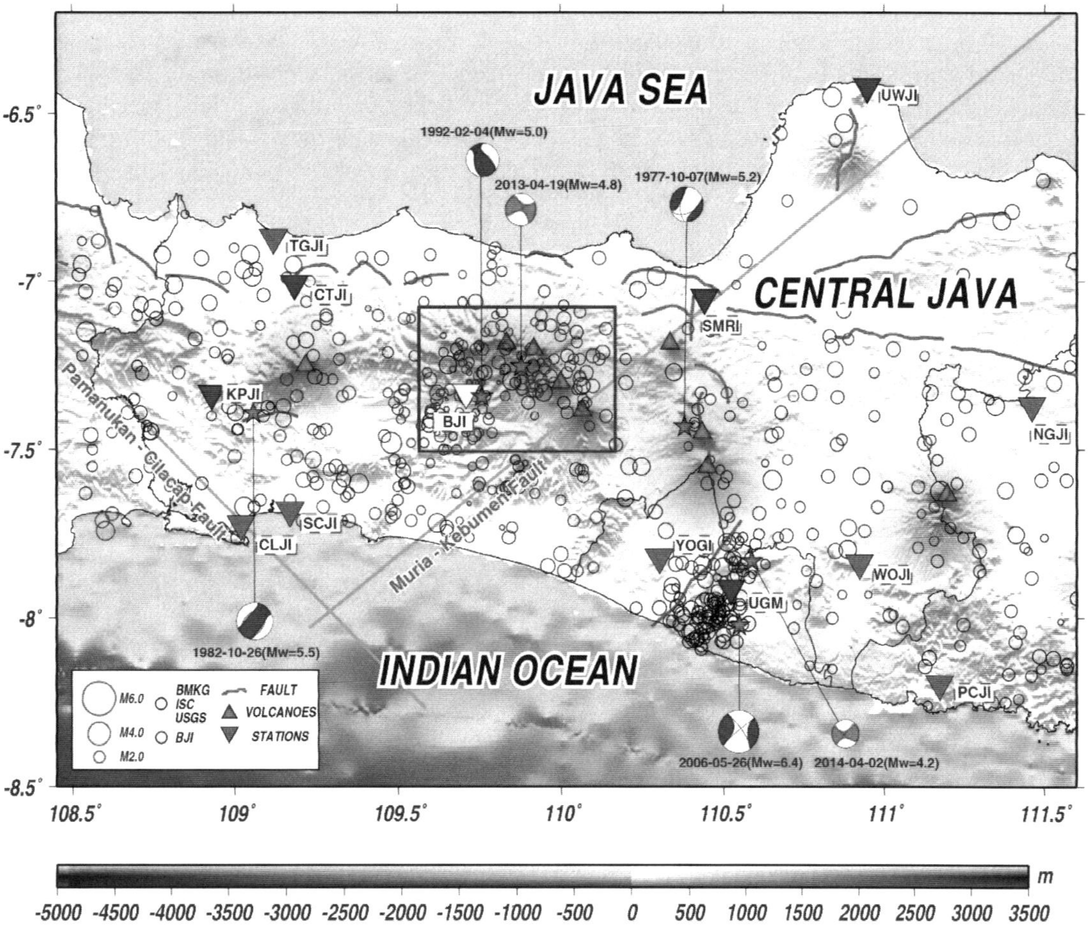

Figure 1. Overview map showing the seismicity, faults, and seismic stations in Central Java. The black open circles indicate combined catalog from ISC-USGS-BMKG catalog while the blue circles recorded by BJI. Blue beach-ball symbols show the earthquake mechanisms listed in GlobalCMT while the red beach-ball from GFZ catalog. Inverted triangle showing us the InaTEWS seismic network, the blue inverted triangle was used to generate mainshock's moment tensor in this study. The yellow one notified the location of BJI sensor. Red triangle shows the volcanoes location. Red lines and orange lines denote mapped faults which confirmed from Earthquake Source and Hazard Map of Indonesia 2017 and proposed in [2,3], respectively.

In purposed to monitoring the aftershocks, BMKG had sent macroseismic survey team and in its enforcement was deployed five local temporary sensors (including three Trillium Compact Seismometer and two Taide Digital Seismograph (TDS)) which covering the damaged area in Kalibening District. Total of 24 aftershocks was recorded up to 26 June 2018 (where 18 events up to 01 May 2018 for the last day of survey). The nine of them were felted by citizens and made a disruption, both at displaced and damaged area.

There are several factors, including the network geometry, available phases, arrival time reading accuracy and knowledge of the crustal structure which controlled the accuracy of absolute hypocenter locations [4,5,6]. In further, the precision of earthquake location in terms of a better, precisely and refined distribution can be further improved by earthquake relocation [6, 7, 8]. Another hand, to investigation the earthquake source processes, the moment tensor is very well suited as a mathematical description of equivalent forces and moments in a point source. Understanding the faulting type of an earthquake to be a valuable and important task by considering the estimation of the moment tensor solution [9].

The purpose of this study is to evaluate the 2018 Kalibening earthquake sequences in lead to calling a revision of earthquake hazard in this study area. First, we relocate the aftershocks, then determine the focal mechanisms both mainshock and aftershocks. From these results, we suggest the deformation scheme in the source region that generated the Kalibening earthquake. Then we show the lacking of InaTEWS seismic network in terms of low area coverages around Banjarnegara Region and its relation with unconfirmed faulting type of the Kalibening earthquake.

2. Data

For studying Kalibening earthquake sequences, we used data from two types of seismic networks. First, to determined the mainshock moment tensor inversion, we had chosen Short Period sensor of InaTEWS seismic network which quite well in recording the signals. There are CTJI, JCJI, KPJI, SMRI, UGM, and UWJI (blue inverted triangle in Fig.1) which we considered as the most suitable sensors due to data availability and to avoid the big residual between observe and synthetic waveform from other sensors (grey inverted triangle).

The second, we used a temporary network to relocated and determined the aftershocks moment tensor inversion. From the total of 24 aftershocks that occurred, we were able to process 14 events (completely shows in Tabel 3). Since the first four came before all the temporary sensors were deployed (before the survey) and another last six events after survey. We set the aftershocks that would be relocated with Event ID (1-14). As a dense temporary network with a density of around 2 km, it can be seen and became quite simple for clearly phase identified even with M 0.8 aftershock. Seismic data that we used were originated from six sites even though in total were eight as the two of them had errors in reading time (YGI1 and YGI2 were unused in this study).

Initially, there were five sites associated with five sensors but for necessity in the macroseismic survey, the three of them were forced to be moved. So that finally became eight sites sensor with different deployed times (detail in Table 1). The spread aftershock distributions at the outside in the south of initial temporary network location became a reason why some of these sensors was moved. Almost all aftershocks were not well recorded by InaTEWS seismic network as its small energy radiations and very shallow depth (for example in Fig.3). Consequently we can not use it as ancillary for relocating and determining the moment tensor inversion. Likewise with the mainshock, we found difficulties in finding the pairing events for relocating.

Table 1. Information of the temporary network during the macroseismic survey for 2018 Kalibening earthquake

No.	Station Code	Sensor Type	Lat. (°S)	Long. (°E)	Location	Deploy Time (d/m h:m UTC)	Status
1.	BJIX	Taide TDL-303 S	7.2178	109.6515	Sidakangen	19 Apr 22:00 – 26 Apr 14:00	Used in this study
2.	KP05	Trillium Compact	7.1939	109.6413	Bedana	20 Apr 10:50 – 21 Apr 10:21	Used in this study
3.	KP03	Trillium Compact	7.2147	109.6332	Sikumpul	20 Apr 09:57 – 22 Apr 13:05	Used in this study
4.	TES2	Trillium Compact	7.1987	109.6563	Karanganyar	20 Apr 02:12 – 23 Apr 13:11	Used in this study
5.	KTSI	Trillium Compact	7.2264	109.6903	Kertosari	21 Apr 15:00 – 23 Apr 13:55	Used in this study
6.	KLBB	Trillium Compact	7.2547	109.6858	Kaligombong	23 Apr 00:00 – 23 Apr 14:36	Used in this study
7.	YGI1	Taide TDL-303 S	7.2183	109.6729	Bakalan	20 Apr 18:00 – 21 Apr 08:00	Unused in this study
8.	YGI2	Taide TDL-303 S	7.2114	109.6800	Kasinoman	21 Apr 14:00 – 25 Apr 00:00	Unused in this study

The combined onshore earthquake catalog (depth < 50 km) from ISC-USGS-BMKG and Banjarnegara Geophysical Station (BJI) alone were used for investigating background seismicity and showing the lacks of InaTEWS seismic network around Banjarnegara Region. This area bounded by the blue rectangle in Fig.1 with the constraint area (109.55°E – 110.25°E and 7.05°S – 7.50°S). The combined catalog (denoted as black open circles) were collected between 1960 and July 2018 while the BJI catalog (blue open circles) since 2004. In avoiding the duplication, we were only selected events from the similar catalog with priority: ISC>USGS>BMKG>BJI. It means we convinced to deleted the catalog from BMKG when ISC catalog presented similar earthquake parameter. We compared that combined catalog to BJI where this catalog was mostly recorded from the single station observations. At this terms, the data used only from 2004 since the initial recording of the BJI catalog digitally. Here we also enclose the moment tensor catalog in Central Java from GlobalCMT and GFZ catalog since 1976 until now.

3. Methods

3.1. Earthquake Relocation

In this study, we used HypoDD to refined aftershocks location. HypoDD is a Fortran computer program package for relocating earthquakes with the double-difference algorithm [6]. We allow the combination use of phase delay times measured from catalog picks and from cross-correlation. Because cross-correlation methods can measure differential phase arrival times with subsample precision for events that are nearby and have similar focal mechanisms [10]. For detail in [7,11]. Solutions are found by iteratively adjusting the vector difference between nearby hypocentral pairs, with the locations and partial derivatives being updated after each iteration [6].

Since the arrival times were picked using Seismic Analysis Code (SAC) for the relocation process, the waveform data must be converted to the same form first (SAC format). The output signal of TDS sensor is in TRC format, we need to convert it to MINISEED before and then to SAC format using MSEED2SAC software. However, Trillium Compact Seismometer has output signal in MINISEED format so it can be converted directly to SAC.

First, we set the origin time as 0 time. The signal was cut 5 s before until 15 s after origin time to simplify phase identification as the raw data had one-hour continuous signals. The bandpass filtered 0.7-8 Hz was applied to all seismograms for better clearly signals. P and S-arrival were picking manually which marked by T1 and T2 as shows in Fig.2. We took the initial aftershocks parameter that generated during the macroseismic survey as part of the input to the program. Total of 160 P-phase pairs and 172 S-phase pairs were used in this processing where linked event pair as 76. We set waveform data 0.1 before – 0.5 s after P-arrival in Vertical Component and S-arrival in Horizontal

Component for cross-correlations. For velocity structure, we used as proposed in [15] from the anisotropic structure beneath Central Java study results (Table 2).

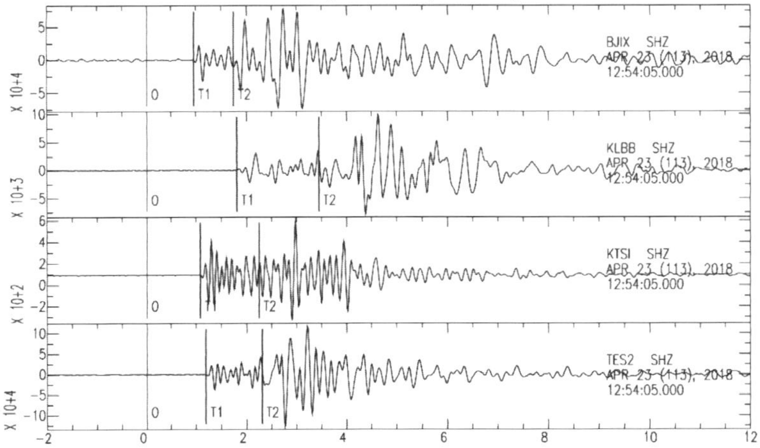

Figure 2. The example of raw waveforms from Event ID 11 recorded in temporary network vertical component (counts/m/s). The straight line with 0 denotes the origin time. T1 and T2 for P-arrival and S-arrival respectively.

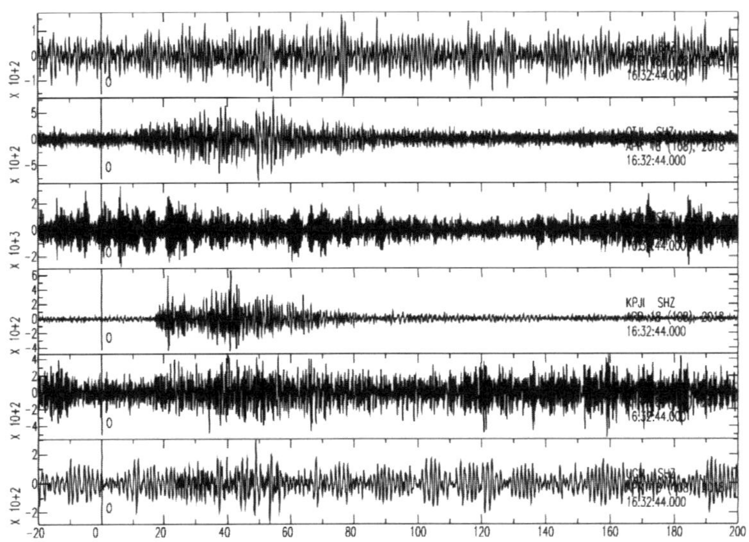

Figure 3. The observed of bandpass filtered 0.7-8 Hz waveforms in vertical component from aftershock Number 2 listed in Tabel 3. This waveform recorded by the nearest SH* InaTEWS network.

3.2. Moment Tensor Inversion

A well-established method and software had applied to perform waveform inversion towards estimate the moment tensor solutions for small and moderate earthquakes. This code constructed by Yuji Yagi [12] where utilize an assumed that horizontal location of centroid can be approximated to the epicenter, and estimated the optimal depth of the centroid and half duration using the grid-search method. Furthermore also in under assumed that a simple source model such as the point source model, in which we assumed the seismic waveform to be radiated from one point to estimate stable moment tensor solution. This method applied the Green's function based on Extended Reflectivity which developed by [13]. The optimum moment tensor solution is reached by best fitting between observation and synthetic waveforms through the inversion process. For more details about the method please refer to [12,14].

In determining the mainshock moment tensor, the waveform data were downloaded from Webdc3 BMKG in FULLSEED format. Then it converted to SAC format using RDSEED software. The instrument response file for each seismometer was removed by deconvolution of the response file. We used 40 s of waveform data with start time 0.5 s before P arrival for inversion. To minimize the effects of low- and high-frequency noise on the measurement of the surface wave dispersion, we used a frequency range of 0.05 – 0.1 Hz based on trial and error. As we unable refined the mainshocks locations, the initial parameter were used forcefully.

After the waveforms were prepared as explained in earthquake relocation section, we removed the instrument response for all channels to generated aftershocks moment tensor inversion. The waveform data that used around 8 s – 12 s with start time 0.1 s before P arrival. In this study, we preferred used the aftershocks parameter resulted from relocation to obtain the optimal position of the centroid of earthquake inversions. Trial and error methods were adjusted to select the fittest bandpass filter in frequency range of 0.28 – 0.38 Hz. This configuration depends on the state that the bigger magnitude and the farther distance should provide longer durations of aftershocks and advisable adjusted with the lower frequency range. The PREM velocity model combining with Koulakov [15,16] were applied for both determining mainshock and aftershocks moment tensor as detailed in Table 2. For mapping, we used Generic Mapping Tools [17]

Table 2. Velocity model from Koulakov in combined with PREM velocity model [15,16]

No.	Vp (km/s)	Vs (km/s)	Density (g/cm^3)	Thickness (km)	Qp	Qs	Vp/Vs
1.	4.30	2.46	1.02	3.0	150.	350.	1.75
2.	4.90	2.80	2.60	5.0	150.	350.	1.75
3.	5.70	3.26	2.60	8.0	150.	350.	1.75
4.	6.90	3.94	2.90	8.0	150.	350.	1.75
5.	7.10	4.06	2.92	53.0	150.	350.	1.75
6.	7.80	4.46	3.30	63.0	150.	350.	1.75

4. Results and Discussion

4.1. Aftershocks Relocation

We had done a double difference relocation combined with cross-correlations to Kalibening aftershocks. As P and S-wave differential times calculated from both catalog picks and cross-correlations, the final relocations have a RMS of 7 ms for catalog pick data and 17 ms for cross-correlation data. After five iterations, the inversion indicates average relative errors of ±19m (horizontally) and ± 20m (vertically). The condition number (CND) for the last iterations was 69 (to avoid over or under damping of the solution, it should between 40 – 80). We successfully refined 13 aftershocks location from the total of 14 that we proposed. Based on the output of that parameters, we assured had obtained better and constrained hypocenters in 3-D projections.

The distribution of relocated aftershocks represents in Table 3 and together mapped with Banjarnegara seismicity shows in Figure 4. In Table 3, the initial locations denote unrelocated parameters while the final locations for relocated aftershocks. Fig.4a represents the location of Kalibening earthquake sequence ruptured in the north of Banjarnegara Region, in the area of Kalibening District. The distribution of relocated aftershocks (blue-black list circles) was more concentrated in SE of Kalibening Basin where heavy damage occurred (marks by the rectangle with blue dashed line) as shown in Fig.4b. The mainshock (red star) seemingly stand a bit apart from relocated aftershocks as we didn't relocate it. Fig.4c we clearly exhibit the relocated aftershocks more patterned in comparing to the unrelocated (yellow-black list circles).

Table 3. Aftershocks detailed information containing initial (unrelocated) to final (relocated) parameter

No.	Event ID	Date (d/m/y)	O. Time (UTC)	Mag	Initial Lat. (°S)	Initial Long. (°E)	Initial Depth (km)	Final Lat. (°S)	Final Long. (°E)	Final Depth (km)	Info.
1.	-	18/04/2018	16:32:46	2.7	7.220	109.670	9.0	-	-	-	Bf. Surv
2.	-	19/04/2018	02:44:24	2.6	7.240	109.649	9.0	-	-	-	Bf. Surv
3.	-	19/04/2018	16:12:28	2.2	7.220	109.640	5.0	-	-	-	Bf. Surv
4.	-	19/04/2018	23:26:23	2.0	7.220	109.660	5.0	-	-	-	Bf. Surv
5.	**1**	20/04/2018	12:24:20	2.1	7.227	109.696	1.0	7.251	109.688	1.909	Dur. Surv
6.	**2**	20/04/2018	21:53:50	2.1	7.238	109.662	1.0	7.237	109.682	4.059	Dur. Surv
7.	**3**	21/04/2018	00:55:15	1.9	7.246	109.642	1.0	7.241	109.663	4.758	Dur. Surv
8.	**4**	21/04/2018	11:19:28	3.4	7.219	109.667	2.0	7.238	109.684	3.532	Dur. Surv
9.	**5**	21/04/2018	14:31:34	2.0	7.240	109.654	1.0	7.233	109.667	2.052	Dur. Surv
10.	**6**	21/04/2018	15:40:08	2.2	7.205	109.678	5.0	7.210	109.689	4.522	Dur. Surv
11.	**7**	21/04/2018	17:55:28	2.0	7.250	109.670	6.0	7.243	109.683	3.885	Dur. Surv
12	**8**	21/04/2018	18:53:05	1.5	7.247	109.671	3.0	7.245	109.676	3.376	Dur. Surv
13.	**9**	21/04/2018	20:58:40	1.4	7.218	109.675	1.0	7.229	109.674	4.608	Dur. Surv
14.	**10**	22/04/2018	20:21:14	0.8	7.238	109.673	4.0	7.231	109.675	3.313	Dur. Surv
15.	**11**	23/04/2018	12:54:05	3.0	7.192	109.707	1.0	7.232	109.692	6.308	Dur. Surv
16.	**12**	23/04/2018	23:00:57	2.8	7.221	109.680	4.0	7.206	109.687	3.261	Dur. Surv
17.	**13**	24/04/2018	10:21:10	1.6	7.220	109.690	3.7	7.235	109.697	4.139	Dur. Surv
18.	**14**	24/04/2018	11:33:28	2.4	7.240	109.700	4.5	-	-	-	Dur. Surv
19.	-	03/05/2018	02:46:26	2.9	7.200	109.690	15.0	-	-	-	Af. Surv
20.	-	07/05/2018	08:29:02	2.2	7.180	109.660	4.0	-	-	-	Af. Surv
21.	-	11/05/2018	17:32:05	1.8	7.020	109.730	7.0	-	-	-	Af. Surv
22.	-	19/05/2018	09:37:15	2.8	7.120	109.710	7.5	-	-	-	Af. Surv
23.	-	23/05/2018	21:55:20	2.6	7.260	109.640	9.0	-	-	-	Af. Surv
24.	-	04/06/2018	13:09:33	2.2	7.220	109.700	5.0	-	-	-	Af. Surv

The connecting line that linking the unrelocated to relocated aftershocks indicates the shifted patterns of aftershocks. In general, events that ruptured in the west, in the northwest and in the northeast of aftershocks centralization were moved to the east, to the southeast, and to the southwest, respectively. In average, the events were moved 0.22 km away where the farthest moved as 0.77 km.

Fig.4d and Fig.4e depict the cross-section of A-B line and C-D line respectively in Fig.4b. With range of 2 – 6 km depth, the relocated aftershocks distribution more vertically oriented as shows in Fig.4d whereas in Fig.4e seems formed a slope which elongated deeper from SW to NE. The shifted patterns of the aftershocks in the A-B and C-D cross-sections respectively represent in Fig.4f and Fig.4g. The connecting line in that two figures overall shows how the unrelocated moved to concentrate lineations of the final aftershocks location in the range of 3 – 5 km depth. Furthermore, event at the upper initial depth (1 km mostly) shifted to the downward otherwise with the event at the lower than 5 km shifted to the upward. However, the shifted vector signifies the uniformity movement in A-B cross-section (Fig.4f) while in C-D cross-section with randomly from SW and NE oriented in the upper of final location initially to the downward in the middle (event concentrated).

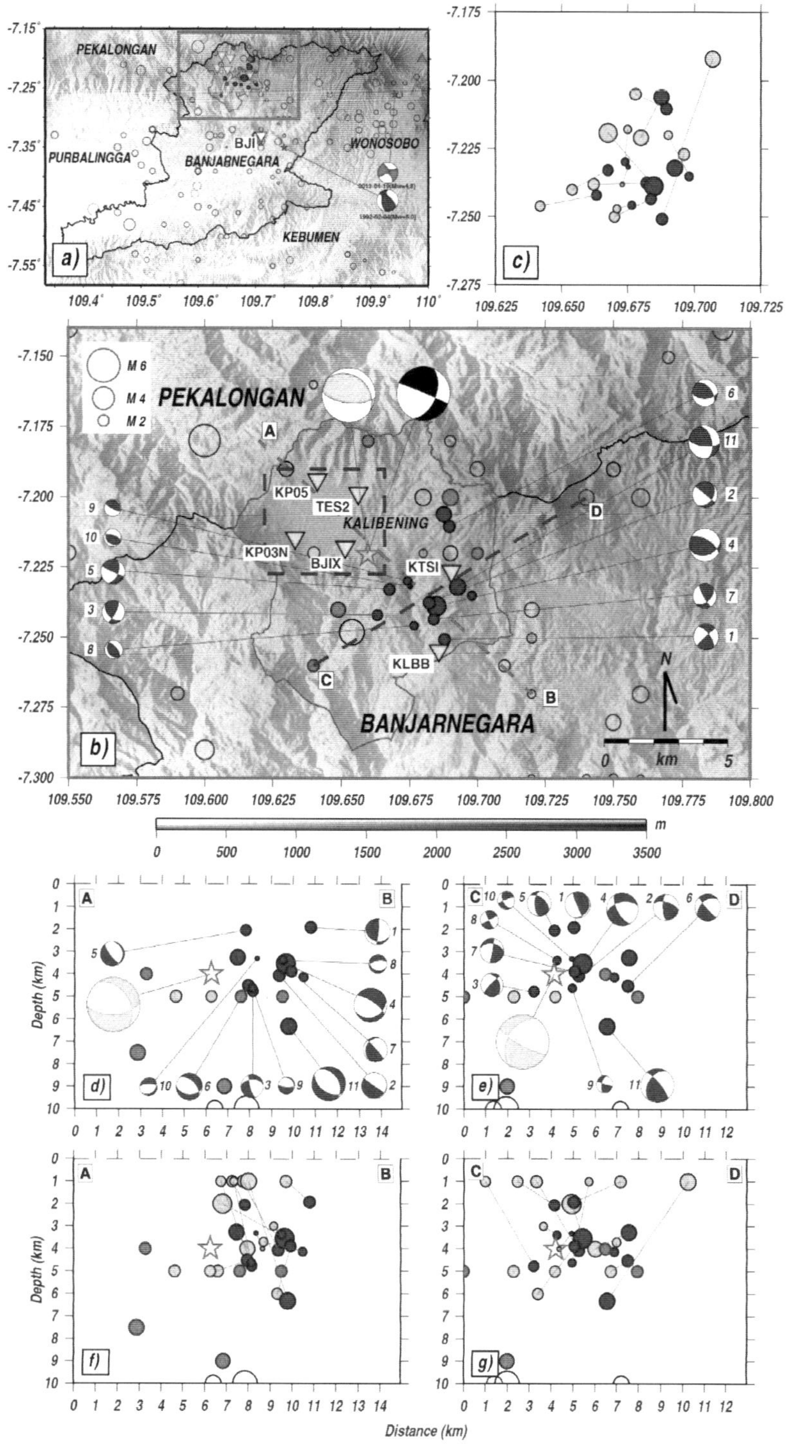

Figure 4. Kalibening Earthquake Seismicity Map **(a)** Location of 18 April 2018 Kalibening earthquake sequence at Banjarnegara seismicity map. **(b)** Zoomed in map from red rectangle in figure 2.(a). Red star, yellow-black list, blue-black list, red-black list, red-blue list, blue open, black open circles denote mainshock, unrelocated aftershocks, relocated aftershocks, felt aftershocks before survey, felt aftershocks after survey, BJI catalog and combined catalog, respectively. Green beach ball, blue beach ball, and black beach ball represent mainshocks, aftershocks moment tensor (the beside number marks event ID) and BMKG results, respectively. Green inverted triangle indicates temporary sensors. **(c)** The shifted patterns of aftershocks where the connecting line linking unrelocated to relocated aftershocks. **(d)** The cross sections of A-B line. **(e)** The cross sections of C-D line. **(f)** The shifted patterns of aftershocks in the cross sections of A-B line **(g)** The shifted patterns of aftershocks in the cross sections of C-D line.

4.2. Moment Tensor Inversion

Our results of mainshock moment tensor inversion show fault plane solution with a thrust faulting mechanism (green beach ball in Fig.4) with the first nodal 95.7/ 64.9/ 75.3 (Strike/ Dip/ Rake) and the second nodal 307.5/ 28.8/ 118.5. Detailed information on the resulting mechanism is listed in Fig.5 and Table 4. We assured this result was good enough as the variance about 0.38.

Table 4. Fault plane solution for mainshock as deduced from moment tensor inversion

Date (d/m/y)	O. Time (UTC)	Mag	Lat. (°S)	Long. (°E)	Depth (km)	Strike (N1)	Dip (N1)	Rake (N1)	Strike (N2)	Dip (N2)	Rake (N2)	Normalized L2 norm (Variance)
18/04/2018	06:28:35	4.4	7.210	109.650	4.0	95.7	64.9	75.3	307.5	28.8	118.5	0.3811

We also had successfully determined a total of 11 Kalibening aftershocks moment tensor inversion (blue beach balls in Fig.4). All of aftershock moment tensor results is listing in Table 5. Even though classified as microearthquakes, we had tried to get the best fitting results as the overall variance under 0.45. At least there are 4 events that we considered as the most suitable parameter to mainshock results (event ID 4, 6, 9 and 11). The detailed information listed in Fig.6. Aftershocks moment tensor solutions in the A-B cross section visible in Fig.4d while in the C-D cross-section in Fig.4e. The fault plane solutions mainly show thrust faulting with slight oblique and observable support the mainshock moment tensor inversion result, even some the aftershocks shows strike-slip mechanisms. By the reason of the small duration which impacted to low waveform fitting quality, we were not given to warrant the aftershock moment tensor solutions and it cannot be considered as the stand-alone conclusion.

There were differences in results where BMKG reported the mainshocks focal mechanisms with strike-slip solutions (black beach ball in Fig.4). It could be happened as BMKG using the first-motion polarities of the P wave and IASP91 velocity model. In here, we keep taking a side with our results as supported by aftershock solutions and also used local velocity structure.

From mainshock moment tensor inversion which supported by aftershock moment tensor inversion results has been explained previously, we focused to the second nodal plane which indicated a thrust faulting mechanisms with 307.5/ 28.8/ 118.5 (Strike/ Dip/ Rake). This parameter seemingly consistent to aftershocks relocations results which the formed lineation trending NW-SE appropriate with Strike = 307.5. Further as shows in Fig.4e, the C-D cross section exhibits aftershocks pattern which elongated deeper and formed a slope from SW to NE approximately fit to Dip = 28.8.

Table 5. Aftershocks fault plane solution and moment tensor elements result

Event ID	Date (d/m/y)	O. Time (UTC)	Mw	Lat. (°S)	Long. (°E)	Depth (km)	Strike	Dip	Rake	Mrr	Mtt	Mff	Mrt	Mrf	Mtf	Mo (Nm)	Variance
1	20/04/18	12:24:20	1.1	7.251	109.688	1.909	321.0	67.7	6.0	-2.14	5.85	-3.71	-3.30	0.07	-0.48	0.5406E+11	0.30
2	20/04/18	21:53:50	1.3	7.237	109.682	4.059	307.2	89.6	-55.5	-0.10	6.14	-6.04	7.54	-4.90	1.12	0.1089E+12	0.44
3	21/04/18	00:55:15	1.2	7.241	109.663	4.758	133.7	45.2	22.8	3.23	3.64	-6.87	4.50	3.91	2.09	0.8649E+11	0.38
4	21/04/18	11:19:28	3.3	7.238	109.684	3.532	300.5	62.8	116.4	1.00	-1.14	0.15	-0.44	0.64	0.13	0.1331E+15	0.32
5	21/04/18	14:31:34	0.8	7.233	109.667	2.052	114.3	87.3	36.5	0.01	1.24	-1.25	1.23	-0.29	1.10	0.2082E+11	0.44
6	21/04/18	15:40:08	1.8	7.210	109.689	4.522	316.5	47.3	126.7	4.69	-3.96	-0.74	1.67	2.60	1.75	0.5490E+12	0.42
7	21/04/18	17:55:28	2.2	7.243	109.683	3.885	150.0	78.8	42.9	0.46	1.22	-1.68	1.40	-0.99	-0.69	0.2338E+13	0.38
8	21/04/18	18:53:05	1.4	7.245	109.676	3.376	321.0	62.5	85.3	1.73	-0.65	-1.08	-0.77	0.77	0.45	0.1883E+12	0.34
9	21/04/18	20:58:40	0.5	7.229	109.674	4.608	324.1	14.7	115.7	4.76	-1.79	-2.97	6.80	-2.90	0.59	0.8271E+10	0.43
10	22/04/18	20:21:14	1.5	7.231	109.675	3.313	291.4	69.4	83.0	0.95	-1.73	0.78	-1.77	0.87	1.71	0.2653E+12	0.46
11	23/04/18	12:54:05	1.2	7.232	109.692	6.308	346.0	50.6	150.9	3.42	-2.16	-1.26	4.58	2.84	5.00	0.7814E+11	0.40

IOP Conf. Series: Earth and Environmental Science **273** (2019) 012018 doi:10.1088/1755-1315/273/1/012018

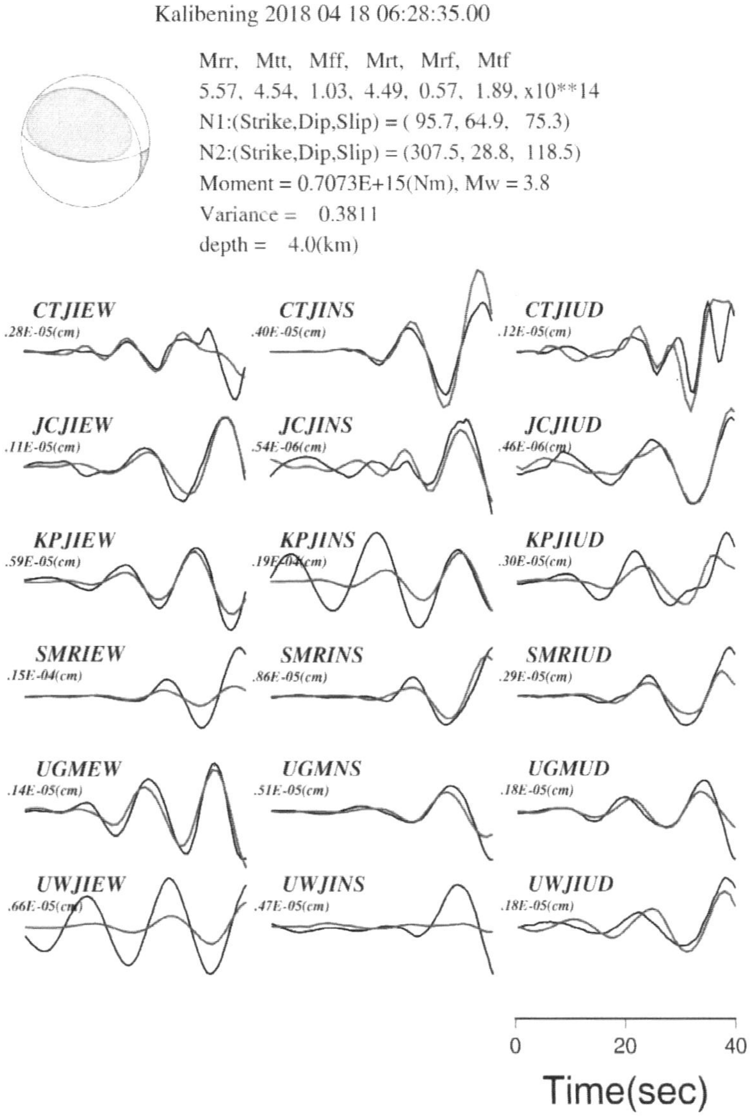

Figure 5. The complete output of moment tensor solution of the mainshock in this study (18 April 2018). The main information of source parameters on the top right-hand graph, the black lines and the red lines are observation waveforms and calculation ones, respectively. The numbers below the station code indicate the maximum amplitude (in unit of cm/sec).

The results from investigating the background seismicity in Banjarnegara Region with constrained area proved there were only 47 events from that combined catalog where 125 events by BJI. It showed there were almost a threefold losses in earthquake cataloging by that combined catalog compared to the closest geophysical station (BJI). Then, that seismicity patterns and the two moment tensor catalog which ruptured on 4 April 1992 (Mw 5.0) and 19 April 2013 (Mw 4.9) (Fig.4) carried us to a big question, why the faulting zone still unconfirmed in that area by Indonesian Seismic Sources and Seismic Hazard Maps 2017. As our suggested, it may be encouraged by the slight confirmed event (combined catalog) unable to leading up the comprehensive study while the BJI catalog less reliable as mostly recorded from single station observation.

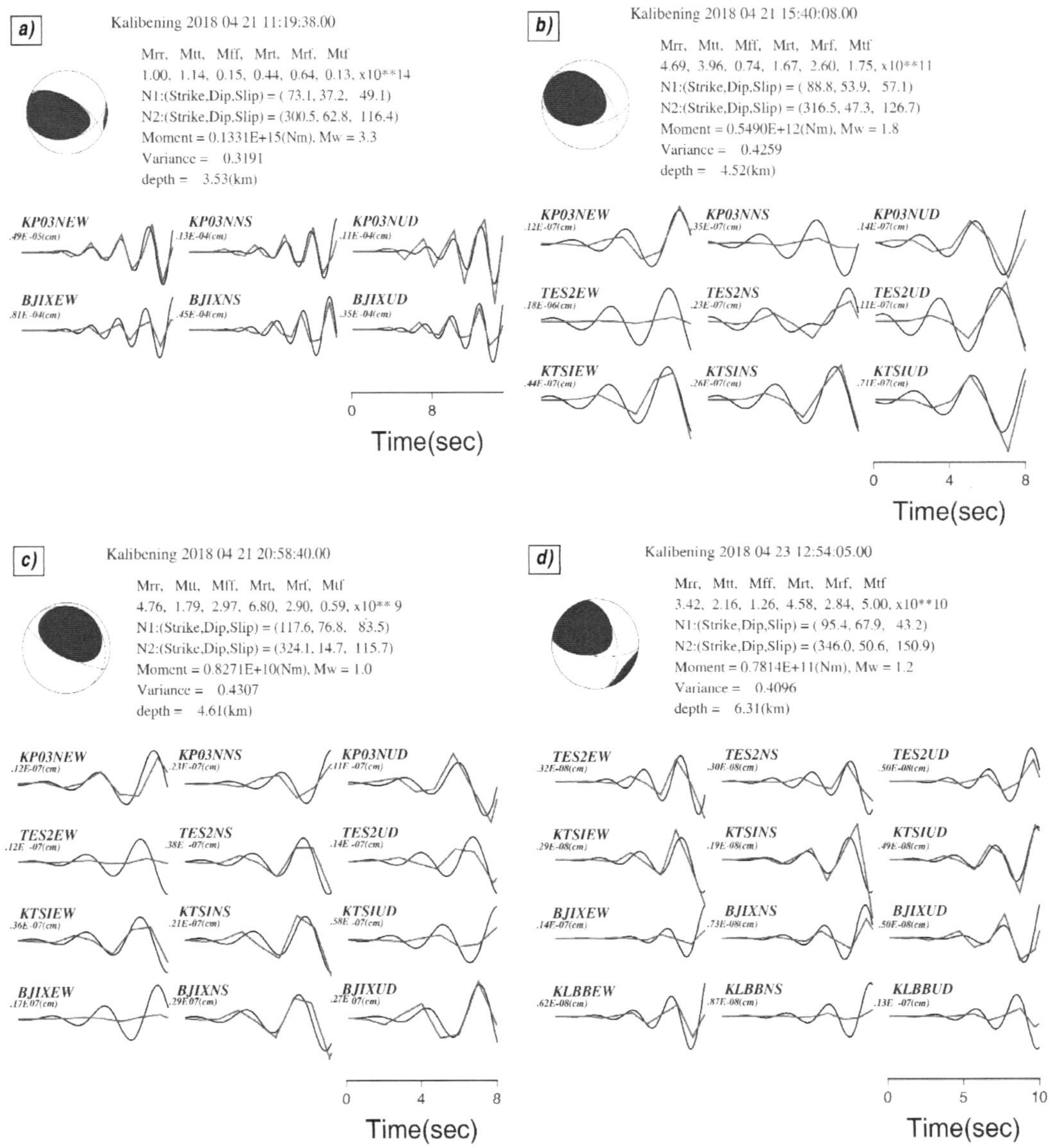

Figure 6. Results of moment tensor inversion for the event ID 4 **(a)**, the event ID 6 **(b)**, the event ID 9 **(c)** and the event ID 11 **(d)** which the most suitable with the mainshock's moment tensor solution. All annotation in this figure is similar within Figure 5.

The low area coverages around Banjarnegara Region were impacted by the sparse and lack of InaTEWS seismic network configurations which based on the geometry is far enough to that area. This situation was worsened since the BJI (yellow inverted triangle in Fig.4) has stopped working at the end of 2015 then the local observation only from TDS recording. The real-time earthquake monitoring in Central Java for InaTEWS not including data from BJI as yet.

The earthquake epicenter mainly ruptured in the tectonic of North Serayu Basin, in the middle between Pamanukan-Cilacap Fault and Muria-Kebumen Fault [2,3]. The systematic approach of lineament analysis to identify active fault onshore Java had been done [18]. This work was performed generally to figure out the neotectonics of Java and supposed active faults exist within North Serayu Basin but it did not even close to 2018 Kalibening earthquake concentration. A suggestion was denoted that Kalibening earthquake occurred in the Kalibening plateau as tectonically depression caused by the existence of dextral strike-slip faulting of Kalibening-Wanayasa. However, based our results we proposed the 2018 Kalibening earthquake generated by a thrust faulting mechanisms.

The previous study [15] even used information from 134 seismic stations covering a region of about 150 km x 200 km for 150 days in Central Java (MERAMEX project) seemingly did not show any significant results in this study area. As six aftershocks still recorded by BJI single station observation (TDS) after survey (Table 3), we assured that the temporary network should be installed for a longer time maybe for one month. In advising for further quantifying active deformation in North Serayu Basin, some comprehensive study like GPS observation [19], a combination of satellite and field-based geology and geomorphology mapping, paleoseismology and moment tensor analysis can be applied [18].

5. Conclusions

The relocation results showed precisely and refined Kalibening aftershocks distribution that formed a lineation trending northwest-southeast. In further, the cross section shows a slope-shaped which elongated deeper from southwest to northeast with a range of 2-6 km depth. We cannot relocate the mainshock as we found difficulties in finding the event pair for relocating.

We suggested the deformation scheme in the source region of Kalibening earthquake with thrust faulting for 307.5/ 28.8/ 118.5 (Strike/ Dip/ Rake) based on moment tensor inversion. This proposed was supported by aftershock moment tensor inversion and seemingly consistent with aftershocks relocation results. The sparse and lacking of InaTEWS seismic network configurations induced the low area coverages in Banjarnegara region with surrounding area and losses in earthquake cataloging.

As unconfirmed faulting by Indonesian Seismic Sources and Seismic Hazard Maps 2017 related to Kalibening earthquake, we suggested an advising for further quantifying active deformation in North Serayu Basin especially in this study area with a comprehensive study. The occurrence of a damaging earthquake in Kalibening called for the revision of earthquake hazard assessment for Central Java province. The research questions are related to the origin and the level of activity of earthquakes in onshore of Central Java.

6. Acknowledgments

We thank Agency for Meteorology Climatology and Geophysics (BMKG) and 2018 Kalibening macroseismic survey team for the earthquake catalog also the seismic waveform data.

7. References

[1] Nasional, P. S. G. (2017). National Center for Earthquake Studies, Indonesian Seismic Sources and Seismic Hazard Maps 2017, Center for Research and Development of Housing and Resettlement, Ministry of Public Works and Human Resettlements (*Peta Sumber dan Bahaya Gempa Indonesia Tahun 2017, Pusat Penelitian Perumahan dan Pemukiman, Kementerian Pekerjaan Umum dan Perumahan Rakyat*). ISBN 978-602-5489-01-3.
[2] Satyana, A. H. (2006, November). New Insight on Tectonics of Central Java, Indonesia and Its Petroleum Implications. In *Proceedings AAPG International Conference and Exhibition, Perth.*

[3] Satyana, A. H. (2007). Central Java, Indonesia–A "Terra Incognita" in petroleum exploration: New considerations on the tectonic evolution and petroleum implications.

[4] Pavlis, G. L. (1986). Appraising earthquake hypocenter location errors: A complete, practical approach for single-event locations. *Bulletin of the Seismological Society of America*, *76*(6), 1699-1717.

[5] Gomberg, J. S., Shedlock, K. M., & Roecker, S. W. (1990). The effect of S-wave arrival times on the accuracy of hypocenter estimation. *Bulletin of the Seismological Society of America*, *80*(6A), 1605-1628.

[6] Waldhauser, F., & Ellsworth, W. L. (2000). A double-difference earthquake location algorithm: Method and application to the northern Hayward fault, California. *Bulletin of the Seismological Society of America*, *90*(6), 1353-1368.

[7] Schaff, D. P., Bokelmann, G. H., Ellsworth, W. L., Zanzerkia, E., Waldhauser, F., & Beroza, G. C. (2004). Optimizing correlation techniques for improved earthquake location. *Bulletin of the Seismological Society of America*, *94*(2), 705-721.

[8] Sipayung, R., Ulfiana, E., & Sianipar, D. (2018, July). The connection of coloumb stresses and aftershock imparted by the 18 August 2012 Mw 6.3 Palu-Koro earthquake. In *AIP Conference Proceedings* (Vol. 1987, No. 1, p. 020037). AIP Publishing.

[9] Abdel-aal, A. A. K., & Yagi, Y. (2017). Earthquake source characterization, moment tensor solutions, and stress field of small-moderate earthquakes occurred in the northern Red Sea Triple Junction. *Geosciences Journal*, *21*(2), 235-251.

[10] Diehl, T., Kraft, T., Kissling, E., & Wiemer, S. (2017). The induced earthquake sequence related to the St. Gallen deep geothermal project (Switzerland): Fault reactivation and fluid interactions imaged by microseismicity. *Journal of Geophysical Research: Solid Earth*, *122*(9), 7272-7290.

[11] Poupinet, G., Ellsworth, W. L., & Frechet, J. (1984). Monitoring velocity variations in the crust using earthquake doublets: An application to the Calaveras Fault, California. *Journal of Geophysical Research: Solid Earth*, *89*(B7), 5719-5731.

[12] Yagi, Y., & Nishimura, N. (2011). Moment tensor inversion of near source seismograms. *Bull. IISEE*, *45*, 133-138.

[13] Kohketsu, K. (1985). The extended reflectivity method for synthetic near-field seismogram. *Journal of Physics of the Earth*, *33*(2), 121-131.

[14] Yagi, Y., & Fukahata, Y. (2011). Introduction of uncertainty of Green's function into waveform inversion for seismic source processes. *Geophysical Journal International*, *186*(2), 711-720.

[15] Koulakov, I., Jakovlev, A., & Luehr, B. G. (2009). Anisotropic structure beneath central Java from local earthquake tomography. *Geochemistry, Geophysics, Geosystems*, *10*(2).

[16] Dziewonski, A. M., & Anderson, D. L. (1981). Preliminary reference Earth model. *Physics of the earth and planetary interiors*, *25*(4), 297-356.

[17] Wessel, P., & Smith, W. H. (1998). New, improved version of Generic Mapping Tools released. *Eos, Transactions American Geophysical Union*, *79*(47), 579-579.

[18] Marliyani, G. I. (2016). *Neotectonics of Java, Indonesia: Crustal deformation in the overriding plate of an orthogonal subduction system*. Arizona State University.

[19] Meilano, I., Abidin, H. Z., Andreas, H., Gumilar, I., Sarsito, D., Rahma Hanifa, R., ... & Fukuda, Y. (2012). Slip rate estimation of the Lembang Fault West Java from geodetic observation. *Journal of Disaster Research*, *7*(1), 12-18.

IOP Conf. Series: Earth and Environmental Science **273** (2019) 012019 doi:10.1088/1755-1315/273/1/012019

Application of C-band Doppler Weather Radar (CDR) for Detecting Volcanic Ash Dispersion of Sinabung Eruption 19 February 2018

I. Irwandi[1,2,3,*], Yopi Ilhamsyah[1], Donaldi S Permana[4], and Melanton P Haloho[5], Budi Prasetyo[5], Isya N. Dana[6]

[1]Tsunami and Disaster Mitigation Research Center (TDMRC), Syiah Kuala University, Jl. Prof. Dr. Ibrahim Hasan, Gampong Pie, Banda Aceh, 23233, Indonesia
[2]Physics Department, Syiah Kuala University, Banda Aceh, 23111, Indonesia
[3]Geophysics Engineering Department, Syiah Kuala University, Indonesia
[4]Center for Research and Development BMKG, Jl. Angkasa I No. 2, Jakarta Pusat, 10720, Indonesia
[5]BMKG Jl. Ngumban Surbakti No. 15, Wilayah I Medan
[6]Center for Volcanology and Geologic Hazards (PVMBG), Bandung, Indonesia

*E-mail: irwandi@unsyiah.ac.id

Abstract. The weather radar is generally employed to forecast the potential of hydrometeorological hazard. In the present paper, we focus on the application of C-band Doppler Weather Radar (CDR) to identify February 19th, 2018 Sinabung volcanic eruption and its dispersion by using a direct product in the form of CMAX (Column Max) retrieved from reflectivity factor (dBz). The distribution of reflectivity factor varies from the center of the eruption to the farther location indicating the dispersion of material size erupted by the Sinabung. Lapili which is larger material size with 48dBz tends to locate in the nearby volcanic area. The medium size of the material like coarse volcanic ash with 33 dBz of reflectivity was floating in the air and carried away by the wind. Meanwhile fine volcanic ash with 10 dBz of reflectivity flies a great distance along with the wind.

1. Introduction

Indonesia is surrounded by active tectonics where many volcanoes are located leading to among countries most at risk of the threats of volcanic eruption [1]. Volcanoes are observed intensively by PVMBG (Indonesian Center for Volcanology and Geological Hazard Mitigation) by deploying volcano monitoring station. In addition, BMKG (Indonesian Agency for Meteorology, Climatology, and Geophysics) is assigned to monitor hydrometeorological hazard by performing the weather radar. Since 2016, as many as 40 weather radars had been installed throughout Indonesia [2].

Weather radar play a major role in providing services associated with weather forecasts, especially short-term weather (near real-time), for aviation services during take-off and landing at the airport, besides it can be utilized to provide early warning to the occurrence of extreme weather such as torrential rain, typhoons, strong winds, and wind shear as well as for analyzing other extreme weather events [2]. Physically, the weather radar works by recording electromagnetic waves reflected by the material, both

IOP Conf. Series: Earth and Environmental Science 273 (2019) 012019 doi:10.1088/1755-1315/273/1/012019

water and volcanic ash [3]. Therefore, weather radar can be employed to monitor the dispersal of the volcanic ash of volcanic eruption as part of disaster mitigation.

Sumatra island located at the boundary margin tectonics have subduction and strike-slip faults known as the Great Sumatra Fault [4][5]. The active faults not only create high seismic [6] but also geologically produce high volcanic hazard zones. For instance, the second largest mountain in North Sumatra, Mount Sinabung is known for never erupting since 1600. But suddenly, Sinabung awake in last 2010. A total of 12 thousand residents must be evacuated to the safe area for three consecutive days from 27 to 29 August 2010 as smoke and lava were released from the mountain. Following no record of eruption since 1600, Sinabung is categorized into volcano type B. As it starts to be active again, Sinabung turns into volcano type A leading its status to raise to be "alert" (*Awas*) [7].

On 19 February 2018 Mount Sinabung returned a relatively large eruption by producing dark gray plume with a high volume of ash that rose to about 16.8 km (55,000 ft). Visibility in five surrounding districts was poor and reduced to only about 5 m. Some residents outside of the evacuation zone were self-evacuated [8]. Since a number of active volcanoes type A are also found in Aceh Province, it then draws TDMRC to study the 19 February 2018 Sinabung volcano eruption as a lesson learned of the volcanic disaster mitigation. A team of TDMRC Unsyiah was involved in the Volcano Eruption Impact Assessment on 22-26 February 2018. The team conducted interviews and field samplings. Raw weather radar data during eruption days are available at BMKG Region I Medan which is is the closest weather radar to Mount Sinabung with a distance of 49.69 km. Based on these data, the present paper discusses the effectiveness of weather radar application on the detection of the dispersal of Sinabung volcanic ash [9].

(a) The Sinabung Volcano in closer view (b) The Sinabung eruption

Figure 1. (a) A closer view of Sinabung from red zone during volcanic eruption impact assessment trip 22-26 February 2018[9] (b) Dark cloud eruption arouse 5000 m height followed by pyroclastic flow, 19 February 2018, at 08.58 local time [10]

2. Method

The weather radar is developed after World War II with the first application in the military field. It is used to mitigate hydrometeorological hazard by observing the precipitation, i.e., rainfall and snow. The range of observation may reach several hundred kilometers depending on the power of emission of the microwaves spectral and the transmitting antenna. The types of weather radar based on the number of antennas are divided into 2 types, e.g., MonoStatic Radar, i.e., Radar where transmitter and receiver are installed in one antenna while another type of Radar is Bistatic Radar where two antennas,e.g., transmitter and receiver are installed separately. At the beginning of the development weather radar in 1960, Non-Doppler Radar is used to detect energy reflectivity from the object only. The next development of radar also includes velocity detection based on Dopler principles (figure 2).

IOP Conf. Series: Earth and Environmental Science **273** (2019) 012019 doi:10.1088/1755-1315/273/1/012019

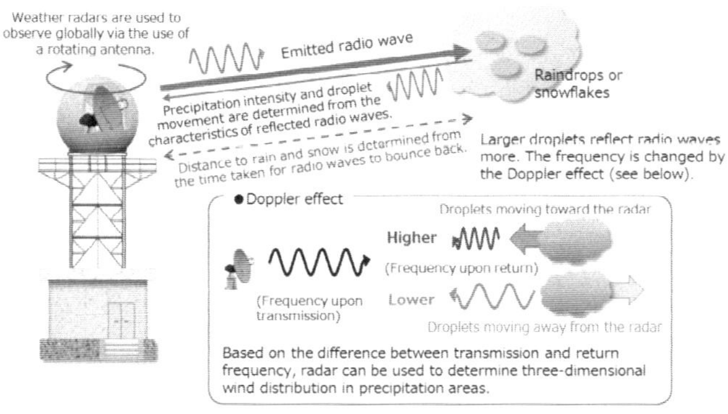

Figure 2.
Cartoons from the Doppler weather radar, the distance to the location of the rain is calculated from the travel time of the electromagnetic wave from the transmitter until it returns to the sensor. Whereas precipitation intensity is calculated from the amplitude of the reflected wave (the strength of echo) https://www.jma.go.jp/jma/en/Activities/radar/radar.html

In the present paper, we utilize radar for detecting volcanic ash dispersion with a case study of Sinabung Eruption February 19th, 2018. We use C-band Doppler Weather Radar (CDR) from BMKG office in Medan. Figure 3a showed the tower of weather radar at BMKG office. The weather radar observation data were obtained during the field trip Volcano Eruption Impact Assessment[9]. In figure 3b, it is shown the map where weather radar was stationed. Mount Sinabung was located 49.69 km from the station that still includes in the coverage area of the weather radar that may reach 200km.

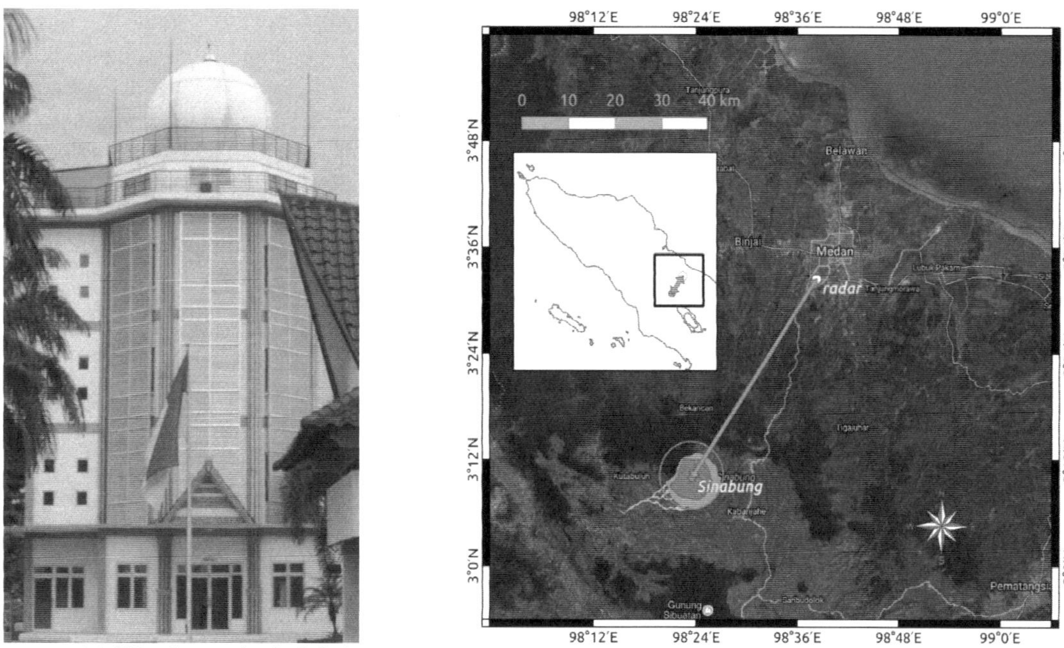

(a) Weather radar in Medan (b) Radar location and Sinabung Volcano.

Figure 3. (a) Weather station is called radar site Medan, located at 3.5398N, 98.6371W with 30 meters above sea level. The BMKG tower was built on December 2009 with 23 meter in height and working on CBAND frequency 1.6 GHz with 1^0 beam width and pulse width 0.8 μs. The radar was made by EEC with the TX Magnetron transmitter type with 250 kW power and Rx Stallo receiver. The lowest radar radiation angle is 0.9 and the highest is 19.5 [11]. (b) The location map of the Weather Radar and Sinabung Volcano is 49.69 km. The map is overlaid with KRB I circle (Volcanic Hazard Zone I) of Volcano Sinabung [12].

Single-polarization Doppler radars can measure horizontally-polarized power echo and Doppler frequency shift can determine ash content and radial velocity. The appearance of volcanic clouds just occurred in a short time from a few minutes to several hours after the eruption, so the initial intensity information when the eruption occurs will be very important to save. Weather radar that has detection capability that is far hundreds of kilometers will be very safe from the danger of pyroclastic which can damage equipment. High spatial resolution within a few hundred kilometers is an advantage of weather radar located on land. Radar that can work 24 hours a day can provide immediate information on the volume of volcanic dust clouds, mass quantities, and eruption height so that it becomes one of the important tools in the mitigation of volcanic disasters [13].

3. Results and Discussion

The data obtained is stored in the volumetric (*.vol) format by the radar system. Then the data is processed using EDGE^{TM} Radar Analysis software to display the CMAX (Column Max) product from the reflectivity factor (dBz). this product will display the maximum value of reflectivity factor in a location from vertical scanning or also known as volume scanning. To facilitate the reading of this product CMAX reflectivity factor is zoomed over focus areas and overlaid with other information in the form of scales, wind directions, and volcanic hazard-prone zones with the help of QGIS [14] software as shown in figure 4.

According to Sinabung Volcano Observation Station PVMBG (Center for Volcanology and Geological Hazards) [10], the erupting started at 08.53 WIB (local time). Three minutes before the event the weather radar did not show any significant reflectivity factor at KRB I, see figure 4a.

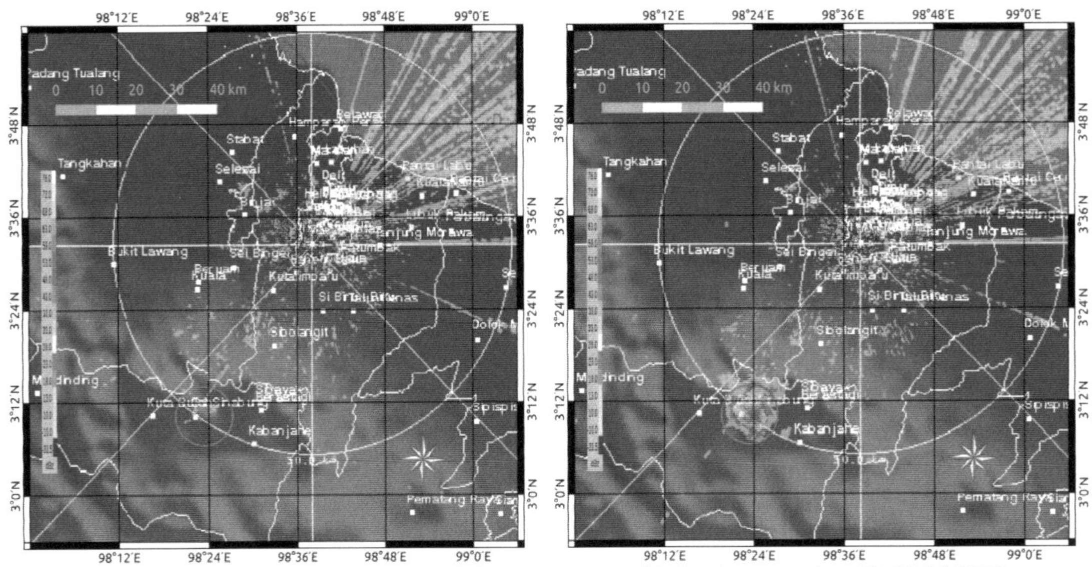

(a) before eruption: 08:50:06 WIB (local time) (b) durring erupting 09:00:06 WIB

Figure 4. CMAX (Column Max) of horizontally-polarized radar reflectivity factor (dBZ), measured by C-band Doppler weather radar Medan site(a) before eruption (b) during eruption

At 09:00 three minutes after the eruption, 48dBz reflectivity factor emerged as shown in figure 4b. This high reflectivity factor rises due to a massive eruption mass concentration such as volcanic bombs (size

IOP Conf. Series: Earth and Environmental Science **273** (2019) 012019 doi:10.1088/1755-1315/273/1/012019

> 64mm) and lapilli (2mm< size <64mm). The high reflectivity factor concentration moves Southeastward because the crater is located slightly to the Southeast from the top of the mountain. The midpoint of the KRB I line is at the top of Sinabung Volcano. As a reference the midpoint of the circle of KRB I is at the top of Sinabung Volcano. At 9:10 a.m. 17 minutes after the eruption, the reflectivity factor dropped to a maximum of 33 dBz, see figure 5a. This happens because particles with large masses (volcanic bombs) including lapilli have fallen down by the gravitational force, while coarse and fine volcanic ash is still in the air. At 09:40 47 minutes after the eruption, reflectivity factor falls to 10dBz and moved 15km to the North, except in the crater area where a concentration point show at 28 dBz as shown in figure 5b. This indicating that all rough volcanic ash has fallen down and could not transport to far. Whereas fine volcanic ash remained in the air and could travel far away by the wind. As a result of field observations during the Volcano Eruption Impact Assessment, we were able to find the volcanic ash dust up to Kotacane and Lhoksemawe [9].

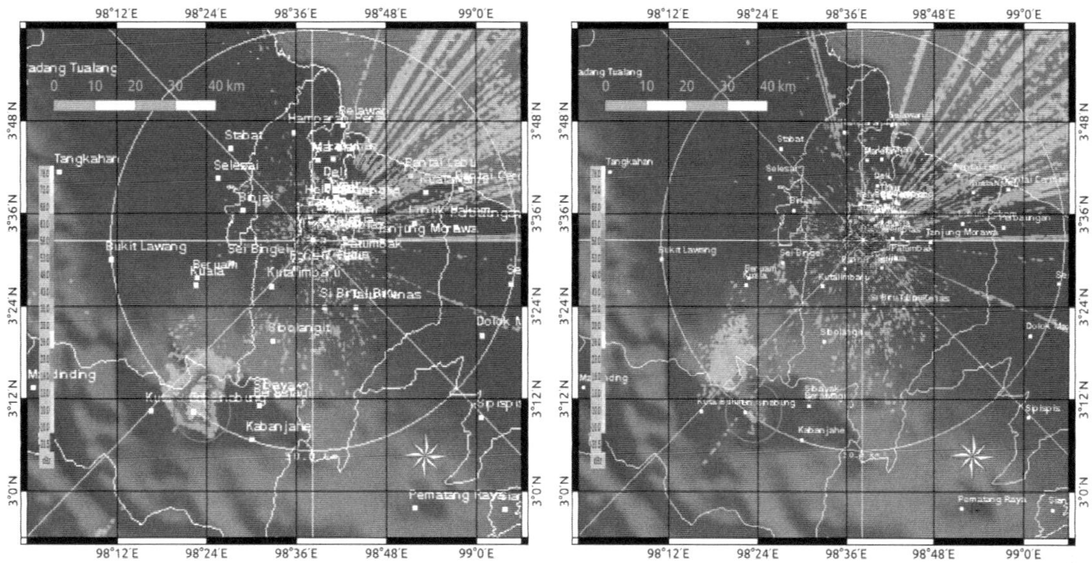

(a) after erupting 09:10:07 WIB (b) 47 minute after erupting 09:40:05

Figure 5. CMAX faktor reflectivity (dBz) after no eruption anymore and lapilli particle fell down dan ash transport by wind (a) 17 minute after eruption (b) 47 minute after eruption

4. Conclusions

Weather radar BMKG region I the C-band Doppler Weather Radar (CDR) field works well and records data during the eruption of Sinabung volcano on February 19, 2018. Observations show that CMAX distribution pattern for volcano eruption can be clearly distinguished from the pattern the distribution produced by clouds or precipitation. Volcano eruption pattern begins suddenly as a high reflectivity factor appears in the middle and then spreads around it with a lower value as increasing distance.

In this article, we only discuss CMAX products from horizontal max projection of reflectivity factors. Actually, EGDE^{TM} has many direct products and derivative products obtained from the weather radar need to be investigated for science and volcanic disaster mitigation processes.

5. References

[1] Loughlin, S.C., Sparks, R.S.J., Brown, S.K., Jenkins, S.F. and Vye-Brown, C. eds., 2015. *Global volcanic hazards and risk*. Cambridge University Press.

[2] Permana, D.S., Hutapea, T.D.F. and Praja, A.S., Fatkhuroyan., Muzayanah, LF (2016). Multi Weather Radar Format Data Processing Using Python-Based Wradlib. Jurnal Meteorologi Dan Geofisika, 17(3), pp.157-164.

[3] Bryan, S., Clarke, A., Vanderkluysen, L., Groppi, C., Paine, S., Bliss, D.W., Aberle, J. and Mauskopf, P., 2017. Measuring Water Vapor and Ash in Volcanic Eruptions With a Millimeter-Wave Radar/Imager. IEEE Transactions on Geoscience and Remote Sensing, 55(6), pp.3177-3185.

[4] Irwandi, Marwan, Muksin and Fashbir, 2017, July. Applications of the VLF-EM method for rapid Sumatran fault identification in Leuser national park, Aceh. In AIP Conference Proceedings (Vol. 1861, No. 1, p. 030050). AIP Publishing.

[5] Tabei, T., Kimata, F., Ito, T., Gunawan, E., Tsutsumi, H., Ohta, Y., Yamashina, T., Soeda, Y., Ismail, N., Irwandi Nurdin and Sugiyanto, D., 2015. Geodetic and geomorphic evaluations of earthquake generation potential of the northern Sumatran fault, Indonesia. In International Symposium on Geodesy for Earthquake and Natural Hazards (GENAH) (pp. 21-28). Springer, Cham.

[6] Irwandi, I., 2017, February. Advantages of Realistic Model Based on computational method: NDSHA versus Standard PSHA. In IOP Conference Series: Earth and Environmental Science (Vol. 56, No. 1, p. 012007). IOP Publishing.

[7] Pallister, J., Wessels, R., Griswold, J., McCausland, W., Kartadinata, N., Gunawan, H., Budianto, A. and Primulyana, S., 2018. Monitoring, forecasting collapse events, and mapping pyroclastic deposits at Sinabung volcano with satellite imagery. Journal of Volcanology and Geothermal Research.

[8] Global Volcanism Program, 2018. Report on Sinabung (Indonesia). In: Sennert, S K (ed.), Weekly Volcanic Activity Report, 14 February-20 February 2018. Smithsonian Institution and US Geological Survey. http://volcano.si.edu/showreport.cfm?doi=GVP.WVAR20180214-261080

[9] Irwandi Irwandi, N. Rizky ,M. Ayub A. H., M. P Haloho SP, Yosi Setiawan, Isya N. Dana, Umar R., Wilfridus FS B., Iskandar, 2018. Laporan Survey Dampak Letusan Gunung Sinabung (Volcano Erruption Impact Assessment) 19 Februari 2018. TDMRC, Unsyiah (unpublished).

[10] Dana,Isya N., Umar Rosadi, Wilfridus F. S. Banggur, Wahidin Akbar, Petrus Tupe Taran, 2017. Laporan Tanggap Darurat Letusan Gunungapi Sinabung, Sumatera Utara Pusat Vulkanologi Dan Mitigasi Bencana Geologi

[11] Ginting, Natal and Simon Sinaga, 2013, Metadata Radar Cuaca Bmkg Untuk Web Wmo Radar Database (WRD). BMKG Wilayah I Medan.

[12] Gunawan, H., A.R. Mulyana, A.Solihin, Pujowarsito and Riyadi. 2014. Volcanic Hazard Map of Sinabung Volcano, North Sumatera Porvince. PVMBG. Bandung

[13] Marzano, F.S., Barbieri, S., Vulpiani, G. and Rose, W.I., 2006. Volcanic ash cloud retrieval by ground-based microwave weather radar. IEEE Transactions on Geoscience and Remote Sensing, 44(11), pp.3235-3246.

[14] Quantum, G.I.S., Development Team (2014).'QGIS Geographic Information System.'Open Source Geospatial Foundation Project.

[15] Lacasse, C., Karlsdóttir, S., Larsen, G., Soosalu, H., Rose, W.I. and Ernst, G.G.J., 2004. Weather radar observations of the Hekla 2000 eruption cloud, Iceland. Bulletin of Volcanology, 66(5), pp.457-473.

[16] Darmawan, Y. and Matondang, C.A., Sensitivitas Radar Cuaca Doppler C-Band (CDR) Terhadap Kejadian Angin Puting Beliung di Kecamatan Siborongborong Tanggal 29 Januari 2013. Prosiding Workshop Radar dan Satelit Cuaca Vol : 1 Desember 2013.

Acknowledgments

I wish to acknowledge assistance from TDMRC for field trip support, BMKG Wilayah I Medan for radar data, BMKG R&D Center for data processing, and PVMBG for geological information.

IOP Conf. Series: Earth and Environmental Science **273** (2019) 012020 doi:10.1088/1755-1315/273/1/012020

Preliminary Results of Horizontal to Vertical Spectral Ratio (HVSR) Across Lembang Fault, Bandung, Indonesia

Zulfakriza, Z. [1*]**, Puspito, N.T.** [1]**, Nugraha, A. D.** [1]**, Pranata, B.**[2,3]**, Rosalia, S.**[4]

[1]Global Geophysics Research Group, FTTM, Institute of Technology Bandung, 40132, Indonesia
[2]Earth Sciences Graduate Program, Bandung Institute of Technology Bandung, 40132, Indonesia
[3]Badan Meteorologi, Klimatologi dan Geofisika, Jakarta, Indonesia
[4]Geophysical Engineering, FTTM, Institute of Technology Bandung, 40132, Indonesia

[*]E-mail: zulfakriza@itb.ac.id

Abstract. Lembang Fault is located in the Bandung basin with is a thick sedimentary of volcanic rocks and the depositional remnants of an ancient lake. Bandung region has a dense population and infrastructure that possible to be vulnerable of seismic hazard. Based on geological studies, Lembang fault is an active fault that possibility to produce a magnitude more than 6.0. Seismic hazard estimation is one of some method to identification of hazard potential due to the Lembang Fault activity. Horizontal to Vertical Spectral Ratio (HVSR) method is one of way to seismic microzonation that identified the dominant frequency (F_0). We deployed 6 seismometers across Lembang Fault area for about 3 months observation. Then we computed the seismic response that corelated with physical properties inside the basin. The variation of seismic susceptibility index around Lembang area indicates the vulnerability of seismic hazard. LEM01 and LEM06 have relatively lower of natural frequency, its about 1.22 (LEM01) and 1.49 (LEM06). The number of natural frequency indicate thick of sediment and vulnerable of seismic hazard.

1. Introduction

Lembang Fault is an active fault which is about 10 km north of the Bandung. The existence of the Lembang Fault provides a potential seismic hazard for the Bandung area and its surroundings. Morphologically, Lembang Fault extends from east to west along 22 km which is divided into 6 segments, namely Cimeta, Cipogor, Cihideung, Gunung Batu, Cikapundang and Batu Lonceng [1]. Based on geodetic observations [2] the slip rate of Lembang Fault is 6 mm / y. At least there were two earthquakes occurred around Lembang Fault. The magnitude M2.9 on July 22, 2011 and M3.3 on August 28, 2011 [2]. M3.3 Earthquake causes damage to several houses in Cisarua District, West Bandung.

In addition to building structure, building damage due to earthquakes is affected by local geological conditions and poorly consolidated rock conditions [3]. Local geological conditions that have impacted on building damage have been assessed using HVRS such as in Yogyakarta [4] which was compare to damaging caused by the 2006 earthquake. [5] shown the regional HVSR calculation in Bandung basin based on 64 seismic stations around Bandung area. Seismic vulnerability index (Kg) values obtained from HVSR calculations provide information about the condition of the structure on the surface banana, the level of solidity of the rock and the thickness of the sediment layer.

Since April to June 2016, ITB installed 6 seismographs (Figure 1) for 3 months closed to the Lembang Fault and its surroundings. The main purpose of installing the seismograph network is to monitor seismicity due to activity of the Lembang Fault. In addition to monitoring, the existence of this seismograph network can be used to understand the local geological conditions around the Lembang Fault by applying HVSR method.

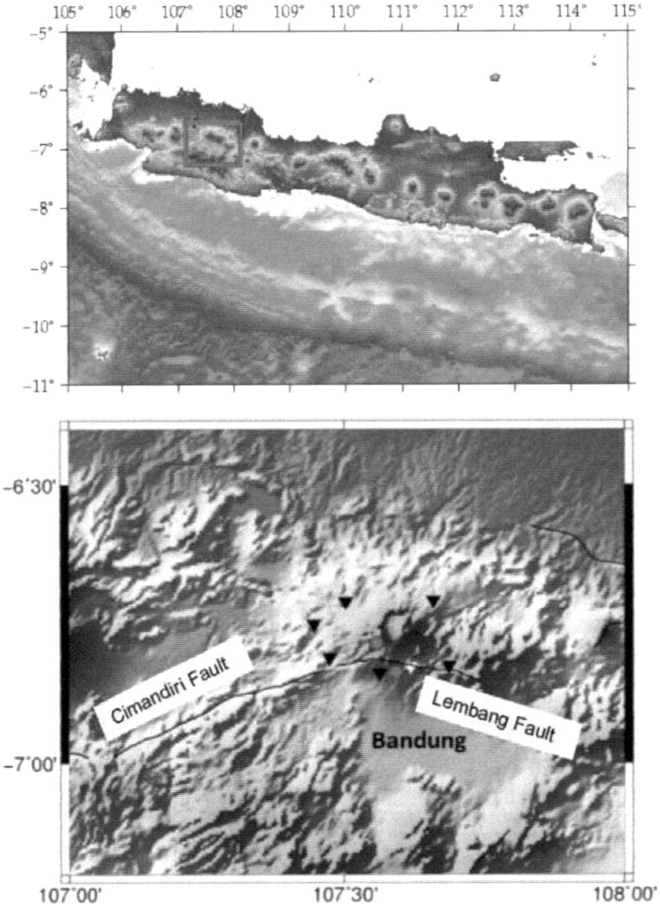

Figure 1. Location of the Bandung basin area. The study area is situated in the central part of West Java, Indonesia about 150 km southeast of Jakarta. Lembang and Cimandiri Fault indicated by black line, ITB seismograph network marked by invert black triangle, BMKG seismograph station marked with invert black triangle, then the two red dots are earthquake events across Lembang Fault.

2. Geological Setting

Geological setting of the Lembang fault is related to evolutionary process of the Bandung basin [6]. Morphological formations of the Bandung region consist of volcano chains. The process of rock formation in the Bandung basin began in the Miocene era (about 20 million years ago), this is indicated by the presence of limestone outcrops in Rajamadala [7].

Morphologically, Bandung area is composed of the Sunda-Tangkuban Perahu volcanic complex. The existence of Mt. Tangkuban Perahu in the North of Bandung which is one of the active volcanoes in the ancient caldera [8]. In the southern part of Mt. Tangkuban Perahu is a relatively flat area but is cut by structures in the east and adjacent to the Lembang city. The expression of continuity provides an illustration of the existence of faults with east-west orientation. [8] mentioned its expression as Lembang Fault (Figure 2).

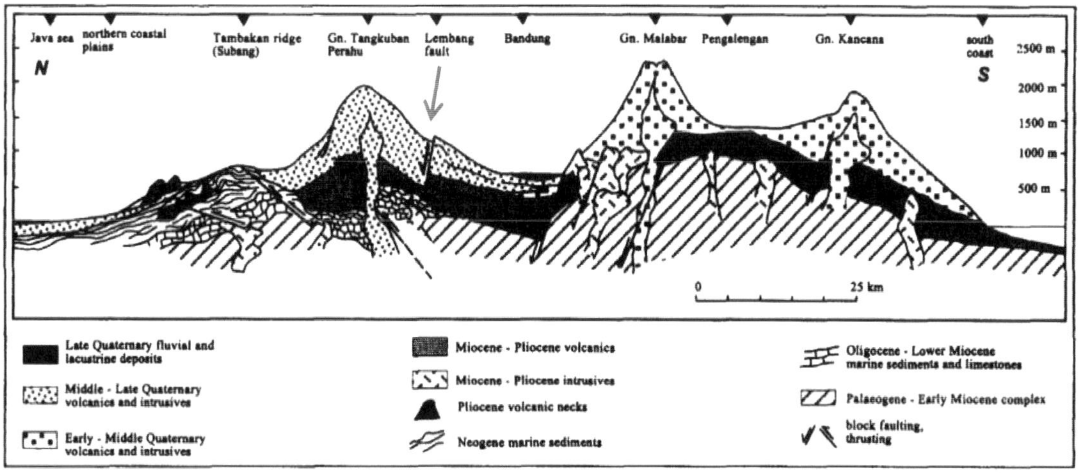

Figure 2. Sketch map vertical cross section of south to north West Java. This figure showing the structural setting of some volcanic chain, Bandung basin dan Lembang Fault that marked by red arrow (modified from [9]).

The continuity of Lembang faults indicates a segmentation of Lembang fault morphology. Morphotectonics are influenced by morphological conditions and tectonic processes that occurred in the past. This condition is caused by the morphology has a spatial dimension and tectonics has a time dimension. Tectonic land forms will express topographic formations that can be used as indicators of active tectonic or tectonic movements. Topographic shapes that have undergone movement displacement can be seen and observed through satellite image that provide morphotectonics features in the form of river flow patterns, hills movement, river deflection, straightness, fault faulting, and river terrace appearance.

3. Data and Method

In this study, we used data from the seismometer network in the Lembang area that recording since April to June 2016 continuously using 6 stations of 3 component broadband seismometer with 250 Hz of sampling rate (Figure 3A). The seismometers distribution covered the Lembang Fault and some area that closed to the Fault, such as Cisarua, Parompong and Lembang City (Figure 1). we used a bunker for seismometer installation with 150 cm depth and mostly the waveform show the signal clearly. The digital recorder of seismometer was connected with a GPS antenna, thus allowing to received coordinate and timing daily. The recorded data is a time series waveform in miniSEED format (Standard Exchange of Earthquake Data).

For HVSR calculation, we used data waveform for 6 hours recording from 21.00 to 03.00. We assumed it would be possible to obtain the recording with low noise due to human and industrial activities (Figure 3B). We used GEOPSY software to calculate f0 with a frequency range of 0.5 - 20 Hz and 50 s time window. The output that will be generated from this processing is an HVSR curve that provides information on the dominant frequency value and amplification.HVSR curve is data segmentation. Segmentation is the process of dividing signals into several parts with different segment widths. In this study, the segment width used is 250 s. this is because in the segment width, it is stable and does not contain noise or transient signal signals. At this value is used so that the variation in the frequency domain becomes more detailed.

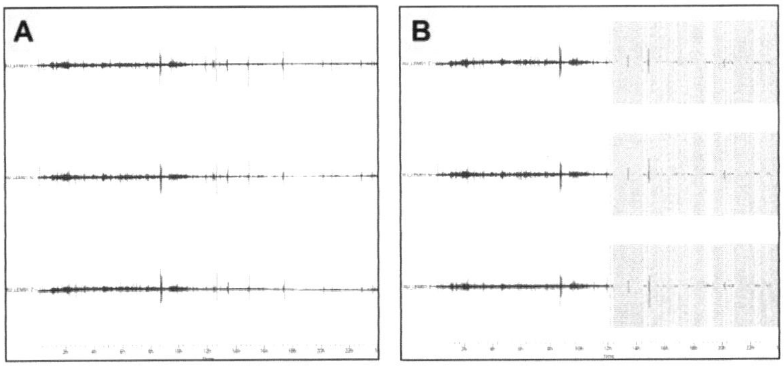

Figure 3. A. Waveform of LEM01 with 3 component that used for HVSR calculation. B. Time window selection 21.00 to 03.00 WIB.

4. Results and Discussion

The spectrum of HVSR can be seen on the curve of frequency and H/V Amplitude. HVSR is a merging curve between horizontal and vertical components. We produced HVSR curve for each point LEM01, LEM02, LEM03, LEM04, LEM05 and LEM06 (Figure 4). The HVSR curve provides information in the form of natural frequency and soil amplification. HVSR is a merging curve between horizontal and vertical components. Extraction of the natural frequency and amplification of the HVSR curve is a reference for calculating the soil vulnerability index. [9] introduced the soil vulnerability index (Kg) equation. The vulnerability index (Kg) value can estimate the distribution of natural frequencies, soil amplification and soil susceptibility levels related to local geological conditions.

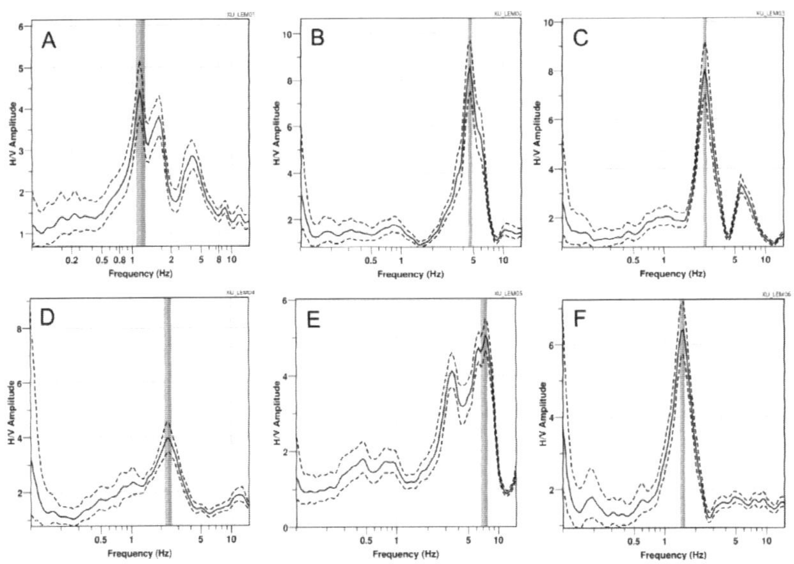

Figure 4. HVRS curve for different station, (A) LEM01 with 1,22 of natural frequency, (B) LEM02 with 4,65 of natural frequency, (C) LEM03 with 2,54 of natural frequency, (D) LEM04 with 2,37 of natural frequency, (E) LEM05 with 7,22 of natural frequency and (F) LEM06 with 1,49 of natural frequency

Based on the natural frequency of each station, the vulnerable zone around the Lembang fault can be identified. Refer to [9], the value of natural frequency is inversely proportional to the vulnerability of the soil. If the natural frequency value is low, thereby the soil vulnerability is higher. In consequence of sedimentary thickness and bedrock deeper. LEM01 and LEM06 have relatively lower of natural frequency, its about 1.22 (LEM01) and 1.49 (LEM06). LEM05 has natural frequency higher than others about 7.22. Figure 5 shows the distribution of natural frequency values for 6 observation stations. The areas with low natural frequencies indicate that has relatively deep bedrock depths and thicker of sediments. LEM06 is close to the Lembang fault line and has a low natural frequency value. It means this area belong to venerability class.

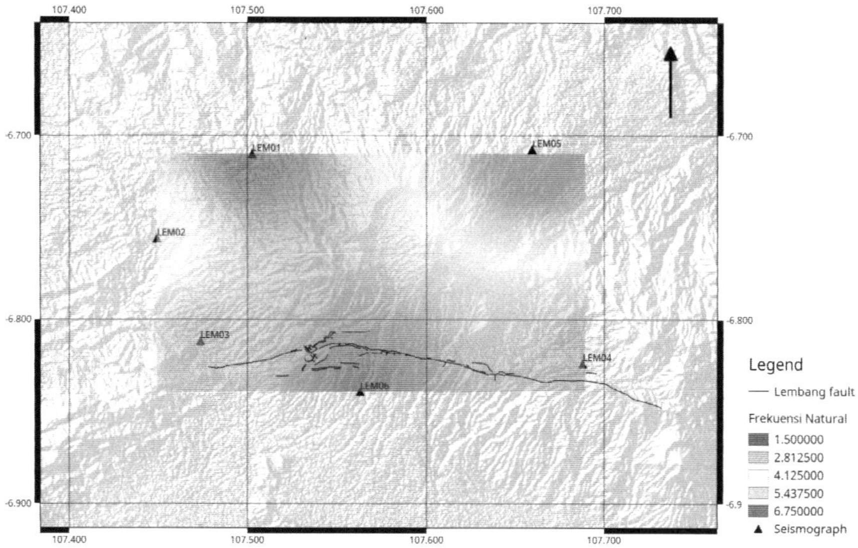

Figure 5. Map of natural frequency (f_0) across Lembang Fault and surrounding area. Black triangle indicate the seismic stations and black line indicate the fault line.

5. Conclusions

We have produced the natural frequency (f_0) to identify the sediment thickness across Lembang fault and surrounding area. Based preliminary result of the study, we found LEM01 and LEM06 have relatively lower of natural frequency, its about 1.22 (LEM01) and 1.49 (LEM06). The number of natural frequency indicate thick of sediment and vulnerable of seismic hazard.

References

[1] PusGeN, 2017. Updating Seismic Hazard Map of Indonesia 2017, Ministry of Public Work of Indonesia.

[2] Meilano, I., Abidin, H.Z., Andreas, H., Gumilar, I., Sarsito, D., Hanifa, R., Rino., Harjono, H., Kato, T., Kimata, F., Fukuda, Y, 2012. Slip Rate Estimation of the Lembang FaultWest Java from Geodetic Observation, Journal of Disaster Research Vol.7 No.1, 2012.

[3] Seed, H.B,. dan Schnabel, P.B., 1972. Soil and Geological effects on site response during earthquakes", Proc. first International Conference on Microzonation for Safer Construction-Research and Application, vol. 1. pp 61-74.

[4] Daryono, Sutikno, Sartohadi, J., Dulbahri, Brotopuspito, K.S. (2009): Local site effect at Bantul Graben based on Microtremor measurements. International Conference Earth Science and Technology. Phonix Hotel, Yogyakarta.

[5] Bayu Pranata, Tedi Yudistira, Erdinc Saygin, Phil R Cummins, Sri Widiyantoro, Budi Brahmantyo, Zulfakriza, 2018. Seismic Microzonation Of Bandung Basin From Microtremor Horizontal-To-Vertical Spectral Ratios (Hvsr) 2018. Seismic Microzonation Of Bandung Basin From Microtremor Horizontal-To-Vertical Spectral Ratios (HVSR), AIP Conf. Proc. 1987, 020004-1–020004-6; https://doi.org/10.1063/1.5047289

[6] Hamilton W (1979) Tectonics of the Indonesian region. US Geol Surv Prof Pap 1078:1–345.

[7] Van Bemmelen, R.W., 1949. The Geology of Indonesia: General Geology of Indonesia and Adjacent Archipelagoes, the East Indies, Inclusive of the British Part of Borneo, the Malay Peninsula, the Philippine Islands, Eastern New Guinea, Christmas Island, and the Andaman and Nicobar Islands. US Government Printing Office

[8] Dam, M. A. C., Suparan, P., Nossin, J. J., Voskuil, R. P. G. A. & G. T. L. Group (1996) A chronology for geomorphological developments in the greater Bandung area, West-Java, Indonesia. J. Southeast Asian Earth Sci. 14(1–2), 101–115.

[9] Nakamura (2000): Clear Identification of Fundamental Idea of Nakamura's Technique and Its Applications. 12WCEE.

Source Mechanism Analysis By Using Tensor Moment Inversion (Study Case : Pidie Jaya Earthquake in 2016 December 7th)

Andrean V H Simanjuntak[1,2,3], **Umar Muksin**[1,3], **Qadariyah**[3], **Yosi Setiawan**[4], **Irwandi**[1,3]

[1] Department of Physics, Syiah Kuala University, Banda Aceh
[2] Mata Ie Geophysical Station of BMKG Aceh, Aceh Besar
[3] Tsunami and Disaster Mitigation Research Center (TDMRC), Banda Aceh
[4] Tuntungan Geophysical Station of BMKG North Sumatera, Deli Serdang

email : andreansimanjuntak@gmail.com

Abstract. The earthquake event with M 6.7, on 7th December 2106 in Pidie Jaya, inflicted heavy casualties and material losses. In 1967, an earthquake event occurred near Pidie Jaya with M 6.1 which caused by Samalanga fault. However, the Pidie Jaya earthquake, although very close, but was not caused by the Samalanga fault activity. The latest, an unkown fault was considered as an active fault and the location of earthquake sources and and a part of the branching of Samalanga fault. It is very interesting to discuss, generally earthquake in Aceh, always happen on Sumatera fault's path. Analysis of earthquake source mechanism is very important to do, to know the fault characteristic and direction. The analysis in this study involves inversion of tensor moments with isola program on each recording component, which will provide a synthetic wave as a comparison. The earthquake mechanism which obtained from the inversion is strike-slip with left-lateral and has the parameter strike 254°, dip 80°, and rake 48° on the first nodal plane as the true nodal. While, the second nodal has the plane parameter is strike 153°, dip 42°, and rake 165°. From these parameters, the first nodal plane that has a strike 153° is considered to be interpreting the new fault line as it is also followed by the direction of the spread of aftershocks. These result indicate that the Samalanga Fault is not the cause the Pidie Jaya earthquake, and shows the real direction of Pidie Jaya fault from South-West to North-East with lef-lateral orientation.

1. Introduction

Interaction between plate tectonics and local faults makes Aceh region vulnerable to the earthquake events. The interaction occurs because of geodynamic process that cause a relative movement in the interface zone between one fault to others. The movement produces energy which can generate a deformations and can make the occurrence of earthquakes. In the Aceh region, earthquake activity oftenly occurs after several major earthquake events happened from subduction zones. The series of earthquakes can provide stress that will trigger some activity of local fault on land and produce new earthquake sources. Aceh region has five local fault that is quite active and has a recurrence time of 10 - 15 years. In the last decade, two faults have been broken at Tripa, and Samalanga-Sipopok fault in 2013.

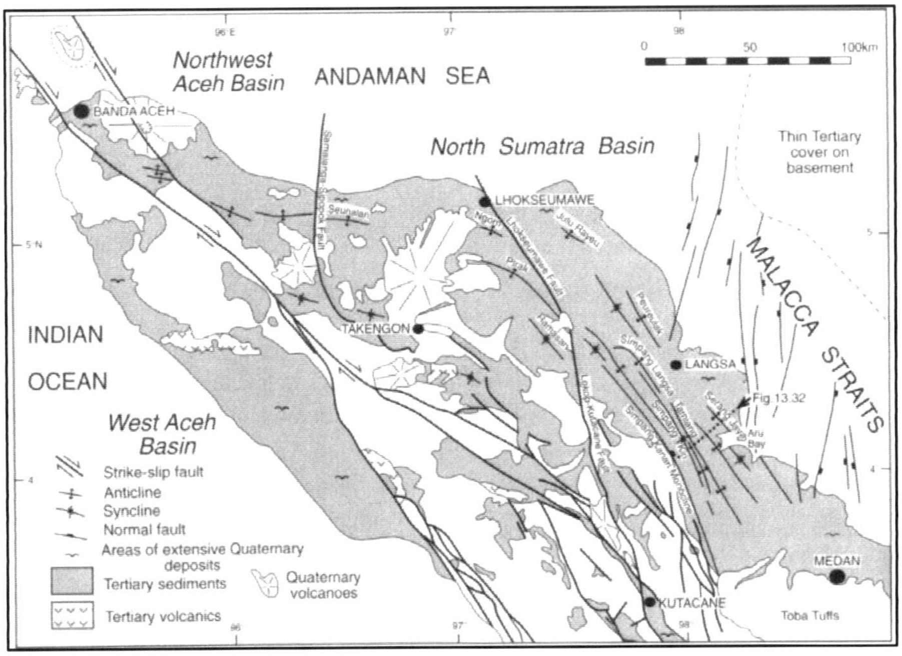

Figure 1. The tectonic map on the basin area at northen part of Sumatera Island [1].

In **Figure 1,** we can see some faults in Aceh region is part of the Sumatran Fault System. The tectonic pattern of Aceh region is controlled by tectonic plates activity from Indian Ocean subduction. The plate is the oceanic plate (Indo-Australian plate), which moves northward with a velocity 6 - 8 cm/year, with larger densities pounding the European continent - Asia (Eurasian plate) and producing subduction. The subduction resulted a series of front-line non-volcanic arches (forearch islands) such as Simeulue, Nias, Batu, Siberut to Enggano Island. In Central Aceh, after Bener Meriah earthquake in 2013. [2] did passive seismic research to found some unknown active faults and imaging them which has right lateral direction.

Fault is a picture of a plane in the earth crust, having side which connected by an asperity (lock). Both sides have interface zones that are mutually attached due to the pressure and friction force on the surface. After each sides still continue to move, the fault plane would be driven gradually by tectonic forces but still firmly sealed. Most of Aceh region is composed by sedimentary and volcanic rocks from the tertiary and quarter periods. Folds are mostly found in the north and east, due to deformation of rock layers resulting from force and pressure from local faults or tectonic plates. Thus, rocks move from the initial position and form the arch. The folds are subjected to horizontal pressure, resulting in anticline and syncline formation. In addition, the Bukit Barisan mountain range with the path and distribution of volcanic constituent rocks are in the middle. The circuit has a straight-line straightness system, and is called the Sumatran Fault that divides Sumatra Island from Semangko Bay to Banda Aceh. The spread of the epicenter on the land of Aceh, always sourced from the alignment zone of the Sumatran Fault System.

However, the Pidie Jaya earthquake is not, but on the north, near the fault location of Samalanga. In the north of Aceh, the fault of the Samalanga is a branch of the Sumatran Fault that aligns northward, and last, the earthquake occurred in 1967 with M 6.1. On December 7, 2016 at 05:03:36 pm, an

IOP Conf. Series: Earth and Environmental Science **273** (2019) 012021 doi:10.1088/1755-1315/273/1/012021

earthquake occurred with M 6.5 located in Pidie Jaya, Aceh, Indonesia. The epicenter coordinates location at 5.25 ° N and 96.24 ° E, precisely on land with depth at 15 km and have a distance 18 kilometers southeast from Sigli, and 2 kilometers north from Meureudu. **Figure 2** shows us the earthquake instensity was felt in Pidie Jaya IV SIG-BMKG (VII-VIII MMI), Lhokseumawe, Lhoksukon, Bireun II SIG-BMKG (IV-V MMI), Banda Aceh, Meulaboh II SIG-BMKG (III-IV MMI). According to the National Disaster Management Agency (BNPB) report, at least 104 people died as a result of this earthquake. The initial analysis, Pidie Jaya earthquake hypocenter is not far from the Samalanga fault. However, the assumption that the fault of Samalanga as the cause of the earthquake is not appropriate. The previous quake's epicenter was due to the Samalanga fault activity. After quality control on advanced analysis, the result is far from that fault. A hypocenter occured in the alignment of an unknown local fault.

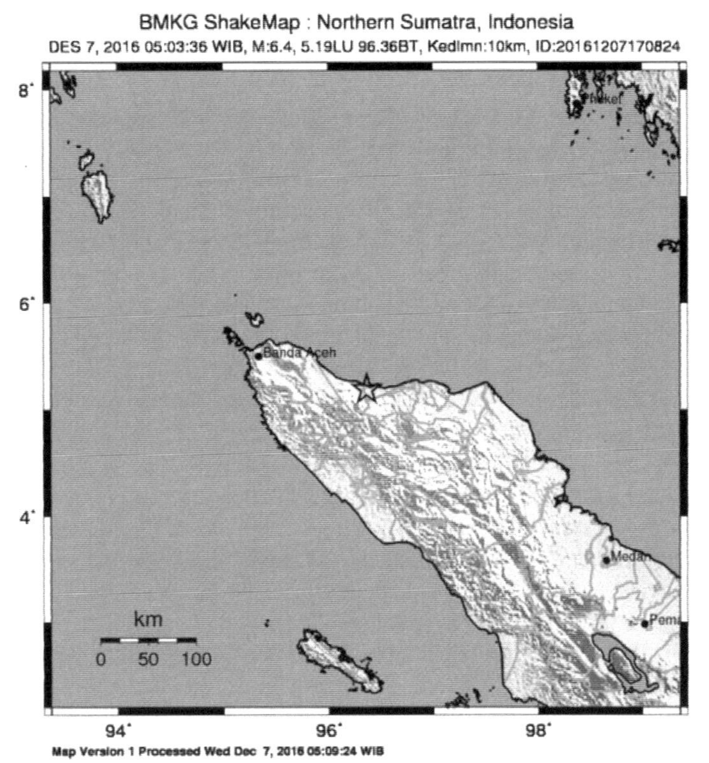

PERCEIVED SHAKING	Not felt	Weak	Light	Moderate	Strong	Very strong	Severe	Violent	Extreme
POTENTIAL DAMAGE	none	none	none	Very light	Light	Moderate	Mod./Heavy	Heavy	Very Heavy
PEAK ACC.(%g)	<0.03	0.3	2.8	6.2	12	22	40	75	>139
PEAK VEL.(cm/s)	*<0.01*	*0.1*	*1.4*	*4.7*	*9.6*	*20*	*41*	*86*	*>178*
INSTRUMENTAL INTENSITY	I	II–III	IV	V	VI	VII	VIII	IX	X+

Scale based upon Worden et al. (2011)

Figure 2. The shakemap of Pidie Jaya earthquake from BMKG with MMI scale.

Parameters that released by USGS provide some information if this earthquake occurred on land and was generated by strike-slip activity in the direction of south-western-trending strike, and unknown type and name of fault. BMKG released the distribution of aftershock which also follows the strike direction on the nodal plane which has a south-northwest trending strike. Therefore, the orientation of the fault, structure and type of fracture in the Pidie Jaya earthquake requires ongoing analysis. It is necessary to map the straightness, type, structure and fault properties, which can later be used to design a mitigation. The analysis in this study aims to model and analyze the source mechanism, and compare it with several other models.

Modeling of source mechanism is done by applying the inversion method to tensor momen. Analysis and modeling of earthquake source mechanism by inversion on moment of tensor conducted by Isola (Isolated Asperities) program and using green function [3]. The concept of green function with discrete wave number method from [4] and [5], to produce synthetic data on displacement waveform. The entire wave is used and the green function is calculated. The moment tensor of sub-events is obtained by minimizing leastsquare from a match between the observation and synthetic signals, while the position and time of the sub-events are optimized through grid-search. Solutions and approaches in Isola use the concept of multi point-source by doing iterative deconvolution techniques, previously performed by [6]. The solution of earthquake source mechanism with Isola has been successful and widely applied by many researchers from Indonesia as well as outside.

Modeling the source mechanism with the Isola will produce the Mw (moment magnitude) value, the moment tensor value on each resultant of the x, y, z, azimuth, plunge, DC (Double Couple), CLVD (Compensated Linear Vector Dipole) components, centroid time and the value of reduction variance. [7] analysed and got a solution for 2009 and 2010 Himalaya earthquake, with applied a frequency band filter for inversion with range 0.02 - 0.11 Hz (2009 earthquake) and 0.08 - 0.20 Hz (2010 earthquake). [8] conducted an analysis to identify source mechanisms on 15 earthquakes with a magnitude range of moments (Mw) 4.1 - 6.4, which occurred in ten major segments located along the Sumatera Fault System. Many researchers get the focal mechanism of earthquake by using the Isola program because it based on GUI Matlab and easy to learn. The tensor moment inversion method in Isola program is very good to modeling the focal mechanism of earthquake which use near-field station.

2. Materials and Method

To do a calculation, the digital waveform of seismogram with 3 components (NS, EW, and UD) is needed and used in this research. Data was taken and downloaded from BMKG (Meteorology Climatology and Geophysics Agency of Indonesia), and also taken from IRIS Wilber3 for regional stations from Thailand and Malaysia. All data selected for Pidie Jaya earthquake on 7 December 2016 with M 6.5 are from IA (Indonesia) with 3 stations, MY (Malaysia) 3 stations, and TM (Thailand) 2 stations. The selected stations are combined between local and regional with seismograph sensor type is broadband (0.000 - 50.000 Hz). In addition, polezero data, velocity model data and earthquake parameter data are required as initial inputs. The waveform data used in the SAC (Seismic Analysis Code) format. Polezero data is used to obtain real earthquake recording by eliminating the instrument response, because the received waveform data is still combined data in the form of convolution between waveform, path, and response instrument.

We used the velocity model used in this paper from global tomography model, given the absence of local model that is suitable with geological and tectonic conditions in the research area. The method which used to calculate the source mechanism solution in the Isola program is doing a inversion on moment tensor in each axis (x, y, z) and iterated by deconvolution [9]. The green function is used to generate synthetic signals based on the structure and model of the earth, in which the signal is considered a waveform model to be seismographically recorded from the signal. The green function is calculated by applying the discrete wavenumber method. Synthetic results from calculations with green functions in the form of displacement data will be compared with observation. The parameters of the earthquake source will be estimated by the inversion method using the model of the green function. With green function, estimation can be done because using synthetic signal. The synthetic signal will be compared with the observation signal, the comparison will show the fitting waveform of the three best components. Good fittings are seen with overlapping overlays.

Accuracy was measured by overlayed observed signals with synthetic signals. To get an overlay form, try and error method must be done by giving good band-pass filter in order to get the fit and best inversion result. The model will has a value of reduction variance (0 - 1) which is a measure of the compatibility of the observed signal and the accuracy of the synthetic signal. The big reduction variance will signify the results more accurate, precise and represent the tectonic structure in there. The measured reduction variance is obtained by providing the appropriate band-pass filter with 4 corner (f1, f2, f3, f4). The inversion used in the ISOLA-GUI program is deviatoric doment densor inversion which calculates the inversion process on the basic components of tensor moments. DC (Double Couple) and CLVD (Compensated Linear Vector Dipole) which used in inversion is the most common and good method [10]. The result of the source mechanism obtained from tensor moment inversion still contains ambiguity. The ambiguity is due to the existence of two nodal fields in earthquake source modeling in the form of strike, dip and rake parameters of both nodal fields. The ambiguity in this case means that the researcher must determine the "true fault plane" of the two fields.

3. Result and Discussion

Observation signal used for green function and inversion from 13 local and regional stations around Aceh province, Malaysia, and Thailand. From 13 stations, only 8 stations that had sufficiently good variant values which obtained from the trial and error process. For visual inspection, we overlayed the comparative results between synthetic and observational signals. Here is the 8 station location which used in this study, and global velocity model.

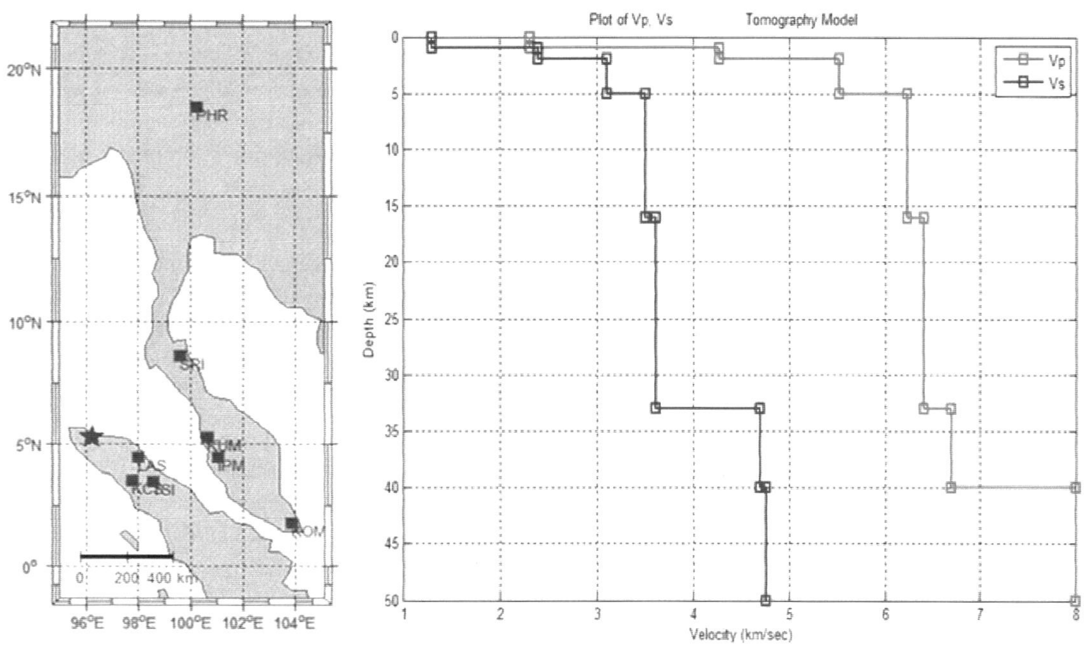

Figure 3. The location of stations which used to do a calculation (left), the global velocity model which used in this research, blue color (Vp) and red color (Vs)(right).

In the processing of Isola, we give the signal duration as an input to do a green function is 409.6 seconds, which is the duration recommended in the Isola program. The 409.6 second duration is used, since the estimation has covered the entire waveform of each station. The velocity model that used in this research is the global velocity model and shown in **Figure 3**. This velocity model describes the current model of global seismic wave velocity (P wave and S wave). Analysis results from BMKG, that has good quality control, was used as an input for hypocentre location information in Isola. Then, we also got the foreshock from BMKG and any references to assement the direction fault, because foreshock distribution will show the true nodal plane and follow the strike of earthquake focal mechanism.

The filter used for velocity data conversion to displacement data in the inversion process which shows in **Figure 4** is 0.01 - 0.025 Hz. This filter was used because accordance with [11], which suggesting that low frequency filters for displacement data are useful to get a good point source solution. The type of inversion used in this model is deviatoric moment tensor. **Figure 4** shows the matching data in displacement between the observed signal and the synthetic signal in the three component of seismogram (NS, EW, Z).

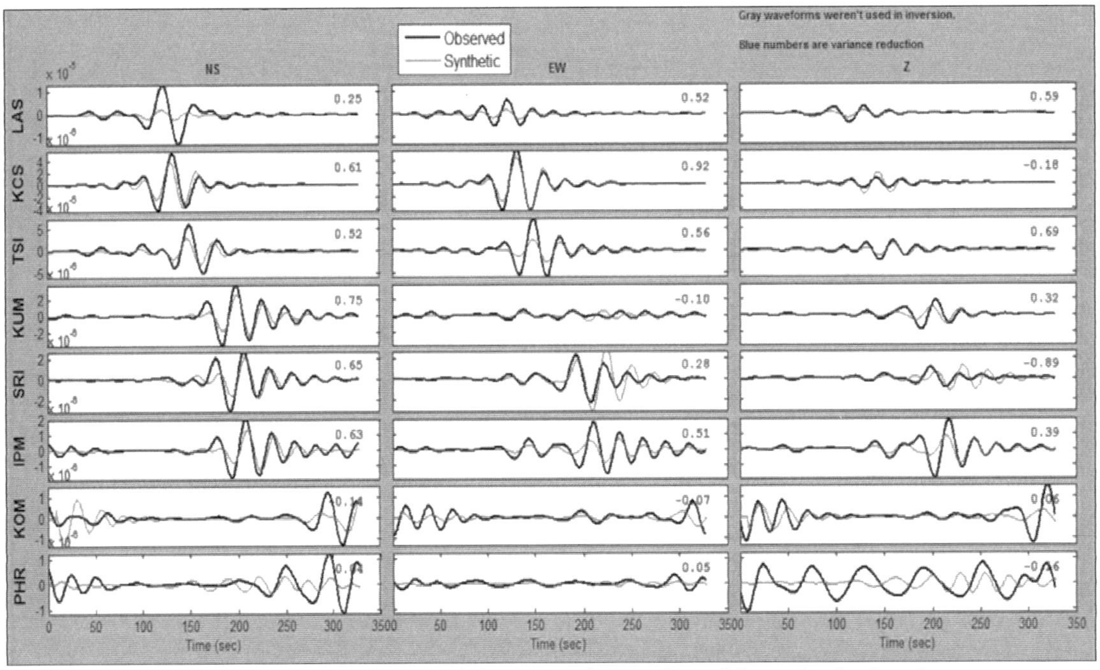

Figure 4. The overlay graph between observation with synthetic signal.

The black signal is an observation signal while the red color signal is a synthetic signal. The value in blue is the value of the variance reduction (0-1). The presented picture is the final result to get a source mechanism model in this research after do a try and error method. This result is obtained directly through Isola's GUI program which runs with Matlab program. Based on Figure 5, there are 8 stations obtained from the inversion on waveform with 3 components. There are 3 stations with initial name LASI, KCSI and TSI with IA code from Indonesia. The value of the match rate between synthetic seismogram and observation of LASI stations on NS, EW and Z components, respectively are 0.25, 0.52 and 0.29. KCSI has a match rate between synthetic seismogram with observations on NS, EW and Z components with value 0.61, 0.92 and -0.18. The level of compatibility between synthetic seismogram and observation of TSI stations on NS, EW and Z components respectively are 0.52, 0.56 and 0.69.

The stations from Malaysia are 3 stations with initial name KUM, IPM and KOM with MY code. The value of the compatibility level between synthetic seismogram and observation of KUM stations on

NS, EW and Z components respectively are 0.75, -0.10 and 0.32. IPM has a good match value between synthetic seismogram and observation on each component of NS, EW and Z, with value 0.63, 0.51 and 0.39. The match value between synthetic seismogram and observation at KOM stations on NS, EW and Z components are -0.14, -0.07 and 0.06, respectively. There are 2 stations with initial name SRIT and PHRA from Thailand was used in this research. The match value between synthetic seismogram and observation at SRIT station on NS, EW and Z component are 0.65, 0.28 and -0.89, respectively. PHRA has good value between synthetic and observations on NS, EW and Z components, the value are 0.04, 0.05 and -0.16.
Based on the reduction variance value, the matching data which shown in Figure 5 generally seems quite relevant, there are only two stations on the Z (vertical) component that still have minus variance values, that is the best value resulting from the trial and error process. The average variance for the four stations used was 0.70 (70%). Based on the value of variance, the results obtained in this study is quite relevant when associated with the actual situation.

4. Focal Mechanism

Earthquakes are caused by fault movement with certain motion as its characters. The fault motion model and the causative character of the earthquake can be identified based on the earthquake moment tensor. This moment tensor is used to describe the direction of earthquake-causing forces. The movement of fault is one of the source parameters cause a vertical deformation in the source area and produce an tectonic earthquake. The fault parameter is also known as the solution of focal mechanism. The focal mechanism is the direction and fault orientation of the tectonic earthquakes. The focal mechanism is divided into four types: strike slip, normal, reverse and oblique.

In general, the angle forming parameters of the faults are strike angle, dip and slip. At the time of the earthquake occurrence, there are two fields that become the reference in determining tensor moment that is auxiliary plane and perpendicular between each fault plane. Based on the results of the strike, dip and rake parameters, we can see the characteristics of the first plane parameters with the strike = 254°, dip = 80° and rake = 48° and the second plane of the strike value = 153°, dip = 42° and rake = 165°. Then, we plotting the fault plane with GMT (Generic Mapping Tool) to obtain the pattern of fault plane formed is strike-slip fault **(Figure 5).**

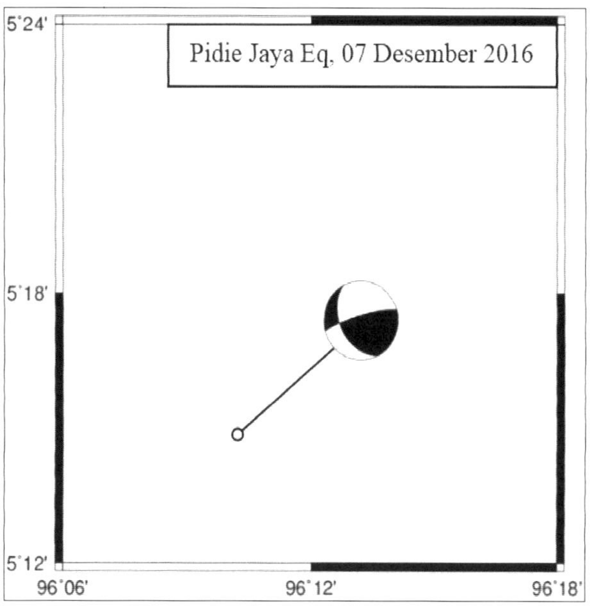

Figure 5. Earthquake Mechanism Solution.

IOP Conf. Series: Earth and Environmental Science **273** (2019) 012021 doi:10.1088/1755-1315/273/1/012021

Based on the strike, dip and rake value, each node field on the beach ball of the earthquake focal mechanism which generated by ISOLA can be compared with the focal mechanism generated by GFZ, GCMT and USGS (**Table 1**). From the data processing using ISOLA-GUI, program and the result of beachball depiction with GMT on the picture shows that there is no significant difference between the processing result using ISOLA with the processing result from GFZ and USGS. So, the results obtained from this data processing can be said quite accurate.

Tabel 1. The difference model of each focal mechanism component.

GFZ	GCMT	USGS	Author
(Strike, dip, slip)	(Strike, dip, slip)	(Strike, dip, rake)	(Strike, dip, rake)
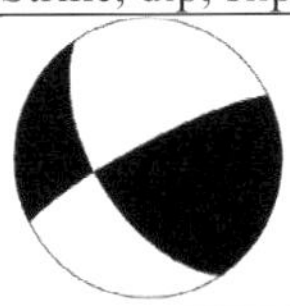	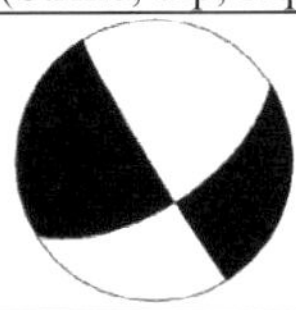	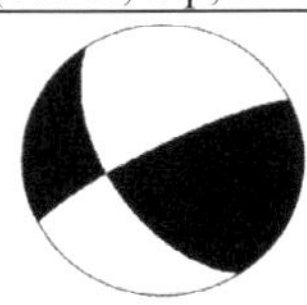	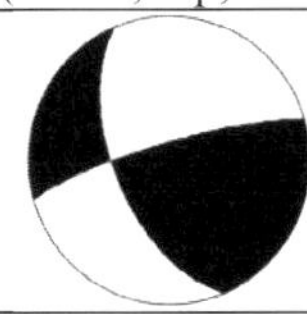
(NP1= 148°, 79°, 170°)	(NP1= 57°, 63°, -4°)	(NP1= 243°, 81°, 33°)	(NP1= 254°, 80°, 48°)
(NP2= 240°, 80°, 11°)	(NP2= 148°, 87°, -153°)	(NP2= 147°, 57°, 170°)	(NP2= 153°, 42°, 165°)

Figure 6 shows us a difference of parameter information like latitude, longitude, depth and magnitude which obtained from GFZ, GCMT, USGS and BMKG data. The blue star shows the epicentre from GFZ, the yellow star shows epicentres from GCMT data, the green star shows the location of epicentre from USGS data and the white star indicates the location of the main quake generated by the author from BMKG station. The red dots indicate the spread of aftershocks from BMKG that occurred on December 7, 2016 in Pidie Jaya District which has direction from North-West (NW) to East-South (ES).

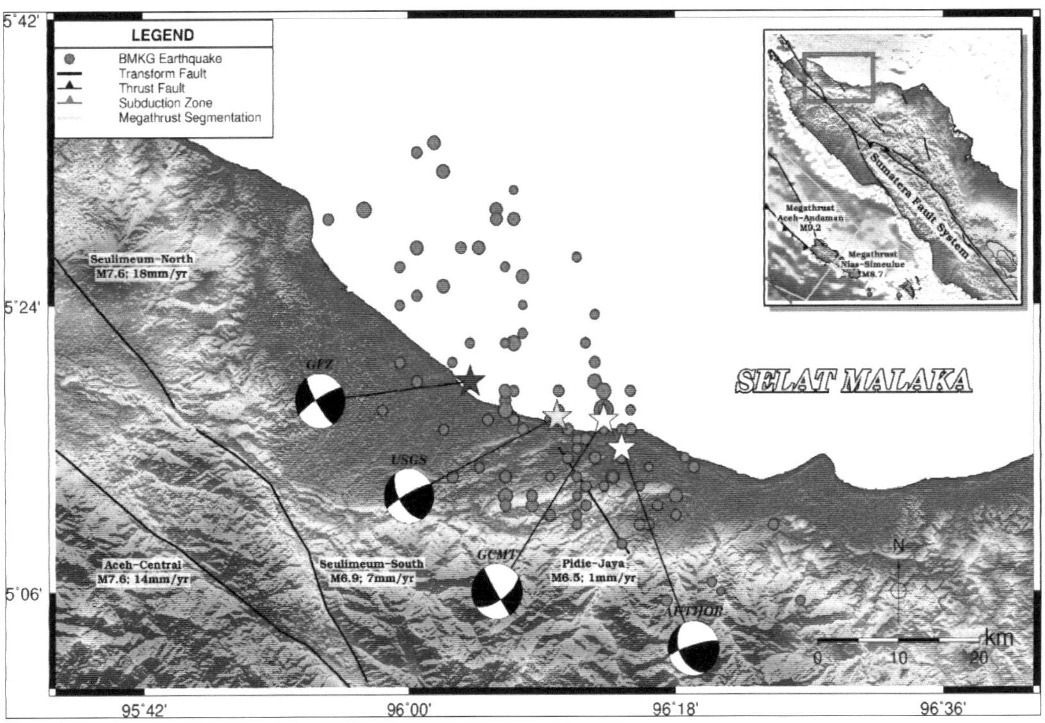

Figure 6. The comparison of focal mechanism solution (Author, GCMT, USGS, and GFZ) with the spreading of foreshock.

Figure 6 shows the result of earthquake source mechanism which obtained from tensor moment inversion process. Based on the source mechanism parameter, Pidie Jaya earthquake has a oblique (strike-slip with vertical movement) fault mechanism solution with two nodal planes. First nodal plane has strike 254°, dip 80°, rake 48° and the second nodal plane has strike 153°, dip 42°, rake 165°. The result of earthquake source parameter still contains the ambiguity because there are two nodal planes in which one of the two nodal fields is the "true fault plane" and the other is the "auxiliary plane". To avoid ambiguity, foreshock analysis is needed to know the direction of straightness fault. The foreshock obtained follows the direction of the first nodal plane with strike direction 254°. From [12], we can see the result has a directional fracture directed South-West (SW) to North-East (NE), and opposite with BMKG result. [12] used a temporary network to analysed the foreshock and local velocity model to get the focal mechanism with Cut and Paste method, than BMKG result which used global velocity model. This result will refer to [12] with has aftershock which follow the focal mechanism strike from South-West (SW) to North-East (NE).

In comparison, the Global Centroid Moment Tensor (GCMT) catalog released the focal mechanism solution of Pidie Jaya earthquake. The solution has the first nodal plane with strike = 57°, dip = 63°, slip = -4°, and second nodal plane with strike = 148°, dip = 87°, slip = -153°. The GCMT Catalog for the mechanism solution is made using seismic data from remote stations (teleseismic), not local or regional data. For parameters of major and subsequent seismic hypocenter locations using localized results from BMKG with 140 epicenter sites. According to [12], the nodal field closest to the hyposenter location is estimated as the "actual fault zone".

5. Conclusions

From this research, we can get the focal mechanism solution of Pidie Jaya Earthquake by combine regional and local station, also successful to analyse this fault with seismology method and approach. From that, we can give an assessment to specify the pidie jaya fault, as we know together which cause of this earthquake. The fault has a strike-slip mechanism with dextral movement and the direction from North-West (NW) to East-South (ES). This result has not much different than others result, that mean we did a good procedural and get the good model too. The novelty in this paper about how to combined the local and regional stations with azimuth < 180°, used global velocity model and got best fitting waveform. Next, we can know how the filter range for all station, and how the waveform characteristic in that field.

References

[1] Barber, A. J., Crow, M. J., & Milsom, J. (Eds.). (2005). *Geological Society of London*.
[2] Muksin, U. et al (2019). *Journal of Asian Earth Sciences*, **171**, 20-27.
[3] Sokos, E. N., & Zahradnik, J. (2008). *Computers & Geosciences*, **34(8)**, 967–977.
[4] Bouchon, M. (1981). *Bulletin of the Seismological Society of America*, **71(4)**, 959–971.
[5] Coutant, O. (1989). *Journal of Geophysical Research: Solid Earth*, **94(B12)**, 17805–17818.
[6] Kikuchi, M., & Kanamori, H. (1991). *Bulletin of the Seismological Society of America*, **81(6)**, 2335–2350.
[7] Kumar, R., Gupta, S. C., & Kumar, A. (2015). *Natural Hazards*, **76(2)**, 1163–1181.
[8] Kasmolan, M., Santosa, B., Lees, J., & Utama, W. (2010). *Open Geosciences*, **2(4)**, 455–474.
[9] Sokos, E. N., & Zahradnik, J. (2008). *Computers & Geosciences*, **34(8)**, 967–977.
[10] Stein, S., & Wysession, M. (2009). *John Wiley & Sons*.
[11] Bormann, P., Baumbach, M., Bock, G., Grosser, H., Choy, G. L., & Boatwright, J. (2002). *IASPEI New Manual of Seismological Observatory Practice*, **1**, 1–94.
[12] Muzli, M., Umar, M., Nugraha, A. D., Bradley, K. E., Widiyantoro, S., Erbas, K., ... & Wei, S. (2018). *Seismological Research Letters*, **89(5)**, 1761-1772

IOP Conf. Series: Earth and Environmental Science **273** (2019) 012022 doi:10.1088/1755-1315/273/1/012022

Multi-frequency effect towards High Conductivity Overburden Responses using Horizontal Coplanar (HCP) coil configuration for Landslide Hazard Instrumentation

Bukhari Bukhari[1,2], Muhammad Syukri Surbakti[1,*], Nazli Ismail[1], Nasrullah Idris[1], Elin Yusibani[1], Muhammad Isa[1], Saumi Syahreza[1,3]

[1]Department of Physics, Faculty of Mathematics and Natural Sciences, Universitas Syiah Kuala, Banda Aceh 23111, Indonesia.
[2]Doctorate Program of Mathematics and Applied Science (DMAS), Postgraduate Program, Universitas Syiah Kuala, Banda Aceh 23111, Indonesia.

[3]Tsunami and Disaster Mitigation Research Center (TDMRC), Universitas Syiah Kuala, Banda Aceh 23111, Indonesia.

*Corresponding author: msyukris@unsyiah.ac.id

Abstract. The instrumentation for the sensing of natural disaster phenomena as like as landslide instrumentation has acquired great importance for the instrumentation community. The novel instrumentation for landslide application has been made and calibrated using Horizontal Coplanar (HCP) coil orientation. The early instrumentation for landslide slopes was gradually equipped with standard and automatic instrumentation consisting of three the piezometric cells, two deep-seated steel wire extensometers, four in clinometric tubes, a rainfall gauge, a snow gauge and an air thermometer. This novel instrumentation works in electromagnetic methods similar with electromagnetic methods in geophysics. HCP coil orientation is a coplanar array where both coils are horizontal. In HCP, the depth of sounding is about 1.5 times of the coils separation which have maximum coupled configuration (HCP, VCP, and Vertical Coaxial configuration). The overburden response had inverted the percent value of maximum negative response then there were three maximum response amplitude peaks. Measurement and data processing were performed using a computer through Delphi program. The diagonal cross sections of the overburden as the centre were the maximum value of inverted response of anomaly positions. In aluminium thin plate overburden, the diagonal cross section point as the overburden centre was the maximum value of inverting value of the highest minimum response amplitude. The measurement graphic in the computer screen with several types of overburdens would be displayed. As the result, that the overburden widths for overburden were determined by subtracted of two turning points positions. The overburden positions were accurate for Type P overburden. The master curve had a high anomaly contrasts with two of positive anomaly peaks. Otherwise, the anomaly positions were between two of the negative anomaly peaks.

1. Introduction

One method that exists in geophysical instrumentation is a method that uses an electromagnetic induction system. The advantage of this method is the operation of measuring instruments that is relatively simple and practical [1]. The advantage if applied to the application of landslide is the penetrating power that is shallow in depth, which is in the order of hundreds of meters in the subsurface. The shallow penetrating power can be regulated not too affected by the material contained below the earth's surface. The Horizontal Coplanar loop configuration (HCP) is relatively lower in noise compared to the Vertical Coplanar (VCP) method. It caused by the atmospheric noises are usually in vertical polarization. The HCP method is also one configuration that results in maximum coupling (McNeill, 1980). The VCP method was selected as a configuration option in this examination.

The electromagnetic induction system has two coils as transmitter and receiver. If there is a change in distance, change in height, and changes in angle between the coils of transmitter and receiver, the receiver receives a change in response [1]. An examination for VCP configuration also has been conducted which had a lower level in the depth penetration and more noise contribution [2]. Changes in response can be used as information on movement or small shifts of land if landslides occur. This electromagnetic method has a weakness, where the material contained below the earth's surface also affect the response. The transmitter emits a signal called the primary electromagnetic field and if there are conductive objects in the subsurface there are eddy currents on the conductive objects. The eddy currents then trigger a secondary electromagnetic field. The secondary electromagnetic field simultaneously with the primary field reach the receiver directly. Fortunately, these primary and secondary electromagnetic fields can be separated so that the signal used is only the primary electromagnetic field while the secondary electromagnetic field is removed. However, the primary electromagnetic field in its propagation to the receiver passess through conductive materials so that this is also undesirable. To overcome this, a pair of frequencies is used.

In this study, several pairs of frequencies is used, which are referred to the three source frequencies (S) and the two of reference frequencies (R). The frequency pairs are 112.5 Hz.(R) & 337.5 Hz. (S), 112.5 Hz. (R) & 1012.5 Hz. (S), 112.5 Hz. (R) & 3037.5 Hz. (S), 337.5 Hz. (R) & 1012.5 Hz. (S), and 337.5 Hz. (R) & 3037.5 Hz. (S). The configuration of the coils from this study is as shown in Figure 1. The effects of the frequencies used as well as the influence of the material on the surface and below the earth's surface, namely overburden and anomalies can be obtained. These data can be used in the design of landslide early warning system instruments. This tool is based on the tools used by GNES instruments that have been used by various applications, some of which are for testing cube anomalies and overburden[1], and metal defect applications [3].

2. Methodology

The HCP method is a variation of the most commonly used Slingram method [4]. This method uses a transmitter and receiver coil arrangement that is placed horizontally at a fixed (co-axial) (dashed line on Figure 1.a) and a plot distance with overburden level and co-axial (dashed line in Figure 1.b).

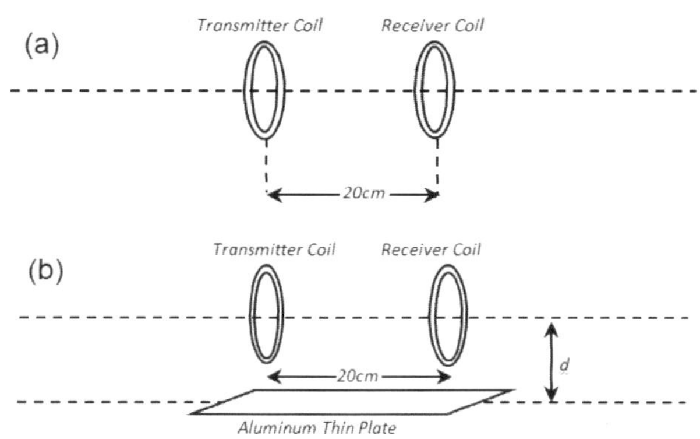

Figure 1. Equipment examination scheme: (a) Without overburden, (b) With overburden.

2.1 The equipment system components for measurement system

The GNES has three components above plus computer-controlled system and automatic data acquisition. The GNES is useful for real time data acquisition that saves the data automatically to the computer. The sensors in GNES system are two coils from the transmitter to transmit the primary field and the receiver to receive the secondary and the primary field simultaneously. The signal conditioning circuits are the band-pass filter, the notch filter, the precision of mathematical operation, and the absolute value function. There are several outputs in the GNES system i.e. two of analog displays, two of digital displays, and the motor stepper system [1].

Table 1. The equipment system for measurement.

The Equipment	Total
GNES Equipment	1 unit
Aluminium sheet Type P (20cm x 20cm), thickness = 0.75mm	1 unit
Aluminium sheet Type Q (36cm x 36cm), thickness = 0.75mm	1 unit
Aluminium sheet Type X, (36cmx6cm), thickness = 0.75mm	1 unit
Aluminium sheet Type Z, (120cmx36cm), thickness = 0.75mm	1 unit

2.2 Physical modeling

The scale factor *n* is important for interpreting the field data using the physical model data. In this physical modelling (*n*=100), the anomaly target and the overburden were a cubical Aluminum and an Aluminum plates respectively. It is assumed that the environment was non-conductive. In the free-air condition which has zero conductivity is representative for this environment. In the zero conductivity, the secondary field from the medium is not presence. In this condition, there is no phase difference between transmitted and received electromagnetic fields.

3. Results and Discussion

3.1 Anomaly response

The HCP coil orientation is a coplanar array where both coils are horizontal. In HCP, the depth of sounding is about 1.5 times of the coils separation which have maximum coupled configuration (HCP, VCP, and Vertical Coaxial configuration). An examination of anomaly response by physical model using thin plate of aluminum with 120cmx36cm of dimension (Type Z), 0.075 cm of thick, dip = 0°, and depth = 2 cm was conducted. The results are described in Figure 2.

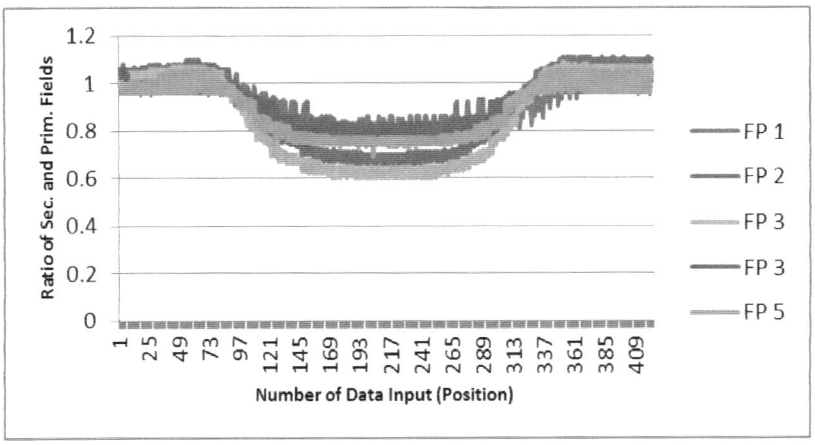

Figure 2. Response of Aluminum thin plate anomaly Type Z.

An online observation through the monitor of computer is can be performed. The secondary and primary fields have the same in quantity when the transmitter or the receiver coils precisely at above the anomaly ($V_S/V_R = 1$). The ratio of secondary and primary field when the anomaly positions placed below the coils is marked by intersections of the graphic with y-axis on y = 1. The anomaly position (AP) is the maximum of secondary field at AP point in the middle of the points of $V_S/V_R = 1$ or a midpoint between the intersections as found Sinha [5].

The anomaly position (AP) of the five pairs of frequency is 70 cm (the number of input data). The transmitter and the receiver coils position when $V_S/V_R = 1$ are 30 cm and 110 cm, respectively. The anomaly contrasts series from the lowest to the highest are in the frequency pairs of FP 1, FP 4, FP 5, FP 2 and FP 3 for the ratio of secondary to primary fields of 0.81833, 0.80583, 0.75417, 0.68667 and 0.6275, respectively

3.2 Overburden response

The overburden response had inverted the percent value of maximum negative response then there were three maximum response amplitude peaks. Measurement and data processing were performed using a computer through Delphi program. The diagonal cross sections of the overburden as the centre were the maximum value of inverted response of anomaly positions.

The diagonal cross section point as the overburden centre was the maximum value of inverting value of the highest minimum response amplitude. The measurement graphic in the computer screen with overburden Type P is displayed in Figure 3. We had the centre position for overburden Type P was 68cm. We had the overburden width for overburden Type P by subtracted two of the turning points positions we found 20 cm. We had the overburden position accurately for Type P. The master curve had a high anomaly contrast. There were two positive anomaly peaks. The anomaly position was between two of negative anomaly peaks (Figure 3).

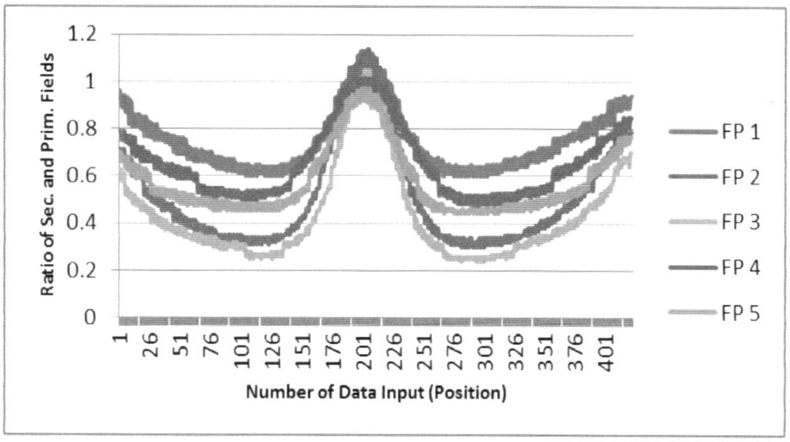

Figure 3. Overburden Type P response by multi-frequency.

The cross sections points of the overburden for five of frequency pairs are 68cm. The ratios of secondary and primary fields on the cross section for Figure 3 start from the higher contrasts to lower contrast are 1.10083 (FP 2), 1.07583 (FP 1), 0.9975 (FP 3), 0.99167 (FP 4) and 0.94583 (FP 5). The turning point of the five frequency pairs are 47 cm and 87 cm. The values of V_S/V_R and diagonal cross section points are given by Table 2.

Table 2. Interpretation of Aluminium thin plate overburden (Figure 3)

Frequency Pairs (FP)	Diagonal Cross Sections Points (cm)	The V_S/V_R Values of The Left Side of Overburden	The Ratio of Secondary to Primary Fields (V_S/V_R) (position)	The V_S/V_R Values of The Right Side of Overburden
FP 1	67	0.653	1.07583	0.66
FP 2	67	0.3767	1.10083	0.37
FP 3	67	0.94	0.9975	0.93
FP 4	67	0.6	0.99167	0.613
FP 5	67	0.5067	0.94583	0.47
Average	67	0.61528	1.07583	0.6086

3.3 Anomaly with overburden

The overburden thickness in the field exploration is often covers the tropical area such as in Malaysia and Indonesia presented as weathering layer. The more of thickness then the more of energy of electromagnetic wave can be absorbed, as the result that the electromagnetic waves become smaller. The electromagnetic wave that reaches the target of below of the surface is also smaller in amplitude. Generally, there are no direct connections between the sizes of the anomaly when the overburdens were present. The response of the overburden is usually dominant.

3.3.1 Aluminium Type X anomaly with Type P overburden

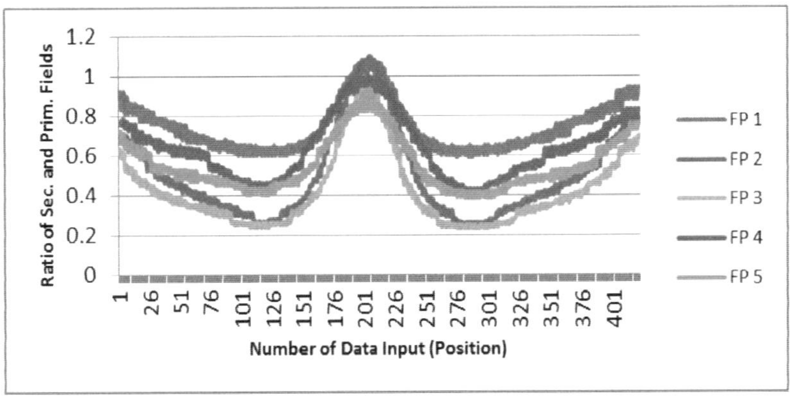

Figure 4. Response of Aluminium Type X with Type P overburden.

The cross sections points of the overburden for five of frequency pairs are 67 cm. The ratios of secondary and primary fields from higher to lower contrasts are 1.0525 (FP 1), 1.04917 (FP 2), 0.97583 (FP 4), 0.925 (FP 3) and 0.87333 (FP 5) as shown in Figure 4. The turning point of the five frequency pairs are 47 cm and 67 cm. The values of V_S/V_R and the diagonal cross section points are given by Table 3.

Table 3. Interpretation of graphic of Figure 4.

Frequency Pairs (FP)	Diagonal Cross Sections Points (cm)	The V_S/V_R Values of The Left Side of Overburden	The Ratio of Secondary to Primary Fields (V_S/V_R) (position)	The V_S/V_R Values of The Right Side of Overburden
FP 1	67	0.5133	1.0525	0.5133
FP 2	67	0.2267	1.04917	0.2467
FP 3	67	0.53	0.925	0.5533
FP 4	67	0.4367	0.97583	0.42
FP 5	67	0.513	0.87333	0.4767
Average	67	0.44394	0.975167	0.442

3.3.2 Aluminium Type Q anomaly with Type P overburden

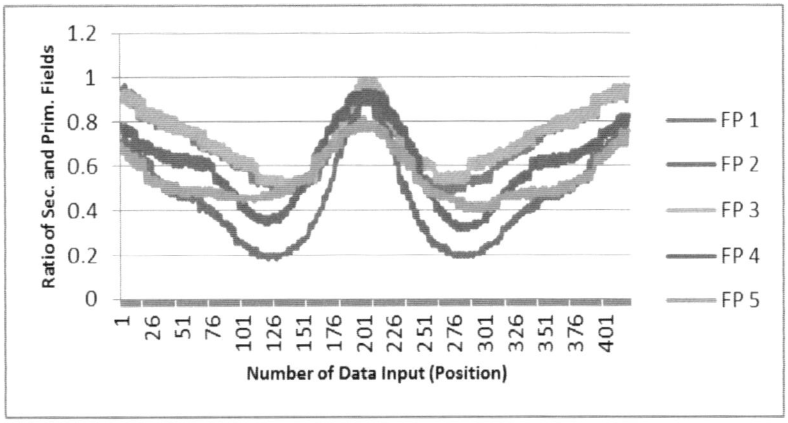

Figure 5. Response of Aluminium Type Q with Type P overburden.

The cross sections points of the overburden for five of frequency pairs are 68 cm (Figure 5). The ratios of secondary and primary fields from higher to lower contrasts are 0.9575 (FP 3), 0.9425 (FP 1), 0.91833 (FP 4), 0.88917 (FP 2) and 0.78583 (FP 5) as shown in Figure 5. The turning point of the five frequency pairs are 47cm and 67cm. The values of V_S/V_R and the diagonal cross section points are given by Table 4.

Table 4. Interpretation of graphic of Figure 4.

Frequency Pairs (FP)	Diagonal Cross Sections Points (cm)	The V_S/V_R Values of The Left Side of Overburden	The Ratio of Secondary to Primary Fields (V_S/V_R) (position)	The V_S/V_R Values of The Right Side of Overburden
FP 1	68	0.5133	0.9425	0.5133
FP 2	68	0.2367	0.88917	0.2333
FP 3	68	0.54	0.9575	0.55
FP 4	68	0.4467	0.91833	0.4
FP 5	68	0.5233	0.78583	0.47
Average	68	0.452	0.898667	0.43332

3.3.3 Aluminium Type Z anomaly with Type P overburden

The cross sections points of the overburden for five of frequency pairs are 68 cm. The ratios of secondary and primary fields on a position points from the higher contrasts to lower contrasts are 0.93167 (FP 1), 0.91667 (FP 4), 0.86583 (FP 2), 0.77333 (FP 5) and 0.711167 (FP 5) as shown in Figure 6.

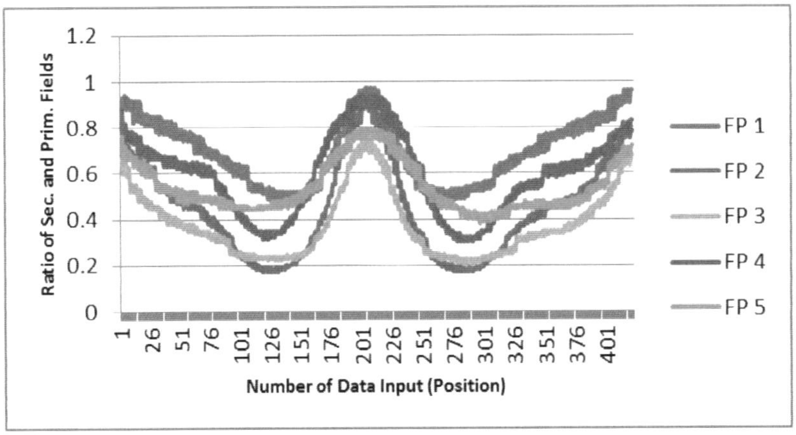

Figure 6. Response of Aluminum Type Z with Type P overburden.

The turning point of the five frequency pairs are 48cm and 88cm. The values of V_S/V_R and the diagonal cross section points are given by Table 5.

Table 5. Interpretation of graphic of Figure 5.

Frequency Pairs (FP)	Diagonal Cross Sections Points	The V_S/V_R Values of The Left Side of Overburden	The Ratio of Secondary to Primary Fields (V_S/V_R) (position)	The V_S/V_R Values of The Right Side of Overburden
FP 1	68	0.5	0.93167	0.51
FP 2	68	0.22	0.86583	0.22
FP 3	68	0.2433	0.71167	0.24
FP 4	68	0.4067	0.91667	0.4
FP 5	68	0.4867	0.77333	0.4833
Average	68	0.37134	0.839833	0.37066

4. Conclusion

Multi-frequency application contribute in the anomaly and the conductive overburden predictions because the veer of the OP and IP components could be predicted and the total responses could be averaged and filtered for this landslide instrumentation. The ratio of secondary and primary field when the anomaly positions placed below the coils is marked by intersections of the graphic with y-axis on y = 1. The anomaly position (AP) is the maximum of secondary field at AP point in the middle of the points of $V_S/V_R = 1$ or a midpoint between the intersections. The diagonal cross sections of the overburden as the centre were the maximum value of inverted response of anomaly positions. The overburden response can inverted the value of maximum negative response then there were three maximum response amplitude peaks. The series shows that the anomaly contrasts are bigger when the different of the reference and the source frequency are bigger. The anomaly position is the mid position of the transmitter and the receiver coils which informed by the maximum of negative response indication. If there is a change in distance, change in height, and changes in angle between the transmitter and receiver, the receiver receives a response change. Changes in response can be used as information on movement or small shifts of land which subsequently collected data can be translated into information if landslides occur.

5. References

[1] Surbakti M S, Jafri M Z M, San L H and Tahrin N M 2011 Overburden Effect Examination Of Cubical Anomaly Using Geometry Normalized Electromagnetic System (GNES) Equipment *Google Scholar*

[2] Surbakti M S, Ismail N, Idris N, Yusibani E, Isa M and Syahreza S 2018 Multi-frequency effect towards High Conductivity Overburden Responses using Vertical Coplanar (VCP) coil configuration for Landslide Instrumentation. In: *Journal of Physics: Conference Series*: IOP Publishing) p 012069

[3] Zakaria Z, Surbakti M S, Syahreza S, Jafri M Z M and Tan K C 2017 Development of Geometry Normalized Electromagnetic System (GNES) instrument for metal defect detection. In: *High-Performance Computing in Geoscience and Remote Sensing VII*: International Society for Optics and Photonics) p 104300J

[4] Frischknecht F 1988 Electromagnetic physical scale modeling *Electromagnetic methods in applied geophysics-theory* 365-441

[5] Sinha A 1980 A study of topographic and misorientation effects in multifrequency electromagnetic soundings *Geoexploration* **18** 111-33

Mathematical Model for Mangrove Protections Toward Nearshore Process

Salmawaty Arif

Mathematics Department, Syiah Kuala University, Jl. Syeh Abdurrauf 3, Darussalam, Banda Aceh, 23111, Indonesia

E-mail: salmawaty@unsyiah.ac.id

Abstract. This research developed a mathematical model to evaluate positive effects of mangroves to nearshore process. Particular purpose of the study was to conduct the analytical calculation for the steady condition of wave energy density and wave heights distributions due to the presence of mangroves. The method used was an expansion of the equations for nearshore quantities from the general condition to the one with the existence of mangroves. First, the wave energy density was calculated analytically by solving the wave energy conservation equation. The analytic solution for energy quantity is a function of position variable, involving longshore currents component and other physical constant parameters. The next step was to insert the dissipation of wave energy flux due to wave breaking into another energy balance equation involving mangroves effects. The mangroves contribution is represented by the dissipation due to the existence of the plants in the water. Other quantities included in the model are the wave group velocity and the dissipation due to bottom friction flux. Then the wave heights distribution concerning the mangroves effect was derived from the wave energy density function governed.

Keywords: mangroves, wave heights, nearshore, mathematical model, analytic solution

1. Introduction

Beach vegetations such as mangroves have been well known providing the natural ability to protect the coastal areas and their communities by reducing the nearshore wave heights and lessen the sediment movements. However, the quantitative approach for investigations to determine the advantages of mangroves and muds in their growing areas towards the nearshore protection and other benefits in a broader context, has rarely been conducted. To fulfil the lack of investigations in this field, this research develop a mathematical model to evaluate the positive effects of mangroves to the nearshore process.

The mathematical modelling approach is an efective way in the effort to know the amount of protection provided by the mangrove plants by reducing coastal process caused by waves, which in turn will become input for the solution to the coastal hazards leading to coastal disaster. Particular purpose of the study was to conduct an analytical calculation of wave energy density and wave heights distributions due to the presence of mangroves. These wave quantity functions may also help as evaluation tools to the results of numerical methods as the commonly selected methods applied for solving problems. The method used was an expansion of the nearshore problem from the general condition to the one with existence of mangroves. The development of mathematical model was conducted, the problem was solved analytically by implementing the mathematical integration methods to solve the differential equation problems. Here the steady case was solved.

2. The model formulation and methods

In this research the wave quantites in the direction parallel to the shore line was assumed to be uniform. The domain of the mathematical model developed is an area in the two dimensional coordinates (x,z), which are the off−onshore with a sloping bottom configuration and vertical directions. The origin of the coordinates was selected at a certain offshore position of the still water surface. Water depth distribution $h(x)$ along the interval was given as the input for the model. First, the wave quantities were governed without the involvement of mangroves. The wave energy density was calculated analytically by solving the wave energy conservation equation. The equation was accompanied by some initial conditions obtained at the origin position. The second step was to insert the dissipation of the wave energy flux due to wave breaking into another energy balance equation involving mangroves effects. The mangroves contribution is represented by the dissipation flux due to the existence of the plants in the water. Other quantities included in the model are the wave group velocity, and the dissipation due to bottom friction flux. Then the wave heights concerning the mangroves effect were derived from the wave energy function governed by using the famous Snel's law for sloping bottom, and also by considering the wave breaking criterion.

2.1. The domain and variables

Consider two dimensions horizontal xz−plane. x refers the on-offshore axis and z be the distance measured upwards. Let $\eta = \eta(x)$ be the mean water level elevation measured above the still water level and $h = h(x)$ is defined as the depth of the water measured from the reference axis to the bottom. If 0 is a reference position selected in the sea and a is the position where bottom slope s starts, let * symbolize the sloping related notation such that for the initial constant, the water depth is written as $h_0{}^* = h_0 + sa$. Let $\theta = \theta(x)$ represents the angle between the incident wave with the positive x-axis, $E = E(x)$ stands for the energy with c is the wave phase velocity, c_{gx} is the group velocity component in x-direction of the group velocity c_g, $H = H(x)$ is the wave height, g is the acceleration of gravity, and ρ is the fluid density. Moreover, let ω be the angular frequency of incident wave, and k be the wave number. The depth averaged flow velocity components in along shore direction is represented by $v = v(x)$.

2.2. The change of wave field due to mangroves

The change of wave field was calculated by using the wave energy conservation equation. Here, the total energy in mangrove plantation area is denoted by E_w with the subscript w stands for waves, the dissipation of wave energy flux due to wave breaking is D_b, the dissipation due to bottom friction flux is D_f, and the dissipation flux due to the plants in water is D_v. The model from [1] for the energy balance is given as

$$\mathrm{d}_x\left(E_w c_g\right) = -D_b - D_f - D_v. \tag{1}$$

In this research, we proposed the analytical forms of the right hand side terms of equation (1), namely the dissipation terms D_b, D_f and D_v. From the calculation for energy density function, the wave heights quantity could be determined by using the relations formulated in linear wave theory as

$$E = 1/8\, \rho g H^2,$$

and the breaking criterion for the breaking constant γ which is defined as

$$H = \gamma(h+\eta).$$

3. The analytic solution

The analytic solution is found by solving the equation (1) analytically by using an integration method to solve the ordinary differential equation.

3.1. The analytic energy density function

The analytic functions of nearshore energy quantity for steady and unsteady flows in the direction perpendicular to the shore line on a sloping bottom up to the wave breaking position have been derived [2]. For steady problem, the one-dimensional energy conservation equation reads

$$\mathrm{d}_x\left(Ec_{gx}\right) + S_{yx}\mathrm{d}_x v = 0, \tag{2}$$

where

$$n = 1/2[1 + 2\,kh\,(\sinh 2\,kh)^{-1}], \quad c = (gh)^{1/2},$$

$$c_{gx} = nc\cos\theta, \qquad S_{yx} = 1/2\,nE\sin 2\theta.$$

S_{yx} is the radiation stress components described in [3]. Recall for a very shallow water, $kh << 1$ such that $n = 1$ and $c_g = c$. For this assumption, we write the equation (2) as a homogeneous first order linear ordinary differential equation:

$$\mathrm{d}_x(Ec\cos\theta) + E\sin\theta\cos\theta\,\mathrm{d}_x v = 0. \tag{3}$$

By Snel's law for a constant K and the subscript 0 refers to offshore values of x-axis, the sloping bottom yields the relation

$$(\sin\theta)/c = (\sin\theta_0)/c_0 = K. \tag{4}$$

Let x_b represents the position of breaking. In this paper, the dissipation of wave energy flux D_b is approximated by the second term of equation (2) at the position of breaking, namely

$$D_b = S_{yx}\mathrm{d}_x v, \text{ for } x = x_b\,. \tag{5}$$

To get the contribution of D_b in the solution of the main equation (1), we utilized equation (5) into equation (2) to be integrated as

$$Ec\cos\theta = -\int D_b\,\mathrm{d}x \tag{6}$$

For the initial constant $E_0^* = E_0 c_0 \cos\theta_0$, the analytic steady solution of equation (2) for the energy between the area of sloping bottom until the breaking wave position x_b is obtained as already given in [2],

$$E(x) = (c\cos\theta)^{-1} E_0^* \exp\left(-Kv\right) \qquad \text{for } a < x < x_b \tag{7}$$

The analytic steady energy solution of equation (7) taken at the breaking position then becomes

$$E(x_b) = (c\cos\theta(x_b))^{-1} E_0^* \exp\left(-Kv(x_b)\right). \tag{8}$$

3.2. The bottom friction energy function

The bottom friction force of Chezy law with the bottom friction coefficient C_f is $\boldsymbol{F} = \rho C_f\,\boldsymbol{u}|\boldsymbol{u}|$. It is interpreted as the force acts in the direction of flow velocity, with the magnitude proportional to the square of the velocity. Following [4], assuming the net current $\boldsymbol{u}$ is small compared to the orbital component, the quadratic form could be approximated by the linear term for the mean flow, namely

$$F = \rho C_f u\left(u_{\text{orbital}}^2 + v_{\text{orbital}}^2\right)^{1/2}$$

$$\approx \rho C_f u u_{\text{orbital}} \tag{9}$$

Since $u_{\text{orbital}} = (\gamma/\pi)\left(g(h+\eta)\right)^{1/2}$, the dissipation of bottom friction flux D_f is then defined as

$$D_f = Fu = (\rho C_f \gamma u^2/\pi)\left(g(h+\eta)\right)^{1/2}. \tag{10}$$

D_f is the transfer of wave energy flux to the turbulent bottom boundary layer, in the absence of vegetation [1]. The cross shore velocity component u in equation (10) is assumed here to be constant and small compared to the longshore component. By defining the surface slope s' as [5]

$$s' = -\,\mathrm{d}_x(h+\eta), \tag{11}$$

the integration result of the bottom friction dissipation term contributes to the main solution of equation (1) is formulated as

$$\int D_f\,\mathrm{dx} = -[2\rho C_f \gamma(h+\eta)][g(h+\eta)]^{1/2}u^2/(3\pi s') + C. \tag{12}$$

3.2.1. The analytic longsore currents component

The analytic steady energy solution (8) involves longshore currents component v above the sloping bottom area which has been derived analytically in [6] adapted from [7]. It was derived by considering forcing terms included in the y−momentum equation for along shore direction. Radiation stress must be balanced by the bottom friction and momentum exchange forces. By taking an empirical constant N as in [5], the longshore currents at the breaking position is written in the form of

$$v(x_b) = A\,v^*\left(\frac{P_1 - 1}{P_1 - P_2}\right), \tag{13}$$

where

$$v^* = 5\pi\gamma(16C_f)^{-1}[g(h+\eta)_b]^{1/2}s'\sin\theta(x_b);$$

$$P = N\pi s'(\gamma C_f)^{-1};\ P_1, P_2 = -3/4 \pm (9/16 + 1/P)^{1/2};\ A = P(1 - 5P/2)^{-1}.$$

3.3. The vegetation dissipation function

The drag force due to the presence of vegetation is defined with the drag coefficient C_d and the area of the plant A_v in the form of $F_d = 1/2\,\rho C_d A_v\, u|u|$. Here the time averaged energy dissipation due to the plants in water D_v is refered to [8] and [9], with the drag coefficient of vegetation is denoted by C_d, the diameter of cylinders of plants is d, area per unit length of each vegetation stand b, and the elevation of the top of the plant relative to the bottom is s_v.

The dissipation of wave energy due to the plants D_v is then defined as [8]

$$D_v = 16\rho C_d d(3\pi k)^{-1}b^{-2}(gk/\omega)^3\left(3k\cosh^3 kh\right)^{-1}\left(\sinh^3 ks_v + 3\sinh ks_v\right)H^3. \tag{14}$$

The contribution of D_v in the solution of the main equation (1) can be written as

$$\int D_v\,\mathrm{dx} = 16\rho C_d d(3\pi k)^{-1}b^{-2}\left(\sinh^3 ks_v + 3\sinh ks_v\right)\int (gk/\omega)^3\left(3k\cosh^3 kh\right)^{-1}H^3\,\mathrm{dx}. \tag{15}$$

Recall that the angular frequency ω is defined as

$$\omega^2 = gk \tanh kh$$

For $kh << 1$, then it is approximated that $\tanh kh \approx kh$ and $\cosh kh \approx 1$ so then $\omega^2 = gk^2 h$. By implementing these relations, and defining the water depth on the sloping bottom as $h = h_0+sa-sx$, the result of equation (15) is obtained, so the wave energy density due to mangroves is formulated as

$$\int D_v \mathrm{d}x = 16\rho C_d d g^{3/2} (9\pi ks)^{-1} b^{-2} \left(\sinh^3 ks_v + 3\sinh ks_v\right) H^3 \int h^{-3/2} \mathrm{d}h \quad (16)$$

or

$$\int D_v \mathrm{d}x = -32\rho C_d d g^{3/2} (9\pi ks)^{-1} b^{-2} h^{-1/2} \left(\sinh^3 ks_v + 3\sinh ks_v\right) H^3 + C. \quad (17)$$

3.4. The analytic energy density of mangrove contribution

By combining the analytic forms of equation (8) for the energy of breaking waves, equation (12) for the turbulent energy of bottom friction, and (17) for the energy loss due to the plants, the final result of analytic energy density due to the presence of mangrove is obtained:

$$E_w = -1/c\left\{E_0^* \exp[-Kv(x_b)] - [2\rho C_f \gamma(h+\eta)][g(h+\eta)]^{1/2} u^2/(3\pi s')\right\}$$
$$+ 1/c\left[32\rho C_d d g^{3/2} (9\pi ks)^{-1} b^{-2} h^{-1/2} \left(\sinh^3 ks_v + 3\sinh ks_v\right) H^3\right] + C, \quad (18)$$

with the form for $v(x_b)$ is given in equation (13). The constant C can be determined by giving some respective quantities at the wave breaking position.

4. Results and Discussion

The analytic function for energy density governed in equation (18) will help the investigations of the mangroves' contribution in nearshore process. The energy contains the contributions of the energy due to breaking waves, bottom frictions, and the plants. The result also shows the role of three kinds of slopes which contribute toward the energy density, namely the bottom slope, the waves surface slope, and the top of mangroves slope.

To utilize the result, the initial conditions and data are needed as the input. Wave celerity, wave angle, wave height, and water depth at a selected initial position offshore are needed to determine the portion of energy at the wave breaking. Water depth distribution and flow velocity component along the cross shore section contribute to the energy due to the bottom friction. Quantities of the plant, particularly the diameter of the trunk, the area per unit length of each vegetation, and the slope of the canopy are needed to determine energy due to the plants.

Further work might be needed to see the variability of value combinations of the slopes for the distribution of wave energy and heights, as well as the contribution of the longshore currents. The wave height distribution due to the presence of mangroves could also be directly calculated from the energy density by using the relation given in the Subsection 2.2. The wave energy density governed from the model can also be utilized for other calculations such as the water level distribution at the shore line.

5. Conclusion

The analytic forms should then help as the tool for the numerical as well as the experimental works in the subject which also could evaluate the performance of the functions obtained at the same time.

6. References

[1] Guannel G P, Ruggiero, Faries J, Arkema K, Pinsky M, Gelfenbaum G, Guerry A and Kim C K 2015 Integrated modeling framework to quantify the coastal protection services supplied by vegetation *J. Geophys. Res. Oceans* **120** 324–45(AGU Publications). DOI: 10.1002/2014JC009821

[2] Arif S 2014 Mathematical study on energy of coastal waves, *Proc. 9th AIWEST_DR (Banda Aceh)* 129–135 (Banda Aceh: Syiah Kuala University Press)

[3] Longuet-Higgins M S and Stewart R W 1964 Radiation stresses in water waves; a physical discussion, with applications *Deep Sea Res.* **11** 529-62 (Pergamon Press Ltd.)

[4] Vreugdenhil C B 1998 *Numerical Methods for Shallow-Water Flow* (Dordrecht: Kluwer Academic Publishers)

[5] Mei C C 1989 The applied dynamics of ocean surface waves *Advanced Series on Ocean Engineering* **1** (World Scientific)

[6] Arif S and Hadi S 2003 Approaching analytic wave driven longshore currents *Mihmi* **9** 1 59–68 (Bandung: Penerbit ITB)

[7] Longuet-Higgins M S 1970 Longshore current generated by oblique incident sea waves *J. Geophysical Res.* **75** 33 6778–89

[8] Dalrymple R A, Kirby J T, Hwang P 1984 Wave diffraction due to areas of energy dissipation *J. Waterway Port Coastal and Ocean Eng.* DOI: 10.1061/(ASCE)0723-950x(1984)110-67

[9] Hendriks J M 2014 Waves dissipation in mangroves *Master Thesis* (Enschede: University of Twente).

Estimation of return period of the destructive earthquake in aceh using the maximum likelihood method

Jumila[1], H Sofyan[1*], L Rahayu[1]

[1] Department of Statistics, Faculty of Mathematics and Natural Science, Syiah Kuala University, Banda Aceh

E-mail: hizir@unsyiah.ac.id

Abstract. Aceh is one of the provinces in Sumatra Island which coincides with the meeting of the Eurasian plate and the Indo-Australian plate. The negative impact of the plate meeting is the occurrence of an earthquake. The earthquakes that occurred in this region have caused various losses including casualties. Therefore, estimates of earthquake events are required for disaster mitigation caused by earthquakes. In this research the period of earthquake re-occurring in Aceh and Surrounding region using the relation between frequency and magnitude of earthquake known as Gutenberg-Richter relationship. The maximum likelihood method is used to determine the seismic parameters obtained from the relationship. The data used ware sourced from the USGS earthquake catalogue (1907-2008) and the BMKG earthquake catalogue (2009-2017) with minimum depth is 80 Km that coincides at 92 ° - 98.5 ° East Longitude and 0 ° - 8 ° North latitude. The obtained magnitude data were converted to the moment magnitude (Mw), then selected data that have a minimum Magnitude (Mw) of 5 which is a destructive earthquake. The results obtained from this study are the period of the destructive earthquake ($Mw \geq 5$) is expected to re-occur in 19.7 years in the sea region between Aceh Barat Daya and Simeulue.

1. Introduction

The province of Aceh is one region that coincides with the meeting of the Eurasian plate with the Indo-Australian plate along the Indian Ocean. In this zone, there have been some major earthquakes that have caused severe damage in some areas around the Indian Ocean, among them the earthquake and tsunami of December 26, 2004 which has a magnitude of 9.1 SR and the earthquake that occurred in Pidie Jaya on December 7, 2016 measuring 6.5 SR.

Research on seismotectonic potential as a precursor of earthquake level in Aceh has been done, among them, analysing the level of seismicity and the period of earthquake re-occurring in Aceh region using the relationship between the frequency with earthquake magnitude, using the Maximum Likelihood method. The study concludes that between 2 to 5 years there will be an earthquake for the 6 Mw scale in Simeulue Island, Northwest Aceh and North Aceh (Andaman)[1]. An earthquake with a greater magnitude generally concentrated in the district of Simeulue, the western part of Aceh Besar and northwest of Sabang[2]. Based on the above background, spatially and temporally, the researcher wants to calculate the period of the earthquake in Aceh region as the initial information for mitigation, so that the impact of the earthquake that will happen can be minimized.

2. Materials and Method

Tectonic earthquakes occur due to the encounter or movement between plates, the active tectonic plates with other tectonic plates have three types of movement, i.e. the two plates are spreading, collision, and mutually (transform) [3].

Aceh Province is one of the areas on the island of Sumatra located at $2^{o} - 6^{o}$ North Latitude and 92° - 98° East Longitude. The island of Sumatra is an island that coincides at the meeting of the Eurasian Plate and the Indo-Australian Plate. The collisions between the plates resulted in a series of non-volcanic island arcs (such as Simeulue Island, Pulau Banyak, Nias Island to Enggano Island) and a series of hillside mountains with volcanic trails in the centre[4]. In addition to being in two active plate meetings, the Aceh region also belongs to the active fault zone of Sumatra. The existence of this fault resulted in frequent earthquakes in the Aceh region with great power and small, both on land and at sea.

2.1 Earthquake Frequency (N) dan Magnitude (M) Relationship
Earthquake frequency-magnitude relationships can be used to see the number of earthquakes occurring in a region, one of them using the Gutenberg-Richter law [5].

$$log\, N(M) = a - bM \quad (1)$$

where:
N = the number of earthquakes
a = a determination that depends on the period, total area and area of observation activity
b = seismotectonic parameter of an area in which in an earthquake occur and depending on the local rock.
M = magnitude
[6]

The value of the tectonic parameters in the research area can be seen by looking at b-value. b-value can show the stress level of the rock. If b-value is low then the stress condition in the area is high (Nuannin, 2006). According to Kulhanek, *et al.* (2018), b-value is said to be low if b-value has a value smaller than the average value of b-value in the entire study area. b-value approximation is correct if the standard deviation of b-value is less than one [7].

2.2 Maximum Likelihood Method
Maximum Likelihood aims to estimate statistics of random variables by maximizing probabilities with existing samples. This method is often used in different applications and studies in various situations[8]. Likelihood's maximum method is better for calculating b-value. Magnitude (M) is considered as a random variable with the Probability Density Function (PDF) for M is as follows [9]:

$$p(M) = b\ln(10)\,\frac{10^{-bM}}{10^{-bM_{min}} - 10^{-bM_{max}}} \quad (2)$$

Where M_{min} is the minimum magnitude and M_{maks} is the maximum magnitude, if $M_{max} > M_{min}$ then equation (2) becomes:

$$p(M) = b\ln(10)\,10^{-b(M-M_{min})} \quad (3)$$

the maximum etic of likelihood in equation (3) has b-value by maximizing the likelihood function, where the equation becomes:

$$b = \frac{1}{\ln(10)(\mu - M_{min})} \quad (4)$$

With $^{1}/_{\ln(10)}$ equal to $\log e$, then the b-value obtained from the maximum of Likelihood is expressed as follows:

$$b = \frac{\log e}{\bar{M} - M_0} \quad (5)$$

Where:
$\bar{M}$ = Average magnitude

M_0 = Minimum magnitude
$log\ e = log\ 2{,}71828 = 0{,}3434$
The error rate or standard deviation of the b-value calculation with the magnitude can be estimated using the equation:

$$\sigma_b = 2{,}30b^2\sqrt{\frac{\sum_{i=1}^{N}(M_i-\mu)^2}{N(N-1)}} \quad (7)$$

In addition, there is also an a-value that states the level of seismicity in a region. The smaller the a-value, the smaller the seismic activity [10].

For a-value can be calculated from the cumulative frequency relationship that is:

$$a = \log N(M \geq M_0) + \log(b \ln 10) + M_0 b \quad (8)$$

In the calculation is used observation value of, which then obtained a-value[1].

2.3 Earthquake return period

To get the average value of earthquake return periods in a region can be calculated using the following equation:

$$P(\Theta) = \frac{1}{N_1(M \geq M_0)} \quad (9)$$

Where Θ is the period of earthquake repeat for magnitude $M \geq M_0$.

2.4 The probability of the occurrence of earthquakes
The probability of earthquakes occurring is the possibility of destructive earthquakes in one area over a period. Assuming the time interval follows the exponential form e^{-NT}, the probability of earthquake occurrence is as follows:

$$P(M,T) = 1 - e^{-N_1(M \geq M_0)T} \quad (10)$$

Where $P(M,T)$ is the probability of an earthquake with magnitude *M* and period *T*.

3. Methodology
The data used is earthquake occurrence data from January 1907 to December 2017. Because the data obtained from Badan Meteorologi Klimatologi and Geofisika (BMKG) only from 2009, then the data from 1907-2008 accessed from the catalogue of earthquakes United States Geological Survey (USGS). The selected data is earthquake data located on longitude 92° - 98.5° East Longitude and latitude 0° - 8° North latitude which coincides in Aceh province and at maximum depth 80 Km, research area was divided into the grid so obtained 832 grids with earthquake incidence as much 9202. The distribution of earthquakes in the region of Aceh and Surrounding in the period 1907 to 2017 can be seen in [11].

The cumulative frequency of earthquake events in the region of Aceh shows that the earthquake increased significantly after the year 2000 where in 2004 there has been a major earthquake that caused the tsunami and there have been many subsequent earthquakes after that. For recalculation periods, the earthquake distribution was grouped by the division of the study area [11].

4. Result
The spatial b-value analysis is done in the research area with the number of earthquakes as much as the occurrence, the value of b-value is obtained and the standard deviation is obtained which means less than one then the b-value approximation is correct. Spatial variation of b-value is determined by dividing the research area located at $92^0 - 98{,}5^0$ East Longitude and $0^0 - 8^0$ West Longitude into 832 parts with grid $0{,}25^0 \times 0{,}25^0$. The result of the analysis shows that b-value ranges from 0,132 - 3,434 with average 0,924. There is 347 grid that has a low b-value that is a region with b-value smaller than average, the grid that has a high b-value is an area with b-value more than average. In addition, 331 grids cannot be analyzed because 4 grids have an average magnitude equal to the

minimum magnitude and 327 grids with no earthquake data. However, for mapping, the area with an earthquake frequency of more than 10 was used, resulting in 73 areas with an average b-value of 0.751. The variation of b-value can be seen in Figure 2.

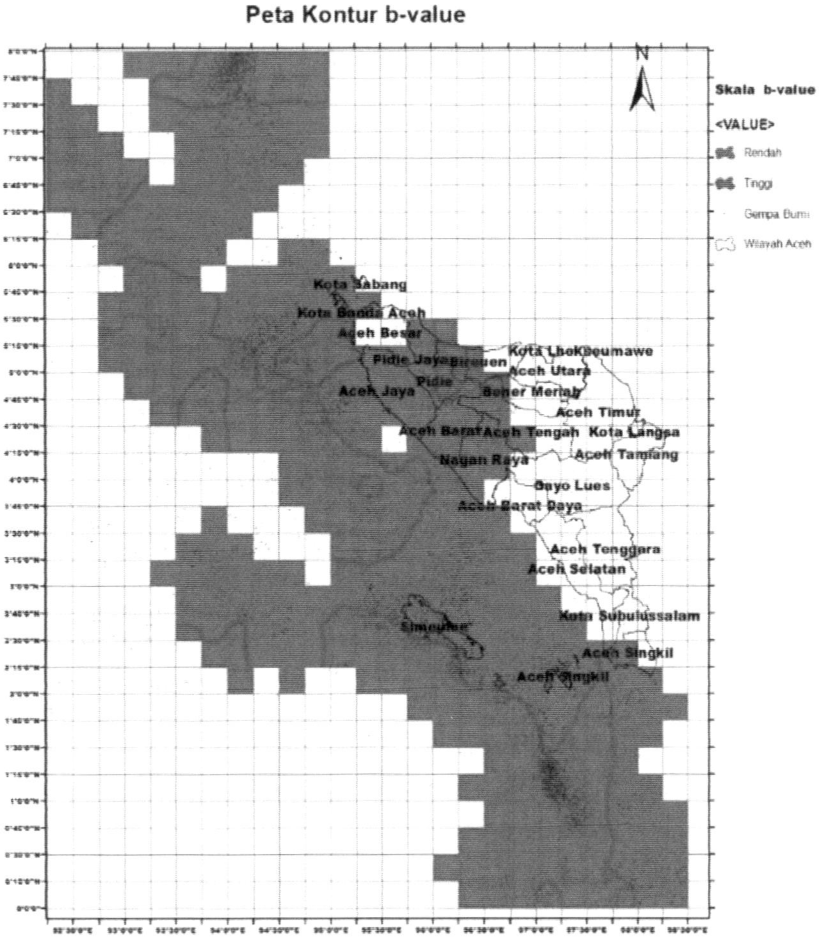

Figure 1. Map of b-value spatial variation in Aceh area

Figure 1 shows that the blue region is a low b-value area, the region is a region with high levels of stress and rare earthquakes, but in case of earthquakes, there will be earthquakes with a large magnitude. It may be due to the buildup of energy in the region and when energy is released will result in large-scale earthquakes. While the red area is a region with high b-value, it indicates that the area is often earthquake with small magnitude due to the earthquake immediate energy released. The level of accuracy of the calculation can be seen based on the standard deviation value of b-value. Overall the standard deviation of b-value is less than 1 which means the calculation of b-value is close to true.

The temporal b-value analysis is done by distributing earthquake occurrence time. Temporarily in the event of a b-value earthquake decreases until a major earthquake occurs. In 2012 and 2016, there has been a major earthquake in Simeulue and Pidie Jaya. In 2011, b-value increased after the earthquake in simeulue in 2010. In addition, b-value also increased in 2017 after the earthquake in Pidie Jaya at the end of 2016.

4.1 Return period and Probability of Earthquake.

The re-occurrence period of earthquakes in Aceh and surrounding areas with $M_w \geq 5$ is illustrated in the following contour maps.

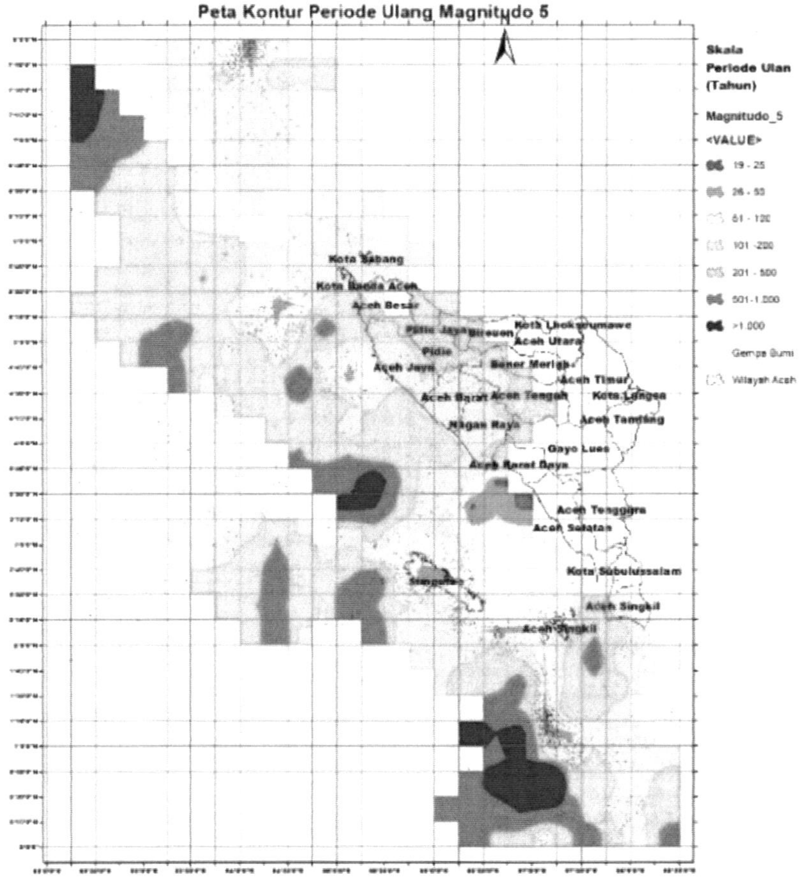

Figure 2. Map of contours of earthquake repeating period

Figure 2. shows that red areas are areas with periods of re-earthquakes in the shortest period of 19 to 25 years, then the occurrence of repeated periods of earthquakes continues increase every change in red, yellow, green to blue. While the white colour indicates the area that there is an earthquake frequency of less than 10. The tipe of return period the earthquake can be seen in Table 1.

Table 1. The period of return period of destructive earthquakes

Grid	Period (years)	Year estimates of return period	Occurrence probability
356	19,7	2037	0,632122
151	21,5	2039	0,632652
125	22,8	2040	0,632452

Table 1 shows that the most rapidly estimated earthquake will occur in the 356th grid that coincides in the southwestern seas of Southwest Aceh District or between Aceh Barat Daya and Simeulue with a repeat period of 19.7 years. The 151 and 125 grids coincide in the southernmost seas of Pulau Banyak island of Aceh Singkil with a return period of approximately 21.5 years and 22.8 years.

The probability of earthquake occurrences with on the grid to 356, 151 and 125 is about 63 percent. The estimated year the earthquake repeat period is calculated after 2017 where the last earthquake event data is used, so earthquakes are expected to occur in the 356th grid by 2037, the 151rd grid by 2039 and the 125th grid by 2040. As for the validity of return period analysis, an analysis of the probability of earthquake occurrence in the near-estimated period will occur the period of the earthquake. The probability of earthquake occurrence can be seen in Table 2.

Table 2. The probability of the occurrence of destructive earthquakes

Grid	Probability				
	5 years	10 years	30 years	50 years	100 years
356	0,224	0,398	0,782	0,921	0,994
151	0,208	0,372	0,753	0,903	0,991
125	0,197	0,355	0,732	0,889	0,988

Table 2. shows that the probability of an earthquake re-occurrence increases with increasing period. It is known that in the 356th grid is estimated in the next 10 years has a probability of earthquake 39.8 percent and in the next 30 years has a probability of 78.2 percent. The results are by the estimated repeat period of the earthquake that occurred within 19.7 years with a probability of occurrence of 63.2 percent (Table 1).

In addition, in Table 2 it can be seen that the greatest probability of earthquakes with an estimated 356th grid within 50 years will occur with a probability of 92.1 percent and within 100 years with a probability of 99.4 percent.

5. Conclusion

Based on estimation of return period of the destructive earthquake in Aceh area using the maximum likelihood method, the fastest destructive earthquake is expected to occur in the sea area between Aceh Barat Daya and Simeulue at 96° - 96.25° East Longitude and 35°-35.25° North Latitude in 19.7 years, it means that the earthquake expected to occur in 2037.

6. Reference

[1] Rahmat Tauladani, "Kajian Seismisitas dan Periode Ulang Gempa Bumi di Aceh," *Skripsi*, 2015.

[2] M. Affan, M. Syukri, L. Wahyuna, and H. Sofyan, "Spatial Statistic Analysis of Earthquakes in Aceh Province Year 1921-2014: Cluster Seismicity," *Aceh Int. J. Sci. Technol.*, vol. 5, no. 2, pp. 54–62, 2016.

[3] D. P. McKenzie and R. L. Parker, "The North Pacific: An example of tectonics on a sphere," *Nature*, vol. 216, no. 5122, pp. 1276–1280, 1967.

[4] D. H. Natawidjaja *et al.*, "Paleogeodetic records of seismic and aseismic subduction from central Sumatran microatolls, Indonesia," *J. Geophys. Res. B Solid Earth*, vol. 109, no. 4, pp. 1–34, 2004.

[5] L. De Arcangelis, C. Godano, and J. R. Grasso, "Statistical physics approach to earthquake occurrence and forecasting," *Phys. Rep.*, 2016.

[6] B. Gutenberg and C. . Richter, "Magnitude and energy of earthquakes," *Annu. Geophys.*, vol. 9, 1956.

[7] S. Happrobo, "Variasi b-value terhadap Ruang dan Waktu sebagai Prekursor Gempa Bumi Daerah Andaman Sumatera," Syiah Kuala Univesity, 2016.

[8] J. P. Wang, Y. Xu, and Y. M. Wu, "Earthquake probability in Taipei based on non-local model with limited local observation: Maximum likelihood estimation," *Soil Dyn. Earthq. Eng.*, vol. 99, no. March 2015, pp. 150–156, 2017.

[9] Q. Han, L. Wang, J. Xu, A. Carpinteri, and G. Lacidogna, "A robust method to estimate the b-value of the magnitude-frequency distribution of earthquakes," *Chaos, Solitons and Fractals*, vol. 81, pp. 103–110, 2015.

IOP Conf. Series: Earth and Environmental Science **273** (2019) 012024 doi:10.1088/1755-1315/273/1/012024

[10] H. Kanamori, "The nature of seismic patterns before large earthquakes. In earthquake prediction: An international review," *Maurice ewing Ser.*, vol. 4, pp. 1–9, 1981.
[11] H. Sofyan, Jumila, and L. Rahayu, "The application of R and ArcGIS software as a learning media to estimate the return period of the destructive earthquake in Aceh areas using maximum likelihood method," *J. Phys.*, vol. 1088, no. 1, 2018.

IOP Conf. Series: Earth and Environmental Science **273** (2019) 012025 doi:10.1088/1755-1315/273/1/012025

Modelling of sea surface temperature by using generalized additive mixed models in risk detection

R A Humaira[1], Miftahuddin[1], and I Setiawan[2]

[1]Department of Statistics, Faculty of Mathematics and Science, Syiah Kuala University
[2]Department of Marine Science, Faculty of Marine and Fisheries, Syiah Kuala University, Banda Aceh, Indonesia

Abstract. Rising temperatures from year by year caused an extreme weather which could result in all kinds of natural disasters. One of methods that can analyze the factors which affect sea surface temperature (SST) is Generalized Additive Mixed Model (GAMM). The GAMM is able to analyze more complex, especially those related to random effect and data which have not normal distribution. The purposes of this study is to construct a model of SST dataset by interacting with several variables, then the factors that affect SST be discovered. The type of data that used in this research is secondary data obtained through the National Oceanic and Atmospheric Administration website. The data has been taken from 2 locations, namely points 4°N90°E and 1.5°N90°E. The data used is daily unitin periods from September 17, 2006 to June 14, 2017. The variables used are SST, wind speed, air temperature, dynamic height, heat content, rainfall, relative humidity, shortwave radiation, sea surface density, and sea surface salinity, and a random effect namely location. Based on the results obtained the best model is the third model with Akaike Information Criterion, Bayesian Information Criterion and log-likelihood values respectively are -986,021, -894,911, and 515,752. Then the factors that significantly affect SST are wind speed (X_1), air temperature (X_2), dynamic height (X_3), heat content (X_4), sea surface density (X_8), and sea surface salinity (X_9). Furthermore, there are several interacting predictor variables that affect SST including interactions between air temperature and dynamic height, dynamic height and sea surface density, wind speed and rainfall, and air temperature with sea surface density.

1. Introduction

Climate change has a significant impact on Indonesia region. The effects of climate change can be felt by people who living in rural, urban and coastal areas. Rising temperatures from year by year caused an extreme weather which could result in all kinds of natural disasters. According to Meteorology, Climatology and Geophysics [1], in the last few years, it's been five times that happened tropical cyclone in indonesia.

The other effects that caused by climate change are drought, forest fires and storms that occured on the high seas. In 2017, one of the districts in Aceh Province, namely Nagan Raya Regency, experienced a peat fires and West Aceh Regency experienced a forest fires, sixty five-acre lot. Not only forest fires, on February 2018 there are thousands of acres of rice fields in several districts in Aceh Province experienced drought thus effect to rice plants being threatened corp failure. There are 8 districts that experienced drought on paddy fields, namely Southeast Aceh, Aceh Tamiang, Lhokseumawe, Langsa, North Aceh, Aceh Besar, Banda Aceh, and Nagan Raya. The most extensive area of wetland that that experienced drought is North Aceh Regency, 2,002 acres and Aceh Besar Regency as vast as 1,882 acres. Then some time ago there was a storm in the waters of the Indian Ocean and The Malacca Strait in the northern tip of Aceh which caused a high level of concern. This is because of two boats have sunk on the Banda Aceh - Sabang route and in the waters of North Aceh [2].

There are several parameters that affect the climate system on the Earth, especially for regions with tropical climates that are very influential in determining the relationship between the atmosphere and oceans that affect climate change. One of the interactions between the atmosphere and the ocean is the interaction between SST and rainfall. the sea is affected by surface currents, solar heat, cloud conditions, and upwelling especially in estuaries and along coastlines. In addition, several meteorological factors also affect SST, such as rainfall, evaporation, humidity, air temperature, wind speed, and the intensity of solar radiation.

According to previous research on SST in the Indian Ocean region at 4°N90°E, [3] the data period used from 2011 to 2015 using Generalized Additive Models method. Based on the optimum model obtained a few significant variables namely wind speed, SSS, shortwave radiation, salinity, humadity, prec. rain, dynamic height, density, temperature water, isoterm, conductivity, current velocity, zonal, and merid.

Additive models are comprehensive, the meaning is able to analyze more complex, especially those related to random effect and data that have not normal distribution. Generalized Additive Model (GAM) is an extension of the ordinary linear regression model but replaces the linear function into an additive function , thus it can be used in cases where there is a nonlinear relationship between the response variable and the predictor variable. Analysis by using the GAM method is not only carried out on the response variable which have not normal distribution, but also it can be used in distributions that are included in the exponential family. However, GAM cannot be used if there are two effects in a model, namely random effects and fixed effects. This study uses a random effect namely location. There are 2 locations that are used as a random effect, these are location A which is the point of data retrieval point 4°N90°E and location B is point 1.5°N90°E. Therefore the Generalized Additive Mixed Model (GAMM) approach is used to detect both effects to analyze SST data by interacting with several variables such as rainfall, wind speed, air humidity, and etc.

2. Literature Review

2.1. Generalized Additive Model

[4] The additive model is the development of a linear model in which the sum predictor component of the smoothing function. Suppose there are a number of data $\{y_i, x_{i1}, x_{i2}, \dots x_{ip}\}_{i=1}^{n}$ where n is the number of observations. The form of the additive model can be explained as follows.

$$\boldsymbol{y} = \boldsymbol{X}^{*}\alpha + \sum_{j=1}^{p} f_j(x_j) + \in \qquad (1)$$

$\boldsymbol{y}$ = vector of response variable
$\boldsymbol{X}^{*}$ = tranpose of fixed effects matrix, start the observation to the *i*
$\boldsymbol{\alpha}$ = vector of fixed effects coeficients
$f_j(x_j)$ = smoothing function of covariate x_j
x_j = predictor variable to the *j*
$\in$ = random error

[5] The generalized Additive Model (GAM) is an extension of the additive model which assumes that the mean of the response variable depends on the additive predictor variable. GAM also assumes that the response variable has an exponential distribution. GAM is also an extension model of Generalized

IOP Conf. Series: Earth and Environmental Science **273** (2019) 012025 doi:10.1088/1755-1315/273/1/012025

Linear Model (GLM) by replacing the linear predictor function $\sum_{j=0}^{p} x_j\beta_j$ to be an additive predictor function of $\sum_{j=0}^{p} f_j x_j$. The general form of the GAM model is defined as follows:

$$g(\mu) = \boldsymbol{X}^{*}\alpha + \sum_{j=1}^{p} f_j(x_j) \quad (2)$$

$g(\mu)$ = link function
$\boldsymbol{X}^{*}$ = tranpose of fixed effects matrix, start the observation to the *i*
$\boldsymbol{\alpha}$ = vector of fixed effects coeficients
$f_j(x_j)$ = smoothing function of covariate x_j
x_j = predictor variable to the *j*

2.2. Generalized Mixed Additive Model (GAMM)
Generalized Additive Mixed Model (GAMM) is used when there is a nonlinear relationship between the response variable and the predictor variable. The use of GAMM for quantitative variable data with the estimated smoothing function used is smoothing spline. GAMM is an extension of the Generalized Linear Mixed Model (GLMM) model where GAMM replaces linear functions in the GLMM to additive functions. The general form of the GAMM model is as follows.

$$g(\mu) = \boldsymbol{X}^{*}\alpha + \sum_{j=1}^{p} f_j(x_j) + \boldsymbol{Zb} \quad (3)$$

$g(\mu)$ = link function
$\boldsymbol{X}^{*}$ = tranpose of fixed effects matrix, start the observation to the *i*
$\boldsymbol{\alpha}$ = vector of fixed effects coeficients
$f_j(x_j)$ = smoothing function of covariate x_j
x_j = predictor variable to the *j*
$\boldsymbol{Z}$ = tranpose of random effects matrix, start the observation to the *i*
$\boldsymbol{b}$ = vector of random effects coeficients

Generalized Additive Mixed Model (GAMM) is an extension of the GAM and GLMM methods, only replacing linear functions into additive functions. The GAMM model is able to identify and explain the effect of random components more effectively and efficiently in a model. The GAMM method includes random effects and fixed effects in the model [5]. Random effects are randomized individual-experimental retrieval of a unit from the population, while the fixed effect (fixed effect) are all related parameters taken from the population in an experiment. A model that is influenced by a fixed effect and a random effect is called a mixed effect [6].

2.3. Link function
[7] The link function g (·) is a function that connects a linear predictor η with the expectation value of the response y, which is μ. The link function or the inverse contact function is determined by the variable category of the response variable as seen in the table 1.

Table 1. Connect function of several distributions

Response Variable	Distribution	Link function	Reverse connection function
Continuous	Normal	Identity	η
	Gamma	Invers	$1/\eta$
		Log	e^{η}
	Invers Gaussian	$\frac{1}{\mu^2}$	$-\frac{1}{\theta^2}$
Proportion	Binomial	Logit	$e^{\eta} = (1 + e^{\eta})$

whole	Poisson	Probit	$\phi(\eta)$
		Logit	e^{η}
Biner	Bernoulli	Probit	$e^{\eta} = (1 + e^{\eta})$

2.4. GAMM parameter estimator

The estimation is the estimated value of the population parameter based on the available data, to estimate the parameters of GAMM, the first step is to describe the probability density function of the exponential family as the response variable as follows:

$$(y;\theta) = exp[a(y)b(\theta) + c(\theta) + d(y)] \tag{4}$$

GAMM uses the maximum likelihood estimator (MLE) approach to obtain the β and estimator parameters b. The estimated value obtained in GAMM is obtained by maximizing the log-likelihood function [6].

2.5. Significant Parameter Test

One test that can be used to test the significance of the β coefficient in the model is a partial test. Partial test (T-test) is used to determine the effect of an individual predictor variable on the response variable. The hypothesis used is

H_0 : $\beta_j = 0$

The j predictor variable does not significantly influence the response variable

H_1 : $\beta_j \neq 0$

The j variable predictor has a significant effect on the response variable

Test statistics used for partial testis:

$$t_{calculate} = \frac{\beta_n}{SE\ (\beta_n)} \tag{5}$$

2.6. Model Selection

2.6.1. Akaike's Information Criterion (AIC)

[8] Choosing the goodness of a model used a criterion in measuring it. One type of criteria used is AIC.

$$AIC = -2LL + (2 \times K) \tag{6}$$

where, LL is the Likelihood log model and K is the number of parameters estimated in the model. The model with the smallest AIC value is the chosen model.

2.6.2. Schwarz's Bayesian Information Criterion (BIC)

BIC is an asymptotic result derived from the assumption that the distribution of data in exponential families. Formula in calculating the BIC value is

$$-2 \cdot \ln p(x|k) \approx BIC = -2 \cdot \ln L + k \ln(n) \tag{7}$$

to determining the optimal model, the model chosen is the model that has the smallest BIC value. This is because BIC improves the function of k. That is, variations that cannot be explained in the response variable and the number of predictor variables increase the BIC value. Therefore, small BIC values explain that fewer predictor variables, more appropriate or both [9].

3. Research methodology

The type of data used in this study is secondary data obtained through the National Oceanic and Atmospheric Administration website [10] at points 4°N90°E and 1.5°N90°E. Data retrieval is through the points 4°N90°E and 1.5°N90°E because this point is closest to Aceh Province. The data used is daily unit in periods from September 17, 2006 to June 14, 2017.

4. Results and Discussion

In this study, before carrying out further analysis, a descriptive statistical analysis was first carried out to obtain an overview of the data as shown in the Table 3 and Table 4.

Table 2. *Summary statistic* data point 4°N90°E (location A)

Variabel	Min	Kuartil 1	Median	Mean	Kuartil III	Max	Range
Y	28,57	29,05	29,27	29,39	29,57	30,99	2,42
X_1	0,40	3,50	5,10	4,98	6,52	11,00	10,60
X_2	26,75	28,17	28,62	28,57	29,01	29,94	3,19
X_3	104,00	116,50	121,30	121,70	125,80	140,80	36,80
X_4	35,08	35,24	35,32	35,32	35,38	35,58	0,50
X_5	0,00	0,00	0,02	0,32	0,18	9,99	13,16
X_6	73,10	78,70	80,20	80,22	81,92	87,70	14,60
X_7	19,75	20,57	21,02	20,98	21,23	21,63	1,88
X_8	24,04	176,23	236,04	214,97	271,05	315,18	291,14
X_9	3,20	3,34	5,10	4,98	6,52	11,00	7,80

Table 3. *Summary statistic* data point1.5°N90°E (locationB)

Variabel	Min	Kuartil 1	Median	Mean	Kuartil III	Max	Range
Y	28,06	28,78	29,03	29,14	29,41	30,60	2,54
X_1	0,60	2,90	4,70	4,59	6,37	10,30	9,70
X_2	26,52	27,91	28,23	28,28	28,75	30,22	3,70
X_3	9,99	115,19	122,51	119,75	127,70	139,37	129,38
X_4	-9,99	35,27	35,36	34,26	35,43	35,59	25,60
X_5	0,00	0,01	0,035	0,453	0,44	9,99	9,99
X_6	73,80	79,62	81,40	81,68	84,05	89,20	15,40
X_7	13,04	161,40	234,71	222,06	273,66	999,99	32,05
X_8	9,99	21,02	21,35	21,18	21,18	21,65	11,66
X_9	33,05	33,66	34,14	34,07	34,40	35,23	2,18

Based on Table 3 for data in location A and Table 4 for the data in location B above, it is known that the minimum, quartile I, median, mean, quartile III, maximum and data range for each variable.

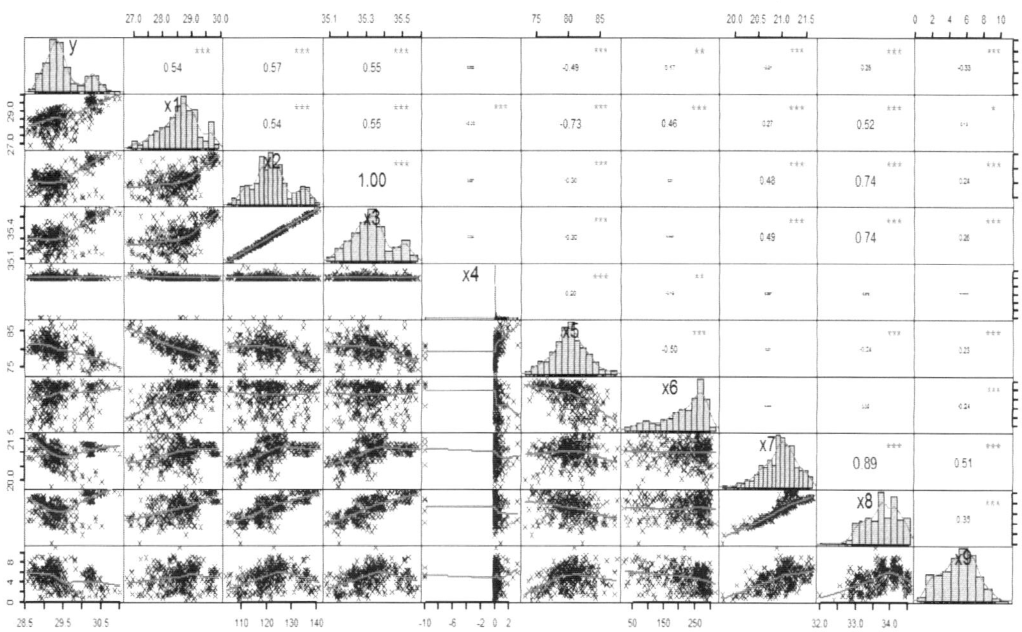

Figure 1. Correlation matrix of SST dataset

The correlation matrix provides information about relationship patterns, forms of distribution, and correlations between both dependent and independent variables and between independent variables. Figure 1 shows that of the nine variables, only one variable had a symmetrical histogram, namely the X_5 (relative humidity). Y (SST), X_2 (dynamic height) and X_3 (heat content) have a bimodal histrogram. For variables X_1 (air temperature), X_6 (wind speed), and X_7 (short wave radiation) have the shape of a histogram sticking to the left. While the variables X_4 (rainfall), X_8 (sea level density), and X_9 (sea surface salinity) are not symmetrical because they have different data that goes up or down.

Based on Figure 1 also known the pattern of the relationship between the response variable (Y) with all predictor variables (X) based on the distribution of data. There are several variables that have a formless distribution pattern so that smoothing is done in the modeling. Then there is one independent variable that has a perfect correlation (strong correlation) that is the correlation between variables X_2 and X_3 with a correlation value of 1. Besides that there are also several variables that are linearly related and have a strong correlation, variables X_7 and X_8, X_3 and X_8, X_2 and X_8 with a correlation value of 0.89; 0.74, and 0.74. In addition to seeing the pattern of relationships and distribution between response variables and explanatory variables, other requirements of using GAMM method are the response variables included in the exponential family. Some distributions included in the exponential family are Normal, Gamma, Inverse Gaussian, Exponential and so on. Here are visualizations and comparison tables of several distributions of data.

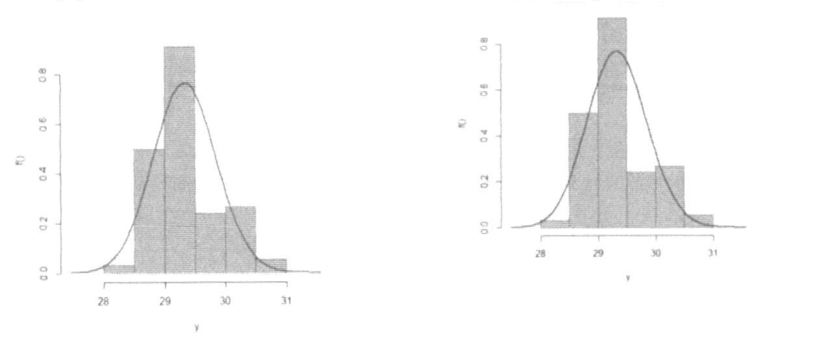

(a). Normal Distribution (b). Eksponensial Distribution

(c). Gamma Distribution (d). *Inverse Gaussian* Distribution

Figure 2. Visualization for the selection of data distribution

Table 4. Comparison of distribution

Distribution	AIC
a. Normal	629,852

b. Eksponensial	3557,64
c. Gamma	626,256
d. *Inverse gaussian*	**624,511**

Based on Figure 2, it can be seen that the distributions (a), (c), and (d) have similar forms of distribution visualization. So that, it is difficult to determine the appropriate distribution. Next in Table 5. can be seen the comparison of data distribution based on the AIC value. From Table 5 it is known that the distribution that matches the data is the Gaussian Inverse distribution because it has the smallest AIC value. Then, based on table 1. for the inverse gaussian distribution, the link function that used is $\frac{1}{\mu^2}$.

4.1. Modeling using GAMM

The response variable of sea surface temperature (Y) is modeled with nine predictor variables namely wind speed (X_1), air temperature (X_2), dynamic height (X_3), heat content (X_4), rainfall (X_5), relative humidity (X_6), short wave radiation (X_7), sea surface density (X_8), and sea surface salinity (X_9). There are several alternative models for modeling sea surface temperature data. Modeling is done by adding locations as random influences, location interactions with each predictor variable, and interactions between predictor variables that have the greatest correlation value. Then the best model is chosen which has the lowest AIC, BIC, and log-likelihood values.Based on the data analysis conducted, 3 GAMM models were obtained as follows.

a. The 1[st]GAMM model is a model with random influence that interacts with all predictor variables with the addition of interactions between variables X_2 and X_3, X_1 and X_5, X_2 and X_8, X_3 and X_8, and X_7 and X_8. Interaction between predictor variables is determined through the correlation value obtained based on Figure 1.

$$g(\mu) = 1{,}160 \times 10^{-3} + f_1(x_1) + f_2(x_2) + f_3(x_3) + f_4(x_4) + f_5(x_5) + f_6(x_6) + f_7(x_7) + f_8(x_8) + f_9(x_9) + x_2x_3 + x_1x_5 + x_2x_8 + x_3x_8 + x_7x_8 + (5.808 \times 10^{-7}) \quad (8)$$

b. The 2[nd]GAMM model is a model with random influence that interacts with all predictorvariables.

$$g(\mu) = 1{,}160 \times 10^{-3} + f_1(x_1) + f_2(x_2) + f_3(x_3) + f_4(x_4) + f_5(x_5) + f_6(x_6) + f_7(x_7) + f_8(x_8) + f_9(x_9) + (1{,}57 \times 10^{-5}) \quad (9)$$

c. The 3[rd]GAMM model is a model with a random influence and interaction between predictor variables are correlated, between variables X_2 and X_3, X_1 and X_5, X_2 and X_8, X_3 and X_8 and X_7 and X_8.

$$g(\mu) = 1{,}163 \times 10^{-3} + f_1(x_1) + f_2(x_2) + f_3(x_3) + f_4(x_4) + f_5(x_5) + f_6(x_6) + f_7(x_7) + f_8(x_8) + f_9(x_9) + x_2x_3 + 2{,}886x_1x_5 + x_2x_8 + x_3x_8 + x_7x_8 + (-7{,}633 \times 10^{-7}) \quad (10)$$

4.2. Optimal Model Selection and Interpretation

Based on the results in Table 6. it is known that the alternative GAMM model that has the AIC, BIC, and the smallest log-likelihood values is the 3[rd]GAMM model with a random influence and the interaction between predictor variables that are strongly correlated between the variables X_2 and X_3, X_1 and X_5, X_2 and X_8, X_3 and X_8, and X_7 and X_8. Based on Table 6. it is known that the model has the AIC, BIC and the smallest log-likelihood is 3rd GAMM model and then the 3rd GAMM model is the best model.The AIC, BIC, and log-likelihood values obtained for 3rd GAMM model are -981,0602; -873,853 and 517,2894.

Table 5. Comparison of models based on AIC, BIC, and Log-likelihood values

IOP Conf. Series: Earth and Environmental Science **273** (2019) 012025 doi:10.1088/1755-1315/273/1/012025

model	AIC	BIC	*Log-likelihood*
1st GAMM	-1652,839	-1357,575	900,1183
2nd GAMM	-1464,454	-1208,145	796,2026
3rd GAMM	**-981,0602**	**-873,853**	**517,2894**

That is known that there are some variables that are not significant, so that the parameter significance test is determined and determine the best model using the backward elimination method.

4.3. Parameter Significance Test

Parameter significance test is conducted to determine whether the predictor variable has a significant effect on the response variable. Here are the hypotheses used.

H_0 : The i-predictor variable has no significant effect on response variable

H_1 : The i-predictorvariable has a significant effect on the response variable

The rejection area used is reject H_0 if the p-value is <alpha (0.05).

Table 6. Parameter significance test for 3rd GAMM model

Parameter	Estimated value	*P-value*
Intersep	1,160 x10^{-3}	<2e-16***
Location B	-7,366 x10^{-7}	0,371
s(X_1)	4,11 x10^{-10}	0,000407***
s(X_2)	5,825 x10^{-12}	8,83x 10^{-5}**
s(X_3)	-1,655 x10^{-10}	7,04 x 10^{-6}***
s(X_4)	5,423 x10^{-12}	0,031796 ·
s(X_8)	8,619 x10^{-4}	<2e-16***
s(X_9)	-8,657 x10^{-11}	<2e-16***
ti(X_2 X_3)	2,354 x10^{-6}	0,030021*
ti(X_1 X_5)	2,691 x10^{-8}	0,057082·
ti(X_2 X_8)	-4,302 x10^{-6}	0,054772·
ti(X_3 X_8)	-2,683 x10^{-4}	0,024302*

Description: significant at α 0,1% (***), 1% (**), 5% (*) and 10% (·)

Based on the results in Table 7. all significant parameters were obtained with different levels of significance. There are 5 predictor variables and 2 significant predictor variable interactions on α 5%, respectively, variables X_1, X_2, X_3, X_8 and X_9 and X_2X_3 and X_3X_8. Then obtained 1 predictor variable and 2 significant predictor variable interactions on α 10%, respectively, X_4 variables and X_1X_5 and X_2X_8 interactions. The following are the AIC, BIC, and log-likelihood values obtained for the significant 3rd GAMM model, which are -986.0211; -894,911 and 515,7519. Here are the best models obtained.

$$g(\mu) = 1{,}163 \times 10^{-3} + f_1(x_1) + f_2(x_2) + f_3(x_3) + f_4(x_4) + f_8(x_8) + f_9(x_9) + x_2x_3 + \\ + x_3x_8 + 2{,}918x_1x_5 + x_2x_8 + (-7{,}633 \times 10^{-7})$$

5. Conclusion

Based on data analysis and discussion can be concluded that:

1. The best model of sea surface temperature data using the generalized additive mixed models approach is the3rd GAMM model. The 3rdGAMM model has the smallest AIC, BIC, and log-likelihood values with values of -986.0211; -894,911 and 515,7519. Following is the construction of the 3rd GAMM model:

$$g(\mu) = 1{,}163 \times 10^{-3} + f_1(x_1) + f_2(x_2) + f_3(x_3) + f_4(x_4) + f_8(x_8) + f_9(x_9) + x_2x_3 + \\ + x_3x_8 + 2{,}918x_1x_5 + x_2x_8 + (-7{,}633 \times 10^{-7})$$

2. Factors that significantly influence (α = 0.5%) on sea surface temperature (Y) are wind speed (X_1), air temperature (X_2), dynamic height (X_3), sea surface density (X_8), and sea surface salinity (X_9). Then there are several interacting predictor variables that affect sea surface temperature including interactions between air temperature with dynamic height and dynamic height with density of sea surface temperature.

References

[1] Badan Meteorologi, Klimatlogi, dan Geofisika 2018 *Iklim.* www.bmkg.go.id Accessed on February 20, 2018 (Indonesia)
[2] Serambi Indonesia 2017 Bencana Perubahan Iklim www.acehtribunnews.com Accessed on April 24, 2018
[3] Ananda S, Ariska R, Humaira R.A, and Miftahuddin 2017 Pemodelan Suhu Permukaan Laut Menggunakan *Generalized Additive models* (GAM) Dalam Efek Waktu. *Seminar Nasional dan Rapat Tahunan (SEMIRATA)* (Indonesia)
[4] Hastie, T and Tibshirani, R 1990 *Generalized Additive Models, Statistical Science* Chapman and Hall Washington USA
[5] Wood S.N 2017 *Generalized Additive Models An Introduction with R* CRC Press, United Kingdom
[6] Jamilatuzzahro, Caraka, R.E, and Herliansyah, R 2018 *Aplikasi Generalized Linear Model pada R.* Innosain Yogyakarta
[7] Codd C.L 2014 A review and comparison of models and estimation methods for multivariate longitudinal data of mixed scale type *Disertasion* The Ohio State University USA
[8] Shadish W.R, Zuur A.F, and Sullivan K.J 2014 Using Generalized Additive (Mixed) Models to Analyze Single Case Designs *Journal of School Psychology* **52**, 149-178
[9] Dziak, John J, Coffman, Donna L, Lanza, Stephanie T, and Li, R 2017 Sensitivity and Specificity of Informaion Criteria *Peerj Preprints*
[10] National Oceanic and Atmospheric Administration 2017 *Global Tropical Moored Buoy Array Pacific Marine Environmental Laboratory.*www.noaa.gov Accessed on February 11, 2018

IOP Conf. Series: Earth and Environmental Science **273** (2019) 012026 doi:10.1088/1755-1315/273/1/012026

Geohazard Identification Using 2-D Resistivity Imaging For Seulawah Agam Volcano, Indonesia

Muhammad Syukri[1], Rosli Saad[2], Marwan[3], Sabrian Tri Anda[3] . R. Safitri[1], Zul Fadhli[3]

[1]Department of Physics, Faculty of Sciences, Syiah Kuala University, Darussalam, Banda Aceh, 23111, Indonesia.
[2]Geophysics Section, School of Physics, University Sains Malaysia, Penang 11800, Malaysia.
[3]Department of Geophysics Engineering, Faculty of Engineering, Syiah Kuala University, Darussalam, Banda Aceh 23111, Indonesia.

Corresponding author: m.syukri@unsyiah.ac.id

Abstract. Common parameter used for delineating geothermal system is resistivity. Hydrothermal alteration of primary minerals depends on temperature distribution and forms various clay minerals with conductivity properties. Clay minerals formed at high temperatures are more resistant than lower temperature. Most geothermal systems within active volcanic regions show low resistivity (1-5 Ohm.m) surrounding an inner core of higher resistivity. 2-D resistivity study conducted at Seulawah Agam active volcano and its vicinity (Aceh) with geothermal sources (Herzt, Ie Jue and Ie Seu'um) flowing from south-east to north-west. Two lines, L1 and L2 were conducted crossing the suspected geothermal flow using ABEM SAS4000 system with 10 m minimum electrode spacing and Pole-Dipole array. Resistivity results indicated two main zones; (i) top layer, <100 m depth with resistivity value of >60 Ohm.m and (ii) cap-rock, 30-100 m depth with resistivity value of 6-50 Ohm.m. Few fractures and low resistivity regions (<5 Ohm.m) were identified in the cap- rock. Hot water and hot mud have low resistivity value; hence the low resistivity regions were interpreted as geothermal flow and it is convincingly suspected to be a geohazard zone. New extension of Seulawah Agam geothermal flow was identified by 2-D resistivity method.

1. Introduction

Resistivity is a common parameter used for delineating geothermal system since it is temperature dependent. Sounding methods (Schlumberger and TEM) are mostly used for shallow exploration, <1 km [1]. In volcanic regions, hydrothermal alteration of primary minerals depends on temperature distribution determined by heat convection (hydrothermal water) and forms various clay minerals (Figure 1). These clay minerals have varying electrical conductivity properties, thus becoming a significant factor in the search for geothermal heat source [2].

IOP Conf. Series: Earth and Environmental Science **273** (2019) 012026 doi:10.1088/1755-1315/273/1/012026

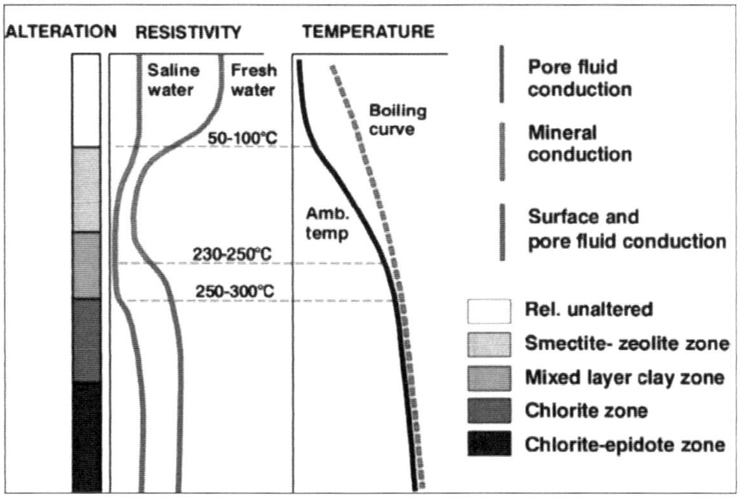

Figure 1. Electrical conduction vs. temperature and hydrothermal alteration [2].

Temperature and pressure change an electrical resistivity, physical properties and other physical parameters of rocks due to changes in chemical reactions. Mineral alteration is a chemical reaction that depends on mineral content, pressure and temperature of hydrothermal fluid. Archie's law is a formula that approximates the resistivity of rocks/formations:

$$r = r_w a f^{-m}$$

where:

r = Bulk resistivity (measured)
r_w = Resistivity of pore fluid
f = Porosity of total volume of rocks/formation
a and m = Empirical parameter (~1) and cementing factor (~2), respectively.

Rocks composed of minerals including water; carry mineral/chemical, physical and electrical properties. Electrical currents transfer in rocks as conduction, electrolytic and surface conductivity. Ionic conduction occurring in rocks contains free ionic exchange as rocks are filled with fluids in pore spaces. Rocks behave as semi-conductors if they contain minerals, which are electrically semi-conducting. Geothermal field is usually associated with volcanic or intrusive sources. The electrical conductivity is associated with electrolytic and surface electrical conduction of thermal water and the presence of clay minerals. Clay minerals formed at high temperatures are more resistant than clay minerals formed as alteration of rocks at a lower temperature. These properties are significant in the interpretation of electrical methods used for exploration of geothermal resources [3].

Most geothermal systems within the active volcanic regions, show common resistivity structure with low resistivity of 1-5 Ohm.m surrounding an inner core of higher resistivity. This increasing resistivity with depth is associated with a change in the conduction mechanism, from interface conduction to electrolyte conduction due to a change in alteration minerals at about 240°C [4]. Resistivity for geothermal area is substantially different and generally lower than the resistivity in areas with colder subsurface temperature [5]. Water penetration in Earth cracks produces a reduction of the bulk resistivity, and makes the fault easy to detect. Figure 2 shows a conceptual model of geothermal system [6].

The purpose of this paper is discussed the resistivity change towards temperature using 2-D resistivity imaging at geothermal fields of an active volcanic region (Seulawah Agam, Aceh, Indonesia), and to relate it and suspect it is a geohazard area, both due to volcanoes and faults in the study area. Most geothermal systems within the active volcanic regions, show common resistivity structure with low resistivity of 1-5 Ohm.m surrounding an inner core of higher resistivity. This increasing resistivity with depth is associated with a change in the conduction mechanism, from interface conduction to electrolyte conduction due to a change in alteration minerals at about 240°C [4]. Resistivity for geothermal area is substantially different and generally lower than the resistivity in areas with colder subsurface temperature [5]. Water penetration in Earth cracks produces a reduction of the bulk resistivity, and makes the fault easy to detect.

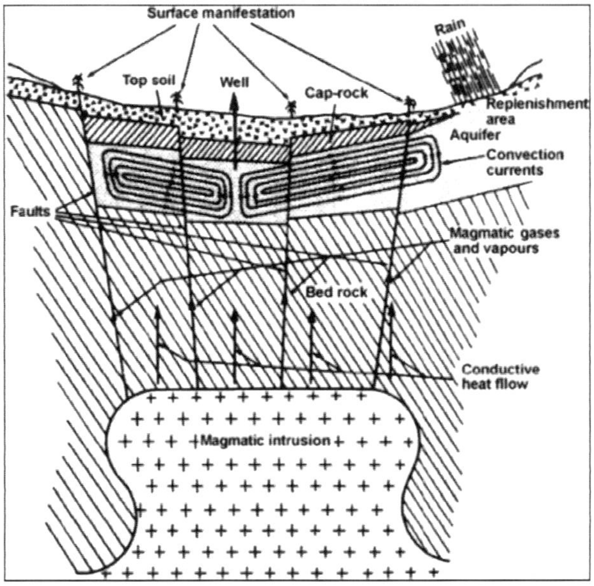

Figure 2. Conceptual model of a geothermal field [6].

2. Geology Settings

Seulawah Agam volcano, one of three active stratovolcanoes in the Aceh province, is located at the northwest end of Sumatra (Figure 2). Seulawah Agam has two craters, van Heutsz (Heszt), the most active crater at an elevation of 714 m on the north side, and Simpago on the south side. The lithology of the study area is dominated by Lam Teuba volcanic composed of andesitic to dacitic volcanic, pumiceous breccia, tuffaceous, calcareous sometimes cross-bedded sandstones, conglomerates, agglomerate, minor mudstones and ash flows which intruded of the Seulimum formation. The Seulimeum formation is composed of tuffaceous and calcareous sandstones, conglomerates and minor mudstones [7]. The geological formation formed a topographic depression, occupied with alluvial flat and low flat-topped hills within Barisan Range. The adjacent Sumatra in the west coast constitutes volcanic arc with many Quaternary volcanoes such as Pulau Weh and Seulawah Agam.

The volcanic belt is extended along the tectonically weak SFS and this tectonic weakness is supposed to have triggered volcanism [8]. In the northern most Sumatra, the Sumatran Fault System (SFS) splits into two major dextral strike-slip faults which are the Seulimeum and the Aceh Fault [7]. Sumatran Fault system occurs from the Middle Miocene and the opening of the Andaman Sea. Pre-Tertiary basements rocks outcrop mainly along the central spine of the Barisan Mountains, which extend the length of the island parallel to the southwest cost.

The area from northeast and southwest is overlain by Tertiary sedimentary and volcanic rocks. From upstream of the Jantho region to downstream of Indrapuri, the Pleistocene coarse-grained partly volcanic sands and gravels form a prominent terrace surface on both sides of the Krueng Aceh. Downstream of Indrapuri, the alluvial deposits can be subdivided into a shallow aquifer system and a deep aquifer system. Upstream of Indrapuri, the alluvial sandy-gravelly deposits in the vicinity of the river courses, the older terrace sand-gravel deposits and the semi-consolidated sandstones are assumed to constitute the main aquifers of the upper part of the Krueng Aceh valley [9].

IOP Conf. Series: Earth and Environmental Science **273** (2019) 012026 doi:10.1088/1755-1315/273/1/012026

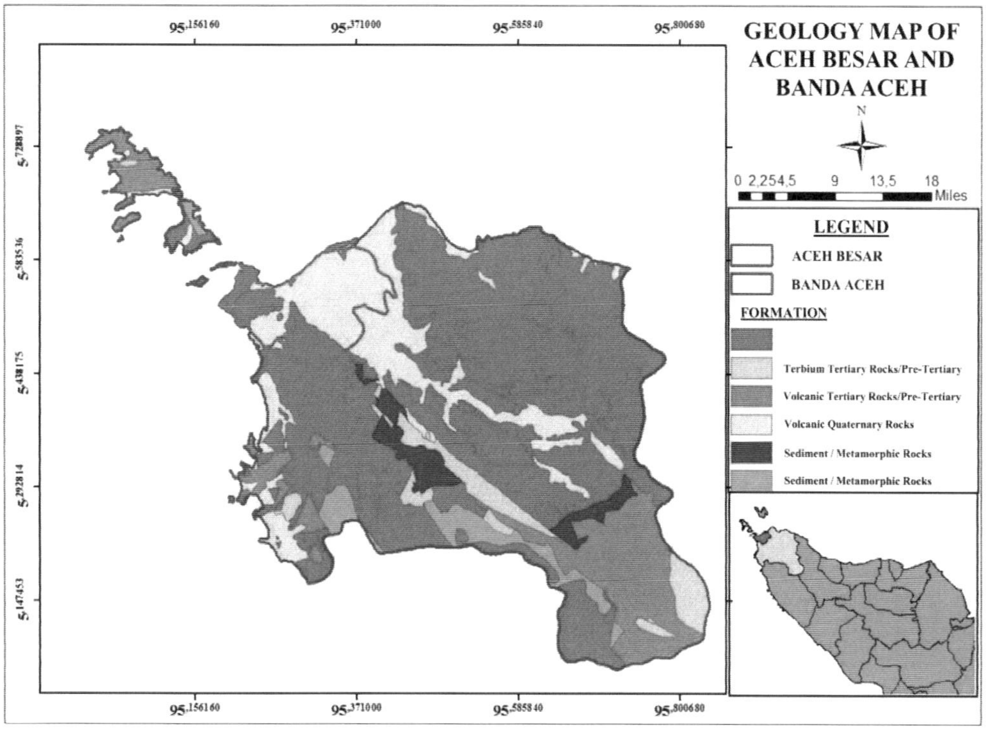

Figure 3. Map of geology of the study area [12]

3. Study Area and Methods

2-D resistivity study was conducted at Seulawah Agam, Aceh, Indonesia and its vicinity. Seulawah Agam is an active volcano with geothermal sources such as Herzt, Ie Jue and Ie Seu'um. Flow of the geothermal is suspected from south-east to north-west from Seulawah Agam. Two survey lines, L1 and L2 were conducted crossing the suspected geothermal flow using ABEM SAS4000 system with 10 m minimum electrode spacing and Pole-Dipole array (Figure 4).

The data acquisition used roll along techniques which make a total length of the survey line is 1200 m. The data processed involve standard processing, standard processing with mathematical data extrapolation and modelled using RES2Dinv software. The data were then outputted into Surfer software for gridding, contouring and final presentation.

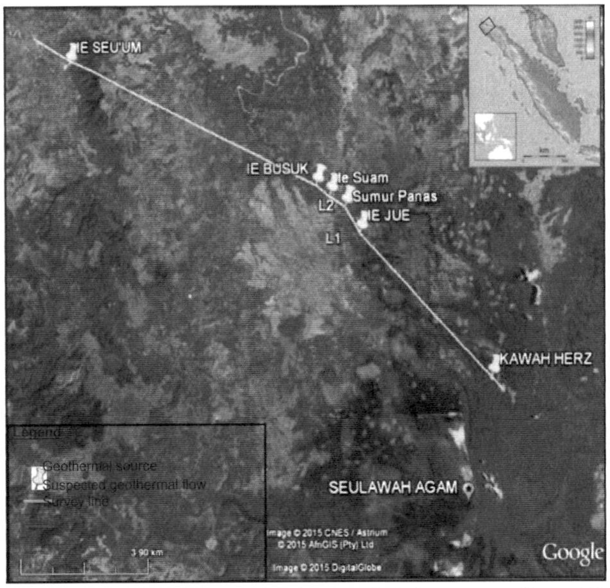

Figure 4. Study area and its vicinity with 2-D resistivity survey lines (Google, 2015).

The 2-D electrical resistivity apply base on an injected electrical current (I) through two metallic electrodes and measuring the potential difference (V) between two other electrodes. The apparent resistivity (r_a) is given by [10]:

$$\rho_\alpha = k\frac{\Delta V}{I}$$

where k is geometric factor.

The measurements provide information about apparent resistivity for a medium whose volume is proportional to the electrode spacing. The data point corresponding to the investigated volume is conventionally represented on a section at a depth equals to the electrode spacing. The apparent resistivity is conventional: it consists of a pseudosection with x in abscissa and z in ordinate (not a real depth). Software such as Res2DInv [11] used to process the data for real resistivity inversion model interpretation.

4. Result and Discussions

2-D resistivity study was conducted at Seulawah Agam, Aceh, Indonesia and its vicinity. Seulawah Agam is an active volcano with geothermal sources such as Herzt, Ie Jue and Ie Seu'um. Flow of the geothermal is suspected from south-east to north-west from Seulawah Agam. Two survey lines, L1 and L2 were conducted crossing the suspected geothermal flow and faults area. Figure 5 and 6 show 2-D resistivity pseudosection of line L1 and L2. Generally, resistivity values of the study area is 0-500 Ohm.m and covered depth of up to 320 m from ground surface. The area is divided into two main zones;
(i) top layer with thickness up to 100 m and resistivity value of >60 Ohm.m and (ii) cap-rock with depth of 30-100 m and resistivity value of 6-50 Ohm.m. Few fractures and low resistivity regions (<5 Ohm.m) were identified in the cap-rock. Hot water and hot mud have low resistivity value; hence the low resistivity regions were interpreted as geothermal flow and it is convincingly suspected to be a geohazard zone. This shows that 2-D resistivity method can be used to identify geothermal flow and faults area as an identity of geohazard area, both volcanic and tectonic hazard. Figure 7 shows new extension of Seulawah Agam geothermal flow identified by 2-D resistivity method.

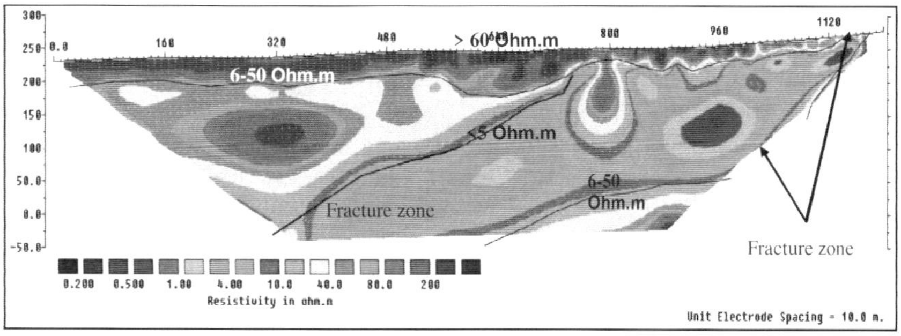

Figure 5. Resistivity pseudosection of line L1.

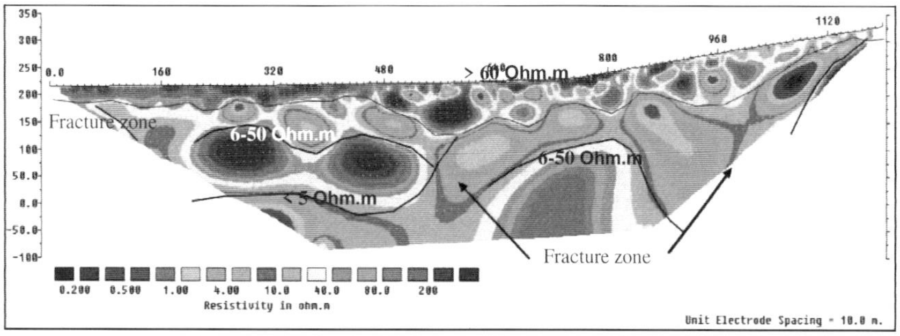

Figure 6. Resistivity pseudosection of line L2.

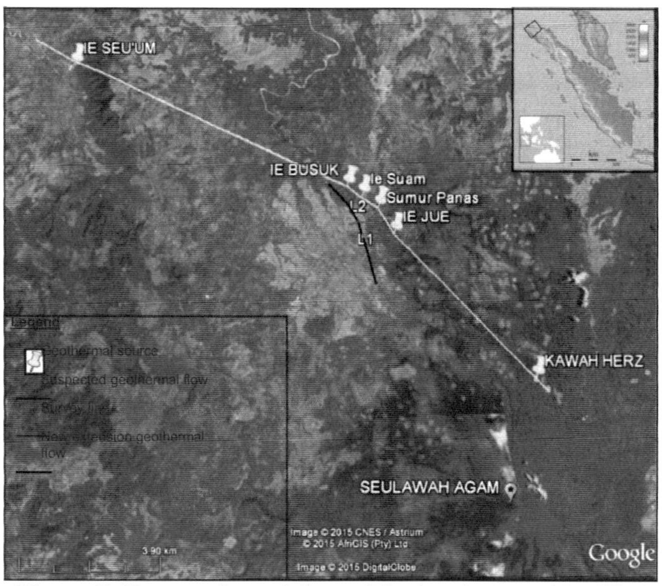

Figure 7. New geothermal flow of Seulawah Agam identified by 2-D resistivity method

IOP Conf. Series: Earth and Environmental Science **273** (2019) 012026 doi:10.1088/1755-1315/273/1/012026

5. Conclusion

Generally the fractures and hydrothermal outflow regions were identified in the cap-rock, and it was detected at depth 50-120 m with resistivity value of <5 Ohm.m, and its identify as geohazard area, both volcanic and tectonic hazard. The fault was identified at distance 300-800 m, and it is suspected that the fault is >150 m deep. This shows that 2-D resistivity method can be used to identify the potential geohazard area. The resistivity profiles indicates the presence of several fracture in some geothermal spots at the study area. These results are also shown from the characteristics of the resulting resistivity values, which characterize the presence of fractures and hydrothermal outflow in the study sites. This can also be an implication of the presence of volcanic hazard in the region. The interpreted resistivity and geothermal data of the Seulawah Agam area have revealed new information and improved the knowledge about the internal structure beneath this area. Detail ground geophysical study such as 2-D resistivity, seismic, gravity is recommended for environmental and safety purposes.

References

[1] Chiragwile, S.A. (2009) Interpretation of resistivity sounding in the Krysuvik high-temperature geothermal area, SW-Iceland, using joint inversion of TEM and MT data: Geological Survey of Tanzania (GST), p.85-123.

[2] Flóvenz, Ó.G. Spangenberg, E. Kulenkampff, J. Árnason, K. Karlsdóttir, R. and Huenges, E. (2005) The role of electrical interface conduction in geothermal exploration: Proceedings of the World Geothermal Congress 2005, Antalya Turkey, CD, p.1-9.

[3] Wright, H.M.N. Cashman, K.V. Gottesfeld, E.H. and Roberts, J.J. (2009) Pore structure of volcanic crusts: Measurements of permeability and electrical conductivity: Earth and Planetary Science Letters, 280, p.93-104.

[4] Hjalmar, E. Knutur, A. and Olafur, G.F. (1994) Resistivity methods in geothermal prospecting in Iceland: Surveys in Geophysics 15: 263-275.

[5] Oskooi, B. Pedersen, L.B. Smirnov, M. Árnason, K. Eysteinsson, H. Manzella, A. and DGP Working Group. (2005) The deep geothermal structure of the Mid-Atlantic Ridge deduced from MT data in SW Iceland: Phys. Earth Planet. Inter., 150, 183-195.

[6] Berktold, A. (1983) Electromagnetic studies in geothermal regions: Geophys. Surv., 6, p.173-200.

[7] Bennet, J. D., Bridge, D. Mc., Cameron, N. R., Djunuddin, A., Ghazali, S. A., Jeffery, D. H., Kartawa,W., Keats,W., Rock, N. M. S., Thomson, S. J. and Whandoyo, R. 1981, Peta Geologi Lembar Banda Aceh, Sumatra (1:250.000). Geologic Map of the Banda Aceh Quadrangle, Sumatra, Lembar (Quadrangle) 0421 (Banda Aceh), pp. 19, Bandung.

[8] Gasparon, M. 2005, Quaternary volcanicity. (In: Barber, A. J. et al. (Eds.): Sumatra, Geology, Resources and Tectonic Evolution), pp. 120 – 130, London.

[9] Ploethner D.,and Siemon B., 2006, Hydrogeological reconnaissance survey in the province Nanggroe Aceh Darussalam, Northern Sumatra, Indonesia – Survey area Banda Aceh / Aceh Besar 2005, Helicopter Project Aceh (HELP ACEH). BGR Report Vol. C-1, Archives No. 0126196 C-1, Hanover.

[10] Dahlin, T. (2001) The development of DC resistivity imaging techniques, Computers & Geosciences 27, p. 1019–1029.

[11] Loke and Barker (1996), Tutorial : 2-D and 3-D electrical imaging surveys.

[12] Syukri, M, Marwan, Safitri, R., Fadhli, Z., Andika, F., and Saad, R., 2017 Groundmagnetic survey used to identify the weathered zone, in Blang Bintang, Aceh, Indonesia, AIWEST-DR 2016, IOP Publishing, IOP Conf. Series: Earth and Environmental Science 56 (2017) 012017.

Acknowledgments

The authors would like to thank the Ministry of Education and Culture, Indonesia for its financial support in the scheme of Decentralized Research Competitive Grants Program 2017, Universiti Sains Malaysia, Penang (Malaysia) Geophysics team, students and staffs of Faculty of Sciences and Faculty of Engineering, Syiah Kuala University (UNSYIAH), Banda Aceh (Indonesia).

IOP Conf. Series: Earth and Environmental Science **273** (2019) 012027 doi:10.1088/1755-1315/273/1/012027

The Correlation Between P-Wave Velocity Value With Pathways Of Geothermal Manifestation In The Area Ie Jue, Aceh Besar As Hazardous Side Study

Asrillah[1,*], D Wahyuni[1], and Umar Muksin[1,2]

[1]Geophysical Engineering Departement, Engineering Faculty, Syiah Kuala University, Darussalam Banda Aceh, 23111, Indonesia, * email: asrillah@unsyiah.ac.id

[2]Department of Physics, Syiah Kuala University, Darussalam Banda Aceh, 23111, Indonesia.

Abstract. Refraction seismic survey has been carried out in the manifestation area of Ie Jue, Seulawah Agam Volcano Region, Seulimuem, Aceh Besar Regency. This research aims to identify weak zones related to channel of geothermal outflow that would bring about a potential hazard for human activity and that goal is obtained by exstracting information from P-wave velocity. The data acquisition was conducted for two seismic lines located near to the geothermal manifestation area of Ie Jue. Both of the seismic lines were set up by linking the 24 geophones to the Seismograph PASI 16S and placing them in three meters spacing of each other. The offsets either left or right were laid out at 35 meters from first and last geophones, so the total spread length of each seismic line is 139 meters. The data recorded by the tool then were processed by using ZONDST2D and finally, the 2D velocity profile was obtained. The results show that the 2D velocity profiles illuminate us to potray and interpret the features figured out. The most valuable information provided by the profile is the existance of such the weak zones of the first and second seismic lines at relatively near-surface depth. These weak zones were suspected as the fractured or low-velocity zones, where their velocity ranges approximately at 0.2 - 1.2 km/sec. These low-velocity layers are strongly predicted as the pathways or channels of the geothermal fluids that appears to the surface. The shallow weak zones are potentially to become the hazard grounds because they are regularly traversed by timber trucks from which the dynamic load will be pushing them down. As consequence, the ground surfaces could gradually become a land subsidence endangering human activity within the vicinity of the area.

1. Background

Aceh is one of the provinces in Indonesia that has geothermal potential which can be explored and developed for alternative energy. One of them is in Aceh Besar Regency which is in the Seulawah Agam volcano. The geothermal manifestation has two important sides. The first side is as an advantageous view, which indicate a potential resource of geothermal energy. However, the second view also has a potential hazard which are channels or pathways of geothermal fluid flow. This

IOP Conf. Series: Earth and Environmental Science **273** (2019) 012027 doi:10.1088/1755-1315/273/1/012027

research focuses on the disadvantageous side. To intensify this concern, refraction seismic method was considered as one of the geophysical methods that can be used to find out more about the shallow structures that are associated with the geothermal manifestation such as outflow, fumaroles etc. The refraction seismic method utilizes the seismic wave propagation from which the subsurface model velocity value can be generated based on the processed seismic data. The value of subsurface model velocity distribution toward depth is then used as the result to interpret the subsurface information of a research interest.

This study focuses on identifying weak zones where the settlements use the manifestation path as their traffic to support their livelihood and those zones could potentially cause the hazard for them. The weak zones are a subsurface layer that is soft and not compact because of heating by manifestation fluid. The weak zones have the characteristics of smaller density and larger porosity values than stronger zones, so that this gives the possibility of potential hazard to be occur such as subsidence, considering that the area of Ie Jue itself is an active area traversed by timber trucks. The dynamic and regular pressure affected by timber trucks on the surface could cause the surface going down and it gradually can raise the occurrence of subsidence in the area, so that this is a research background was conducted.

2. Study Literature

2.1 Geological setting of the study area

This research area is located in geothermal manifestation area of Ie Jue, Lamteuba Village, Seulimum Sub-district, Aceh Besar Regency. Geothermal manifestation of Ie Jue is situated in Seulawah Agam Volcano area. Based on geological map in Figure 1, informed that the areas of Seulawah Agam Volcano are dominated by rocks of lava member (Qvlt) and volcanic rock (QTvt) and consist of andesite volcanic rock to dacite, breccia, pumice, tuff, agglomerate, and ash flow [1]. The Seulawah Agam volcano has some near subsurface manifestations, for example Ceumpaga crater, located in the Southside, Heutz Crater, located in the Northside, manifestation of hot spring in the Ie Su'um Krueng Raya area, and hot spring manifestation in Ie Jue Lamteuba. There are some manifestations in geothermal area Ie Jue, like warm ground, steaming ground, hot springs, fumarole, and mud puddle [3].

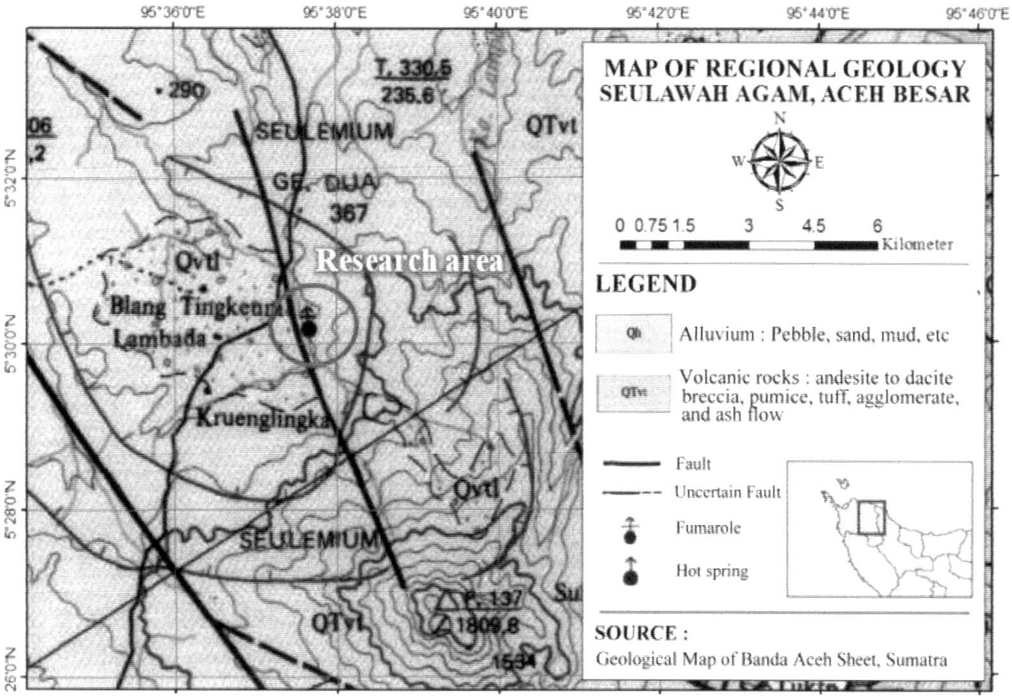

Figure 1. Geological map of the research area.

2.2. Weak zone

Weak zone is a layer of soil or rock that is soft and not compact. Weak zone can be identified by a variety of geophysical methods, one of which is by using seismic investigation. Refraction seismic and 2D resistivity survey are two geophysical methods that are considered suitable for obtaining information about the characteristics of rocks for the weak zone. According to [7], a zone classified as soft layer if it includes the following types of characteristics such as, zones that contain weaker material; tension fracture zones; weathered zones; shear zones and fault zones.

Based on geophysical research using seismic methods and examining of rock samples collected at field in rock mechanics laboratories at the Lulea University of Technology, [8] concluded that zones, where P wave velocity values of layer lower than 4000 m /s, can be grouped as weak zones. These weak zones are characterized by contrast of velocity between weak zones and surrounding rocks. Weak zones have the characteristics of small density and large porosity values compared to their surrounding rocks.

In addition to measurements using the refraction seismic method, the 2D resistivity measurement is also one of the geophysical methods that can be used to identify weak zones. This method provides different resistivity values of each rock body or stratum. The difference in resistance of each contrast layer will affect how the electrical current flow through different layers in the soil or rock. According to the geophysical research using resistivity method, zones that have low resistivity values are identified as weak zones

2.3. Seismic wave

Seismic wave can be distinguished based on the way of their propagation, where the earth media are strongly control their direction propagation. Based on that view, they are classified as body wave that propagate through the earth's interior and some are propagating through the earth's surface called surface wave [11]. The body wave are the wave that propagate in the elastic medium and the direction of its propagation throughout all parts of the earth. They have a higher frequency than surface wave.

Based on the motion of particles in the media and their direction of propagation, they can be distinguished into P and S wave.
Historically, in seismic survey it only P wave is used because it facilitates survey techniques in two ways. Firstly, the ease of data acquisition because it does not require a three-component detector (the detector records the motion of particles vertically). Secondly, a higher P wave velocity makes it always the first time to reach a detector (geophone), making it easier to recognize [4]. Some P wave velocity values in various rock types are summarized in Table 1.

Table 1. P wave velocity [2].

Materials	Velocity (ms^{-1})
Topsoil	0.2 – 0.6
Weathered layered	0.3 – 0.9
Alluvium	0.5 – 2.0
Clay	1.0 – 2.5
Sandstone and shale	2.0 – 4.5
Limestone	2.0 – 6.0
Granite	5.0 – 6.0
Metamorphic rock	3.5 – 7.0
Basalt	5.4 – 6.4

The velocity of seismic wave propagation in subsurface rock layers is different each other, this difference is influenced by several physical properties of the rock layer. According to [10], there are several factors that influence the velocity of seismic wave propagation in subsurface rock layers, such as rock lithology, density, porosity, depth, pressure, and temperature.

2.4. Seismic method

For active seismic technique, it is divided into two methods, that are refraction and reflection seismic. Refraction seismic is used to investigate shallow subsurface of geological structure. While, reflection seismic is applied for imaging of deep geological structure. Refraction seismic method is considered as suitable method that used to identify the shallow subsurface features of geothermal area. It utilizes concept of seismic waves that refracted by boundary layers. The survey of refraction seismic aims to identify shallow subsurface of geological structure, to determine the depth of layer subsurface, and to specify rocks layer by analyzing seismic wave velocities that propagate in rocks layer [11].
The refraction seismic method utilizes the time taken by the wave to propagate on the rock from the position of the seismic source to the receiver at a certain distance. In this method, the waves that appear after the first signal (first break) are ignored, because the refraction seismic waves travel fastest compared to other waves except at relatively close to offset so that the first arrival of waves recorded at receiver plays the important role. The P wave velocity is greater than the S wave velocity by then the arrival time of the P wave is used in the refraction seismic method. Parameters of distance and time of wave propagation are associated with fast wave propagation in the medium. The magnitude of wave propagation velocity is controlled by a group of physical constants in the material known as elasticity parameters [6].

IOP Conf. Series: Earth and Environmental Science **273** (2019) 012027 doi:10.1088/1755-1315/273/1/012027

2.5. *Method of seismic tomography*

In this research, tomographic modeling method is being used. The tomographic method is a special technique that can be used to provide an inner profile of a solid object without cutting or slicing it. This is done by taking measurements outside the object from various directions (which are called making projections), then reconstructing them.

In addition to this, the tomographic refraction method begins with making the initial model and then continuing forward refraction raytracing through modeling. The next stage is comparing the travel time of the calculation with travel time of the measurement and then trying or modifying the initial model to minimize the difference (misfit) between the travel time of the calculation and observation. By repeating these processes until the time of calculation and measurement reaches a minimum value (approaching zero), and obtaining misfit or Root-Mean Squares (RMS) error is at the lowest value, the velocity model presents a predicted subsurface features. The main goal of raytracing technique is to search the minimum travel time between source and receiver for each pair of recipient sources [5].

3. Method

This research was conducted in Lamteuba Village, Seulimuem Sub-district, Aceh Besar. The approaching methodology performed in this research was the refraction seismic method. Refraction seismic survey is used to verify the P-wave velocities of different strata from the subsurface geothermal manifestation by which the probably existed hazard of ground or surface can be analyzed. The data acquisition were conducted for two seismic lines. These two seismic lines oriented from the Southwest-Northeast, where the coordinate of the initial and final positions of the first line is 5°30'23.51"N / 95°37'40.57"E and 5°30'27.34"N / 95°37'42.80"E, where the second line is 5°30'23.22"N / 95°37'39.37"E and 5°30'25.16"N / 95°37'43.21"E. Both lines were intersected with each other intending to strengthen prediction of subsurface manifestations. These lines were positioned close to the surface manifestations (hot steam) which are located within radius of more and less 3 meters.

Table 2. Research equipment.

Tools dan materials	Quantity
Seismograph PASI 16S-24P	1 unit
Power Supply 12 Volt	1 unit
Battery connecting cable	1 unit
Sledge-hammer 5 Kg	1 unit
Iron plate	1 unit
Geophone trigger 10Hz	1 unit
Trigger connecting cable	1 set
Geophone 10 Hz	24 pieces
Geophone connecting cable	2 set
Global Positioning System (GPS)	1 unit
Gauge	2 unit
Software ZONDST2D	1(*sharedware*)

This research was conducted by utilizing PASI 16S-24P Seismograph equipment. All of the seismic lines were set with the same layout geometry such as three meters spacing of each. The left and right offsets were placed at 35 meters, so that the total spread length of each line is 139 meters. The data generated from the tool were then processed by using ZONDST2D to generate the profile of P-wave velocity describing the distribution of P-wave velocity in the subsurface. Based on the velocity profiles and their values, the correlation between the values of P-wave velocity and pathway or channel of geothermal manifestations for particular subsurface features would be analyzed. To meet the goal, the equipment used in this research include:

The procedure carried out in this study consists of 3 stages, include:

3.1. Data acquisition
The first stage was a preliminary survey. This preliminary survey aims to review several locations that are used as locations for research data acquisition, observe the geological surface conditions at the research locations and also for planning of layout data acquiring. After determining the field design, the next step is data acquisition. The travel time obtained is then used to modeling the wave velocity of the subsurface layer based on the depth of the layer at the research location.

3.2. Refraction seismic data processing
The subsurface model was obtained from the data generated during measurements in the field, so that data processing is needed to obtain the seismic wave velocity value and followed by analyzing then interpreting the velocity profile. Data processing and 2D modeling were carried out with four main stages including such as data input and geometry measurements, first break picking and band-pass filtering, review of wave propagation travel time and inversion process with refraction seismic tomography method (non-linear least square).

3.3. Data interpretation
This interpretation stage was done after obtaining a model 2D of data analysis, then comparing with the geological data of the actual location that has previously been studied and also with other (secondary) supporting data, such as previous study in this research area. The data interpretation was also performed by means of qualitative interpretations based on the seismic wave velocity models derived from the distribution of the subsurface layer of the research location. The value of seismic wave velocity found in the model 2D was then matched with the reference table of seismic wave velocity values as shown in Table 1, so that it can be identify the shallow subsurface weak zone structure in the geothermal manifestation.

4. Result and discussion

4.1. Line 1
The first seismic line was an expansion of path that crosses directly into the streamlet, so that end-shot 1 was very close to the streamlet. Determination of the first seismic line was based on the possibility of existed a weak zone that could be clearly seen due to the outflow of manifestation flows directly into the streamlet. The line marked by red color was positioned intersecting with the second line colored by yellow sketched out in figure 2. That is an intention to reinforce the prediction of existence a weak zone. The first line is a measurement trajectory that was located close enough to one of geothermal manifestations where its distance was relatively 3 meters to the outflow. The type of outflow is hot steam-water mixture.

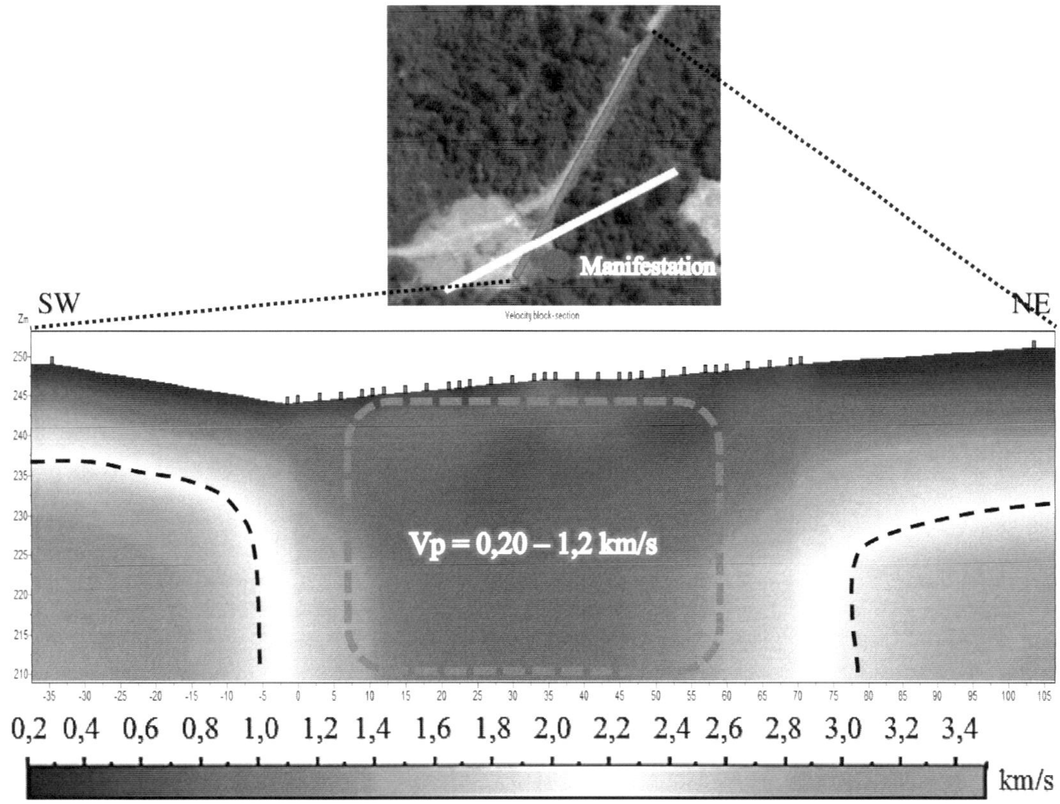

Figure 2. The 2D velocity profile of the first line.

The 2D velocity profile of the first line shows that there was a weak zone that could be identified. The weak zone was located at distances of around 15-50 meters as shown in the dashed red line in Figure 2. The weak zone has P wave velocity value that ranges from 0.2 km/s - 1.2 km/s and it was identified as topsoil or totally complete weathered soil. This estimation of weak zone based on the velocity interval referred to P wave velocity values lower than 4000 m/s (4 km/s) that was an approach value studied by [8]. The weak zone that was found from the first line was suggested as a fracture zone associated with the geothermal system in the Ie Jue geothermal field, Seulawah Agam, Seulimum, Aceh Besar, where the fracture zone was assumed as the path of the fluid exits to the surface.

4.2. line 2

The second line was an intersected line to the first line at the point where it was closed to streamlet flow. Determination of the second line was also based on the possibility identification of a weak zone intended to verify or validate to the first line result. The center point of the second line was placed close to the streamlet its self. Same as first line, the second line was also a measurement line that was close enough to the hot steam-water mixture which was ± 3 meters away. Based on the results of data processing from the second line, it indicates as a weak zone at a distant range of 10 - 65 meters as shown in the marked box of red line in Figure 3. This weak zone also has a P-wave velocity value ranging from 0.2 km/s - 1.2 km/s and it was similarly predicted as completely changed to surface soil. This could be caused by the influence of temperature that involves in geo-chemical process of weathering. When the temperature in a rock layer increases, the rocks or soils would experience an expansion that affects porosity enlargement and shrinks density [10]. The weak zone produced from the second line was strongly indicated as a weak zone associated to the the weak zone of the first line. This zone was also expected as the pathway of the manifestation.

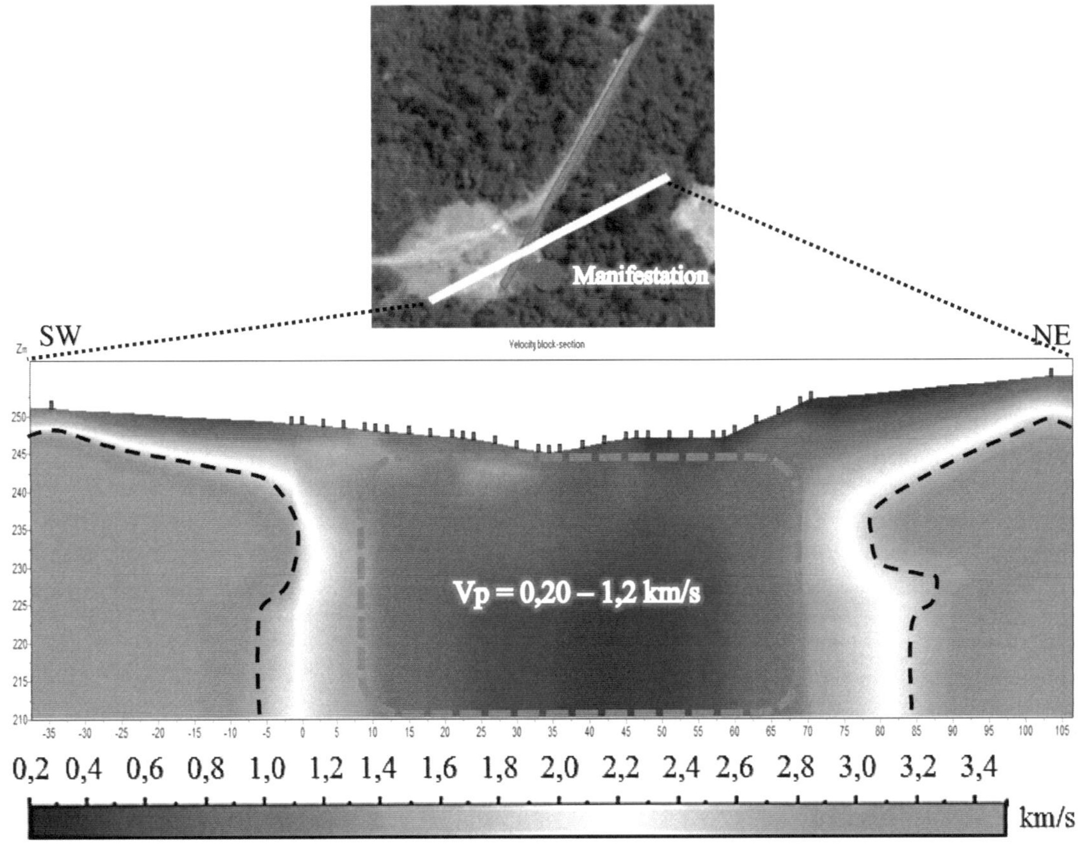

Figure 3. The 2D velocity profile of the second line.

The predicted weak zones obtained from these two seismic lines were suggested as the manifestation pathways from which fluids of have reached to surface. The pathway would gradually to be weaken because of continuous heating from the boiling water and gas fluids, so that the porosity of the layer would be even larger. Enlargement of the porosity causes the layer under the surface not to be compact. This unconsolidated zone could potentially raise a hazard. As previously explained, the Ie Jue area itself is an active area that has traversed by timber trucks. This routine traffic becomes the dynamic loads pressuring from the surface to few meters of subsurface where the outflow channels are situated. As consequence, these expected weak zones could potentially produce a decline of bearing capacity. Finally, the land surface results a subsidence in the vicinity of the area. Figure 4 shows the results of the interpretation of the weak zone profiles association from the first and second lines where the lines intersect of each other and quite close to the surface manifestation of A1.

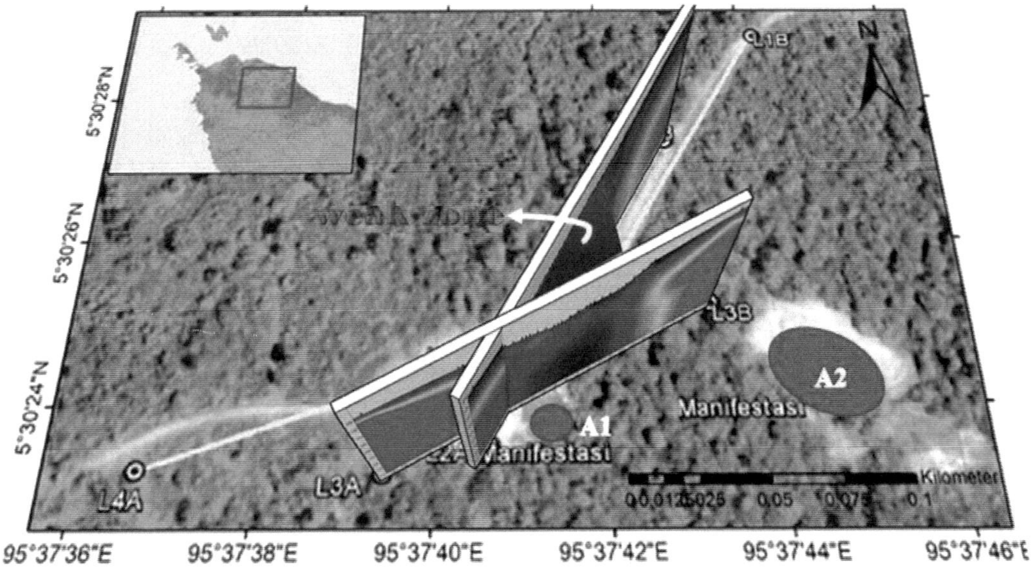

Figure 4. The intersection of the 2D velocity profile combining of weak zones of the first and second line

5. Conclusion

The result of this research shown that for both measured lines there are a strong estimation of relationship between P-wave velocity values with shallow structure as a pathway of fluid flow originated from geothermal source. The near surface structure consists of weak zones. This assumption derived from a relatively low P-wave velocity value. For the first line, the position of weak zone laterally is at a distance of 15-50 meters and vertically at 35 meters depth, while for the second line its position horizontally and vertically is at 10 - 65 meters distance and 35 meters respectively. These structures of weak zone are strongly predicted as the low-velocity zones or low-velocity layers (LVZ) and being a dominantly weathered zones. They are predicted as the channels for fluids from which geothermal manifestation have emerged to the surface. The low P-wave velocity values (0.2 - 1.2 km/sec) may reliable evidence indicating the manifestation pathways of geothermal. These pathways are frequently passed by timber trucks and motorcycles from which producing a dynamic load that pressures the ground overlaying the weak zones. This daily traffics will possibly produce a subsidence endangering for human activity.

6. References

[1] Bennett, J.D., Cameron, N.R. dan Bridge, D.McC. 1981. *Peta Geologi Lembar Banda Aceh, Sumatera*. Bandung : Pusat Penelitian dan Pengembangan Geologi.
[2] Burger, H. R. 1992. Exploration Geophysics of The Shallow Subsurface. New Jersey : Prentice Hall.
[3] Ismail, N., Marwan, dan Wangsa, A. 2018. Interpretasi Kuantitatif Data Anomali Gravitasi di Kawasan Panas Bumi Seulawah Agam, Aceh Besar. Journal of Aceh Physics Society, Vol. 7, No. 1, Hal. 6-12.
[4] Kearey, P., Brooks, M., and Hill, I. 2002. An Introduction to Geophysical Exploration 3rd .ed. United Kingdom : Blackwell Science Publishing.
[5]. Munadi, S. 1992. Mengenal Tomografi Seismik. LPI, No.3/1992, Hal : 239-248. Indonesia.: Lemigas.

IOP Conf. Series: Earth and Environmental Science **273** (2019) 012027 doi:10.1088/1755-1315/273/1/012027

[6]. Nurdiyanto, B., E. Hartanto, D. Ngadmanto, B. Sunardi, dan P. Susilanto. 2011. Penentuan Tingkat Kekerasan Batuan Menggunakan Metode Seismik Refraksi. Jurnal Meteorologi dan Geofisika, Vol. 12, No. 3, Hal. 211-220.
[7] Palmstrom, A. 1995. On Faults And Weakness Zones, Appendix 2 In: Rmi – A Rock Mass Characterization System For Rock Engineering Purposes. Norway : Oslo University.
[8] Renstrom, V. 2016. The Influence of Weakness Zones on The Tunnel Stability Based on Investigations in Bodötunnelen. Sweden : Lulea University of Technology.
[9] Reynolds, M.J. 1997. An Introduction to Applied and Environmental Geophysics. United Kingdom : John Wiley & Sons, Ltd, Publication.
[10] Sheriff, R.E. dan Geldart, L.P. 1995. Exploration Seismology. Cambridge : University Press.
[11] Telford, M.W., Geldart, L.P., Sheriff, R.E., and Keys, D.A. 1976. Applied Geophysics, New York: Cambridge University Press.

Acknowledgments
The author would like to thank for the co-authors in preparing and completing this research and the extended appreciation is delivered to the technicians of the Near-Surface Geophysics Laboratory and students who assisted the data acquisition.

Are we ready for the next strike?

W L Nabigha[1], B Setiawan[2], D R Q Rabbani[3], S M Alhabsyi[4]

[1] SMP Negeri 1 Banda Aceh, Jalan Prof. A. Majid Ibrahim 1, Banda Aceh, 23116, INDONESIA
[2] Program Study of Geological Engineering, Jalan Tgk. Syech Abdurrauf 7, Banda Aceh, 23111, INDONESIA
[3] SMA Negeri 4 Banda Aceh, Jalan T. Panglima Nyak Makam 19, Banda Aceh, 23127, INDONESIA
[4] Dinas Pekerjaan Umum dan Penataan Ruang Kota Banda Aceh, Jalan Prof. Ali Hasyimi, Gp. Pango Raya, Banda Aceh, 23119, INDONESIA

Email: wan.nabigha@gmail.com

Abstract. A tsunami's tragedy happened at Banda Aceh 14 years ago, but most of the city dwellers don't grasp it clearly the hazardous of it. An un-adequate risk perception may be a contributing factor of this un-adequate preparedness. In general observation, we can make a hypothesis that many people in Banda Aceh are unprepared for the next strike. This study is to examine this hypothesis by studying the preparedness of students at SMP 1 Banda Aceh to the natural disaster of tsunami. An interview survey was carried out to randomly selected students to obtain data on their state of emergency preparedness and to ascertain their attitudes toward disaster. Thirty-one (31) students were interviewed. All the respondents know that the school is located at past tsunami's inundated area. About >75% respondents, who participated in the survey, have at least an emergency preparedness plan in place, and approximately 81% of these students believe that they know about contingency plan. However, based on the respondents' response to explain their action if any strong earthquakes strike, only 10% of the respondents have, truly, applied a contingency plan. It is recommended that a contingency plan campaign must be carried out in Banda Aceh.

1. Introduction

Natural disasters occur throughout the world, however the losses to disaster in developing countries are generally much greater than in developed countries in terms of gross domestic product (GDP) percentage or government revenues. The 2004 tsunami is estimated to cause the Province of Aceh loss its capital stock equivalent to 97 percent of their GDP [1].

In term of human perception, most of the Banda Aceh's city dwellers know about the tsunami's tragedy that happen at Banda Aceh 14 years ago. However, the authors believe, that most of the city dwellers don't grasp clearly the hazardous of this tsunami disaster. Tsunami is a rare event, but it will happen again someday [2]. And the question is 'Are we ready for the next strike? Many countries have been damaged or affected by the 2004 Indian Ocean tsunami [3] including our country in this specific city of Banda Aceh [4]. The tsunami was triggered by the second largest recorded seismic ([5]; [6]). This historical tsunami disaster just made some nations to be more aware for the next disaster by creating a contingency plan, instead of just an emergency operation plan. A contingency plan is a specific plan designed to take an escape action due to the possible future disaster event including tsunami. This contingency is prepared to help the people to respond effectively to a significant peril phenomenon [7]. Considering it's a substantial disaster, contingency plan is important for people as it allows people to continue on with their lives and return to their normal daily operations as quickly as possible after an unforeseen event occurs. An example of a contingency plan for this tsunami disaster is conceivably going to higher buildings nearby. In the case of Banda Aceh, tsunami contingency plan is crucial as the city was damage badly during the 2004 tsunami. A contingency plan is more than emergency plan [8]. The differences between the contingency plan and the emergency plan is outlined in Table 1.

Table 1. Differences between contingency planning and emergency operations planning [8]

Aspect	**Contingency plan**	**Emergency Operations Plan**
Relation to emergency event	Before	During
Scope of plan	Global or scenario based	Both strategic and specific
Partners involved	All likely partners	Operational and implementing partners only
Focus	Developing agreed scenarios	Effective and rapid response
Relationships	Developing	Utilizing
Planning Style	Consultative	Directive and consultative
Allocation	Roles	Responsibilities
Time frame	Floating, uncertain	Fixed, immediate

This study aims to investigate the preparedness of the Banda Aceh's dwellers by examining students at the inundated school by the 2004 tsunami. The result of this study is important to carry out further campaigns such as tsunami disaster education, contingency plan promotion, tsunami drill activities or disaster related regulations in Banda Aceh.

2. Tsunami 2004 damage map of Banda Aceh and the location of the school

Figure 1 shows the tsunami 2004 damage map of Banda Aceh and the surrounding areas [9]. The location of this study, which is SMPN 1 Banda Aceh, is also overlaid on the figure (squared black). The Figure 1 demonstrates that the school location is at devastated area. This suggested that the area of the school was highly inundated during the 2004 tsunami. Others documentation by the Japan Times [10] and Tempo [11] suggested devastated damages at the areas nearby the school, as shown in Figure 2.

3. Methodology

A simple questionnaire was developed for this study. The questionnaire includes personal data (i.e. name, address, and contact number), respondent's basic understanding about tsunami disaster & contingency plan, and respondent preparedness. This questionnaire was used for the survey. A simple statistical analysis is employed to conclude the results from the questionnaire.

In addition to the developed questionnaire, respondents were asked to identify five independent variables associated with their preparedness: (1) school location related to the past tsunami event; (2) knowledge of contingency plan; (3) state of personal contingency plan; (4) family contingency plan; and (5) their action if any strong earthquake strike. Using a select framework, the survey data were examined to determine the current state of the student's disaster preparedness. Additionally, the combined effect of risk perception and action on preparedness were also be examined.

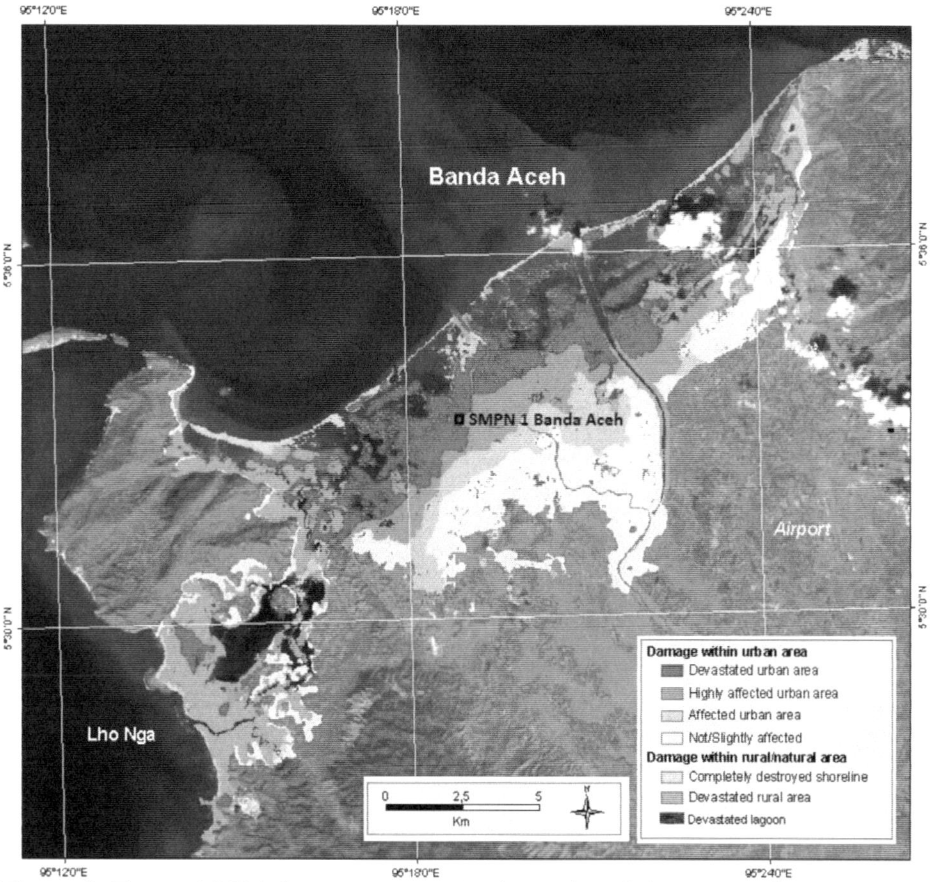

Figure 1. Tsunami 2004 damage map at Banda Aceh and the surrounding areas [9] incorporated the school location (SMPN 1 Banda Aceh-Indonesia).

(a)

(b)

Figure 2. Damages at nearby the school a documentation by the (a) Japan Times [10] and (b) Tempo [11]

4. Results and discussions

Currently, 31 respondents have been interviewed (Table 2). According to the survey, some students at Banda Aceh do have a contingency plan for the next strike, some students decided on the spot, and the remainders don't even have a neither emergency nor contingency plan at all. When explaining their plan, they used words such as run, go, high buildings, safe areas/locations. Some respondents are also mentioning about some places i.e. Neusu, Mata Ie, Lambaro which are unaffected areas by the tsunami 2004. They all aware that they are located an inundated tsunami area. Based on the data as shown in Table 2, a further analysis was carried out to identify whether a contingency plan have been put in a least family level. The results are presented in Table 3.

Table 2. Results of the interview

No.	Name	Do you know that our school is at tsunami inundated area?		Do you know contingency plan?		Is any contingency plan in your family?		What will you do if any strong earthquake strike?
		Yes	No	Yes	No	Yes	No	
1	Respondent 01	1	0	1	0	1	0	Run to high building, no wait
2	Respondent 02	1	0	0	1	0	1	Go out of the house, go to open space
3	Respondent 03	1	0	1	0	1	0	Go to the Tsunami Building[a]
4	Respondent 04	1	0	1	0	1	0	Run to the school gate
5	Respondent 05	1	0	1	0	1	0	Run to a higher building
6	Respondent 06	1	0	1	0	1	0	Run to go to Neusu[b]
7	Respondent 07	1	0	1	0	1	0	Go to Museum Tsunami or higher building
8	Respondent 08	1	0	1	0	1	0	Go to Masjid Raya[c] or higher building
9	Respondent 09	1	0	1	0	0	1	Go to open field or go to higher building
10	Respondent 10	1	0	1	0	1	0	Go to open field or go to higher building
11	Respondent 11	1	0	1	0	0	1	Run to Lambaro[b], as the place is free flood
12	Respondent 12	1	0	1	0	1	0	Go to Mata Ie[b]
13	Respondent 13	1	0	1	0	1	0	Go to higher building
14	Respondent 14	1	0	1	0	1	0	Go to higher building/place
15	Respondent 15	1	0	1	0	0	1	Run to Blang Padang[d]
16	Respondent 16	1	0	1	0	1	0	Go to Museum Tsunami or higher building
17	Respondent 17	1	0	1	0	1	0	Go to Masjid Raya[c]
18	Respondent 18	1	0	1	0	1	0	Go to Museum Tsunami[a]
19	Respondent 19	1	0	1	0	1	0	Following the school's order
20	Respondent 20	1	0	1	0	1	0	Following the school's order
21	Respondent 21	1	0	1	0	1	0	Go to Museum Tsunami[a]
22	Respondent 22	1	0	1	0	1	0	Go to Neusu[b] and go to cousin's house
23	Respondent 23	1	0	0	1	0	1	Nothing
24	Respondent 24	1	0	1	0	0	1	Nothing
25	Respondent 25	1	0	1	0	1	0	Go to places with many people
26	Respondent 26	1	0	1	0	1	0	Go to Pasar Aceh[d]
27	Respondent 27	1	0	1	0	1	0	Go to Museum Tsunami[a]
28	Respondent 28	1	0	1	0	1	0	Go to Museum Tsunami[a]
29	Respondent 29	1	0	1	0	1	0	Go to Museum Tsunami[a]
30	Respondent 30	1	0	1	0	1	1	Nothing
31	Respondent 31	1	0	1	0	1	0	Walk home via Neusu[b]

[a] Tsunami Building or Museum Tsunami is one of escape buildings in Banda Aceh.
[b] Neusu, Lambaro, Mata Ie are unaffected tsunami places during the past 2004 event.
[c] Masjid Raya is considered as safe places during the past 2004 event.
[d] Blang Padang and Pasar Aceh are considered as unsafe places during the past 2004 event.

Table 3. Interpreted results

No.	Name	What will you do if any strong earthquake strike?	Emergency plan	Contingency plan	No plan
1	Respondent 01	Run to high building, no wait	x		
2	Respondent 02	Go out of the house, go to open space	x		
3	Respondent 03	Go to the tsunami building	x		
4	Respondent 04	Run to the school gate	x		
5	Respondent 05	Run to a higher building	x		
6	Respondent 06	Run to go to Neusu		x	
7	Respondent 07	Go to Museum Tsunami or higher building	x		
8	Respondent 08	Go to Masjid Raya or higher building	x		
9	Respondent 09	If earthq. go to open field, if tsunami go to higher building	x		
10	Respondent 10	If earthq. go to open field, if tsunami go to higher building	x		
11	Respondent 11	Run to Lambaro, as the place is free flood	x		
12	Respondent 12	Go to Mata Ie	x		
13	Respondent 13	Go to higher building	x		
14	Respondent 14	Go to higher building/place	x		
15	Respondent 15	Run to Blang Padang	x		
16	Respondent 16	Go to Museum Tsunami or higher building	x		
17	Respondent 17	Go to masjid	x		
18	Respondent 18	Go to Museum Tsunami	x		
19	Respondent 19	Following the school's order	x		
20	Respondent 20	Following the school's order	x		
21	Respondent 21	Go to Museum Tsunami	x		
22	Respondent 22	Go to Neusu and go to cousin's house		x	
23	Respondent 23	Nothing			x
24	Respondent 24	Nothing			x
25	Respondent 25	Go to places with many people			x
26	Respondent 26	Go to pasar Aceh			x
27	Respondent 27	Go to Museum Tsunami	x		
28	Respondent 28	Go to Museum Tsunami	x		
29	Respondent 29	Go to Museum Tsunami	x		
30	Respondent 30	Nothing			x
31	Respondent 31	Walk home via Neusu		x	

The Table 3 is developed based on the descriptive evaluation of the respondent actions if any seismic strike. The classification is based on the [8] as shown in Table 1. This analysis results suggest that most of respondents (74%) have an emergency plan and only 10% of the respondents do applied a (family) contingency plan. The shocking results are 16% of the respondents with no plan at all, even though they know that they are living at the tsunami's affected area.

5. Conclusion

Tsunami disaster will happen again someday. Both emergency and contingency plan play a huge role to save life especially for Banda Aceh city's dweller. This study attempts to investigates the Banda Aceh citizen's preparedness due to tsunami disaster by interviewing students at tsunami 2004 affected school. At the current stage, various responses have been observed. The result of this study shows further campaigns in the tsunami disaster preparedness are essential to be warranted. It is also recommended that further studies should be conducted to community preparedness to broaden the researcher's survey base and strengthen statistical reliability.

6. References

[1] IEG, Hazards of Nature, Risk to Development: An IEG Evaluation of World Bank Assistance for Natural Disasters, The World Bank, Washington DC3 (2006)
[2] J. Goff, C. Chague-Goff, Progress in Physical Geography.38 (2014) 218–240
[3] B. Poisson, M. Garcin, R. Pedreros, Geophys. J. Int. 177 (2009) 1080–1090
[4] S. Syamsidik, T. Tursina, A. Meutia, M. Al'ala, M. Fahmi, E. Meilianda, J. Earthq. Tsunami. 11 (2017) 19 pp.
[5] R. Bilham, Science, 308 (2005) 1127–1133.
[6] B. Setiawan, W. Kusuma, (2013) 8 pp. http://scholarsmine.mst.edu/icchge/7icchge/session04/5
[7] IFRC, (2012) 68 pp. http://www.ifrc.org/PageFiles/40825/1220900-CPG%202012-EN-LR.pdf
[8] Handbook for Emergencies - Second Edition (UNHCR, 1999, 414 p.)
[9] STETIT, 2005) Indonesia - Sumatra - Banda Aceh area: Damage map. https://reliefweb.int/map/indonesia/indonesia-sumatra-banda-aceh-area-damage-map accessed on 10 September 2018
[10] The Japan Times (2018) 2004 tsunami survivors recall how mosques stood firm https://www.japantimes.co.jp/news/2014/12/24/world/2004-tsunami-survivors-recall-how-mosques-stood-firm/#.W5XYRM4zaM8 accessed on 9 September 2018
[11] Tempo.co (2018) Institut Prancis Gelar Pameran Gempa dan Tsunami. https://seleb.tempo.co/read/607762/institut-prancis-gelar-pameran-gempa-dan-tsunami accessed on 9 September 2018.

Knowledge level of earthquake and tsunami disaster at disaster preparedness school and non-disaster preparedness school in Kuta Raja sub-district of Banda Aceh

Teuku Hasan Basri[1], Nuraini[2] and Sitti Raisa[3]

[1]Department of Physics Education, Universitas Samudra, Langsa, Indonesia
[2]Department of Biology Education, Universitas Samudra, Langsa, Indonesia
[3]Department of Geography Education, UniversitasSyiah Kuala, Indonesia

Abstract. Banda Aceh is one of the worst affected areas in the earthquake and tsunami on 26 December 2004, which claimed more than 75% of casualties and also flattened almost all the buildings in this region.After the Tsunami disaster, various international and local agencies, as well as the government, together with tried to rebuild the affected areas includingBanda Aceh.The Tsunami and Disaster Mitigation Research Center (TDMRC) of Syiah Kuala University has begun to build Disaster Preparedness School (SSB) the program that directly touching the main stakeholders in creating disaster-resistant communities.Kuta Raja which is one of the sub-districts in Banda Acehthat is also vulnerable to earthquake and tsunami disaster so it becomes one of the focus in Disaster Preparedness School program.This research was conducted to know what level of Earthquake disaster knowledge and Tsunami in Disaster Preparedness School and Non-Disaster Preparedness School students in Kuta Raja Sub-district of Banda Aceh. The population in this study amounted to 4 elementary schools (2 SSB and 2 non-SSB). Because the population is above 100, the sample is taken 20% from 196, so the total is 40 students for SSB School and 40 students for Non-SSB School.To achieve the research objectives, the method used is a descriptive method with data collection questionnaire technique. Data analysis technique used is calculated asthe percentage by using a simple statistic formula.The results showed that the number of students of the Disaster Preparedness School (SSB) responded to the questionnaire with the correct answer percentage of 46.3% (less), while the Non-SSB students had the correct answer percentage of 66.3% (enough).

Keywords: knowledge level, earthquake, tsunami, school disaster preparedness

1. Introduction

The territory of Indonesia has geographical, geological, hydrological and demographic conditions that allow disasters to occur, both caused by natural factors, non-natural factors,and human factors. A natural disaster occurs when an extreme geological, meteorological, or hydrological event exceeds the ability of a communityto cope with that event [1]. The disaster can cause human deaths, environmental damage, property losses, and psychological impactsand more complex mental health disorders [2]. Indonesia is a disaster-prone area which is shown by the increasing number of disasters every year. Disasters such as tsunamis, earthquakes, landslides, floods, hurricanes, volcanic eruptions, and industrial accidents are often a serious threat to Indonesian citizens.

Banda Aceh as the Capital of Aceh Province has geographic, hydrological and demographic conditions that are prone to disasters. The mega-tsunami generated by the megathrust earthquake in the Aceh-Andaman subduction zone (9.2 Mw) in 2004 was the largest tsunami disaster in the 20th Century which caused the death of more than 200,000 people [3]. Banda Aceh was one of the areas worst affected by the earthquake and tsunami on December 26,2004, which claimed more than 75% of the fatalities and also flattened almost all the buildings in the region and also affected the economic and social life of the people of Banda Aceh.This disaster caused prolonged trauma for the people of Banda Aceh so that it needed a special strategy to anticipate disasters involving the entire community including students at the school.

Education is one of the effective means to reduce disaster risk by incorporating subject matter on natural disasters for every student at all levels, especially in schools in disaster risk areas [4]. Tsunami and Disaster Mitigation Research Center (TDMRC) of Syiah Kuala University has initiated a Disaster Preparedness School (SSB) program. SSB is an activity to build school capacity through strengthening knowledge and attitudes, school policies, emergency response plans, school early warning systems, and resource mobilization based on existing school capacity, especially in anticipating the risk of natural disasters, including earthquakes and tsunamis. Nevertheless, the program to date is still limited to the Disaster Preparedness School (SSB)[5]. In fact, the understanding of anticipating the risk of natural disasters is important for all students in the school. The purpose of this study was to determine the level of Earthquake and Tsunami Disaster Knowledge in School Students of Disaster Preparedness and Non-Disaster Preparedness Schools in Kuta Raja District, Banda Aceh.

2. Methods

The study was conducted in Kuta Raja Subdistrict, Banda Aceh, from December 2016 to January 2017. This study used a descriptive method involving 8 schools consisting of 4 (four) Disaster Preparedness School (SSB) and 4 (four) Non-Disaster Preparedness Schools (Non-SSB) in Kuta Raja Subdistrict, Banda Aceh. The total sample is 40 students for SSB Schools and 40 students for Non-SSB Schools. Interviews are the implementation of the questionnaire verbally and privately with each sample member. This technique is an effort to collect information by submitting a number of written questions to be answered in writing by the respondent. The questions submitted in the questionnaire must lead to the purpose of the research and verification.

2.1 Data Analysis

The percentage level of knowledge is used as a simple statistical formula as follows [6].

$$P= \frac{F}{N} \times 100\%$$

Where:

P = Percentage of knowledge level
F = The item frequency is correct
N = Total number of questions

Table.1 Knowledge Level Scale

No	Value Percentage	Category
1.	76-100%	Good
2.	56-75 %	Enough
3.	40-55 %	Less
4.	< 40 %.	Not good

3. Results And Discussion

3.1 Respondents identity

The number of elementary school students in Kuta Raja sub-district is 5,349 students, with 9,131 male and 7,730 female.[7] However, because this study uses a sampling model, the authors sampled 80 students, namely 40 students from SSB schools and 40 students from non-SSB schools to be respondents in this study. For more details about the sex of students who become respondents can be seen in Table 2 below.

Table 2.The identity of Respondents by Sex at SSB schools

No	Sex	Frequency	Percentage (%)
1.	Male	20	50 %
2.	Female	20	50 %
Total		40	100 %

Table 2 above showed that (50%) ofthe respondents are male and (50%) female. In this study, there was a balance of the number of male and female respondents with status as students attending SDN 17 Banda Aceh and SDN 70 Banda Aceh in Kuta Raja Subdistrict.

Table 3. The identity of Respondents by Sex at Non-SSB schools

No	Sex	Frequency	Percentage (%)
1.	Male	20	50 %
2.	Female	20	50 %
	Total	40	100 %

Table 3 above showed that (50%) ofthe respondents are male and (50%) female. In this study, there was a balance between the number of male and female respondents with status as students attending SDN01 Banda Aceh and SDN08 Banda Aceh in Kuta Raja Subdistrict.

3.2 Sources of Disaster Information on Earthquake and Tsunami

Information sources include anything that can be used to help each person to learn and display competencies that can be in the form of messages, people, materials, tools, settings,and techniques [8]. Sources of information are defined as everything and by which someone learns something [9]. In the learning process the components of learning resources may be used singly or in combination, both planned learning resources and learning resources that are utilized. So that it can be said that the source of information is all sources in the form of data, people and certain forms that can be used by students in learning, both separately and in combination so as to facilitate students in achieving learning goals or achieving certain competencies.Information source for earthquake and tsunami disaster for SSB and Non-SSB studentsare shown in Table 4.

Table 4. Information source for earthquake and tsunami disaster

No	Source of Information	Percentage (%)	
		SSBStudents	Non-SSB Students
1.	School	60	42.5
2.	Parents	30	50
3.	News Paper	0	5
4.	Peers	5	0
5.	Social media	5	2.5

Based on Table 3 we know that students obtain information sources for knowledge of Earthquake and Tsunami disasters as much as 60% come from schools, followed by parents, peers, and social media about 30%, 5%, and 5%respectively. This shows that most of the sources of disaster information for SSB students are obtained from school and only a few come from other sources.Meanwhile, 50% of the information sources for earthquake and tsunami disaster were from parents, 42.5% came from school, 5% came from print media, and as much as 2.5% came from social media. This shows that half (50%) of all non-SSB school students get sources of information from parents and others get information from other sources.

3.3 The level of students' Earthquake and Tsunami Disaster Knowledge
The concept of knowledge in the context of disaster shows that the successof gaining knowledge lies in the relationship between knowledge management and disaster resilience. Where, the knowledge possessed can motivate someone in taking appropriate actions to save lives [10]. The level of students' Earthquake and Tsunami Disaster Knowledge can be seen in Table 5.

Table 5. The level of students' Earthquake and Tsunami Disaster Knowledge of SSB and Non-SSB students

No.	School	Percentage correct answer (%)	Level of students' Knowledge
1.	SSB	46.3	Less
2.	Non-SSB	66.3	Enough

Based on Table 5 we know that the frequency of correct answers for SSB students is 6.95 with a percentage of 46.3% and the frequency of correct answers for Non-SSB students is 9.95 with a percentage of 66.3%. Thus the level of knowledge of SSB students is in the range of 40-55%, which is classified as less, while the level of knowledge of non-SSB students includes in the range of 56-75%, which is classified as Enough.

Thus it can be analyzed that the SSB program run in schools has not been able to make students have the knowledge of the disaster of the Earthquake and Tsunami in a Good level. Even the results of the study indicate that the level of knowledge of disaster in the Earthquake and Tsunami students are still in the Less. The socialization carried out by TDMRC and the programs run by schools require evaluation, especially school is one of the important stakeholders in increasing students' knowledge.

3.4 Application of disaster alert programs in schools
Tsunami and Disaster Mitigation Research Center (TDMRC) Syiah Kuala University has implemented disaster management programs in schools, especially in Banda Aceh. SSB in Banda Aceh cover all levels of education ranging from elementary schools to high schools, disaster prepared schools are moved to schools located right in Tsunami-prone areas such as Meuraxa District, Kuta Raja District, Baiturrahman District, and other Districts in Tsunami-prone areas. With this program, it is expected that school-aged children in Banda Aceh will gain knowledge about the disaster so that the knowledge gained can be applied and of course the simulation must be carried out continuously. Simulation programs are taught to be more interactive using various methods such as role-playing, discussion,and other methods. Programs invested in schools are included in extracurricular activities. Although there are similar activities related to disaster education held in schools, there are significant differences in how to formally enter curriculum-based disaster themes and include extracurricular-based disaster themes [11].

The results of an interview conducted with the Head of SDN 17 Banda Aceh revealed that SDN 17 Banda Aceh had a disaster alert program. Banda Aceh as one of the disaster-prone areas of the Earthquake and Tsunami should be school-age children to get knowledge about the disaster. In addition, the school has also attempted to bring disaster preparedness values closer together with several activities such as simulations. However, disaster alert programs in schools have not been implemented optimally and are limited to training and socialization activities. He hopes that in the future there will be government involvement in developing disaster alert programs in schools.The results of an interview conducted with ateacherin SDN 70 who was the teacher in charge of the SSB program at the school, he said that the disaster preparedness program was not very running because from the TDMRC also did not convey too much the sustainability of the program that would be continued by the school. After the socialization and training of students and teachers, the teacher only gave back to students who did not participate in the socialization and simulation [12].

Disaster prevention can be done in schools using local wisdom models from the local community. This will be more effective if carried out sustainably through the curriculum in schools [13]. A local wisdom-based curriculum will explain the relationship between humans and the natural and cultural environment around them [14]. The curriculum for natural disasters is important in disaster-prone countries including Indonesia so students have knowledge about disaster early on [15].

4. Conclusion

The results showed that the number of students of the Disaster Preparedness School (SSB) responded to the questionnaire with the correct answer percentage of 46.3%, while the Non-SSB students had the correct answer percentage of 66.3%. So the level of knowledge of SSB students in Kuta Raja sub-districts is still classified as Lessand Non-SSB students' knowledge level that is enough.The SSB program that has not been implemented optimally in schools and the lack of sources of information is a factor causing the students' low knowledge of the earthquake and tsunami disaster.

5. References

[1] Lindell, MK and CS Prater, 2003. Assessing Community Impacts of Natural Disasters. *Natural Hazards Review* 4 (4): p. 176-185.
[2] Honesti, L and N Djali. 2012. Pendidikan Kebencanaan Di Sekolah – Sekolah Di Indonesia Berdasarkan Beberapa Sudut Pandang Disiplin Ilmu Pengetahuan. *Jurnal Momentum*12 (1): p. 51-56
[3] Natawidjaja, DH. 2015. Mega-tsunami cycles Aceh Andaman region in historical context. *Ris.Geo.Tam* 25 (1): p. 49 – 62.
[4] Suarmika, PE and EG Utama. 2017. Pendidikan mitigasi bencana di Sekolah Dasar (sebuah kajian analisis etnopedagogi). *Jurnal Pendidikan Dasar Indonesia* 2(2): p. 18 – 24
[5] Darmawan, S. 2012. *Elastisitas dan Getaran Terintegrasi Bencana*. Bandung: Alfabeta.
[6] Arikunto, Suharsimi. 2013. *Prosedur Penelitian Suatu pendekatan Praktik*. Jakarta:Rineka Cipta.
[7] BPS, 2016. *Kecamatan Kuta Raja Dalam Angka 2015*. Banda Aceh.
[8] Hayati, D. *Pilihan Tindakan Penanggulangan Bencana*. Jakarta: Rosda.
[9] Susetyo, H. 2006. *Penanganan dan Pendidikan Bencana*. Jakarta: Gema Insani.
[10] Suryani, O. R. 2016. Pengetahuan yang Menyelamatkan. *Smong News*. 07 Nopember 2016.
[11] Jaelani, A. Pedoman Kesiapsiagaan Menghadapi Gempa Bumi. Yogyakarta: Bhuana Ilmu.
[12] Nurchayat, A. Nuray. 2013. *Perbedaan Kesiapsiagaan Bencana Antara Kelompok Siswa SD yang Dikelola Pedagogik dan Andragogi*. Skripsi. Jakarta: Universitas Gadjah Mada.
[13] Astuti, SI and Sudaryono. 2010. Peran Sekolah Dalam Pembelajaran Mitigasi Bencana. *Jurnal Dialog Penang gulangan Bencana* 1 (1): p. 30-42
[14] Desfandi, M. 2014. Urgensi kurikulum pendidikan kebencanaan berbasis kearifan lokal di Indonesia. *Sosio Didaktika* 1 (2): p. 191-198
[15] Prasetyo, D. 2013. *Kesiapsigaan Bencana dan Multifungsi Bencana*. Jakarta: Bumi Aksara.

An overview of the down-scaling values of physical and social vulnerability parameters for tsunami disaster at the city level – a case study of Banda Aceh

A S Sugiarti[1], E Meilianda[2,3], E Fatimah[3]

[1]Graduate Program of Disasters Science, Syiah Kuala University, Banda Aceh, 23111,
[2]Tsunami and Disaster Mitigation Research Center (TDMRC)Syiah Kuala University, Banda Aceh, 23233, Indonesia
[3]Civil Engineering Department, Faculty of Engineering Syiah Kuala University, Banda Aceh, 23111, Indonesia

*Corresponding Author : annisasrisugiarti@gmail.com

Abstract. The Indian Ocean earthquake and tsunami occurred on December 26, 2004, has been one of the deadliest disasters in modern history. Banda Aceh City is one of the severely affected cities by the disaster. Many of the casualties were due to lack of knowledge in the community about the disaster and preparedness in facing the disaster. Now, after more than 10 years since the deadly disaster, the Indonesian government has created disaster risk maps for the provincial level and utilizing some of the provincial level parameters to create a district- or city-level risk maps. The purpose of this study was to analyze the dynamics of physical and social vulnerabilities at the district level over the last 10 years and to evaluate the sensitivity of the disaster vulnerability values by down-scaling the parameter values to the district or city level. This study is using Geographic Information System (GIS), which spatial data are encompassing the infrastructure, housings and the growth of population at risk up to the year 2017. The results reveal the remarkable distribution of risk levels at each investigated village. The physical vulnerability is valued moderately in Banda Aceh city about 54% whereby Gampong Laksana at Kuta Alam sub-district are among the highest vulnerability areas (0.933). As for the social vulnerability is valued high in Banda Aceh city, the highest vulnerable areas are located in Punge Ujong at Meuraxa sub-district (0.833). Overall, the vulnerability parameter value for a tsunami disaster in Banda Aceh city remains high, particularly at those villages located very close to the sea. More mitigation effort should, therefore, be reinforced at those highly risky coastal areas.

Keywords: tsunami, Banda Aceh, GIS, risk mapping, vulnerability

1. Background.

Indonesia is a vulnerable country to disasters both from natural events and those caused by human activities. The process of natural disasters occurrence can take place suddenly or progress slowly closely related to geographical conditions, geology, climate and other factors. The earthquake and tsunami disaster on December 26, 2004 was a deadly disaster in modern history, especially in Aceh. The death caused from the disaster reached more than thousand people and destroyed infrastructures and coastal areas, especially on the west coast of Aceh. The large number of casualties due to lack of community knowledge about the disaster and preparedness in the face of disasters. The city of Banda Aceh, which is directly adjacent to the sea, makes this region vulnerable to the tsunami disasters. Vulnerability has a dynamic characteristic in spatial-temporal because it is influenced by the number of populations, property, and environmental conditions in areas with high threats. The dynamics of vulnerability in an area that has the potential for disaster threats within a certain period of times determines the level of disaster risk in the region. To minimize these conditions, the efforts are needed to socialize knowledge and potential disasters in their respective neighbour-hoods.

Knowledge about disasters is one part of disaster mitigation. One product that results from disaster knowledge is a disaster risk map. Disasters risk map can be used as a tool to increase a disaster knowledge and awareness. Indonesia already has a disasters management agency (BNPB), disasters management law Nr. 24/2007 and disasters guidelines for risk assessment that is Head of BNPB regulation Nr. 2/2012. In the guideline there are various type of disasters such as for earthquake, tsunami, floods, etc.

Each province and district in Indonesia should have a disaster risk map to find out the potential disasters in their respective regions. For making a risk map it requires valid data so that it can represent the actual conditions that occur in the field. Aceh itself already has a provincial-level disaster risk map called Aceh Disasters Risk Mapping (ADRM) created by TDMRC in 2011. The ADRM contains 11 potential disasters in Aceh including abrasions, storm surges, floods, epidemics, earthquakes, volcanoes, forest fires, droughts, landslides and tsunamis. The risk map is based on the Head of BNPB (Indonesia Agency for Disasters Management) regulation Nr. 2/2012 concerning general guidelines for disaster risk assessment. Banda Aceh city is one of the cities that are prioritized to have a risk map because Banda Aceh city is ranked fifth as tsunami disasters prone area based on the Indonesian Disaster Risk Index book [2].

Vulnerability is a condition that is determined by physical, social, economic and environmental factors or processes that can increase vulnerability in a community [3]. Based on these four factors, physical vulnerability and social vulnerability are the most influential in the disaster, especially the tsunami disaster. The previous research about tsunami disaster risk in the city of Banda Aceh has been conducted in 2014 with the result the city of Banda Aceh was in low level of vulnerability [4]. So, the purpose of this study was to study the dynamics of physical vulnerability and social vulnerability at the district/city level after more than 10 years of the tsunami by taking research locations in villages in Banda Aceh. Villages that are chosen as the object of the research are the only villages that affected by the tsunami threats.

2. Method

This is a semi-qualitative research that takes place in the city of Banda Aceh which is divided into nine sub-districts. The method in making the physical vulnerability and social vulnerability maps of the tsunami disaster depends on the use of geographic information systems (GIS) and the stages of data collection based on secondary data obtained from Statistic Center Agency (BPS) of Banda Aceh and Regional Development Planning Agency (BAPPEDA). The data for population, poverty, disability, the number place of worship, school and health facility was obtained from BPS. For administrative map was obtained from BAPPEDA.

2.1. Physical vulnerability

Factors that affecting physical vulnerability generally refer to the weaknesses of the location and the environment that has a vulnerability to the disasters. The vulnerability is related to threats so that risks can change linearly as a function of threat and vulnerability parameters if they are interrelated [4]. Indicators of physical vulnerability include the density of houses, the density of worship buildings, the density of school buildings and the density of health buildings. After obtaining the building density from each indicator of physical vulnerability, it is necessary to define an index classification of each indicator. Index classifications used in the physical vulnerability are shown in table 1 as follows

Table 1. Physical vulnerability parameters [5]

Component/Indicator	Index Classification (Score)			Total Weight
	Low (0.333)	**Moderate (0.666)**	**High (1.000)**	
House Density	<150 per km^2	150 – 1000 per km^2	> 1000 per km^2	40%
Place of Worship Density	< 5 per km^2	5 – 15 per km^2	> 15 per km^2	20%
School Density	< 2 per km^2	2 – 6 per km^2	> 6 per km^2	20%
Health Facility Density	< 3 per km^2	3 – 9 per km^2	> 9 per km^2	20%

2.2. Social Vulnerability

Social vulnerability is a function of the presence of victims of disaster events. The considerable parameters used in social vulnerability are the lives of individuals, communities and people behaviour. Indicators of social vulnerability include population density, sex ratio, ratio of poverty, ratio of disabled people and ratio of age groups. The next step is to define index classification with units on the four indicators of social vulnerability. Table 2 shows index classification used in social vulnerability

Table 2. Parameters of social vulnerability [5]

Indicator		Index Classification (Score)			Total Weight
		Low (0.333)	**Moderate (0.666)**	**High (1.000)**	
Population Density		<500 people/km^2	500-4000 people/km^2	>4000 people/km^2	60%
Vulnerability Group	Sex Ratio	>80%	80-60%	<60%	10%
	Poverty Ratio	<20%	20-40%	>40%	10%
	Disability Ratio	<20%	20-40%	>40%	10%
	Age Group Ratio	<20%	20-40%	>40%	10%

3. Results and Discussion

3.1. Physical Vulnerability

Based on the results of research, the highest house density in the city of Banda Aceh is in Bitai (Jaya Baru sub-district) with 11935.05 houses per km2 and the lowest is in Rukoh (Syiah Kuala sub-district) with 88.53 houses per km^2. The level of house density in the city of Banda Aceh is shown in figure 1.

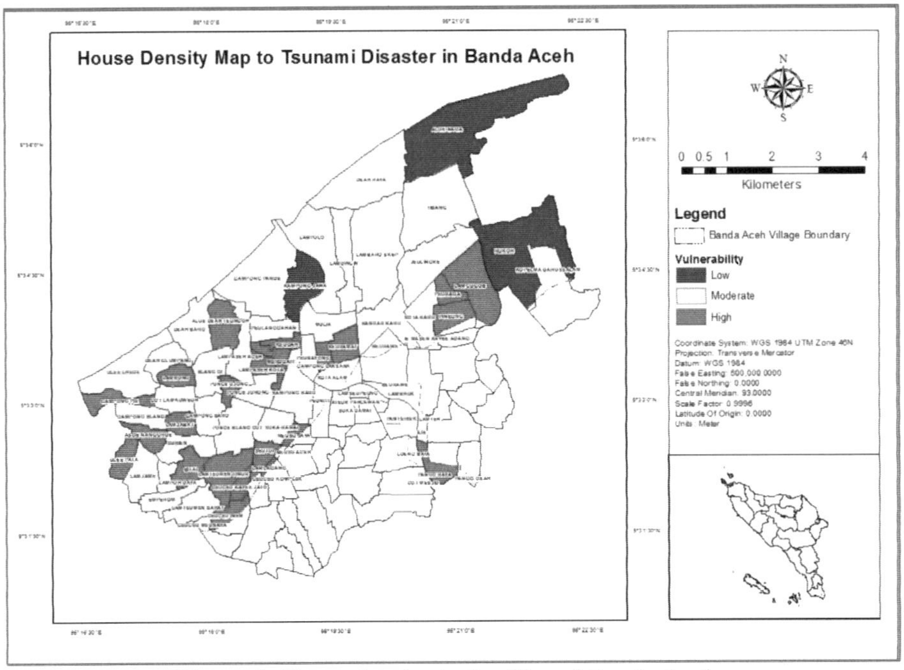

Figure 1. House density in the City of Banda Aceh

The density of worship buildings included in the calculation such as mosques, mushalla, churches, temples, and monasteries which found in the villages of the city. The level of the place of worship density is shown in figure 2. The results showed that the highest density of worship buildings in the City of Banda Aceh was in Gampong Laksana (Kuta Alam sub-district) with 30.01 unit per km^2 and the lowest was in Pango Raya (Ulee Kareng sub-district) with 0.

IOP Conf. Series: Earth and Environmental Science **273** (2019) 012030 doi:10.1088/1755-1315/273/1/012030

School buildings are one indicator of physical vulnerability that must be considered. The City of Banda Aceh itself has 236 school buildings scattered into 9 sub-districts in each village in the sub-district. The highest density of school buildings in the City of Banda Aceh is at Ateuk Pahlawan (Baiturrahman sub-district) with 32.92 units per km^2 and the lowest is in Bitai (Jaya Baru sub-district) with 0 because the area does not have a school building. Figure 3 shown the level of school building density of Banda Aceh in the object area.

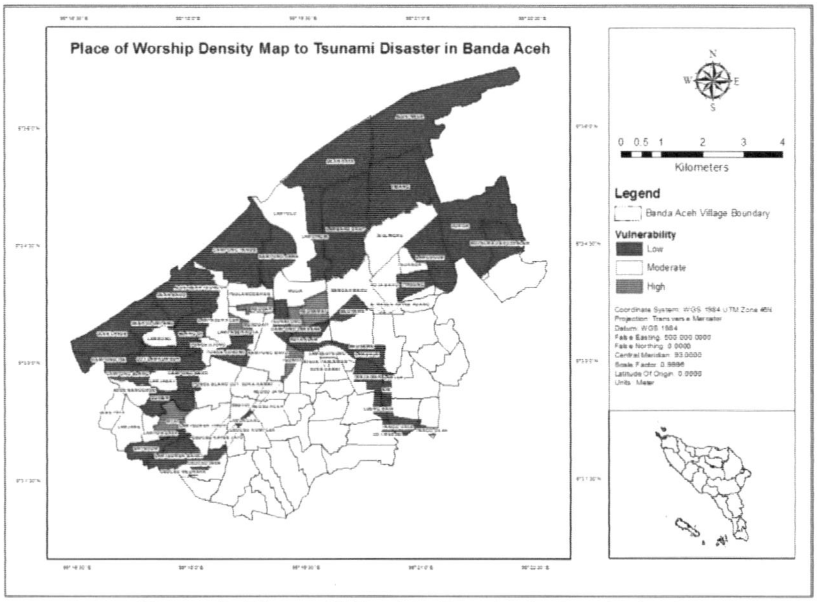

Figure 2. Place of Worship Density in Banda Aceh

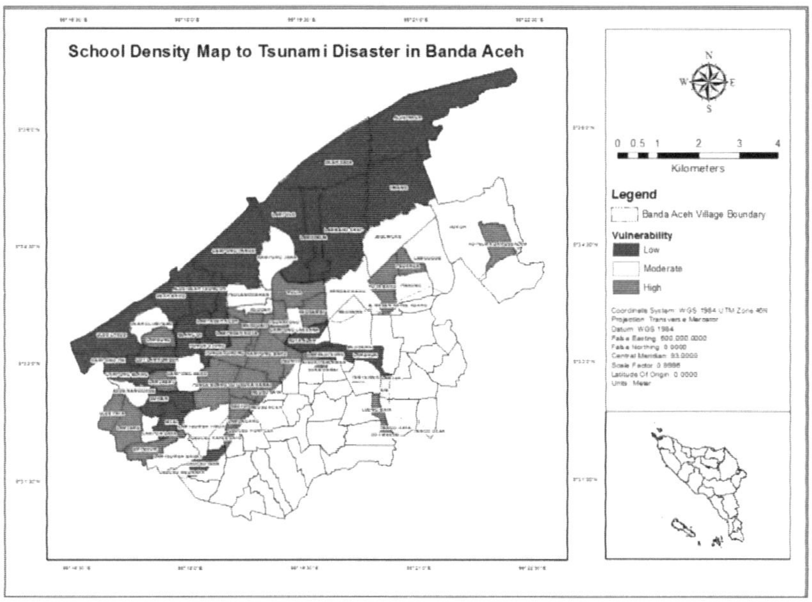

Figure 3. School building density in Banda Aceh city

Similar to place of worship and school buildings, the location of a health facility also needed to be considered. Health facilities are important during a disaster and it categorizes as a critical facility because in the health facility usually there are many vulnerable groups, especially sick people who might be become the victims during the disasters. In Banda Aceh city, there are 112 health facility such as hospitals, maternity hospital, health-center, and others. So the calculations were carried out in

IOP Conf. Series: Earth and Environmental Science **273** (2019) 012030 doi:10.1088/1755-1315/273/1/012030

the areas where the health facility are located in the vulnerable of tsunami disasters area. The vulnerability level of health facility density was shown in Figure 4.

In Banda Aceh city, the highest density of health facility was in Gampong Laksana (Kuta Alam sub-district) with 35.01 unit per km^2 and the lowest one was in Keudah (Kuta Raja sub-district) with 0 because there were no health buildings in the area.

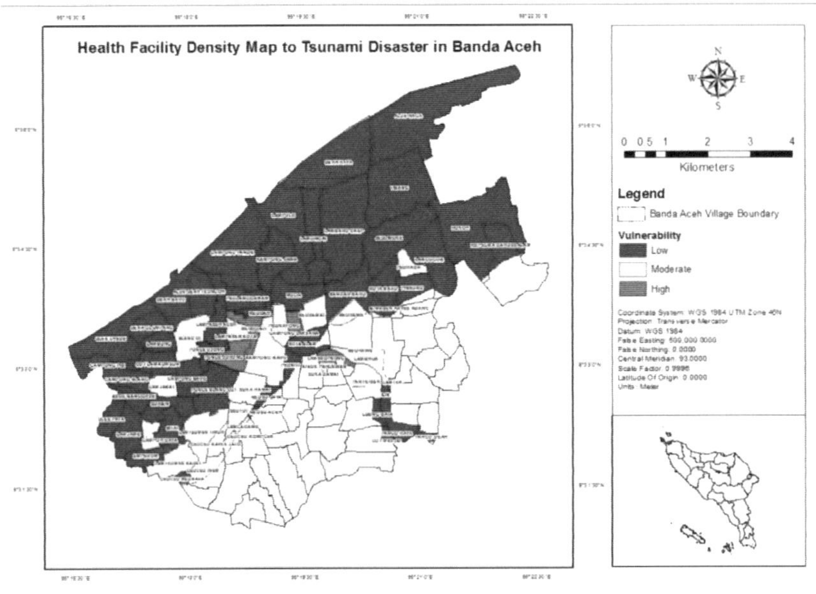

Figure 4. Health facility density in Banda Aceh city

The physical vulnerability which is obtained from the four indicators in the Banda Aceh city was shown in Figure 5 as follows. Based on the figure, it can be seen that the physical vulnerability in the Banda Aceh city were in moderate level of vulnerability. The level of a high category is 44% of the area, 54% of the area is in the medium category and 2% of the area is in a low category. Alue Naga (Syiah Kuala sub-district) is a village in the City of Banda Aceh which is categorized as the lowest one and Gampong Laksana (Kuta Alam sub-district) is a village in the city that is the highest one of physical vulnerability. The score of total physical vulnerability at Alue Naga (Syiah Kuala sub-district) is 0.333 while the score at Gampong Laksana (Kuta Alam sub-district) is 0.933.

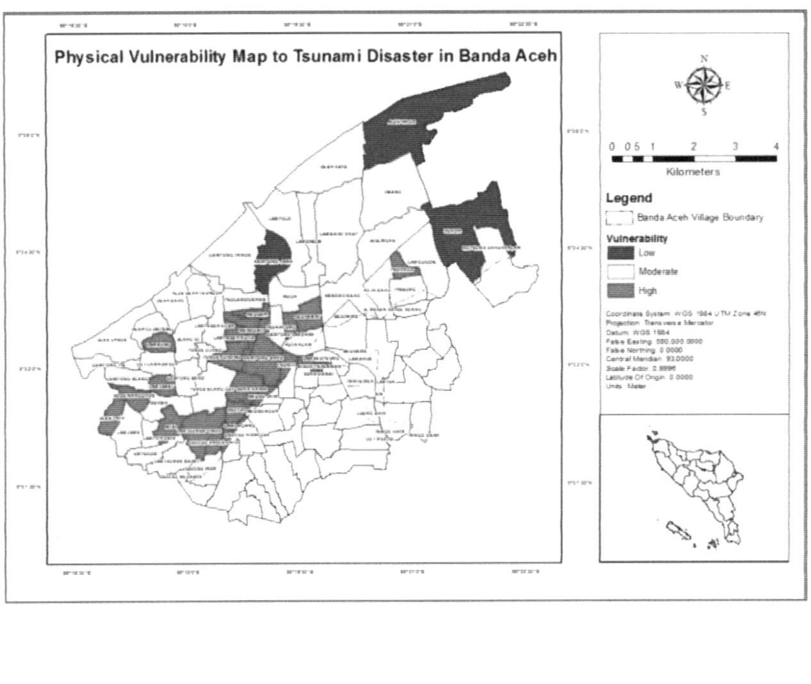

IOP Conf. Series: Earth and Environmental Science **273** (2019) 012030 doi:10.1088/1755-1315/273/1/012030

Figure 5. Map of Physical Vulnerability in Banda Aceh city

3.2. Social Vulnerability

There are five indicators should be considered in defining social vulnerability maps for the District / City level. The indicators are population density, sex ratio, poverty ratio, the ratio of disability, and the ratio of age groups.

Based on the calculation that is shown in Figure 6 the highest population density of Banda Aceh is at Gampong Laksana (Kuta Alam sub-district) with 25095 people / km^2 and the lowest one is in Gampong Pande (Kuta Raja sub-district) with 359 people / km^2.

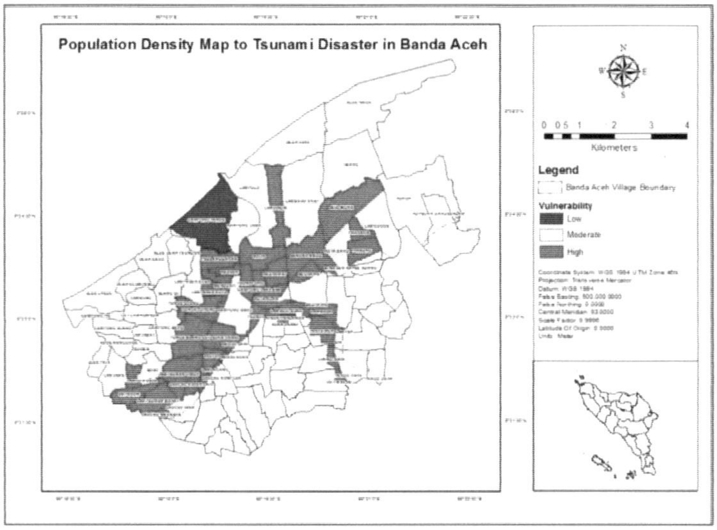

Figure 6. Population Density in Banda Aceh city

The population density in Banda Aceh city is categories as the medium category with almost all villages close to the beach were in the medium category except for Gampong Pande (Kuta Raja sub-district) which falls into the low category.

Almost all the sex ratio of the City of Banda Aceh has a low level of social vulnerability by the tsunami disaster. Only one village that has considered as moderate level. The lowest sex ratio (percentage male: female) in the City of Banda Aceh was in Gampong Pande (Kuta Raja sub-district) which is categorized as moderate level and the highest sex ratio was in Lamseupeung (Lueng Bata sub-district) in a low-level category as it shown in Figure 7.

IOP Conf. Series: Earth and Environmental Science 273 (2019) 012030 doi:10.1088/1755-1315/273/1/012030

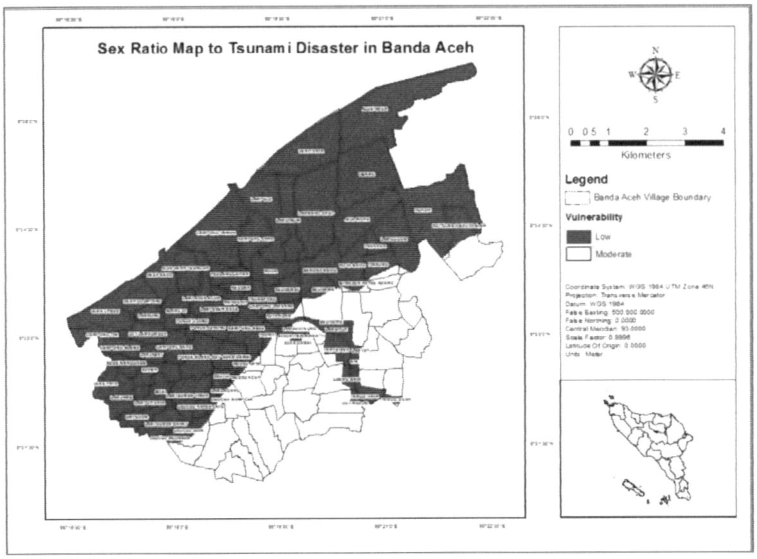

Figure 7. Sex ratio in Banda Aceh city

The poverty ratio data describes the level of poverty that exists in an area. In the City of Banda Aceh, the lowest poverty ratio was in Gampong Pande (Kuta Raja sub-district) and the highest one was in Punge Ujong (Meuraxa sub-district). Figure 8 shows the poverty ratio of social vulnerability in the city.

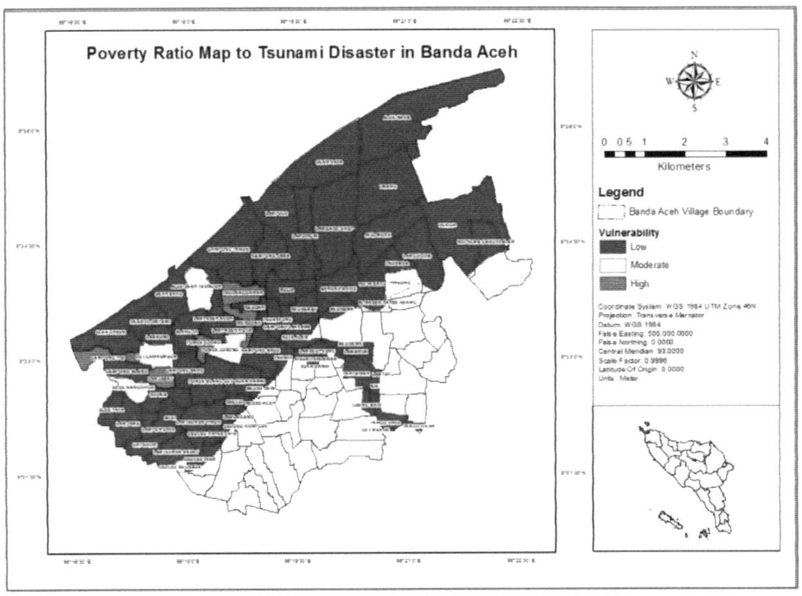

Figure 8. Poverty ratio in Banda Aceh city

In Banda Aceh city, all of the villages were in the lowest disability ratio. The calculation based on the total number of disability divided by the hazard area in km^2. The disability ratio map is shown in Figure 9.

IOP Conf. Series: Earth and Environmental Science **273** (2019) 012030 doi:10.1088/1755-1315/273/1/012030

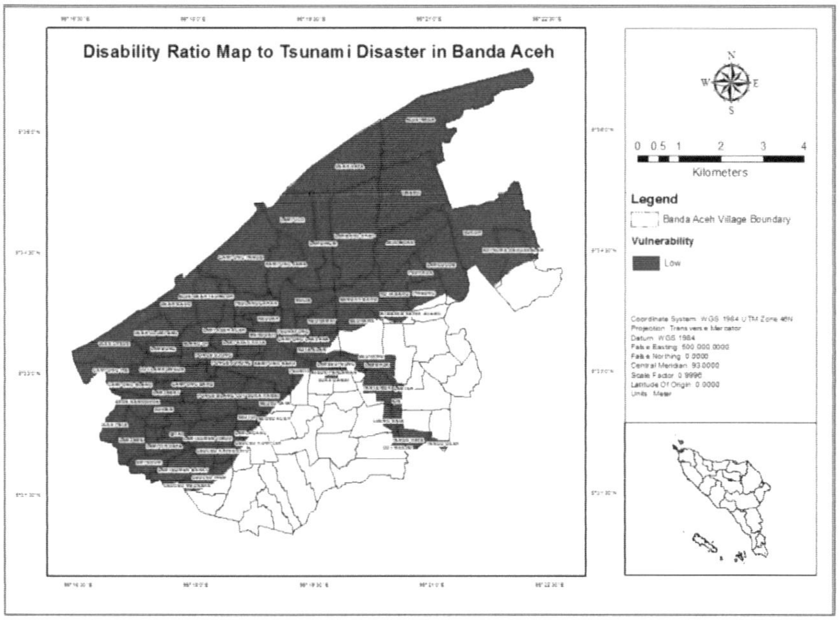

Figure 9. Disability Ratio in Banda Aceh city

The ratio of age group data in defining social vulnerability maps is classified based on the age of (0-4) years and the elderly over 65 years. The calculation is based on the ratio of the total number of the (0-4) and the elderly over 65 years divided by the hazard area in km^2. The highest age group ratio in Banda Aceh city is at Kopelma Darussalam (Syiah Kuala sub - district) and the lowest one at Lamjabat (Meuraxa sub-district) as shown in Figure 10.

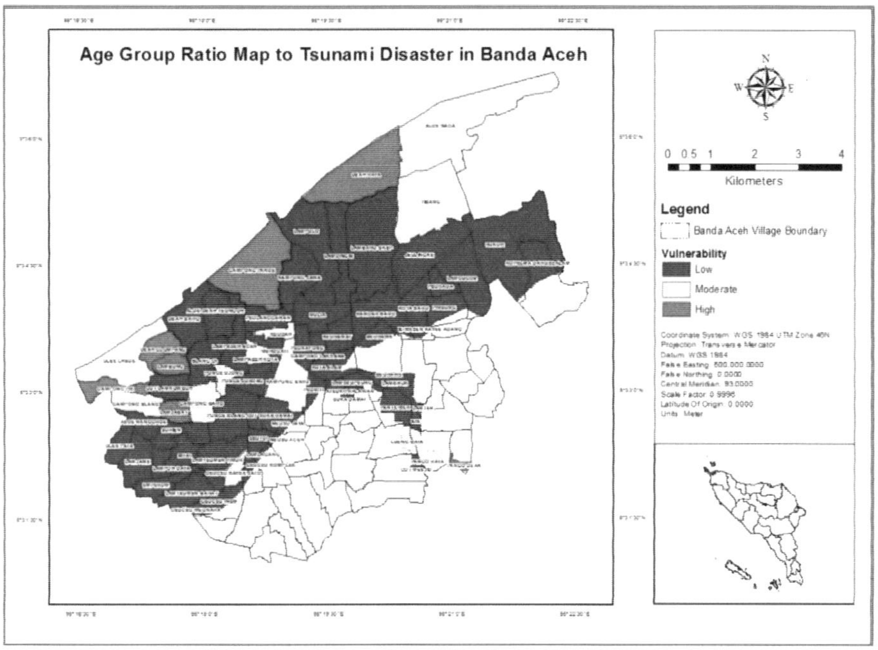

Figure 10. Age Group Ratio in Banda Aceh

Overall with five indicators, Banda Aceh city were in high social vulnerability with 43% of its area was in the medium category and 57% of its area were in the high category. The village with the lowest social vulnerability in Banda Aceh is in Gampong Pande (Kuta Raja sub-district) and the highest one is in Punge Ujong (Meuraxa sub-district). The map of social vulnerability to tsunami disaster is shown in Figure 11.

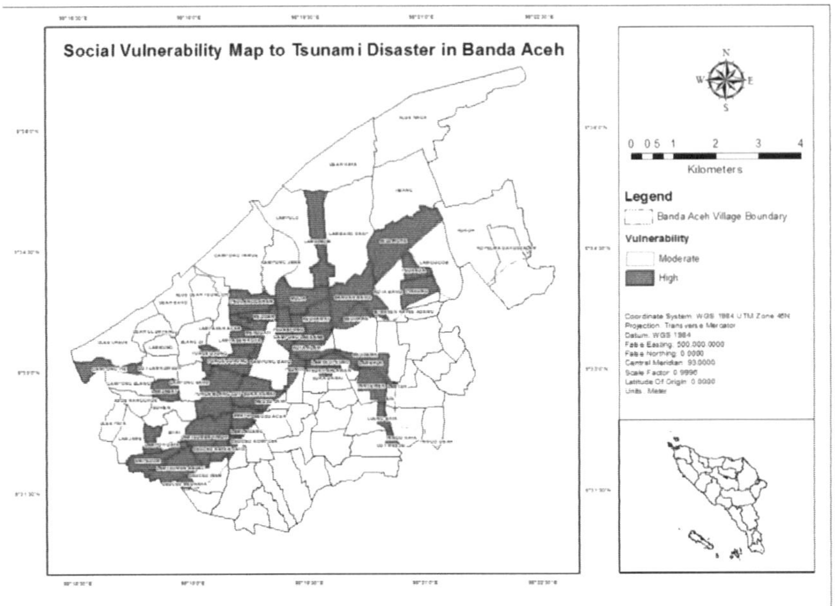

Figure 11. Social Vulnerability in Banda Aceh

3.3. Vulnerability level in the City of Banda Aceh

In the physical vulnerability of the District / City level, the most influential indicators are the house density and the school density. The density of building like house density and school density have an important role. Because the higher building density in the area, the more vulnerable that area will be. Whereas in the social vulnerability the most influential ones are the population density and the age group ratio. These two indicators are relatively dominant for the effect of the disaster. And also the indicators might have the most effective in causing on the number of victims of the disaster. Women, children and the elderly are as vulnerable groups as components which are included in these two influential indicators. These groups should be take into account in the management process of evacuation.

4. Conclusion

Based on the results, it is indicated that 54% of the area in the City of Banda Aceh was included in the category of moderate physical vulnerability. The lowest physical vulnerability is found at Alue Naga village (0.333) and in the contrast of that, the highest one is at Gampong Laksana (0.933). The most influential indicators in determining the highest level of physical vulnerability are the density of houses and the density of school buildings.

In social vulnerability, it is found that 57% of the area were categorized as a high social vulnerability. It means that the number of the total score is still quite high. Punge Ujong (0.833) is the village with the highest level of social vulnerability in Banda Aceh city. The most influential indicator of social vulnerability in the Banda Aceh city are population density and age group ratio.

5. References

[1] BNPB 2013 Indeks Risiko Bencana Indonesia (IRBI) Tahun 2013 *Direktorat Pengurangan Risiko Bencana Deputi Bidang Pencegahan dan Kesiapsiagaan : Sentul*

[2] Republik Indonesia Pedoman Umum Pengkajian Risiko Bencana *Perka BNPB No.2 Tahun 2012*

[3] Løvholt F, Setiadi N J, Birkman J, Harbitz C, Bach C, Fernando N, Kaiser G and Nadim F 2014 Tsunami risk reduction – are we better prepare today than in 2004? *International Journal of Disasters Risk Reduction vol 10* pp 127-142

[4] Fauziah, E. Fatimah, Syamsidik 2014 Penilaian Tingkat Risiko Bencana Tsunami Untuk Kawasan Kota Banda Aceh Berdasarkan Skenario Tsunami Desember 2004 *Jurnal Teknik Sipil Universitas Syiah Kuala Vol 3* pp 145- 56

IOP Conf. Series: Earth and Environmental Science **273** (2019) 012030 doi:10.1088/1755-1315/273/1/012030

[5] BNPB and JICA 2015 Petunjuk Teknis Penyusunan Peta Ancaman dan Risiko Bencana untuk Tingkat Kabupaten/Kota

Acknowledgments
Special thanks to Dedy Alfian for his assistance to conduct this research.

IOP Conf. Series: Earth and Environmental Science **273** (2019) 012031 doi:10.1088/1755-1315/273/1/012031

Dhamar Wulan and Menak Jinggo War; Mythological Interpretation on The Perception of Disaster Risk at Raung Eruption

A Purnomo*[1] and B Kurniawan[1]

[1]State University of Malang, Malang City, Indonesia

*agus.purnomo.fis@um.ac.id

Abstract. The response in the face of disaster provided by the community depends on many factors, one of which is the perception they build by the knowledge, experience, or inheritance of its predecessor. In the case of people who have little contact with information technology, perceptions will be strongly influenced by the inheritance of their predecessors. As in dealing with disasters, they will use the knowledge gained from the experience or knowledge of previous generations. This article aims to interpret the public perception of Bondowoso Regency about the risk of volcanic eruption disaster. Using a qualitative approach with 64 subjects representing three districts namely, Tlogosari, Sempol, and Sumber Wringin. Interview techniques are used for data collection which is then analyzed by four-stage interactive analysis model; (1) Data collection, (2) Data reduction, (3) Data Display, and (4) Verification. Residents from the eight villages believe that the flow of lava (lava flow) in the area they live in is the former blood flow from the fight Dhamar Wulan and Menak Jinggo. So there is no disaster threat in their neighborhood. This knowledge is passed on to the next generation. When confidence in the environment mixes with information from authorities, they still believe that the environment in which they are raised will not pose a catastrophic threat. They are just natural routines that have been experienced by previous generations.

1. Introduction

Already since the beginning of civilization, several groups of humans lived in the Volcanic region [1]. Fertile land and water resources are the main attraction for the community. Volcanoes that are still active in Indonesia are 129 and 70 of them are categorized as very threatening [2].

The community around the slopes of the Raung Volcano has long lived there. They have knowledge and ways to deal with the environment for their survival (local wisdom) [3], including the volcanic eruption disaster. Their response in the face of events is formed from the experience of knowledge, understanding, and the meaning of each event, phenomenon, hope, and problems that occur around it [4].

Responses when disasters are affected by perceptions of disaster [5] [6]. Individual experiences in dealing with disasters will increase perceptions of disasters and become motivations for actions or actions when disasters occur [7] [8] [9]. Other influential factors are socio-economic and demographic characteristics. Individuals in each social group will receive information and evaluate disasters and prepare themselves to face disasters with different resources.

Based on etymology, the risk in Italian comes from the word rischio, risico, or Greek rhizikon, which means root or stone that is uprooted from the ground. The etymology has no meaning until in the 16th century the word risk was meaningful, namely at the time of the emergence of industrialization in Europe, where the meaning was to take the challenge to get success. Similar to China, the risk is interpreted as two currencies: danger and opportunity.

At present in Indonesia, in various regional languages, the word risk has been accepted as part of everyday life. But it turns out it is not easy to get the equivalent of risk words in regional languages. It is easier to find equivalent words of danger or threat in Bugis, Acehnese, Javanese, Sundanese, Minang or Malay languages for example. This is interesting to study further, whether people in various parts of Indonesia are like a shadow of disaster risk reduction scientists and workers who use narrative and risk words as an effort to make social change. Another interesting question is how important is the meaning of risk in the Indonesian context. The risk seems to be a new term for us.

So there are many factors that influence readiness in dealing with disasters such as age, sex. Socio-economic characteristics such as income and education obtained both formally, informal and/or non-formal. On this rational basis, the purpose of this article is to discuss how the influence of the myth of Dhamar Wulan and Menak Jinggo war on people's perception of the threat of Raung Volcano disaster.

2. Methods

This research was conducted using a qualitative approach. A qualitative approach is chosen to get a picture of the profound empirical reality of a social phenomenon. The interview technique used is a very important key in this study [10]. The location of the study was conducted in Bondowoso district, East Java Province. The research subjects were 64 people representing three sub-districts namely, Tlogosari, Sempol, and Sumber Wringin. The results of the data collected are then analyzed in four stages, namely; (1) data collection, (2) data reduction, (3) data presentation, and (4) conclusion drawing [11] [12] [13]. Data collected in the form of interviews, direct observation, documentation, and literature review.

The research location is in three Districts of Bondowoso Regency. Namely Sumber Wringin, Sempol, and Tlogosari Districts. The three sub-districts were directly affected by the Raung Volcano eruption primary disaster. Primary disasters that may occur in the eruption of Mount Raung include pyroclastic flows (hot clouds, pyroclastic falls in the form of rocks (incandescent) and ash dry ash in the type of ash dry. from 10km from the crater, here is a map of the research location.

Figure 1. Settlements in the KRB I Pakisan Village, Tlogosari District

Some villages in the Disaster Prone Area (in Indonesia is *Kawasan Rawan Bencana* -KRB) I region, namely in Sumber Wringin Subdistrict, are Sepanas Hamlet and Toll Roads in Rejo Agung Village, Darungan Hamlet in Sumber Wringin Village, and South Brick Village in Tegal Jati Village. In Tlogosari Subdistrict, there is the Brambang Hamlet, Pakisan Village. And in Sempol District there is Jampit Hamlet in Jampit Village.

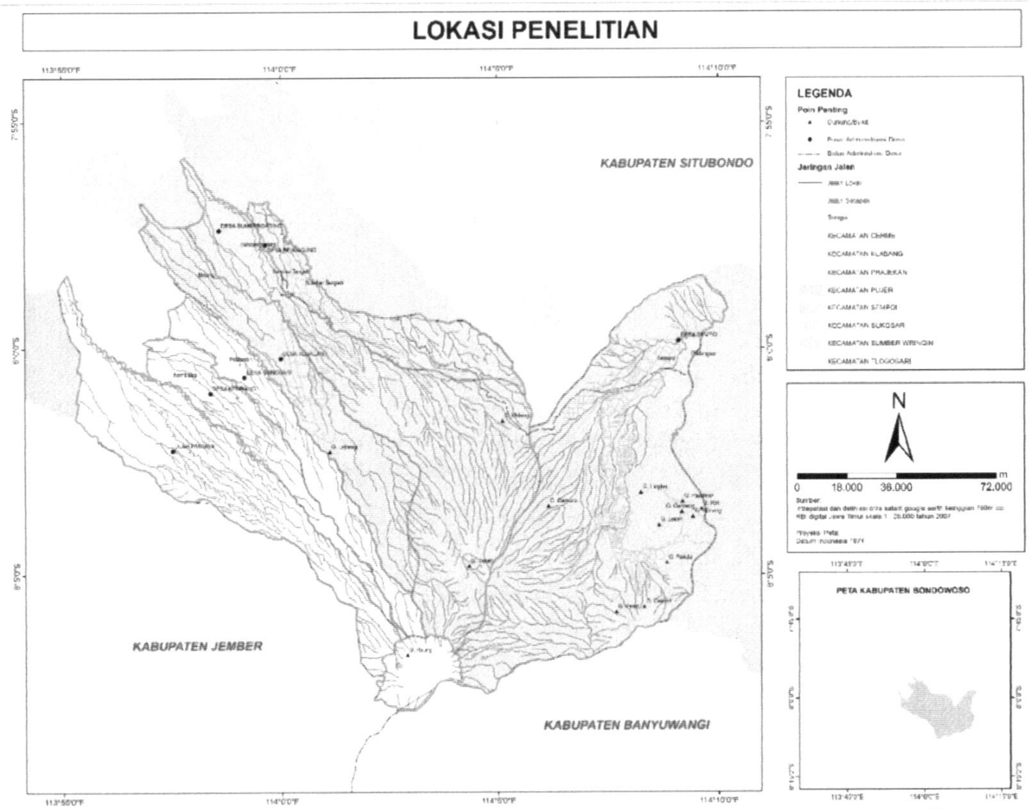

Figure 2. Map of Research Location

Of the three existing sub-districts, Sumber Wringin District has a settlement with the location closest to the crater of the Raung Volcano, which is in Sepanas village. This hamlet is the most severely affected area of eruption in 2015. This area is the Perhutani (Perhutani is mean Business Indonesia Properties or *Badan Usaha Milik Negara* - BUMN, that have the duty and authority to carry out planning, management, exploitation and protection of forests in their working areas) region with the function of the Bondowoso Regency research forest. Most of its citizens are workers in Perhutani with a profit-sharing system. This means they get permission to process and plant forest land with annual crops (coffee) while caring for rubber trees.

Figure 3. Access to the main road in Rejo Agung Village, Sumber Wringin District

IOP Conf. Series: Earth and Environmental Science **273** (2019) 012031 doi:10.1088/1755-1315/273/1/012031

Road access to eight villages in the three sub-districts is 40% severely damaged. Many roads with asphalt surfaces have all been stripped off (30%) and the rest are still in the form of land (10%) which inhibits the rate of the vehicle if an evacuation is carried out. For two-wheeled motor vehicles, the average maximum speed that can be reached is 40km /jm and four wheels are around 20-25km/jm.

From the existing location, eight villages from three sub-districts are available in accessibility and quite difficult to evacuate. The BPBD keeps the posts together to facilitate picking up residents in each village. Improvement of accessibility facilities is an urgent matter that needs to be done. In addition to supporting safety aspects, it will also encourage economic growth. But because most of the area is Perhutani, the improvement of these facilities requires gradual permission from several agencies.

3. Finding and Discussion

Communities around Mount Raung build knowledge about the threat of eruption from the eruption experience in 1956. Many residents consider that the Raung eruption only occurs inside the crater and does not pose a threat. In Bondowoso, only a few people wanted to evacuate during the eruption in 2015. The reason was that the location they were living at now was owned by Perhutani, so they did not object to leaving it. People under the age of 40 who did not experience an eruption in 1956 only gained knowledge from information provided by the Regional Disaster Service Management (in Indonesia is *Badan Penanggulangan Bencana Daerah* -BPBD). When trying to prepare for evacuation, the previous generation provided information that "Eruptions will not be harmful, we used to erupt in the 1950s. Take it easy!".

In Bondowoso Regency, the community interpreted Mount Raung as a place of life that would not harm them. Some of the closest hamlets belong to Perhutani's work area. They work to manage coffee and rubber plants and have a safer place in the lower area. When there is an appeal to evacuate they do not object because they feel there is nothing to fight for or protect.

During the socialization with residents in Gunosari Village, Tlogosari Subdistrict, Bondowoso, the BPBD provided evidence that their area included the lava flow area. But the community interpreted the lava flow as blood flow due to the battle of Dhamar Wulan and Menak Jinggo. This myth became a belief when mingling with the experience of the preceding generation in the Raung eruption in 1956. They believe they do not need to evacuate because the eruption of Raung is an eruption in the crater and will not come out. The myths circulating in society are theoretically expressed as a subconscious feeling of humans [14]. The unconscious feeling is then felt together by the people who are passed down from generation to generation. In the communication of Javanese people (including the communities around Mount Raung), the myth has always been an important consumption to protect the social environment and the natural environment around them. This is because myth also has a function as a control force rather than human life [15] [16]. The result is the maintenance of the social behavior of the Javanese people both towards fellow humans and the universe. This means that the myths believed by Javanese people give birth to certain mindsets for them [17] including their perception of the eruption of Mount Raung which is associated with the myth of Dhamar Wulan and Menak Jinggo.

Picture 4. Lava Flow in Kali Asat, Gunosari Village, Tlogosari District

Knowledge and beliefs in the form of this myth are widely known and get a lot of support, especially in rural areas [18]. Knowledge combined with aspects of experience will build people's perceptions of their environment. This perception is built from traditional sources of information/knowledge. In communities with younger age groups, perception is a combination of knowledge and information from previous generations. This is a modern perception.

Longer life means more experience in dealing with disaster threats from the living environment so as to build perceptions of disasters [5] [6] [7] [8] [9]. This experience is transmitted to the next generation through learning in the event of a disaster. However, if the occurrence of a disaster is not felt to have a significant effect on his life, this new generation will build the perception that disaster is a routine activity in the environment around his residence. This perception is stronger when all their livelihoods depend on the surrounding environment without diversification. They will assume that disasters that occur are part of their lives that have two sides (dualism). That is as a source of livelihood (fertile land) and disasters (damaging livelihoods).

The construction of these two perceptions is getting stronger with actions and information from the government that are too excessive. The local government asks residents in affected areas to evacuate in the area that has been prepared. It should be noted, that residents who are asked to evacuate have the main job as farmers in the garden or breeders. So they feel it is important to protect the assets that are the source of their livelihood. When they feel confident and agree to evacuate, what they get is a false warning sign. Eruption is not as informed by the government. The erosion that occurred was only a burst of ash and was no worse than what had already happened. The damage caused is only temporary in garden plants due to volcanic ash cover.

Residents in Jampit and Rejo Agung Villages experienced a 50% and 80% decline in coffee and rubber production. Coffee plants whose leaves are exposed to ash do not want to flower and produce fruit the following season (one year). While rubber plants experience a production decline of up to 80% because the leaves are covered with ash and dry. The impact is if the plant is forced to remain tapped (the sap is taken) it will die. Therefore farmers do not wiretap until 6 months after the eruption.

The losses caused by the residents of these two villages are not too large. That's because they only manage Perhutani's land with an agreement that 50% of the harvest is owned by the residents and the rest is handed over to Perhutani. Ownership factors greatly influence the response when a disaster occurs. They do not mind being moved at any time because currently they only occupy official homes to look after the gardens. They already have a house in a safer area.

Disturbed sources of livelihood make residents in the two villages have to make decisions to find temporary sources of livelihood. Livestock is assets that are easy to be used as temporary sources of livelihood. But in the event of a disaster, sometimes livestock is one of the variables that makes people reluctant to evacuate. The need to protect livestock mixed with the belief that they live in a safe area builds the perception of residents around active volcanoes that it is difficult to evacuate. Even when in refugee there are some residents who want to enter the danger zone just to see the condition of their livestock.

Consideration of citizens to survive because of more rational reasons. Myths born without logos can be rational when getting a lot of support. These rational reasons ultimately make individuals make decisions as a response to the surrounding environment [19].

Changing the mindset that exists in the community in the face of the threat of catastrophic volcanic eruptions in Bondowoso District needs to study their sources of livelihood. Apart from the aspect of education there needs to be the identification and planning to suppress the negative impacts of this disaster on the character of the community. The aspects of vulnerable livelihoods must be one of the government's attention. The following is the identification of aspects of livelihoods that have vulnerabilities in the event of a disaster.

Table 1 Identification of Vulnerability Aspects and Recommendations for Local Governments

No	Aspect	Vulnerability	Recommendation
1	Physical	Agricultural land is very vulnerable to damage when there is ash rain	Land restoration after a disaster occurs and giving new seeds after the land is ready for planting
2	Economy	Damaged agricultural land results in economic activity stopping, residents will divert livelihoods in the form of livestock savings. The reasons for protecting livestock also have an impact on the decision to evacuate during a disaster.	The government provides shelter and feed locations for livestock

Perception in disaster risk reduction efforts is an important approach point. Mitigation planners must look at local values in the surrounding community so that mitigation efforts do not become a responsive effort in dealing with disasters. These activities must be part of the community (risk society) by integrating into people's lives so that mitigation becomes an anticipatory activity.

The Risk Society does not produce a community that is able to manage risk, but it is emphasized that this community is able to do so because it has fulfilled certain conditions, including the ideal 'risk communication'. The prerequisite is implied in the definition of risk communication, which is an exchange of information and opinions, as well as building effective dialogue between the various parties involved in reviewing, minimizing and managing risks, together with those (communities) affected by the risk.

However, for the sake of effective risk communication, the community needs to have a critical attitude towards reality, ask questions and question science itself, as a form or characteristic of modernity. Dialogue and democratic processes in discussing risk information must be fulfilled. These prerequisites are clearly not fulfilled if we are still trapped in a false news and information octopus (hoax), just reading the big titles without looking further at the contents of the news, believing it, even participating in disseminating it. This is what literary weakness means. When we do not have the preconditions of a critical attitude as a 'society risk' or a society that is familiar with risk, then we need to look further at the causes. Is the level of literacy or reading power of the community weak, the level of education that is still low, distrust of modern science, or so forth. With risk communication still having problems, it is difficult for us to hope that people can take ideal disaster risk reduction measures as imagined.

In the case that occurred in Raung, the government chose to take on a scare role so that the community was always ready. But this is a bad value for the government if what is informed does not occur. Public trust in the government will be lost so that it will be difficult when a disaster occurs in the future. It is important for the government to consider aspects of myth or local value as one approach that can accommodate risk communication. As happened in the Puailiggoubat newspaper in Mentawai. Three years after the 2004 tsunami, the media still discussed the role of the community in restoring all damaged aspects (Figure 5).

Local values in the form of increasing religiosity will have an impact on reducing the potential for disasters in the future and are also found in many Indonesian people, especially Jawa [20]. The story of the war of Dhamarwulan and Menak Jinggo is a very thick and inherent history in the eastern part of East Java (Banyuwangi, Jember, Bondowoso, and Situbondo). The belief in the value of the cult from the surrounding environment provides a different view in dealing with disasters. They will understand disasters as part of the natural routines around them [21], or usually referred to as the domestication of disasters.

From this fact, the literacy of the community against disasters is not empty at all. It's just that we as academics view that literacy is incompatible with modern literacy so that preventive measures in the form of efforts to frighten become the most rational choice when facing disasters. If examined further the problems of the community are distrust of modern science. With the still problematic risk communication, it is difficult for us to hope that the public can take ideal disaster risk reduction measures as imagined.

Figure 5. Puailiggoubat Newspaper December 2007
Source issuer/puailiggoubat/docs/ok

4. Conclusion

Local myths are public literacy that has been built before modern knowledge exists. Local myths build more religiosity about the value of the environmental cults in which they live. In disaster risk reduction efforts academics sometimes tend to nullify the role of local knowledge with modern knowledge in the form of "scaring" the community with excessive threats about the impact of disasters. This effort is indeed quite effective for preventive actions so that people will stay away from the danger zone. It's just that if the threat delivered does not occur, it will have an impact on how people face disasters in the future. Experience and knowledge of disasters that exist make them not believe in modern knowledge. Therefore, there needs to be a special room for local knowledge for academics in providing understanding to the community regarding disaster risk reduction efforts as an effort to approach a more populist knowledge.

References

[1] Howes, Dale Dominey. 2004. Perceptions on Hazard an Risk Santorini. Journal of Volcanology and Geothermal Research 137 (2004). P 285-310.
[2] Zamroni, M. Imam. 2011. Islam dan Kearifan Lokal dalam Penanggulangan Bencana di Jawa. Jurnal Penanggulangan Bencana Vol 2 No 1Tahun 2011. Hal 1-10
[3] Mishra, S., and D. Suar. 2007. Do lessons people learn determine disaster cognition and preparedness? Psychology & Developing Societies 19:143-159. http://dx.doi.org/10.1177/097133360701900201
[4] Vazquez, Esperanza lopez and Marvan, Maria Lusia. 2012. Volcanic Risk Perception, Locus of Control, Stress and Coping Responses of People Living Near the Popocatepetl Volcano in Mexico. Journal of Risk Analysis and Crisis Response, Vol 2 No 1 May 2012. P 3-12
[5] Sattler, D. N., C. F. Kaiser, and J. B. Hittner. 2000. Disaster preparedness: relationships among prior experience, personal characteristics, and distress. Journal of Applied Social Psychology 30:1396-1420. http://dx.doi.org/10.1111/j.1559-1816.2000.tb02527.x
[6] Miceli, R., I. Sotgiu, and M. Settanni. 2008. Disaster preparedness and perception of flood risk: a study in an alpine valley in Italy. Journal of Environmental Psychology 28:164-173. http://dx.doi.org/10.1016/j.jenvp.2007.10.006

[7] Russell, L. A., J. D. Goltz, and L. B. Bourque. 1995. Preparedness and hazard mitigation actions before and after two earthquakes. Environment and Behavior 27:744-770. http://dx.doi.org/10.1177/0013916595276002
[8] Lindell, M. K., and R. W. Perry. 2000. Household adjustment to earthquake hazard: a review of research. Environment and Behavior 32:461-501. http://dx.doi.org/10.1177/00139160021972621
[9] Tekeli-Yesil, S., N. Dedeoglu, M. Tanner, C. Braun Fahrlaender, and B. Obrist. 2010. Individual preparedness and mitigation actions for a predicted earthquake in Istanbul. Disasters 34:910-930. http://dx.doi.org/10.1111/j.1467-7717.2010.01175.x
[10] Moleong, J., Lexy. 2014, Metodologi Penelitian Kualitatif, Bandung: PT. Remaja Rosdakarya.
[11] Creswell, J. W., & Clark, V. L. P. (2017). Designing and conducting mixed methods research. Sage publications.
[12] Fatchan, H. A. (2011). Metode Penelitian Kualitatif. Surabaya: Jenggala Pustaka Utama.
[13] Denzin, N. K. (2017). The research act: A theoretical introduction to sociological methods. Routledge.
[14] Jung, C.G., 1998. Jung on mythology. Princeton University Press.
[15] Wilkinson, P. and Philip, N., 2007. Mythology. Dorling Kindersley Ltd.
[16] Audifax. 2005. Mite Harry Potter: Psikosemiotika dan Misteri Simbol di Balik Kisah Harry Potter. Yogyakarta: Jalasutra
[17] Angeline, M., 2015. Mitos dan Budaya. Humaniora, 6(2), pp.190-200.
[18] Dove, Michael R. dan Bambang Hudayana. 2011. The view from the volcano: an appreciation of the work of Piers Blaikie. www.elsevier.com/locate/geoforum
[19] Ritzer, George and Douglas J. Goodman, 2010, Teori Sosiologi Klasik Sampai Perkembangan Mutakhir, Yogyakarta: Kreasi Wacana
[20] Adisaputri Gianisa, Loic Le De, (2018) "The role of religious beliefs and practices in disaster: The case study of 2009 earthquake in Padang city, Indonesia", Disaster Prevention and Management, Vol. 27 Issue: 1, pp.74-86, https://doi.org/10.1108/DPM-10-2017-0238
[21] Bachri S, et al. (2015). The calamity of eruptions, or an eruption of benefits? Mt. Bromo human–volcano system a case study of an open-risk perception. Published by Copernicus Publications on behalf of the European Geosciences Union. Nat. Hazards Earth Syst. Sci., 15, 277–290, 2015

Constructing Volcanic Hazard Video Toward Student Comprehension of Kelud Volcano Disaster Prone Area And Its Environment Characteristics

L Y Irawan[1*], Sumarmi[1], S Bachri[1], A R O Sari[2], A Wahyudi[2], I K Ayuni[2], S Lailunnahar[2], and S D Dirgantara[2]

[1]Lecturer at Department Geography, Faculty of Social Science, Universitas Negeri Malang, Jalan Semarang 5, Malang, Indonesia
[2]Student at Department Geography, Faculty of Social Science, Universitas Negeri Malang, Jalan Semarang 5, Malang, Indonesia

*E-mail: listyo.fis@um.ac.id

Abstract. This paper aim to presents several procedural proccess to develop learning material toward volcanic disaster with case study Kelud Volcano in East Java. Last eruption of Kelud volcano occured on 13th February 2014. This eruption phase has a big impact in large area especially in disaster prone area in Blitar, Kediri, and Malang. Volcanic pyroclastic material and lahar became serious hazardous problem. Some people who live in disaster area actually have not any experience coupling with volcanic disaster. In volcanic disaster management one most important thing is preparing people to reduce disaster risk. To achieve this purpose people especially student need to understand or have comprehension toward hazard, vulnerability, and risk of Kelud Volcano. Hence, need an effective media to reach student comprehension by using video. The video was developed by using design and development research. This video present series of Kelud volcanic hazard, vulnerability, people capacity to reduce disaster risk. The video is used to student in the senior high school. The result of using could increasing students understanding/comprehension toward Kelud volcano its environmental characteristics, hazard, vulnerability, and how to enhance people capacity toward disaster risk reduction. The final finding shows that learning material in form of video is used full to increasing student knowledge, comprehension, and capacity in the efforts of disaster risk reduction.

1. Introduction

Indonesia lies among 129 active volcanoes in the world. These condition has consequences due to a large of volcanic hazards that widespread in Indonesia. Kelud Volcano is one of an active volcano in East Java Province, Indonesia. This volcano types belong to Stratovolcano. The Stratovolcano has explosive eruption characteristics.

Kelud Volcano Eruption in 2014 had a serious damage and loss impacts in several disaster-prone areas. Damage to residents' settlements, paralysis of the agricultural industry, airport closure, and damage to various public facilities is a clear example of the losses experienced during the eruption period. On the other hand, eruption events have several positive influence on the paradigm shift of thinking and acting on the threat of eruption of the Kelud Volcano [1]

Disaster risk reduction education for high school students who are vulnerable to volcanic disasters has not fully accommodated students' learning needs. In general, disaster education in high school had limited resources that explain the environmental conditions of the student's residence and surroundings. The characteristics of hazard threats should be understood by students in a disaster-prone area.

As an ongoing education and learning effort a model in disaster risk reduction education (DRRE) has been widely developed. [2] describes several concepts, namely: disaster risk reduction education (DRRE), education for sustainable development (ESD), climate change education (CCE) and environmental education (EE) which are carried out in various areas and contexts. The 2006 UNISDR campaigned for disaster risk reduction starting from school. This aims to propose integration between disaster risk reduction and the curriculum that is implemented in schools.

Based on the development of disaster risk reduction education trends and student learning needs in disaster-prone areas, the disaster material content should be contextually arranged with the actual conditions in the field. The development of learning design is expected bridging student learning need. Student learning need is a conditions of fully understood about surrounding environment.

Development of learning media for disaster risk reduction must be in line with the basic competencies that are applied in schools, especially in 11th grade of senior high school. Basic competencies in 11th grade of senior high school is suitable to develop of DRRE learning models. One kind of tool that develop in DRRE is video. In this case the video content are characteristic of geographical place/location; its potential hazard; people and their surrounding environment vulnerability; their capacity in order to reduce impact of disaster risk.

Nowadays, learning materials in disaster is currently limited and lack of appropriate learning media. Learning video media is a tool for presenting audio and visuals. This media contains concepts, principles, procedures, and application theory to help student understand their environment [3]. Learning media is helpful for direct information from teachers to students. This function are stimulate students' thoughts, feelings, attention, and interests. Learning process becomes more efficient after using learning media/video. Through good media visualization will make it easier for students to understand about 30% [4].

The learning about natural disasters, especially the eruption of Kelud Volcano, becomes very important. This is important because it can reduce the impact of casualties caused by the eruption of the Kelud Volcano. Students can carry out their own disaster mitigation activities by knowing the vulnerable areas of Kelud, characteristics of eruptions at Kelud Volcano, and disaster mitigation activities that can be carried out.

Geography learning is an attempt to create a learning process that can facilitate students to be able to understand geosphere phenomena, identify geosphere phenomena, and formulate and draw conclusions from a geosphere phenomenon. The geosphere phenomenon in question is a problem that occurs in the environment.

School is a learning institutions where students learn about cultural values, religious values, traditional-modern knowledge, including knowledge about disaster. According to John Dewey in Nasution explained the school as the most effective medium for reconstructing and improving society through education. Therefore, schools in disaster-prone areas must have sufficient preparedness to deal with natural disasters around them. Schools also have a role in improving students' preparedness in dealing with disasters. The purpose of disaster learning in schools is to build a culture of alertness and a safe culture in schools and build resilience in the face of disasters.

Urgency in studying disaster material is a constraint, such as limitations of learning media. Conceptual learning media is needed in daily learning. The available learning media currently using old media that is not contextual. The introduction of the characteristics of the region's environment prone to volcanic eruption and disaster risk reduction must be contained in the geography learning media. The availability of geography learning media about disaster risk is useful in students' understanding of the environment, and the risk of disaster. Kelud Volcano, and how to mitigate disasters that must be done if the eruption of the Kelud volcano is repeated. The availability of learning media does not exist in schools makes students less aware of the characteristics of disasters in their environment.

The media is a link and tool for learning activities. Learning media is used as a tool to channel information from teachers to students and stimulate students' thoughts, feelings, attention, and interests so that the learning process becomes more efficient. By using the media, the teacher can present learning material in audio visual form such as moving pictures coupled with the writing and material presented seem more interesting. Thus the delivery of material by the teacher becomes easier. Especially in the basic competencies of disaster mitigation.

2. Method

The early stage of design and development learning materials based on national curriculum. The national curriculum on geography subject divided into several basic competencies. In the matter of disaster mitigation there is basic competencies in senior high school.

To achieve the goals of learning media development, Operational efforts are carried out with the development of design and development research [5]. In this design there are three stages including:

2.1 Learning Theory And Research;

The first stage of Learning Theory and Research is a stage of identification and analysis of student learning needs. Identification and analysis of needs are included disaster mitigation material. Disaster mitigation materials taught in high school are not contextual. Student learning needs should accommodate the characteristics of the disaster location. Learning material is not only disaster mitigation but also disaster risk reduction. In the material developed disaster risk reduction is aimed at reducing the risk of volcanic disasters.

2.2 Instructional and Instructional Design Theory and Research;

The second stage is based on developing basic competencies in schools. The operational stages include: (1) designing learning objectives, (2) introducing learning material content, (3) preparing learning media in accordance with learning material (in this case the media used is video), (4) preparing learning evaluations and assessments.

2.3 Communication Theory and Research

The third stage is the implementation of disaster risk reduction learning in the classroom. The implementation includes: the use of video media, evaluation aims to test students' understanding and to know the effectiveness of using video media.

The first stage of testing is used by evaluating through subjective tests. Subjective tests include: geographical characteristics of volcanic regions; history of volcanic eruptions; areas prone to volcanic disasters; factors that influence the risk of volcanic disasters; and efforts to reduce the risk of volcanic disasters. The second stage of effectiveness testing is done by the teacher and student responses.

Based on the first testing phase, quantitative data is obtained in the form of student scores. Whereas from the second stage the teacher and student responses obtained a qualitative description of the effectiveness of the development and application of video media.

3. Results and Discussion

The preparation of learning video media is based on the need for disaster risk reduction learning materials in high school. Material needs cover the breadth and depth of the material. The broad aspects of learning materials for disaster risk reduction include: Characteristics of Kelud Volcano, Disaster Prone Areas, Hazards of Kelud Volcanoes, Vulnerability, Risk, and Disaster Mitigation Efforts. While the depth aspect of the material is the detail of organizing the material that has been prepared previously.

The development of students' cognitive level becomes a reference in developing learning materials. The cognitive level of high school students will develop rapidly if supported by learning media that is appropriate to their age development. In addition to the cognitive aspects of learning videos must also pay attention to the affective aspects of students. The affective aspects of students relate to the acceptance of students' attitudes in disaster risk reduction.

Learning videos cannot yet be used extensively before being tested. The test was used to determine the accuracy of the material presented in the video, the level of students' understanding, the response of students and teachers to the video. The test was conducted in schools located in Kelud Volcano eruption prone areas.

Experimental of videos carried out through the stages of delivering learning materials using learning videos. Totally 38 students were involved in video experimental learning product activities. First stage before use the video there is a pre-test. Pre test is used to determine the level of students' understanding of volcanic disasters. The questions given at the pre test are the same as the post test questions. The scope of the question includes: 1) Geographical characteristics of Kelud Volcano; 2) History of Kelud Volcano eruption; 3) Kelud Disaster Prone Areas; 4) Hazard, vulnerability and risk analysis of Kelud Volcano; 5) Efforts to reduce the risk of Kelud volcano eruption disaster.

The pre test results showed an average student score of 58 from a scale of 100. The average pre test score showed that students' knowledge of disaster risk was in sufficient classification. Classification of students level of knowledge regarding volcanic disaster material can be shown in Table 1 as follows:

Table 1. Classification of Students' Level of Knowledge in Volcanic Disaster Material

Range Score	Classification of Student Knowledge Level
0-20	Very limited knowledge
21-40	limited knowledge
41-60	Adequate knowledge
61-80	Well knowledge
81-100	Very well knowledge

The second stage of the Experimental was the delivery of learning material on the risk of Kelud volcano disaster using learning videos. Learning videos contain: 1) Type of Kelud volcano; 2) Types and characteristics of natural disasters; 3) Impact of Kelud Volcano eruption disaster; 4) Disaster management cycle; 5) Institutions/agencies that play a role in the disaster management of Kelud Volcano.

The third stage is the post test. The evaluation questions used in the post test are the same as the evaluation questions at the pre test. The post test results showed an increase in the average score of students which is equal to 73. This means there is a gain score of 15 points from the pre test.

The fourth stage is the response of students and teachers shown in the student and teacher response questionnaire. The indicators used in the student-teacher response questionnaire can be seen in Table 2.

Table 2. Student-Teacher Responds on Media

No	Variables of Student-Teacher Responds on Media	1	2	3	4	5
1	Approprieteness of learning material with its indicator					
2	Role of media in order to achieve student comphrehension					
3	Approprieteness of learning contect and indicator					
4	Material depth					
5	Approprieteness of learning and its environment					
6	Accuration of fact and data in learning material					
7	Approprieteness material content and context					
8	Clarity of learning material					
9	Appearance of materials systematics					
10	Level of student comprehension on learning material					
11	Role of media accordance to understand					
12	Role of pictures to enhance knowledge					
13	Role of task and assignment in increasing student comprehension					
14	Student couriosity toward learning material					

IOP Conf. Series: Earth and Environmental Science **273** (2019) 012032 doi:10.1088/1755-1315/273/1/012032

15	Level of learning material sophisticated

Remark:
Score 1 Very limited; 2 Limited; 3 Adequate; 4 Well; 5 Very Well
Weight 1.3
Minimum value 20; maximum value 100

As a final result Student-Teacher Responds on Media get score 90.62. It shown this media is valuable and can be used to volcanic disaster risk reduction learning material. Media video is suitable for a subject that describes a certain process, several phases of demonstration, and concepts related to everyday life. In the cognitive domain, students obtain information in the form of a real condition from the actual environmental conditions and processes that work in it. On the other hand, video media presents a real condition without students field observation.

Video development process should be consider on several lesson criteria. These criteria could be distinguish into several phases as follows:

1. Type of material

Selection of materials based on the complexity of the content that can be shown on the incidence of Kelud Volcano eruption. On the other hand, preparation of instructional video media useful as supplementary materials are not available in textbooks geography. So as well as students gain a real condition of the processes that occur on Kelud disaster prone area. Based on these explanations, it can be determined in lesson learning materials include:

a. Type of Kelud volcano
b. Types and characteristics of natural disasters
c. Impact of Kelud Volcano eruption
d. Disaster management cycle
e. Institutions / agencies involved in disaster management Kelud Volcano

2. Video presentation format

Learning videos prioritize the presentation of clarity and breadth of material. Video formats that are suitable for learning disaster risk reduction include: narrative explanation, interview presentation and presenter monologue. In the design of this development use a combination format. The combined format chosen for the material characteristics disaster risk reduction can be optimally by using a variety of formats. Narrative format is used to describe the processes of volcano. Interview format used to describe the threat of volcano hazard and its impact. Presenter format used to describe the general conditions in the field [7].

3. Video duration

In the preparation of the video media to be effective when presented for 20-40 minutes. It is based on the ability in students memory and concentration skills. They have limitation in understanding material. Short video duration provides benefits compared to movies. Some materials capture from video can be shown as follows.

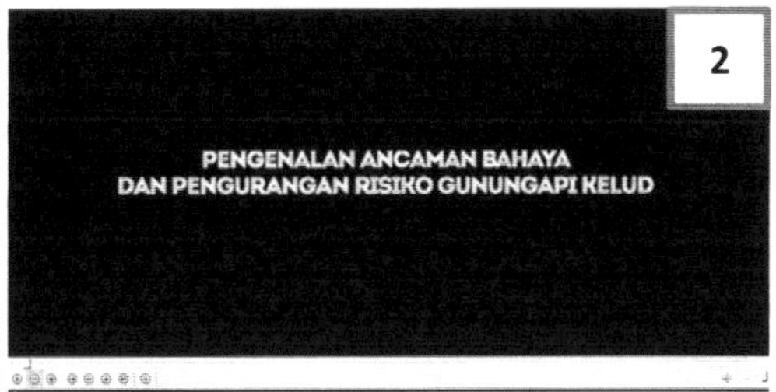

Figure 1. Video opening; **Figure 2.** Introduction of Learning Material on video; **Figure 3.** Aerial view Post Kelud Eruption on 2014; **Figure 4.** Opening speech; **Figure 5.** Opening speech by presenter; **Figure 6.** Kelud Volcanic Disaster Explaination by Head of Volcano Observation ; **Figure 7.** Crater Lanscape after 2014 eruption

4. Concluding Remarks

Video media is suitable for learning material that is describing a particular process, a demonstration plot, a concept or describing something. In the cognitive domain, students obtain information in the form of a real picture of the field conditions and processes. Video can present a real picture for students without doing field observations.

5. References

[1] Sartohadi J and Pratiwi E S 2014 *Bunga Rampai Penelitian: Pengelolaan Bencana pada Kegunungapian Kelud pada Periode Krisis Erupsi 2014* Yogyakarta PUSTAKA Pelajar
[2] Shaw R Shiwaku K Takeuchi Y 2011 *Disaster Education* Emerald Group Publishing Limited
[3] Riyana, C 2007 Pedoman Pengembangan Media Video Jakarta: P3AI UPI
[4] Anderson, H.M. Dale's Cone of Experience http://www.queensu.ca/teachingandlearning/modules/ active/documents/Dales_Cone_of_Experience_summary.pdf
[5] Richey, R. C. & Klein, J. D. (2007). Design and development research. NJ: Lawrence Erlbaum Inc.
[6] Russell Gazda M.S. & Michael G. Flemister Ph.D. 1999 *Design and Production of Video for Instructional Multimedia: Psychological Implications and Proposed Guidelines* Journal of Visual Literacy, 19:1, 85-98, DOI: 10.1080/23796529.1999.11674546

[7] Ellis R Childs M 1999 *The Effectiveness of Video as a Learning Tool in On-line Multimedia Modules*, Journal of Educational Media, 24:3, 217-223, DOI:10.1080/1358165990240305

Acknowledgments
Authors wishing to acknowledge assistance or encouragement from colleagues and students in Department of Geography Universitas Negeri Malang who accompanied this research patiently. Special thanks dedicated to Prof. Dr. Sumarmi, M.Pd, Syamsul Bachri, Ph.D who gave important advice in this research.

An effect of tsunami to hotel occupancy: A case of Phuket, Thailand

J Tang[1], N Leelawat[2,3,4,*], A Suppasri[5] and F Imamura[5]

[1] School of Management Technoogy, Sirindhorn International Institute of Technology, Thammasat University, 131 Moo 5 Tiwanont Road, Bangkadi, Pathum Thani 12000 Thailand
[2] Department of Industrial Engineering, Faculty of Engineering, Chulalongkorn University, Phayathai Road, Pathumwan, Bangkok 10330 Thailand

[3] Disaster and Risk Management Information Systems Research Group, Chulalongkorn University, Phayathai Road, Pathumwan, Bangkok 10330 Thailand
[4] Risk and Disaster Management Program, Graduate School, Chulalongkorn University, Phayathai Road, Pathumwan, Bangkok 10330 Thailand

[5] International Research Institute of Disaster Science, Tohoku University, 468-1 Aramaki-Aza Aoba, Aoba-ku, Sendai 980-0845 Japan

[*]Corresponding author, natt.l@chula.ac.th, n.leelawat@gmail.com

Abstract. As a popular city for global tourists, more than one third of the Phuket GPP depends on the hotel and restaurant sector. The 2004 Indian Ocean Tsunami caused a serious physical damage to hotels and a high cancellation of bookings in the post disaster period, which, in turn, led to most THB 3 billion drop of the hotel and restaurant sector income in 2005 in Phuket. In addition, many tourism facilities in Phuket suffered bankruptcy or a slump, because they started their business again immediately after the reconstruction phase, while tourists did not really come back. This paper is an initiative study of our research project to evaluate the effects of the 2004 Indian Ocean Tsunami on Hotel Occupancy, with a focus on the speed for the physical reconstruction and tourism market recovery. We collected the physical reconstruction period and the hotel occupancy recovery data of six hotels that located in the Patong Beach of Phuket from 2003 to 2017. Through calculation, the average of damage rate (i.e., unavailability of the rooms in this study) is 55.89%, with an average speed for the physical reconstruction to rebuild the rooms about 6.07% of the overall rooms per month. The quotient of these two values indicates that the reconstruction period is around 9.2 months. However, the average period for reconstruction and tourism recovery is about 20 months. There is an about 11 months' delay between that hotel and resorts have finished the reconstruction and the tourists start to visit Phuket as usual. Therefore, the local government, local tourist authority, and hotel entrepreneurs can consider these findings in order to prepare the recovery plan for tourism industry in the future tsunami.

Keywords: 2004 Indian Ocean Tsunami, Hotel Occupancy, Patong Beach, Tourist Industry, Tourist Business Recovery

1. Introduction

Nowadays, tourism contributes to more than 9% of Thai GDP (Gross Domestic Product) [1]. Phuket has the third-highest GPP (Gross Reginal and Provincial Product) among Thailand's provinces, and its

IOP Conf. Series: Earth and Environmental Science **273** (2019) 012033 doi:10.1088/1755-1315/273/1/012033

economy heavily depends on this industry [2]. Especially, the hotel and restaurant sector accounted for more than one third of the Phuket GPP [2].
Back on the 26th December 2004, there was an earthquake in Sumatra Island in Indonesia with a magnitude of 9.1 [3] generating a massive tsunami that claimed lot of lives in 15 different countries with a maximum height approximately 50 meters [4]. In Thailand, it not only resulted in the death of at least 8,000 people [5], but also the devastation of many facilities, such as resorts, roads, households, telecommunication systems, pipeline systems and electricity facilities [6]. Phuket is considered to be the one of the most affected provinces in the 2004 Indian Ocean Tsunami in the south of Thailand, which includes Phang-nga, Krabi, Trang, Ranong, Satun and Phuket. It led to almost THB 3 billion drop of the hotel and restaurant sector income in 2005 [7], because of the serious facility damage and a high cancellation of bookings in the post disaster period. According to Airport of Thailand, the total international passenger arrivals at the Phuket internal airport in 2005 dropped almost 50%, comparing to 2004.

Many tourism facilities in Phuket suffered bankruptcy or a slump, because they started their business again immediately after the reconstruction phase. However, tourists did not really come back to the area for a long time [8]. In other words, the key issue of the tourism industry in Phuket in the post tsunami recovery was not the reconstruction of the tourism facilities, but how to bring back the visitors who have been scared away by the tsunami [9]. Nevertheless, there is a big gap between the anticipated period for the rebuilding of the tourism infrastructure and the real one. Although more than half of the hotels in Phuket were able to re-open a week after the tsunami, the period estimated by the government officials varied from weeks to a year. This would result in a delay in the tourism promotion. Moreover, a good anticipation the average speed for the physical reconstruction and tourism market recovery could help hotels to decide a proper time for their reopening. Patong beach was the beach in Phuket that got the worst hit by this tsunami. Based on a case study of Patong beach in Phuket, this paper tends to study the effects of the 2004 Indian Ocean Tsunami on Hotel Occupancy, with a focus on the speed for the physical reconstruction and tourism market recovery.

2. Background

The 2004 Indian Ocean Tsunami caused a huge damage and affected many different areas in Thailand. Table 1 summaries its economy impacts on three main sectors: social, production, and infrastructure sectors in detail. Out of the total damage and losses 85,747 million Baht, 83.9% (about 71,972 million Baht) is caused by the tourism. As shown in Table 2, the damage and losses of tourism facilities is about 14,648.3 million Baht. The remaining 80% (about 57,324 million Baht) are revenue losses because of the unavailability of rooms and the slow pace recovery of the tourists to arrive in the affected areas. Following Phang Nga, The damage and losses of tourism facilities in Phuket is the second largest in Thailand, around 4,103 million baht (*see* Table 2).

Table 1. Impacts of the 2014 Indian Ocean Tsunami to Thailand.

Sectors	Sub-Sectors	Description	Damage & Losses (Million Baht)
Social	Population	Death of people	--
	Housing	Damage and losses of housing units	866
	Education	Damage and losses of schools	340
	Health	Damage and losses of health facilities	115
Production	Agriculture	Losses of standing crops, together with the reductions in productivity	376
	Livestock	Losses of livestock numbers	18
	Fisheries	Damage and losses of fishing fleets & gears, together with the reductions in productivity	6,481
	Industry	Subsequent losses caused by the losses in agriculture and fishery	2,182
	Commerce	Subsequent losses caused by the losses in agriculture and fishery	1,479
	Tourism	Damage and losses of tourism facilities, together with the drop-in tourist arrivals, and further leads to losses of revenue	71,972
Infrastructure	Water Supply	Damage of water supply systems, together with the loss of revenue	153
	Electricity	Damage of electrical subsystems, together with the loss of revenue	557
	Transportation& communication	Damage and losses of roads, bridges, drainage works, airports and tele-communication facilities	632
	Others	Damage and losses of piers, dams and retaining walls, buildings and others	576

Note. Source [6].

Table 2. Estimated replacement and repair value of tourism facilities of the 2014 Indian Ocean Tsunami (Million Baht).

	Hotels and Resorts				Other Commerce				Total
	#	Bldgs	Goods	Damage	#	Bldgs	Goods	Damage	
Krabi	131	3,256.9	473.7	3,730.6	984	13.4	3.6	17.0	3,747.6
Trang	4	150.0	24.0	174.0	142	1.3	0.2	1.5	1,75.5
Phang Nga	93	5,953.0	340.2	6,283.1	858	20.4	4.9	25.3	6.616.1
Phuket	100	3,170.9	637.8	6,590.8	2,858	367.6	103.3	470.9	4,103.0
Ranong	2	4.4	0.6	5.0	62	0.8	0.2	1.0	6.1
Satun									
Total	328	12,535.2	1,597.8	14,132.6	4,904	403.5	112.2	515.7	14,648.3

Note. Source [6]

3. Background

This study is a pilot study of our research project. In the beginning of 2018, we collected the occupation rate and the availability of room for service data in the post tsunami period from 2003 to 2017 for the hotels existed before the tsunami and remained their operation until now, through face-to-face interviews with managers. Among them, we received responses from six hotels that located in the

Patong Beach of Phuket to allow us to conduct the data collection. As shown in Table 3, two of them are 3-star hotels, and another four are 4-star hotels. Most of them are located very close (less than 100 meters) to the Patong Beach. Only one hotel is around 500 meters away, where it is just at the boundary of area that was heavily damaged by the 2004 Indian Ocean Tsunami. The majority of tourists during that period came from Europe, Austria, China and Japan. Only four hotels were selected, because other two hotels could not provide the confidential data of their damage and the reconstruction rate.

Table 3. Descriptive information of the six samples sections, subsections and subsubsections.

	Hotel Star	No. of Rooms	Nationality of main customers	Location	
				Distance from Patong Beach (m)	Direction in Patong Beach
A	★★★	72	European	50	South
B	★★★	63	European & Australian	50	South
C	★★★★	403	--	100	South
D	★★★★	182	Australian, Chinese	100	South
E	★★★★	350	Australian, Chinese, Japanese	500*	Middle
F	★★★★	71	European, USA, Chinese	50	Middle

Note. *Out of the heavily damaged area during the 2004 Indian Ocean Tsunami.

4. Results

Table 4 shows the average occupation rate before and after the 2004 Indian Ocean Tsunami of the six samples. Before the Tsunami in Dec. 2004, the occupancy rates of all six hotels were above 80%. However, it immediately dropped 30% to 50% in Dec. 2004 after the Tsunami happened. As most of hotels did not provide us the monthly occupation rate in the normal year, so only the yearly occupation rate has been shown in Table 4. In all of our six samples, the occupation rate from before 2014 dismember, and after 2008 for each hotels in our sample are similar, so these periods are treated as Normal situation. It means during these periods, the occupation rate is not affected by the 2014 Indian Ocean Tsunami. Then, the average and standard deviation of the occupation rate in each hotel are calculated. In Table 4, only the values that are highlighted in the bold and italic font are significant different from the average occupation rate in the Normal situation based on the t-test. There are strongly affected by the Tsunami. The impacted periods are several months to 3 years differently on each hotel, according to the data shown in Table 4.

As shown in Table 5, there are only four hotels provide the room ratio that available for service before and after the 2004 Tsunami. Other two hotels did not provide us the confidential data, so we excluded them in Table 5. Table 5 shows that the Tsunami all the hotel rooms are restored by 2005. The availability of rooms directly after the Tsunami varies from only 24% to 95% of rooms in our sample. If we recalculated them based on the room number, on average in 2005, there are only 392 rooms out of 888 rooms that available on service. If the damage rate is defined as the unavailability of the room for service in this research, through calculation, the average damage rate is 55.89%. Among these hotels, three of them decided to close the hotel, and to reopen their business after the reconstruction, and the other one decided to continue its business even during its reconstruction.

The reconstruction of the hotel room is comparatively fast, even the longest one in our sample spent only around a year, and the shortest one spent just 2 weeks (see Table 5). The reconstruction speed for each month is recalculated based on the reconstruction period in Table 5, with the assumption that the reconstruction speed of each hotel is constant (suggested by hotel managers). Weighted by the room number of each hotel, the average speed for the physical reconstruction is about 6.07% per month. This reflects 6.07% restoration of rooms on average per month. The quotient of the average damage rate and the average speed for physical reconstruction indicates that the reconstruction period is around 9.2 months. If we calculated the recovery period of the occupation rate of these 4 hotels that shown in Table 4, the average recovery period for their hotel business is 20 months.

Table 4. Average occupation rate of the six samples (2004~2017).

Hotel	Before Dec. 2004	Dec. 2004	2005	2006	2007	Average in Normal (before Dec. 2004, 2008~2017)	SD in Normal (before Dec. 2004, 2008~2017)
A	90.0%	40.0%*	70.0%*	80.0%*	90.0%	90.0%	0.0%
B	92.0%	60.0%*	98.0%	90.0%	90.0%	90.4%	1.1%
C	82.0%	38.0%*	38.0%*	77.0%	94.0%	83.1%	4.4%
D	90.0%	40.0%*	30.0%*	40.0%*	50.0%*	90.0%	0.0%
E	80.0%*	30.0%*	35.0%*	40.0%*	78.0%*	89.1%	2.9%
F	80.0%	50.0%*	80.0%	89.0%	80.0%	83.4%	5.6%

Note. * significant at 0.05 (One-tail t-distribution, $df = 10$).

Table 5. Availability of room for service (2004~2017).

Hotel	No. of Rooms	2003	Before Dec. 2004	Dec. 2004	2006	2007~2017	Closed Business for Reconstruction Period (Year)
A	72	100%	100%	95%	100%	100%	0.05
B	63	100%	100%	82%	100%	100%	0.25
C	403	100%	100%	24%	100%	100%	0.90
E	250	100%	100%	50%	100%	100%	0
Total No. of Room	888	888	888	392	880	880	--

Note. Availability of room for service of Hotels D and F are not available.

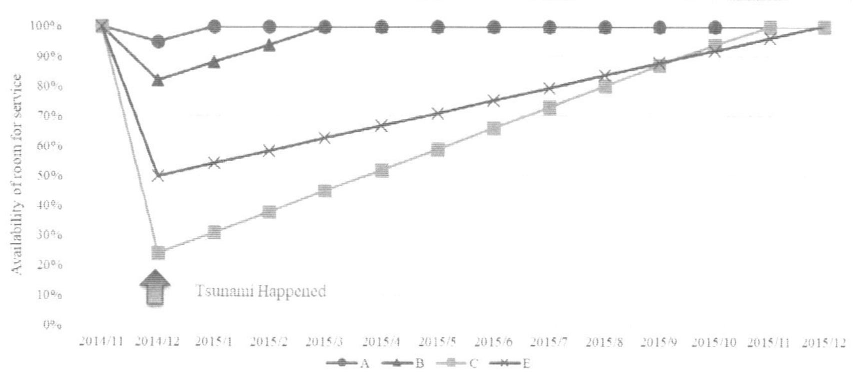

Figure 1. Reconstruction of the hotel room (2004~2005).
Note. This figure is drown based on the linear assumption of reconstruction rate.

5. Conclusions

As mentioned before, through this case study, we found out that the average of damage rate at Patong beach in the Indian Ocean Tsunami is 55.89% of the hotel rooms. On average, the hotels spent around 9.2 months for the reconstruction, which means the average speed for the physical reconstruction about 6.07% per month. However, on the other hand, it took around 20 months to boom up the occupancy rate, which is much longer than the reconstruction period of the rooms. It indicates that many hotels suffered a long period with a low income, after they had spent a big amount of money for the reconstruction. The reason is that the speed of tourists came back to Phuket is much slower than the local people expected. We can image how serious was this problem to the small business. Just as Karatani mentioned in his research [8], many family business in Phuket were bankrupt not because of the damage from the Tsunami, but the slow recovery speed of the tourist.

There are two ways to solve this problem. One is to shorten the recovery time for the tourism industry. In other words, it is to motivate the tourists to come back faster. Combining Table 3 and Table 4, it shows that there is a slower recovery pace of the average occupation rate of hotels (Hotel D and E in the samples) that a part of their major customers is Chinese and Japanese. Meanwhile, most of managers that we have interviewed for this research agree that the Asian tourists came back to Phuket much slower than European and Australian. How to motivate Chinese and Japanese to come back to the Thailand after the disaster will be an important topic for the Thai tourism industry. Meanwhile, when another disaster happened, if the local government knows the average damage rate of hotels, the approximately reconstruction rate could be estimated, based on the Phuket experience that shown in this study. Therefore, it can help the tourism officers to create the tourism promotion strategy and campaign to encourage the tourists to come back right after most of the tourist facilities have been recovered.

The other solution is to provide a forecasting of the recovery speed of the tourism after the disaster to support the business owner to decide a "right time" to reopen the property. This research suggests the average occupation rate of the hotel to be one of the potential index for the tourism recovery, especially in Phuket. It is because of the high ratio of the hotel and restaurant income among the overall tourism income in Phuket. As a result, the overall market can generate the optimized market situation.

Since this is a pilot study, more data is required in order to understand the pattern and behaviour of the tourists' hotel booking. Next, the scales of each hotel are not exactly similar. This factor should be taken into account for the future study. In addition, it is necessary to take into account that the findings are relied on the construction technologies and the government reconstruction plans at that time; as well as the tourist's attitude which is possible to be the effect of the media and information channels at that time. Lastly, it has to be interpreted that 'tsunami' was the first-time experience of Thai government and Thai people at that time. The future tsunami, if any, may cause the different tourist market recovery pattern as the country has prepared the tsunami warning system following the lesson learned in 2004 [5]. Moreover, is it possible that the hotels can learn from some hotels in Japan in which they were changed to be the temporary shelter after the 2011 Great East Japan tsunami [10].

6. References

[1] World Travel & Tourism Council 2017 *Travel & Tourism Economic Impact 2017 Thailand* Retrieved from: https://www.wttc.org/-/media/files/reports/economic-impact-research/countries-2017/thailand2017.pdf
[2] C9 Hotelworks 2015 *Phuket 2015 Economic Overview* Retrieved from: https://www.c9hotelworks.com/downloads/phuket-2015-economic-overview-2015-08.pdf
[3] Henderson J C 2007 Corporate social responsibility and tourism: Hotel companies in Phuket, Thailand, after the Indian Ocean tsunami *Int. J. Hospitality Manag.* **26**(1) 228-239
[4] Suppasri A, Koshimura S, Imai K, Mas E, Gokon H, Muhari A and Imamura F 2012 Damage characteristic and field survey of the 2011 Great East Japan Tsunami in Miyagi Prefecture *Coastal Eng. J.* **54**(01)

[5] Leelawat N, Suppasri A and Imamura F 2015 The tsunami warning system in Thailand: A part of the reconstruction process after the 2004 Indian Ocean tsunami *Post-Tsunami Hazard: Reconstruction and Restoration (Advances in Natural and Technological Hazards Research vol 44)* eds V Santiago-Fandiño, Y A Kontar and Y Kaneda (Cham: Springer International Publishing) chapter 8 pp. 111-119
[6] Asian Disaster Preparedness Center 2007 *The economic impact of the 26 December 2004 earthquake & Indian Ocean tsunami in Thailand*
[7] CB Richard Ellis (Thailand) Co., Ltd. 2008 *MarketView Phuket* Retrieved from: https://www.cbre.co.th/media/research_lang_file/452/q2_08_phuket_market_view.pdf
[8] Karatani Y 2008 Tourism Industry Losses and Recovery Process from the Indian Ocean Tsunami: A Case of the Affected Tourist Destination in Southern Thailand *Proc. of the 14 th World Conf. on Earthquake Engineering*
[9] Nidhiprabha B 2007 *Adjustment and recovery in Thailand two years after the tsunami (No. 72).* ADB Institute Discussion Papers
[10] Leelawat N 2018 Hotel-evacuation-shelter (in Thai) *TPA News vol 259* ed S Kan-uam (Bangkok: Technology Promotion Association (Thailand-Japan)) pp. 43-44

Acknowledgments
This study has been partly supported by the SIIT Young Researcher Award (PI: Dr. Jing Tang) of the Sirindhorn International Institute of Technology; the Ratchadaphiseksomphot Endowment Fund (760003-CC) (PI: Prof. Dr. Suched Likitlersuang) of Chulalongkorn University; and the Japan-ASEAN Science Technology and Innovation Platform Research Network (JASTIP-Net 2017 No. 16) (PI: Dr. Natt Leelawat). Also, the authors would like to thank Mr. Noppawan Chotiwan and Mr. Thanavit Prakitachakul for their help in data collection, as well as the support from all the samples.

Tsunami and Earthquake Symbols and Signs on the Children's Semiotic Understanding

Moriyanti[1], N M Ismail[2], F Yanti[3], E A Rahma[4], R Setyana[5]

[1]Study Program of English Education, Faculty of Teacher Training and Education, Iskandarmuda University, Surien, Banda Aceh 23234, Indonesia
[2]English Department, Syiah Kuala University, Darussalam, Banda Aceh 23111, Indonesia

[3]Life Science Management Department, HTW Berlin University of Applied Sciences, Berlin 10318, Germany
[4]Faculty of Fisheries and Marine, Teuku Umar University, Meulaboh 23615, Indonesia
5Economics Management Department, Faculty of Economics, Teuku Umar University, Meulaboh 23615, Indonesia

Corresponding Author: mori.yanti@yahoo.com

Abstract. This qualitative research aims at describing the coastal children's knowledge and perspective about the Tsunami and earthquake's symbols and signs. The use and purpose of the signs on the warning board is very important for children on the coastal area as an understanding of disaster preparedness. Consequently, the researchers chose 43 respondents with 8-13 years old who live at the coast of Kampung Jawa, Banda Aceh to respond questionnaire and interview. The questionnaire set contained the verbal and non-verbal symbols and signs about tsunami and earthquake which was formulated in the form open-ended questions. It was found that only 12 children who could answer 80% of the questions correctly from the questionnaire. This implies that their understanding and background knowledge about the tsunami and earthquake's symbols and signs are very low level even though the symbols and signs are put on their enviroment. Moreover, during interview process, children had wrong interpretation about the signs and symbols and did not know the function because they never get information about the symbols, the way to evacuate and preparation before the disaster happened from their school and enviroment. Consequently, it is suggested that stakeholder should conduct practices for disaster education that can improve their disaster awareness.

Keywords: Tsunami and earthquake; disaster; symbols; the study of signs; semiotics

1. Introduction

The coastal areas of Aceh have been struck by such catastrophic ocean waves named Tsunami in 2004. There had been a lot of damages—both regarding the physical and non-physical issues undergone by the Acehnese people, specifically speaking for people in Banda Aceh. The attack of a natural disaster can be sudden and recovery may take years and even decades.[1] Consequently, a lot of international countries have also done their favor to help the impacts of the disaster. The favors included physical necessities such as food, building reconstruction, and other staple needs for survival; while the non-physical helps have been seen in the manifestation of psychological guidance, educational supports, and until today, there are continuous research on the Tsunami and earthquakes that can be taken into account as the pro activeness of the disastrous experience in the past. In fact, linguists have also taken a stand to contribute to this research area in attempt to minimize the effect of the disasters in the future. The linguists' concern being discussed here is about the use of signs in the warning boards for Acehnese people—the semiotics. Chandler noted that the value of a sign depends on its relations with other signs within the system.[2] Someone cannot get involved in a semiotic study of how meaning is delivered in text and cultural practice without adopting a philosophical standpoint in relation to nature of sign, illustration and reality. As a result, there are mutual relationship between signs and meaning in real life. In Banda Aceh, several standardized symbols and signs have been designed to be put around the cautious areas in the coastal areas. The symbols and signs are clearly stated in the government rules; Badan Nasional Penanggulangan Bencana (BNPB) no: 07, 2015[3] and Disaster Preparedness Manual from Badan Nasional Penanggulangan Bencana (BNPB), 2017.[4] This study intended to seek the information whether Acehnese children who live around the coastal areas are well-informed of these symbols and signs and their perspective toward tsunami and earthquake signs in daily life.

2. Literature Review

A sign is defined based on various types of things—a shading, a motion, a wink, a question, a scientific condition, and so on—that stands for an option that is other than itself. The word "red" qualifies as a sign since it does not remain for the sounds r-e-d that involve it, but instead for a specific sort of shading and different things. All things considered, the term semeiotics (spelled along these lines) was instituted by Hippocrates (460-377 BC) as the exploration of manifestations and symptoms. The side effect, Hippocrates guaranteed, was referred to as semeion—which means "check" or "sign in Greek". Disentangling what a side effect rely on, how it shows itself physically, and why it is demonstrative of specific illnesses or conditions is the quintessence of medicinal determination. These days, while the objective of semiotics today is to research something very extraordinary (a sign, for example, red), it in any case has held a similar fundamental technique for request.[5]

Danesi further urges that the thing to which a sign alludes is referred to, sensibly, as the referent.[5] There are two lunds of referents: (1) a solid referent, for example, the creature assigned by the word feline, and (2) a unique referent, for example, the "bright ideas" assigned by—commonly—a light bulb. The second one is something that can be appeared to exist in reality – for example, a "feline" can be demonstrated by basically indicating one. The former is fanciful and cannot be shown by just indicating it-how might you point to a "bright idea" inside somebody's mind? Signs enable us to allude to things and thoughts, despite the fact that they probably will not be physically present for our faculties to witness. When we say or hear the word feline, the picture of the creature being referred to comes immediately to mind, regardless of whether the genuine creature is not around for us to see with our sense. In addition, semiotics has the representation which is the forms, object which is the actualization, and interpretant which is the value of the symbols.[6]

In a similar line of satirize contention forever guaranteeing semiosis, Chang addresses on the semiotic ramifications of an ongoing worldwide occasion, in actuality a calamity, that has influenced where it occurred as well as where it did not happen, either; to be specific, the quake and tidal waves in South Asia that asserted almost a quarter million of lives, a significant number of whom were from this remote Nordic piece of the bewildered world.[7] At that point, what are fiascos? How are they encoded as they are decoded? Why semioticians did not give careful consideration to it, why there has not been a semiotics of calamity. For all our enthusiasm in globalization, for what reason has not there been a semiotics of worldwide disaster?

The calamity has constrained us to ponder the semiosis of a system of frameworks, both regular and social. The polysystem of that charitable quake and its ensuing executioner ocean floods is intricate to the point that it justifies inside and out semiotic investigation. It is a system comprising of numerous logical and folkloric translations of arrangement of changes among famous, indexical and representative signs, one understanding offering ascend to another; revealed creature and human reactions to semiotic activities and responses, where mankind is isolated into national, ethnic and social gatherings, some of which, similar to those minorities in the Andaman archipelagos, have been up to this point covered in puzzle.[8] The outcomes are sound geophysical readings, observer stories of different purposes of perspectives, a considerable measure of which are semantically overcoded by religion and superstition, for example, creatures sparing human lives, and so forth. Tourism, media, altruistic associations, debacle alleviation gatherings, flotsam and jetsam, bodies – all these had a part to play in shaping this hopeless mixture of enormous scale.[7]

Take for instance the Tsunami disaster, the innate older folks in the Andaman ocean locale saw the approaching tide and alarmed his kin to escape and look for higher safe houses. His alarm is undoubtedly etymological or conceivably paralingual, given the direness existing apart from everything else. Here the interpretant is no uncertainty the Peircian 'propensity', however propensity is now semantically intervened or, at any rate, when the message was conveyed, the normal signs must be meant phonetic signs. Clarke contends that before the experience is conveyed, the quick view of regular issue can be free from dialect intercession. Unconvinced by Peirce, he proposes the qualification between common sign and sensibly more crude 'natsign', the last being a 'short-go' sign, which is delivered not for correspondence and can be deciphered without surmising.[9]

Indonesia is one of the archipelagic country which has tropical climate. Besides, the country has many active volcano and mountain that are potentially occurred earthquake, tsunami, and landslides.[10] The need of disaster socialization is very essential for people as one of disaster risk reduction. One of the efforts is introducing disaster education at schools. School preparedness of disaster risk reduction can be done by prevention, protection, mitigation, response and recovery activities (REMS).[11] Those terms can be defined in the following: prevention can be done with the action of schools to prevent of the actual incident; protection is the progress actions of school to take the safeguard of the school, students, and staff from emergency event; mitigation is the action of schools to eliminate and reduce of injury and poverty damage; response is the action of putting the schools as a place to respond the emergency events and provide the fast need for students and staff in particular; and recovery is the action of cooperation with community to restore educational programming, the physical environment, business operation, social, emotional, and behavioral health.

Schools are expected to be familiar with mitigation efforts in the areas of safety, security and emergency help and it can be done by drills. For instance, the drills conducted to prepare emergency for community from earthquake. According to Zein, the frequent disaster occurrence in Indonesia both natural and non-natural disaster has given impact for the community so that they have had experience about it.[12] Based on the reason, it becomes the important issues in socialization efforts related to disaster migration. Each country has different types of disasters that also need the different handling. Therefore, Indonesian people must have built the awareness of potential disaster will occur. It makes the community will more prepared to disaster will happen. Based on disaster risk reduction research in the Philippines, Mamon, *et.al* argue that there is the integration of the concepts about the disaster preparedness, awareness, mitigation, prevention, and adaptation towards the knowledge and understanding of students on disaster risk reduction.[13] Concerning to the relation between education and mitigation, it is undeniable to the statement that the school as an educational for the citizens. Hence, education and public awareness toward disaster mitigation reduction have correlation each other. School can consider to put the learning material about disaster mitigation to build awareness of disaster migitation reduction.

According to Tuladhar *et.al*, the most of the students in Nepal do not have specific knowledge and have anxiety for various disasters.[14] All variety of disasters can occur any time in their region. The disaster information, disaster knowledge, disaster readiness, disaster awareness, disaster adaptation, and disaster risk perception of the students were evaluated regarding their exposure to past disasters. The result represents that previous experience of students to disaster is not contributed to obtaining the knowledge of disaster risk reduction. Therefore, disaster education for young people with disaster readiness behaviours and disaster adaptation are necessary tools for disaster risk reduction.

3. Research Methods

The approach used in this study was merely qualitative. There were 43 respondents involved in this study. They were children aged between 8-13 years old. These children are homeless and they are school drop-outs who live in a coastal area named Kampung Jawa, Peunayong, Banda Aceh, Indonesia. Kampung Jawa is one of coastal area in Banda Aceh that happened tsunami and earthquake. The research was conducted in May 8th – 13th, 2018 at *Taman Edukasi Anak Pemulung* Place and the research instruments were questionnaire and interview. The respondents were given a questionnaire set contained the verbal and non-verbal symbols and signs about tsunami and earthquake which was formulated in the form open-ended questions. They were asked to answer the questionnaire set based on researchers' guidance.

There are seven types of questionnaire questions described in the table as follows:

Table 1. Type and Total of Questions in Questionnaire

Questions Types	Total of Questions
Symbols/ Signs	10
Safety Color	5
Action Code for Earthquake	20
Action Code for Tsunami	5
Written Signs	4
Word/ Phrase/ Sentence	3

As they finished filling it out, the data were analyzed using percentage to depict their understanding about the symbols and signs. After gaining the result of questionnaire, researcher interviewed respondents to clarify the questionnaire data about their perspective on tsunami and earthquake symbols and signs for their daily life. The symbols and signs were adopted from Disaster Preparedness Manual and government rule. Then, from the percentage, discussions and conclusions over the study as a whole was settled descriptively.

4. Results and Discussion

The percentange of respondents' understanding of symbols and signs can be seen in the table as follows:

Table 2. Percentage of Respondents' Understanding of Symbols and Signs

Students who can answer correctly	Thourough understanding of all type questions (%)
12	80%
14	50%
11	30%
4	20%
2	5%

From the questionnaire data, it was found that only 12 of the respondents who could answer 80% of the questions correctly. This shows that their understanding very low level although the symbols and signs are put on their coastal area. Delicado, *et.al* refer to his case studies in Portugal that highlight the need to involve children as active participants and have a vital role to play in disaster risk prevention and mitigation.[15] In reality, children who live at coastal area of *Kampung Jawa* still lack of knowledge about disaster education, tsunami and earthquake symbols and signs.

Based on the result of questionnaire about children's understanding on tsunami and earthquake symbols and signs, children got confused about the meaning of symbols and signs. The knowledge of symbols and signs is needed to extend the understanding of the disaster's impacts and to reduce the hazard of disaster. Providing this knowledge is a child-centered mitigation and preparedness program to reduce disaster impacts. In brief, Reid has summarized that there is a nearby correspondence between original perception and natural language.[16] The signs in unique discernment are sensations, of which nature has given us in such an awesome assortment, suited to the assortment of the things meant by them. Nature has built up a genuine association between the signs and the things meant; and nature has additionally showed us the translation of the signs—so that, past to encounter, the sign recommends the thing connoted, and make its conviction.

In addition, to clarify the result of questionnaire, researchers did interview for respondents and they responded enthusiastically. In fact, there are several signs on the road to show the direction of evacuation for citizen when the disaster happened. The respondents thought that the purpose of signs to give the direction for transportation on the road. As a result, they have the wrong interpretation about the signs. They are also difficult to respond the interview questions about the location that we can find the symbols and signs because they rarely pay attention the symbols and signs in daily life. They do not realize the function and benefit of symbol and signs for their safety and preparedness. Understanding the symbols and signs of the cautions is so important for disaster-effect reduction.[12]

Moreover, several verbal signs are also difficult for them to comprehend because there are some vocabularies that they are not familiar for their language. They are difficult to read because most of them are school drop-outs and lack of education. In interview process, they also responded that they never get information about the ways to evacuation when the disaster happened and steps to prepare something before the disaster happened. Marskole, *et al.* claims that the scope of disaster learning is unsatisfactory and there is a tremendous need of providing knowledge to children and youth.[17] They are the social strength who can help perform and reduce the risk of disasters. Effective training and purposive programmes are to be timely organized to create the future generation aware. Khorram-Manesh states that there is a children's and youth's need for self-motivation education in the field of emergency and disaster management.[18] The purpose of this education is to enhance emergency preparedness and individual skills, increase the understanding and advocate for young people to grab responsibility in emergency situations and gain the concept of the psychological effects of a disaster.

5. Conclusions

From the results narrated above, it can be concluded that the children who live in the coastal area of Banda Aceh are still lacking knowledge about the Tsunami and earthquake symbols and signs. Gibbs *et al.* claim that participatory approaches can participate children and young people in community preparedness, response, and recovery processes in a supported way to encourage a self-efficacy and capability that can prevent the injured effects to disasters.[19] It is important for them to understand this as an entire perception to live with—regarding the fact that they spend their time in the on the agitated catastrophic area on the daily basis. Additionally, it is important for the stakeholders who are directly involved with the disaster mitigation counseling to include the counseling about symbols and signs along with other disaster awareness topics during any counseling programs. Mitigation education is actions taken to prevent or minimize the risk to life, social and economic activities, and natural resources from natural hazards. The guidance to reduce the impacts of a natural disaster on communities is widespread awareness, education, preparedness, prediction, and warning systems. Children must be made aware of what natural hazards they are likely to take on their own communities. They should know previously what specific preparations to do before a disaster, what to do during an earthquake, tsunami, or another likely disaster, and what actions to take in its aftermath. Preparedness is the manner of turning awareness of the natural hazards and risks encountered by a community into actions that renew or enhance its capability to respond and recover from disasters. Further perspective that can be drawn based on the data found above is the fact that all signs and symbols around the cautious areas need to be renewed as most of them are now hard to see clearly.

IOP Conf. Series: Earth and Environmental Science **273** (2019) 012034 doi:10.1088/1755-1315/273/1/012034

6. Appendices
Appendix A Questionnaire Set Questions

NAME : ________________________
nama
AGE : ________________________
umur
ADDRESS : ________________________
alamat

1. What is the kind of disaster below?
Apa jenis bencana di bawah ini?

a.____________________

b. ________________

c.______________________

d. ________________
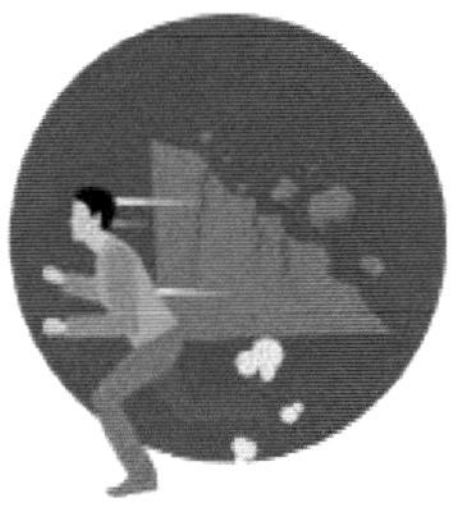
e.______________________
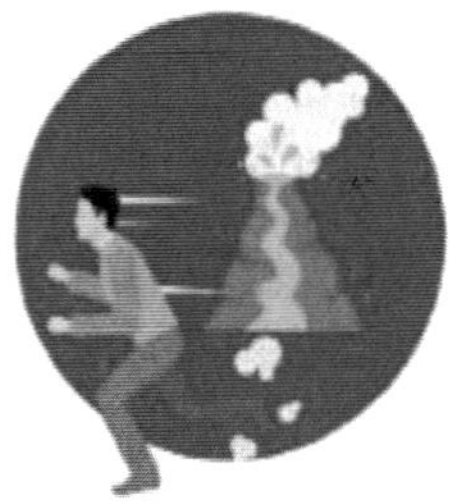
f. ______________

2. What is the meaning of Symbols and Sign below?
Apa maksud dari tanda dan simbol di bawah ini?

a. _____________ b. ___________________

IOP Conf. Series: Earth and Environmental Science **273** (2019) 012034 doi:10.1088/1755-1315/273/1/012034

c.  ____________________

d.  ____________________

e.  ____________________

f.  ____________________

g.  ____________________

h.  ____________________

i.  ____________________

j.  ____________________

3. What is the meaning of Safety Color below?
Apa maksud dari warna keselamatan di bawah ini?

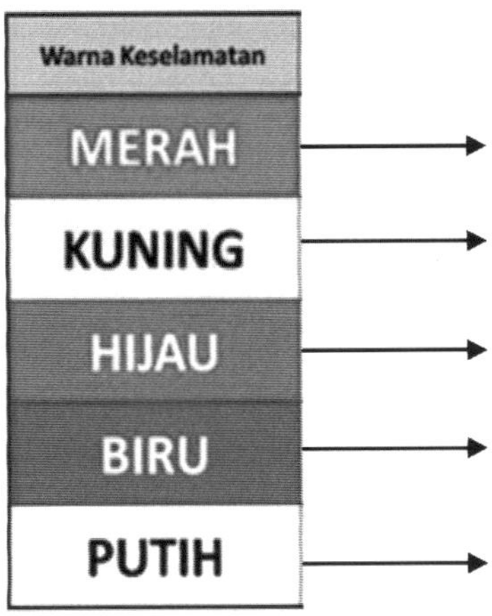

4. What is the meaning of Action Code for Earthquake below?
Apa maksud dari kode tindakan gempa bumi di bawah ini?

IOP Conf. Series: Earth and Environmental Science **273** (2019) 012034 doi:10.1088/1755-1315/273/1/012034

a.  ________________

b.  ________________

c.  ________________

d.  ________________

e.  ________________

f. 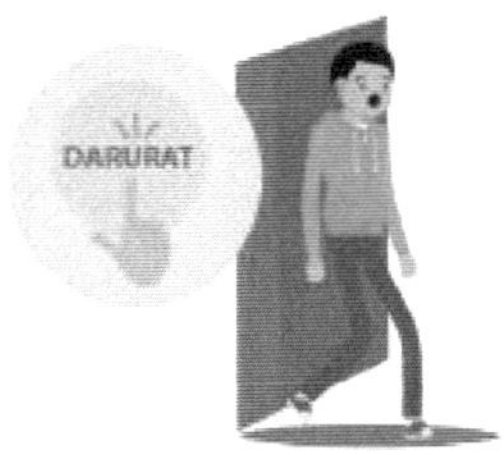

____

g.  ________________

h.  ________________

i.  ________________

j.  ________________

k.  ________________

l. 

____

IOP Conf. Series: Earth and Environmental Science **273** (2019) 012034 doi:10.1088/1755-1315/273/1/012034

m.  ____________________

n.  ____________

o. 

____

p.  ____________

q.  ____________________

r.  ______________

s.  ________________

t.  ____________

5. What is the meaning of Action Code for Tsunami below?
Apa maksud dari kode tindakan tsunami di bawah ini?

u.  ________________

v.  ______________

w.  ________________ x. 

____

y. 

____

6. What is the meaning of Written Sign below?
Apa maksud dari tanda yang tertulis di bawah ini?

a. 

________________ b. 

____

c. 

____________ d. 

____

7. What is the meaning of word/ phrase/ sentence below?
Apa maksud dari kata/ frasa/ kalimat di bawah ini?

Word:
kata
a. Smong =

Frasa:
phrase
a. ie raya =

Kalimat:
Sentence
a. Plueng ie raya =

Appendix B Interview Questions

1. What is the function of tsunami and earthquake symbols and signs?
Apa fungsi dari simbol dan tanda tsunami dan gempa bumi tersebut?

2. What is the benefit for us to know the function of tsunami and earthquake symbols and signs?
Apa manfaat bagi kita untuk mengetahui fungsi dari simbol dan tanda tsunami dan gempa bumi?

3. Where is the tsunami and earthquake symbols and signs put on?
Dimana simbol dan tanda tsunami dan gempa bumi sering di pasang?

4. Have you ever gotten information about tsunami and earthquake symbols and signs?
Apakah kamu pernah mendapatkan info mengenai simbol dan tanda tsunami dan gempa bumi?

5. Have you ever gotten information about evacuation and disaster education from your environment and school?
Apakah kamu pernah mendapatkan informasi tentang evakuasi dan pendidikan kebencanaan di lingkungan dan sekolahmu?

7. References

[1] UNESCO 2014 *Stay safe and be Prepared: a Teacher's Guide to Disaster Risk Reduction. Paris: The United Nations Educational, Scientific, and Cultural Organization*. Available from: http://unesdoc.unesco.org/images/0022/002289/228963e.pdf (Accessed 15 July 2018)
[2] Chandler D 2007 *Semiotics the Basics* (New York: Routledge)
[3] Badan Nasional Penanggulangan Bencana 2015 *Rambu dan Papan Informasi Bencana* (Jakarta: Perpu Kemenkumham RI)
[4] Badan Nasional Penanggulangan Bencana 2017 *Membangun Kesadaran, Kewaspadaan dan Kesiapsiagaan dalam Menghadapi Bencana* (Jakarta: Direktorat Kesiapsiagaan Deputi Bidang Pencegahan dan Kesiapsiagaan)
[5] Danesi M 2004 *Messages, Signs, and Meanings* (Toronto: Canadian Scholars' Press Inc.)
[6] Morentin J M 2011 *Semiotics of the Edges*. Available from: www.semioticmethodology.com/Edges.pdf (Accessed 10 August 2018)
[7] Chang H 2005 Disaster semiotics: An alternative 'global semiotics'? *Proceeding of the Int. Semiotics Institute, Seminar 2: Global Semiotics* 11-19 June
[8] Greenway H D S 2005 Please don't rescue us. Just leave us alone *Int. Herald Tribune 19th.* January 7
[9] Clarke Jr. D S 1987 *Principles of Semiotics* (London: Routledge & Kegan Paul)
[10] Kastolani W and Mainaki R 2018 Does educational mitigation needs to be introduced in school? *SHS Web of Conference EDP Science.* **42** 1–5
[11] Readiness and Emergency Management for School (REMS) Technical Assistance (TA) Center. Available from: https://rems.ed.gov/Docs/Mitigation_Fact_Sheet_508C.pd (Accessed on August 20, 2018)
[12] Zein M 2010 *Community Based Approach to Flood Hazard and Vulnerability Assessment in Flood Prone Area: A Case Study in Kelurahan Sewu* (Surakarta: Disertasi Universitas Gajah Mada)
[13] Mamon M A C, Suba R A V and Son Jr. I L 2018 Disaster risk teduction knowledge of grade 11 students: impact of senior high school disaster education in the Philippines *Int. J. Health System and Disaster Management.* **5** 69–74
[14] Tuladhar G, Yatabe R, Dahal R K and Bhandary, N P 2015 Knowledge of disaster risk reduction among school students in Nepal *Geoenvironmental Disasters*. **2** 1-11
[15] Delicado A, Rowland J, Fonseca S, Almeida A N, Schmidt L and Ribeiro, A S 2017 Children in disaster risk reduction in Portugal: policies, education, and (non) participation *Int. J. Disaster Risk Sci*. **8** 246–57
[16] Reid T 1994 *The Works of Thomas Reid* (Edinburgh: Maclachlan and Steward)
[17] Marskole P, Mishra A, Kumar P, Gaur P, Aharwar P, Patidar P and Shejwar P 2018 A study to assess awareness on disaster management among school going children in Gwalior *Int. J. Community Medicine and Public Health*. **5** 1371–75
[18] Khorram-Manesh A 2017 Youth are our future assets in emergency and disaster management *Bull. Emerg and Trauma: Trauma Research Center, Shiraz University of Medical Sciences*. **5** 1-3
[19] Gibbs L, Pietro M D, Harris A, Ireton G, Mordech S, Roberts M, Sinclair J, Wraith R and Sugimoto Y 2014 Core principles for a community-based approach to supporting child disaster recovery *Aus. J. Emergency Management*. **29** 18-24

IOP Conf. Series: Earth and Environmental Science **273** (2019) 012034 doi:10.1088/1755-1315/273/1/012034

Acknowledgments
The Authors' deepest thanks goes to the leader of *Taman Edukasi Anak Pemulung* at Kp. Jawa, Banda Aceh, Maulidar Yusuf and young children at the coast of Kp. Jawa, Banda Aceh for their help and support in order to get the realistic data for this research.

IOP Conf. Series: Earth and Environmental Science **273** (2019) 012035 doi:10.1088/1755-1315/273/1/012035

Evaluating Disaster Instructional Material Questions in Geography Textbook: Using Taxonomy of Spatial Thinking to Support Disaster Preparedness

S Ridha[1,2*], S Utaya[3], S Bachri[3], B Handoyo[3]

[1]Doctoral Student of Department of Geography Education, Postgraduate of State University of Malang Indonesia, Jalan Semarang No. 5 Malang,
[2]Department of Geography Education Sekolah Tinggi Keguruan dan Ilmu Pendidikan Al-Washliyah Banda Aceh, Indonesia, Jalan Al-washliyah No.1 Banda Aceh
[3]Department of Geography Education, Faculty of Social Science, State University of Malang Indonesia, Jalan Semarang No. 5 Malang

*E-mail: syahrul.ridha.1607219@students.um.ac.id

Abstract. This paper discusses whether instructional material questions about disaster in geography textbooks contain three components of spatial thinking, namely: 1) concepts of space, 2) tools of representation, and 3) processes of reasoning. The taxonomy of spatial thinking is used to evaluate the instructional material questions about disaster in geography textbook of senior high school in Indonesia. A survey was conducted to select geography textbooks that were most dominantly used in senior high school. Four geography textbooks with curriculum 2013 were chosen and evaluated. These textbooks are used as learning resources by majority of teachers and standardized by the National Education Standards Agency (BSNP).Then, the books were reviewed in instructional material questions about disaster and the questions were evaluated. Evaluation is done by coding technique using taxonomy of spatial thinking. The evaluation result showed that spatial concepts of instructional material questions about disaster were still classified as low, even 73 percent of questions was designed without using concepts of space (non-spatial). Besides, the majority of representation tools were not used. From 189 questions evaluated, only 27 questions or 14 percent used tools of representation. In addition, the majority of reasoning processes was at the average level (processing) or 47 percent. The processes of reasoning should be run with student's activities to obtain high level cognitive processes, for example by create maps and designing an area for disaster evacuation. In brief, from spatial perspective, this study revealed that instructional material questions about disaster in geography textbooks have not supported preparedness in facing disaster. Therefore, teachers and textbook writers need to develop instructional material questions about disaster from spatial perspective to support disaster preparedness.

Keywords: evaluating, disaster instructional material, questions, spatial thinking, disaster preparedness.

1. Introduction

Spatial thinking is a process of cognitive ability [1]. Spatial thinking is essential part in geography education because it can help improve the ability to represent, plan, analyze and design an area and have an ability to connect between physical and human being [2]. Spatial thinking consists of three components, namely concepts of space, using tools of representation, and processes of reasoning [3], [4]. Those three components have a relationship with one another. In learning geography, spatial thinking is the main foundation in geography skills [5]. This is because each component of spatial thinking influences each other. In addition, spatial thinking is important to improve understanding of location, distribution, interrelation of geosphere phenomena and the use of geospatial technologies (GST), such as GIS, remote sensing and GPS (global positioning systems) [1]. The result of previous study evaluating questions in geography textbooks in senior high school indicated that the questions are dominated by relatively low spatial concept questions, the spatial representation provided in geography textbooks is not combined with activities to obtain higher cognitive processes [3]. The result of another study showed that questions in geography textbooks at university have higher spatial level than questions in geography textbooks in senior high school [6]. The questions designed at university stimulate students to think in higher level and lead to problem solving.

Based on the result of those studies, the purpose of this study is to analyze the content of components of spatial thinking in instructional material questions about disaster, namely: concepts of space, tools of representation, and processes of reasoning. Therefore, it is important to evaluate instructional material questions about disaster with spatial thinking perspective to support disaster preparedness in senior high school students. Disaster preparedness must be measured in learning geography in senior high school. The measurement aims to find out to what extent students have been prepared for disaster in school. As a result, it is necessary to know whether questions in instructional material about disaster include three components of spatial thinking or not? This question can be answered by evaluating instructional material questions about disaster in geography textbook in senior high school using taxonomy of spatial thinking. Thus, instructional material questions about disaster can be identified from spatial perspective to know their lacks. Table 1 below shows taxonomy of spatial thinking used to evaluate instructional material questions about disaster.

Table 1. Taxonomy of spatial thinking

Components of spatial thinking		Taxonomy
Category	Subcategory	
1. Concepts of space	Non-spatial	-
	Spatial primitives	Place-specific, Identity, Location, Magnitude
	Simple spatial	Distance, Direction, Connection & Linkage, Movement, Transition, Boundary, Region, ShapeReference, Frame, Arrangement Adjacency, Enclosure
	Complex spatial	Distribution, Pattern, Dispersion & Clustering, Density, Diffusion Dominance, Hierarchy & Network, Spatial Association, Overlay, Layer, Gradient, Profile Relief, Scale, Map Projection, Buffer
2. Using tools of representation	Use	Map, Diagram, Chart, Graph, Photo
	Non-Use	-
3. Processes of reasoning	Input	Name, Define, List, Identify, Recognize, Recite, Recall, Observe, Describe, Select, Complete, Count, Match
	Processing	Explain, Analyze, State causality, Compare, Contrast, Distinguish, Classify, Categorize Organize, Summarize, Synthesize, Infer, Make analogies, Exemplify, Experiment, Sequence
	Output	Evaluate, Judge, Predict, Forecast, Hypothesize, Speculate, Plan, Create, Design, Invent, Imagine, Generalize, Build a model, Apply a principle, Complex

Source: [Jo I, Bednarz S W. and Metoyer S. 2010. Selecting and Designing Questions to Facilitate Spatial Thinking,Geogr. Teach., vol. 7, no. 2, pp. 49–55.]

The first component of spatial thinking is concepts of space. Concepts of space is a form of knowledge to understand location, distance, pattern, affordability, morphology, association, spatial relationship of geosphere phenomenon. To understand spatial concepts, concepts of space are divided

IOP Conf. Series: Earth and Environmental Science **273** (2019) 012035 doi:10.1088/1755-1315/273/1/012035

into four subcategories, namely non-spatial, spatial primitives, simple-spatial, and complex-spatial [7]. Questions categorized as non-spatial are questions that do not contain the component of spatial thinking. For example, how many people come from Pidie District in Banda Aceh City? Spatial primitives are the lowest spatial concepts that use the concept of location, place-specific, identity, and magnitude in the questions [6]. For example, what province is located between East Java Province and West Nusa Tenggara Province (NTB)? This question can be identified as a specific place between East Java and NTB. Simple-spatial is a higher concept of space than spatial primitives based on concepts and distribution, distance, direction, connection and linkage, movement, transition, boundary, region, shape, reference frame, arrangement, adjacency, and enclosure [6]. For example, in what climate regions can the tundra-forest be found? Complex-spatial is the highest spatial concept based on spatial distribution [8], including distribution, pattern, dispersion and clustering, density, diffusion, dominance, hierarchy and network, spatial association, overlay, layer, gradient, profile, relief, scale, map projection, and buffer [9]. For example, where is the best place in Banda Aceh to build a tsunami evacuation tower with consideration of population distribution and land available on an urban planning map? This question uses the concept of spatial distribution and association. This will be evaluated as complex-spatial.

The second component of spatial thinking is using tools to represent information (using tools of representation). The tools include maps, diagrams and graphs [8]. The third component is processes of reasoning. Processes of reasoning is an activity of reasoning on objects mapped. Reasoning is an activity of interpreting information contained in a map to be used by students as an information, such as distribution of animals and plants in Indonesia. Interpretation is done by mentioning, explaining, and analyzing objects on the map. The reasoning process is divided into three categories, namely input, processing and output, in which these three subcategories have different taxonomies [6].

Disaster preparedness not only needs the ability of organization, but also needs spatial ability as basic foundation in making decision to deal with disaster [9]. The important goal of learning disaster preparedness from spatial perspective is the provision of spatial concepts that students need when they face disaster. Thus, instructional material questions are very important to contain spatial thinking components to support disaster preparedness.

There are three key elements of spatial thinking that can be used in disaster preparedness [9], namely: 1) concepts of space, students understand about spaces and directions that are around them, such as point, region, and distance, in which it becomes an important object in building spatial concepts . For example, when a disaster occurs, students know where they have to run, where the shelter buildings are located. 2) Knowing about representation of space, a map becomes a space representation tool that must be understood, both two and three dimensions of region. Students must be able to connect an abstract representation on the map with the real world. For example, colour on a map, students must be able to recognize their environment, such as where the danger zone is, how wide is the submerged area and how much the rainfall occurs. These questions can be described on the map to represent the actual environmental conditions. 3) Knowing about processes of reasoning, knowing about reasoning process leads to problem solving through spatial perspective using various cognitive skills and knowledge. For example, how to determine a good evacuation route during tsunami disaster and where the safe meeting place is when an earthquake occurs. These questions require good reasoning skills to understand. Thus, it has to be taught to students.

Students need to know those three elements of spatial thinking as a knowledge that must be provided for disaster preparedness. The knowledge can be implicated through the design of spatial thinking instructional material questions about disaster, namely using the component of spatial thinking in the questions. The questions can be combined into textbooks as an student's exercise in the form of assignment questions, multiple choices, and essays. Thus, students can train themselves to face disasters with spatial abilities.

2. Method

This study evaluates instructional material questions about disaster in geography textbooks used in high school. The textbooks was chosen by conducting a survey in all senior high schools in Banda Aceh in order to select the dominant geography textbooks used in teaching-learning process. Then, the dominant books used were reviewed based on instructional material questions about disaster. The survey result showed that there are four textbooks mostly used as a source of learning in learning geography. The four books are: 1) Geografi Untuk SMA/MA Kelas XI Berdasarkan Kurikulum 2013 Edisi Revisi [10], 2) Jelajah Dunia Geografi SMA/MA Kelas XI Kurikulum 2013 Revisi Kelompok Peminatan [11], 3) Mengkaji Ilmu Geografi Untuk Kelas XI SMA dan MA Kurikulum 2013 Edisi Revisi [12], and 4) Buku Siswa Geografi Untuk SMA/MA XI Peminatan Ilmu-ilmu Sosial [13]. Besides, these books were also standardized by the National Education Standards Agency (BSNP). Data collection was done by coding technique. Coding is used to identify the use of three components of spatial thinking in instructional material questions about disaster. Then, coding results are showed in percentage of relative frequency table.

The questions evaluated were assignments and practice questions at the end of books' chapter. The questions of assignment were in the form of individual and group assignments, while the questions of practice were in the form of multiple choice and essay. Table 2 below shows the number and location of questions evaluated.

Table 2. Location of questions in instructional material about disaster

Questions Location	Instructional Material A	Instructional Material B	Instructional Material C	Instructional Material D	Total
Assignment	17 (23%)	15 (27%)	2 (9%)	10 (25%)	44 (23%)
Multiple Choice	26 (36%)	29 (53%)	10 (48%)	12 (30%)	77 (41%)
Essay	30 (41%)	11 (20%)	9 (43%)	18 (45%)	68 (36%)
Total	73 (100%)	55 (100%)	21 (100%)	40 (100%)	189 (100%)

Coding is specifically focused on component of spatial thinking (Figure 1): 1) classifying questions into spatial concepts, namely non-spatial, spatial primitives, simple-spatial, and complex-spatial. 2) determining the use of representation tools on questions, namely use and non-use. 3) classifying cognitive processes on questions, namely input, processing and output [3]. Table 3 below shows an example of coding questions using taxonomy of spatial thinking.

Table 3. Example of questions coding

Questions	Categories		
	Concepts of space	Tools of representation	Processes of reasoning
Non-spatial thinking questions - List examples of senior high school student participation in natural disaster management through local wisdom and modern technology.	Non-spatial	Non-use	Input
Spatialthinkingquestions - Create a map/ plan of disaster evacuation route in the area where you live. Determine the gathering location (assembly point) and the fastest road to reach the evacuation location.	Complex spatial	Use	Output

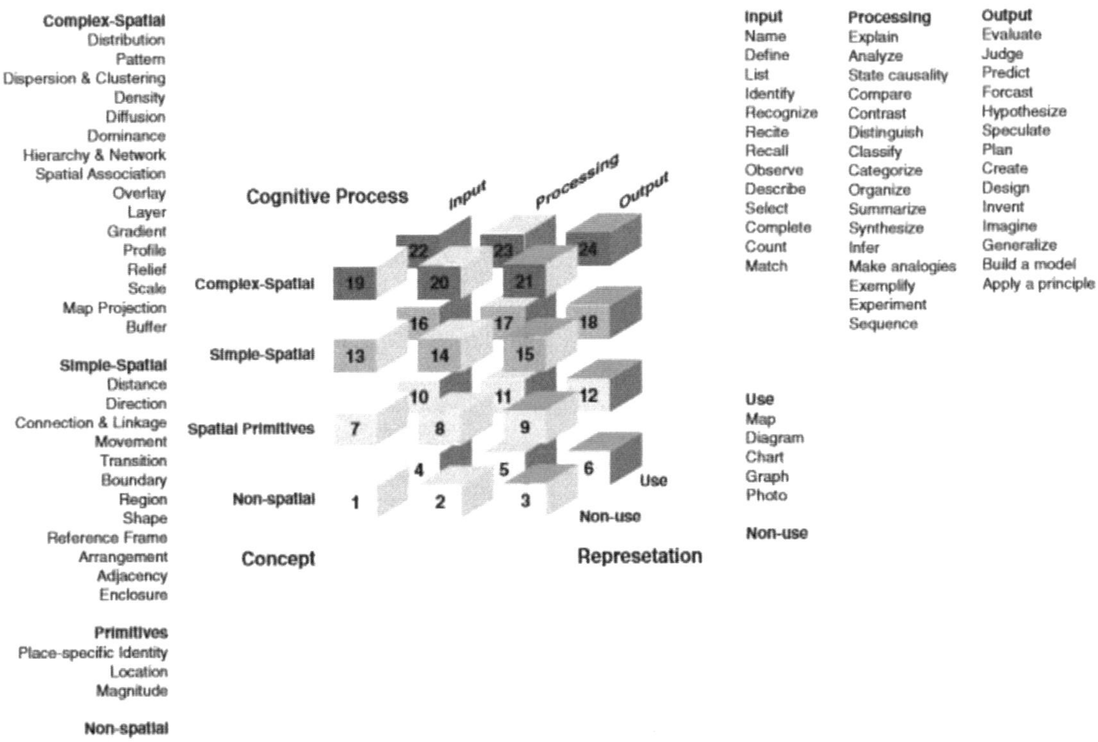

Figure 1.Taxonomy of spatial thinking used in evaluating questions in instructional material about Disaster. Sources: [Jo I and Bednarz S W. 2009. Evaluating Geography Textbook Questions from a Spatial Perspective: Using Concepts of Space, Tools of Representation, and Cognitive Processes to Evaluate Spatiality, J. Geogr., vol. 108, no. 1, pp. 4–13.]

3. Results and Discussion

Spatial thinking in this study is defined as one form of thinking that combines some cognitive skills [14]. The combination includes three components, namely concepts of space, using tools of representation and processes of reasoning. These three components should be used in a complex way at high level, for example, concepts of space (complex-spatial), tools of representation (use) and processes of reasoning (output). Furthermore, the result of instructional material questions evaluation about disaster is showed by following three components of spatial thinking. Those three components are explained as follows.

Concepts of space

The spatial concept contained in 189 instructional material questions is non-spatial as much as 73 percent. While, the other spatial concepts used are follows: spatial primitives with 12 percent, simple-spatial with 8 percent and complex-spatial with 7 percent. The imbalance in the use of spatial concepts in these questions is caused by the preparation of questions not based on spatial thinking. Table 4 below shows the percentage of spatial concepts in instructional material questions about disaster.

Table 4. Percentage of spatial concepts

Subcategory	Disaster Instructional material A	Disaster Instructional material B	Disaster Instructional material C	Disaster Instructional material D	Total
Non-spatial	60 (82%)	41 (74%)	16 (76%)	20 (50%)	137 (73%)
Spatial primitives	5 (7%)	3 (6%)	1 (5%)	14 (35%)	23 (12%)
Simple-spatial	6 (8%)	4 (7%)	3 (14%)	3 (7,5%)	16 (8%)
Complex-spatial	2 (3%)	7 (13%)	1 (5%)	3 (7,5%)	13 (7%)
Total	73 (100%)	55 (100%)	21 (100%)	40 (100%)	189 (100%)

The spatial concept used in instructional material questions about disaster should be at high level, namely complex-spatial. This complex-spatial level aims to make students able to improve disaster preparedness. It can be seen from the ability to understand the distribution of disaster-prone locations, the ability of planning disaster evacuation routes with overlay technique andthe ability to save themselves by going to the safest building to seek protection. In addition, textbook writers and teachers must concentrate on the sustainability of development of instructional material questions about disaster at high spatial level.

Using tools of representation
The use of tools means using map, diagram, chart, graph, and photo [15]. The majority of representation tools is not used in instructional material questions about disaster. From 189 questions evaluated, there are 162 questions or 86 percent not using tools of representation and only 27 questions or 14 percent use tools of representation. This number shows that there is a lack of instructional material questions about disaster that must be corrected. Table 5 below shows the percentage of the use of representation tools in instructional material questions about disaster.

Table 5. Percentage of using tools of representation

Subcategory	Disaster Instructional material A	Disaster Instructional material B	Disaster Instructional material C	Disaster Instructional material D	Total
Use	6 (8%)	10 (18%)	1 (5%)	10 (25%)	27 (14%)
Non-use	67 (92%)	45 (82%)	20 (95%)	30 (75%)	162 (86%)
Total	73 (100%)	55 (100%)	21 (100%)	40 (100%)	189 (100%)

Using tools of representation in disaster preparedness is important to be considered in compiling instructional material questions about disaster. It can help students to understand visualization of images on map, both two-dimensional and three-dimensional maps with the real world. Besides, it can also help students to interact with maps by combining the tools of representation into instructional material questions about disaster [16]. For example, by asking a few questions, where is the heat/danger zone? How many areas are flooded and where? How is the weather forecast for the coming days? [9].

Processes of reasoning
Spatial thinking is the ability of cognitive skills that require complex reasoning [17]. Giving information through high-level cognitive processes becomes one of the reasoning complexity measure, for example, the student's activities for creating thematic map and plan in using land of region. Besides, it can be seen from the way students determine the best evacuation route during tsunami and flood disaster. However, the evaluation results hows that the majority of cognitive process in instructional material questions about disaster is average at 47 percent in the processing subcategory, low-level reasoning at 39 percent in the input subcategory, and high-level of reasoning only at 14 percent. Only 27 questions out of 189 direct students to high-level reasoning or output which includes

creating, planning, and predicting about disaster on people's lives. Table 6 below shows the percentage of reasoning processes in instructional material questions about disaster.

Table6. Percentage of processes of reasoning

Subcategory	Disaster Instructional material A	Disaster Instructional material B	Disaster Instructional material C	Disaster Instructional material D	Total
Input	31 (42%)	25(45%)	9 (43%)	9 (22%)	74 (39%)
Processing	40 (55%)	14 (26%)	11 (52%)	23 (58%)	88 (47%)
Output	2 (3%)	16 (29%)	1 (5%)	8 (20%)	27 (14%)
Total	73 (100%)	55 (100%)	21 (100%)	40 (100%)	189 (100%)

Complex reasoning is needed by each student for disaster preparedness, especially for students who live in disaster-prone locations, such as beaches, earthquake, landslide and flood prone areas. It must be supported by education systems, such as curriculum support, teaching and learning material in schools are based on spatial thinking in disaster learning. In addition, teaching materials should contain questions with complex reasoning in input, processing and output. These three levels of questions must be integrated into instructional material questions about disaster.

A good question about disaster is a question containing component of spatial thinking in high-level, covering complex-spatial, using tools of representation and processes of reasoning that can trigger high-level cognitive (output). The complex-spatial level is a process of thinking about spatial distribution. For example, where is the best place in Lampuuk Beach, Aceh to build a coastal disaster-prevention forest by considering the distribution of population and land available on the map? Then, create a new map for tsunami disaster as a plan to minimize its effect. This question shows that the concept of spatial distribution and association is used to minimize tsunami damage. In addition, in the process of learning students create map related to a plan for minimization of tsunami damage. Thus, questions with complex-spatial components and using tools of representation and output can train students for disaster preparedness from a spatial perspective.

4. Conclusion

Spatial thinking is one of the disaster preparedness benchmarks in disaster education because everyone is absolutely thinking about the space where he is, where to run and where to shelter when facing a disaster. Therefore, spatial thinking needs to be integrated in learning, including by designing instructional material questions about disaster with spatial thinking perspective. Generally, evaluation results show that the spatial concept of instructional material questions about disaster is still classified as low, even 73 percent of questions is designed without using spatial concepts (non-spatial). Besides, the majority of representation tools were not used. From 189 questions evaluated, only 27 questions or 14 percent used representation tools. Furthermore, the majority reasoning process was at average level (processing) or 47 percent. Therefore, reasoning process should be managed with student activities to obtain high level of cognitive processes, for example by creating map and designing an area for disaster evacuation. Good instructional material questions about disaster is questions that contains complex spatial thinking components, including using complex-spatial with high-level concept, interpretation tools and reasoning processes that can trigger high cognitive levels of students. The question that only asks students to memorize information is not enough to facilitate student's spatial thinking skills and disaster preparedness. Therefore, instructional material questions about disaster which contain three components of spatial thinking needs to be developed in geography textbook.

References

[1] Gersmehl P J. 2018. Teaching geography. New York: Guilford Press.
[2] Heffron S G and Downs R M, Eds. 2012. Geography for life: National Geography Standards, 2nd. ed. Washington, DC: National Council for Geographic Education.

[3] Jo I and Bednarz S W. 2009. Evaluating Geography Textbook Questions from a Spatial Perspective: Using Concepts of Space, Tools of Representation, and Cognitive Processes to Evaluate Spatiality, J. Geogr., vol. 108, no. 1, pp. 4–13.
[4] Lee J and Bednarz R. 2012. Components of Spatial Thinking: Evidence from a Spatial Thinking Ability Test, J. Geogr., vol. 111, no. 1, pp. 15–26.
[5] Jo I, Bednarz S W, and Metoyer S. 2010. Selecting and Designing Questions to Facilitate Spatial Thinking,Geogr. Teach., vol. 7, no. 2, pp. 49–55.
[6] Scholz M A, Huynh N T, Brysch C P, and Scholz R W. 2014. An Evaluation of University World Geography Textbook Questions for Components of Spatial Thinking, J. Geogr., vol. 113, no. 5, pp. 208–219.
[7] Lee J, Jo I, Xuan X, and Zhou W. 2018. Geography preservice teachers' disposition toward teaching spatial thinking through geography: a comparison between China and Korea, Int. Res. Geogr. Environ. Educ., vol. 27, no. 2, pp. 135–148.
[8] Jo I, Klein A, Bednarz R. S, and Bednarz S W, 2012. An exploration of spatial thinking in introductory GIS courses," in Teaching Geographic Information Science and Technology in Higher Education, Unwin D J, Foote K E, Tate N J, and DiBiase D, Eds. Oxford: Waley-Blackwell, pp. 211–229.
[9] Fuhrmann S et al. 2008. Teaching Disaster Preparedness in Geographic Education, J. Geogr., vol. 107, no. 3, pp. 112–120.
[10] Sindhu Y. P. 2017. Geografi Untuk SMA/MA Kelas XI Berdasarkan Kurikulum 2013 Edisi Revisi. Jakarta: Erlangga.
[11] Tika P, Amin, and E. P. Rahayu. 2017. Jelajah Dunia Geografi SMA/MA Kelas XI Kurikulum 2013 Revisi Kelompok Peminatan. Jakarta: Bailmu.
[12] Sugianto and D. Endarto. 2017. Mengkaji Ilmu Geografi Untuk Kelas XI SMA dan MA Kurikulum 2013 Edisi Revisi. Surakarta: Platinum.
[13] Arifin A. 2017. Buku Siswa Geografi Untuk SMA/MA XI Peminatan Ilmu-ilmu Sosial. Surakarta: Mediatama.
[14] National Research Council, Ed. 2006. Learning to think spatially. Washington, D.C: National Academies Press.
[15] Metoyer S and Bednarz R. 2017. Spatial Thinking Assists Geographic Thinking: Evidence from a Study Exploring the Effects of Geospatial Technology, J. Geogr., vol. 116, no. 1, pp. 20–33.
[16] Gillen L, Skryzhevska L, Henry M C, and Green J. 2010. Map Interpretation Instruction in Introductory Textbooks: A Preliminary Investigation, J. Geogr., vol. 109, no. 5, pp. 181–189.
[17] Jo I and Bednarz S W. 2011. Textbook Questions to Support Spatial Thinking: Differences in Spatiality by Question Location, J. Geogr., vol. 110, no. 2, pp. 70–80.

Acknowledgement
The authors would like to thank Lembaga Pengelolaan Dana Pendidikan/LPDP (Indonesia Endowment Fund for Education), The Ministry of Finance, the Republic of Indonesia for providing the scholarship for Syahrul Ridha.

Community Resilience: Lesson Learnt from Disaster Survivors in Yogyakarta Province Indonesia

Z Akbar[1]

[1]Department of Psychology, Universitas Negeri Jakarta, Jalan Rawamangun Muka, Jakarta Timur 13220, Indonesia

Abstract. This article describes community resilience in disaster survivors in Yogyakarta Province. Model of community resilience applied in Cangkringan, an affected districts by the Merapi eruption disaster in 2010. The aims of this research are, first, to clarify the definition of community resilience in disaster studies; second, to understand how the community resilience process in affected people, and finally to interpret the disaster as a 'window of opportunity' for sustainable trajectories. Participants were drawn from the inhabitants of Cangkringan districts Yogyakarta Province. It illustrates the concept model of 'community resilience', which is widely explored from community perspectives. This research also seeks to understand how the community resilience process can be enacted in alternative social practices. It found that number of themes emerged in community resilience integrated model includes partnership and institutional, education and engagement, available resources (personal and culture values, psychological strength), also risk and vulnerability groups in society.

Keywords: Community Resillience, Survivors, Natural Disaster

1. Introduction

Indonesia is a hazard-prone country as it is situated at the meeting point of three active plates in the world: the Indo-Australian plate in the south, the Euro-Asian plate in the north, and the Pacific plate in the east. The three plates are moving and thrusting towards one another in such a way making the area prone to natural disasters such as volcanoes, earthquakes, tsunamis. The movement of the plates also causes the area to become a tectonically and volcanically active region [2]. Thus, natural disasters occur almost every year in Indonesia, some of which hit Yogyakarta Province in October 2010 when volcanic eruption of Mount Merapi, caused many casualties and extensive property damage.

Mount Merapi eruption occurring between October and November 2010 which the impact was suffered by those residing Yogyakarta province and Central Java province. The eruption was the most powerful and the worst after its last eruption in 1870, causing a lot of casualties of both lives and properties. It was considered the worst because it had forced a total of 32 villages with a population of more than 70.000 people to leave their homes. Based on the statistical data from the National Agency for Disaster Management (BNPB), the total number of casualties caused by the eruption included 277 from Yogyakarta province and 109 from Central Java province (National Agency for Disaster Management, 2010). The data also recorded there were a total number of 2,527 refugees in Central Java province and 12,839 refugees in Yogyakarta province [2].

The 2010 Mount Merapi caused physical damages in almost all sectors including residential, infrastructure, social, and economic. It destroyed many of physical infrastructures such as houses, public places, schools, and worship places as well as the installation of water and electricity. The National Agency for Disaster Management recorded more than 2,636 units of houses in Yogyakarta province were heavily damaged, 156 houses were slightly damaged, and 632 houses were partially damaged. There were a total 3,424 of houses in Yogyakarta province damaged as the impact of Mount Merapi eruption. Data in Central Java province recorded that a total of 174 houses were heavily damaged, slightly damaged, and partially damaged. Based on a financial study, the National Agency for Disaster Management reported the loss caused by the damage was estimated to be 3.62 trillion rupiah, consisting of 1.69 trillion rupiah (4.64%) in economic production sector, 707.42 billion rupiah (19.50%) in infrastructure, 626.65 billion rupiah (17.27%) in houses, 408.75 billion rupiah (13.22%) in cross sector, and 122.47 billion (3.38%) in social sector [2].

Disasters cause a lot of damage and changes to people who become victims. Efforts to build or increase the potential for community resilience to overcome the impact of the changes brought about by the disaster will be very important. Potential resistance in the community is called resilience in the community. Community resilience is an attempt to tackle, confront and mitigate the stress posed by the traumatic accident of disaster [3]. Community resilience development efforts in post-disaster is rated to have not yet become a focus of concentration until now, although it will be very important to be developed in community.

Preliminary research done by Akbar [1] found some psychological aspects that become a protective factor that affects the long-term psychological state after the disaster on the survivors in Yogyakarta province. These factors include social support, spirituality, effective coping and disaster experience. Psychological factors can be further developed into psychological skills that can then be built into a strength to improve community resilience.

The researches carried out about the community in the context of disaster in Yogyakarta province, among others, research by Rakhman and Kuswardhani [10] concerning the empowerment of local wisdom as a resilient community model to face the disaster found that the values of local wisdom as *nrimo, ngaruhke*, and mutual cooperation can be empowered as capital of community resilience in the face of the earthquake in Yogyakarta. Suharta, Prasetya, and Tohani [12] found a conceptual model of disaster awareness community integrated with the Community Learning Center as a unit of non-formal education in the community, especially at the village level. From these researches it can be seen that the focus of the model of post-disaster community available is only limited on models that are conceptual, the absence of an integrated model because it is only specific to certain aspects such as the educational aspect, and also has not facilitated other aspects that are considered important in the community, for example the effort to facilitate risk and vulnerability groups based on personal / psychological strength in the community.

This paper investigates community resilience model by a group of affected people in Cangkringan sub-district, Sleman district, Yogyakarta province. Community resilience is an approach to the case of people to investigate the community-led reaction to the Merapi eruption disaster. Community resilience encompasses individual preparedness as well as establishing a supportive social context in communities to withstand and recover from disasters.

2. Purpose

The aims are, first, to clarify the definition of community resilience in disaster studies; second, to understand how the community resilience process in affected people, and finally to interpret the disaster as a 'window of opportunity' for sustainable trajectories [5]. The research questions were the following:

a. How does people in Cangkringan district in affected area enact community resilience processes towards sustainable trajectories?
b. How can this case study contribute to an understanding of community resilience?
c. What model of community resilience that implemented in this area?

3. Methods

Researcher followed two research strategy to identify community resilience practices employed at the local level. The first phase focused on documentary sources to identify the actions taken at the community and state level in response. A second phase of the research involved interviews to triangulate the documentary sources with the practices of people who experienced the disaster. In identifying the

IOP Conf. Series: Earth and Environmental Science **273** (2019) 012036 doi:10.1088/1755-1315/273/1/012036

research participants for this study researcher employed a snowball technique to seek interviewees with firsthand experiences of enduring Merapi eruption in the past.

The instrument used in this research is two namely survey and interview. The survey was conducted to determine the community needs and circumstances of the post-disaster in Yogyakarta province. The second instrument is the interview. Interview is used to collect data on people's perceptions of readiness, capability, and the local potential as an initial effort for the community resilience development of post-disaster. The results of these surveys and interviews will be baseline data in the research, and then become important information in the research. The respondents of this research consists of representatives from each group of citizens that includes fathers, mothers, adolescents, and the elderly. The number of respondents in this research is 150 people comprising 60% male and 40% female.

4. Results

Communities have an important role in the process of disaster resilience. Community resilience is a response to the disaster, which focuses on balance in community between intervention policy of formal institutions that are top-down up to facilitating the needs of the expectations of the grassroots. Community resilience not only is a top-down policy rules that are formal, but also emphasizes the processes that are bottom-up of autonomous process based on abilities, skills, and behavior action raised by the people of the region [6, 8]. Community resilience rooted in community's capacity to improve the social infrastructure, the initiative of the community, as well as to manage human resources and local community resources [11]. Community resilience is a process and not an outcome [7].

Disasters can cause the appearance of a spontaneous reaction and voluntarian applying solidarity independently. Even if people do not have enough resources to deal with disasters or not formally authorized coming from institutions, collective action that comes from local resources is a way that can be done to overcome the effects of the disaster. Community resilience is a process that comes from grassroots and spontaneous reactions in overcoming the effects of traumatic events e.g. disaster [13].

Here are the data obtained based on interviews conducted in the community: The number of people in Kaliadem village around 100-500 residents with a percentage of 60% male and 40% female. The language used in everyday life is Javanese and Indonesian.

In the community agree that there are those who rated responsible for disaster preparedness / emergency response:

Responses	Amount	Percentage
Yes	90	60%
None	45	30%
Do not know	15	10%

Parties who interact/relate/communicate/responsible for disaster preparedness/emergency response in the community is:

Responses	Amount	Percentage
The Department of Health (national or local level, health centers or community health workers)	52	35%
Civil Protection	30	20%
National Disaster Management Office	23	15
United Nations	0	0
NGO	37	25%
I do not know	0	0
Other	8	5%

IOP Conf. Series: Earth and Environmental Science **273** (2019) 012036 doi:10.1088/1755-1315/273/1/012036

Community access to information on the Internet:

Responses	Amount	Percentage
Many people have access to the internet	45	30%
Few people have access to the internet	105	70%
No one has access to the internet	0	0

Natural disaster that often occurs in Kaliadem Village community is:

Responses	Amount	Percentage
Eruption of Mount Merapi	150	100%

Health problems experienced by people when a disaster occurs are:

Responses	Amount	Percentage
Severe injury	0	0
Epidemic of a disease	45	30%
Damage to health clinics	0	0
Contamination of water/water shortages	45	30%
Lack of food	45	30%
Other	15	10%

Groups in community that are more affected / influence of the disaster are:

Responses	Amount	Percentage
Parents	52	35%
People with disabilities	16	10%
Children	52	35%
Woman	30	20%

People's knowledge of presence/absence of a disaster plan:

Responses	Amount	Percentage
Yes	30	20
No	105	70
Do not know	15	10

Challenges faced by the community in response to the disaster or what the consequences are faced with the presence of disaster include: must follow the instructions of the government, thinking about jobs, the state of livestock, thinking of the future post-disaster life, continue to keep together with other citizens. In community, in the development of disaster management planning the following parties have a very important role, namely: Village Head (Kepala Dukuh), RT, RW, Volunteers, and community leaders who participate in BNPB.

Outside the community, in an effort to prepare and implement a disaster plan are the following parties:

Responses	Amount	Percentage
National government	75	50%
Agencies of the United Nations	0	0

The groups of volunteers	75	50%

Does Disaster response plan have specific components of the health aspects?

Responses	Amount	Percentage
Yes	147	98%
No	3	2%
Do not know	0	0

The contents of disaster planning, namely: to teach to always be alert, disaster response training, coordination is done from the government to the village head is forwarded to the community.

The community knows regarding information material about health, disaster preparedness and disaster response that comes from national authorities (government):

Responses	Amount	Percentage
Yes	150	100%
No	0	0
Do not know	0	0

Information is available in a language that is easily understood and can be used in the community:

Responses	Amount	Percentage
Yes	150	100%
No	0	0
Do not know	0	0

Such information helps the community to prepare and respond to disasters?

Responses	Amount	Percentage
Yes	150	100%
No	0	0
Do not know	0	0

Media used by the community in receiving information regarding a disaster or emergency situation:

Responses	Amount	Percentage
Newspaper	0	0
Family and friends	30	20%
Television, radio	90	60%
Internet	15	10%
Other	15	10%

Method of the first warning or other disaster-related information distributed or disseminated to the community is through: announced through mosques, through the traditional tool that facilitates direct communication with Village Head to be distributed to the community, via the siren sounded, the village head forwarded to be delivered to all citizens. Examples of how communities prepare for or respond to disasters namely preparing important papers, preparing transportation, knowing the evacuation routes, infrastructure, preparing a full citizen vigilance. Preparedness and disaster response efforts of the community namely people already know the assembly point and ready for further evacuated. After the information to the community and they are ready to wait for orders thus after rising status to be alert the community is ready evacuated in quick time.

Are there any laws, policies or regulations related to disaster management in the country of Indonesia?

Responses	Amount	Percentage
Yes	45	30%
No	0	0
Do not know	105	70%

Does the community have their own policies or regulations related to disaster preparedness and response?

Responses	Amount	Percentage
Yes	105	70%
No	45	30%
Do not know	0	0

Disaster issues faced by people who are not covered by the national policy or laws or community among others concerning effort to recognize the signs of nature arising such as the air temperature rises, the animals of the forest go down, already visible hotspots at the top of Mountain.

Next steps recommended to reduce the health risks in emergencies and to improve disaster preparedness are to use masks for health, provision of decent evacuation facilities, the special room for special groups (infants, children, and the elderly), preparation of personal medicine. The presence of health service posts which participate in the rescue of disaster management for example through a mobile clinic.

Who is considered to be responsible for initiating the planning process?

Responses	Amount	Percentage
Members of the community	15	10%
Community leader	120	80%
Municipal authorities or national authorities (BPBD and BNPB)	15	10%
Special NGO	0	0

The parties should be involved in planning the next steps to improve disaster preparedness and to reduce health risks such as government, community leaders, NGOs, volunteers, and the School Department.

Toughness education to face disaster in community: There has been available but considered not effective, it should have been made the flow of information and coordination with residents who have a vehicle, making it easier to evacuate residents when a disaster occurs, disaster simulation can be often done. Special handling at the time of the disaster occurs to risky community group (elderly, children, people with disabilities): There has not been available, so far this special group is evacuated beforehand, placement in refugee camps together with other residents. The hope is that the risky people should be addressed specifically, accompanied, served all the needs, given the special handling as well

The values that exist in the community or local wisdom related to disaster prevention/ preparedness, namely: *pasrah lan tawakkal* (surrendering everything to God), feeling to be always protected by the God, togetherness with other citizens (*urip ora urip manganese ora manganese sing penting bareng*).

This research found the conceptual model of community resilience in post-disaster community namely partnership and institutional, education and engagement, available resources, and risk and vulnerability:

Domain Areas	Topics
Partnership and Institutional	Formal and informal relations between relevant institutions and community-based organizations (neighborhood associations, faith-based organization, business, etc)
Education and Engagement	Preparedness education level, the activities carried on disaster preparedness and community resilience, perspective on the level of neighbor-to-neighbor reliance and voluntarian capacity, perspectives on the ability of community resilience and the ability to educate the community
Available Resources	The values of personal / psychological strength that will be a force in the community, the resources availability of the various parties can be a force in the community, among others, family, school and other community organizations
Risk and Vulnerability	Activities relating to disaster preparedness, communication and response given, training and education, and community preparedness against vulnerable and risky groups such as children with special needs, the elderly, patients with mental disorders

The results of this research can be summarized as follows:

Domain Areas	Research Result
Partnership and Institutional	• People agree that there are the most responsible parties for disaster preparedness • Party responsible for disaster preparedness / emergency response in the community is the Ministry of Health • Parties outside the community are responsible for seeking to prepare and implement a disaster plan are national governments and volunteer groups • Community leaders are rated as the party responsible for initiating the planning process
Educational and Engagement	• Community considers the lack of knowledge about disaster management plan • Community needs to know about information material about health, disaster preparedness and disaster response comes from national authorities (government) • Community considers that disaster plan must have certain components in particular on the health aspects
Available Resources	• Only a few people who have access to the Internet • The available information is expected in the language that is easily understood and can be used in the community • The available information can help people to prepare for disaster response • Television and radio become media which has important role in receiving information related to a disaster or emergency situation

	• Community is not aware of any laws, policies or regulations related to disaster management in Indonesia • Community has its own policies or regulations related to disaster preparedness and response
Risk and Vulnerability	• Health problems experienced by the community when a disaster occurs are an outbreak of disease, contamination of water / lack of water and food shortages • Groups in community are more affected by the disaster are parents and children

Disaster resilience should not be interpreted as an outcome but as a process. Community resilience can be seen as ‘a process linking a set of networked adaptive capacities to a positive trajectory of functioning and adaptation in constituent populations after a disturbance. It should consider sustainability the forefront of all community planning efforts in society. Community resilience defines as a process generated by grassroots groups that assume the responsibility for organising themselves in a collaborative, spontaneous and autonomous way, exploiting local skills, knowledge and resources, and using the disaster as an opportunity to ameliorate the social dynamics of the community and to enhance its sustainability [4]. The following section shows how the main elements identified in community resilience model in Cangkringan district Yogyakarta Province includes partnership and institutional relate to community based organizations, education and engagement, available resources (personal and culture values, psychological strength), also risk and vulnerability groups in society.

This is also in line with several community resilience programs implemented in other countries. Research conducted by Fois and Forino [5] described the development of post-disaster community resilience in the villages in the Pescomaggiore area, Italy after an earthquake L'Aquila in 2009. The occurred disaster becomes an opportunity for the development of community resilience in communities in the area. Studies of Colten, Grimsmore, and Simms [4] found that people in the coastal areas of Louisiana American who are as victims of cyclones and also the oil spill on the coast develop a model of community resilience after the disaster event. The development model of community resilience is based on the values of the local community and local knowledge, activities that enhance community resilience, external programs related to disaster mitigation, and post-disaster recovery programs.

5. Conclusion

A number of themes emerged in community resilience integrated model includes partnership and institutional relate to community based organizations, education and engagement, available resources (personal and culture values, psychological strength), also risk and vulnerability groups in society.

6. References

[1] Akbar Z 2016 Post-Traumatic Growth: Long Term Psychological Conditions Among Disaster Survivors in Yogyakarta – Indonesia *Dissertation* University of Leipzig, Germany

[2] BNPB 2010 *National Disaster Management Plan 2010-2014* Jakarta Indonesia

[3] Chandra A, Williams M, Plough A, Stayton A, Wells K, Horta M and Tang J 2013 Getting Actionable About Community Resilience: The Los Angeles County Community Disaster Resilience Project *American Journal of Public Health* e1-e9

[4] Colten C, Grimsmore A and Simms J 2015 Oil Spills And Community Resilience: Uneven Impacts And Protection In Historical Perspective *Geographical Review* **105** 4 391-407

IOP Conf. Series: Earth and Environmental Science **273** (2019) 012036 doi:10.1088/1755-1315/273/1/012036

[5] Fois F and Forino G 2014 The Self-Built Ecovillage In L'Aquila, Italy: Community Resilience As A Grassroots Response To Environmental Schock *Disasters* **38** 4 719-739

[6] Lewis J and Kelman I 2010 Places, People, And Perpetuity: Community Capacities in Ecologies of Catastrophe *ACME: An International E-Journal for Critical Geographies* **9** 2 191-220

[7] Manyena S 2006 The Concept of Resilience Revisited *Disasters* **30** 4 433-450

[8] Murphy B 2007 Locating Social Capital in Resilient Community-Level Emergency Management *Natural Hazards* **41** 2 297-315

[9] National Agency for Disaster Management 2010 *Merapi Eruption Disaster in Yogyakarta and Central Java Province* Jakarta, Indonesia: National Agency for Disaster Management

[10] Rakhman A and Kuswardani I 2012 Studi Kasus Gempa Bumi Yogyakarta 2006: Pemberdayaan Kearifan Lokal Sebagai Modal Masyarakat Tangguh Menghadapi Bencana *Prosiding* Seminar Nasional Aplikasi Sains dan Teknologi (SNAST) Periode III 3 November 2012 Yogyakarta.

[11] Sefyang G 2010 Community Action for Sustainable Housing: Building A Low-Carbon Future *Energy Policy* **38** 12 7624-7633.

[12] Suharta R, Prasetyo I and Tohani E 2014 Pengembangan Komunitas Sadar Bencana Terintegrasi "PKBM" Di Kabupaten Bantul Dan Sleman *Laporan Penelitian Terapan*. Fakultas Ilmu Pendidikan Universitas Negeri Yogyakarta.

[13] Zhou H 2010 Resilience to Natural Hazards: A Geographic Perspective *Natural Hazards* **53** 1 21-41.

Factors Affecting Post-Disaster Housing Reconstruction on Preconstruction Stage in Pidie Jaya Regency Indonesia

Nurul Malahayati[1,*], Munirwansyah[2], Mochammad Afifuddin[3] and Syamsidik[4]

[1,2,3,4] Faculty of Engineering, Syiah Kuala University , Jl. Syech Abdurauf No. 7 Kopelma Darussalam, Banda Aceh, Aceh 23111, Indonesia

*E-mail: malahayatijj@unsyiah.ac.id

Abstract. Pidie Jaya regency is one of the regencies located in Aceh Province, Indonesia. On 7 December 2016 an earthquake rocked the Pidie Jaya with a magnitude of 6.5 on the Richter scale on coordinates 5.25 ° North and 96.24 ° East latitude with an estimated in-depth of 15 km. The impact of the disaster caused a lot of damage to infrastructure, especially public housing. The Reconstruction of disaster victim housing in Pidie Jaya District uses community-based methods designed with a community organizing strategy. In the early stages of reconstruction many problems arose between the government, the management consultant, and the community beneficiaries resulting in a delay in the construction housing implementation. This period is called preconstruction, where the activities carried out focused on the preparatory activities to rebuild housing of disaster victims. Questionnaire surveys and interviews were given to stakeholders from the government, the management consultants and the community involved in housing reconstruction. Data were analyzed using statistical software. The results of this study have identified that the dominant factors influencing implementation in pre-construction are starting at the moment (1) recruitment of facilitators; (2) program training for the community; (3) formation of community organizations; (4) program socialization and coordination; (5) inventory, verification and validation of damage data; (6) design and funding. The dominant factors in the pre-construction are interrelated with each other because the problems that occur at the beginning of the pre-construction period, result in delays in the implementation of the construction of the houses. Identifying specific factors that create problems that affect the success of the project in this study is the basis for practitioners, stakeholders and the community in managing a post-disaster housing project, so that it can anticipate the occurrence of similar things in the implementation of reconstruction in the future.

Keywords: Preconstruction, post disaster housing reconstruction, community–based

1. Introduction

On December 7, 2016 an earthquake rocked Aceh Province, precisely in Pidie Jaya, Bireun and Sigli Regencies. The enormous impact occurred in Pidie Jaya District, especially in the housing sector. Damaged houses recorded as an earthquake rose in Pidie Jaya Regency amounted to 17,673, of which 2,202 were severely damaged, 4,542 were moderately damaged, and 10,929 were slightly damaged [1]. The reconstruction phase for disaster housing in Pidie Jaya regency in using the community based and coordinating with Management consultants and the Post Earthquake Disaster Rehabilitation and Reconstruction Coordination Team under the Regional Disaster Management Agency (Badan

Penanggulangan Bencana Daerah/BPBD) of Pidie Jaya Regency. The approach in post-disaster housing reconstruction uses community-based methods, where people become owners, supervisors even as implementers of their own house reconstruction projects (Jha et al., 2010., Silva, 2010., Ophiyandri, et al., 2010). In the early stages of reconstruction many problems arose between the government, management consultants and beneficiary communities. At this time it was called pre-construction where the activities carried out focused on the preparation to rebuild the housing of the disaster victims. Activities carried out during the pre-construction period, namely the procurement of construction management services, facilitator recruitment and training, socialization of the reconstruction program, damaged validation, formation and training of Community Groups (Kelompok Masyarakat/Pokmas), and the initial disbursement process. The problem in this study was the implementation of delayed housing construction for 11 months which was calculated since the completion of the emergency response transition period until the construction of housing was carried out. The construction of new disaster victims housing took place in February 2018, This should be a serious concern by the government considering that the community as victims affected by the disaster cannot be left too long in temporary homes or in damaged housing conditions. The purpose of the research is to get the factors that influence the pre-construction stage, thus impacting the postponement of post-disaster housing construction in Pidie Jaya District. Identifying obstacles that can delay the reconstruction program can help improve the reconstruction process in the future.

2. Literature Review

Based on Aceh Governor Regulation number 147 of 2016 concerning Action Plans for Post-Disaster Earthquake Rehabilitation and Reconstruction in Pidie, Pidie Jaya and Bireun Districts 2017-2019, that the earthquake that struck Aceh on December 7, 2016 at 05:02:36 WIB, shallow earthquake, 10 km, with a magnitude of 6.5 MW centered in Pidie District and had an impact on Pidie and Bireuen Regencies. This disaster resulted in fatalities and caused tremendous damage to community housing, infrastructure and socio-economic communities. As a result of the impact of the earthquake, the Governor of Aceh declared the provincial disaster response status for fourteen days, dated from 7 to 20 December 2016, through the Disaster Emergency Response Statement Number 39 / PER / 2016. The government also sets emergency transition status from December 21, 2017 to March 20, 2017. After this period the rehabilitation and reconstruction period began in the community housing sector although during the emergency transition status the rehabilitation and reconstruction phase has begun in other infrastructure sectors.

Based on the Regulation of the Head of National Disaster Management Agency (Badan Nasional Penanggulang Bencana/BNPB) No.11 of 2008 concerning the Post Disaster Rehabilitation and Reconstruction Guidelines stated that reconstruction is the formulation of policies and efforts as well as well planned concrete steps, consistent and sustainable way to permanently rebuild all infrastructure, facilities and institutional systems, both at the government and community levels. The objective of the reconstruction is to rebuild in the long term permanently some or all physical and non-physical facilities and infrastructure, along with all institutional systems and services damaged by the disaster, so that conditions are restored and their functions run well and the community can be better protected from various disaster threats [5].

The Indonesian government in the community-based housing sector rehabilitation and reconstruction regulations is designed with community organizing strategies and relies on initiatives and community initiatives (participatory development) without abandoning local wisdom. Especially for Community Direct Assistance (Bantuan Langsung Masyarakat/BLM) implemented through community empowerment in Community Groups (Pokmas) [6]. Community-based construction projects pose many risks in their implementation. Critical risk has been identified into 6 categories, namely project initiation, community contribution and participation, budget and finance, skilled personnel, material procurement and technical supervision and quality control. To manage these risks, it is necessary to plan, assess and mitigate in the future [7]. In Indonesia this community-based method was first applied in Aceh Province after the 2004 earthquake and tsunami disaster and was considered successful in its implementation so that it is also applied to other regions such as Nias,

Yogyakarta, Central Java and West Sumatra. Insufficient funding for reconstruction contributes to the slow progress of reconstruction in the housing sector. After almost four years after the disaster, housing reconstruction is still ongoing in West Sumatra [8].

Activities carried out at the time of community-based housing reconstruction in the original location (on-site) post-earthquake disaster in Pidie Jaya District in 2017 consists of program socialization, verification and validation of house damage data, finalization of data on recipients of housing assistance and formation community based (Pokmas), Pokmas training, identification of needs / technical surveys, preparation of Housing Construction Technical Document (Dokumen Teknis Pembangunan Perumahan/ DTPP), verification of technical documents & determination of the funding for Housing Assistance (Bantuan Dana Rumah/BDR), submission of funds disbursement, implementation of housing construction and accountability of funds [9]. The pre-construction period referred to in this study is from the program's socialization activities up to the submission of funds disbursement, while the implementation of the house and the accountability of funds are included in the construction of the house.

Traditionally, pre-construction is the initial planning and engineering carried out by the project team so that the project can run according to the agreement in the contract, such as defined project scope, schedule, costs, methods of implementation, identifying potential problems as early as possible with the most efficient use of resources. Pre-construction efforts finally help the owner determine whether the project can work or not. As for post-disaster projects in Indonesia, pre-construction is carried out in accordance with local government policies and after the emergency response period ends. The most common measure of the success of a post-disaster reconstruction project is the speed at which a damaged housing can be rebuilt [10].

Many important risks of managing community-based construction projects are in the pre-construction phase, such as unconfirmed sources of funding, lack of technical facilities, lack of consensus, lack of cooperation, non-conformance with standard specifications, inability to recruit skilled workers, unavailability of skilled workers, labor incompetent, a long tender process, and a lack of scheduled work [11]. In line with this, many problems with implementing community-based approaches in post-disaster contexts have also been found in the pre-construction phase, such as inappropriate policies regarding reconstruction strategies, lack of understanding of the meaning of community-based programs, coordination and communication problems, and lack of database systems [7, 12, 13]. There are 12 most dominant factors of 32 Critical Success Factors (CSFs) in the pre-construction phase of community-based post-disaster housing that have been implemented in Indonesia, namely transparency and accountability, followed by appropriate reconstruction policies / strategies, understanding community-based methods, gathering trust from the community, facilitator capacity, good coordination and communication, adequate funding availability, implementing capacity, significant level of community participation / control, involvement of all community members, involvement of all community members, success in identifying damage, and government support [14]. Success factors can be indicated by identifying specific factors that create problems or areas that have risks that affect the success of the project.

3. Methodology

The method of data collection was done by collecting secondary data from the BPBD of Pidie Jaya Regency, namely documents relating to the reconstruction of the post-disaster housing sector in Pidie Jaya Regency. Primary data consists of respondent data and data on factors that influence the slow start of housing construction during the pre-construction period. In this study the statements in questionaire are limited to the activities included in the implementation of pre-construction namely starting from program socialization to the community, conducting verification & validation of data on the level of damage to houses, formation and training of community group, design and funding. The data was obtained from dissemination of questionaire to respondents consisting of 10 people from BPBD, 10 people from management consultants consisting of 3 leaders and 7 facilitators, and 30 people from community group. Determination of respondents using the snowball method where several respondents told potential respondent researchers. The distribution and collection of questionnaires is done by visiting respondents directly to the research location that lasts for 3 months, starting in April 2018 and ending July 2018. All respondents were those directly involved in the reconstruction of houses after the earthquake in Pidie Jaya District. The questionnaire is conducted in closely manner with a Likert rating scale, namely (1) very bad; (2) Bad; (3) medium; (4) Good; (5) very good.

Data processing method uses the following steps:

1. Characteristics of respondents were analyzed using frequency distribution, namely grouping data or summary data tables that show the frequency / number of items / objects in each class that exists.
2. Data resulting from questionnaires about the factors of activities carried out during the pre-construction period. The activity is based on the perceptions of respondents with rating scales from very well done to very poor rating scales. These factors will find the value of validity by using multiple correlation formulas to determine whether or not the questionnaire is valid.
3. Next, a reliability test is conducted on the questionnaire to determine whether the questionnaire can be used more than once, at least by the same respondents who will produce consistent data so that questions can be reabeled to be submitted.
4. After the questionnaire is distributed, the data from the respondents' answers are calculated using the mean.
5. The next stage is the results of the study used for descriptive analysis of the results obtained to describe the conditions of pre-construction influenced by several factors in its implementation so that the construction of houses after the disaster was identified to be late.
6. Guided interviews with selected respondents consist of 2 people from BPBD Pidie Jaya Regency, 2 people from Management Consultants and 2 people from the community to obtain further information about the results that have been obtained from the factors of activities that affect the implementation of post-disaster housing pre-service in Pidie Jaya Regency.

4. Analysis and discussion

The results of secondary data analysis and interviews with the timing of the pre-construction of post-disaster housing in Pidie Jaya Regency can be seen in the figure below. Figure 1 explains that the rehabilitation and reconstruction period for housing starts since the end of the emergency response transition period, which is March 20, 2018. At this stage BPBD identified the level of damage to the house in collaboration with a team from Syiah Kuala University.

After the emergency transition period ended, housing rehabilitation and reconstruction activities were carried out by the BPBD of Pidie Jaya Regency, one of which was the procurement of a Regional Management Consultant conducted in May 2018. As many as 2 Regional Management Consultants who signed a contract with BPBD in Pidie Jaya Regency which began in August 2017 and ended in July 2018. One of the tasks of the regional management consultant is to recruit facilitators, pre-service training and mobilization of facilitators and then the facilitator conducts program socialization and coordination activities up to 40% of the phase I funding disbursement. The pre-construction period is very long, roughly ± 11 months, which is calculated from the end of the emergency response transition period until the construction of the house begins. The period of construction of the house began in February 2018 after the completion of the phase I application for disbursement of funds. The Rehabilitation and Reconstruction period will end in 2019 based on regulations set by the Aceh Province government.

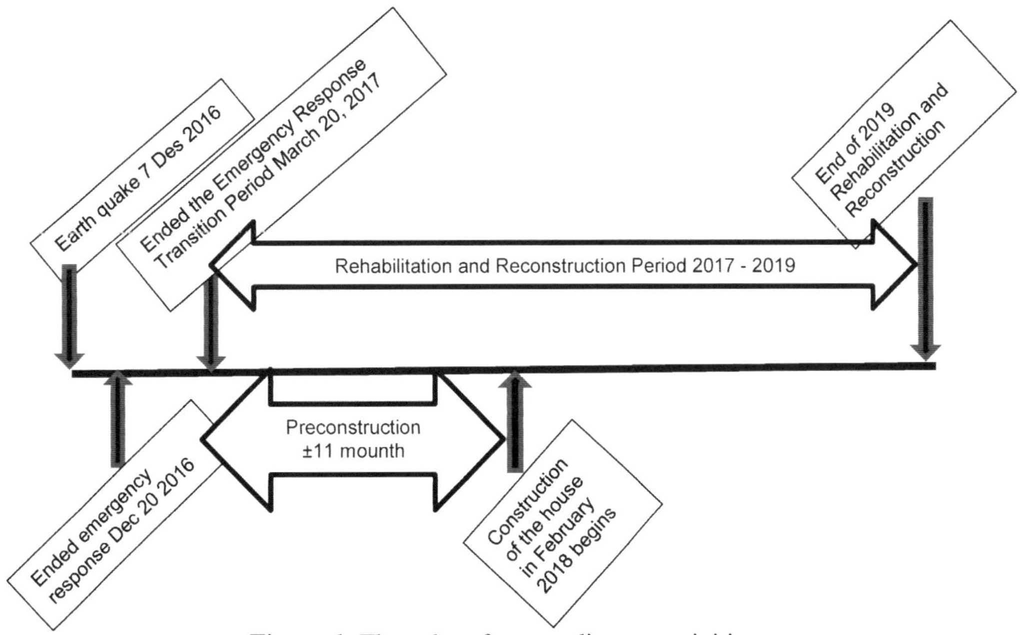

Figure 1. Flow chart for post-disaster activities

As many as 50 respondents who filled out the questionnaire in this study, the characteristics of respondents can be seen in the table below. Based on the characteristics of the respondents it can be explained that the respondents from the government and management consultants generally have a Bachelor degree and from the community there are those who have very low education, namely elementary school. Six respondents were identified from the community who had no experience in the construction sector. In general, Pokmas are not experienced in the construction sector, but in the characteristics of these respondents only identify that if the respondent knows little about how to build a house, then the experience will be between 1 and 5 years. Full description is shown in Table 1.

Table 1. Characteristics of Respondents

No	Demographics	Total	Percentage
1.	Group of respondents		
	a. Government (BPBD)	10	20%
	b. Management Consultant		
	Team Leader	3	6%
	Facilitator	7	14%
	c. Community		
	Head of Community Group (Pokmas)	4	8%
	Member of Pokmas	26	52%
2.	Age of Respondents		
	31 - 40 years old	18	36%
	41 - 50 year old	22	44%
	>51 years old	10	20%
3.	Education Attained		
	Elementary School	7	14%
	Junior High School	9	18%
	Senior High School	9	18%
	Diploma	1	2%
	Undergraduate	24	48%

4.	Experience in construction		
	0 Year (Never)	6	12%
	1-5 Years	29	58%
	5-10 Years	4	8%
	11-15 Years	8	16%
	>15 Years	3	6%

The results of data processing using statistical methods found that there were 26 factors of activities carried out during the pre-construction period. The activity is based on the perceptions of respondents with rating scales from very well done to very poor rating scales. The ranking sequence of these factors can be seen in the table below.

Table 2. Mean Score

Code	Factors		Mean Score	Rank
Socialization and Program Coordination				
P1	BPBD & facilitators coordinate at the District / Village level		4,180	4
P2	BPBD & facilitator socialize to the community		4,160	5
P3	BPBD & facilitator build togetherness, solidarity and volunteerism		3,540	23
P4	BPBD & facilitator communicate well with the community		3,620	22
P5	All communities participate in socialization & coordination		3,740	19
P6	Program socialization and coordination in accordance with the planned time		3,940	13
		Mean	3,863	
Inventory, verification and validation of damage data				
P7	BPBD conducts inventory & identification of damage		3,920	14
P8	BPBD & facilitator verify & validate damage data		4,040	8
P9	Communities participate in identification, verification & validation		3,900	15
P10	BPBD & facilitator conduct verification & validation fairly		4,020	9
P11	Inventory, verification & validation according to the planned time		3,860	16
P12	BPBD & facilitator validate data on time		3,760	18
P13	Finalizing the data until the District Head decree is issued for the recipient of the house on time		3,700	20
		Mean	3,886	
Formation of community organizations				
P14	Selection of Head of Pokmas according to skills in the field of construction		3,640	21
P15	Establishment of Pokmas according to the planned time		3,960	12
P16	The selection of Pokmas and members is conducted openly		3,780	17
		Mean	3,793	
Design and funding				
P17	The facilitator makes the design of the house on time as planned		4,200	3
P18	The facilitator makes the budget plan of the house on time as planned		4,340	1
P19	The design plan and bugget plan are consulted with the community		3,340	24
P20	Transparency and accountability to the community in setting funding		4,100	6

	targets			
P21	The government provides funding in accordance with the results of the validation of damage		4,240	2
P22	Disbursement of phase 1 (40%) funds on time		4,080	7
		Mean	4,050	
Program training for the community				
P23	The facilitator trains the Pokmas		4,000	10
P24	The training program is absorbed by the Pokmas		3,220	26
P25	The training results can empower Pokmas		3,260	25
P26	Training according to the specified time		3,980	11
		Mean	3,615	

From these 26 factors, the very low factor was the program training for the community or Pokmas, the design plan and bugget plan consulted with the community, building togetherness, solidarity and volunteerism, communication, election of the Pokmas chairman, and finalizing the data until the decree from the recipient. The factors with the lowest mean value were identified that the activity was carried out with a moderate level in the Likert scale or meant that the activity had not been fully carried out very well and there must have been a very influential cause in the implementation of the post-disaster housing construction.

The results of the analysis were then asked again with several respondents in the form of guided interviews to find accurate information. From the results of interviews with respondents that from the procurement activities of the Regional Management Consultants included in the pre-construction activities carried out by the BPBD and those activities that proceeded as planned. The regional management consultant consists of 2 companies whose task is to coordinate with the parties involved in the implementation of Rehabilitation and Reconstruction, identifying and formulating activities, compiling general guidelines and showing technical implementation, preparing and controlling all activities and identifying and facilitating resolution of problems that occur in the implementation of Housing Sector Rehabilitation and Reconstruction. The results of the analysis of interviews with respondents found that many led to delays in the implementation of housing construction starting from the activity of recruiting facilitators, resulting in delayed activities. The scope of questions in guided interviews starts from the recruitment of facilitators to design and funding activities as outlined below.

Recruitment of facilitators

The recruitment of facilitators is the duty and responsibility of the Regional Management Consultant so that all regulations and procedures have been determined by the consultant but still coordinate with the BPPD of Pidei Jaya Regency. However, in its implementation it was impressed by the public that in the recruitment process it was carried out in a non-transparent manner such as excessive requirements, a non-transparent recruitment process including graduation announcements was only conducted through personal Short Message Service (SMS). There was dissatisfaction from the community towards the recruitment process which resulted in delays in issuing decrees on the determination of facilitators and subsequent activities such as training and facilitating facilitators too late too. From this fact, it is necessary to implement a better human resource procurement management system in the future so that it can reduce problems as has happened in Pidie Jaya District.

Program socialization and coordination

Program socialization and coordination were held together with BPBD and facilitators at the Subdistrict, village and community levels, but not all beneficiaries were present at the event so that after the program socialization and coordination program many people were dissatisfied and visited the BPBD of Pidie Jaya Regency, and not all can be serviced by BPBD, so it seems that communication between BPBD / facilitator and community is carried out with very limited time.

Inventory, verification and validation of damage data
The damaged house validation carried out by the facilitator experienced an increase in implementation time because the community was not satisfied with the results of the initial identification of house damage that had been carried out during the emergency response and transition. The damaged house validation carried out by the facilitator experienced an increase in implementation time because the community was not satisfied with the results of the initial identification of house damage that had been carried out during the emergency response and transition. This happens because there are some inaccurate data during the initial identification period and also due to a lack of knowledge of the category of damage to houses by the community. Program socialization to the community is very necessary so that it increases the knowledge and confidence that the results of identification and validation that have been done are correct. The number of houses that must be validated has increased from the original plan so that the number of Technical Facilitators that have been recruited is no longer sufficient as planned in the contract between the BPBD of Pidie Jaya Regency and the Regional Management Consultants. Validation of the level of damage to the house was carried out for a long time, resulting in delays in public testing activities and the issuance of Regent Decrees for recipients of houses, formation and training of Pokmas, preparation of Housing Development Technical Documents, design and budget plan.

Formation and training of Pokmas
The facilitator facilitates the community in choosing Pokmas leaders and members. Generally the leader of the Pokmas chosen is by deliberation between the community and its members without being seen from their expertise and skills in managing housing construction. The formation and training of the Pokmas is partially delayed due to the increase in the number of Pokmas from the original plan because of the additional validation of the level of damage to the house. The addition resulted in insufficient training funds planned. Pokmas training does not run effectively and efficiently so it is feared that it will affect the non-empowerment of the community in carrying out the process of building houses.

Design and funding
The task of the facilitator is to design and calculate the cost requirements based on the level of damage. The cost for a heavily damaged house is funded in the amount of Rp. 85,000,000. / House and a moderately damaged house of Rp. 20,000,000. / House. For houses categorized as heavily damaged, built with 2 reconstruction methods, the first method will be built from foundation work to finishing using 36 house type designs and uniform technical specs if identified all old buildings cannot function. While if a part of the house has been identified as functioning it will be heavily damaged, it will be designed according to the needs of the community but adjusted to the funding and technical specs that have been set. However, for the latter category in reality many people were dissatisfied with the design made by the facilitator and it was stated that the design made by the facilitator was not fully consulted with the community. Often changes in design based on the wishes of the people must be made a policy about its own technical specs. Not all community desires about the design can be accommodated by the facilitator and the average design results based on the results of the analysis of the field survey conducted by the facilitator not based on the wishes of the community so that the timing of the design and budget plan is exactly as planned.

5. Conclusion

The results of this study have identified that specific factors that create problems that affect the success of housing reconstruction at the pre-construction stage, namely (1) recruitment of facilitators; (2) program training for the community; (3) formation of community organizations; (4) program socialization and coordination; (5) inventory, verification and validation of damage data; (6) design and funding. The dominant factors in the pre-construction are interrelated with each other because the problems that occurred at the beginning of the pre-construction period resulted in unavoidable delays in the construction of houses during reconstruction. The results of the identification of these factors form the basis for practitioners, stakeholders and the community in managing post-disaster housing projects so that they can anticipate similar things in the implementation of reconstruction in the future.

6. References

[1] Anonim. (2017). Pedoman Umum Penyelenggaraaan Rehabilitasi dan Rekonstruksi Sektor Permukiman Berbasis Masyarakat (*Insitu*) Pasca Bencana Gempa Bumi di Kabupaten Pidie Jaya.
[2] Jha, A.K., Barenstein, J.D., Phelps, P.M., Pittet, D. and Sena, S. (2010). Safer Homes, Stronger Communities: A Handbook for Reconstructing after Natural Disasters, The World Bank, and Washington, DC.
[3] Silva, J. (2010). Lessons from Aceh: Key Considerations in Post-disaster ReconstructionPractical Action, Rugby.
[4] Ophiyandri, T., Amaratunga, D. and Pathirage, C. (2010). "Community based post-disaster housing reconstruction: Indonesian perspective", Proceeding of CIB World Congress, Salford, 10-13 May 2010.
[5] Anonim. (2008). Peraturan Kepala Badan Nasional Penanggulangan Bencana (BNPB) No. 11 Tahun 2008 Tentang Pedoman Rehabilitasi dan Rekonstruksi Pasca Bencana.
[6] Anonim. (2013). Peraturan Kepala Badan Nasional Penanggulangan Bencana (BNPB) No. 3 Tahun 2013, Petunjuk Teknis Pelaksanaan Anggaran Kegiatan Rehabilitasi dan Rekonstruksi Wilayah Pascabencana
[7] Ganapati, N. E., & Ganapati, S. (2008). Enabling participatory planning after disasters: A case Study of the World Bank's housing reconstruction in Turkey. *Journal of the American Planning Association*, *75*(1), 41-59.
[8] Pranoto, S. (2011). *Lessons learned: pembelajaran rehab rekon pasca gempa di Sumatera Barat, 30 September 2009: building back better*, Gagak Jingga.
[9] Anonim. (2017). Petunjuk Teknis Rehabilitasi dan Rekonstruksi Sektor Permukiman Berbasis Masyarakat (Insitu) Pasca Bencana Gempa Bumi Di Kabupaten Pidie Jaya., BPBD Kabupaten Pidie Jaya
[10] Anthopoulos, L. G., Kostavara, E., & Pantouvakis, J. P. (2013). *An effective disaster recovery model for construction projects*. Procedia-Social and Behavioral Sciences, 74, 21-30.
[11] Manelele, I. and Muya, M. (2008). "Risk identification on community-based construction projects in Zambia", Journal of Engineering, Design and Technology, Vol. 6 No. 2, pp. 145-161.
[12] Dercon, B. and Kusumawijaya, M. (2007). "Two years of settlement recovery in Aceh and Nias: what should the planners have learned?", paper presented at 43rd ISOCARP Congress, Antwerp, Belgium, 19-23 September.
[13] Davidson, C.H., Johnson, C., Lizarralde, G., Dikmen, N. and Sliwinski, A. (2007). "Truths and myths about community participation in post-disaster housing projects", Habitat International, Vol. 31, pp. 100-115.
[14] Ophiyandri, T., Amaratunga, D., Pathirage, C., & Keraminiyage, K. (2013). Critical success factors for community-based post-disaster housing reconstruction projects in the pre-construction stage in Indonesia. *International Journal of Disaster Resilience in the Built Environment*, *4*(2), 236-249. Dorman L I 1975 *Variations of Galactic Cosmic Rays* (Moscow: Moscow State University Press) p 103.

Acknowledgments
Authors wishing to acknowledge assistance or encouragement from colleagues, special work by technical staff or financial support from organizations should do so in an unnumbered Acknowledgments section immediately following the last numbered section of the paper.

Disaster preparedness in the red zone schools at 13 years post tsunami 2004

S Yusuf[1], **Asniar**[1], **Rachmalia**[1, 2,*] and **S S Susanti**[1]

[1] Faculty of Nursing, Syiah Kuala University, Jl. Tgk. Tanoh Abee, Darussalam, Banda Aceh, 23111, Indonesia

[2] Tsunami and Disaster Mitigation Research Center (TDMRC), Jl.Prof. Dr. Ibrahim Hasan, Meuraxa, Banda Aceh, 23233, Indonesia.

[*] Corresponding author email : rachmalia@tdmrc.org

Abstract. Based on the Indonesian Disaster Risk Index, 75% schools in Indonesia are located in the medium to high-risk disaster-prone area. However, many educational institutions in Indonesia especially in Aceh Province, still consider disaster preparedness as inessential issues and tend to be ignored. This study aims to evaluate the preparedness of the high risk school communities after 13 years of post- tsunami 2014; related to policy parameters, knowledge, emergency planning, disaster warning systems and resource mobilization capacity. This is a descriptive explorative study using questionnaires to collect data from 80 elementary school teachers and 265 elementary students. A semi structured interviews with eight elementary school principals was also conducted to enrich the research findings. The study findings show that the school preparedness is in the low category (total index value 20.50); school teacher preparedness is in the medium category (total index value 61.79) and student preparedness is in the low category (total index value 52.12). The preparedness of the school community (a composite index of school institutions, teachers and students) is also in low category (total index value 43.09). The result of the study indicated that disaster preparedness programs in most elementary schools have not been well implemented. The lack of sustainability of the disaster preparedness school program and its monitoring and evaluation might contribute to this condition.

Keyword: preparedness, disaster, earthquake, tsunami, elementary school children.

1. Introduction

The province of Aceh is one of the regions in Indonesia that is vulnerable to earthquakes and tsunamis due to its closest location to the epicenter of the 2004 Indian Ocean earthquake. Therefore 16 of the 23 districts/cities in Aceh are in high-risk classification or in the red zone of earthquake disaster and tsunami [1]. Based on the map of Disaster Risk Index, district of Aceh Besar located in the red zone of earthquake and tsunami disaster [1]. Reducing disaster risk is not only the responsibility of the government, but also needs the participation of the entire community, including the participation of educational institutions. The participation of educational institution should not be ignored since its potential as an effective, dynamic and sustainable information transfer strategy in building the knowledge capacity to improve community preparedness for disasters [2].

Nevertheless, only few of educational institutions considers disaster preparedness as an important issue. Based on the results of a survey conducted by [3], 19 schools in the city of Banda Aceh, the sustainability of many disaster preparedness schools or *Sekolah Siaga Bencana* (SSB) was halted due to the mutation of school principals. This condition affected the discontinuation of SSB policy, which was also contributed from the lack of support from government and non-government institutions and funding problems. In addition, the results of the survey also show that only 15.79% of schools have a high level of preparedness for policy and direction parameters. As for emergency planning, early warning and resources mobilize capacity, only 47.37% have high preparedness levels. This is a worrisome condition as children are one of the most vulnerable groups to suffer from the impact of disasters due to their weak physical condition and their inability of to make decisions; especially in the ocassion when they became separated from their parents during the disaster [4].

Preparing for children to face future disasters could be conducted by increasing the ability of children to make decisions. Thus, if there is a threat to their safety, the child is able to take appropriate actions. Therefore, disaster education is considerably needed in schools, particularly in terms of the individual, family and community education to help the children developing disaster preparedness culture [5]. Vulnerabilities relate to the level of hazard preparedness [4]. Thus, vulnerability can be reduced by increasing the capacity. Capacity building might be achieved by providing education and training about disaster knowledge in schools [6].

Disaster risk reduction strategies in schools could be carried out in the form of structural and non- structural manner, in order to establish a culture of preparedness and safety [7]. With the purpose to emphasize the importance of disaster risk reduction, the government of Aceh Besar district issued a *Qanun* (sharia law) on regional disaster management, in which mentions the appeals for customary institutions, non- governmental organizations, religious institutions, business institutions and communities in order to play role in disaster management. Furthermore, the Sendai Framework for Disaster Risk Reduction / SFDRR (2015-2030) in *Priority for Action 4*: also emphasizes the importance to strengthen preparedness to deal with disasters [8].

Considering the importance of disaster preparedness, this study aimed at evaluating the preparedness of the school community in the red zone of the earthquake and tsunami disaster Peukan Bada Aceh Besar community health center working areas. This research will focus on studying the five parameters of the earthquake and tsunami disaster preparedness including:
1) Disaster preparedness policy (PS), 2) Disaster knowledge (K), 3) Emergency planning (EP), 4) Disaster warning system (WS), and 5) Resource mobilization capacity (RMC) [9] [10].

2. Material and method

This research employed an explorative descriptive design with qualitative and quantitative data collection. The study was conducted in eight elementary schools located in the red zone of the earthquake and tsunami disaster prone area. The quantitative data was collected using a questionnaires developed by [10]. The questionnaires was filled by eight school principals, 80 teachers and 265 students at fourth and fifth grade of elementary school located in Peukan Bada subs-district. The sample size was determined based on [11] table, using a power of 0.8; the value of $\alpha = 0.05\%$; and effect size 0.2; which resulted 241 samples. To avoid the lack of response rate, the researchers added 10% of the sample size to 265 people. Probability random sampling technique was performed by a systematic random sampling.

In addition to collecting quantitative data, this study also collected qualitative data using a semi-structured interview method and windshield survey. The semi-structured interviews were conducted to collect qualitative data related to disaster preparedness programs in schools. Whereas windshield survey were conducted to observe the surrounding conditions of the schools which are relevant to the earthquake and tsunami disaster preparedness. The interview guide was prepared by the researchers by addressing the input, process and output aspects in carrying out the disaster preparedness program [9]. While the windshield guide was developed based on the components described in the table of *Learning about the Community on Foot* as part of the *Community as Partner Model* [12].

The study was conducted by distributing two type of questionnaires which both implied several parameters including policy, knowledge, emergency planning, disaster warning system, and resource mobilization capacity pertinent to earth quake and tsunami disaster. The first type of questionnaire is designed to be filled by the elementary school principals, while the second one is to be distributed for the teachers and students of the elementary schools. The data collected by those questionnaires will be used to estimate the parameter index as follows.

IOP Conf. Series: Earth and Environmental Science **273** (2019) 012038 doi:10.1088/1755-1315/273/1/012038

$$\text{Parameter index} = \frac{\text{Total of parameter's real score}}{\text{Parameter's maximum score}} \times 100$$

Furthermore, the index value is interpreted based on the following categories: a) high preparedness (value index 79.50 - 100), b) moderate preparedness (index values 55.00 - 79.49) and c) low preparedness (index value <55).

3. Result

The questionnaires that distributed to the school institutions achieved 100% response rate and another one distributed to students achieved a response rate of 96.23%. Whereas from 80 questionnaires distributed to elementary school teachers, the response rate was 98.75%. Quantitative data collection using a questionnaire identified the school distance from the beach is less than 500 meters (50%). Most of the building walls type are 100% concrete/brick. The condition of the walls on the lower floors of a multi-stores school building was half solid and has enough windows/ventilation (66.7%). Most respondents did not know the condition of the school buildings (62.5 %) and only 12.5% school buildings that follow earthquake resistant building standards and consider the influence of a tsunami load was only 12.5%.

3.1. The school preparedness

The results of the earthquake and tsunami disaster preparedness evaluation at eight elementary schools in Aceh Besar District can be seen in the figure below. The parameters used in this study were 1) Disaster preparedness policy (PS), 2) Disaster knowledge (K), 3) Emergency planning (EP), 4) Disaster warning system (WS), and 5) Resource mobilization capacity (RMC).

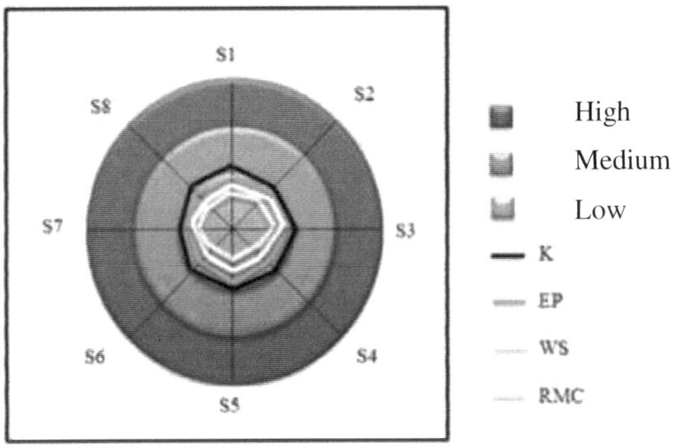

Figure 1. The school preparedness index

Table 1. The percentage of school preparedness parameters

	Low		Medium		High	
Parameter	Number Of Schools	%	Number of Schools	%	Number of Schools	%
PS	8	100	-	-	-	-
EP	8	100	-	-	-	-
WS	7	87.5	1	12.5	-	-
RMC	8	100	-	-	-	-
School Preparedness	8	100	-	-	-	-

The evaluation results show school preparedness is in the "low" category in the policy parameters with an index value of 19, emergency planning with an index value of 22, disaster warning system with an index value of 23 and resource mobilization capacity with an index value of 19. Therefore, it can be concluded that school preparedness for the earthquake and tsunami disaster is in the "medium" category with an index value of 20.50.

3.2. The teacher preparedness

The results of the teacher preparedness evaluation for the earthquake and tsunami disaster in 79 teachers from 8 elementary schools in Aceh Besar District can be seen in the figure below.

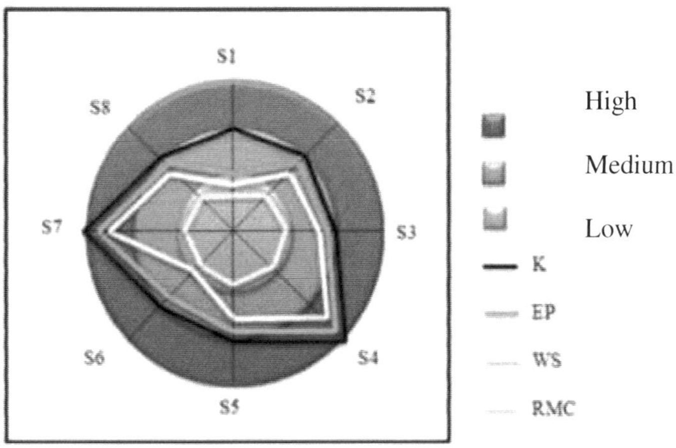

Figure 2. The teacher preparedness index

Table 2. The percentage of teacher preparedness parameters

Parameter	Low		Medium		High	
	Number Of Teachers	%	Number of Teachers	%	Number of Teachers	%
KAP	5	6.3	33	41.8	41	51,9
EP	24	30.4	31	39.2	24	30.4
WS	28	35.4	24	30.4	27	34.2
RMC	79	100	-	-	-	-
Teacher Preparedness	24	30.4	51	64.6	4	5.1

The evaluation results show teacher preparedness in the "medium" category in three parameters, such as: Knowledge (77.77) index value, emergency planning (66.21) and disaster warning system (64.48). However, it was in the "low" category on resource mobilization capacity parameters with an index value of 38.69. The study concluded the teacher preparedness for the earthquake and tsunami disaster is in the "medium" category with an index value of 61.79.

3.3. The student preparedness

The results of the student preparedness evaluation for the earthquake and tsunami disaster in 255 students in 8 elementary schools in Aceh Besar District can be seen in the figure below.

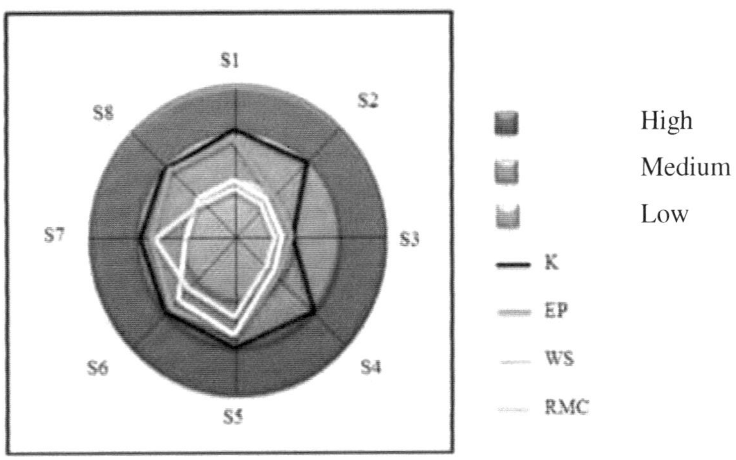

Figure 3. The student preparedness index

Table 3. The percentage of student preparedness parameters

	Low		Medium		High	
Parameter	Number Of Students	%	Number of Schools	%	Number of Schools	%
K	83	32.5	145	56.9	27	10.6
EP	95	37.3	127	49.8	33	12.9
WS	182	71.4	58	22.7	15	5.9
RMC	151	59.2	87	34.1	17	6.7
Student Preparedness	140	54.9	104	40.8	11	4.3

The evaluation results show student preparedness in terms of the knowledge parameter was in the "medium" category with index value (61.61) and disaster emergency planning (57.97). However, the disaster warning system parameter was in the "low" category (46.56) and resource mobilization capacity (42.35). Thus, the study concluded that student preparedness for the earthquake and tsunami disaster is in the "low" category with an index value of 52.12.

3.4. The school community preparedness (SCP)
The results of the school community preparedness evaluation in 8 elementary schools in Peukan Bada Aceh Besar described in the figure below.

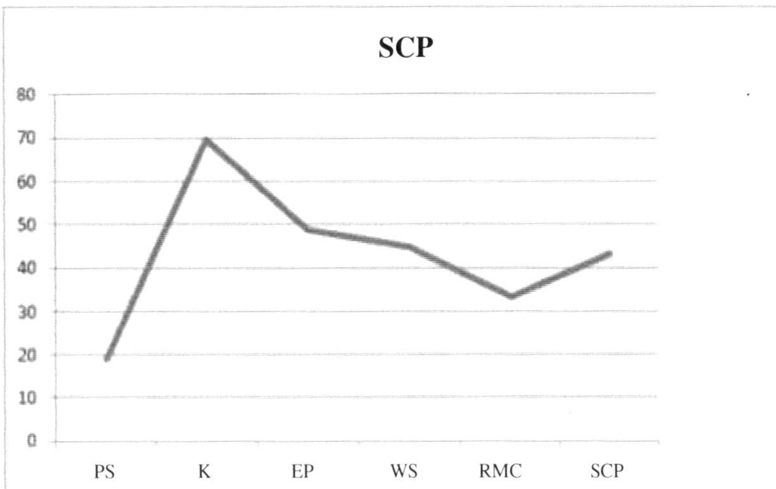

Figure 4. The school community preparedness index

Table 4. The percentage of school community preparedness parameters

Parameter	Low		Medium		High	
	Number of Teachers	%	Number of Teachers	%	Number of Teachers	%
School	8	100	-	-	-	-
Teacher	24	30.4	51	64.6	4	5.1
Student	140	54.9	104	40.8	11	4.3

The evaluation results show that the preparedness of the school community was in the "low" category in the policy parameters with index value of 19. Disaster emergency planning (48.73), disaster warning system (44.68) and resource mobilization capacty (33.35). However, the knowledge parameter was in the "medium" category with the index value of 69.69. The study concluded that the school community preparedness for the earthquake and tsunami disaster is in the "low" category with an index value of 43, 09.

From the results of the windshield survey at eight elementary schools in the Peukan Bada showed that all the schools did not have the evacuation route mark and disaster gathering point. In addition, there were no banners or posters related to earthquake and tsunami disaster preparedness. This result was in contrast with the interviews with the stakeholder of community health center and schools' principals about the earthquake and tsunami preparedness.

The interviews with stakeholders in the community health center found that the disaster preparedness program at had not optimally implemented. Whereas disaster preparedness was very important since the the location of the community health center was in the costal area and prone to earthquakes and tsunamis. The monitoring and evaluation of the disaster preparedness program at the community health center should be conducted in order to perceived some barriers and challenges in running the disaster preparedness program. The community health center has never provided disaster-related promotive and preventive services to the elementary school children; since officers in charge of the disaster preparedness at the community health center have never been trained to make programs. Therefore, most of them had lack of knowldge in commencing promotive and preventive efforts towards disaster preparedness. For this reason, the head of the community health center felt that it was better for the officers whom responsible for disaster preparedness programs to be trained in order to establish and implement the programs based on the guidelines on disaster preparednes.

From the results of the interviews with all eight schools'principals found that all schools did not have the disaster preparedness program. Therefore, some teachers delivered the knowledge about earthquake and tsunami disasters to students according to their abilities, despite the implementation of the programmes was inadequate and unstructured. To this point, the schools did not have any guideline on the implementation of the earthquake and tsunami preparedness program. Nevertheless, before 2010 many governmeny and non-government institutions visit the school regarding the disaster preparedness. However, none of them conducted or following up any plans for disaster preparedness program sustainability. Acording to the school principals, the schools were ready to implement disaster preparedness programs under the programmes' guidance to school institutions, teachers and students. Based on the interview it was obtained about the obstacles affecting the implementation of disaster preparedness programs in schools, such as the absence of good cross-

IOP Conf. Series: Earth and Environmental Science **273** (2019) 012038 doi:10.1088/1755-1315/273/1/012038

program and cross-sectoral collaboration with related sections. In the future, the authorities need to facilitate good cooperation between cross-programs and cross-sectors.

Besides establishing cross-program and cross-sectoral cooperation, it is necessary to manage education that emphasizes school independence and creativity in order to improve earthquake and tsunami disaster preparedness in schools. The innovative learning strategies was one of the strategies which emphasize on actively involving students' participation in disaster preparedness program.

4. Discussion

4.1. The school preparedness

The result showed that school preparedness for the earthquake and tsunami disaster was in the "low" category. These results identified the school institutions categorized as "unprepared" to deal with the earthquake and tsunami disasters. Based on the 2013 Aceh earthquake and tsunami risk map, 100% of elementary schools in Peukan Bada are in the red zone of the earthquake and tsunami disaster prone area. Therefore, it is necessary to strengthen school capacity to support teachers and students' preparedness in improving earthquake and tsunami disaster preparedness through School-Based Management [3]. [13] also strongly recommends reactivating disaster preparedness which supported by the local government policies to ensure the implementation of preparedness throughout the schools. Moreover, according to [14] realizing the importance of disaster safe education unit, it is necessary for the schools to undertake disaster resilient education efforts through the pillars of safe school facilities. The activities include the placement of school locations in disaster safe areas with the design of school buildings in accordance with the building rules and standards. In addition, these efforts must also be accompanied by the support of education sector policies and planning also local disaster management plans [14].

4.2. The teacher preparedness

From the study result showed that the teacher preparedness for the earthquake and tsunami disaster was in the "medium" category. These results identified teachers as "almost ready" to face the earthquake and tsunami disasters based on 3 parameters; such as knowledge, emergency response plans and disaster warning systems. While the resource mobilization parameters of all teachers were in "unprepared" condition. The teacher is the biggest source of information for the students to obtain knowledge of the earthquake and tsunami disasters in schools. If teacher resource mobilization is ready, the teacher will understand and be able to convey information about the earthquake and tsunami disaster preparedness to the school community. In accordance with the results of the research conducted by [15] showing the higher the knowledge of psychosocial disaster preparedness of the teachers, the better the knowledge of children who learn from them. In order to improve the resource mobilization, teachers need to be provide the education, trainings and workshops in order to be capable in simulating the earthquake and tsunami disasters in the schools also to strengthen the other 3 parameters. Thus, the high disaster preparedness could be achieved.

4.3. The student preparedness

The result of the study identified the students' preparedness for the earthquake and tsunami disaster was in the "low" category. This result identifies students as "unprepared" in facing the earthquake and tsunami disasters. The children are vulnerable group to the earthquake and tsunami disasters [16]. Most children also spend of their time in the schools; therefore, they need be asured to be always ready to face the earthquake and tsunami disasters. To understand the disaster-resilient, education need to be implemented in comprehensive efforts which is child-centered. Understanding the children specific needs in dealing with disasters is important, by actively engaging the children according to their capacities and interests [14]. This opinion is in accordance with the results of research by [17] which provides strong empirical evidence that education shows positive externalities to disaster risk reduction. [18] also found that knowledge and information received will be processed to produce a decision to act towards disasters.

4.4. The school community preparedness
The result of the study showed that the school community's preparedness for the earthquake and tsunami disasters was in the "low" category. This result identified the school community as "unprepared" to face the earthquake and tsunami disaster. Disaster events are unpredictable and can occur during the school hours. Therefore, it is necessary to develop a school institutional capacity to improve the preparedness through the School-Based Management (SBM). Thus, the schools are expected to have continuous independence in order to ensure sustainability of programs that aim to increase resilience to earthquake and tsunami disasters in schools and communities [3]. This opinion is also supported by the research conducted by [19], that school-based disaster management can facilitate the schools to regularly check and verify various disaster management tasks and to improve their ability to disaster preparedness.

5. Conclusion
The study showed that the school and student preparedness were in a low category, while teachers were in the medium category. This condition affected the preparedness of the school community in the Peukan Bada as in the low category. This conditions needs to be anticipated by various efforts to improve the preparedness that aimed at children, teachers, school administrators and other relevant parties.

The improvement of children's preparedness to deal with the disasters could be conducted by improving the ability of children to make decisions and folowed by disaster education in schools in a sustainable manner. Decision-making ability in children is very necessary since the children are vulnerable groups. Therefore, the need to educated in order take appropriate action towards disasters. These efforts could be achieved by incorporating disaster preparedness education into formal and informal education; thus, disaster education will be sustainable. Sustainable disaster education is considered essential and very effective also dynamic information transfer strategy. To ensure the sustainability of disaster education, the role of teachers, principals and related parties will be needed optimally. With the active role of all parties, it is expected that the disaster preparedness of the school community will be improved. Therefore, the disaster impact will be reduced in the event of disasters.

References

[1] IRBI 2013 *(IRBI) Indeks risiko bencana Indonesia* (Sentul: Direktorat Pengurangan Risiko Bencana. Deputi Bidang Pencegahan dan Keselamatan Provinsi Jawa Barat)
[2] CDE 2011 *A framework of school-based disaster preparedness* (Jakarta: Consortium for disaster education)
[3] Oktari R S, Kemala I D, Rachmalia, and Husna H 2015 Strengthening institutional capacity of school to enhance community resilience against disaster *Symp. Nasional Mitigasi Bencana Tsunami 2015,* Banda Aceh, 21-22 Desember 2015.
[4] Forum Keperawatan Bencana and Japanese Red Cross Society 2009 *Keperawatan bencana*, edisi pertama (Banda Aceh: Palang Merah Indonesia)
[5] Shaw R, Kobayashi K S H, and Kobayashi M 2004 Linking experience, education, perception and earthquake preparedness *Disaster Prev. Manag.* **13** 39–49
[6] Natawidjaja D H 2007 Gempa bumi dan tsunami di Sumatra dan upaya untuk mengembangkan lingkungan hidup yang aman dari bencana alam *Geospasial.menlh KLH report* (Jakarta)
[7] MENDIKNAS 2010 *Undang-Undang Nomor 24 Tahun 2007 dan Peraturan Pemerintah Nomor 21 Tahun 2008* (Jakarta) p 1–2
[8] UNISDR 2015 *Sendai framework for disaster risk reduction 2015 - 2030*, 1st ed (Switzerland)

[9] Triyono and Koswara A 2017 *Panduan monitoring dan evaluasi sekolah siaga bencana*, Januari 2011 (Jakarta: LIPI Press)
[10] LIPI-UNESCO/ISDR 2006 *Kajian kesiapsiagaan masyarakat dalam mengantisipasi bencana gempa bumi dan tsunami* (Jakarta: LIPI Press)
[11] Cohen J 1998 *Statistical power analysis for the behavioral sciences*, 2nd ed (New York: Lawrence Erbaum Associates)
[12] Anderson E T and McFarlane J 2006 *Buku ajar keperawatan komunitas teori dan praktik*, edisi ketiga (Jakarta: EGC)
[13] Sakurai A, Bisri M B F, Oda T, Oktari R S, Murayama Y, Nizammudin and Affan M 2017 Exploring minimum essentials for sustainable school disaster preparedness : A case of elementary schools in Banda Aceh City , Indonesia *Int. J. of Disaster Risk Reduct.* **29** 73-83
[14] Amri A 2017 *Pendidikan tangguh bencana: mewujudkan satuan pendidikan aman bencana di Indonesia.*(Jakarta: Kementerian Pendidikan dan Kebudayaan)
[15] Elangovan A R and Kasi S 2015 Psychosocial disaster preparedness for school children by teachers *Int. J. of Disaster Risk Reduct.* **12** 119–124
[16] BNPB 2012 *Pedoman penerapan sekolah/madrasah aman dari bencana.* (Indonesia: Peraturan Kepala Badan Nasional Penanggulangan Bencana) pp 1–40
[17] Hoffmann R and Muttarak R 2017 Learn from the past, prepare for the future: impacts of education and experience on disaster preparedness in the Philippines and Thailand *World Development* **96** 32–51
[18] Azwar S 2011 *Sikap Manusia: Teori dan pengukurannya* 15ed ed (Yogjakarta: Pustaka Pelajar).
[19] Wang J J 2016 Study on the context of school-based disaster management *Int. J. Disaster Risk Reduct.* **19** 224–234

IOP Conf. Series: Earth and Environmental Science **273** (2019) 012039 doi:10.1088/1755-1315/273/1/012039

Transforming the Archives of the Great East Japan Earthquake into Global Natural Disaster Archives

A. Shibayama[1], S. Boret[1]

[1]International Research Institute of Disaster Science, Tohoku University, Aoba 468-1, Aramaki, Aoba-ku, Sendai, JAPAN

Abstract. Soon after the devastation brought by Great East Japan Earthquake (GEJE), a variety of Japanese agencies and organizations began to build digital archives of the disaster. Seven years onwards, over fifty digital archives collect and release millions of earthquake records. These earthquake disaster digital archive created for the Great East Japan Earthquake have the potential to lead the way for more general measures and countermeasures to reduce the impact of future 'natural' disasters. [1] We also believe that the records and knowledge of 'natural' disasters must be systematized and shared worldwide. As a first step, this research sheds light on examples and subjects of the digital archive of GEJE and report on basic knowledge that serves for the development of digital archive of 'natural' disasters' in the future.

1. Introduction

The Great East Japan Earthquake that occurred at 14:46 on March 11, 2011, was the largest ever recorded in Japan with a magnitude 9. The epicentre was located off the cost of northeast Japan and caused a massive tremor in the Tohoku region. The observation of seismic reported intensity between 1 and 2 in the distant Kyushu island. Literally, the whole of Japan shook that day. Within an hour of the earthquake, a great tsunami struck across the Tohoku region. The residents of many coastal cities and villages evacuated but the people and the buildings left behind were washed away. The Fukushima Daiichi Nuclear Power Station located in Okuma Town and Futaba Town in Fukushima Prefecture was hit by the tsunami causing serious damage to the facility and the worst outcome possible: its meltdown. The widespread evacuation began in the municipalities neighbouring the nuclear power plants from one morning to the next. Some people cannot return to their homeland seven years on. Among the 19,620 precious lives lost the Great East Japan Earthquake, 2,569 people are still counted as missing. Even after seven years, the search for missing persons is being held in the coastal areas of the Tohoku region months after months.

Mirroring the magnitude of the catastrophe, a tremendous effort was observed among the various agencies and organizations that started to put together digital archives to preserve the records of the Great East Japan Earthquake occurred. Seven years since the earthquake, more than 50 digital archives

[1] The authors understand that there is no such thing as a 'natural' disaster. All disasters are unequivocally the result of social processes as well as environmental hazards. For the sake of intelligibility, we use this sentence to refer to all disasters which are tied or induced by natural hazards (i.e. earthquake, tsunami, typhoon, hurricane, drought, storm, flood, and so forth).

are currently operating in Japan. Together, they gathered and manage millions of earthquake records. Based on our observation and contribution to this effort, we argue that the digital archive built after the Great East Japan Earthquake will serve to establish measures and countermeasures to protect human populations from future 'natural' disasters. Also, we believe in the necessity of preserving the records and knowledge of natural disasters that will occur in the future worldwide. To this end, this research intends to discuss examples and subjects of the digital archive of the Great East Japan Earthquake to report on basic knowledge necessary for the development of digital archives of natural disasters in the future.

2. About the disaster archive of the Great East Japan Earthquake

2.1. About the Disaster Archives

The discussion about the 'disaster archives' in this paper includes disaster-related knowledge and teachings, as well as the collection, preservation, and utilization of these records. A term similar to disaster archives is 'disaster big data'. Disaster big data refers to the body of information necessary for disaster prevention and mitigation, and emergency treatment and restoration from the time immediately following the disaster. The main difference between disaster archives and disaster big data is the respective nature of the data handled. Disaster archives manage digital records including SNS, photographic and video footage, as well as analogue documents including record magazines and publications, and handles relics and physical remains. In contrast, disaster big data deals only with digital records and is a component of disaster archives. The latter handles all information from disaster big data that focuses on data from the time immediately following the disaster, as well as records from prior to the disaster to the revival.

The concept of disaster archives was not recently created. In fact, we can refer to the collections in libraries and museums of disaster-related documents such as books and record magazines as disaster archives. Following the trend in recent years that has seen the evolution of the internet and an increase in digital material, the move toward material storage to digital archives is becoming mainstream, and the institutions implementing this are beginning to exceed the frame of libraries and museums.

2.2. Earthquake Disaster Archive Trends in the Great East Japan Earthquake

Following the GEJE, a Reconstruction Design of Seven Principles was proposed at the Reconstruction Design Council on June 25, 2011. Principle one states: "We shall record the disaster for eternity, including through the creation of memorial forests and monuments, and we shall have the disaster scientifically analysed by a broad range of scholars to draw lessons that will be shared with the world and passed down to posterity." [1] This principle saw the construction of a number of disaster archives. Examples include disaster archives built by the library group centred around the National Diet Library[2] and disaster-stricken areas, research institutions such as Tohoku Gakuin University[3], Tohoku University[4], Iwaki Meisei University[5], Japan Society of Civil Engineers[6], and the National Research Institute for Earth Science[7] and Disaster Resilience, media such as the Japan Broadcasting Corporation (NHK) [8] and Kahoku Shimpo Publishing Co. [9], private enterprises such as Google[10] or Yahoo! JAPAN[11], municipalities such as Tagajo City in Miyagi Prefecture[12], Miyagi Prefectural Library[13], Koriyama City in Fukushima Prefecture[14], Kuji city, Noda village, and Fudai village in Iwate Prefecture[15], and Iwate Prefecture[16]. Contrasting with previous practice, the present framework for disaster archives takes into consideration a number of factors. These include the scale and severity of the damages caused by the disaster, the challenges for a single system or institution to collect all earthquake-related records, people's desire to transmit lessons to future generations beyond Japan, and the large number of earthquake records that were created due to the spread of digital technology.

Phases of the disaster archives process can be divided into collecting, organizing, preserving, and exhibiting materials in a contents holder in each institution as well as a portal site operating system that allows cross-searching between contents holders. Typical portal sites include the National Diet Library's

IOP Conf. Series: Earth and Environmental Science **273** (2019) 012039 doi:10.1088/1755-1315/273/1/012039

'Hinagiku' [2]and Harvard University Reischauer Institute of Japanese Studies 'Japan Disasters Digital Archive' [17]. The majority of others are contents holders only. The National Diet Library's 'Hinagiku' has collaborated with dozens of institutions and, to date, the number of registered disaster-related documents of the Great East Japan Earthquake is approximately one million. Registered contents include materials from immediately after the earthquake to the recovery and reconstruction, including photographs, testimonies, visual and audio recordings, administrative sentences; over half of the items are photographic records. These disaster-related data serve disaster prevention education and reconstruction tourism, publications, media, and research materials. Finally, we note that there are three main types of disaster archives depending on their modes of construction: 1) disaster archives using conventional means to gather material; 2) disaster archives storing large amounts of earthquake records, and 3) disaster archives passing on disaster-related documents to future generations.

Table1. Major disaster archive of the Great East Japan Earthquake (2017)

Date	Site Name (Names of institutions)	Number of contents
June, 2011	East Japan Earthquake Picture Project (Yahoo! JAPAN)	64,993
June, 2011	Memories for the Future (Google)	62,662
June, 2011	Center for Remembering 3.11 (Sendai mediatheque)	1,232
September, 2011	Michinoku Shinrokuden (Tohoku university)	125,531
March, 2012	Great East Japan Earthquake Archives (NHK)	1,803
March, 2012	Archives of agricultural, fishery and forestry cooperative organizations' activities for the restoration from the Great East Japan Earthquake (Norinchukin Research Institute Co., Ltd.)	4,251
April, 2012	Memories of Hamadoori: Disaster Archive for the future (Iwaki Meisei University)	299
September,2012	Remembering 3/11(FNN)	69
March, 2013	KAHOKU SHIMPO DISASTERS ARCHIVE (Kahoku Shimpo Publishing Co.)	127,098
March, 2013	The Great East Japan Earthquake Archives Fukushima (Keio University)	18,795
May, 2013	Remembering 3.11 (Tohoku Gakuin)	8,786
January, 2014	Red Cross Nuclear Disaster Resource Center Digital Archives (Japanese Red Cross Society)	1,054
March, 2014	Tagajo kenbunoku (Tagajo city in Miyagi prefecture)	24,519
April, 2014	Aomori Disaster Archive (Hachinohe City, Misawa City, Oirase Town and Hashikami Town in Aomori Prefecture)	93,514
June, 2014	Fukushima Nuclear Accident Archive (Japan Atomic Energy Agency)	112,824
March, 2015	Kuji-Noda-Fudai Disaster Archive (Kuji City, Noda Village and Fudai Village in Iwate Prefecture)	127,415
April, 2015	Koriyama Disaster Archive (Koriyama City, Tomioka Town, Futaba Town and Kawauchi Village in Fukushima Prefecture)	49,901
June, 2015	Great East Japan Earthquake Archive Miyagi (Miyagi Prefectural Library)	222,291
July, 2015	Urayasu Disaster Archive (Urayasu City in Chiba Prefecture)	34,241
February, 2016	Radiation Medical Science Center for the Fukushima Health Management Survey digital archive (Fukushima Medical University Radiation Medical Science Center)	1,881
March, 2017	Iwate Earthquake Tsunami Archive "Hope" (Iwate prefecture)	236,498

3. Problems of disaster archives in Japan

The fifty archives of the Great East Japan Earthquake contain more than four million points of data about the Great East Japan Earthquake. We know of no comparable case throughout the world, where, at most, ten archives had been devoted to a single 'natural' disaster. Despite these encouraging figures, the totality of four million points are not organized using a unified scheme of metadata and therefore cannot be easily and systematically retrieved by the archive end-users. Although this problem includes issues peculiar to Japan, others may commonly arise during the future construction of a disaster archive somewhere else in the world. In this section, we reveal the problems encountered by the disaster archive of the Great East Japan Earthquake and ways that may remedy to the problems of disaster archiving.

3.1. Disaster Archive Guidelines

Before the Great East Japan Earthquake, there were no guidelines on disaster archives in Japan. Therefore, each agency established its own standards and its own policy for the establishment of its disaster archive. In March 2013, the Japanese Ministry of Internal Affairs and Communications formulated the "Guidelines for Construction and Operation of Digital Archive related to the Great East Japan Earthquake Disaster."[18] Disaster archives are now being established according to these standards. Nevertheless, the Ministry of Public Management, Home Affairs, Posts and Telecommunications guidelines did not have provisions such as the definition of earthquake data, the classification criteria of documents, the lexical control of material keywords, and the disclosure criteria. Therefore, each archival platform remains different and problems to unify document retrieval. Much data remain inaccessible to the public. The authors think that it is necessary to rethink the guidelines of disaster archives and to improve the legislation.

3.2. Metadata schemes

Among digital archives of disasters in Japan, a few sites designed their own metadata schemes. Most archives use the scheme the Great East Japan Earthquake Archive Metadata Schema of the National Diet Library [19] or some additional items. The metadata schema of the National Diet Library independently expanded the international metadata standard Dublin Core to include corresponding items of the Japanese language and disasters. The metadata of the National Diet Library is made up of over 300 items in terms of coverage and convenience. However, many items such as GIS information cannot be stored. Likewise, the solving of multilingual correspondence is still at an impasse. While it can be a reference for the future archives of natural disasters, the metadata schema of the National Diet Library requires revisions to meet the needs of future databases worldwide.

3.3. Vocabulary Control in Disaster Archives

Items of metadata include titles, creation dates, publication dates, file formats, keywords, document classifications, location information, descriptions of document contents, and modification history. Among them, keywords allowing access to the materials need particular attention. These terms are generally selected among the commonly used vocabulary. However, this process of selection must depend on the knowledge and vocabulary of the user and the country of origin. In Japanese, the same word can be written in three different scripts (i.e. kanji, hiragana, or katakana). For example, the word of rubble pronounced 'gareki' can be written 瓦礫(kanji), がれき(hiragana) and ガレキ(katakana). Even when using English keywords, the same item can be described using several words as in the case of car, motor vehicle, auto car, automobile, and motorcar. We observe further variations between the vocabulary used in the United Kingdom and the USA. The simple solution is to build a thesaurus. However, we predict that the task would meet further challenge from technical and scientific vocabularies, as they also find many equivalents. For instance, the word 'tsunami' which, originally Japanese, can be described with the words tidal wave, seismic sea wave, and hydraulic bore. In other words, the plurality and complexity terminologies related to natural disaster make it the creation of a unified and universal thesaurus challenging albeit necessary.

3.4. The information about the locations of the material

IOP Conf. Series: Earth and Environmental Science **273** (2019) 012039 doi:10.1088/1755-1315/273/1/012039

For most disaster-related materials, the location information includes the name and address of the places of the object of the data. This information can be said to be useful for screening and analysing elements in many ways. However, common problems are that the location of items of past disasters might now have a completely different name and address, several places might exist for the same address, or else regional names of an area may be changed after being used for a long time. Some of these issues may be resolved by using geocodes for a given period. However, there periods cannot be defined based on some algorithm and need to be carefully considered. For instance, after the Tohoku earthquake, some areas could no longer be used or inhabited because of the area has been considered prone to tsunami and inhabitable. Some of the facilities located in these zones have been rebuilt in inland while conserving its original name. As a result, the same place may have different locations in the past, the present and possibly the future. In such instance, you cannot establish a single geocode for a building name.

With regards to location information, an additional complexity concerns the archiving of photos. Currently, pictures taken with a smartphone have their location automatically saved. The information corresponds to the location of the photographer, not the location of the scene appearing on the snapshot. Within the metadata schema of the disaster archives of the National Diet Library, there are options that allow you to access both the location of the camera and the location of the scene. However, it seems that the latter is rarely provided. The reason is that this information has to be identified and entered manually. Although it is too early to say, this issue could be resolved using automatic identification technology and artificial intelligence.

3.5. The cost of building disaster archive

The investment required by the archives of the Great East Japan Earthquake is considerable. According to our research, the cost for a single item represents approximately $13.30 or 1,500 Japanese yen. This amount includes the expenses for the collection, the receipt, the handling right or acquisition for public disclosure and secondary usage, mass categorisation and organisation, the creation of metadata, paper materials, the registration in the database, the construction cost of the system for publishing the contents and the operating and maintenance expenses. The most significant among all seems to be the classification of hundreds of materials and the attribution of metadata. In order to develop future disaster archives, we must envisage the reduction of expenses via the use of artificial intelligence and automation.

4. Conclusion

This paper examined the evolutions and some of the general issues facing the digital archives of the Great East Japan Earthquake. We also discussed the future of disaster archive and pointing out to some of the avenues or solutions to allow for better archiving of natural disaster worldwide. To develop a more general scheme for natural disaster archive, we believe that useful lessons can be drawn from the archival activities that surround the Tohoku earthquake. For instance, we mentioned that the necessity of building a thesaurus of natural disasters, create a metadata schema for natural disaster, and introduce AI for the classification and organization of information. We will continue to draw lessons from the archives of the Great East Japan Earthquake and hope to lead the way towards the internalization of disaster archives.

5. References

[1] The Reconstruction Design Council in response to the Great East Japan Earthquake (25 June 2011), Towards Reconstruction "Hope beyond the Disaster", http://www.cas.go.jp/jp/fukkou/english/pdf/report20110625.pdf (Accessed, 1 August 2018)

[2] National Diet Library, NDL Great East Japan Earthquake Archive "Hinagiku", http://kn.ndl.go.jp/node?language=en (Accessed, 1 August 2018)

[3] Tohoku Gakuin, Remembering 3.11(in Japanese only), http://archive311.tohoku-gakuin.jp/(Accessed, 1 August 2018)

[4] Tohoku university, Michinoku-Shinrokuden (in Japanese only), http://search.shinrokuden.irides.tohoku.ac.jp/(Accessed, 1 August 2018)

[5] Iwaki Meisei university, Memories of Hamadoori : Disaster Archive for the future, http://hamadoori-kioku.revive-iwaki.net/ (Accessed, 1 August 2018)
[6] Japan Society of Civil Engineers, Great East Japan Earthquake Archive(in Japanese only), http://www.jsce.or.jp/library/eq20110311/index.shtml (Accessed, 1 August 2018)
[7] National Research Institute for Earth Science, 311 Marugoto Archive(in Japanese only), http://311archives.jp/ (Accessed, 1 August 2018)
[8] NHK, Remembering 3.11 Great East Japan Earthquake Archive, http://www9.nhk.or.jp/311shogen/en/ (Accessed, 1 August 2018)
[9] Kahoku Shimpo Publishing Co., Kahoku Shimpo Disaster Archive, http://kahoku-archive.shinrokuden.irides.tohoku.ac.jp/ (Accessed, 1 August 2018)
[10] Google, Memories for the future, https://www.miraikioku.com/en/ (Accessed, 1 August 2018)
[11] Yahoo! JAPAN, East Japan Earthquake Picture Project, http://archive.shinsai.yahoo.co.jp/ (Accessed, 1 August 2018)
[12] Tagajo City in Miyagi Prefecture, Tagajo Kenbunoku(in Japanese only), http://tagajo.irides.tohoku.ac.jp/ (Accessed, 1 August 2018)
[13] Miyagi Prefectural Library, Great East Japan Earthquake Archive Miyagi, https://kioku.library.pref.miyagi.jp/index.php/en/ (Accessed, 1 August 2018)
[14] Koriyama City in Fukushima Prefecture, Koriyama Disaster Archibe(in Japanese only), http://shinsai.koriyama-archive.jp/ (Accessed, 1 August 2018)
[15] Kuji city Noda village, and Fudai village in Iwate Prefecture, Kuji-Noda-Fudai Disaster Archive (in Japanese), http://knf-archive.city.kuji.iwate.jp/ (Accessed, 1 August 2018)
[16] Iwate Prefecture, Iwate Earthquake Tsunami Archive "Hope", http://iwate-archive.pref.iwate.jp/ (Accessed, 1 August 2018)
[17] Harvard University Reishauer Institute of Japanese Studies, Japan Disasters Digital Archive, http://www.jdarchive.org/en/home (Accessed, 1 August 2018)
[18] Ministry of Internal Affairs and Communications, Guidelines for Construction and Operation of Digital Archive related to the Great East Japan Earthquake Disaster (in Japanese only), http://www.soumu.go.jp/menu_seisaku/ictseisaku/ictriyou/02ryutsu02_03000114.html (Accessed, 1 August 2018)
[19] National Diet Library, The National Diet Library Great East Japan Earthquake Archive Metadata Schema, http://kn.ndl.go.jp/static/metadata?language=en (Accessed, 1 August 2018)

IOP Conf. Series: Earth and Environmental Science **273** (2019) 012040 doi:10.1088/1755-1315/273/1/012040

Communicating Risk in Enhancing Disaster Preparedness: A Pragmatic Example of Disaster Risk Communication Approach from the Case of *Smong* Story

Alfi Rahman[1,2] and Khairul Munadi[2]

[1] Faculty of Social and Political Sciences, Syiah Kuala University, Darussalam, Banda Aceh, 23211, Indonesia, alfi.rahman@unsyiah.ac.id
[2]Tsunami and Disaster Mitigation Research Center (TDMRC), Syiah Kuala University, Banda Aceh, 23233, Indonesia

Abstract. Communicating risk has been widely accepted as a key strategy in mitigating the the impact of disasters. The concepts and practices of risk communication have been evolving since the 1980s due to the complexity and diversity of risk, developments in the communication sciences, and the complex nature of stakeholders and media. Failure to communicate risk may lead to an increase in the number of fatalities in future events. In this study, disaster risk communication (DRC) is proposed based on Lasswell's communication model and seen as an interactive process of exchanging information and opinions about risk among individuals, groups, and institutions to help them reduce uncertainty and undertake appropriate decisions and actions in the event of disaster. This study also describes the case study of the *Smong* story, which was successfully used to convey a simple tsunami risk message and help people make appropriate decisions, undertake appropriate actions, and save lives during the 2004 Indian Ocean tsunami. It involves multiple factors, such as messages about the nature of risk and other messages, not strictly about risk, that express concerns, opinions, or reactions to risk or to legal or institutional arrangements for risk management in strengthening community preparedness before, during, and after disasters.

Keywords: Disaster Risk Communication; Smong Story; Community Disaster Preparedness; Tsunami Preparedness

1. Introduction

The frequency at which disasters have occurred in recent years has had significant impact, with numerous lives lost as well as serious damage to infrastructure, the economy, livelihoods, etc. Disasters have affected more than 200 million people [1] worldwide, and this number will most likely increase in the future. Before the 2004 Indian Ocean tsunami, disasters were considered unavoidable acts of God or punishment for people's sins. People accepted disaster more passively. Since the tsunami, Disaster Risk Reduction (DRR) efforts have been constructing and promoting a comprehensive approach to disaster management; however, disasters are still perceived as unavoidable events that humans are powerless against and must accept as their destinies.

Before the 2004 Indian Ocean tsunami, disaster studies tended to focus on the crisis and emergency stages. Since then, it has been realized that disaster studies should also consider comprehensive factors such as population growth, urbanization, environmental degradation, climate change, etc.[2].

IOP Conf. Series: Earth and Environmental Science **273** (2019) 012040 doi:10.1088/1755-1315/273/1/012040

The high number of fatalities in the 2004 Indian Ocean tsunami was attributed in part to a lack of preparedness in communicating risk [1]. For example, the tsunami revealed weaknesses in communicating risk and warning systems that could have saved people's lives, while the 2011 Japan tsunami revealed flaws in perceptions and preparations [3] when most people thought they were already situated in safe places.

The worst effects of disasters have been discussed widely in literature, and efforts have been made to prepare communities by introducing structural and non-structural mitigation efforts. In recent years, research into risk communication as related to natural hazards and disasters has also attracted increased attention. Researchers have suggested communicating risk through community preparedness as a more effective means of dealing with the effects of disaster [4,5]. However, most DRC strategies have been constructed based on a limited assumption that views the public as passive and commonly perceives the message as being one-way, without any chance for response.

The Sendai Framework for Disaster Risk Reduction (SFDRR) 2015–2030 highlighted the importance of communicating risk in encouraging disaster preparedness for effective response [5]. More specifically, there are correlations between risk communication and efforts to increase community disaster preparedness. For example, Comfort et al. (2010) define community disaster resilience as a result of communities' capacity to effectively communicate and manage risk as well as respond to and recover from the disaster [6]. As such, an important component that should be considered in achieving disaster community resilience is how disaster risk communication efforts can contribute to community risk preparedness.

Recently, disaster risk communication has been widely accepted as a key strategy for reducing the worst effects of disasters. The concepts and practices of risk communication have been evolving since the 1980s due to the complexity and diversity of different risks, developments in communication science, the complex nature of stakeholders, increased media technology, and abundance of information that can easily be accessed by individuals [7]. Failure to communicate risk may increase vulnerability.

The 2004 Indian Ocean tsunami was a turning point for Indonesia in reducing risk. Indonesia's disaster paradigm changed into a comprehensive and preventive approach [6]. Three years following the tsunami, the Indonesian government established a regulatory and institutional framework for DRR through the Law 24/2007; this was followed by the establishment of the National and Local Disaster Management Agency [7,8]. The Indonesian disaster paradigm started moving from emergency response to a more comprehensive DRR approach. Since then, the Disaster Management Agency has taken a leading role in reducing risk.

Information related to disaster risk that is commonly received by communities is generated by the Local Disaster Management Agency (*Badan Penanggulangan Bencana Daerah*, BPBD), Non-Governmental Organizations (NGOs), and academia. In fact, there are many types of risk information that should reach various audiences. Efforts to reduce the worst impacts of disasters have been introduced in many ways by researchers and practitioners in the field of DRR. The efforts produced vary based on their creators' backgrounds. For example, disaster education researchers and practitioners suggest that disaster education should be implemented through formal and non-formal education at every level, and that structural mitigation is a main factor in helping communities be more prepared to deal with future disasters.

Communication efforts have been conducted, studied, and improved over many years to help planners include strategic factors that contribute to community resilience based on all available resources [4].

In this study, disaster risk communication (DRC) is defined as an interactive process of exchanging information and opinions about risk among individuals, groups, and institutions [7] to help them reduce uncertainty and make appropriate decisions and take appropriate actions in the event of a disaster [9]. It involves multiple factors, such as messages about the nature of risk and other messages, not strictly about risk, that express concerns, opinions, or reactions to risk messages or legal/institutional arrangements for risk management. These strategies also strengthen community preparedness.

IOP Conf. Series: Earth and Environmental Science **273** (2019) 012040 doi:10.1088/1755-1315/273/1/012040

2. Purpose

The purpose of this study is to introduce a pragmatic conceptual framework of DRC for community disaster preparedness. This study also describes a case study that was successfully used to convey a simple tsunami risk message and helped people make appropriate decisions and take appropriate actions to save lives during the 2004 Indian Ocean tsunami.

3. Methods

This research is a qualitative study which begins with a review of disaster risk communication discourses. It uses data from literature reviews, interviews, and a case study. The framework is constructed on the understanding and analysis of various recent discourses related to DRC in enhancing community disaster preparedness. This paper also presents as its case study the *Smong* story from Simeulue Island. Interviews were conducted in September 2017 and January 2018 and based on a parallel research study by the authors on disaster risk reduction in Simeulue Island.

4. The Importance of Disaster Risk Communication

Disaster studies now hold that human perception, knowledge, and values can contribute to disaster preparedness and are fundamental in increasing community disaster resilience. Individuals, groups, or communities make choices whether or not to spend money on mitigation measures despite knowing the potential implications of not making enough effort. Ignorance of the effect of disaster is also a major problem that can occur. This situation can occur due to the lack of quality information on disasters that can help them make appropriate decisions and undertake appropriate actions.

From this point of view, Disaster Risk Communication (DRC) takes on an important role. The main purpose of DRC is to increase the quality of future decisions and actions in the event of a disaster. When discussing DRC, it should be accepted as an interactive and dynamic process in which information and opinions are exchanged between individuals, groups, and institutions. This situation involves multiple factors, not only risk, but also include culture, mental attitudes, opinions, and reactions to risk messages [10,11]. The current approach to encouraging preparedness is ineffective, and a new method of communicating risk is needed.

DRC is believed to help people see the big picture of risk through community preparedness [12]. To optimize the effectiveness of disaster preparedness, a plan should be put in place to manage the flow of risk messages. DRC programs should be integrated as much as possible into larger, comprehensive efforts that address and influence multiple elements.

Disaster risk management is not entirely new to DRR efforts. Academics and practitioners have long recognized the benefits of such DRR efforts as disaster education, as well as the use of local resources for DRR; this may include, for example, using local and indigenous knowledge for DRR and disaster risk assessment (see Figure 1).

Unfortunately, most DRR efforts seem separate from each other. It is difficult to see links between these efforts. In fact, studies have shown that most communities lack the capacity to prepare for future disasters. In other words, DRR efforts have been improved but, in the process, communities have lost still more of their capacity to mitigate disaster [13]. Communities have their own styles of risk management, based on the resources that they can potentially have [14,15].

The main function of DRC is to extract information or messages from major DRR efforts and deliver them to target audiences (see Figure 1). DRC is a means of extracting risk messages from many resources, framing them into DRC plans, and spreading them through communities.

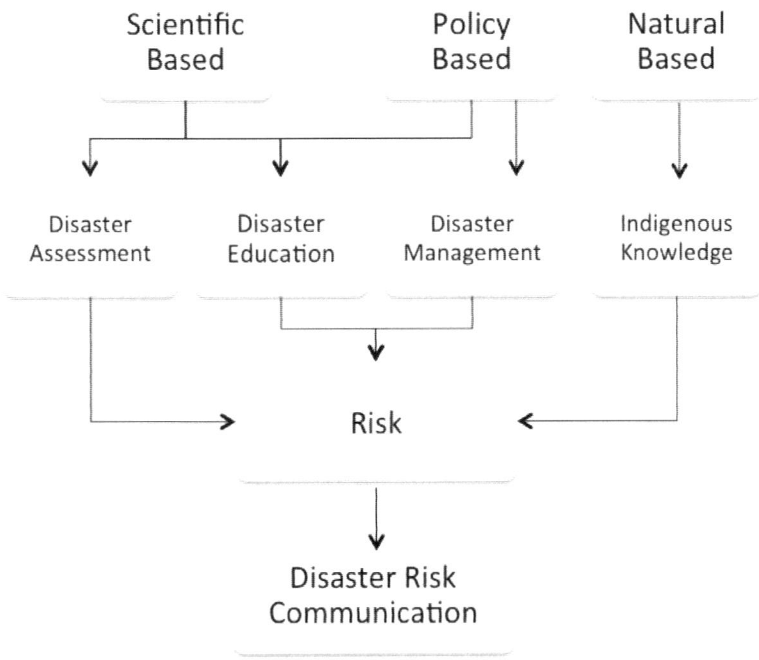

Figure 1. Major DRR efforts and their relationship with disaster risk communication

In achieving public disaster preparedness, Coppola (2009) emphasizes the importance of three stages of DRC: (1) early planning, (2) developing campaign strategies, and (3) implementing and evaluating campaigns [4]. This suggests that DRC plans rely on comprehensive planning, which is the foundation for formulating the flow of risk messages.

The DRC framework of this study is used to describe the process of developing a relationship between audiences and communicators and delivering specific messages created by the communicator. Communication science has great potential to advance disaster preparedness efforts.

5. Results

Lasswell's communication model theory is commonly known as "who says what in which channel with what effect?", and is considered by researchers in the field of communication [16] as an adaptive model that explains the flow of communication. Lasswell's communication model also includes consideration of a variety of factors for determining the impact of communication.

So, the DRC framework in this study is constructed based on Lasswell's communication model for several reasons: (1) interactive; this model gives communicators the opportunity to create risk messages based on risk planning, (2) multiple messages; it allows a communicator to create multiple risk messages, since the available information is interpreted in more than one way according to appropriate applications and channels, (3) multiple channels and specific target audiences; multiple media can reach multiple target audiences with very specific messages, (4) concern for the effects of the process, which could give the opportunity for the DRC plan to set goals.

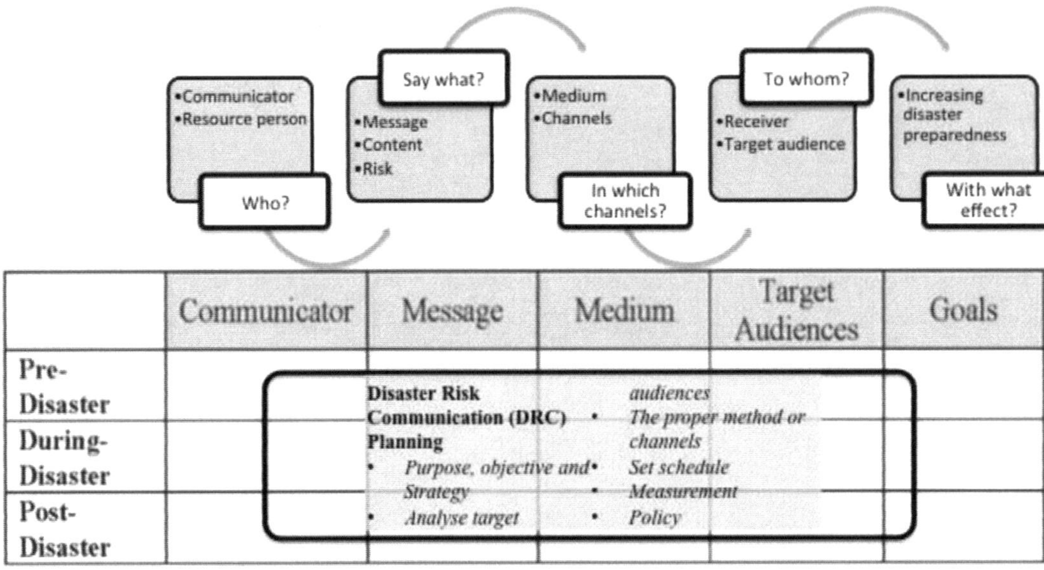

Figure 2. The disaster risk communication framework

The DRC framework of this study needs a certain amount of planning, based on the flow of a message from the sender to the target audience. DRC framework proposes and involves planning and strategy development, and is divided into five elements (see Figure 2), as follows:

- The first is to identify the role of the sender with the authority to create risk messages. The sender or communicator should also be concerned with the planning strategy development. The communicator has the authority to identify target audiences and set objectives based on specific target audiences. A series of assessments should also be done objectively, such as by a disaster management agency that is mandated to formulate risk messages.
- The second concerns the risk message or risk content developed from various resources.
- The third concerns the channels or media that transfer the risk messages to the target audiences
- The fourth concerns to the target audiences who receive the risk messages to facilitate them making the right decisions and taking the right actions before, during, and after disaster
- The fifth and the final element focuses on measuring the impact of the DRC plan on resilience

The model is also concerned with the various channels chosen to deliver the risk message to different target audiences. The effects of the DRC are broken into three main categories: (1) cognitive (person's belief or opinions about risk), (2) effective (person's feelings about risk), (3) behavioral (person's behavior related to risk) [4].

To plan DRC, this framework provides several steps that should be considered: (1) determine the purpose and objectives of DRC; (2) analyze the target audience; (3) develop the DRC messages; (4) determine the proper media or communication channels; (5) set a schedule; and (6) put all of these pieces together into a comprehensive plan.

A DRC framework helps the public focus its efforts and ensure all are involved in assessing, communicating, managing, and integrating risk throughout the entire range of disaster risk reduction efforts. This framework also concerns the impact of globalization, which could affect public perception and knowledge by enabling access to varied and unlimited information. People should also have the skill to select related information [17].

6. The Case of the *Smong* Story in Transmitting Tsunami Risk Messages and Saving Lives during the 2004 Tsunami in the Simeulue Island, Aceh, Indonesia

Simeulue Island is located in southwestern Aceh and has been hit by tsunamis in 1907, 2004 and 2005 [3,19]. Because of this situation, the Simeuluean people have created their own knowledge related to tsunamis, called *Smong*, as a way for dealing with these disasters. An example of the traditional implementation of disaster risk communication from the case of the *Smong* story could be seen in Table 1.

Table 1. A pragmatic example of disaster risk communication from the case of the *Smong* Story in saving lives during the 2004 Indian Ocean tsunami

Disaster Phase	Who? Communicator	Say what? Message	In which channels? Medium	To whom? Receiver or Target Audiences	With what effect?
Pre-disaster	• Person who experienced the 1907 tsunami • Person who received the *Smong* story from the person who experienced the 1907 tsunami	• Story of the 1907 tsunami • A big earthquake followed by receding seawater, the seawater reached the land	*Nafi-nafi* (Storytelling)	• Children • Adults	Increasing the tsunami preparedness
During disaster	• Person who received the *Smong* story from the person who experienced the 1907 tsunami	• *Smong* story content in the *Nafi-nafi*	• Older persons took the role of delivering the tsunami warnings	• Messages delivered through the community action	• Saving lives • Three people were killed
Post-disaster	• Disaster Management Agency • BPBD, NGOs (international, national, and local level) • Related stakeholders. • Community groups • Local leaders	Success of the *Smong* story in saving lives has inspired others.	Encouraging them to reach other channels such as books, films, pamphlets Other traditional channels such as *Nandong*	All community members and related stakeholders, institutions, etc.	Increasing awareness that the *Smong* story should be passed to the next generation.

Source: Rahman et. al (2017, 2018) [13,18] and interview results (November 2017 and January 2018)

The *Smong* story successfully saved many Simeuluean people from the 2004 Indian Ocean tsunami. Only three of the island's 70,000 residents were reportedly killed in the tsunami [18,20]. *Smong* knowledge is contained in traditional oral stories, known as *Nafi-nafi*. The success of the

Smong story's risk message content is clear from the appropriate decisions and actions of the Simeuluean people during the 2004 Indian Ocean tsunami [13,18,21–23].

The *Smong* story community contains common knowledge on tsunami risks that existed before the 1907 tsunami but failed to save lives at the time. Since then, the *Smong* story has resonated through the Simeuluean community, and it worked well in 2004.

The *Smong* content in *Nafi-nafi* has been recognized by most Simeuluean people. It tells about: (1) tsunami phenomenon; if the big earthquake occurs, following by the receding of seawater, the seawater will reach the land, (2) the action should be taken when seeing this phenomenon: run away from coastal areas or flee to higher places/mountains, (3) transmission of the story across generations.

According to a former Simeulue Regency leader, the simple message of the *Smong* story has been accepted and maintained through the common memory of the Simeuluean people. He added that, because the 1907 tsunami killed more than 50% of Simeulue's population at the time, it made the *Smong* story resonate in the Simeuluean people's collective memory.

However, many scholars argue that we need to be aware of and guard against the potential erosion of indigenous community knowledge. For example, during the longtime conflict between the Free Aceh Movement (FAM) and Indonesian Armed Forces, the *Smong* story was transmitted through traditional channels and reinforced by the earthquakes that occurred frequently.

7. Conclusion

This study forwards a pragmatic conceptual framework of DRC for enhancing disaster preparedness. This framework's underlying assumption is that DRC can increase community disaster preparedness. It should be integrated through all DRR efforts. The critical function of DRC is planning, implementing, exercising, and evaluating efforts to help people make appropriate decisions and take appropriate actions during disasters, and should be combined with other strategies.

The *Smong* story proves that a clear and simple message can contribute to increased community tsunami preparedness. It shows that a simple message and the right traditional media can help the community make appropriate decisions and take appropriate actions when disaster occurs. The next stage of this study is to apply the framework in a practical situation to ensure the flow of risk messages from sender to target audience.

8. References

[1] United Nations 2015 *GAR: Global Assessment Report on Disaster Risk Reduction 2015*

[2] Hilhorst D 2013 Disaster, conflict and society in crisis *Everyday Politics of Crisis Response* (Routledge London)

[3] Bryant E 2014 *Tsunami: the underrated hazard* (Springer)

[4] Coppola D P and Maloney E K 2009 *Communicating emergency preparedness: Strategies for creating a disaster resilient public* (Auerbach Publications)

[5] Cheng D, Claessens M, Gascoigne T, Metcalfe J, Schiele B and Shi S 2008 Communicating science in social contexts *New Model. new Pract. (Berlin, 2008)*

[6] Djalante R, Garschagen M, Thomalla F and Shaw R 2017 Introduction: Disaster Risk Reduction in Indonesia: Progress, Challenges, and Issues (Springer, Cham) pp 1–17

[7] BNPB 2015 *Indonesia: national progress report on the implementation of the Hyogo Framework for Action (2013–2015)* (Jakarta)

[8] Indonesia U-U R 2007 Nomor 24 Tahun 2007 tentang Penanggulangan Bencana

[9] Eisenman D P, Cordasco K M, Asch S, Golden J F and Glik D 2007 Disaster planning and risk communication with vulnerable communities: lessons from Hurricane Katrina *Am. J. Public Health* **97** S109–15

[10] Kasperson R E, Renn O, Slovic P, Brown H S, Emel J, Goble R, Kasperson J X and Ratick S 1988 The social amplification of risk: A conceptual framework *Risk Anal.* **8** 177–87

[11] Morgan M G, Fischhoff B, Bostrom A and Atman C J 2002 *Risk communication: A mental models approach* (Cambridge University Press)

[12] Veil S R, Buehner T and Palenchar M J 2011 A work-in-process literature review:

Incorporating social media in risk and crisis communication *J. contingencies Cris. Manag.* **19** 110–22

[13] Rahman A, Sakurai A and Munadi K 2017 Indigenous knowledge management to enhance community resilience to tsunami risk: lessons learned from Smong traditions in Simeulue island, Indonesia *IOP Conf. Ser. Earth Environ. Sci.* **56** 012018

[14] Hiwasaki L, Luna E, Syamsidik and Shaw R 2014 Process for integrating local and indigenous knowledge with science for hydro-meteorological disaster risk reduction and climate change adaptation in coastal and small island communities *Int. J. disaster risk Reduct.* **10** 15–27

[15] Sillitoe P and Marzano M 2009 Future of indigenous knowledge research in development *Futures* **41** 13–23

[16] Lasswell H D 1948 The structure and function of communication in society *Commun. ideas* **37** 215–28

[17] Peters E, Hibbard J, Slovic P and Dieckmann N 2007 Numeracy skill and the communication, comprehension, and use of risk-benefit information *Health Aff.* **26** 741–8

[18] Rahman A, Sakurai A and Munadi K 2018 The analysis of the development of the Smong story on the 1907 and 2004 Indian Ocean tsunamis in strengthening the Simeulue island community's resilience *Int. J. Disaster Risk Reduct.* **29** 13–23

[19] Whitlow K F 2008 The 2004 and 1861 Tsunami Deposits on Simeulue Island, Western Sumatra

[20] KATADATA 2016 Jumlah Penduduk di Kabupaten Simeulue, Aceh 2001 - 2013

[21] McAdoo B G, Dengler L, Prasetya G and Titov V 2006 Smong: How an oral history saved thousands on Indonesia's Simeulue Island during the December 2004 and March 2005 tsunamis *Earthq. Spectra* **22** 661–9

[22] Syafwina 2014 Recognizing Indigenous Knowledge for Disaster Management: Smong, Early Warning System from Simeulue Island, Aceh *Procedia Environ. Sci.* **20** 573–82

[23] Sanny T A 2007 *The Smong Wave from Simeulue: Awakening Strategic Development of Regency of Simeulue* (Simeulue: Pemerintah Kabupaten Simeulue)

Disaster Risk Communication Issues and Challenges: Lessons Learned From the Disaster Management Agency of Banda Aceh City

Shylvana Adella[1,*], Muhammad Fauzan Nur[1], Nazriatun Nisa[1], Arief Tirtana Putra[1], Fira Rizki Ramadhana[1], Fitri Maiyani[1], Khairul Munadi[1,2], and Alfi Rahman[1,2,3]

[1]Department of Disaster Science, Postgraduate Program, Syiah Kuala University, Darussalam Banda Aceh, Indonesia
[2]Tsunami and Disaster Mitigation Research Center (TDMRC), Syiah Kuala University, Banda Aceh, Indonesia.
[3]Faculty of Social and Political Sciences, Syiah Kuala University, Banda Aceh, Indonesia

*email: ishyladella@gmail.com

Abstract. The *Badan Penanggulan Bencana Daerah* (BPBD or Disaster Management Agency) of Banda Aceh City was established through Banda Aceh bylaw (or *Qanun*) No. 3 of 2011 following the significant 2004 Indian Ocean tsunami. Meanwhile, as Banda Aceh's population has grown, an increasing number of people have moved to and decided to stay in areas once affected by the tsunami. Given this background, disaster risk communication is indispensable. It is important for the BPBD of Banda Aceh City to disseminate risk messages throughout the community. Disaster risk communication is informed by the communication targets set by the BPBD of Banda Aceh City. The purpose of this study is to explore the issues and challenges in implementing disaster risk communication in the Banda Aceh community and across various sectors. A qualitative research design is applied to gain an insight into the disaster risk communication of the BPBD of Banda Aceh City. Data were collected through interviews and document analysis. The results of this study show that the BPBD of Banda Aceh City has carried out disaster risk communication in accordance with its primary mandated tasks and functions. However, there have been several challenges and issues, namely limited funding and a lack of technology and disaster knowledge. Consequently, the BPBD of Banda Aceh City has been unable to efficiently realize its disaster risk communication goals.

Keywords*:* Disaster Risk Communication, Disaster Management Agency, BPBD of Banda Aceh City

1. Introduction

In December 2004, a catastrophic earthquake that measured 9.2 on the moment magnitude scale struck west of Sumatra Island, Indonesia [1]. The earthquake triggered a calamitous tsunami, and Aceh was the most damaged area. The cataclysmic disaster caused 127,720 fatalities, with a further 93,285 people missing and 635,384 becoming internally displaced persons (IDPs) [2]. Banda Aceh was the area most affected by the tsunami in Aceh; more than 60% of Banda Aceh was damaged by the 2004 tsunami, with 64,461 people dead or missing and 2,830 injured [3].

Today, the population of Banda Aceh is growing rapidly, with an annual growth rate of 1,65% since 2009; this rate is predicted to increase by 2020 [4]. In some tsunami-prone sub-districts of Banda Aceh, such as Jaya Baru and Meuraxa, the population growth has been 0.76% and 1.20%, respectively, since 2012 [5]. Although the area has high risk of disaster, people have chosen to live in these sub-districts due to economic reasons, namely lower housing prices.

After the 2004 Indian Ocean tsunami, most people decided to return to tsunami-affected area, indicating the ineffectiveness of relocation and spatial regulation in Banda Aceh. Denis Stanley and Chang Seng (2013) indicate that conflict and rapid decentralization could be factors hindering institutional changes for dealing with disaster risk. However, a major disaster can be an important trigger for change and reorganization [6]. Since the 2004 tsunami, many countries have reformulated their understanding of natural disasters so that people are more prepared to face similar situations in the future. Indonesia, for example, established disaster management agencies at every level of government (national, provincial, and district), with the purpose of implementing more comprehensive and effective disaster risk reduction (DRR) and disaster risk communication (DRC). In each district of Indonesia, there is a *Badan Penanggulan Bencana Daerah* (BPBD or Regional Disaster Management Agency), while at the national level the BNPB (*Badan Nasional Penaggulangan Bencana* or National Disaster Management Agency) directly reports to the President and coordinates the development of disaster risk reduction plans. At the provincial and district level, these same duties are conducted by the BPBD [7].

According to Regulation of the Ministry of Domestic Affairs (or PERMENDAGRI) No. 46 of 2008 [8] concerning the Guidelines and Working Procedures of BPBD in Indonesia, BPBD has three main functions: coordinating, commanding, and implementing disaster management at the district level. The regulation states that the BPBD is to be chaired ex-officio by the Regional Secretary and accountable to the Mayor of the city—in this case, the Mayor of Banda Aceh. Research conducted by Yuniarti (2016) [9] found that disaster management is not only concerned with physical infrastructure, but also requires synergic cooperation among stakeholders and related service providers to overcome problems using an appropriate communication approach. Therefore, the need for skilled communicators within the organization is necessary to strategically maintain and define its position, particularly when facing crisis-related criticism, threats, and uncertainties [10].

The BPBD of Banda Aceh was established in 2013 through Qanun (local bylaw) No. 3 of 2011 [11]. Communication is a crucial and important component that contributes to or helps avert disaster situations. Lindell and Perry, as cited in Rodrigeuz (2004), argues that one important function of risk communication is to explicitly or implicitly promote appropriate protective behavior by those at whom the information is directed. The primary goal of disaster communication is to inform people or communities about disaster risks and protect them from impending hazards [12]. The ultimate objective is to reduce the lives lost and number of injuries [12]. One of BPBD's tasks is to disseminate effectively disaster risks among the community to help them make appropriate decisions and take appropriate actions.

Risk communication is a dynamic and interactive process that involves the exchange of information between individuals, groups, and communities [13]. Risk communication can contain dramatic elements, problematic, solutions, and action/reaction that arise from the various human characters in coping with disaster [13].

DRC has expanded to the understanding of risk, particularly technical and scientific, with the increased public pressure for more information [10]. This form of communication has become increasingly ubiquitous and almost routine as a form of public messaging [10]. The reason why DRC is important is that natural hazards have caused thousands of fatalities and billions of dollars of damage to infrastructure and property, which could be prevented if communities were informed about risk and took preventive action before the hazards occurred. The disaster risk information provided by the authorities is critical, because it will help communities create perceptions of disaster risk and mitigate impact [14].

According to the research conducted by Eisenman (2007) [15] people who lack disaster experiences or are newly settled in disaster-prone areas are vulnerable groups owing to their lack of disaster knowledge. Wachinger et al. (2013) [16] believes that previous experience with natural hazards and trust in government/expert capacity have the most substantial impact on risk perception. If people trust the government, they are more likely to take warnings seriously and act accordingly. However, such responsiveness depends on their perception of their agency to engage in effective protective actions and on the strength of their belief that personal responsibilities can be delegated to public emergency management. These issues need to be taken into account when developing communication processes and participatory activities.

Many BPBDs have limited personnel, logistical capacities, equipment and funding [18]. Many have problems, such as a lack of physical offices, and many of their staff are inadequately trained in key aspects of disaster management and communication. Commonly, BPBDs remain focused on disaster response rather than mitigation. Effective disaster management requires cooperation between governments and the various institutions that comprise them. Regional authorities often possess local knowledge about previous disasters and logistics that are indispensable in disaster prevention, mitigation, and response. However, effective disaster management also requires a strong central authority with the power to take control where necessary. Without local and central government cooperation and coordination, this jurisdictional overlap and gap could lead to both central and regional governments seeking to impose different policies and directives. This can cause, at worst, confusion leading to possible institutional paralysis or, at best, inefficient disaster management and avoidable deaths and property loss [17].

The BPBD of Banda Aceh City faces similar issues and challenges. For example, in terms of funding, the BPBD of Banda Aceh City only gets 0.6% of the APBK (district budget) [18]. As population growth has increased over the years, many of the newcomers coming to Banda Aceh have had less information about the hazards surrounding them. Aside from that, the BPBD of Banda Aceh City has also faced rejection from the community; for example, landowners may break disaster signs, fearing that they will decrease the price of land.

It is important to highlight the issues and challenges faced by the BPBD of Banda Aceh City in communicating disaster risk. Involved stakeholders are believed to be potential resources that can be used to ensure the disaster management process and its stages can reach the target audiences. Therefore, coordination and cooperation need to be developed into a proper process. Leaders who can mobilize and coordinate when dealing with potential disaster are likewise required. As such, the regional head can be the key to successful disaster management.

2. Purposes

The purpose of this study is to highlight the disaster risk communication issues and challenges faced by BPBD of Banda Aceh City after the 2004 Indian Ocean tsunami in providing and disseminating risk messages to its citizens.

3. Methods

This study is qualitative research. Data were collected through semi-structured interviews with the division heads and staff of the BPBD of Banda Aceh, then transcribed and analyzed using content analysis to identify the main issues and challenges faced by the BPBD of Banda Aceh City.

4. Results

The interviews were held on September 10, 2018, at the offices of the BPBD of Banda Aceh City. The interview results highlighted some relevant information, such as the agency's primary tasks and functions, and also focuses on its implementation of disaster risk communication issues and challenges.

4.1 The Primary Tasks and Functions of the BPBD of Banda Aceh City

The staff of the BPBD of Banda Aceh City routinely report the implementation of disaster management activities to the mayor of Banda Aceh, with reports made to the Regional Secretary—as the ex officio head of the BPBD of Banda Aceh—after the activities are finished. "Reporting to the Regional Secretary as the Head of BPBD of Banda Aceh is carried out after the activity takes place, both verbally and in the [written] report", said the Head of the Rehabilitation and Reconstruction Division. Furthermore, "the Chief Executive of the BPBD of Banda Aceh is actively communicating the outcomes of the activity with the Regional Secretary for evaluation and future agendas". This is in accordance with Article 8 of *Qanun* No. 3 [11] which stipulates that the BPBD must be accountable and report directly to the mayor of Banda Aceh at least once a month. Communication of activity reports between the BPBD and the leader/mayor is also appropriate, based on Yunarti's finding [9] that social cooperation must also be considered when handling disasters.

The BPBD of Banda Aceh City has 44 staff (38 civil servants and 6 non-civil servants). This is in accordance with the statement of a member of the Field Management of State Property Board, "There are many staff here who are civil servants. Also, we always ask the Regional Civil Service Agency for more staff to fill the shortcomings of our office".

As the leading agency in cases of disaster, the BPBD of Banda Aceh City should ensure that its staff receive appropriate and relevant training in disaster management. Several staff have attended such training, which is held by the BNPB. According to the Section Chief of Emergency Affairs, "Every year we send some staff from the BPBD of Banda Aceh City to attend the training held by BNPB". This is in accordance with Article 18, Paragraph 1, and Article 20, Paragraph 1, of the *Qanun* of the Province of Aceh [11] which states that the implementing elements of BPBD must have the expertise, knowledge, and integrity needed for disaster management, including cross-sectoral coordination functions.

In its cross-sectoral coordination function, the BPBD of Banda Aceh City is actively coordinating with other units and services. This was stated by the head of the Emergency and Logistics Division: "On September 5, 2018, we implemented a Table Talk Exercise (TTE) that was coordinated by the Chief Executive of BPBD of Banda Aceh. The TTE was attended by several agencies and services, such as the police, health services, social services, and Indonesian Inter-Citizen Radio or RAPI (*Radio Antar Penduduk Indonesia*)." The cross-sectoral coordination function is regulated through Articles 11, 12 and 13 of the Regulation of the Ministry of Domestic Affairs [8] concerning the implementation, coordination, and command of BPBD functions in Indonesia.

4.2. Issues and Challenge from in Disaster Risk Communication by the BPBD of Banda Aceh City

Based on its primary task and function, the BPBD of Banda Aceh has three functions: implementation, coordination, and command.

4.2.1. Implementation Function

Based on interviews, the implementation function of the BPBD of Banda Aceh City involved the dissemination of disaster materials to schools, villages, and other agencies. Many benefits were provided by school-based DRR programs and for evaluating challenges in disaster risk management. As such, the Indonesian government must integrate DRR into the education sector, such as by integrating DRR in the school curriculum and providing teachers with DRR training and preparedness in schools [19].

"We have been continuing to socialize disaster preparedness and mitigation programs in many schools based on our initiatives and in collaboration with other sectors or services," said the head of the Emergency and Logistics Office. Furthermore, the BPBD of Banda Aceh City collaborated with BRI (*Bank Rakyat Indonesia*) to socialize disaster education and conduct earthquake evacuation simulations for their staff. Nevertheless, this is not a routine activity, but only held occasionally, if there is an invitation or if funding is available. The purpose of this activity is to build people's knowledge of disaster risk, which is in accordance with Deverell's (2012) [20] research concept of knowledge management within an organization, which is expected to change people's behavior so they can live with the threat of disaster.

4.2.1.1. Communication Target
The BPBD of Banda Aceh City has had various targets for disaster risk communication, reaching from elementary school students, the general public, and the disabled community. The head of the Prevention and Preparedness Division said, "disaster risk communication programs are targeted at various communities, for example, the establishment of disaster community resilience (*Desa Tangguh Bencana*, DESTANA) in several villages located in tsunami-prone areas". However, most DRC activities have been funded by non-government organizations or through cross-sectoral collaboration. As such, the BPBD of Banda Aceh City functions as an executor, not as an implementer. Furthermore, the disaster risk communication conducted by the BPBD of Banda Aceh City has not effectively reached the newcomers who are living with existing communities, a situation that could contribute to increased vulnerability [15]. This condition could also situate them in difficult positions when disaster occurs.

4.2.1.2. Strategy
The BPBD of Banda Aceh City is well aware of the potential hazards that are spread throughout Banda Aceh. Therefore, it has applied different strategies for disaster risk communication. As the Head of the Emergency and Logistics Division said, "before conducting disaster risk communication programs or other activities, the BPBD of Banda Aceh City always forms a team that will formulate the right strategies for reaching the targeted audiences". Disaster risk communication strategies have been formulated in several programs, including a series of workshops, disaster drills, schools, local communities, etc. The BPBD of Banda Aceh City has also conducted disaster risk communication programs in conjunction with the mass media, but extending only to coverage of BPBD activities. According to Asteria (2016) [13], the media's role in communicating disaster risk is important, as it is a benchmark of institutions such as BPBD's success in tackling disasters in the region. The mass media is used to provide information related to risks and hazards, and the ways to deal with them. The BPBD of Banda Aceh City has also promoted its activities so that the public can know of its programs. Social media also can be used in emergency situations, as proven by Alexander (2014) [21]. It is important for the BPBD of Banda Aceh City to use social media that are popular with the residents of Banda Aceh, such as Instagram and Facebook, in disseminating disaster risk messages and information.

4.2.2. Function of Coordination
In accordance with *Qanun* of Aceh Province No. 3 of 2011 [11], the BPBD of Banda Aceh City also has a coordinating function. Interviews found that the BPBD of Banda Aceh City has carried out this function through contingency plans, which were formed to cope with floods that involve various related agencies. The BPBD of Banda Aceh City has yet to specifically create a contingency plan for earthquakes or tsunamis, as it continues to use the contingency plan of the Aceh Disaster Management Agency or *Badan Penanggulangan Bencana Aceh* (BPBA) and BNPB. The head of the Rehabilitation and Reconstruction Division said, "The coordinating function of the BPBD of Banda Aceh City is carried out by coordinating contingency plans with other agencies. What has been completed is the flood contingency plan, which was prepared in 2016 for Baiturrahman and Lueng Bata sub-districts. Other agencies have adjusted their respective expertise". As such, according to some staff, during a

disaster or emergency response period, the BPBD of Banda Aceh City communicates the emergency to the Regional Secretary, who then returns the mandate directly to the BPBD of Banda Aceh City. In this case, the BPBD of Banda Aceh City has collected cross-sectoral trust to ensure it remains trustworthy institution in disaster management. This is in accordance with the research of [16] that disaster agencies must get public trust to ensure that disaster risk communication can influence the public mindset.

4.2.2.1. Communication Targets
The communication target of the BPBD of Banda Aceh City's coordinating function is disaster-related agencies, especially in Banda Aceh. These include the Social Service Agency, Health Service Agency, Search and Rescue (SAR), Regional Development Planning Agency (BAPPEDA), and other services. The communication process is conducted at the BAPPEDA office every week and involves representatives from each relevant agency.

4.2.2.2. Strategy
The BPBD of Banda Aceh City applies a coordination strategy by creating contingency plans that involve related agencies and organizations. As already conveyed by the head of the Emergency and Logistics Division, "cooperation with other agencies is still carried out using contingency plans and Medium-Term Development Plans (*Rencana Pembangunan Jangka Menengah,* RPJM) so that each agency conducts its activities as related to their respective fields".

4.2.3. Function of Command
In accordance with the Regulation of Ministry of Home Affairs [8] and *Qanun* No. 3 of 2011 of Banda Aceh [11], the BPBD of Banda Aceh City has been appointed as an agency for disaster management in Banda Aceh. The command function can be completed because the BPBD of Banda Aceh City is headed by the Regional Secretary, who can lead all related agencies if a disaster occurs in Banda Aceh.

4.2.3.1. Communication Target
In performing disaster risk communication, communication targets are all of the existing agencies and organizations coordinated by the BPBD of Banda Aceh City.

4.2.4. Challenges
In carrying out disaster risk communication activities in Banda Aceh, the BPBD of Banda Aceh City has several problems and challenges, including funding, technology, and knowledge capacity.

4.2.4.1. Lack of Funds
The lack of funding is the BPBD of Banda Aceh City's biggest issue in implementing its programs. It has also influenced such strategic programs as capacity building. This issue is also the reason why some programs are only carried out until the first stage, i.e. socialization stage, and not until more advanced stages such as simulations and drills. The new institutional arrangement for DM (Disaster Management) financing guarantees the availability of funds as part of the National Action Plan for DRR; however, there is still an over-emphasis on spending for post-disaster situations. Additionally, there is a need to improve the procedures and mechanisms for the equitable and timely allocation of funds and resources at the local level [6].

The head of Rehabilitation and Reconstruction Division said, "the Japanese government provided us with seismographs, but they cannot be installed because we lack the funds to hire experts". Research conducted by Oktari (2017) [18] states that the BPBD of Banda Aceh City only has limited funding through the Banda Aceh regional budget (*Anggaran Pengeluaran dan Belanja Kota*), and so it cannot perform maximally, including in its disaster risk communication programs.

4.2.4.2. Lack of Equipment

The main problem with equipment mentioned by the BPBD of Banda Aceh City in the context of completing disaster risk communication is the lack of appropriate devices. As the Head of the Emergency and Logistics Division said, "If a disaster occurs, we can't use a telephone or cellular phone to communicate with other agencies or regions; only Handy Talkie (HT) radios work during such cataclysmic events. However, our agency does not have HT Radio frequency. Because we lack the tools and frequency, it will be a big problem if a disaster occurs. We must borrow the radio frequency from the BPBA". According to Barbara's (2005) research [10], good disaster risk communication requires good technology in order to call public attention to the threat of disaster. Institutions such as BPBD should have technologically advanced communication equipment for disseminating disaster information.

4.2.4.3. The Limited Capacity of Staffs

In some cases, the local government of Banda Aceh has hired staff who are not aligned with the BPBD's needs. For example, the BPBD of Banda Aceh City needs an IT expert, but the staff assigned to the agency lack skills. As one staff member of the Field Management of State Property said, "Sometimes, the assigned staff do not meet our needs. They are transferred to the BPBD of Banda Aceh City from another agency and lack knowledge about disaster. The expert staff of the BPBD of Banda Aceh City have even been transferred to another agency." The head of the Emergency and Logistics Division added, "Since the BPBD of Banda Aceh parted with the firefighting department, we have had staff shortages when responding to emergencies, for example, expert staff that can set up tents quickly".

5. Conclusions

The BPBD of Banda Aceh City has conducted disaster risk communication through *Qanun* of Banda Aceh No. 3 of 2011 [11]. Communication is carried out in accordance with the agency's functions, namely coordinating, commanding, and implementing disaster management in the region. Communication is carried out through different forms and strategies, in accordance with the pertinent communication targets. However, there are several issues and challenges, such as limited funds and equipment, as well as a lack of staff capacity in implementing more comprehensive disaster risk communication.

References

[1] Ghobarah A, Saatcioglu M and Nistor I 2006 The impact of the 26 December 2004 earthquake and tsunami on structures and infrastructure *Eng. Struct.* **28** 312–26

[2] BRR 2009 *BRR Buku 0: Tsunami*

[3] Meisl C S, Safaie S, Elwood K J, Gupta R and Kowsari R 2006 Housing reconstruction in Northern Sumatra after the December 2004 Great Sumatra earthquake and tsunami *Earthq. Spectra* **22**

[4] Achmad A, Hasyim S, Dahlan B and Aulia D N 2015 Modeling of urban growth in tsunami-prone city using logistic regression: Analysis of Banda Aceh, Indonesia *Appl. Geogr.* **62** 237–46

[5] Bappeda Kota Banda Aceh 2017 *Statistik Banda Aceh 2017* (Banda Aceh: Bappeda Kota Banda Aceh)

[6] Chang Seng D S 2013 Tsunami resilience: Multi-level institutional arrangements, architectures and system of governance for disaster risk preparedness in Indonesia *Environ. Sci. Policy*

[7] Djalante R and Thomalla F 2012 Disaster risk reduction and climate change adaptation in Indonesia: Institutional challenges and opportunities for integration *Int. J. Disaster Resil. Built Environ.* **3** 166–80

[8] Menteri Dalam negeri P 2008 PEDOMAN ORGANISASI DAN TATA KERJA BADAN PENANGGULANGAN BENCANA DAERAH
[9] Yuniarti T 2016 Kepemimpinan dan pengelolaan modal sosial dalam penanggulangan bencana banjir *Makna* **3** 94–128
[10] Barbara R and W. Seeger M 2005 Crisis and Emergency Risk Communication as an Integrative Model *J. Health Commun.* **10** 43–55
[11] Walikota Banda Aceh 2011 Qanun No. 3 Tahun 2011 Kota Banda Aceh: Tentang Susunan Organisasi dan Tata Kerja Badan Penanggulangan Bencana Daerah Kota Banda Aceh
[12] Rodriguez H, Diaz W and Aguirre B 2004 *Communicating Risk and Warnings: An Integrated and Interdisciplinary Research Approach*
[13] Asteria D 2016 Optimalisasi Komunikasi Bencana di Media assa Sebagai Pendukung Manajemen Bencana *J. Komun. Ikat. Sarj. Komun. Indones.* **01** 1–11
[14] Rodríguez H, Quarantelli E L and Dynes R R 2007 *Handbook of Disaster Research*
[15] Eisenman D P, Cordasco K M, Asch S, Golden J F and Glik D 2007 Disaster planning and risk communication with vulnerable communities: lessons from Hurricane Katrina. *Am. J. Public Health* **97 Suppl 1** 109–15
[16] Wachinger G, Renn O, Begg C and Kuhlicke C 2013 The risk perception paradox-implications for governance and communication of natural hazards *Risk Anal.* **33** 1049–65
[17] Butt S 2014 Disaster management law in Indonesia: From response to preparedness? *Asia-Pacific Disaster Management: Comparative and Socio-Legal Perspectives*
[18] Oktari R S, Fahlevi H and Irawati W 2017 Disaster budgeting of Banda Aceh's local government: Trends and analysis of post-tsunami Aceh 2004 *IOP Conference Series: Earth and Environmental Science* vol 56
[19] Amri A, Bird D K, Ronan K, Haynes K and Towers B 2017 Disaster risk reduction education in Indonesia: Challenges and recommendations for scaling up *Nat. Hazards Earth Syst. Sci.* **17** 595–612
[20] Deverell E 2012 Investigating the Roots of Crisis Management Studies and Outlining Future Trajectories for the Field\r\n *J. Homel. Secur. Emerg. Manag.* **9** Article 24
[21] Alexander D E 2014 Social Media in Disaster Risk Reduction and Crisis Management *Sci. Eng. Ethics* **20** 717–33

IOP Conf. Series: Earth and Environmental Science **273** (2019) 012042 doi:10.1088/1755-1315/273/1/012042

Braille Monopoly Game as a Tool of Disaster Mitigation Education for Visual Disabilities

D Maharani[1], M Afifuddin[2,*], D A Putri[3], K Setiawan [1], and S A Ansari[4]

[1] Civil Engineering Department Student, Engineering Faculty, Syiah Kuala University, Banda Aceh, Indonesia
[2]Lecturer Civil Engineering Department, Engineering Faculty, Syiah Kuala University, Banda Aceh, Indonesia
[3]Mathematics Education Department, Teacher Training and Education Faculty student, Syiah Kuala University, Banda Aceh, Indonesia

[4]Medical Education Department, Medical Faculty student, Syiah Kuala University, Banda Aceh, Indonesia

[*] Corresponding author: m.afifuddin@unsyiah.ac.id

Abstract. Disaster mitigation is a series of efforts to reduce disaster risk, through physical development and capacity building to face disaster threats. This effort is important, especially in Aceh Province, which is a disaster-prone area. People with visual disabilities are vulnerable to disaster threat. Special facilities are needed to educate them about disaster mitigation. This program designs "Monopoli Bencana (Monca)" as a disaster mitigation learning media for people with visual disabilities. Monca aims to increase their awareness, knowledge and understanding related to disaster mitigation. Program implementation uses the one shoot pretest-posttest method. This program covers three important things, starting with socialization, direct discussion, and the practice of playing Monca. The target of this program is the people with visual disabilities in the UPTD Rumoh Seujahtera Beujroh Meukarya (RSBM). This program has three final results, namely: 1) Monca as an educational game equipped with braille: 2) Disaster mitigation pocket book printed in braille; 3) There is an increase in knowledge and awareness of the importance of disaster mitigation for at least 11 children who participated in the program.

1. Introduction

Aceh Province is one of the disaster prone areas. Stretching an area of 5,677,081 ha, Aceh Province is crossed by at least two active fault lines, from around 1900 km of the Sumatra Fault system [1]. As a result, Aceh is vulnerable to repeated earthquakes, landslides, and tsunamis [2]. Structuring buildings in Aceh that have not implemented the disaster preparedness principle also increases the likelihood of casualties if at any time a disaster occurs. Such conditions make it difficult for Aceh to be free from disaster threats.

Disaster mitigation education becomes an irreplaceable need [3]. Disaster mitigation is not only operational standards and procedures for handling disasters, but also an overall effort to minimize losses and casualties caused by disasters. Disaster mitigation education should be obtained by all groups, not only adults, but also children. In addition, people living in disaster-prone areas, as well as people with disabilities should be the main target of disaster mitigation education due to their vulnerability.

One community group that is difficult to get access to disaster mitigation education is people with visual disabilities [4]. They are a group of people with an inability to see partially or fully, so they cannot receive education normally. Need special approaches and facilities to educate them about disaster mitigation [5]. This program aims to introduce disaster mitigation to people with disabilities and increase public awareness about the importance of disaster mitigation education for all people. The introduction of mitigation is pursued through socialization and game practice. The game is chosen because it can present material in interesting and fun way.

2. Methodology

Based on its final results, this program uses research and technological development methods. However, the main purpose of this program emphasizes the educational aspect. So, one shoot pre-test and post-test is selected as the implementation of the media. Therefore, this program involves two important stages, namely the design of the media and the implementation of media through socialization.

2.1. Design of the Media

The design of the media refers to the identification of needs and solutions. Identification is explored during the survey directly to locations before of the implementation of media. The identification steps are shown in Figure 1. After the needs and solutions are obtained, the design of the playing media is carried out according to the steps shown in Figure 2.

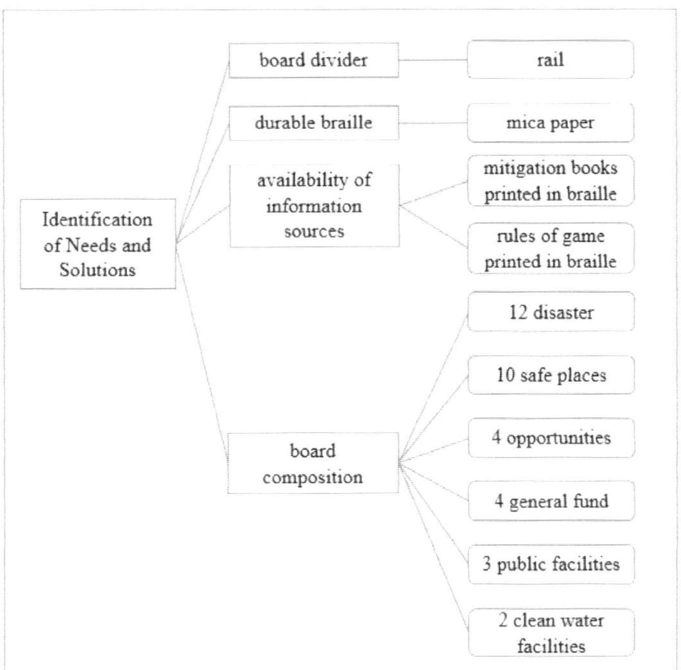

Figure 1. Identification of needs and solution.

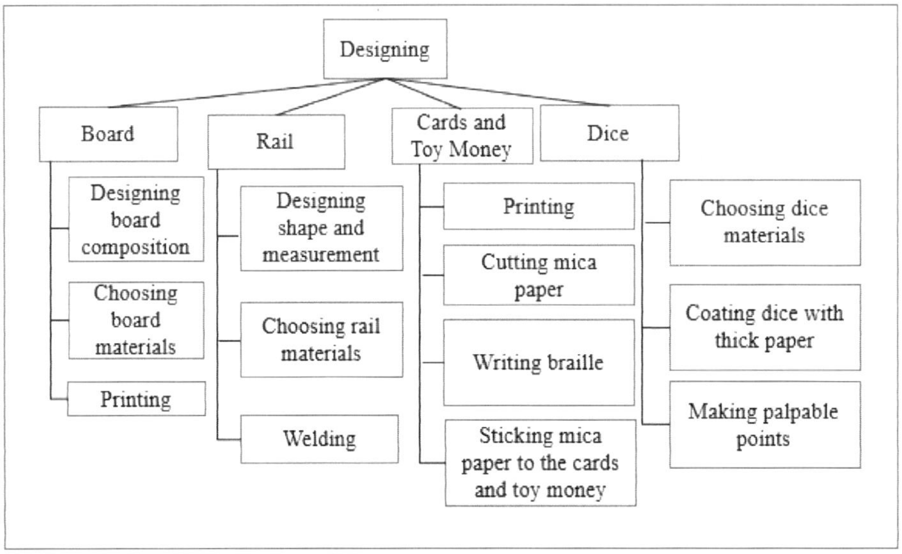

Figure 2. Designing entire media.

The media consists of 36 boxes containing 12 disasters along with anticipation and countermeasures, general fund boxes and opportunities that equipped with braille. Each box is given instructions about disasters which are found, orders to buy certain items, or evacuate to the certain location. It also introduces important items, such as clean water, logistics, first aid kits, masks, and fire extinguisher tanks. The philosophy of sharing is also introduced through a shared spending system and humanitarian donations. This media is designed to present practical, effective and enjoyable disaster mitigation learning for people with visual disabilities. The prototype of media is presented in Figure 3.

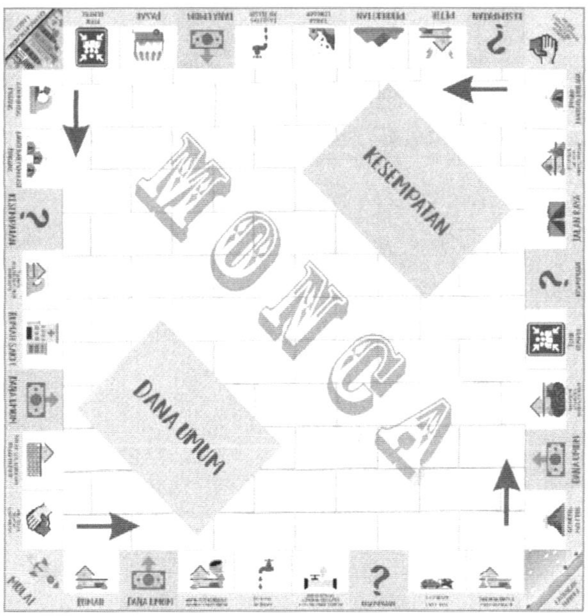

Figure 3. Monca's prototype.

The game can be played by four people in every round. Each player may have an assistant. The game lasts for approximately 30 minutes, or according to the agreement. After the time is up, the player is declared defeated if he is in a disaster box or hospital. At start time, each player is provided with Rp.790 worth of money with details 8 sheets of Rp.5, 8 sheets of Rp.10, 6 sheets of Rp.20, 6 sheets of Rp.25, 4 sheets of Rp.50, and 2 sheets of Rp. 100. Each player is also provided with 2 points of clean water, and 1 point of logistics. If the player stops in the disaster box, the player must find a safe point and put his pawn as soon as possible, or the player must enter the hospital and pass one turn. On the next turn, the player returns to the original disaster box and continues the game from that point.

2.2. Socialization

The socialization was held at the UPTD Rumoh Seujahtera Beujroh Meukarya (RSBM), Ladong, Mesjid Raya, Aceh Besar. The targets of this program are children (under the age of 18) with visual disabilities who live there. The socialization lasted for seven sessions, with 90 until 120 minutes per session.

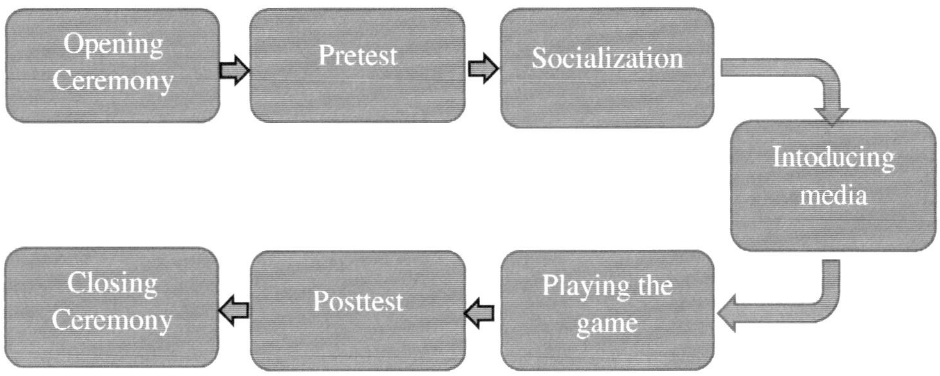

Figure 4. Rundown of implementation media trough msocialization.

As shown in Figure 4. The socialization activity began with an official opening by the head of the UPTD RSBM, followed by a pre-test to measure participants' initial knowledge about disaster mitigation. After that, socialization of disaster mitigation was carried out, as a basic understanding that is important for them. Strengthening participants' understanding is done through game sessions. So that participants will be aware of being a disaster resilient community. After socialization and playing the game, post test was conducted to measure participants' final knowledge of disaster mitigation. At the end of the program, all participants received an appreciation for their cooperation during the activity.

3. Result

3.1. Monca as Disaster Mitigation Learning Media

The details of the media that have been produced consist of a Monca ± 70 cm × 70 cm game board equipped with ± 73 cm × 73 cm playing rails. Media are equipped with six-edged dice, sliding pawns, toy money, cards, and game guides that are printed in braille. The information that printed on the board, card, and toy money is coated with mica paper that has been written in braille. The Monca game package can be used as a disaster mitigation education media for visual disability. The Monca circuit that has been assembled can be seen in Figure 5.

IOP Conf. Series: Earth and Environmental Science **273** (2019) 012042 doi:10.1088/1755-1315/273/1/012042

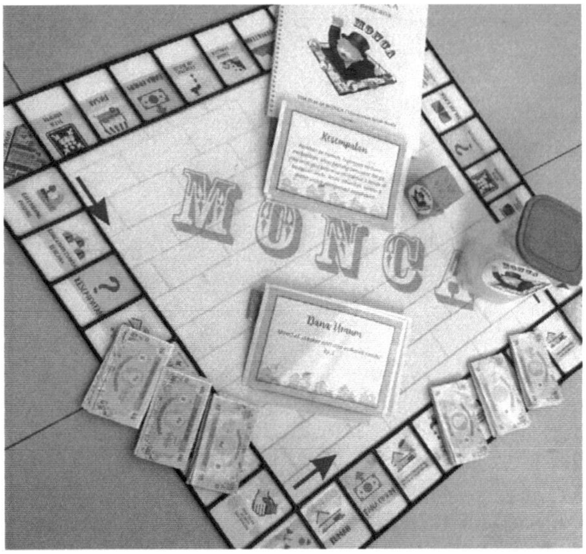

Figure 5. Entire media

3.2. Disaster Mitigation Books and Monca Play Rules in Braille

Disaster mitigation books printed in braille are summaries of the "Buku Saku Tanggap, Tangkas, Tangguh" National Disaster Management Agency (BNBP) 2012. The book contains a brief description of disasters, including causes, signs, and mitigation or risk reduction efforts. The disasters described included traffic accidents, floods, tsunamis, tidal waves, fires, earthquakes, landslides and volcanic eruptions. The availability of these reading materials can trigger participants' enthusiasm to improve their skills in mastering braille. It is also make them easier to obtain any information's about disaster mitigation. Books and rules of the game are presented in Figure 6.

Figure 6. Mitigation books and rules of game printed in braille

3.3. Knowledge Improvement by The Participants

Knowledge improvement of the participants was measured by conducting pre-test and post-test through interviews. The results of the pre-test showed that seven out of eleven children who participated in the program did not have sufficient knowledge about disaster mitigation. After that, socialization about disaster mitigation and utilization of Monca were carried out to facilitate them to remember the material given. Then, evaluation of mitigation knowledge is carried out through post-test. At the time of the post-test, ten of the eleven children could reiterate the things taught during socialization and games. There was an increase in knowledge of disaster mitigation for about 54% of the participants. It shows that the implementation of this program can increase participants' knowledge about disaster mitigation. The pre-test and post-test results conducted at the UPTD RSBM can be seen in Figure 7.

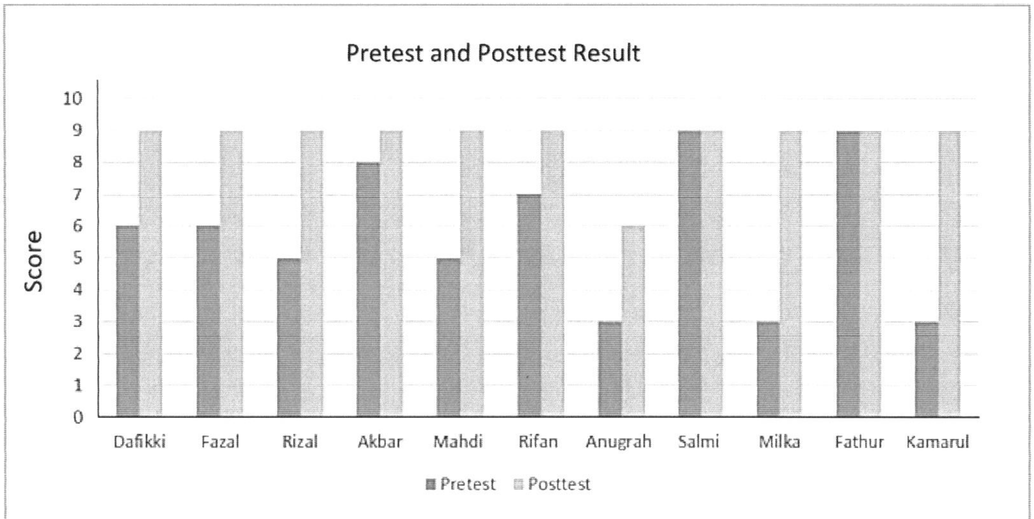

Figure 7. Result of pretest and posttest

4. Conclusions

After the entire series of activities carried out, the following conclusions are obtained:

- The Monca game tool is suitable to be used as a medium to socialize disaster mitigation.
- The disaster mitigation book that printed in braille can be used again by visual disabilities after the socialization session was held. The game and the books trigger the enthusiasm of participants to improve their skills in mastering braille.
- There is an increase in the knowledge of participants of the activity. The matters conveyed during the socialization session and the game can be delivered again by the participants at the post-test.

5. References

[1] Sugiyanto D, Ismail N, Zulfakriza Z and Meilano I 2015 Analisa Deformasi Permukaan Patahan Aktif Segmen Seulimum dan Segmen Aceh. *Proc. of the Seminar on the Results of the TDMRC-Unsyiah Disaster Research* (Banda Aceh: TDMRC-Unsyiah)

[2] Nasaruddin, Munadi K and Yuliansyah D 2011 Sistem Informasi Multi Ancaman Bencana Alam di Aceh *J. Seminar Nasional Informatika* **5** pp 124–32.

[3] Suhardjo D. 2010. *Arti Penting Pendidikan Mitigasi Bencana Dalam Mengurangi Resiko Bencana*. (Yogyakarta: Universitas Islam Indonesia)

[4] Ratih P. 2012. *Keterlibatan Penyandang Disabilitas Dalam Penanggulangan Bencana.* (Yogyakarta: Balai Besar Penelitian dan Pengembangan Pelayanan Kesejahteraan Sosial (B2P3KS) Yogyakarta Kementrian Sosial RI)
[5] Br. Damanik H, and Susanti SS, 2017 *Kesiapsiagaan Bencana Tsunami Pada Tuna Netra.* (Banda Aceh: Keperawatan Univ Syiah Kuala) pp 1–10.

Acknowledgements
Special thanks to Kemenristekdikti for funding this program through the Student Creativity Program 2017/2018. We also express our deep appreciation for the Academic Community of Syiah Kuala University as well as all administrators and residents of the UPTD Rumoh Seujahtera Beujroh Meukarya who have actively participated and supported this program.

Media Engagement in Communicating Disaster Risk After the 2004 Indian Ocean Tsunami: A Perspective from Aceh

Fakhruradhi[1*], Norma Susanti RM[1], Agung Pratama[1], Rina Fitri[1], Siti Raudhiah[1], Zahrina[1], Alfi Rahman[2,3] and Khairul Munadi[2]

[1]Department of Disaster Science, Postgraduate Program, Syiah Kuala University, Darussalam Banda Aceh, Indonesia
[2]Tsunami and Disaster Mitigation Research Center (TDMRC), Syiah Kuala University, Banda Aceh, 23233, Indonesia
[3]Faculty of Social and Political Sciences, Syiah Kuala University, Darussalam, Banda Aceh, 23211, Indonesia

*E-mail: fakhrurradhi11@gmail.com

Abstract. The conflict that lasted in Aceh for approximately 32 years, followed by the tsunami of 2004, had a significant impact by causing numerous casualties as well as material and nonmaterial losses. As a result, the media in Aceh recognized the need to be more active in reporting news about disasters and disaster mitigation. The role of the media as a liaison between the government and the community has been regulated by law, including as related to the publication of disaster mitigation efforts. The purpose of this study is to determine the role and involvement of the media in supporting disaster risk reduction and communicating disaster risks through the Acehnese community. This study used descriptive qualitative methods to obtain an actual picture of the involvement and role of the media in Aceh in communicating disaster risk. The results of this study indicate that the media in Aceh have not played a significant role in providing disaster risk information. Disaster programs trended and presented in the media mostly during the emergency stage, being less common after the disaster event.

Keywords: disaster risk communication; disaster; media; tsunami Aceh

1. Introduction

Aceh experienced a lengthy armed conflict in which the Free Aceh Movement (*Gerakan Aceh Merdeka*, GAM) demanded Aceh's independence from the Unitary State of the Republic of Indonesia (*Negara Kesatuan Republik Indonesia*, NKRI). To secure the situation, the Government of Indonesia established various operations, including the Military Operations Area (*Daerah Operasi Militer*, DOM; 1989–1998), the Military Emergency Status (2001–2002) and Civil Emergency Status (2003–2004). This conflict destroyed public facilities and infrastructure, as well as individual property, while residents became victims of violence and human rights violations. The conflict also contributed to destroying the customs and cultural institutions of the Acehnese people [1].

In this difficult situation, Aceh experienced twin natural disasters in December 2004: an earthquake and tsunami. Massive damage was also caused in Thailand, Sri Lanka, and other countries around the Indian Ocean. The destruction caused by this natural disaster compounded that caused by the conflict, resulting in 127,720 deaths in Aceh and 93,285 people being reported missing [2]. The community lacked knowledge of tsunamis, even though tsunamis had occurred in several other regions in Indonesia. As stated by [3] in a paper that collected data on tsunami events in eastern Indonesia and data from other research results, in a ten-year period, Indonesia experienced 6 tsunamis [4]. Aceh, meanwhile, had experienced a tsunami in 1907 on Simeulue Island [5].

The information presented from the 2004 Indian Ocean tsunami should be a source of knowledge for disaster education and DRR efforts. During the conflict, the media in Aceh reported more about the conflict and related topics than the tsunamis that occurred in other parts of Indonesia. On the other hand, media restriction due to the conflict situation also contributed to the media's lack of coverage of disasters. As such, it was not surprising that the information presented by the media seemed very monotonous.

Promoting physical reconstruction and improving the role of the media after the 2004 Indian Ocean tsunami required a lot of support from various parties, including international assistance. Security guarantees were needed for international representatives. The Government of Indonesia (GoI) and GAM agreed to sign a Memorandum of Understanding (MoU) on August 15, 2005. Under this momentum, Aceh's security condition slowly began to recover, and people outside Aceh became involved in reconstruction and peace building.

Since that time, there has been an increase in the number of media and information related to disaster and peace. Donors have also funded the development of media to spread information through the Acehnese community. During the rehabilitation and reconstruction phase, most media presented a significant amount of disaster information, including the recovery process and disaster education. This situation was a window and opportunity for the Acehnese people to refine disaster knowledge.

Law no. 24 of 2007 concerning Disaster Management [6] specifies that communities have the right to information related to disasters as part of disaster risk communication. For this reason, the role of the media is very important, as it plays an intermediary/liaison function between the source of information and the recipients of information. This reflects the argument by Morgan et al. (2007) [7] that "risk communication is the exchange of information and views on risk and all related parties, between the experts/risk assessors, managers, and recipients or other related parties, aiming to provide correct, relevant and accurate information clearly and easily understood by groups that are targeted in the risk communication". The role of the media was also mentioned in Law No. 40 of 1999 concerning the Press [8], Article 2 of which asserts that the media has a role in conveying information to and educating communities; this also includes disaster risk communication. In a disaster, the media can be used to distribute disaster information, even as an Early Warning System (EWS). Ideally, this role can be carried out by the media either independently or in collaboration with other parties. Therefore, in the delivery of disaster information, it is important to ensure the media gives proportional attention to other topics as it carries out its role in fulfilling the public's right to know, as stated in Article 6 of Law No. 40/1999.

"The media has an important role in providing education and early warnings related to natural disasters by gathering information about the needs of victims of natural disasters for government organizations and volunteers and [by] being a potential channel for discussion and feedback from the community about natural disasters [9]".

The study conducted by Asteria (2016) [10] describes three important approaches that can be used in disaster journalism as follows:

- The impact and chronology of disasters, and also to educating the public about the occurrence of disasters and reporting the facts that could be related to mitigating the impact of future disasters.
- Must reflect the values of social humanism by revealing accurate data and facts, because the information conveyed also becomes material for social education and could impact victims and non-victims.
- Report and broadcast contents should not have potential for sensitive impact on victims or make victims discouraged by the reality experienced.

Furthermore, it was stated by Asteria (2016) [10] that the media under study had not provided detailed and clear knowledge about disasters, and could not provide effective early warning to citizens, and thus it needed to be improved with an emphasis on humanity to ensure that the mass media can support disaster preparedness and disaster education, thereby building the community's capacity to face future disasters.

2. Purpose

The purpose of this study is to determine the role/involvement of the media and analyze its existence and role in supporting disaster education and risk communication in Aceh following the 2004 Indian Ocean tsunami. It is expected that the results of this study will be used as a reference by the government and media in contributing and managing disaster risk information for disaster risk reduction efforts.

3. Methods

This research is qualitative research, exploring social phenomena or reality. The qualitative approach was used in this study to understand the roles and involvement of the media in Aceh after the 2004 Indian Ocean tsunami. Data were collected through semi-structured interviews, which used open questions to provide and develop on-going conversations. The open questions gave informants the freedom to answer questions as they want, because no answer choices are provided [11].

This research was conducted in Banda Aceh between April and May 2018. The resource of this study were media that represented the government (electronic) and private (print, online, electronic) media. The government media were the Televisi Republik Indonesia (TVRI) of Aceh and the Radio Republik Indonesia (RRI) of Aceh, while the private media included print media (Serambi Indonesia), television (Aceh TV and iNews), radio (Djati FM, and Serambi FM) and online media (Serambi Indonesia).

The research went through three major stages, namely:

- Preparation: the formulation of a list of questions used as an interview instrument, covering the information of programs/rubrics related to disasters, involved elements (such as persons who influenced disaster programs), funding support, and community responses to represented media.
- Data collection: Data collection was carried out using an interview and observation approach. Interviews were conducted by contacting informants. The interview results were written on the interview transcript sheet. Observations of resource media were used to see rubrics and programs and to review the news coverage of disaster issues.
- The analysis and report writing; all data were collected, discussed and analyzed based on intended purposes.

4. Results

Representatives from eight media were interviewed; one was from government media, while the rest were from private media. Research identified numerous examples of the media communicating risk to promote disaster awareness after the 2004 Indian Ocean tsunami. Surprisingly, a negative trend was found, in which media engagement in covering disasters and disaster risks decreased. The interview results showed that most of the media gave frequent coverage to disaster topics only during the recovery phase. Disaster programs were only limited to reporting mitigation and the reconstruction process. After the end of the BRR contract in Aceh, media engagement in communicating disaster risks also decreased. For instance, media such as Serambi FM, RRI, and Djati FM had special disaster programs, but these concluded mostly by the end of 2011. Therefore, from 2012 until 2018, these media focused mainly on reporting disaster events and response times.

However, each media involved in this research has media programs related to disasters. The media programs are described as follows

Table 1. Media Programs Related to Disasters

No	Media	Types of Program		
		News	Tentative	Advertisement
1	TVRI Aceh	Aceh News	Interactive dialogue	PSA (Public Service Advertisements, per BPBA request)
2	RRI Aceh	Pro 3 dan Pro 4	Interactive dialogue	Routine PSA, Weather Information
3	Serambi Indonesia	Serambi Indonesia Rubric	Opinion and reporting Citizen Reporter, *Droe keu droe,* Column Culture	BPBA Advertisement
4	Serambi FM	Salam Serambi	Talk show	PSA in collaboration with BPBA
5	Serambi News	News (Nanggroe)	Opinion and reporting Citizen Reporter, *Droe keu droe,* Column Culture	-
6	Djati FM	Breakfast	Talk show Momentum/event, public dialogue, emergency radio training in providing information	-
7	TV Aceh	Aceh Today, All about Aceh	Morning Coffee, Parliamentary	-
8	iNews TV	iNews Aceh	*Peunegah Aceh*	-

.

4.1. Disaster programs/funnels

First, both types of media have funnels for reporting disasters. Although the topic of disaster was recognized as a media concern, it also covered other types of information. This means that the media do not have specific programs related to disasters. All media have two of three funnels; only the government media and one private media (Serambi Indonesia) have an advertising funnel. This means that the media are popular among Acehnese. As a result, the Aceh Disaster Management Agency (Badan Penanggulangan Bencana Aceh, BPBA) has been interested in cooperating with these media.

The news funnel is mostly used for reporting up-to-date news. The information contained in this funnel also matches the needs of society, including—for example—economics, diplomacy, lifestyle, education, disaster, and so on. Most media do not have special rubrics for disasters or report disasters as regular news. Information is presented in various forms, which can be ascertained through direct coverage from the disaster sites, as proclaimed by news shows. For electronic media such as television and radio, interview programs are conducted directly by inviting interviewees to the studio, while print media present the news by drafting it first before publishing.

Second, tentative programs have been organized only by electronic and online media. Some programs, as mentioned above, drove this funnel. This included, for example, a live program inviting viewers/listeners to contribute directly in the form of questions and responses. However, this dialogue program also did not include regular disaster programming. Dialogue mainly focused on the trending topics of the week or month. Third, advertising: advertisements are used to inform society about disaster prevention and forecasts.

Materials and programs owned by electronic and printed media could be distinguished by their form of presentation. For example, RRI's electronic media conveyed disaster information that focused more on the response or emergency phase. One forms is PSA, which provides information related to disaster mitigation and awareness in different regions based on the risks in said regions. In this case, TVRI has a government-supported talk show and interactive dialogue program related to disaster issues.

Meanwhile, print media such as Serambi Indonesia often also provide opinion rubrics that give students, experts, and even community members the opportunity to spread and share their opinions about disaster. These opinions and articles are not included regularly, but incidentally, depending on the sources obtained by the authors.

4.2. Program Resources

In running disaster news and programs, all media are involved in the selection and interviewing of informants. Experts are commonly used as resource persons by the media and invited to speak on their disaster programs (e.g. Disaster Management Agency, Department of Social Affairs). The media have also involved legislature (to look at approved policies and budget allocations), academics, and representatives of the Centre for Disaster Studies and NGOs. The role of a communicator here is to provide disaster risk information in the form of data, study results, experiences, etc.

The public is more confident and interested in watching or listening to programs that have news or impressions sourced from disaster institutions. The media are also open to involving other competent parties in delivering disaster risk messages, such as local leaders who have local knowledge about dealing with disasters, such as the Simeuluean people who were saved from the 2004 Indian Ocean tsunami—only three of the island's 70,000 residents were reported killed.

However, the involvement of experts is not only the main choice, as the main focus of the media programs is to attract public attention and ensure public preparedness for future disasters. While the influences of experts in the interview process support the preparations handled by journalists, informants also come from local communities who experienced disaster. Representatives of vulnerable groups such as women and disabled groups could also be involved as resource persons. It is important to know the common problems that are faced by such vulnerable groups during disasters.

4.3. Factors Causing the Absence of Special Programs

The absence of special programs in the media was influenced by many factors, including the kinds and levels of community needs for disaster information, the availability of funding, and the level of creativity in presenting information.

4.3.1. The Kinds and Level of Community Needs on Disaster Information. In general, to ensure the effectiveness of disaster programs, the media should also be concerned about listeners/readers/viewers' point of view; that is, the media should know the kind and level of disaster information needed by audiences. Audiences' points of view provide the media with input for improving disaster programs, which could be a strong reason for providing special space or rubrics. Disaster programs could include:

- Interactive dialogue,
- Disaster rubric for readers' opinion,
- Comment columns,
- Social media platform.

However, such special programs have not optimally sought listeners/readers/viewers contributions. This is also indicated by the fact that disaster risk information has not yet resulted in community awareness for reducing risk, especially the information presented in the form of disaster education. Community interest in disaster increases significantly when disasters occur, both within or without Aceh. This situation may likely be caused by the media's inconsistent presentation of disaster events, which is more intensive during the emergency or response phase and continues through the rehabilitation and reconstruction process. Limited feedback from viewers/readers will undermine the media rating.

4.3.2. The Availability of Funding. In addition to the level of community needs for disaster information, funding availability is also a factor that influence the media's designing of special disaster programs. Based on interview results, it can be concluded that both government and private media have no special budget for regular disaster programs. In print media such as Serambi Indonesia, the presence of a special program was found to depend on the situation, such as the occurrence of disaster and the sponsorship of parties interested in developing a special program on disaster topics. This is due to operational costs that are not budgeted by media management. Programs with low ratings will not able to boost circulation, thereby affecting media income. This most prominently affects private media.

Government media, meanwhile, should not benefit financially. It has the responsibility to provide disaster risk information, and as such its budget must be allocated to support the planned programs. However, so far, the budget allocation for such programs is still very limited. This was conveyed by an informant from RRI, who stated that "there is no special budget to invite resource persons (in an interactive dialogue), but the speaker is present on a voluntary basis". This is similarly true at TVRI, which still runs disaster programs situationally.

4.3.3. Presentation/Packaging Information Creativity. In print and online media, which gain their power through presentation and writing, disaster information seems monotonous and old-fashioned, with a rigid and visually uninteresting writing style. It has been made less attractive to the broader public, especially youths.

In electronic media, the style of presentation is similarly not innovative. The selection of resource persons is mostly aimed at certain groups; because of the hosts' lack of knowledge, they are unable to dig deeper and make the information interesting. Coupled with the use of resource persons or experts who are not easily understood by ordinary persons—the main target of disaster information. Moreover, information in not presented by balancing the scientific and religious approach, and as such it has not been able to motivate the Acehnese community to change its attitudes and paradigms, which hold that disaster is destiny and as a result informs a more passive and fatalistic approach to it. Without a balanced approach, it is impossible to create a new paradigm where people can actively reduce disaster risks.

5. A Good Initiative

The Disaster Care Journalist Forum (*Forum Jurnalis Peduli Bencana*, FJPB) was established in Aceh as recognition of the media's need to be concerned with disaster risk information. This initiative was carried out by the members of this forum to provide disaster risk information to the public. The FJPB carries out the role of broadcasting disaster risk information related to:

- The ability to adapt to disaster
- The ability to save lives when disasters occurs
- The ability to take the role on disaster risk reduction efforts

For example, Djati FM Radio has worked together with other forum members to frequently conduct media preparedness training. This training targeted representatives of local communities, the RAPI (Indonesian Inter-Population Radio) community, students, and regional BPBD. This training was intended to inform communities and related stakeholders about the disaster risks surrounding them. Local radio broadcasters were also partners in disseminating disaster risk messages and helping communities to convey their problems to the government.

6. Conclusions

Based on the above findings, media engagement in supporting disaster risk reduction and communicating disaster risk in Aceh can be summarized as follows:

- All of the media in Aceh following the 2004 Indian Ocean tsunami have started programs related to disaster, but most depend on the level of leadership and media concern for disaster issues,
- Even though there is increased awareness of the essential role of the media in disseminating risk messages through community perceptions, it seems to focus only on when disasters occur,
- The lack of journalists' capacity to present disaster programs is also a main issue,
- The allocation of funding is the media's main problem in running disaster programs.

.

7. References

[1] Konflik T, Sosial-politik R, Desentralisasi E, Ilmu F, Politik I, Universitas F and Zainal S 2016 Transformasi Konflik Aceh dan Relasi Sosial- Politik di Era Desentralisasi *J. Masy.* **21** 81–108
[2] BRR, MDF U 2012 *TSUNAMI Habis Bencana Terbitlah Terang*
[3] Arkwright D 2008 Potensi Kejadian Tsunami Di Perairan Timur Indonesia
[4] LATIEF H 2004 Tsunami Aceh 2004
[5] Hilman D, Penelitian N P and Lipi G 2015 Siklus Mega-Tsunami Di Wilayah Aceh-Andaman Dalam Konteks Sejarah Mega-tsunami cycles Aceh Andaman region in historical context **25** 49–62
[6] Presiden Republik Indonesia 2014 Undang-undang Republik Indonesia Nomor 24 Tahun 2007 tentang Penanggulangan Bencana *Igarss 2014* 1–5
[7] Morgan G, Fischhoff B, Bostrom A and Atman C J 2002 *Risk Communication A Mentals Approach* (Cambridge University Press)
[8] Indonesia P 1999 Undang-Undang Republik Indonesia Nomor 40 Tahun 1999 tentang Pers 1–11
[9] Ahmad F, Mustaffa N, Rahim S a, Basri F K H, Ahmad A L and Aziz J 2011 Confronting Environmental Risk via Communication *Innov. J. Public Sect. Innov. J.* **16** 1–13
[10] Asteria D 2016 Optimalisasi Komunikasi Bencana di Media assa Sebagai Pendukung Manajemen Bencana *J. Komun. Ikat. Sarj. Komun. Indones.* **01** 1–11
[11] Pujihastuti I 2010 Prinsip Penulisan Quisioner Penelitian *J. Agribisnis dan Pengemb. Wil.* **2** 43–56

Archiving Disaster Remains: The Case of *"Sasanao Factory"* in Yuriage Village, Natori City, Miyagi Prefecture

S Sato[1], A Suppasri[1], S P Boret[1], M Nakagawa[2] and F IMAMURA[1]

[1] International Research Institute of Disaster Science (IRIDeS), Tohoku University, Sendai, 980-8572, Japan
[2] Ishinomaki Future Support Association, Ishinomaki, 986-0822, Japan

Abstract. In the areas affected by the 2011 Great East Japan tsunami, some damaged buildings were proposed to be preserved as "Disaster Remain" to share this experience and educating future generations. However, many these potential buildings were eventually demolished. This study aims to provide an alternative and eventually a consensus among communities that faced conflicts over the preservation or destruction of such disaster buildings. The methodology consists of 3D building modelling, tsunami numerical simulation and recording tsunami evacuation behaviour with interview survey. In sum, this practical study of "Sasanao Factory" demonstrates that the archiving of disaster remain can offer a compromise between those who wish to preserve disaster ruins, here the owner of the building, and those who want or need their destruction, here local government.

1. Introduction

In the areas affected by the 2011 Great East Japan tsunami, some damaged buildings were proposed to be preserved as "Disaster Remain" to share this experience and educating future generations [1]. However, many these potential buildings were eventually demolished because of the budget provided by the Japanese government for the reconstruction of the coastal areas covers only the preservation of one building per one municipality [2] after receiving the agreement of local residents [3]. For those buildings that must be demolished, an original solution is to archive them as "Disaster Remain."

This study first aims to introduce a whole new process of archiving disaster remain using digital technology. Secondly, it investigates the effectiveness and problems of implementing this approach in the short term/technically speaking. We expected that this approach will provide an alternative and eventually a consensus among communities that faced conflicts over the preservation or destruction of such disaster buildings.

2. Methods

During this study, we digitally archived the "Sasanao Factory" building which used to manufacture boiled fish pastes in Yuriage area, Natori City in Miyagi prefecture. This operation was realized after the demolition of the building as ordered by the local government of Natori City. The method consists of making a 3D building model of the building using UAV scanning, tsunami numerical simulation and recording experience of the stakeholder using interview survey.

IOP Conf. Series: Earth and Environmental Science **273** (2019) 012044 doi:10.1088/1755-1315/273/1/012044

2.1. 3D building model of the building

Vertical direction (Nadir) images of the building and oblique images of its surroundings and circumferential position surrounding the building were taken in total of 671 images using UAV (DJI Inspire1) and autopilot software DJI Ground Station and Litch for DJI drones (Figure 1). In addition, 1,127 pictures of the surroundings and the interior of the building were taken by digital single lens reflex (Nikon D 700). In order to create a 3D model by combining successive images with Context Capture using SfM (Surface from Motion) software, it is desirable to take a series of images from a distant view position to a position close to the photographing field angle by a digital single lens reflex camera when photographing with UAV. Finally, laser measurement using a handheld laser scanner ZEB - REVO and movie shooting with a set of ZEB – CAM were also conducted (Figure 2).

Figure 1. Images shooting using UAV

Figure 2. Laser measurement using ZEB-REVO

2.2. Tsunami numerical simulation

An important parameter for tsunami simulation is the initial seafloor deformation that is computed from the fault parameters. This study estimated the initial water level based on a rectangular fault model and assumed that the change in the sea surface is the same as the seafloor deformation. The rectangular faults are assumed to have a fast movement toward the sea surface that is only a vertical displacement [4]. The fault parameter of the 2011 earthquake based on the Tohoku University model is used [5]. To obtain the tsunami inundations, a numerical tsunami simulation was conducted with the TUNAMI-N2 model [6]. The TUNAMI-N2 model uses the nonlinear theory of the shallow water equation, which is solved using a leap-frog staggered-grid finite difference scheme with the bottom friction represented by Manning's roughness coefficient. The simulation was run on six computational domains using a nesting grid system, from the largest region (1,215 m resolution) to the smallest region (5 m resolution) along the coastline of Yuriage village using the fixed ratio of the two domains as one-third (Figure 3 left). Initial water level calculated using the fault parameters are shown in Fig. xx1 right. This study also considered tsunami mitigation effect from the reconstruction (elevated land, elevated road and new seawall) as shown in Figure 4.

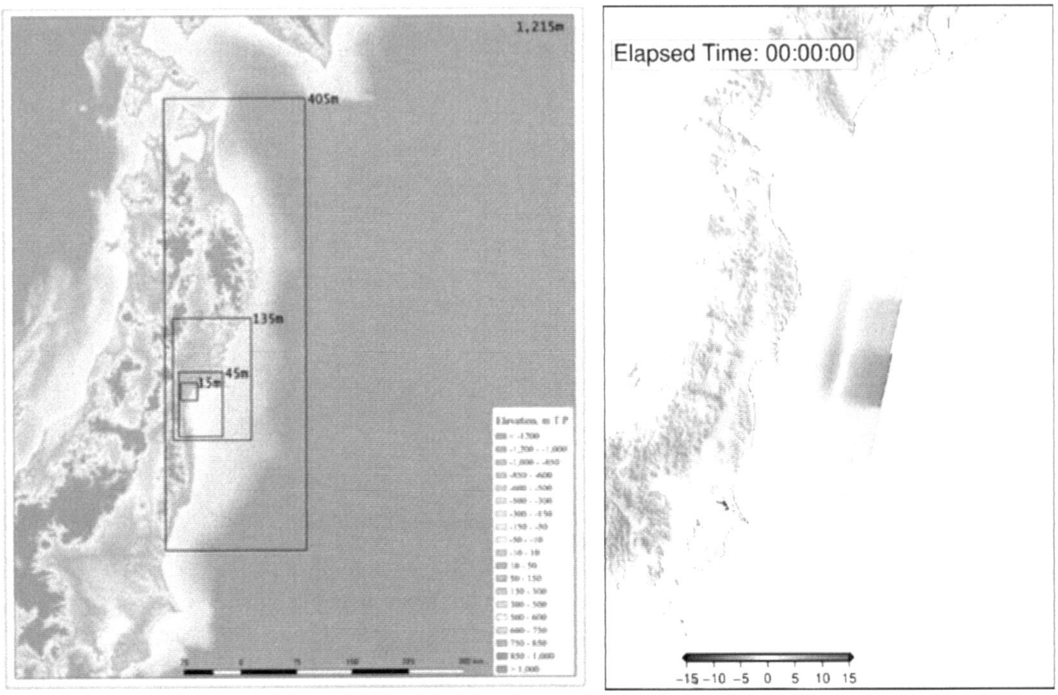

Figure 3. Left: Bathymetry and topography data in nesting grid system and Right: Initial water level computed using the proposed fault parameters

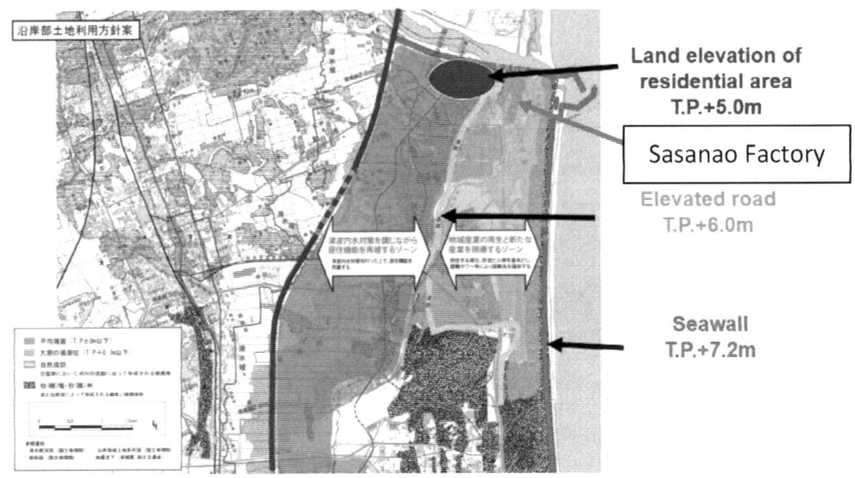

Figure 4. Location of Sasanao factory and reconstruction plan of Yuriage village

2.3. Interview survey

In order to record and to describe the tsunami evacuation and the other first response of the building owner (The company's CEO), we conducted detailed interviews, spending a total of 2 hours to obtain information on just before the earthquake and the company recovery process for about one year after the disaster, starting from immediately after the quake. The interviews were conducted in a semi-structured approach based on the development of talks by interviewee so that the process of the talks would not be hindered and the interviewees would be urged to speak further. Special consideration was given so that detailed information on experiences of individual interviewees could be obtained in addition to common premise-based questions by the interviewer. The data were made to a transcript.

IOP Conf. Series: Earth and Environmental Science **273** (2019) 012044 doi:10.1088/1755-1315/273/1/012044

Then, the transcript was compiled the reading material to learn the tsunami behaviour, his tsunami evacuation and the building damaging process.

3. Results

3.1. 3D building model of the building

As benefit of using SfM software, it became possible to produce 3D models from continuous images (Figure 5, Figure 6). Context Capture, one of SfM software was used to photograph the Sasanao factory using UAV and digital single lens reflex camera and combined the images. 1,762 images were combined from a total of 1,818 images, and a 3D model was produced from the point cloud (set of points having coordinates) data.

Figure 5. Construction of 3D building model by SfM Software

Figure 6. 3D building model of Sasanao Factory

By using free viewing software Actute 3D Viewer and embedding it in a web site, 3D model can be browsed easily while rotating, zooming in and out. Wireframe and texture can be also switched on and off (Figure 7, Figure 8).

IOP Conf. Series: Earth and Environmental Science **273** (2019) 012044 doi:10.1088/1755-1315/273/1/012044

Figure 7. 3D building model (facade)

Figure 8. 3D building model (inside)

In addition, moving images taken by the handy laser scanner ZEB-REVO and the set ZEB-CAM is cut out by cutting 1,376 consecutive still images every 0.5 seconds. Based on the point cloud shape data obtained from the laser scanner, 3D model was created by combining images taken with three cameras of UAV, digital single lens reflex camera, and ZEB - CAM. Although it was expected that the 3D model with the laser survey shape data was expected to have high building details but some data loss in the shadow of the laser such as rooftops, detail reproduction of the internal image etc (Figure 9, Figure 10). This work gave example in preserving and digitalizing the disaster remains using UAV and digital single lens reflex images that can be transferred to future generations.

Figure 9. Comparison of 3D model (Left: with Scanner Data, Right: without Scanner Data)

Figure 10. Comparison of 3D model (Left: with Scanner Data, Right: without Scanner Data)

3.2. Tsunami numerical simulation

The 2011 tsunami inundation heights and depths were collected by the 2011 Tohoku Earthquake Tsunami Joint Survey Group [7]. The numerical tsunami model results were verified using a geometric mean (K) and geometric standard deviation (κ) proposed by Aida [8]. These values refer to the deviations or variances from the proportion of observed and computed data. The simulation results give value of $K = 1.03$ and $\kappa = 1.31$ which can be classified as "good agreement" based on a guideline proposed by Japan Society of Civil Engineers (JSCE). Figure 11 shows snapshots of tsunami inundating Yuriage village at 65, 69 and 82 min after the earthquake. Figure 12 shows comparison of simulated tsunami flow depth between before and after reconstruction. This simulation result illustrated that tsunami arrived Yuriage village around 67 min. It can be seen that residential area was not inundated after the reconstruction with the same 2011 tsunami-like event and inundation area was much decreased.

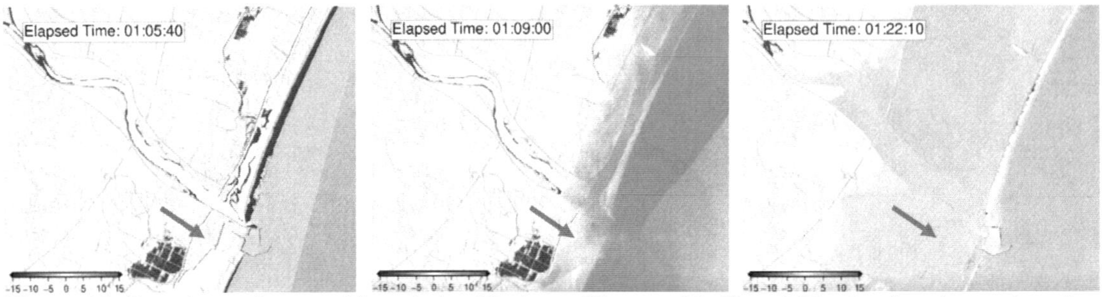

Figure 11. Snapshots of tsunami inundating Yuriage village at 65, 69 and 82 min after the earthquake

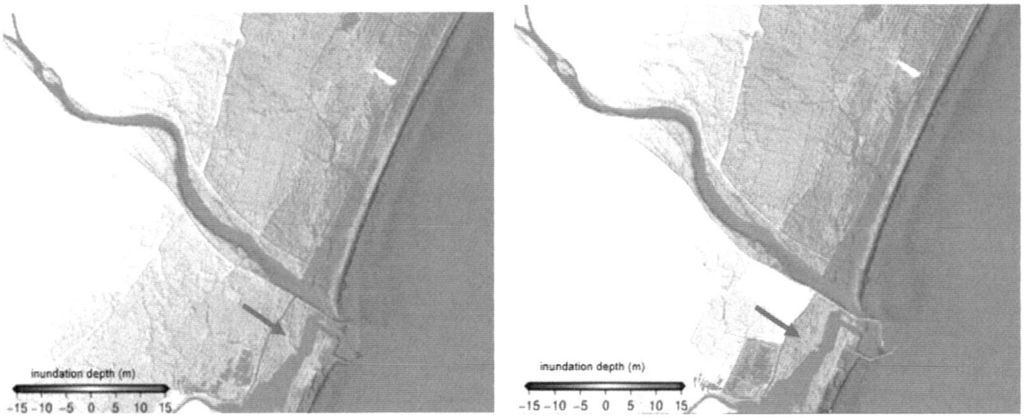

Figure 12. Comparison of the maximum flow depth between before reconstruction (at the time of the 2011 tsunami) and after reconstruction

3.3. Interview survey

The results of interview survey are summarized as follows.

- He took a lunch break at home about 200 m from Sasanao Factory in Yuriage village before the quake.
- After the earthquake, he moved to the factory to inspect the factory building and employees.
- He did not order evacuation to employee immediately because he did not have awareness of tsunami coming after earthquake at the coast.
- In the meantime, water level of "Teizan-bori" moat became lower.
- He was worried about tsunami coming to see the phenomena which is backwash before tsunami.
- His elder sun order evacuation to employees. CEO's wife, the sun and others moved to Yuriage elementary school building which is three stories to inland by cars. Fortunately, they didn't run into a traffic jam.
- He stayed the Factory building because he thought that the building is not destroyed by tsunami attack and wanted to see the tsunami behaviour.
- Then, went upstairs and took pictures of tsunami and situation of around the building with his cell phone camera. However,

Figure 13. Situation of interview survey

4. Conclusion

The 3D building model of the Sasanao factory was completed using commercially available UAV (Unmanned Aerial Vehicle). This model was then shown to the owner of the factory, who shared his impressions and thoughts. Unstructured interview survey based on direct interview with the owner was made. The owner was asked to narrate his own experience of the day of 2011 Great East Japan Earthquake and Tsunami including tsunami related information such as the maximum flow depth and arrival time. The recording of the owner's experience was then transcribed. In addition, a tsunami numerical simulation followed by a tsunami CG was created using the eyewitness's information of the owner. Both transcription and tsunami CG were shown to the owner to receive his feedback and carry out further corrections.

Digital data of such disaster remains can be utilized in various ways, such as publishing in a form that anyone can publicly view, creating a 3D movie in combination with the experience stage of the disaster, and reproducing the 360 degrees' view model. Although methods of digitalizing and preserving such precious disaster remains are still during a trial and error stage that reflected advancement of technology of that period. However, the cost is chapter than maintenance cost of the real structures. Therefore, this work shows a good example of a modern way in digitalizing and preserving disaster remains to transfer disaster experience to future generations. Such method can be applied to some tsunami disaster remains such as electric generation ship in Banda Aceh and fishing boats and patrol boat in Thailand [9].

Tsunami numerical simulation was also performed to reproduce tsunami characteristics such as flow depth and arrival time at the whole Yuriage village and verified by testimony the eyewitness (owner of Sasanao Factory) such as tsunami trace at his factory and tsunami arrival time at the village. The numerical was also conducted to illustrate effectiveness of the reconstruction plan in the future.

The outcomes of the research were finally compiled and donated to the owner in the form of a written document and a digital model of the building in replacement of the ruin to be demolished. In addition, these archived contents (3D building model, transcription and tsunami CG) will be exhibited at a new disaster museum to be built by the city. In sum, this practical study demonstrates that the archiving of disaster remain can offer a compromise between those who wish to preserve disaster ruins, here the owner of the building, and those who want or need their destruction, here local government. We hope that this small experiment may give way to large project in Japan and beyond.

Reference

[1] Sato S 2017 The Great East Japan Disaster-Present situation of disaster remains *Shinsai-gaku* **11** Tohoku Gakuin University 146-161
[2] Reconstruction Agency 2013 Support for the preservation of disaster remains , http://www.reconstruction.go.jp/topics/m13/11/20131115_press_sinsaiikou.pdf
[3] Sato S, Kawashi S, Imamura F 2018 The Establishing Process of Disaster Remains Affected by the 2011 Great East Japan Earthquake and Tsunami Disaster in Kesennuma City *Proceedings of the annual conference of the Institute of Social Safety Science* **7** 81-86

[4] Okada, Y 1985 Surface deformation due to shear and tensile faults in a half-space. *Bull. Seismol. Soc. Am.* **75**, 1135–1154.
[5] Imamura, F, Koshimura S, Murashima Y, Akita Y, Shintani Y 2012 The Tsunami Source Model of the 2011 Tohoku Earthquake (ver.1.2). Available online: http://www.tsunami.civil.tohoku.ac.jp/hokusai3/J/events/tohoku_2011/model/dcrc_ver1.2.pdf (accessed on 25 April 2012).
[6] Imamura, F 1995 Review of tsunami simulation with a finite difference method. In Long-Wave Runup Models *World Scientific Pub Co Inc*. 25–42.
[7] The 2011 Tohoku Earthquake Tsunami Joint Survey (TTJS) Group. 2011 Nationwide field survey of the 2011 off the Pacific coast of Tohoku earthquake tsunami. *J. Jpn. Soc. Civ. Eng.* **67**, 63–66.
[8] Aida, I 1978 Reliability of a tsunami source model derived from fault parameters *J. Phys. Earth* **26**, 57-73.
[9] Suppasri A, Goto K, Muhari A 2015 Ranasinghe, P., Riyaz, M., Affan, M., Mas, E., Yasuda, M. and Imamura, F A decade after the 2004 Indian Ocean tsunami - *The progress in disaster preparedness and future challenges in Indonesia,* Sri Lanka, Thailand and the Maldives, Pure and Applied Geophysics, **172**(**12**), 3313-3341

Acknowledgments
This research was supported by JSPS program for "Responding to Real Society", Top-setting program to advanced cutting-edge humanities and social sciences research, "Modelling and Implementation of Effective and Sustainable Disaster Tradition Methods" (PI: Shosuke Sato) and Natori City.

Fiqh Anticipation: Mitigation Concept Based on Islamic Law

Mizaj Iskandar

Law Department, Sharia and Law Faculty, Ar-Raniry State Islamic University, Banda Aceh, Indonesia

Abstract. Disasters as social phenomena that occur when a community suffers from disruption and loss due to natural processes or technological accidents. Based on this definition, earthquake, tsunami and other disasters are not limited to natural disasters, it can also be a social disaster if the natural disaster is directly in contact with humans and causes material and immaterial losses. Indonesia as a country with the largest Muslim population in the world cannot be separated from Islam in various aspects of life, including the issue of disaster mitigation. for Indonesian Muslims, religious entities are perceived to be bigger than state entities. Therefore, the mitigation approach of science must be harmonized with religious teachings. With the result that natural disaster mitigation is widely accepted by the community. Contemporary Islamic scholars, especially those who have been involved in statistics and other related sciences, will dare to state that future estimates based on honest and correct statistical calculations can be regarded as reality, because according to scientific calculations it is relatively certain to occur. Therefore *Fiqh* anticipation as one of the mitigation management in Islam can take the form in the following three steps; (1) Reducing or avoiding the impact of losses from natural disaster hazards; (2) Ensure fast arrival of assistance to victims; (3) Achieve a quick and effective recovery of society.

Keywords: Islam, fiqh anticipation, natural, disaster, mitigation.

1. Introduction

The natural disasters of the earthquake and tsunami that occurred in Palu, Donggala, Sigi and Parigi Moutong added to the long list of state failures in anticipating potential natural disasters that come along with people's lives. Indonesia is very familiar with natural disasters. Various natural disasters which are based on geographical, geological and hydrological conditions should encourage Indonesia to build community resilience in the face of natural disasters.

The territory of Indonesia is the largest archipelago in the world. Geographically, Indonesia located between the continents of Asia, Australia, the Indian and Pacific Oceans it has 17,508 islands [1]. Although Indonesia has natural wealth and the beauty of islands, it is also important to realize that the archipelago has 129 volcanoes active, or known as the ring of fire, and located at the confluence of three world active tectonic plates: Indo-Australian, Eurasian and Pacific Plate [2].

Ring of fire and being at the meeting of three tectonic plates puts the archipelago in the potential to threaten natural disasters. On the other hand, its position in the tropics and hydrological conditions trigger other natural disasters, such as tornado, extreme rain, floods, landslides and drought [3].

There are many examples of major natural disasters that have occurred in Indonesia such as earthquakes, tsunamis and volcanic eruptions with huge casualties and property. Mount Krakatau was erupted in 1883, and then the earthquake that followed the tsunami in Aceh on December 26, 2004 with a total of 130.000 people died. The 8.7-magnitude earthquake in Nias caused the death of approximately

900 people, the Yogyakarta earthquake killed 300 people, the Padang earthquake in 2009 with the death toll of 500 people. The Mentawai earthquake in October 2010 caused 509 people to die, the Lombok earthquake killed 563 peoples and the last earthquake and tsunami in Central Sulawesi which according to the latest data released by The National Agency for Disaster Countermeasure (BNPB) has resulted in 1,571 deaths and that is not including those missing [4]. All that does not include natural disasters caused by flash floods in Wasior which claimed up to hundreds of people.

Based on the Asia-Pacific Report 2010 data compiled by the United Nations Economic and Social Council (ECOSOC) for the Asia Pacific region, since 1980 to 2009 natural disasters in Indonesia have recorded huge losses of 191,164 people killed (second place in Asia Pacific), 17 million people the victim (ranked ninth) with economic losses reaching USD. 22.582 billion. In 2010, losses reached Rp. 5 trillion [5]. Facing the threat of material and mental losses of natural disasters, it is considered important to realign the way to view natural disasters and the efforts that can be done in order to deal with these threats presented in an Islamic perspective in the following explanation.

The role of religion in post-disaster such as reducing psychological its impact, religious-based organizations relief and assistance have been widely acknowledged. In fact, every disaster always causes the feeling of worry, uncertainty, sadness and distress of the survivors especially those who lost their lovely family members [6]. Religious-based relief organizations have been playing important role both in assisting survivors socially and psychologically in building their life. There has been acknowledged that Holy Book contains explanation about the meaning of human sacrifice, life, death and life after death then connects to natural disaster events.

On the other hand, the increase in unpredictable natural disasters events for a decade has led to put the disaster preparedness as a central issue in disaster management. Despite such efforts to encourage people to take preparedness action, previous studies on natural disaster is that people neglect to take preparation for such disaster events [7]. The question is that can the religious teaching contribute to motivate people to be more prepared for disaster as the dominance of religious views that disaster is caused by God's wrath?

2. Methodology of Study

In order to extract the most suitable and related verses from Quran as well as important Hadith from accessible and translated Hadith compilation (for this study: Bukhari and Tirmidhi) keywords that are related to disaster issues were researched.

The difficulty of this methodology was that the occurrence of these keywords was very low and that verses found did not give sufficient information that would be useful in the context of fiqh anticipation as mitigation concept based on Islamic law. Therefor the keywords to be researched were extended, which proved to be necessary shift in methodology to come to appropriate results.

The next step consisted of compiling both Quranic verses and Hadith and the attempt to comment each keyword. The comments however are based on the experience and the knowledge of the author and try to connect the verses and Hadith to the context of disaster preparedness and response. The translated of Quran used here is *The Meaning of Holy Quran* by Abdullah Yusuf Ali [8].

3. Qur'an and Disaster

In the Qur'an, there are three terminologies related to disasters, namely: *muṣībah* (misfortune/disaster), *balā'* (tremendous trial), *'azāb* (penalty/punishment) [9]. While *fitnah* (tumult) is part of the disaster itself. All disasters that afflict unbelievers are called *'azāb*. Like the flood of Noah, the survivors of the disaster were only believers who followed the teachings of Noah. Infidels to Prophet Lot destroyed but people who pious survived. Those who do not believe to the Prophet Shelah are afflicted with a terrible plague, but strangely, a believer even though his house is next to him is not affected by the disease, while the infidels are destroyed by a terrible disease. Moreover, the plague that attacked all eaters of camel meat of both the Prophet Shelah and the Prophet Jethro (*Syuaib*) were all affected by the virus, but those who did not eat were not infected with the virus. Abraha's army was destroyed because punished by God with the stone which Ababil's bird threw, but nothing happens around. So, the *'azāb* is intended for people who are disobedience and unbelief to God [10].

Whereas *muṣībah* aims to test the faith of a Muslim. But *muṣībah* is take more massive level, because *muṣībah* is no deal whit religion, nation, language, color and sex. Everyone could be hit by *muṣībah.* Therefore, in Qur'an chapter 2 verse 155 God explains that *muṣībah* takes the form of fear (*khawf*), hunger (*jauw'*), some loss in goods (*naqṣ min al-amwāl*), some loss in lives (*naqṣ min al-anfus*) and loss some fruits (*naqṣ min al-tsamarāt*). all of these are forms of testing from God (*wa lanabluwannakum*).

While *balā'* are more individual and mechanical in nature. Such as when God gave an individual test to the Prophet Abraham to slaughter his son Ishmael, whose test was mechanically aimed at improving the spiritual quality of Prophet Abraham. That's why in Qur'an chapter 37 verse 106 God calls the test as *al-balā' al-mubīn* (individual tremendous trial) [11].

From above explanation I conclude that natural disasters have attacked Indonesia are not punishments from God but ordinary natural calamities mentioned in Qur'an as *muṣībah* (disaster). Disaster must be regarded as ordinary natural events. Comes anytime, anywhere, without having to believe and do things that are mythical. There are several attitudes that must be considered in responding to disasters. Firstly, how to improve our insight in the face of disaster including disaster preparedness. second, do not understand the disaster as a curse, punishment or penalty (*'azāb*) from God, but it's only natural phenomena. Third, do not mythologize natural disaster, because its polluting the purity of faith. And fourth, in Islamic view, disaster has a positive dimension, namely to pay off our sins in the day afterlife. In this case The prophet Muhammad said *"Never a believer is stricken with a discomfort, an illness, an anxiety, a grief or mental worry or even the pricking of a thorn but Allah will expiate his sins on account of his patience"*

4. Natural disaster and Islamic teaching

Smith and Petley defines disasters as social phenomena that occur when a community suffers from disruption and loss due to natural processes or technological accidents [12]. Based on this definition, it can be understood that natural disasters are actually social phenomena due to natural events. Not all natural events such as earthquakes or landslides can be called natural disasters. But when in contact with humans and cause loss of property and soul, that is what is called a natural disaster. Based on this definition, earthquake, tsunami and other disasters are not limited to natural disasters, it can also be a social disaster if the natural disaster is directly in contact with humans and causes material and immaterial losses.

Indonesia as a country with the largest Muslim population in the world cannot be separated from Islam in various aspects of life, including the issue of disaster mitigation. for Indonesian Muslims, religious entities are perceived to be bigger than state entities [13]. This at least shows how disaster victims shout for help from their gods not to their country or government. Therefore, the mitigation solution for Indonesian Muslims does not only focus on science, but even the mitigation approach of science must be harmonized with religious teachings. With the result that natural disaster mitigation is widely accepted by the community.

In Islam, natural disasters that hit humans are *qadha* from Allah (god's predestination) [14]. However, behind the god's predestination there is a natural phenomenon that can be digested. Including efforts to avoid it before a natural disaster occurs. In the event of a natural disaster there is a domain that is in the power of human beings and that is beyond human control [15]. All efforts that can minimize and even avoid the dangers and risks of natural disasters are in the domain of human power. Natural events that produce natural disasters cannot be prevented or eliminated. But all efforts to avoid interaction between natural events that cause natural disasters with humans, this is included in the efforts of management and mitigation of natural disasters in fiqh disaster countermeasures [16].

Save our own self or others from danger is included in the category of endeavor (effort) that must be done. Umar bin Khattab was said to have avoided an area that was plagued by an epidemic. When asked whether such acts do not mean avoiding destiny because being infected is a destiny that if God has determined it, then humans will not be able to avoid it. Then 'Umar replied that he was avoiding a destiny to lead another destiny [17]. This story illustrates that fatalism in a natural disaster situation is an attitude that is not justified by Islam. God forbids believers to despair when natural disasters occur and requires to be positive thinking (*husn al-dhann*) for Allah's help. In this case God said: *wa la tai'asu min rauhillah innahu la yaiasah min rauhillah ila qaumul kafirun* (no one despairs of relief from Allah except the disbelieving people) (Quran, 12:87).

In the same spirit even though in a different context, the Prophet Muhammad peace be upon him requires his people to fulfill their safety rights, for example the prohibition on fasting along time and states that the human body and eyes have the right to rest in order to stay healthy [18]. In a broader sense, this means the obligation to save human life.

Technically, efforts to manage natural disasters in Islam are not much different from the many methods that have been implemented today across the world. But [19]the difference in view at the source of the creator of natural disasters, namely with the provision of God, resulted in a slight difference in the initial steps when a natural disaster occurred, as exemplified by the Messenger of Allah and the caliph afterwards, Umar bin Khattab [20].

Once there was an earthquake in Medina. The prophet then put his hands on the ground and said, *"Calm down ... the time has not yet come for you"*. Then, Muhammad turned to the companions and said, *"God is rebuke us with this disaster"* [21]. Apparently, Umar bin Khattab recalled the incident. When an earthquake occurred during the Caliphate, he said to the people of Medina, *"O Man, what is this? How fast are you doing (from immorality to Allah)?*. Caliph Umar bin Abdul Aziz also did not remain silent when an earthquake occurred during his leadership. He immediately sent a letter to all the guardians of the country. He writes: *"In fact this earthquake was a warning from Allah to His servants"* [22].

The initial step taken when a natural disaster occurs is to repent while remembering what disobedience is done which caused God to send the disaster to human being [23]. It also serves as the guardian of the consciousness and spiritual condition of the community, especially those in natural disaster-prone areas to always maintain adherence to religious teachings within the scope of individuals and society, because natural disasters possible to come at any time and destroy everyone in the area.

5. Fikih Anticipation as a Model of Mitigation in Islam

Fiqh is Islamic jurisprudence. *Fiqh* is often described as the human understanding of the divine Islamic law as revealed in the Quran and the Sunnah (the teachings and practices of the Islamic prophet Muhammad) [24]. In *fiqh* there is a situation called emergency (*al-dharurah*). If this situation occurs (afflicts someone), then he is given permission to commit prohibited actions that are not normally done in normal circumstances [25]. Emergencies in general are always associated with saving lives. A frequent example, in a state of extreme hunger, which is thought to be life threatening, a person or group of people may eat *haram* food to avoid death.

The books *Qawa'id al-Fiqh al-Kulliyyah* that I read, all of them stated that this great principle (*qaidah kulliah*) was accepted and practiced in all schools of thought in Islamic law (mazhab) [26]. The argument for that is some Qur'anic verses and the Prophetic traditions (*hadits*) that are understood inductively. One of them, which is relatively very popular is hadith *"la dharara wa la dhirara"* (there is no harm/ do not hurt yourself and others) [25]. Islamic scholars, when explaining this principle includes an important requirement, that the prohibited acts in an effort to save themselves or others, should not be an act that would harm others, moreover to the level of endangering the lives of others [27].

An extreme example that is often used, in a ship accident (sinking at sea), passengers may fight over life jacket as long as they have not been controlled by someone. If the life jacket has been controlled by someone, then other people may no longer grab it, because the struggle is strongly suspected to threaten the life of the previous life jacket holder [24]. According to the ulamas (Islamic scholars), the lives of both people (holder and robber of life jacket) at the same level must be protected by *fiqh* [28]. Thus there is no right for the second person to take it from the first person. Thus, there is no right for the second person (robber) to take it from the first person (holder).

Although the ulamas agreed on the basic principle, they were still discussing about emergency category in *fiqh* [29]. The majority of Islamic scholars agreed that an emergency actually encompasses everything that threatens the five main things that must be protected (*al-dharuriyyat al-khamsah*), namely protection of religion, life, reason, descent (self-esteem), and property [30]. But in the descriptions, they seem to limit themselves only to saving lives. Most books do not give examples for the protection of religion, reason, descent (self-esteem) and property [31].

A condition called emergency if truly threatening life, for example being lost in the forest, falling into a cave, displaced on a rural island, severe famine season, suffering from illness (which is strongly suspected will cause death) and so on. This life-threatening situation must be truly real, already in sight, not just a possibility or a guess. If someone is afflicted with a disease that is difficult to heal, but not to the level of life threatening, then the situation is not considered an emergency. This situation is considered only to the level of need (*hajiyyat*) only [32].

The problem that now arises, what if the alleged disaster, loss of life or mental and physical disability that is expected to appear is not really real. For example, today's scientists have been able to estimate areas prone to earthquakes, even though they have not been able to estimate when exactly an earthquake will occur [33]. But they dare to ensure that at one time, in the future, there will be an earthquake that is very likely to be very powerful that will take a lot of casualties. The scholars agreed that the prediction of the impending disaster must be anticipated [34], for example by making houses that are earthquake resistant or not making houses in earthquake-prone areas.

Contemporary Islamic scholars, especially those who have been involved in statistics and other related sciences, will dare to state that future estimates based on honest and correct statistical calculations can be regarded as reality, because according to scientific calculations it is relatively certain to occur [35]. According to scientists the results of statistical calculations that meet the requirements, will be considered as *sunnatullah*, because almost never wrong [29].

If the results of the study of statistics can be accepted as a reality or even *sunnatullah* in predict something that has not happened in the future. Therefore *Fiqh* anticipation as one of the mitigation management in Islam can take the form in the following three steps [20]:

1. Reducing or avoiding the impact of losses from natural disaster hazards.
2. Ensure fast arrival of assistance to victims.
3. Achieve a quick and effective recovery of society.

Management of natural disasters aims to reduce or avoid potential losses from hazards caused by natural disasters, ensure prompt and appropriate assistance to victims of natural disasters, and carry out a fast and effective recovery process [36]. The natural disaster management cycle describes an ongoing process in which religious teachings and civil society plan to reduce the impact of natural disasters react when and after natural disasters, and take steps to recover after natural disasters occur [35]. Appropriate action at all stages in this cycle results in better preparedness, better awareness, and will reduce the level of vulnerability to natural disasters in the period the next repetition of this cycle [20]. A complete natural disaster management cycle includes aspects of public policy preparation and good planning in modifying the causes of natural disasters or reducing the impact of natural disasters on people, property and infrastructure [37].

Mitigation and preparedness phases are carried out to anticipate natural disasters. A sustainable development perspective plays a key role in contributing to community mitigation and preparation to effectively deal with natural disasters [17]. When natural disasters occur, those involved in natural disaster management are involved in the rapid response phase and long-term recovery [13]. The four management phases of natural disasters described here are not always carried out in a fixed order. Often the phase in this cycle overlaps and the duration of each stage depends largely on the severity of the natural disaster [38]:

1. Mitigation. Minimizing the impact of natural disasters. Example: building standards and zoning prone to natural disasters; vulnerability analysis; public education.
2. Preparedness. Planning to respond to natural disasters. Example: preparedness plan; emergency training; prediction and early warning system.
3. Response. Efforts to minimize the hazards created by natural disasters. Example: search and rescue; emergency assistance.
4. Recovery. Normalization of people's lives. Example: temporary housing; grant; medical treatment

6. Conclusion

Islam should take a central role in efforts to avoid the impact of natural disasters or minimize them. Since before natural disasters, during the emergency response period, until the recovery period and life return to normal.

In essence, disaster is something that cannot be separated from life. This view provides direction that disasters must be managed thoroughly both before, during and after a disaster. Therefore, a special management is needed to handle it. At present, in the disaster management, there have been several patterns of shifting views, namely from being responsive to being preventive, from government affairs to community participation, from the responsibilities of several sectors to the responsibilities of various sectors, as well as from the pattern of dealing with impacts to reducing risk.

In the Qur'an chapter 3 verse 200 it is said that the person who believes to always be in a state of alert before a dangerous thing will happen, *"O you who believe, be patient and strengthen your patience and stay prepared and fear Allah so that you are lucky"*. Furthermore, in chapter 6 verse 131 God Says *"That is because your Lord does not destroy cities with persecution, while the population is in a state of neglect"*. The Qur'an advocates for a populated area and has a government to have standby planning which leads to readiness and ability to estimate, reduce impacts, deal effectively and recover from impacts, and if possible prevent disasters themselves.

Chapter 59 verse 18 is an important part of the Qur'an which emphasizes the concept of planning in all fields. Prominent scholars such as Imam Al-Ghazali interpret the verse as an order to humans to improve, increase faith, and devotion to god through a process of life that must not be the same as the previous life. Al-Ghazali also affirmed the word pay attention to where humans must pay attention to every act that has been done, and prepare themselves (plan) to always do the best for tomorrow (Al-Ghazali: 1999).

The concept of planning in chapter 59 verse 18 is the same subject as the guidelines for the preparation of contingency plans issued by the The National Agency for Disaster Countermeasure (BNPB) in 2011. There are five aspects that can be seen from the similarity of the planning concept, namely: First, planning must involves the process of determining the desired future state (impact analysis), second, the desired future situation compared to the present reality so that the gap can be seen (gap analysis). Third, to close the gap needs to be done an effort that can be done with various endeavors and alternatives (emergency scenario). Fourth, it is necessary to choose a good alternative, in this case includes effectiveness and efficiency (allocation of tasks and resources). Fifth, the alternatives that have been chosen should be detailed to be able to provide guidance and guidelines in decision making and policy (synchronization and harmonization). The features of the mitigation concept in fiqh are not limited to prevention and alertness in dealing with natural disasters, but also offers a spiritual solution when the impact of the disaster concerns a community.

References

[1] Adiyoso W, Kanegae H. The 3 rd International Conference on *Sustainable Future for Human Security The preliminary study of the role of islamic teaching in the disaster risk reduction (a qualitative case study of Banda Aceh , Indonesia).* Procedia Environ Sci [Internet]. 2013;17:918–27. Available from: http://dx.doi.org/10.1016/j.proenv.2013.02.110

[2] Fearnley CJ, Bird DK, Mcguire WJ. *Advances in Volcanology Observing the Volcano World* [Internet]. Fearnley CJ, editor. Switzerland: Springer Open; 2018. 754 p. Available from: https://link.springer.com/content/pdf/10.1007%2F978-3-319-44097-2.pdf

[3] Mclaughlin L. *Hard Lessons Learned.* Asian Ethnol [Internet]. 2016;75(1):105–37. Available from: http://asianethnology.org/downloads/ae/pdf/AE 75-1 McLaughlin.pdf

[4] Adisaputri G. *The role of religious beliefs and practices in disaster : The case study of 2009 earthquake in Padang City , Indonesia The role of religious beliefs and practices in disaster : The case study of 2009 earthquake in Padang City , Indonesia* [Internet]. Auckland University of Technology; 2016. Available from: https://core.ac.uk/download/pdf/80334097.pdf

[5] Gignoux J, Menéndez M. *Benefit in the wake of disaster: Long-run effects of earthquakes on welfare in rural Indonesia.* J Dev Econ [Internet]. 2015;118(January):1–58. Available from: http://dx.doi.org/10.1016/j.jdeveco.2015.08.004

[6] Tzeferakos GA, Douzenis AI. *Islam, mental health and law: A general overview. Ann Gen Psychiatry* [Internet]. 2017;16(1):1–6. Available from: https://annals-general-psychiatry.biomedcentral.com/articles/10.1186/s12991-017-0150-6

[7] Mor NS, Dardeck KL. *Mitigation of Posttraumatic Stress Symptoms From Chronic Terror Attacks on Southern Israel.* J Soc Behav Heal Sci [Internet]. 2017;11(1):47–64. Available from: http://scholarworks.waldenu.edu/cgi/viewcontent.cgi?article=1232&context=jsbhs

[8] Al-amer R, Ramjan L, Glew P, Darwish M. *Language translation challenges with Arabic speakers participating in qualitative research studies.* Int J Nurs Stud [Internet]. 2016;54(February):150–7. Available from: http://dx.doi.org/10.1016/j.ijnurstu.2015.04.010

[9] Iskandar M. *Dayah Darussalam Network and Dayah Awakening in Aceh.* Budapest Int Res Critics Inst [Internet]. 2018;I(3):13–22. Available from: http://bircu-journal.com/index.php/birci/article/view/27

[10] Iskandar M. *Sarf al-Zakat fi al-Masalih al-Amat ('Iadat al-Fikr 'an Sarf Zakat bi Siyaqi al-Ma'ashirat.* Media Syari'ah [Internet]. 2014;XVI(1):25–48. Available from: http://jurnal.ar-raniry.ac.id/index.php/medsyar/article/view/1798

[11] Ghofur A, Susilo S. *Maslaha as the philosophical, political, and legal basis on the Islamic banking legislation in Indonesia.* Glob J Al-Thaqafah [Internet]. 2017;7(1):7–17. Available from: https://doi.org/10.1186/s12991-017-0150-6

[12] Ganguli AC, Harmon JP. *An Introduction to the Special Issue on Pollinators.* J Psychol Christ [Internet]. 2018;33(3):3. Available from: http://linkinghub.elsevier.com/retrieve/pii/S0190052811500276

[13] Iskandar M. *Nalar Konstitusi Dalam Wacana Reformulasi GBHN.* J Justisia [Internet]. 2016;99:391. Available from: http://jurnal.ar-raniry.ac.id/index.php/Justisia/article/view/2561

[14] Science P, Makassar UB. *FLOOD DISASTER , LOCAL BELIEF AND ISLAM-SUFISM.* el Harakah [Internet]. 2017;1:41–53. Available from:

https://search.proquest.com/openview/dbf9dc72ec5c69e42f467ff2ee9c8493/1?pq-origsite=gscholar&cbl=2049068
[15] Zakavi AA, Marzband R. *The Relationship between Human Performance and Natural Disasters Based on Islamic Teachings*. J Islam Stud Cult [Internet]. 2016;4(1):103–14. Available from: https://doi.org/10.15640/jisc.v4n1a13
[16] Azhari Ramli, Muhammad Zen, Munauwar Mustafa RZY. *Strategic Planning and Organizational Performance in the Public Sector, A Study of Local Goverment in Aceh, Indonesia*.pdf. Medwell Journals. 2017;12(6):902–6.
[17] McGeehan KM. *Religious Narratives and Their Implications for Coping, Recovery and Disaster Risk Reduction*. Disasters [Internet]. 2017;41(2):24. Available from: http://asianethnology.org/downloads/ae/pdf/AE 75-1 McLaughlin.pdf
[18] Marzband R, Hosein S, Karnami H, Republic I. *The orodental healthcare based on Islamic Traditions (Hadith) Assistant Professor and Scientific Mission Member of Islamic Thought* , at Department of. Int J Med Res Heal Sci [Internet]. 2016;5(2):36–41. Available from: https://www.ijmrhs.com/medical-research/the-orodental-healthcare-based-on-islamic-traditions-hadith.pdf
[19] Road M. *Spirituality in Resilience Processes in International Contexts : An Introduction*. J Psychol Theol [Internet]. 2016;44(2):107–8. Available from: https://journals.sagepub.com/doi/abs/10.1177/009164711604400201?journalCode=ptja
[20] Sonja A. Al Amry. *Disaster Preparedness in Islam* [Internet]. Second. Islamabad: Evaluation Report Cyclone YEMYIN and Floods Respnse; 2008. 1-18 p. Available from: http://www.ndma.gov.pk/Publications/Disaster Preparedness in Islma.pdf
[21] Burge SR. Myth, *Meaning and the Order of Words: Reading Hadith Collections with Northrop Frye and the Development of Compilation Criticism*. Islam Christ Relations. 2016;27(2):213–28.
[22] Schlehe J. *Anthropology of religion: Disasters and the representations of tradition and modernity. Religion* [Internet]. 2010;40(2):112–20. Available from: http://dx.doi.org/10.1016/j.religion.2009.12.004
[23] Hidayat A, Priatna H. *Environmental Management of the Islamic Perspective*. Atantis Press [Internet]. 2018;261(Icie):116–8. Available from: http://jfatwa.usim.edu.my/index.php/jfatwa/article/view/136/103
[24] Iskandar M. *HAM Dalam Prespektif Islam*. Media Syari'ah [Internet]. 2017;19(1):111–26. Available from: http://jurnal.ar-raniry.ac.id/index.php/medsyar/article/view/2017
[25] Siti Fariza M.F, Hasanah A.K, Irwan M.S AA. *Aplikasi Kaedah Fiqh Al-Darurag Tuqaddar Biqadariha Dalam Agihan Zakat Untuk Mangsa Banjir*. J Fatwa Manag Res [Internet]. 2018;8(79):79–90. Available from: http://hrmars.com/hrmars_papers/Conventional_Disaster_Management.pdf
[26] Thalib P. *Application of Qowaid Fiqhiyyah in contemporary Islamic Law*. Yuridika [Internet]. 2016;31(1):108–20. Available from: http://e-journal.unair.ac.id/index.php/YDK/article/view/1958
[27] Bachri AS. *Pengaruh al-Qawaid al-Usuliyyah dan Fiqhiyyah Terhadap Perbedaan Pendapat Dalam Fikih*. Et-Tijarie [Internet]. 2018;5(2):66–89. Available from: http://journal.trunojoyo.ac.id/ettijarie/article/view/4586/3197
[28] Mueller DM. *Sharia Law and the Politics of " Faith Control " in Brunei Darussalam Dynamics of Socio-Legal Change in a Southeast Asian Sultanate*. Int Asianforum [Internet]. 2015;46(3–4):313–45. Available from: https://crossasia-journals.ub.uni-heidelberg.de/index.php/iaf/article/view/3728/3830
[29] Azhar Rosly S. *Shariah parameters reconsidered*. Int J Islam Middle East Financ Manag. 2010;
[30] Opwis F. *New Trends in Islamic Legal Theory : Maqāṣid al-Sharīʿa as a New Source of Law* ? K

Brill Nv [Internet]. 2017;57:7–32. Available from: https://brill.com/view/journals/wdi/57/1/article-p7_3.xml
[31] Hakim ML. *Pergeseran Paradigma Maqasid Al Syari'ah Dari Klasik Sampai Kotemporer*. Al-Manahij [Internet]. 2016;10(1):1–16. Available from: http://ejournal.iainpurwokerto.ac.id/index.php/almanahij/article/view/913
[32] Rahimin Affandi Abdul Rahim, Paizah Ismail, Idris Awang. *Pendekatan Baru Maqasid al-Shariah Dalam Pengajian Syariah di Malaysia*: Satu Analisis. J Fiqh. 2006;
[33] Entwistle DN, Moroney SK, Aten J. *Integrative Reflections on Disasters, Suffering, and the Practice of Spiritual and Emotional Care*. J Psychol Theol [Internet]. 2018;46(1):67–81. Available from: https://journals.sagepub.com/doi/abs/10.1177/0091647117750658
[34] Iskandar M. *Mitigasi dalam prespektif islam*. Prosiding, Simp Nas Mitigasi Bencana Tsunami [Internet]. 2015;1(3):21–2. Available from: http://tdmrc.unsyiah.ac.id/id/wp-content/uploads/2016/02/20.-MITIGASI-DALAM-PRESPEKTIF-ISLAM.pdf
[35] Gaillard JC, Texier P. *Religions, natural hazards, and disasters: An introduction. Religion* [Internet]. 2010;40(2):81–4. Available from: http://dx.doi.org/10.1016/j.religion.2009.12.001
[36] Sun L, Deng Y, Qi W. *Two impact pathways from religious belief to public disaster response : Findings from a literature review*. Int J Disaster Risk Reduct [Internet]. 2018;27(August 2017):588–95. Available from: https://doi.org/10.1016/j.ijdrr.2017.10.004
[37] Wilkinson O. *'Faith can come in, but not religion': secularity and its effects on the disaster response to Typhoon Haiyan*. Disasters [Internet]. 2018;42(3):459–74. Available from: https://onlinelibrary.wiley.com/doi/abs/10.1111/disa.12258
[38] Guarnacci U. Joining the dots: *Social networks and community resilience in post-conflict, post-disaster Indonesia*. Int J Disaster Risk Reduct [Internet]. 2016;16:180–91. Available from: http://dx.doi.org/10.1016/j.ijdrr.2016.03.001.

Spatial analysis of post tsunami 2004 coastline changes in Lampuuk, Aceh Besar

Iswana Adela[1], Gartika Setya Nugraha[1], Muhammad Irham[2*], Saumi Syahreza[3,4]

[1]Dept. of Geological Engineering, Faculty of Engineering, Syiah Kuala University, Banda Aceh Indonesia 23111
[2]Faculty of Marine and Fisheries, Syiah Kuala University, Banda Aceh Indonesia 23111
[3]Physics Department of Mathematics and Science Faculty of Syiah Kuala University, Banda Aceh - Indonesia 23111
[4]Tsunami and Disaster Mitigation Research Center (TDMRC) of Syiah Kuala University, Banda Aceh - Indonesia

[*]Corresponding author: irham@unsyiah.ac.id

Abstract. Based on the geological records, the tsunami in 2004 has resulted in coastline shifts along the west coast of Aceh. One of the beaches in the western region that undergoes a major coastline change is Lampuuk beach. This study aims to determine the shoreline changes occurring at Lampuuk beach. The study of coastline changes was done by utilizing remote sensing using satellite imagery and geographic information system. Geographic information systems were used to analyze spatial data obtained from multitemporal satellite imagery before and after the tsunami disaster. The results show that coastline changes occur very significantly during the 2004 tsunami and persisted until 2005, but after 2010 the coastline was back to normal. During 2001–2005 there was a reduction in coastal plains caused by the 2004 tsunami waves. In 2005-2018 the sedimentation process at Lampuuk Beach was more dominant, assisting to adjust the shape of the coastal profile as before the 2004 tsunami. The results also show that the Lampuuk area was more sedimentary than erosion.

1. Introduction

Aceh is an area located at the end of the Sumatra fault known as the great Sumatra Fault (fault of Semangko) and vulnerable to the threat of geological disaster [1], one of which is the tsunami disaster. The highly vulnerable area of the tsunami lies along the West Coast of Aceh bordering the Indian Ocean [2]. Based on the tsunami geological records that occurred in 2004 has resulted in coastline shifts along the west coast of Aceh. One of the beaches in the western region that changed the shoreline is Lampuuk beach. This beach is located in Meunasah Mosque Village, Lho'nga District, Aceh Besar District.

The shoreline change is a process that occurs through various processes and occurs continuously [3]. Shoreline changes can be affected by coastal abrasion and accretion [4-5]. Coastal abrasion is a condition in which the coastline progresses towards the land while coastal accretion is a condition where the coastline is increasingly retreating towards the ocean [6]. the depression and abrasion that occur on the coast can be affected by ocean waves, sediment deposition and transportation [7].

Shoreline changes can also be caused by tsunami waves. The tsunami that occurred in 2004 was a tsunami that was very severe and very influential to the changes in coastline that occurred along the West Coast of Aceh. Tsunami waves are waves with very large energy and the waveform extends from the sea floor to the surface so that the tsunami waves produce enormous energy that can cause shoreline changes [2].

Not only by the tsunami, the shoreline changes are also caused by the dynamics of the sea such as waves, tides, currents and geological conditions (geomorphology) and geographical location [8-9]. The process of marine dynamics results in coastal abrasion and accretion while the geological conditions that cause coastline changes are coastal sedimentation processes [10-11]. Sediments are the main ingredients for topographic and coastal bathymetry so that the mechanism of erosion, sediment transport and sedimentation leads to shoreline changes [12]. Therefore, the shoreline changes that occur at the beach of the lights need to be studied because the beach is a tourist beach so that spatial analysis is needed to determine the current status of the coastline. The purpose of this research is to study how the condition of the shoreline changes that occur before and after the tsunami in Lampuuk Beach.

2. The methods of the research

Data obtained directly from the research location was processed using ArcMap. Spatial data analysis of shoreline changes was done in several stages: image collecting, georeferencing, on-screen digitization and image overlay. The imagery to be used in this study was a multitemporal satellite imagery obtained free of charge from the Google Earth Pro software. By using ArcMap shape file, georeferenced was made to adjust the coordinates of the field where the data was taken. While on-screen digitization was done to differentiate the shoreline on each image so that the shoreline changes can be seen. Finally, the overlay was made to see the shoreline difference in each image by combining the image data recorded in different years.

The image used in this research was satellite imagery obtained for free from the Google Earth Pro software. Satellite imagery used is satellite imagery in 2001, 2005, 2010, 2011, 2013, 2015, 2017 and 2018. Geometric correction was done using ArcMap by activating georeferencing tools. Georeferencing was carried out by providing a coordinate system at several reference points at the research location. The provision of coordinate systems in the image according to the coordinate system used was 1984 WGS datum zone 46 N.

3. The Results and discussions

Changes in the coastline that occurred due to the influence of the tsunami waves in 2004 can be seen in Figure 1. In 2001 the coastline was closer to the shore while in 2005 there was an abrasion because of the influence of the 2004 tsunami wave so that the coastline progressed towards land. Abrasion caused by the 2004 tsunami caused the land area to decrease by 118668.4 m^2. The reduction in land area caused by the tsunami wave was very large because Lampuuk Beach had a very gentle slope (0-2%) so it was unable to withstand tsunami waves and triggered a large abrasion. Very gentle coastal slope conditions (0-2%) can cause high waves when tsunami waves approach the coastline. High tsunami waves are caused by wavelengths that are shortened due to the effect of reduced wave velocities as they approach sloping coastlines [13].

Table 1 shows that during the year of 2005 - 2010 there was an abrasion of 4681.92 m^2 and accretion of 44779.36 m^2 causing the addition of coastal area of 40097.44 m^2. It can be seen that the impact of coastline changes by the 2004 tsunami disaster at Lampuuk beach only had a temporary (non-permanent) impact. This evident can be seen from the analysis of shoreline changes in 2010 where the coast began to reside and rebalance its position as it was before the 2004 tsunami. Nevertheless, some changes can not be resumed as originally caused by marine dynamics and coastal topography [14].

IOP Conf. Series: Earth and Environmental Science **273** (2019) 012046 doi:10.1088/1755-1315/273/1/012046

Table 1. Abrasion and accretion at Lampuuk coast in 2005 to 2018

	Year							Average/ years (m²)
	2005 - 2010	2010 - 2011	2011 - 2013	2013 - 2015	2015 - 2017	2017 - 2018	2005- 2018	
Abrasion (m^2)	4681.92	362.12	2440.59	6236.05	2764.37	1836.79	-	2617.40
Accresion (m^2)	44779.36	14584.84	5040.83	2588.94	11655.19	4812.71	64968.47	21204.33

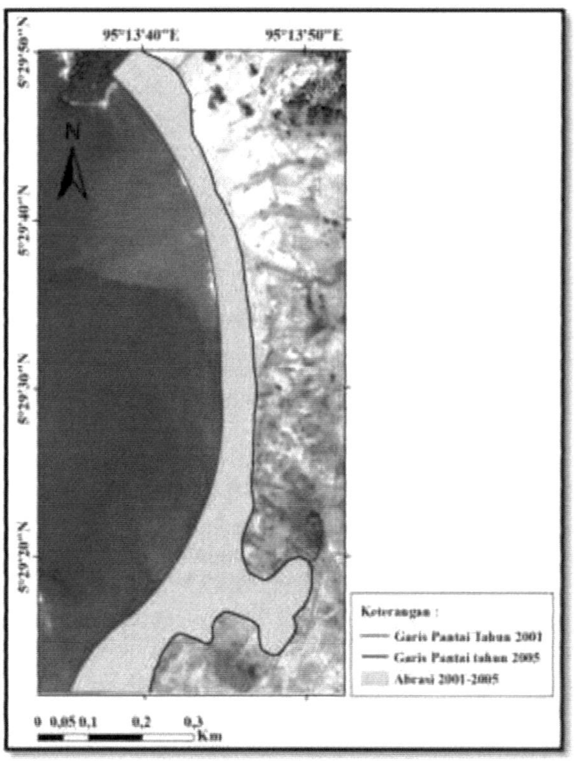

Figure 1. The spatial condition of coastline changes from 2001 - 2005.

The influence of the large waves caused by the 2004 tsunami has caused the coastal plains to experience a large erosion, which has changed the shape of the coastal profile. The influence of the normal wave coming and the process of deposition of coastal sediments after the tsunami can help adjust the shape of the beach profile back to its original shape. Sediments deposited not too far from the beach because of the current carried by the tsunami wave can be transported back to shore by the influence of normal waves so as to help adjust the shape of the beach profile so that it can return to its original profile [15].

Adding sediment to the coastal plains causes the coastline to experience more sea-going swings. Sediments transported to the coastal plain are carried by currents and waves so that marine sediments are transported and deposited along the coastline. Sediments transported to Lampuuk Beach can also come from the Lampuuk Beach cliff which is continuously eroded and eroded by ocean waves (Figure 2). The cliff at the end of Lampuuk Beach consists of a unit of carbonate limestone with mineral calcite making it can be eroded by water. The presence of chemical reactions from seawater and rainwater to carbonate rocks causes a continuous erosion of cliffs and causes the addition of sediment on the coastal plain [16].

Figure 2. Lampuuk Beach cliffs are eroded by ocean waves.

Figure 3 shows the state of Lampuuk shoreline in 2005 (after tsunami affected) and its change in 2010. By looking at the comparison of changes occurring during 2005 to 2010, it can be concluded that coastal erosion caused by tsunamis in the eastern region of eastern Lapuuk beach for five years has been re-sedimented so that the changes change from erosion to sedimentation.

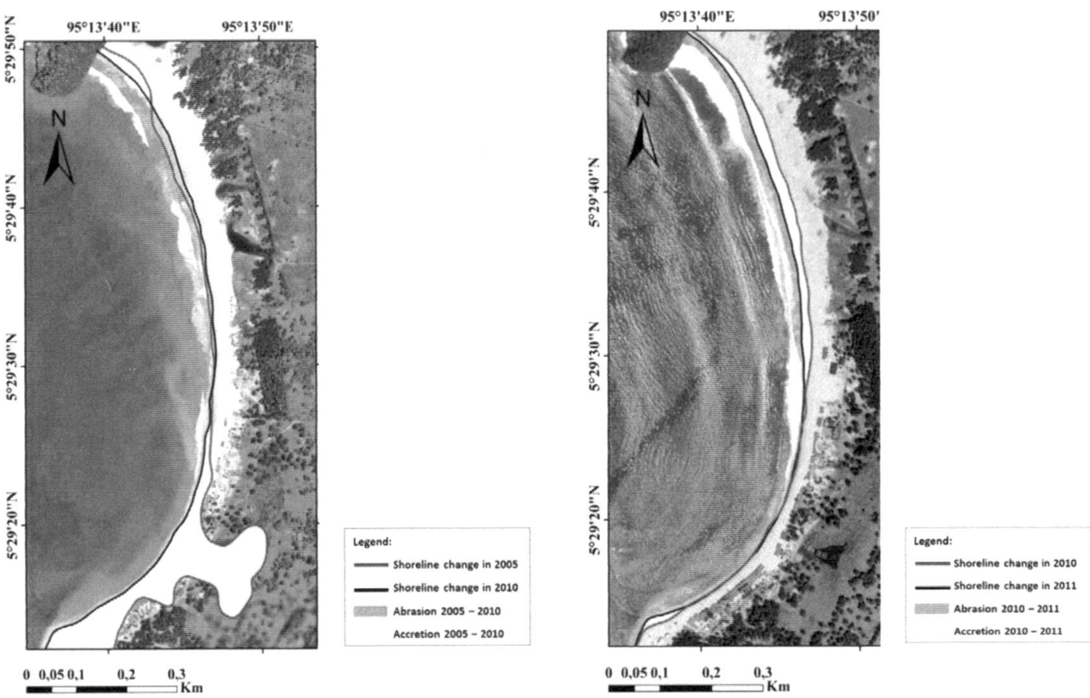

Figure 3. Shoreline change in 2005 after the tsunami 2004 compared to the condition in 2010.

Table 2 shows the addition and reduction of the Lampuuk coastline. The average abrasion / year that occurred during the last 14 years at Lampuuk beach is 2617.41 m2 and the average accretion that happened annually in Lampuuk beach is 21204.34 m2. This indicates that the addition of coastal areas that occur in Lampuuk beach is greater than the reduction of coastal areas that occur. The addition of coastal areas due to the sedimentation process causes the shoreline shifted towards the sea.

Table 2. addition and reduction of the Lampuuk coastline area

Year	addition and reduction (m^2)	Dominant change
2005 - 2010	+40097.44	Sedimentation
2010 - 2011	+14222.72	Sedimentation
2011 - 2013	+2600.24	Sedimentation
2013 - 2015	-3647.11	Erosion
2015 - 2017	+8890.81	Sedimentation
2017 - 2018	+2975.91	Sedimentation
2005-2018	+64968.47	Sedimentation

The shoreline changes that occurred at Lampuuk beach were enormous during the 2004 tsunami and persisted until 2005. These changes are gradually recovered by the presence of sedimentation. Until 2010, the shoreline changes no longer significantly changed due to Lampuu beach topography and sea dynamics processes. The dominance of coastline changes in Lampuuk is more sedimentation than erosion.

4. Conclusion

The shoreline changes that occurred at Lampuuk beach were enormous during the 2004 tsunami and persisted until 2005. These changes are gradually recovered by the presence of sedimentation. Until 2010, the shoreline changes no longer significantly changed due to Lampuu beach topography and sea dynamics processes. The dominance of coastline changes in Lampuuk is more sedimentation than erosion.

References

[1] Natawidjaja, D. H., Bradley, K., Daryono, M. R., Aribowo, S. and Herrin, J. 2017. Late Quaternary eruption of the Ranau Caldera and new geological slip rates of the Sumatran Fault Zone in Southern Sumatra, Indonesia. Geosci. Lett. (2017) 4:21

[2] Subandono,D.B. 2005. Tsunami. *Buku Ilmiah Populer* (Popular book of science). Bogor.

[3] Irham, I., Fadla, Y., and Setiawan, I. (2018). The spatial distribution of suspended sediment analysis along Krueng Cut River, Banda Aceh. IOP Conference Series: Earth and Environmental Science 106 (1), 1 – 6.

[4] Irham, M. and Setiawan, I. (2018). The numerical model of the sediment distribution patern at Lampulo National Fisheries Port. IOP Conference Series: Earth and Environmental Science 106 (1), 1 – 6.

[5] Pethick, 1997, An Introduction to Coastal Geomorphology. Edward Arnold a Division of Holder and Stougthon, London.

[6] Irham, M. and Setiawan, I. 2017. The study of flow resulting from wave on lhonga beach, Aceh Besar, OmniAkuatika, 13(1), 5-12, 1858-3873.

[7] Irham, M., Miswar, E., Ilhamsyah, Y., and Setiawan, I. (2018). The nortern tidal dynamic of Aceh waters: A 3D numerical model. IOP Conference Series: Material Science and Engineering (MSE) 352 (1), 1 – 6.

[8] Setiawan, I. and Irham, M. 2018. Wave Trajectory Study on the Coast of Lhoknga, Aceh Besar, Indonesia: A Numerical Model Approach. Civil Engineering Dimention. 20(1), 30 -34.

[9] Setiawan, I., Rizal, S., Haditiar, Y., Purnawan, S., Yuni, SM., and Irham, M. (2018). Study of current circulation in the Northern Waters of Aceh. IOP Conference Series: Earth and Environmental Science 176 (1), 1 – 6.

[10] Eliot, M., 2016: Coastal sediments, beaches and other soft shores. CoastAdapt Information Manual 8, National Climate Change Adaptation Research Facility, Gold Coast.

[11] van Lanckera, V., Lanckneusa, J., Hearnc, S., Hoekstrad, P., Levoye, F., Milesf, J., Moerkerkea, G., Monforte, O. and Whitehousec, R. 2004. Coastal and nearshore morphology, bedforms and sediment transport pathways at Teignmouth (UK). Continental Shelf Research 24 (2004) 1171–1202.

[12] Irham et al, 2017. The Spatial Distribution of Bed Sediments at Krueng Cut Estuary: The Sieve Analysis Approach

[13] Zhao, X., Chen, Y., Huang, Z., Hu, Z. and Gao, Y. 2017. A numerical study of tsunami wave impact and run-up on coastal cliffs using a CIP-based model. Nat. Hazards Earth Syst. Sci., 17, 641–655.

[14] Narayan, P., Sharma, M. and Maheshwari, B. K. 2005. Effects of medu and coastal topography on the damage pattern during the recent Indian Ocean tsunami along the coast of Tamil Nadu. Science of Tsunami Hazards 23(2), 1 – 9.

[15] Keevil, G. M., Peakall, J. and Best, J. L. 2008. The formation and dynamics of sediment waves. Marine and River Dune Dynamics. 179 – 183.

[16] Masselink, G., Earlie, C. and Russel, P. 2017. The role of beach morphology on coastal cliff erosion under extreme waves.: Coastal cliff erosion: beach morphology and extreme waves. Earth Surface Processes and Landforms. 43, 1213 – 1228.

Acknowledgments

We thank you very much to Syiah Kuala University through the facility support at Geological Engineering Laboratory during the research.

IOP Conf. Series: Earth and Environmental Science **273** (2019) 012047 doi:10.1088/1755-1315/273/1/012047

Drinking Water Bags Based on Chitosan Forward Osmosis Membranes for Emergency Drinking Water Supply

Saiful[1,*], Ulfa Riana[1], Marlina[1], Muliadi Ramli[1], and Nizar Mahmud[2]

[1]Department of Chemistry, Faculty of Natural Science, Syiah Kuala University, Jalan T. Tanoh Abbe No. 3, Darussalam, Banda Aceh 23111, Indonesia
[2]Department of Mechanical Engineering, Faculty of Engineering, Syiah Kuala University

[*]Corresponding author: saiful@unsyiah.ac.id

Abstract. A fast and simple method and technologies in drinking water supply are strongly needed in an emergency situation. One alternative method is based on membrane forwards osmosis (FO) technology. In this study, chitosan-based FO membranes have been developed into drinking water bags used as purification media for dirty water and seawater. The chitosan membrane used for the assembly of drinking water bags has a thickness of 0.043 mm, the porosity of 30.3%, the tensile strength of 28.83 kgf /mm^2, swelling degree of 43.5% and elongation of 7.16%. The drinking water bags are made from a combination of polypropylene plastic (PP) and aluminum foil plastic with the interface of the FO membrane inside. The drinking water bag can be applied to purify dirty water and seawater into energy drinking water, which can be used for drinking water supply in an emergency situation. Energy drinks water is produced from the FO process using a variety of draw solution, specifically glucose, fructose, and sucrose. The highest drinking water flux was obtained by using 3M sucrose concentration as a draw solution. The clean water fluxes for dirty water and seawater samples were 5.25 L/m^{2}hour and 4.25 L/m^{2}hour respectively. The parameters drinking water quality test are proved that pH, total dissolved solids (TDS), salinity, electrical conductivity, heavy metals and the content of Escherichia coli bacteria are agreed with drinking water quality standards based on PERMENKES regulations No. 492/MENKES/Per/IV/2010. The FO drinking water bag based chitosan membrane has the potential to be used as an alternative solution for energy drinking water supply in an emergency.

Keywords: drinking water, drinking water bag, chitosan membrane, forward osmosis membrane, emergencies.

1. Introduction

The clean water crisis, especially drinking water in various parts of the world, has attracted the interest of researchers to conduct studies and develop methods to obtain clean water and drinking water from various types of available water sources. Community access to clean water, especially drinking water is minimal both to meet local and global needs. The availability of surface water and soil that is abundant cannot be used directly as clean water or drinking water [1]. Water in nature that can be used directly for drinking water is less than 1%, this is because most of the available water is in the form of seawater (97%), glacier (2%), river and brackish which cannot be directly consumed as a quality drinking water [2]. Besides, polluted water sources that are not suitable for consumption are also the reason for the lack of availability of clean water that can be directly drunk, while the need for water continues to increase to meet human consumption [3]. The unavailability of drinking water can cause dehydration, and metabolic processes are disrupted, attacked by various diseases and even lead to death. Drinking water is a critical basic need and must be available even in emergencies situation [4].

Emergencies are defined as conditions that threaten and disrupt people's lives and livelihoods caused by natural or human factors to create emergency conditions in almost every aspect of life.

Indonesia is a country that is very vulnerable to emergencies. Geographically, Indonesia is a country with a high potential for disaster (hazard potential). The movement of the Eurasian, Australian and Pacific Ocean basins has caused earthquakes, tsunamis, active volcanoes, landslides, floods and others [5,6]. Aftershock requires readiness and preparedness in dealing with any disasters and handling during emergencies, one of them is the availability and quality of clean water, especially drinking water. Drinking water becomes difficult to obtain because of damage and contamination of clean water sources in disaster areas and the difficulty of access to isolated disaster locations. Therefore, the development of methods and technology for drinking water supply in an emergency situation that is easier, faster and simpler is needed. Innovation and technology development in the provision of clean water, especially drinking water, still needs to be developed further, especially to respond to emergencies that are very likely to occur, for example in disaster areas, water crisis areas, sea trips, backpackers and other emergencies. One of them is by procuring drinking water bags based on membrane technology through forward osmosis (FO) membrane process [7].

FO Membrane is an alternative method that is being developed to improve the efficiency of polluted water purification processes, brackish water, and seawater into clean water, especially drinking water. In the FO membrane method, water is purified by moving from a low concentration solution to a solution with high concentration through a semipermeable membrane due to an osmosis gradient [8]. The advantage of this method is that it does not require external pressure (hydraulic), the possibility of fouling on small membranes and the possibility of low contamination and does not require electrical energy in its application so that it is simpler, cheaper and practical to use [7]. In this study, a chitosan-based FO membrane has been developed as a filter bag for drinking water in an emergency. Saiful *et.al* [9] has reported that chitosan membrane is one of FO membranes that can be applied to brackish filtration into energy drinking water. The quality of the energy of drinking water product has met the drinking water criteria according to the Indonesian Government's drinking water regulations. In this research, further development of drinking water bag membrane FO made from chitosan which has been carried out for purification of dirty water and seawater. The FO testing process is carried out using a variety of sugar solutions as a draw solution, namely glucose, sucrose, and fructose. Energy water products from the FO process are analyzed for quality as drinking water related to the main parameters of drinking water quality, namely heavy metal, pH, salinity, TDS, electrical conductivity and E. coli bacteria content.

2. Material and Method

The equipment used in this study were a magnetic stirrer, oven, beaker, measuring cup, Erlenmeyer, ceramic, duct tape, glue, scissors and sealing machine. The chemicals used are chitosan, sodium hydroxide, acetic acid, dimethylformamide, distilled water, glucose, fructose and sucrose and Polypropylene (PP) plastic and aluminum foil plastic.

2.1. Membrane preparation

Dissolved 3 grams of chitosan with 100 mL of 1% acetic acid (v / v), added 100 mL of 10% (v / v) dimethylformamide solution as an additive. Erlenmeyer is closed and stirred using a magnetic stirrer for ± 24 hours at room temperature; a solution of the chitosan polymer is obtained. Then the solution is printed as a thin membrane on a ceramic at a thickness of 50 μm. The solvent is allowed to evaporate using an oven at a temperature of ± 30 ° C until the membrane dries and detaches from the ceramic. After that the membrane was washed with 1% NaOH (b/v), rinsed with distilled water until neutral and dried at room temperature. The chitosan membrane obtained was used for FO testing (Saiful et al., 2018[b]).

2.2. Drinking water bags preparation

Drinking water bags are made using PP plastic and aluminum foil plastic. The two plastics are cut in a rectangular shape with the same size, then between the two are placed the chitosan membrane that has been made beforehand to form 3 layers. Next, the three plastics are joined together through lamination on all the edges of the plastic using a sealing machine to form a bag of drinking water and installed a lid on the front and back side.

2.3. Draw solution preparation

The draw solutions used in the study were glucose, fructose, and sucrose with concentration 3M dissolved by using distilled water. The draw solutions were fill in into the water bags for testing FO processes.

2.4. Use of drinking water bags

The drinking water bags were made as reported in the previous study by Saiful *et al.* (2018[a]). The drinking water bags filled 100 mL of draw solution via the front lid and then feed solution of dirty water or seawater filled 200 mL from the back lid. During the FO process, the draw solution will attract the water in the feed solution through a semipermeable FO membrane. Therefore most of the water will move from the feed solution to the draw solution. The FO process is carried out for 1 hour.

2.5. FO water product analysis

Water quality FO products were analyzed by measuring several water quality parameters, specifically pH, TDS, conductivity, salinity, heavy metal and *E.coli* bacteria. pH was analyzed by using pH meter and salinity meter was used for salinity analysis. TDS and conductivity were analyzed by using conductivity meter, therefore analysis of heavy metals including Hg, As, Mn, Zn, Cu, Cr, Fe, and Cd by using Atomic Absorption Spectrophotometry (AAS), and analysis of E.coli content using the MPN (Most Probable Number) method.

3. Result And Discussion

3.1. Membrane preparation and characterization

The process of making FO membrane was carried out by dissolving chitosan in 1% (v / v) acetic acid and adding 10% DMF. The addition of DMF can increase porosity because DMF is a porogen additive compound, which is able to create greater space between polymer chains, causing more membrane pores to form and can facilitate more water to move through the membrane (Saiful *et al.*, 2017).

Chitosan membrane is physically transparent, thin and strong with a tight and porous structure [9]. The FO chitosan membrane has a thickness of 0.043 mm, the porosity of 30.3%, swelling degree of 43.5%. Mechanical characteristic FO membrane demonstrated the tensile strength of 28.83 kgf / mm2 and percent elongation of 7.16%. SEM picture in Figure 1 shows that the membrane has an asymmetry structure, where the structure and pore sizes of the membrane are not uniform. Morphology shows that there are differences in structure between the top and bottom surfaces. The membrane morphology on the top surface is tight and flat, while the lower surface is rough. Most of all the forward osmosis membrane has an asymmetric structure [10]. In the cross-section of the membrane illustrated that the membrane has a well-connected structure and is free of macrovoid. In this study, the desired membrane for FO testing application is a tight and porous membrane structure.

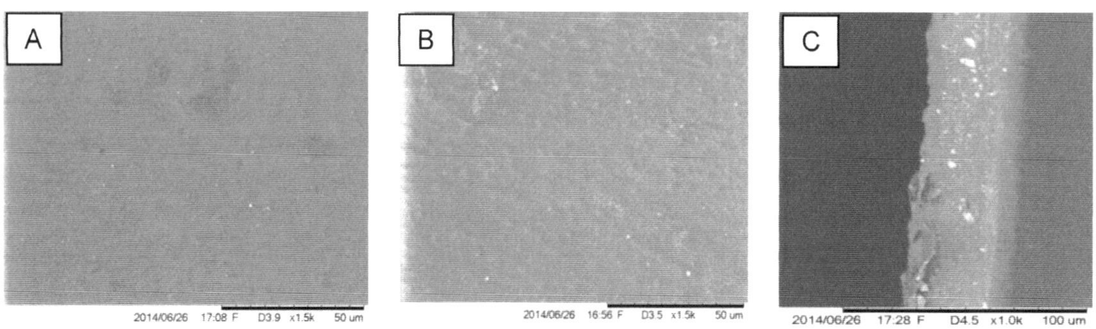

Figure 1. Morphology of 10% chitosan membrane DMF characterized by SEM with 2500x magnification (a) top surface (b) bottom surface (c) cross section.

3.2. Use of drinking water bags

The drinking water bag produced has a total volume capacity of ± 400 mL, which consists of two sides with a volume of ± 200 mL each, which is separated by an FO chitosan membrane. The front part is used to introduce a draw solution in the form of glucose, fructose or sucrose. Then the back part of the water bag is filled in the feed solutions specifically dirty water or seawater. The shape of the drinking water bag is shown in Figure 2.

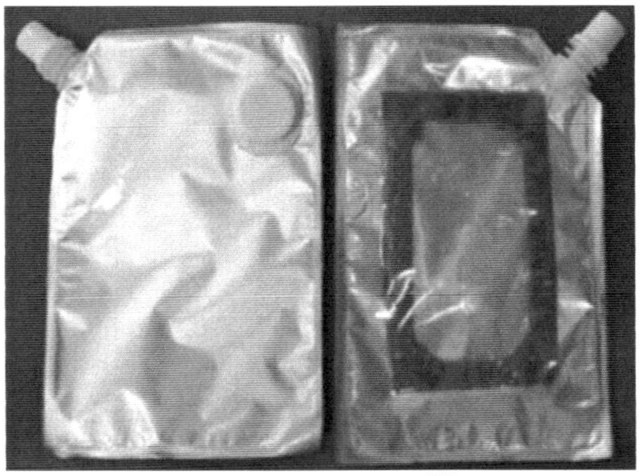

(a) (b)

Figure 2. The shape of the drinking water bag (a) front looks and (b) back looks

The use of drinking water bags for the FO process was carried out by inserting 200 mL of the sample as a feed solution through the back cover of the drinking water bag. Then added 100 mL of a draw solution specifically of sucrose, fructose, or sucrose with a concentration of 3M from the lid of the front of the bag. The FO process was performed for 1 hour with an effective surface area of 41.25 cm^2 membrane. Within 1 hour FO processes the volume of permeate product is close to the maximum volume of the water bag capacity. Besides, based on previous research, the first hour of the FO process is the optimum time with the highest flux value, while after that the flux starts to decrease to a constant [9]. The amount of water flux-product for each draw solution is shown in Figure 3.

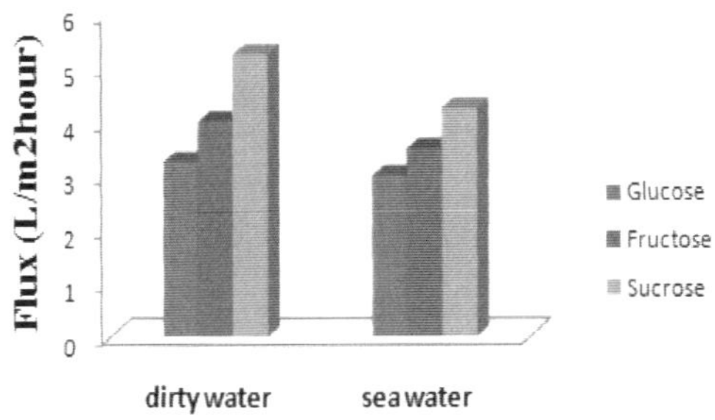

Figure 3. The clean water fluxes of FO chitosan membrane within 1-hour process for dirty water and seawater as feed solution

Based on figure 3, it can be seen that sucrose as draw solution was obtained as the highest clean water fluxes in both seawater and dirty water. The maximum clean water flux generated from 3M sucrose as draw solution for dirty water samples is 5.25 L/m^2h, while for seawater the resulting flux value is 4.25 L/m^2h. Sucrose produces a relatively higher clean water flux because it has a higher osmotic pressure and solubility compared to fructose and glucose, allowing the increased volume of water moving. Dirty water samples produce higher flux than seawater due to the total dissolved solutes in dirty water is lower than seawater, to facilitate more water and free to move towards the draw solution.

3.3. Water quality analysis of FO products

3.3.1. pH, salinity, conductivity and TDS analysis.

The feed solution used is dirty water taken in the Limpok, Aceh Besar village and seawater from the Alue Naga area, Aceh Besar. The water quality of the FO product was measured after 1 hour of the FO processes. Based on Figure 4. Illustrated that the results of pH analysis for all draw solutions were not a significant change in pH towards the draw solution. The pH of the pH of the water obtained from ranged from 7.46 to 7.72. This value is still in the range of pH requirements for drinking water according to PERMENKES No: 492/Menkes/Per/IV/2010, the pH allowed for drinking water is 6.5 - 8.5. According to pH analysis demonstrated that chitosan FO membrane could be applied for producing drinking from dirty water and seawater.

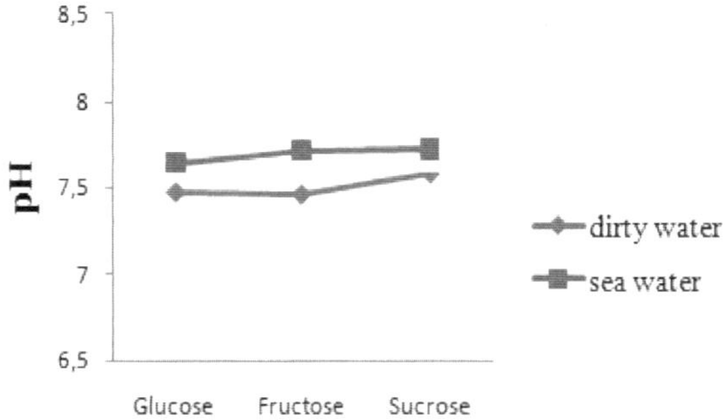

Figure 4. The results quality analysis of the FO drinking water products with pH parameters using a solution of glucose, fructose and sucrose

Another drinking water quality parameters specifically salinity, conductivity, and TDS for FO water product are shown in Figure 5. The Figure 5, it is proved that all the results of the measurement parameter are meet to drinking water quality standard. It means that the FO membrane is able to hold the substance or species dissolved in the feed solution so that it does not move towards the draw solution, while the one that is attracted to and able to pass through the semipermeable membrane is only water.

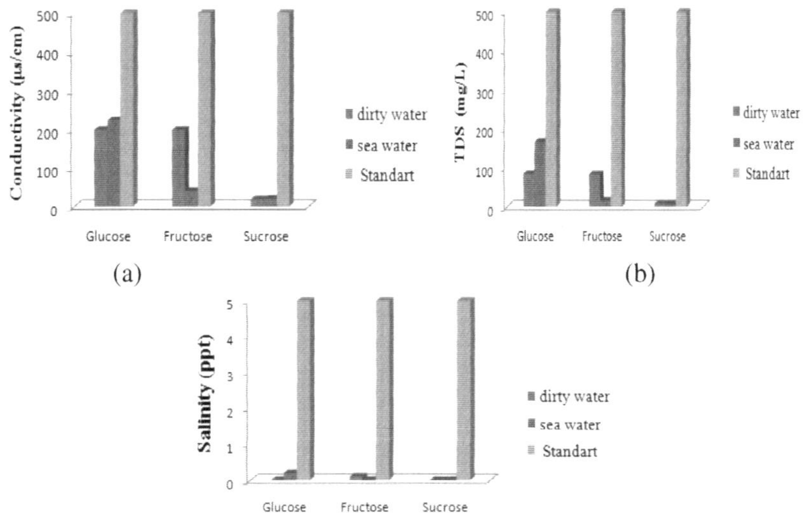

(c)

Figure 5. The analysis results of quality the FO drinking water energy product against the parameters: (a) conductivity (b) *Total Dissolved Solid* (TDS), (c) salinity.

The value of water salinity for FO water products also demonstrates that chitosan FO membrane can block all salt ion from the feed solution, where the salinity values obtained ranged from 0 - 0.2 ppt. This value confirms that the water produced is still classified as fresh water and is below the maximum salinity limit for fresh water which is commonly used 5 ppt. TDS values for FO water products ranged from 8-168 mg/L from the initial TDS of 427mg/L for dirty water and 48350 mg/L for seawater. Then the conductivity of the FO product ranges from 18.8–223 mg/L from an initial conductivity of 1016 μ s/cm for dirty water and 67800 μ s/cm for seawater. Overall TDS values and conductivity of all draw solutions after the FO process was permissible because the value was below 500 mg/L. The content of the TDS agreed with PERMENKES regulation No. 492/Menkes/Per/IV/2010 that is the maximum allowable TDS standard is 500 mg/L.

3.3.2. Metal contents analysis

Analysis of the content of heavy metals in feed water and FO product water was measured using Atomic Absorption Spectrophotometer (AAS). Where the types of metals measured are Hg, As, Mn, Zn, Cr, Fe, Cu, and Cd. All of these metals are chemical parameters requirement in standard drinking water analysis according to PERMENKES No 492/MENKES/PER/IV/2010. The results of the heavy metal analysis are shown in Table 1.

Table 1. Heavy metal analysis of the drinking water FO products.

No	Parameter	Unit	Analysis Result				Standard
			Dirty water sample	FO product	Seawater sample	FO product	
1	Mercury (Hg)	mg/L	<0,001	<0,001	<0,001	<0,001	0,001
2	Arsenic (As)	mg/L	<0,003	<0,003	<0,003	<0,003	0,01
3	Mangan (Mn)	mg/L	<0,002	<0,002	<0,002	<0,002	0,4
4	Zinc (Zn)	mg/L	<0,01	<0,01	0,8207	0,0488	3
5	Copper (Cu)	mg/L	<0,002	<0,0026	<0,0052	<0,0026	2
6	Cromium (Cr)	mg/L	<0,002	<0,002	<0,002	<0,002	0,05
7	Iron(Fe)	mg/L	<0,009	<0,009	<0,009	<0,009	0,2
8	Cadmium(Cd)	mg/L	<0,002	<0,002	<0,002	<0,002	0,003

The analysis results of heavy metals from samples from both dirty water and seawater indicate that the samples used do not contain metals with high concentrations, where the concentration of small metals is below the AAS detection limit. Then the water from the FO product did not show any excess metal content. Based on Table 1. It is described that all types of metals measured are still below the drinking water quality standard. However, based on previous studies reported chitosan-based FO membrane is able to block the heavy metal from passing through the FO membrane. Chitosan FO Membrane is able to filter the smallest size salt ion, and this means that it is automatically can block all bigger size heavy metal ions do not pass through the membrane towards the water of the FO product [9,11].

3.3.3. Bacterial analysis

Analysis of the presence of bacterial content in water from the FO process was carried out using the Most Probable Number (MPN) method. The samples analyzed consisted of seawater and dirty water as well as water from FO products. The results of the E. coli analysis are shown in Figure 6. Where in dirty water samples containing 43/100 mL of E.coli and seawater samples containing 23/100 mL of E.coli. In contrast, the water from FO products from both samples was negative containing E. coli (0/100mL). The FO membrane is proving to be able to filter E.coli from the sample because the membrane pore size is smaller than the size of the bacteria so that the FO product is free of E. coli.

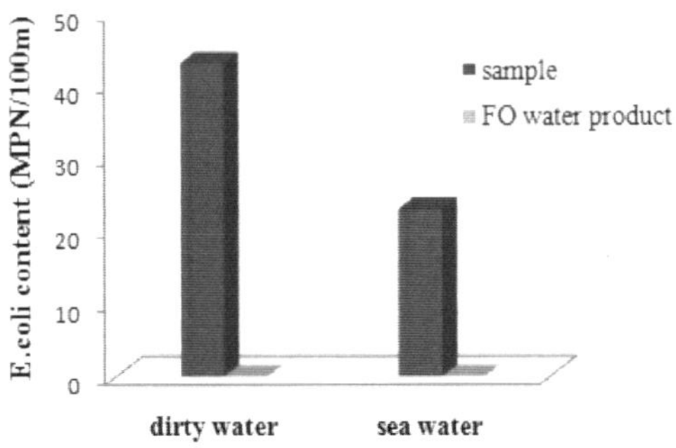

Figure 6. The analysis result E.coli content of FO water product

The type of membrane used as a filter in drinking water bags is a dense membrane with a pore size that is very small and dense. The pore size of the dense membrane has a pore <0,0001 μm in diameter while the size of bacteria is 0.2 to 1 μm, and the virus is between 0.02 to 0.4 μm. FO membrane is a membrane whose ability to filter out dissolved and suspended species in water including bacteria and monovalent ions [12,13].

4. Conclusion

Chitosan FO Membrane can be developed into ready-to-drink filter bags, where maximum flux is obtained with 3M sucrose concentration as draw solutions with a flux value of 5.25 J/m^{2}h for dirty water feed solution and 4.25 J /m^{2}h for seawater feed solution. The results of the analysis of drinking water quality parameters in the term of pH, TDS, salinity, conductivity, metal and E.coly bacteria content showed that the quality of water produced by FO process meets drinking water quality standards according to PERMENKES regulations No. 492/MENKES/Per/IV/2010. The FO membrane-based drinking water bag can be used as an alternative solution for energy drinking water supply in an emergency situation.

5. References

[1] Jincai Su, Sui Zhang, Ming-Ming Ling, Tai-Shung Chung 2012 *Forward Osmosis: an Emerging Technology for Sustainable Supply of Clean Wate, Department of Chemical & Biomolecular Engineering* (National University of Singapore)

[2] Budiyono and Sumardino, S. 2013 *Teknik Pengolahan Air* (Graha Ilmu, Semarang)

[3] Su, J., Zhang, S., Ming, M.L. and Shung, T.C. 2011 *J. Membr Sci* 376:214–224

[4] Haryoto, R. dan Hidayat, W. 2007 *Penyediaan Air Siap Minum Tanggap Darurat Bencana Alam, Pusat Pengkajian dan Penerapan Teknologi Lingkungan* (BNPD.*JAI*. 3(1))

[5] Tejakusuma IG., 2005 *Alami, Vol. 10 No.2, 19-21*

[6] KLH, 2007 *Analisis Potensi Rawan Bencana Alam di Papua dan Muluku (Tanah longsor-Banjir-Gempa Bumi-Tsunami)* (Deputy for Development of Technical Facilities and Capacity Building, Office of the Environment, Jakarta)

[7] Cormick, J., Pellegrino, F., Mantovani, G., And Sarti. 2008 .*Journal of Membrane Science*. 325: 467-478.

[8] Cath, T.Y., Childress, A.E., and Elimelech, M. 2006 . *Journal of Membrane Science*, 281: 70-87.

[9] Saiful, Ulfa Riana, Marlina, Muliadi Ramli, and Nizar Mahmud 2018 *Aceh International Journal of Science and Technology, Accepted for publication*

[10] Tang W dan How Yong Ng 2008 *Desalination*, 224, 1–3, 143-153

[11] Saiful, Marlina, Muliadi. R, Nizar. M dan Maizar. 2017 *Proceeding ISC 2017* UNPAD-Bandung : 33-37.

[12] Padaki, M., Arun, M.I., Jenifer, F. and K.Narayan, P. 2011 *Desalination*. 280: 419-423.

[13] Shakeri, A., Hasan, S. and Masoud, R. 2017 *Carbohydrate Polymers*. 174: 658–68.

Acknowledgments
We want to acknowledge the financial support of this work through the Ministry of Research, Technology and Higher Education of the Republic of Indonesia has financed this research through Competitive Grant Research.

IOP Conf. Series: Earth and Environmental Science **273** (2019) 012048 doi:10.1088/1755-1315/273/1/012048

Spatial Autocorrelation of Earthquake Magnitudes in Tripa Fault, Aceh Province, Indonesia

H SOFYAN [1*], L RAHAYU [1], E LUSIANI [1]

[1] Statistics Department, Faculty of Mathematics and Natural Science, Syiah Kuala University, Darussalam, Banda Aceh, 23111, Indonesia

*e-mail: hizir@unsyiah.ac.id

Abstract. The position of Aceh coincides with the fault zone along the Bukit Barisan called Sumatran Fault. The Tripa Fault is marked by the mountainous area and the trajectory of the Kuala Tripa and Meureubo Rivers along 180 km. Tripa faults are categorized as active faults indicated by the occurrence of several destructive earthquakes in the region. One effort that can be done to reduce the negative impacts caused by the earthquake disaster is to make predictions by mapping the occurrence of earthquakes. The purpose of this study was to determine the relationship between the incidence of earthquakes in the Tripa Fault region of Aceh Province. The data used in this study are earthquake occurrence parameters in the Tripa Fault region from 1990-2017. The results obtained show that there is a global spatial autocorrelation between the magnitude of the occurrence of earthquakes in the region around the Tripa Fault in Aceh Province with several magnitudes of earthquake events having local spatial autocorrelation.

1. Introduction

The Sumatra fault stretching from Aceh to the Sunda Strait. One part of the Sumatra Fault is a Tripa fault. Tripa fault is categorized as active fault. Tripa Fault is characterized by mountainous areas and trajectories of large river along the 180 km. The big river is Kuala Tripa River and Meureubo River. Some earthquakes caused by Tripa Fault movements have considerable and destructive power. Some earthquakes caused by Tripa fault movements have considerable and destructive power. At the center of the segment precisely in the Gayo Lues region in 1990 an earthquake occurred with magnitude 6.1.

Efforts in disaster management (disaster mitigation) to minimize losses caused by earthquakes can be done by mapping in the form of maps prone to earthquake disasters [1]. A spatial earthquake research in Aceh province years 1921-2014 was undertaken by Affan in 2016, the purpose of the research is to detect groups and look for spatial patterns during the period 1921-2014 using GIS. The methods used are Average Nearest Neighbor, Moran Global Index, The Getis-Ord General G, Anselin Local Moran Index, Getis-Ord Gi* and Kernel Density Estimation. One result of the study shows that earthquakes are grouped from the southwest to the northern part of the province [2].

The background above is the reason why the author performs data analysis to see the magnitude of the occurrence of earthquakes in relation to the magnitude of other earthquake events in an area and see the return period of earthquakes in the Aceh region, especially around the Tripa fault. This research is expected to be able to provide disaster information in increasing public awareness, especially those living in the area around the Tripa Fault in Aceh Province.

2. Data and Methods

2.1. Data

The data used in this study is secondary data, namely the occurrence of earthquakes in the Tripa Fault region of the Province of Aceh with variables of longitude, latitude and earthquake magnitude. Earthquake occurrence data used began from January 1990 to December 2017. The total occurrence of earthquakes that occurred in that time span was 598 events. The magnitude of the earthquake occurrence data starts from $M = 1.8$ to $M = 6.7$. Because the Meteorology, Climatology and Geophysics Agency (BMKG) only has an earthquake data repository from 2009, data from 1990 to 2008 were obtained from the United States Geological Survey (USGS) repository. The variables used in this study are longitude points (x), latitude points (y) and earthquake strength variables denoted by (m). The research area is only limited to the area around the Tripa Fault in Aceh Province.

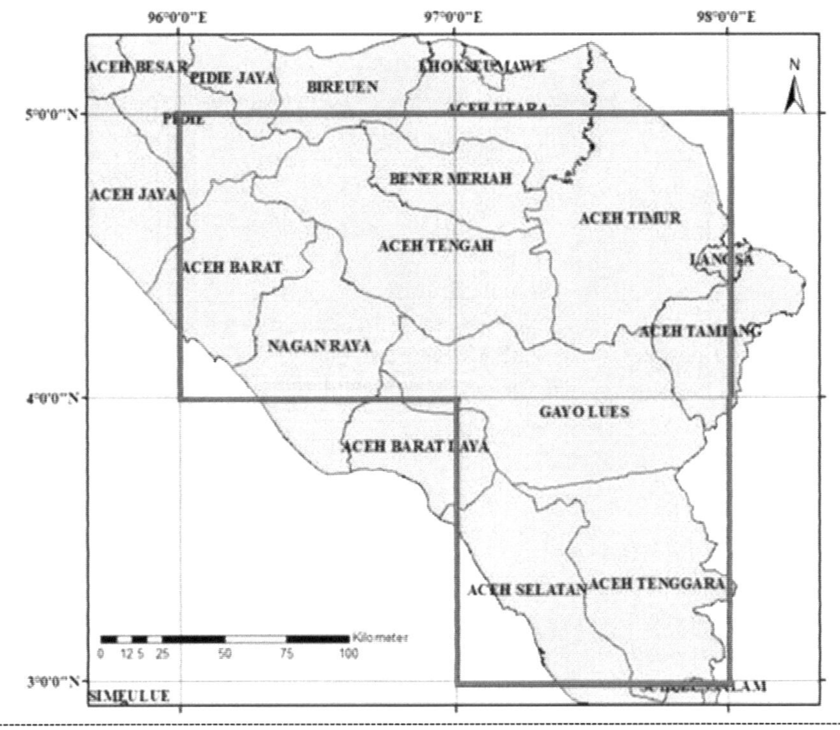

Figure 1. The research area.

2.2. Data Analysis Methods

Data analysis begins by combining data from both sources into one dataset. Then the time of the earthquake occurrence from USGS was converted to Waktu Indonesia Barat (WIB) and the magnitude of *Mw* so that the data from the USGS and BMKG had the same type of time and magnitude. Then the data is selected based on the magnitude of $Mw \geq 3$ so that 322 data remain. Identification of the pattern of earthquake magnitude distribution is done by spatial analysis, namely the global and local spatial autocorrelation test using *R studio 1.0.143* dan *ArcMap 10.4.1*.

Spatial object geometry information is entered into several forms, namely points, lines and polygons. Spatial analysis of earthquake magnitude uses point objects. Point data consists of a collection of event location points with spatial coordinates *x* (longitude) and *y* (latitude point) in an area [3]. The Moran index can be used to see the relation of point objects and distribution patterns from point data. The spatial autocorrelation test was tested in two ways, namely global testing and local testing [4].

The global Moran index can be calculated using the following formula:

$$I = \frac{n}{S_0} \cdot \frac{\sum_i^n \sum_j^n w_{i,j}(y_i - \bar{y})(y_j - \bar{y})}{\sum_i^n (y_i - \bar{y})^2} \tag{2.1}$$

where

y_i = observation *y* at the *i* location.
y_j = observation *y* at the *j* location.
$\bar{y}$ = average value of *y* in all observations.
n = total number of locations.
w = spatial weighting matrix.
S_0 = factor scale or the sum of all elements of the *W* value.
i = longitude.
j = latitude [5].

The Moran Index is used to test spatial dependency or autocorrelation between observations or locations. The hypothesis used is:
$H_0 : I = 0$ (there is no spatial autocorrelation)
$H_a : I \neq 0$ (there is spatial autocorrelation)
Decision making rejects H_0 if $p - value \leq \alpha$, with a value of $\alpha = 0{,}05$ [6].

After testing the spatial autocorrelation globally, it is necessary to test the validity of the results using a local spatial autocorrelation test. The local spatial autocorrelation test was carried out with the Local Indicators of Spatial Association (LISA). LISA is an exploratory analysis method for spatial data that is able to detect spatial relationships at the local level and their influence globally. LISA can be calculated by the equation (2.2).

$$I_i = \frac{(x_i - \bar{x})}{\sqrt{\sum_{i=1}^n \frac{(x_i - \bar{x})^2}{n-1}}} \sum_j w_{ij} \frac{(x_j - \bar{x})}{\sqrt{\sum_{j=1}^n \frac{(x_j - \bar{x})^2}{n-1}}} \tag{2.2}$$

Where $i = 1, 2, \ldots, n$ with *n* is the number of observation locations, $\mathbf{w}_{ij}$ is an element of the spatial weighting matrix that has been standardized in rows, $\bar{x}$ is the average value of x_i, x_i is the value of observation at the *i* location and x_j is the value of observation at the *j* location. The hypothesis used for the LISA test is:
H_0: $I_i = 0$ (there is no spatial autocorrelation at the *i* location)
H_1: $I_i \neq 0$ (there is spatial autocorrelation at the *i* location)

The test statistics used in the LISA test are the same as those on the Moran's index. However, calculation of variance from LISA is done by equation (2.3).

$$\text{var (I)} = \frac{w_i^{(2)}(n-b_2)}{n-1} + 2w_{i(kh)} \frac{(b_2-n)}{(n-1)(n-2)} - \frac{w_i^2}{(n-1)^2} \tag{2.3}$$

Where $w_i^{(2)} = \sum_{j \neq i} w_{ij}^2$, $w_{i(kh)} = \sum_{k \neq i} \sum_{h \neq i} w_{ik} w_{ih}$, $w_i^2 = \left(\sum_{j \neq i} w_{ij}\right)^2$, $b_2 = \frac{m_4}{m_2^2}$, $m_2 = \sum_i z_i^2 / n$, $m_4 = \sum_i z_i^4 / n$, and z_i = standardized observation values. Decision making is done if *p-value* $\leq \alpha$ (0.05) then reject H_0, which means there is spatial autocorrelation at the *i* location [7].

The Moran index coefficient ranges between -1 and +1. If the Moran index is negative it indicates that the data distribution pattern is spreading. Whereas if the autocorrelation is positive, the data distribution pattern is in groups [8].

The weighting matrix used in this study uses distance weights. Distance weight is an inverse distance matrix formed based on Euclidean distance. If given two locations with coordinates (x_i, y_i) and (x_j, y_j), then the Euclidean distance from that location is:

$$d_{ij} = \sqrt{(x_i - x_j)^2 + (y_i - y_j)^2} \tag{2.4}$$

The elements of distance weight matrix are defined as follows.

$$W_{ij}^{(\ell)} = \begin{cases} \dfrac{\dfrac{1}{1 + d_{ij}^{(\ell)}}}{\sum_{j=1}^{n} \dfrac{1}{1 + d_{ij}^{(\ell)}}}, & j \text{ is neighbor } i \text{ at the } \ell \text{ lag} \\ 0, & \text{others} \end{cases}$$

With d_{ij} is the Euclidean distance and W_{ij} is the weighting matrix element [9].

3. Result and Discussion

3.1. Global spatial autocorrelation test

The global spatial autocorrelation test is conducted to determine the relationship between the magnitude of the earthquake occurrence in the region around the Tripa Fault in Aceh Province. The global spatial autocorrelation test uses the following hypothesis:
H_0 : $I = 0$ (there is no spatial autocorrelation)
H_a : $I \neq 0$ (there is spatial autocorrelation)

```
> moran.test(data$m, listw=dist.mat.inve, alternative="two.sided")

        Moran I test under randomisation

data:  data$m
weights: dist.mat.inve

Moran I statistic standard deviate = -91.762, p-value < 2.2e-16
alternative hypothesis: two.sided
sample estimates:
Moran I statistic       Expectation          Variance
    -1.530901e-01     -3.115265e-03      2.671255e-06
```

Figure 2. The result of the global spatial autocorrelation test.

Based on the results of testing the global spatial autocorrelation in the area around the Tripa Fault (Figure 2), the resulting p-value is $2{,}2 \times 10^{-16}$. Because *p-value* < α (0,05) then the decision rejects the null hypothesis. The decision shows that there is a spatial relationship between the magnitude of earthquake events for all study areas located around the Tripa Fault of the Province of Aceh. Because the Moran index value obtained is -0,153, then there is a spatial relationship between magnitudes which is negative in the Tripa Fault in Aceh Province, so that it can be concluded that the occurrence of earthquakes adjacent to the area around the Tripa Fault in Aceh Province has a different magnitude. In addition, the negative Moran index also identifies that the magnitude of earthquake events of the same size are spread or not adjacent.

3.2. Local indicators partial autocorrelation test

As the validity of the global spatial autocorrelation test which states that there is a relationship between the magnitude of the occurrence of earthquakes, the spatial autocorrelation test is carried out locally. This test is carried out to see which earthquake points are affected by other earthquake points. Testing is done using the LISA Test. Visually the results are as shown in Figure 3.

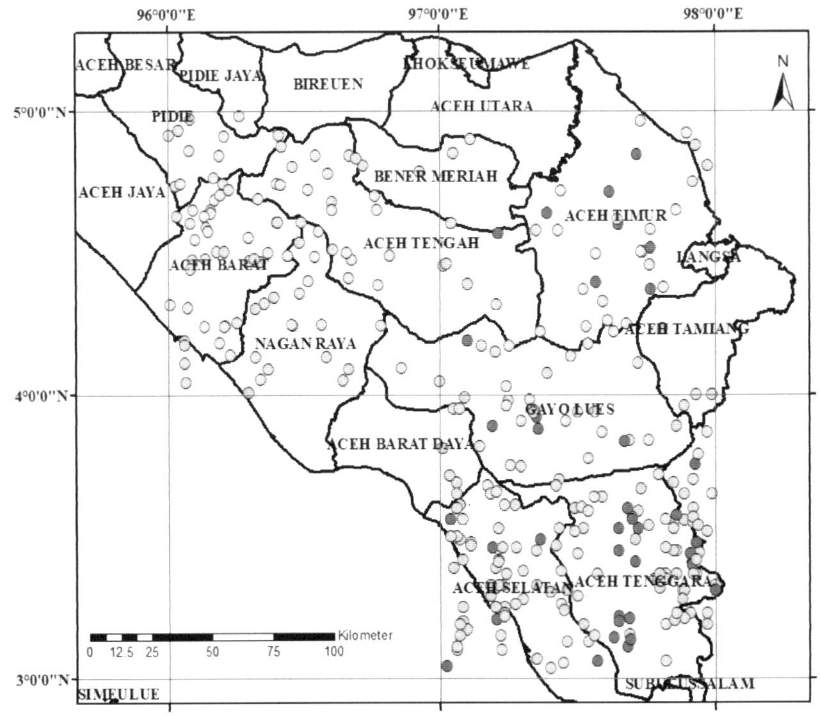

Figure 3. Visualizing local spatial autocorrelation testing result.

Locally, as much as 37 earthquakes magnitude have *p-value* less than $\alpha = 0.05$. 37 magnitudes have local spatial autocorrelation. This means that from 322 magnitudes earthquake occurrence around the Tripa fault with magnitude ≥ 3 there are 37 magnitudes earthquake occurrence that are really affected by the magnitude of adjacent earthquake. Because the Local Moran index of 37 earthquake magnitudes is positive, then the earthquakes have the similar magnitude as its neighbors.

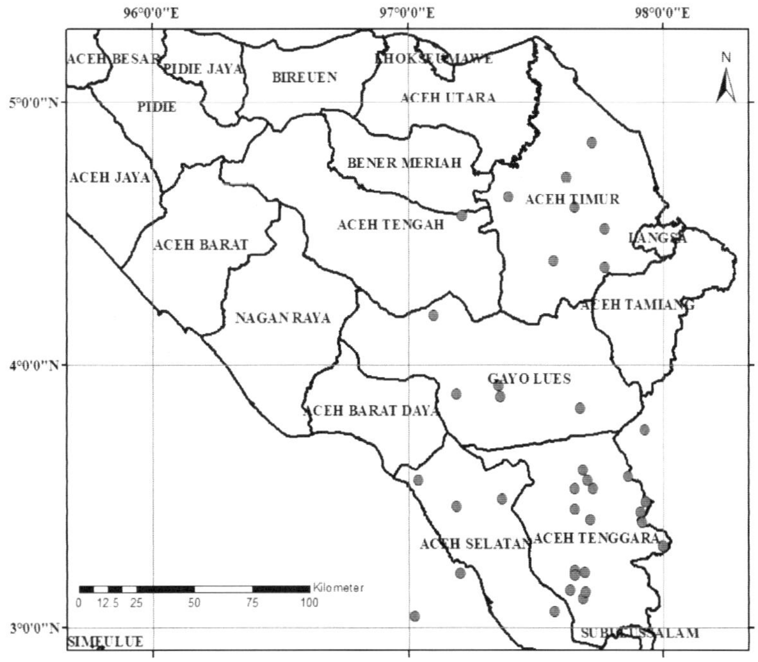

Figure 4. LISA test result with *p-value* ≤ 0.05.

Based on Figure 4. it can be seen that the earthquake magnitude points marked in red have *p-value* ≤ 0.05 or have spatial autocorrelation, which means that the occurrence of earthquakes with the same magnitude is in an adjacent location, while the occurrence of earthquakes with different magnitudes is in a spread location.

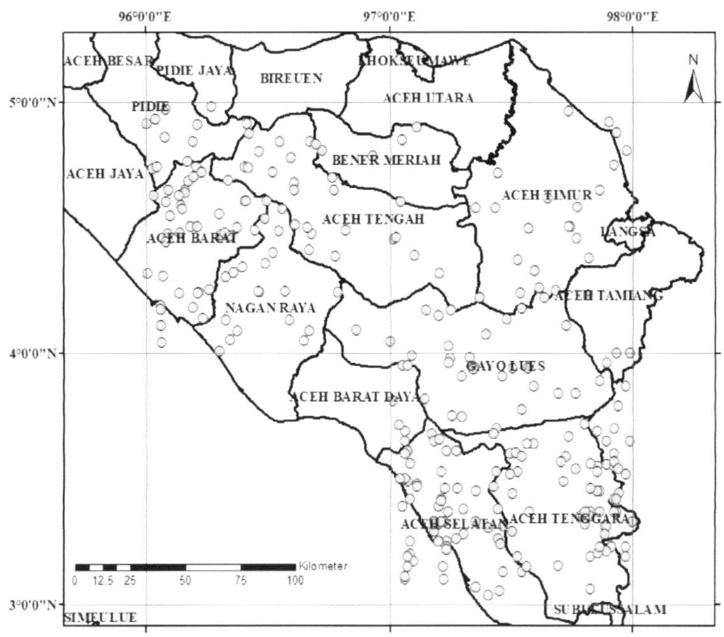

Figure 5. LISA test result with *p-value* > 0.05.

Based on Figure 5. it can be seen that the magnitude of the earthquake marked with a green dot have *p-value* > 0.05 or there is no spatial autocorrelation, which means that the occurrence of earthquakes with the same magnitude is at a spread location, while earthquakes with different magnitudes are in adjacent locations.

4. Conclusions

Based on the results, the conclusions this research are: there is a relationship between the earthquake magnitudes globally in the Tripa fault region, the negative Global Moran index indicates that any earthquake occurrence of an adjacent area around Tripa Fault in the Aceh Province has a difference magnitude, in other words the similar earthquake magnitudes have spread distribution or not close together. Locally, 37 earthquake magnitudes are actually affected by the magnitude of the adjacent earthquake magnitude and have a magnitude that tends to be similar to the neighbors.

References

[1] Djauhari N 2006 *Geologi Lingkungan* (Yogyakarta: Graha Ilmu)
[2] Affan M, Syukri M, Wahyuna L and Sofyan H 2016 *Aceh Int. J. Sci. Technol.* **5(2)** 54-62
[3] Eddy P 2014 *Sistem Informasi Geografis* (Bandung:Informatika)
[4] Mauricio S 2016 *Lecture 1 : Introduction to Spatial Econometric* (Chile: Universidad Catolica del Norte)
[5] Aminuddin A 2017 *AJIE - Asian Journal of Innovation and Entrepreneurship* **2(2)** 90-109
[6] Harmes, Bambang J, Ernan R and Baba B 2017 *Journal of Regional and Rural Development Planning* **1(2)** 192–201
[7] Luc A 1995 Local Indicators of Spatial Association—LISA *Geographical Analysis* **27(2)** 93–115
[8] Jay L. dan David S W 2001 *Statistical Analysis with Arcview GIS* (New York: John Willey & Sons. Inc)

IOP Conf. Series: Earth and Environmental Science **273** (2019) 012049 doi:10.1088/1755-1315/273/1/012049

GIS Application in Mapping of Forest and Land Fires and Mitigation in Banjarbaru City, South Kalimantan

Fikhriyah Khairunnisa[1] & Indah Ratna Rosalena[1]

[1]Department of Geography, Faculty of Mathematic and Natural Sciences
University of Indonesia

E-mail: fikhriyah.khairunnisa@ui.ac.id

Abstract. Forest and land fires is an important issue to be concerned due to its impact to the environment, especially in Banjarbaru City, South Kalimantan. Since the city is used as government area and mostly located in peatland, the prevention is needed to overcome the fires disaster. It can be done by using Geographic Information System (GIS). This study aims to mapping and analyzing the susceptible area of forest and land fires and its mitigation to overcome it. the level of susceptible of forest and land fires is the possibility of fires disaster which is done by GIS to define the priority area of mitigation. The modelling processed to obtained the susceptible area of forest and land fires by using GIS application. Geographic Information System (GIS) applied by using in ArcMap 10.2 facilitates the coating process between fires-causing factors. Physical environmental factors and human activities are two main factors of forest and land fires disaster. Factors used in modelling are land type, altitude and landuse. Each of them has score and percentage due to its risk to the fires disaster. From the modelling of fires-causing factors in Arc. Map 10.2 has been generated the susceptible area of forest and land fires. There are five level of susceptible area, such as not susceptible, low, moderate, high, and very high. Banjarbaru has five district, namely Lianganggang, Landasanulin, Banjarbaru Utara, Banjarbaru Selatan, and Cempaka. The very high area is around 15% area of Banjarbaru City and located in north and west area of Lianganggang district, north area of Landasanulin District and in the west of Banjarbaru Utara district. The high area has about 32% area of Banjarbaru City which is located in South Lianganggang district, southwest and east of Landasanulin district, in the center and west area of Cempaka district and also in north and east part of Banjarbaru Utara. The moderate level of forest and land fires disaster has about 17% area of this city that located in the middle of Lianganggang, southeast of Landasanulin, and south and west part of Cempaka district. And the low level has 32% area of the city and mostly located in cempaka, Banjarbaru utara dan Banjarbaru Selatan district. The last, not susceptible area is only about 0,13% in Cempaka district. The mitigation mostly located in Landasanulin dan Lianganggang district by making canals and boreholes. Local people are also encouraged to do quick response action by giving them water machine to extinguish the fires if it suddenly happened around them while waiting the helicopter from BNPB coming to help. Banjarbaru City, South Kalimantan Selatan is susceptible enough of forest and land fires disaster, especially in the west of Banjarbaru city. The susceptible area of forest and land fires is useful to assist in monitoring and determining priority areas in fires disaster mitigation. Besides, it useful to improve the level of local people awareness to not do something at risk of causing fires.

1. Introduction

Forest and land fires in Indonesia have become an international concern especially since forest fires that occurred in the 80s [1]. The causes of forest and land fires are more caused by human activities than natural processes [2]. In the annual report of WorldLife Fund, a non-governmental organization (NGO) that focuses on environmental issues states that Kalimantan Forests are believed to be shrinking by 75 percent by 2020 if the rate of deforestation is not stopped. Of the approximately 74 million hectares of forests owned by Kalimantan, only 71% were left in 2005. While the number in 2015 shrank to 55%. If the rate of deforestation does not change, Kalimantan is believed to be losing 6 million hectares of forest by 2020, meaning that only less than one third of the forest area is left [6].

The city of Banjarbaru is a city that part of its territory stands on peat land [8]. Forest and land fires have become not only in Indonesia but also in the Southeast Asian region which affects various aspects of life such as disturbances in daily life, transportation barriers, environmental damage, poverty, economy and health problems [9]. Forest and land fires can be a particular threat to this city because of course there are activity centers, such as government offices, transportation facilities in the form of airports, and markets as a place of trading activity.

According to Minister of Environment Regulation No.10 of 2010 concerning the Mechanism of Pollution Prevention and / or Environmental Damage Related to Forest and / or Land Fires (State Secretariat, 2010 in eco mapilata 2013), it was explained that prevention of forest and land fires can be done by providing data and information include location/area of fire and areas prone to forest and land fires. Location / area of ex-fire can be identified using remote sensing technology [7].

From this study, researchers utilize spatial data by processing using applications based on Geographic Information System (GIS), namely ArcMap 10.4 to obtain areas that are susceptible to fires, especially on peatlands. Thus, land fires disasters can be monitored in real-time combined with field surveys. In addition, the ability of GIS can also be maximized as information on how to handle forest and land fires, which during this time the action was only carried out at the time of the fires, but there was no pre and post-event prevention. As a result, the impact will be greater due to the absence of prevention. In this study the author aims to map susceptible area to forest fires and analyze their distribution. The problems studied were hotspot distribution, forest fires hazard mapping and mitigation carried out in dealing with cases of forest fires.

2. Study Area and Data

2.1. Study Area

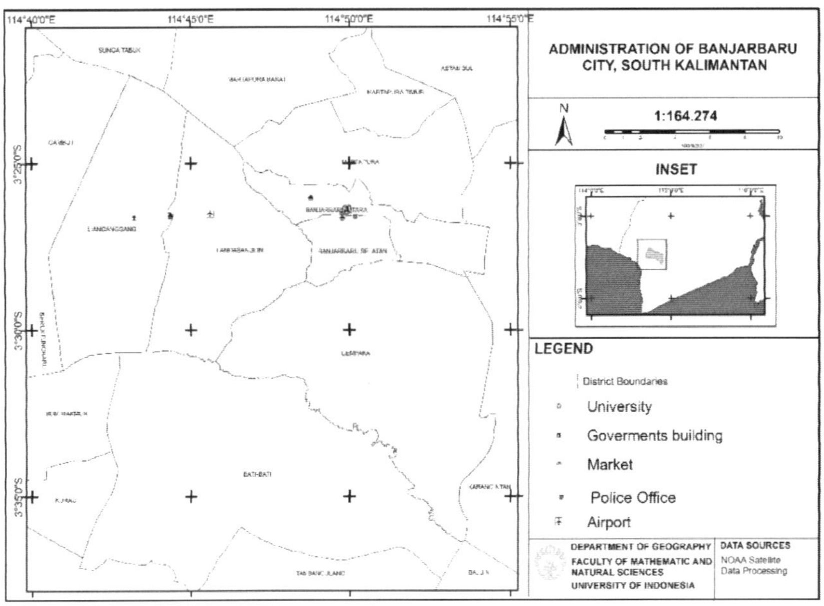

Figure 1. *Study Area*

IOP Conf. Series: Earth and Environmental Science **273** (2019) 012049 doi:10.1088/1755-1315/273/1/012049

Banjarbaru City is located between 3°25'40 "to 3°28'37" South Latitude and 114°41'22 "to 114°54'25" East Longitude. The city of Banjarbaru is at an altitude of 0–500 m above sea level, with a height of 0-7 m (33.49%), 7–25 m (48.46%), 25–100 m (15.15%), 100– 250 m (2.55%) and 250-500 m (0.35%). The area of Banjarbaru City is 371.38 km² divided into 5 (five) Subdistricts and 20 (twenty) Kelurahan, with boundaries where the north is bordered by Martapura District, Banjar Regency, bordering on the south by Tanah Laut Regency, on the west bordering Gambut Subdistrict, Banjar Regency, and to the east bordering Karang Intan District, Banjar Regency [10].

Banjarbaru, South Kalimantan has a protected forest area of 900 hectares, in the Sub-district of Liang/Ulin and grounding [1]. At 2015 when disaster struck the smog in South Kalimantan, Banjarbaru contributed to the scale of forest fires and the breadth of reach 500 hectares more. The incidence of forest fires have caused an impact in various aspects of both ecological, social, economic, and political [8].

2.2. Data Selection

The data required in the processing of this technical module based on updates to the forest and land fires susceptible map in South Sumatra Province the year 2015 of the Biodiversity and Climate Change Project (BIOCLIME), including:

a. Data Administration of Banjarbaru (Geospatial Information Agency).
b. Land cover Data for the year 2007 (from Geospatial Information Agency).
c. Elevation Data/DEM SRTM DATA.
d. Map of soil types (from the Ministry of Environment and Forestry).
e. Hotspots in 2017 (from NOAA Satellite).

3. Metodology

3.1 Workflow

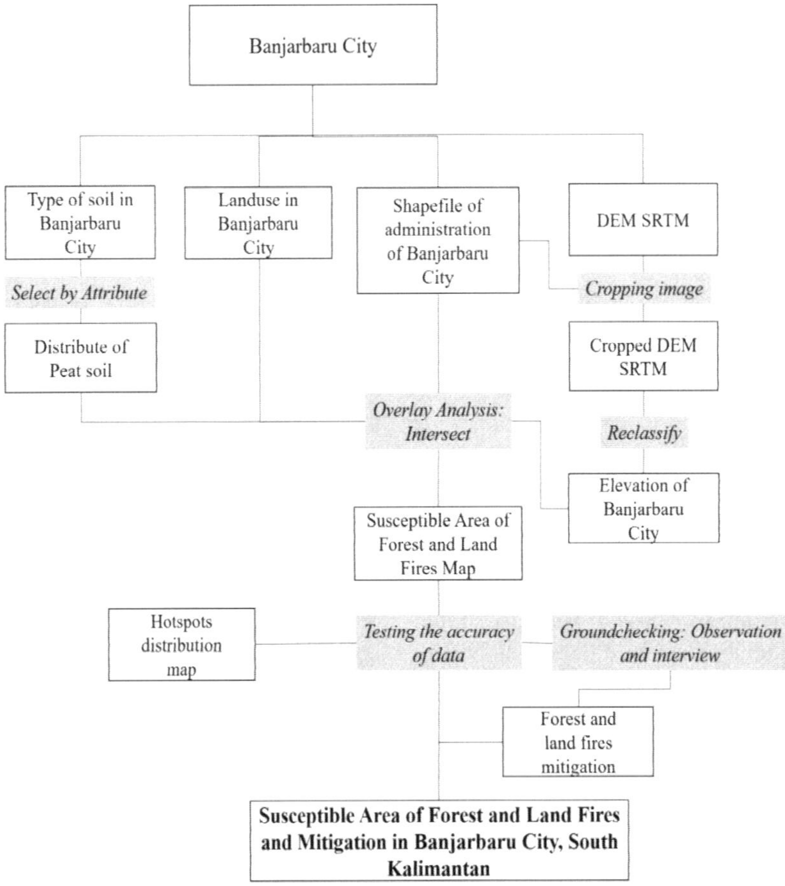

Figure 2. *Workflow of the research*

The research is to determine the level of fire susceptibility in the city of Banjarbaru using three variables: elevation, land use, and soil type. Data on soil types is indicated by the distribution of peat soil. Elevation variables were obtained from Digital Elevation Model (DEM) SRTM data which processed using ArcMap 10.4 to be able to display the required the classification of elevation in the form of elevation of under 25 meters and above 25 meters. In addition, to requiring data relating to research variables, it is also needed data in the form of administrative shapefile of Banjarbaru City. This shapefile in spatial data processing serves to show the boundaries of the study area and for cropping data based on the study area. The modelling process is performed with GIS software. At this stage data processing is still done in ArcMap 10.4. The overlay type used is Intersect. Then, we do weighted overlay process to giving score of each variable. The processing result show the mapping of forest and land fire susceptibility. After the map is obtained, it is tested with groundchecking and interviews to find out how conditions are in the field.

3.2 Data Processing

Based on the 2015 technical module for updating the hazard map for forest and land fires in South Sumatra Province from the Biodiversity and Climate Change Project (BIOCLIME), the matrix for susceptible forest fire areas is determined below:

Table 1. The matrix of susceptible area of fires.

Factors	Percentage	Classification	Score
Landuse	40%	Forest of industry plants	5
		Plantation	3
		Settlements	2
		Dry land agriculture	3
		Rice fields	2
		Temporary land farming	2
		Moor/fields	3
Type of soil	30%	Peat	5
		Non-peat	1
Elevation	30%	0-25 meters	5
		25-500 meters	1
		>500 meters	1

$$\text{The Susceptible Area} = 0{,}4 \text{ x landuse} + 0{,}3 \text{ x elevation} + 0{,}3 \text{ x soil type} \quad (1)$$

The results of the spatial analysis of the susceptibility map, based on the technical module for updating forest and land fire hazard maps in South Sumatra Province in 2015 from the Biodiversity and Climate Change Project (BIOCLIME) will be classified into 5 classes, based on the above formula, namely:

Table 2. Classification of susceptibility class scores.

Class of susceptibility	Score
Not susceptible area	0-0,9
Low	1-1,9
Moderate	2-2,9
High	3-3,9
Very high	4-4,9

4. Results and Discussion

4.1 Elevation in Banjarbaru City

The elevation used as climate factor to define the susceptible area of forest and land fires. Elevation affected to the land temperature, precipitation and sunlight intensity. The higher elevation makes the temperature lower, the precipitation higher, and sunlight intensity lower. The lower elevation makes the temperature higher, the precipitation lower, and sunlight intensity higher which can drive to forest and land fires disaster happened.

The elevation data obtained from SRTM, and processed by GIS application. Elevation divided into 2 classification, under 25 meters and above 25 meters. The area that has elevation under 25 meters is very susceptible to forest and land fires, and viceversa. As it impact to the disaster, elevation contributes 30% to forest and land fires.

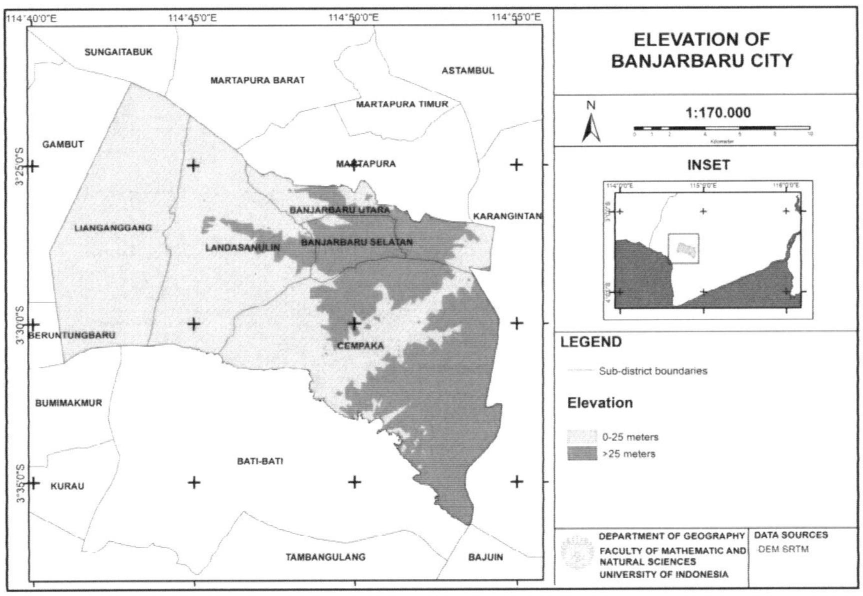

Figure 3 Elevation in Banjarbaru City, South Kalimantan.

Banjarbaru city mostly located at low elevation, especially in the west area of Banjarbaru city. In Lianganggang district, all of the area has elevation 0-25 meters. In Landasanulin district, the elevation under 25 meters is on the west, south and north of the district, meanwhile on the east area of Landasanulin has elevation above 25 meters. In Banjarbaru Selatan district, most of the area is above 25 meter. So does Banjarbaru Utara district, only the west and east area that has elevation under 25 meters. And in Cempaka district from the east to southeast area and the northwest area are above 25 meters, the west to southwest and the center area are under 25 meters. Based on the elevation, the west area of Banjarbaru city is more susceptible to forest and land fires.

4.2 Type of Soil in Banjarbaru City

Type of soil is the second variable as physical environmental factor that contributes 30% to forest and land fires. The data of type soil obtained from Ministry of Environment and Forestry. The composition of type of soil map classified according to analysis needs by distinguishing between mineral soil types and peat. The peat map component is the key information in developing a fires map because peat becomes a material that is flammable in the dry season. Peat soil is irreversible drying and when long dry season, it will be difficult to tie the water back so it is susceptible to burning. Peat soil has highest score, meanwhile non-peat soil has the lowest score to forest and land fires mapping.

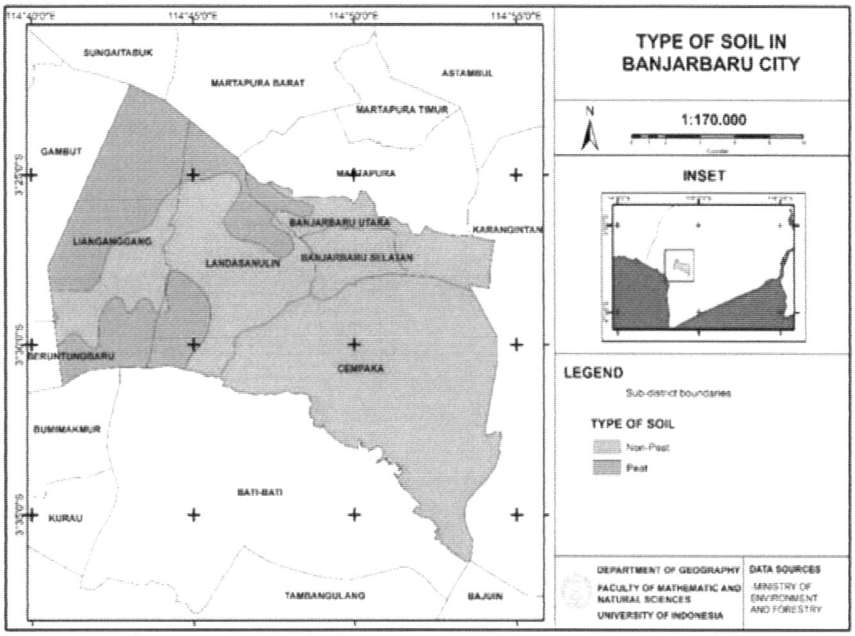

Figure 4. Type of soil in Banjarbaru City, South Kalimantan.

In Banjarbaru City, the peatland is mostly located in Lianganggang district and Landasanulin, and also the west area of Banjarbaru Utara District. Non Peat soil type located on west to east area of Lianganggang district, most of the area in Landasanulin district from west to east and the southeast area of the district, most of the Banjarbaru Utara district, except the western part of the district, all of Banjarbaru Selatan district, and also all of Cempaka district. From the type of soil which classified into peat soil and non-peat soil, Lianganggang and Landasanulin district are more susceptible than other district in Banjarbaru city.

4.3 Landuse in Banjarbaru City

Human factor in the field was represented spatially by land use. Land use greatly affect forest and land fires. Some land coverings are included in the fires-susceptible ecosystem, which is an open area with covered shrubs, litter and twigs both on land and in peat swamps, plantations, especially in peat swamps, agriculture, especially in peat swamps, and good secondary forests on land and in peat swamps. Land use data obtained from the Geospatial Information Agency was then processed to determine the distribution of land use in the study area by using Arc. GIS 10.2.

The weighting of landuse is carried out based on the sensitivity of the type of vegetation concerned to the occurrence of fires. A weight value to a very sensitive type of vegetation , which is very flammable is higher than the landuse that difficult to burn.

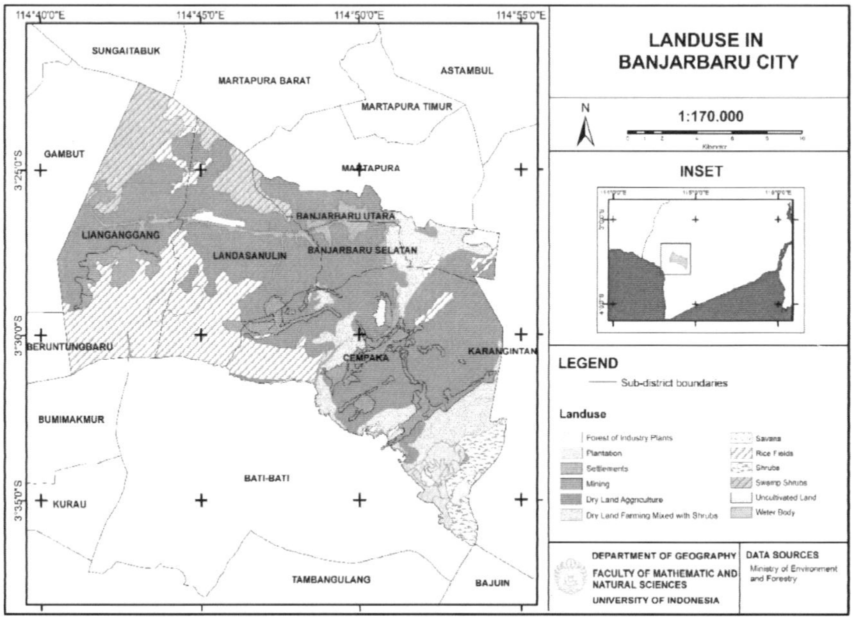

Figure 5. Landuse in Banjarbaru City, South Kalimantan.

Landuse in Banjarbaru city dominated by dry land agriculcure. Every district has area of dry land agriculture. This landuse has area 14703,9 hectare in Banjarbaru city. The second largest landuse is rice field. It has area about 6602,95 hectare mostly located in Landasanulin and Lianganggang district. Then, landuse of dry land farming mixed with shrubs has 3372,38 hectare area that located in Cempaka district and North and South Banjarbaru district. Mining area mostly located in Cempaka and Landasanulin district has area 2530,04 hectare. Settlement located near boundary of North and South Banjarbaru district, Lianganggang and also Landasanulin district. It has area 1062,05 hectare. Another landuse in Banjarbaru city has area under 1000 hectare. Due to its impact to forest and land fires and reflected as human factor, landuse contributes 40% to the disaster of forest and land fires.

4.4 Susceptible Area of Forest and Land Fires in Banjarbaru City

Forest and land fires have various impacts on human life. Syumanda (2003) stated that the impacts of forest fires are (1) social, cultural and economic impacts, (2) impacts on ecology and environmental damage, (3) impacts on relations between countries and (4) impacts on transportation and tourism. Forest fires often cause smoke disasters that can disrupt the activities and health of the surrounding community. From this study, researchers utilized spatial data by processing using GIS applications to obtain areas that are susceptible to forest fires.

From the results of the overlay of the three variables, types of soil, altitude, and land use are found to be susceptible to forest fires. Susceptible areas of forest fires are divided into 5 classes of non-susceptible area, low, moderate, high, and very high.

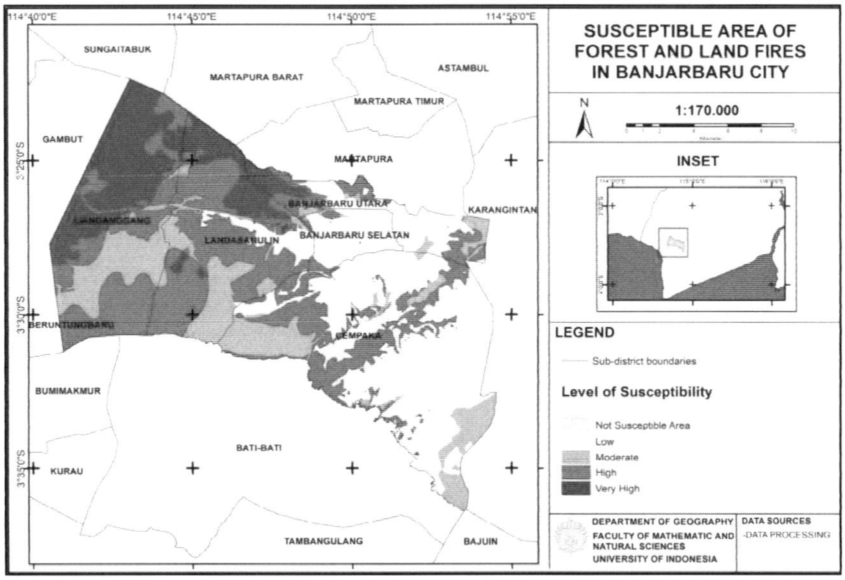

Figure 6. Susceptible area of forest and land fires in Banjarbaru City, South Kalimantan.

From processing result, the map of susceptible area of forest and land fires showing that the most susceptible area in Banjarbaru city is in Lianganggang district. It has large area of high and very high of susceptible. The second suceptible district is in Landasanulin district. And low susceptible of forest and land fires are Cempaka, North and South Banjarbaru district. The very high area is around 15% area of Banjarbaru City and located in north and west area of Lianganggang district, north area of Landasanulin District and in the west of Banjarbaru Utara district. The high area has about 32% area of Banjarbaru City which is located in South Lianganggang district, southwest and east of Landasanulin district, in the center and west area of Cempaka district and also in north and east part of Banjarbaru Utara. The moderate level of forest and land fires disaster has about 17% area of this city that located in the middle of Lianganggang, southeast of Landasanulin, and south and west part of Cempaka district. And the low level has 32% area of the city and mostly located in cempaka, Banjarbaru utara dan Banjarbaru Selatan district. The last, not susceptible area is only about 0,13% in Cempaka district.

Table 3. Area of susceptibility of forest and land fires in Banjarbaru City.

District	Level of Susceptiblility Area (Ha)					
	Not Susceptible	Low	Moderate	High	Very High	Area
Lianganggang	0,00	0,00	1.632,76	3.027,80	3.550,56	8211,12
Landasanulin	0,00	599,18	1.456,18	3.885,71	1.303,23	7244,30
Banjarbaru Utara	0,00	1.270,60	335,75	795,22	356,94	2758,51
Banjarbaru Selatan	0,00	1.332,28	34,32	135,21	0,00	1501,81
Cempaka	42,56	8.130,15	2.098,16	2.768,25	0,00	13039,12
Total	42,56	11332,21	5557,17	10612,19	5210,73	32754,86

Lianganggang district is the most susceptible area of all the district in Banjarbaru. The very high area is about 3550,56 Ha, which more than one third of the total district area. The high susceptible is 3207,80 Ha which also more than one third of the district area (table 3). Liangganggang are located in elevation under 25 meters, more than half of the area is peat land, and the the majority of landuse are rice field, swamp shrub, and dry agriculture. In Landasanulin district the area of each level of susceptibility are shown in table 3, where the high susceptible area is dominate in Lianganggang. In Lianganggang has similar landuse with Lianganggang and more than half area is under 25 meters and also have peatland in the north area which make the north area is very high susceptible and also located in southwest. While, Cempaka, Banjarbaru Utara and Banjarbaru Selatan are dominating with low susceptible.

4.5 Point of Interest in Banjarbaru City

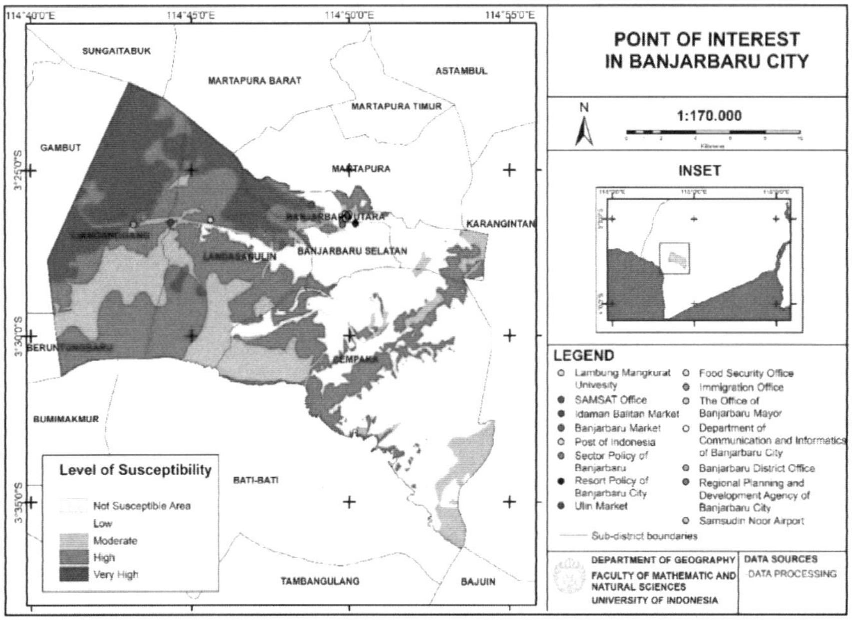

Figure 7. Point of Interest in Banjarbaru

As a government area, Banjarbaru city has several important Point of Interest such as Airport, Goverment buildings, University, and markets. For that reason, Banjarbaru is prioritize to restoration programs. The result show that important point such as university and airport are located in high susceptible area. Meanwhile, markets and goverment area are in moderate and low susceptible area

4.6 Hotspot Distribution in Banjarbaru City

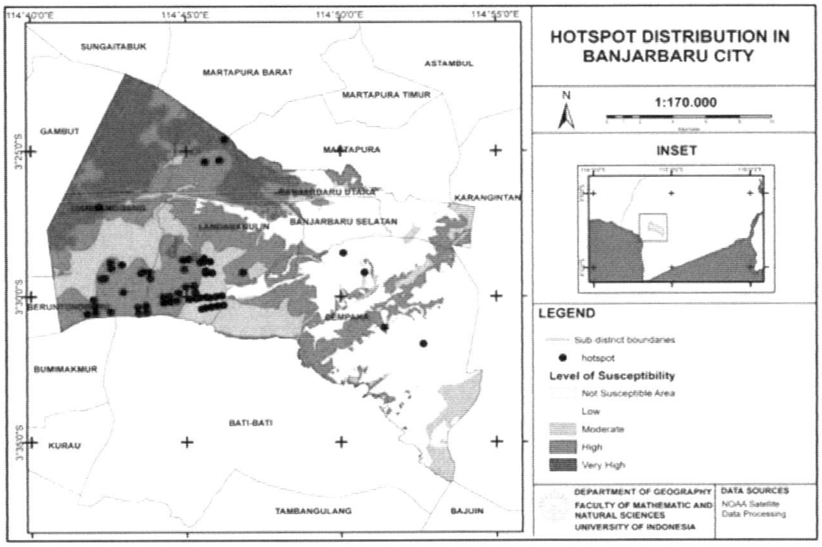

Figure 8. Hotspot distribution in Banjarbaru City

In the context of forest fires, GIS has been used to make efforts to predict forest fires by estimating flammable areas through fires-susceptible maps combined with the distribution of hotspots. Hotspot data is obtained from NOAA satellites. With the distribution of hotspots, it will strengthen the results of mapping susceptibility of fires areas in addition to the influence of the 3 previous physical factors, namely the type of land, height and land use. Hotspot points are abundant on peat soils and on bush land use, dry land farming and fields so that it has high potential to become a hotspot for fires fires.

The distribution of hotspot in 2017 in Banjarbaru city mostly located in South Lianganggang and Landasanulin district. The highest amount of hotspot is in high susceptible area. It has 36 hotspot. The second one is in moderate area in Landasanulin district has about 18 hotspot. Hotspot trigger the forest and land fire, but from the field study, the society said the human activity also cause the forest and land fire, such as improper burning of land and throwing cigarette butts carelessly which can also trigger the forest and land fire. The very high area has only 2 hotspot but the area is located in swamp shrub, peatland and the elevation is under 25 meters. Then, the result show that the location is very high susceptible eventhough there is only 2 hotspot.

Table 4. Hotspots distribution in Banjarbaru City in each level of susceptibility

Suscebtibility Level	*Amount of Hotspot*
Not Susceptible	*0*
Low	*3*
Moderate	*18*
High	*36*
Very High	*2*

After knowing hotspot distribution that can trigger to forest and land fires, we must take action quickly when a fire suddenly appears. Based on the results of the field survey, residents living in the area admitted that fires occur frequently each year. So that they are equipped with individual water machines at each house to take quick action in handling fires that can suddenly appear around their homes. Given that the area is peat and dry land farming. They said that during the dry season the area was very susceptible to fires, even fires could appear suddenly from the middle of fields or bushes and grazed until they grew bigger. To handle this, the water machine is used to prevent the fires from getting bigger while waiting for the fires extinguisher to help from the helicopter sent by the BNPB of Banjarbaru city.

4.7 Mitigation in Banjarbaru City

Forest and land fires mapping can be used as a reference to determine disaster mitigation measures to minimize the impact of the disaster. From the results of field observations and surveys, forest fires disaster mitigation carried out is as follows:

4.7.1 Making Channels

(a) (b)

Figure 8a and 8b. Channels for mitigation

Banjarbaru City is a priority for peatland restoration programs in South Kalimantan because airports are in the city so that they are severely affected by fires on peatlands. The peat restoration program that has been implemented anticipates forest and land fires in Banjarbaru, one of which is the construction of canals around the peat area. The canalization system to prevent the area from burning and also serves as a water source that can be used to water hotspots in the surrounding area.

4.7.2 Drilling Wells

Another forest fires disaster mitigation program is the construction of boreholes. Drilling wells are made up of 50 bore wells in Banjarbaru City. Drilling wells will be very useful during the dry season because the area of peatland will experience drought and flammability, wetting can be through water sources (Head of Peat Restoration Agency, Nazir Foead in Banjarbaru, 2017).

5. Conclusion

From the research, GIS helps to mapping the susceptible area of forest and land fires in Banjarbaru City. As a result, banjarbaru Selatan is susceptible enough of forest and land fires disaster, especially in the west of Banjarbaru city. The distribution of hotspot drive to the forest and land fires. The susceptible area of forest and land fires is useful to assist in monitoring and determining priority areas in fires disaster mitigation by making canals and drilling wells. It also helps to show the center of human activity places in each level of susceptibility. So, it useful to improve the level of local people awareness to not do something at risk of causing fires.

6. References

[1] Akbar, Acep dan Faidil, Sukhyar. 2007.Kebakaran Hutan dan Lahan Rawa Gambut: Penyebab Faktor Pendukung Dan Alternatif Pengelolaannya. *Balai Penelitian Kehutanan Banjarbaru.*

[2] Latifah, R. Pamungkas, A. 2013. Identifikasi Faktor-Faktor Kerentanan Terhadap

[3] Bencana Kebakaran Hutan dan Lahan di Kecamatan Liang Anggang Kota Banjarbaru. *Jurnal Teknik Pomits Vol. 2, No. 2, (2013)*

[4] Nugroho, Dudy dkk. 2015. Modul Pemutakhiran Peta Rawan Kebakaran Hutan dan

[5] Lahan dan di Provinsi Sumatera Selatan Tahun 2015. *Biodiversity and Climate Change Project (BIOCLIME).*

[6] Rizal, Yose. 2017. Banjarbaru Jadi Prioritas Restorasi Gambut. Kalsel Antaranews (accessed from https://kalsel.antaranews.com/berita/43891/banjarbaru-jadi-prioritas-restorasi-gambut accessed on March 28th 2018)

[7] Solichin, L. Tarigan, P. Kimman, B. Firman, dan R. Bagyono. 2007.*Pemetaan Daerah Rawan Kebakaran.*

[8] Syaraffudin. 2017.*Berkah dan Bencana Lahan Gambut Banjarbaru.* Kalsel: ProKalsel. (http://kalsel.prokal.co/read/news/12048-berkah-dan-bencanalahangambutbanjarbaru.html accessed on February 15th 2018).

[9] Syaufina L. 2008. Kebakaran Hutan dan Lahan di Indonesia: Perilaku Api, Penyebab, dan Dampak Kebakaran. Malang: Bayu Media.

[10] Humas dan Protokol Banjarbaru. 2018. Geografis. https://www.humas.banjarbarukota.go.id/tentang-banjarbaru/geografis/ accessed on November 29th 2018.

Acknowledgments

The paper helped by our lecturer. We also thank to government people, because of the informations that they gave help us to make the paper done. The society is also so friendly to give information for us. We hope the paper will usefull for them.

IOP Conf. Series: Earth and Environmental Science **273** (2019) 012050 doi:10.1088/1755-1315/273/1/012050

Conceptual Design of Mobile Application for Post-disaster Rapid Assessment of Damaged Houses

I Setyawati[1], K Munadi[2,3,*], Y Idris[2,4], Syamsidik[2,4], S Muchallil[3]

[1] Post Graduate of Disaster Science Program, Syiah Kuala University, Banda Aceh, Indonesia
[2] Tsunami and Disaster Mitigation Research Center (TDMRC), Syiah Kuala University, Banda Aceh, Indonesia,
[3] Electrical and Computer Engineering Department, Syiah Kuala University, Banda Aceh, Indonesia

[4] Civil Engineering Department, Syiah Kuala University, Banda Aceh, Indonesia

*Corresponding author: munadi@tdmrc.org

Abstract. The current rapid assessment of post-earthquake damaged houses has been done manually. The determination of the damage level is also very subjective. Moreover, the available time during a rapid assessment and limited number of enumerators lead to ineffective decisions for responses and assistances. There are several application had appeared trying to answer the problems, but still they have many types of data that need to be input and even need a lot of time to process. The system design to be simple, more user-friendly and time efficient to be operated during the assessment. This study is aimed to develop a conceptual design of a system to collect data of houses damaged by earthquake using mobile application which propose to work both online and offline. The system will consists of five components: mobile device, user, data center (web server and database), internet, and data processing center (administrator and manager to verify and analyze the data). The application design to be integrated with Google Maps containing concise data entry to facilitate users (trained users and common users) and to shorten the time of data collection. Final collected data will be presented through the website and perform descriptive and spatial analysis.

1. Introduction

In the time of a disaster occurs there are two urgent things that must be done immediately. The first is emergency response which is rescue and handling of disaster victims activities. The latter is rapid assessment which aims to assess the area affected by the disaster and the needs of victims. Rapid assessment is conducted to assess the disaster-affected area and the needs of disaster victims. Activity during rapid assessment is collecting the basic data needed later in recovery phase, and also make a priority list for responses. The aspects evaluate including location, damages, and sources so that the severity of damage can be determined [1].

The current rapid assessment especially in assessing damages has been done by manually [2]. Surveyors have to fill a rapid assessment form then gathered and inputted to the database. This activity takes time and energy while the available time usually only two weeks constraints for the surveyor to work fast. This situation can be worsened by the lack of resources, neither human resources nor supporting facilities to mobilize. Due to these limitations, the data obtained becomes less accurate and affecting the decision that would be made.

IOP Conf. Series: Earth and Environmental Science **273** (2019) 012050 doi:10.1088/1755-1315/273/1/012050

Based on Pidie Jaya earthquake on last December 2016, where rapid assessment activities were disrupted by the limitation of sources that led to unoptimum data from rapid assessment process. This phenomena gave an impact related to the unavailability of the accurate data which produced the unoptimized countermeasure. Based on that circumstances, it would be very helpful if there is a system that can accommodate the building assessment which efficient and accurate through the smartphone technology. The use of smartphone utilization in the development of this system by considering the ease and practicality of the user, in this case, the surveyor.

The development of this application focused on Android-based smartphone. Android smartphones currently lead the world in terms of their users. According to Gartner, Inc. (a research and advisory company), smartphone users reach 86.1% worldwide in the first quarter of 2017 [3]. Thus, maximum number of users can utilize the application if Android smartphone is used as basic of application development [4]. There are also two applications that had been developed previously in Italy, named Earthquake Damage Assessment Manager (EDAM) [5] and Clarinspect in New Zealand. Both applications have been relatively adequate for building data collection after the earthquake. But the amount of data that must be filled to the form are time-consuming, especially when the number of enumerator is limited, while the time span during the rapid assessment period relatively short (three days to two weeks).

This study focus on data collection of damaged houses affected by earthquake. The application has two working mode; online and offline. It can help surveyor to collect damages data with or without internet connection. The form designs to be simple and brief to ease the data gathering process and shorten the time. This Android-based data collection application is expected to facilitate the surveyor in the field. In addition, the spatial analysis also can be performed with map-based due to google map API services integration.

2. Mobile Application on field assessment

Mobile apps utilization keep increasing due to the increasing of smartphone users. The application purpose has vary from simple game to scientific tools. Todays application were supported Google maps API which location based to specified location and described it visualy intuitive. The Application has been utilized for navigation, tourism, etc. The availability of taking coordinates from google maps API services and geotagging photo based app would be highly useful for helping field assessment to collect location based data. There are few existing mobile apps that built for field assessment due to earthquake impacts. These apps analized and compared for collecting information in designing conceptual framework for mobile app of Building Assessment due to Earthquake impact. They collected data such as photos, videos, damaged description and location. The application that developed in China [2] are only working with the internet connection (online). On the other site, there are two more application that developed in Italy [5] and Hawaii [6] which can operate both offline and online. But the requirement for the rapid assessment in Indonesia need to adopt the system and framework requirement which is legally standared by Indonesian Government (PUPR). The application that we propose is the application that can work both offline and online with easy user-interface to be utilize with broader users. The application will focus on house damage assessment to help the process in time efficiency and data accuracy.

IOP Conf. Series: Earth and Environmental Science **273** (2019) 012050 doi:10.1088/1755-1315/273/1/012050

Tabel 1. Building assessment app in some countries.

Mobile App.	**(Xu et al)**	**EDAM**	**MERCI**	**Proposed App.**
Country	China	Italy	Hawaii	Indonesia
Type of survey	Damaged buildings caused by earthquake	Damaged buildings caused by earthquake	Damaged buildings caused by earthquake	Damaged houses caused by earthquake
Input	Data, photo, video	Data, photo, video, and sketch	Text, photo, video	Data photo
Mode	Online	Online/ Offline	Online/ Offline	Online/ Offline
Output	- Map - Descriptive analysis	- Map - Damage evaluation		- Map - Descriptive analysis

3. System Requirements

The Apps that proposed is to help the process of damaged houses assessment to be performed faster than usual using mobile apps. The limitation of time after the earthquake occurred mostly disturbing the accuracy of the houses damaged assessment for calculating loss. In less than two weeks the data should be gathered and analyzed to determined the type of the damaged and the fund that government should spend to help the victims rebuild, or repair their houses. Based on experience of Pidie Jaya earthquake (December 7th, 2016) in where data collection was using the form from Ministry of Public Work and Civil Houses (Kemen PUPR). The Ministry determined the type of the building damaged into two only classifications; moderate and heavy damaged. This two categories not really represented the real condition which happened in Aceh Province.

According to guidebook released by Ministry of Public Work in 2013, the damaged categories were redetermined into three classes; light, moderate and heavy [7]. Light damaged is described as nonstructural damaged which only need minor repairment and can still be inhabited. The moderate were described as wall damaged, and also light structural damaged that can be repaired to be reinhabited. The last, heavy damaged that translated into high volume of structural damaged that needs to be demolished and rebuild to be re-inhabited. The application needed the Google Map API service to work as location-based app. Beside the API services, the WebGIS also will be used to utilized the spatial data for more informative result visualization purpose. Mainly we have a few instruments for collecting data with the mobile application which are; Mobile device (android), Users (data collectors), data (photos, coordinate and text), internet connection, and server (offline and web).

3.1 Google Map API services

The apps that run in Android for damaged houses assessment is location based. The location-based is to the data classification by region using photo with geotagging features. The android app utilize google map API services to adopt the location-base application. The apps will help the data gathering in providing the coordinates of the location both with geotagging, or tagging manually offline in the maps that provided.

3.2 WebGIS

Since the android apps is highly utilize for the data gathering speed up, the visualization were later processed in Web page, in form of WebGIS. The data were stored and analyzed in a server-based and visualized trough the GIS Web page.

4. Conceptual Design of Rapid Assessment Mobile App.

The Application built with user-friendly interface. Working both offline and online this proposed mobile application had a slightly improved advantaged over similar rapid assessment application (Xu et al., 2016). The system consists of five components (Figure 1):

1. Mobile device, such as mobile phone, tablet, other devices with Android-based, have GPS feature and can be connected to the internet.
2. User, is a data collector or surveyor which divided into trained user and common user.
3. Internet network.
4. Data center, consist of web server and database.
5. Data processing center, where the manager verified and analyze the data.

In the users side the components divided into two catageries which are trained users and common users. Both use the mobile device to collect data from the field. The input were transported to web server which are both data center and data processing center through internet. The simple flow of the apps can be describe with input from both users from the field through the internet and continued to be stored and validated via web server.

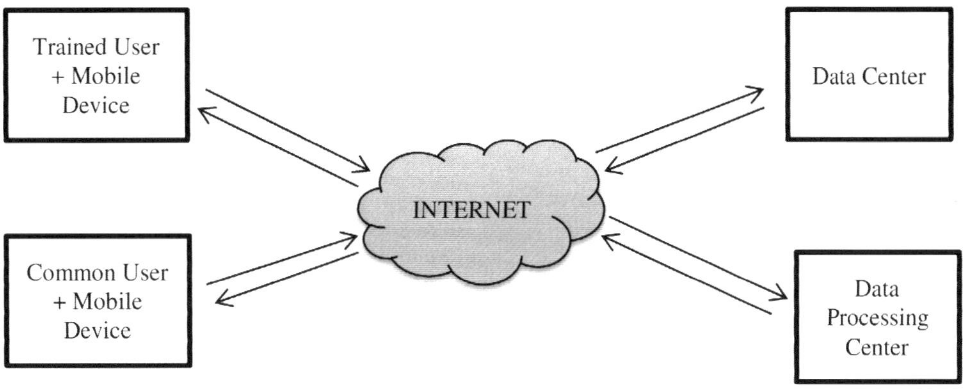

Figure 1. Design of system.

The flow of data collection on the mobile application started with launching the application and setting up the project. Data collection will obtained with coordinate marking, geotagging photo taking, and building condition descripting. The data will be saved in the mobile device in offline mode, and continue to be upload to the server when the internet is available. The data which has been gathered by the users will be verified and validated by the manager. Validated data will be analyzed by managers. The result represented in descriptive and spatial visualization through web for more useful and informative result.

In general, there are four actors which play a role in the apps, which are; trained users, general users, admin, and manager. Both trained and general users were performed as data collectors for damage houses which continue with uploading the data to the server. The manager had higher authority in data section. Manager is able to direct and manage trained users, manage the data (verification and validation), also analyze the data. the highest authority were taken by admin, where they can manage the whole projects and managers.

The system developed is on request by open and close setting. This means that the system is opened to be used if there is a request, in case of earthquake occurs. Furthermore, when the limit time for collecting data has been terminated, then this system will be closed again. This setting is aims to avoid the unwanted data, for example, the data that is submitted out of specified time. This arrangement also intend to focus data on an event in a particular place at a certain time.

4.1 Smartphone based App design

As mention before, the app is developed to be easy to used, simple and brief to fulfill the need of efficiency and effectiveness in collecting damaged houses data during rapid assessment period. Users directed to fill the simple form with brief explanation. Therefore users can gather as many houses data as possible. Users in the App was categorized into common and trained users. The app interface will contained of direct command to fill owners info (name, address and house coordinate) and continued with building types, part that damaged with photos and descriptions included (figure 2).

First user is specified to trained users while the second is common users for more general usage and term. Both common and trained users will be directed to gather the data about the house, ie; house ownership, address, type of houses, and take some pictures. Furthermore, trained users will gather more detailed and technical data for determining the level of damage on the houses. The operation were performed by answering for 3 houses categories; brick wall, structural brick wall, and wooden houses. Structural brick wall were assess three essential part; wall and top gavel, structural part, and roof. Wooden houses have less essential part to be assessed which are walls and frame, and roof. The assessment performed by answering descriptive question about the condition of the building. All the data analyzed in comprehensive resulting the level of damage which divided into 3 damage levels; heavy, moderate, and light. The three of them has their own value per each question. The level of the damages will determine based on their percentage of the total value. Here are the classification of damage by percentage value:

a. Light damage: <25% of total value,
b. Moderate damage: 25%> total value <50%,
c. Heavy damage: >50% of total value.

Meanwhile, the common users are not directed to collect technical data regarding damage houses condition. Location coordinates will be automatically recorded when collecting the data, specifically photos (geotagging). However, in order to anticipate improper or shifting coordinates, columns are also provided to fill the coordinates manualy.

4.2 Web-based App design

Web is used to visualized the collecting data and to analyze data. Web-based application is only for users at the level of administrator and manager. Administrator have access to manages (add, view, edit, and delete) managers and projects. Meanwhile, managers have features to manage the trained users, data (validation and verification), and analyze the data. Valid data can be analyzed and presented both descriptive and spatial analysis (Figure 3).

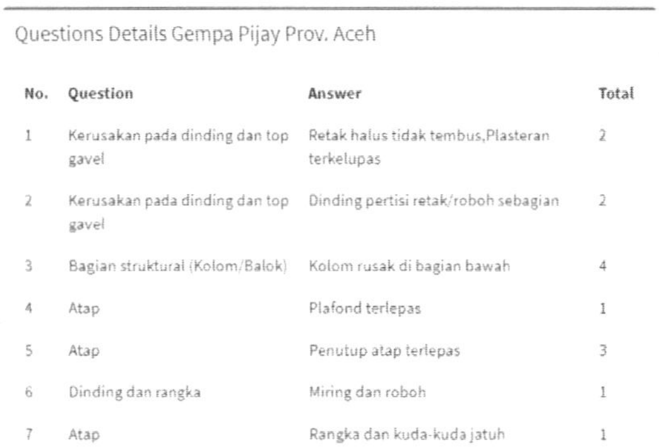
Questions Details Gempa Pijay Prov. Aceh

No.	Question	Answer	Total
1	Kerusakan pada dinding dan top gavel	Retak halus tidak tembus,Plasteran terkelupas	2
2	Kerusakan pada dinding dan top gavel	Dinding pertisi retak/roboh sebagian	2
3	Bagian struktural (Kolom/Balok)	Kolom rusak di bagian bawah	4
4	Atap	Plafond terlepas	1
5	Atap	Penutup atap terlepas	3
6	Dinding dan rangka	Miring dan roboh	1
7	Atap	Rangka dan kuda-kuda jatuh	1

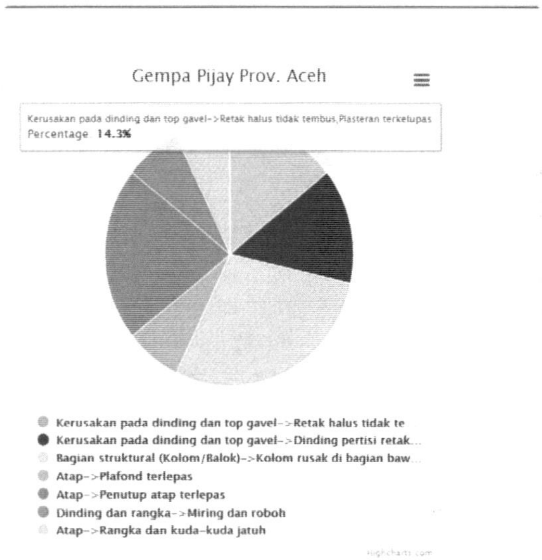

Figure 3. Data Analysis through web.

5. Discussion

The utilization of mobile apps had became a necessity in the digital era. The needs of massive amount of data while remained accurate in a short time frame had produced the mobile apps as one of the most effective solution. The apps would be capable of handling both the efficiency and the integration of the data of damaged houses rapid assessment. The development of the mobile application conceptual design were developed using the waterfall software development method. The method capable of giving the ease to the developer in maintaining every steps clearly and well organized to help checking with the mistaken steps.

The concept of the house damaged rapid assessment mobile application will ease the beneficiery (goverment) in performing the rapid assesment right after the disaster, especially earthquake. The concept were adopting the 3 classification of the houses damaged class which are light, moderate and heavy which refere to former assessment in Aceh. The concept also planned the apps to work 2 time faster then manual. This achievement will be gained with merging the data collecting and data store process through the app integration with internet. The data also become easier to be re-check and analize while being validate first

6. Conclusion

The needs of time efficiency which coupled with data accuracy of data gathering during emergency response had force the way of rapid assessment process to be modernized through mobile era. Mobile application gave more time efficiency and controlled accuracy by two level users. Both offline and online capability also add more function and flexibility to match the blackout that usually happened after the earthquake occurred. The use of the app is very promising for future utilization with proper development.

References

[1] Arii, M. (2013) 'Rapid assessment in disasters', *Japan Medical Association Journal*, 56(1), pp. 19–24.

[2] Xu, Z. *et al.* (2016) 'A System for Post-Earthquake Investigation of Building Damage Based on Smart Phones'. The 16th International Conference on Computing in Civil and Building Engineering, Osaka-Japan.

[3] Gartner (2017) 'Gartner Says Worldwide Sales of Smartphones Grew 9 Percent in First Quarter of 2017', *Gartner Newsroom*. Available at: https://www.gartner.com/newsroom/id/3725117 (Accessed: 25 January 2018).

[4] Suwal, D. *et al.* (2015) 'Geofencing post-disaster scenario using android app'. International Workshop on Strengthening Opportunity for Professional Development and Spatial Data Infrastructure Development, Kathmandu-Nepal.

[5] Cimellaro, G. P. *et al.* (2014) 'Rapid building damage assessment system using mobile phone technology', *Earthquake Engineering and Engineering Vibration*, 13(3), pp. 519–533. doi: 10.1007/s11803-014-0259-4.

[6] Kim, K., Pant, P. and Yamashita, E. (2015) 'Disasters, Drones, and Crowd-Sourced Damage Assessment'. International Conference on Computer in Urban Planning and Urban Management, Cambridge-USA.

[7] Permen PU Nomor 29/PRT/M/2006 tentang Pedoman Persyaratan Teknis Bangunan Gedung.

Acknowledgments

We would like to extend our gratitude for Tsunami and Disaster Mitigation Research Center (TDMRC) Syiah Kuala University for funding the application development.

IOP Conf. Series: Earth and Environmental Science **273** (2019) 012051 doi:10.1088/1755-1315/273/1/012051

WEBGIS Application for Planning the Tsunami Evacuation Route

Nizamuddin[1], W A Aldika[1], Ardiansyah[2], Hizir[3], and Muzailin[1]

[1] Informatics Department, Faculty of Mathematics and Life Sciences, Syiah Kuala University, Darussalam, Banda Aceh, 23111, Indonesia

[2] Electrical and Computer Engineering Department, Faculty of Engineering, Syiah Kuala University, Darussalam, Banda Aceh, 23111, Indonesia

[3] Statistics Department, Faculty of Mathematics and Life Sciences, Syiah Kuala University, Darussalam, Banda Aceh, 23111, Indonesia

Abstract. The tsunami that occurred on December 26, 2004 has resulted in massive casualties in Banda Aceh City. As one of the steps to avoid the huge casualty if the tsunami repeats itself, a number of Escape Buildings have been built in the city. In spite of the availability of the Tsunami Escape Buildings, the number of casualties can be reduced only if everyone in the tsunami-prone areas knows the nearest Escape Building from his or her position and the best route to reach the building. The purpose of this research was to build a GIS web application that can help determine the nearest Escape Building from a user-selected location and the best route to reach it. This application utilizes a web service that implements the Dijkstra algorithm to find the closest facility and the best route to the facility. The web GIS application was built using the open source API. The web GIS application developed in this research helps users find the nearest Escape Building and the best route to reach it without the need of any GIS software. The application is easy to use because it provides a user-friendly interface via a web browser.

1. Introduction

The earthquake and tsunami disaster which occurred in Aceh on 26 December 2004 took a large number of casualties, especially in the city of Banda Aceh and surrounding areas. Many people were killed, missing or injured when the tsunami wave hit the city because they had not reached safe places before the wave inundated the city. After this earthquake and tsunami disaster, several countries in the world provided assistance to the Indonesian government to rebuild Aceh. One form of this assistance was the construction of escape buildings by Japanese government. The escape buildings are intended to be places for evacuation whenever the earthquake and tsunami disaster threatens the safety of citizens again. The escape buildings were constructed in 4 villages, namely Ulee Lheue, Lambung, Deah Geulumpang, and Alue Deah Teungoh Village.

Finding the route to an escape building may be easy for people living in each village where the building is located, but that might not be the case for other people. Since the escape buildings are located in separate locations, anyone who happens to be around Ulee Lheue, Lambung, Deah Geulumpang, and Alue Deah Teungoh village when a tsunami alert is issued has to decide to which escape building he or she must go. The person must go to the nearest escape building from his or her current location and use the fastest route to reach the building. This is not easy for someone who is not familiar with the area, so an application or a tool which can help people find the closest escape building and the route to reach it from their location is necessary.

A web-based Geographic Information System (GIS), or commonly known as Web GIS, can be a solution to the above problem. Web GIS is a system that consists of servers that provide geographic data access to clients or commonly referred to as Web GIS applications [1]. GIS and Dijkstra's shortest path algorithm were used, for example, by Inoguchi *et. al.* who develop a Web-based supporting application for individual evacuation plans [11]. Their application let a user choose a starting location, and then, pick the first and the second evacuation places. The application showed the shortest path to reach the first and the second evacuation places. Another application, which was developed by Schuster and Gomez, allowed a user to calculate an escape route out of a location within a hazard zone at the touch of a button [12]. The web GIS application prototype described in this paper is different from those applications. The user only needs to let the application to detect his or her location, and then it will find the nearest escape building and the shortest route to reach it from the location of the user. The location of user, the closest escape building and the shortest route is displayed on a digital map. The application was developed to help people evacuate to the nearest escape building.

2. Methods

1.1. Data

The data needed in developing the web-based GIS application were road geospatial data and coordinates of escape building location. The coordinates data were converted into point data which show location of escape buildings on the digital map. The road data were used to generate road vertex point data which define the road network of Banda Aceh. This road network data are the most important data for the application because the nearest escape building and the shortest route to the building data are determined based on the road network data.

1.2. System Architecture

The Web GIS application was built as a client-server application as shown in Figure 1. The system architecture of the developed Web GIS application was a modification of the common Web GIS system architecture. As a distributed information system, a Web GIS commonly consists of a Web Application Server as the server and a Web Browser as the client [4]. The Server-side system of the developed Web GIS has three components, namely a Web Server, a GIS Server, and a DBMS. The Web Server receives a request from clients and sends the response in form of web pages which contains an online map. The web server uses the online map from the OpenStreetMap as the base map for the Web GIS application. The OpenStreetMap, which was launched on 2004, hosts a free editable map database of the world [5]. The local GIS Server retrieves the geospatial data of escape buildings and the shortest route from the Database Management System (DBMS) and publishes them as Web Map Services (WMS), which is the most widely adopted and popular specification for geospatial data exchange. This specification allows a GIS Server to provide compiled map images which may contain both vector and raster data [6]. The images can be in the form of PNG, JPEG, and GIF. Web Server consumes the service and adds the escape buildings along with the route geospatial data to the base map to produce the requested online map.

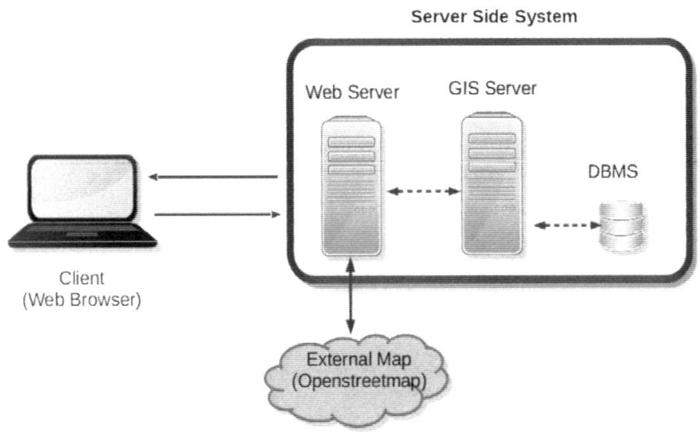

Figure 1. The architecture of the Web application

1.3. Web GIS Development

The online map display of the Web GIS application as created using the OpenLayer API. OpenLayer is a Javascript library for requesting and visualizing geospatial data via the Internet, especially the vector or raster data which have been published as WMS [7]. Every time the client will move to a different area or enlarge an area on the map, the client will give a new request to the web server [8]. The Web GIS application used Geoserver as the GIS Server which published geospatial data for web map applications. The Geoserver uses geospatial data of the type vector and provides Web Map Service (WMS) [9].

The WMS service acts as a media link between the Web Server and the Geoserver which functions as the GIS Server. A web browser sends a pair of location coordinates to the web server, which passes it to the geoserver via the WMS service. The geoserver then executes some previously built functions to calculate the distance to each escape building from the location determined by the coordinate pair and to find the minimum of the calculated distances. Output of these functions are location of the nearest escape building and the shortest route to reach it. These outputs are sent by the Geoserver via the published WMS, and the Web Server consumed the WMS to generate a map which is sent to the client later.

The functions which are used for finding the nearest escape building and the shortest routes were built using the PL/pgSQL script from the PostGIS and pgRouting library, which was installed as an extension of the PostgreSQL Database Management System (DBMS). PostgreSQL is an open source database that is licensed by BSD. This DBMS has a PL/pgSQL procedural language that makes it easy to develop functions for running SQL commands on the database [10]. PostgreSQL can also process spatial data through PostGIS. PostGIS is a geographical extension of the PostgreSQL database, which can store geographic data as tables in the database as well as process the complex geographic data and visually explore them [11]. pgRouting is an additional function library in PostgreSQL/PostGIS which provides some functions for checking the structure of a road network to find the shortest path to the destination [12].

3. Results and Discussion

The Web GIS application for planning tsunami evacuation routes have a simple Graphical User Interface (GUI) as shown in Figure 2. The GUI consists of a map of Banda Aceh City which shows streets and buildings in the city and three buttons for finding and displaying the shortest route to the nearest escape building. The map also shows several special icon markers which mark the location of 4 escape buildings in the city. Users can zoom in / out by clicking the plus ('+') and minus ('-') near the top left corner of the map to get the detail or broad view of the area.

Each of the three buttons below the map has different function. If a user wants to find the closest escape building from his or her current location, the user has to click the "Locate Me" button to acquire the coordinate pair of his or her current location. After that, the user can click the "Show Route" button. When the "Show Route" button is clicked, the client web browser will send the location coordinate to the server. The server will use the coordinate for finding the closest escape building and its shortest route from the location of the user. The shortest route will be sent by the server to the client as a WMS. The client web browser will overlay the route on the base map and show it in the map. Figure 2 show the shortest route to the nearest escape building as a dark line on the map. The user can clear the point location and the route by clicking the "Clear" button.

The Web GIS application as tested for several user's locations, and the application was able to show the nearest escape building dan the shortest route to the building from the tested user's locations. Since the escape building are concentrated in the Meuraxa District at present, the Web GIS application can only be used within the district. The result will not be logical if the application is used in other area because it will direct the user to an escape building in the Meuraxa District although the building is actually far from the location of the user. Actually, the result is an indication that escape buildings are needed in other parts of Banda Aceh city which are located near the coast.

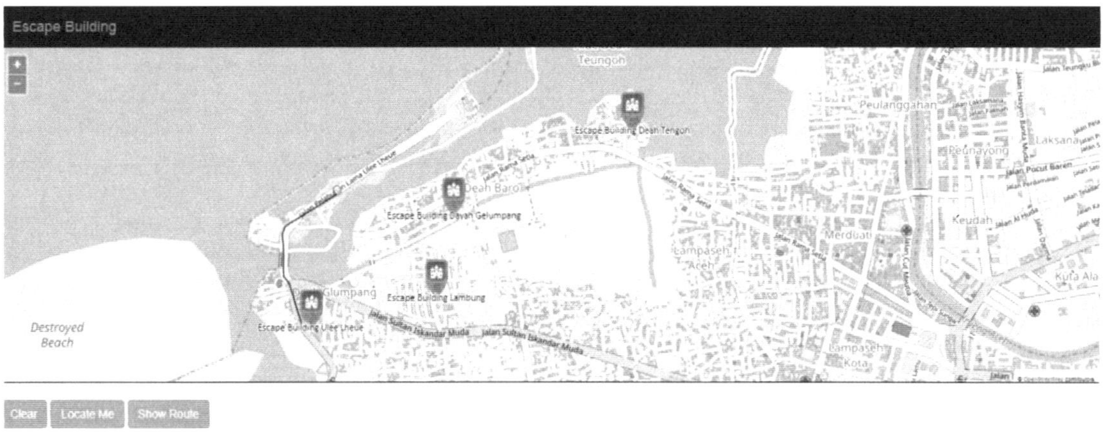

Figure 2. The Graphical User Interface (GUI) of the Web application

4. Conclusions

Based on the results of research, it can be concluded that a WebGIS application for planning the tsunami evacuation route can be built by using the OpenLayer API to display online map, Geosever as a server to provide Web Map Service (WMS), and PL/pgSQL scripts based on the PostGIS and pgRouting library as functions which were used for finding the nearest escape building and the shortest routes. In order for the WebGIS application to function effectively, information on more places than can be used as a temporary place for tsunami evacuation is needed. The places can be either new escape buildings or other tall buildings which are suitable or the tsunami evacuation.

References

[1] H.Cormen, T., Leiserson, C., L.Rivest, R. and Stein, C., 2009. Introduction to Algorithm. Cambridge: The MIT Press.

[2] Schuster M, Gomez C (2013) Evacuation routing out of tsunami hazard zones. In GI Forum 2013: creating the GISociety, Berlin, pp 206–215.

[3] M. Inoguchi, T. Sekikawa, and K. Tamura, "Developing a Web-Based Supporting Application for Individual Evacuation Plans Through Hazard Risk and Geographical Analyses," *J. Disaster Res.*, Vol.12, No.1, pp. 6-16, 2017.
[4] Lacovella, S., 2014. GeoServer Cookbook. Packt Publishing Ltd.
[5] Jokar Arsanjani J., Zipf A., Mooney P., Helbich M. (2015) An Introduction to OpenStreetMap in Geographic Information Science: Experiences, Research, and Applications. In: Jokar Arsanjani J., Zipf A., Mooney P., Helbich M. (eds) OpenStreetMap in GIScience. Lecture Notes in Geoinformation and Cartography. Springer, Cham
[6] Michaelis C.D., Ames D.P. (2017) Web Feature Service (WFS) and Web Map Service (WMS). In: Shekhar S., Xiong H., Zhou X. (eds) Encyclopedia of GIS. Springer, Cham
[7] Gratier, T., Spencer, P. and Hazzard, E., 2011. OpenLayers 2.10 Beginner's Guide. Packt Publishing Ltd.
[8] Sample, J.T., Shaw, K., Tu, S. and Abdelguerfi, M., 2008. Geospatial Services and Applications for the Internet. Springer Science & Business Media.
[9] Fu, P. and Sun, J., 2010. Web GIS: Principles and Applications. Esri Press.
[10] Pilz, J., 2008. Interfacing Geostatstics and GIS. Springer Science & Business Media.
[11] Marquez, A., 2015. PostGIS Essentials. Packt Publishing Ltd.
[12] pgRouting Community, pgRouting. http://pgRouting.org/. Accessed on 5 November 2015.

IOP Conf. Series: Earth and Environmental Science **273** (2019) 012052 doi:10.1088/1755-1315/273/1/012052

Mercury (II) Ions Assessment as a Toxic Waste Hazard in Solution Based on Imagery Data for a Part of Environmental Disaster Management

Zarlaida Fitri[1,2], Muhammad Adlim[2], Muhammad Syukri Surbakti[3], Ahmad Fairuz Omar[4], Fakhry Ananda Sijabat[3], Saumi Syahreza[3,5]

[1]Graduate School of Mathematics and Applied Science, Universitas Syiah Kuala, Darussalam, Banda Aceh 23111, Indonesia.
[2]Chemistry Department, FKIP, Universitas Syiah Kuala, Darussalam, Banda Aceh 23111, Indonesia.

[3]Physics Department, Faculty of Mathematics and Natural Sciences, Universitas Syiah Kuala, Darussalam, Banda Aceh 23111, Indonesia
[4]School of Physics, Universiti Sains Malaysia, Minden, Penang 11800, Malaysia.
[5]Tsunami and Disaster Mitigation Research Center (TDMRC), Universitas Syiah Kuala, Banda Aceh 23111, Indonesia.

Corresponding author: msyukris@unsyiah.ac.id

Abstract. It has been made and calibrated a device to detect dissolved mercury (II) ions as a toxic hazard using Arduino Uno and color sensor TCS3200. The purpose of this design was to realize a device for detecting dissolved mercury (II) ions as heavy metal pollution based on the relation of soluble mercury ion concentration to the liquid from the imagery data. Conjoint natural and technological disasters like the Mercury may pose tremendous risks to regions which are unprepared for such disasters. The results could be an environmental disaster, particularly when these occur in a highly populated area. Controlling the levels of dissolved mercury ions on a regular basis is a part of the environmental disaster management. The device works by converting any light by the photodiode optical sensor when the output is such a magnitude of an electrical current which proportional to the received light color level. The currents that are processed using the Arduino IDE software to extract the RGB values of the dissolved mercury (II) ions image data. Dissolved mercury ions generally do not have the necessary color, therefore, biocatalyst as reagents to generate color in a solution containing the mercury and this study using the Silver Nanoparticles. After reacted with a solution of known concentration of mercury in the value of the solution then carried out on a sample data collection to determine the RGB color values in the solution. In general, the device can be used as a color calibrator applied to the detection of dissolved mercury (II) ions by the obtained image data.

1. Introduction

Mercury is considered one of the most harmful metals found in the environment as a toxic hazard [1]. Combining natural and technological disasters such as the Mercury may pose tremendous risks to regions, which are unprepared for such disasters. The results could be an environmental disaster, particularly when these occur in a highly populated area. Controlling the levels of dissolved mercury ions on a regular basis is a part of the environmental disaster management [2].

Environmental pollution is a serious problem for developing countries like Indonesia. One is pollution by heavy metal ions. Mercury (Hg) is one of the most dangerous heavy metals. In nature, we can see the mercury in three forms: elemental (Hgvapor0toxic), inorganic mercury (Hg22+ and Hg2+) and organic mercury (methyl mercury, ethyl mercury and phenyl mercury). Hgions2+ dissolved in water is pollutant mercury is the most common and most stable. This means that the Hg2+ may accumulate in vital organs of humans through the food chain and can cause damage to the brain, nervous system, kidneys, heart and endocrine system[2].

Currently, there are a lot of measuring instruments used to detect mercury such as Atomic Absorption Spectroscopy (AAS), Inductive Coupled Plasma-Mass Spectrometry (ICP-MS) and Atomic Fluorescence Spectrometry (AFS), however, these instruments are high cost, and requires special skills to be able to operate, relatively long time and not portable [3]. Therefore, it is very necessary means to detect the dissolved mercury ions with a fast, simple, economical, effective and portable.

One promising method is detection based on color or colorimetric sensors due to the simplicity of its operation, the process of rapid, high sensitivity, effective and easily observed with the naked eye. One example of a colorimetric sensor that is widely used today is the TCS3200 color sensor. TCS3200 color sensor with Arduino-Uno can also be used to detect any number of colors. The sensor changes color in the form of a stream and converted into a frequency signal. Values obtained from the frequency of the color sensor readings are processed by using a microcontroller and displayed on a computer or screen.

The basic principles of color are grouped into three basic colors, namely red, green, and blue (RGB). The color is referred to as primary colors. The combination of the two primary colors will produce a new color or referred to as a secondary color, for example, green and red color mixing will produce a yellow color. Primary colors are red (like blood), blue (like the sky or the sea) and yellow (such as egg yolk). The secondary color is the color of the merger of two primary colors with the composition of 1: 1. Tertiary colors are the result of merging the primary color with a secondary color or the merging of the three colors either two primary and secondary colors as well as the incorporation of other secondary colors [4].

Arduino-Uno is a board microcontroller based ATmega328, this is a single-board microcontroller that is open source support, derived from the Wiring platform, designed to facilitate the use of electronics in various fields. Hardware is an Atmel AVR processor and software it has its own programming language [5]. Arduino can be programmed using the Arduino IDE software. The microcontroller of ATmega328 on the Arduino Uno is bootloader that allows uploading the generated code without having to use an external hardware programmer. Arduino software can be installed on a variety of operating systems (OS) such as Linux, Mac OS, and Windows.

Mercury as an element or the ion in solution is toxic or hazardous materials. Therefore, the waste containing mercury with all forms is also a toxic waste. Mercury contained in the waste disposal in public waters or modified by the activities of microorganisms into components of metal mercury (CH3-Hg) which have toxic properties and a strong holding capacity in addition to its high solubility. This resulted in mercury accumulated through the process of bioaccumulation and biomagnification in the body tissues so that the levels of mercury can reach levels dangerous for the lives of aquatic animals as well as human health, eating the aquatic animals [6].

2. Material and Methods

All chemicals used in the present study were of the highest purity (p.a). Chitosan of medium molecular weight and sodium borohydride purchased from Aldrich (USA). Silver nitrate, acetic acid, methanol, salt of additional cation tested and all other chemicals bought from Merck (Germany). Characterization of size and shape of the synthesized silver nanoparticles was carried out using Transmission Electron Microscopy (TEM) recorded on JEOL JEM-1400 operating at 100 kV [7].

2.1 Synthesis of Silver Nanoparticles

Colloidal silver was prepared by reduction of AgNO3 using sodium borohydride as our reported method with slight modification [8]. All glassware was washed with a mixture of nitric acid and hydrochloric acid (1:3) before use. Distilled water was used throughout the experiment. Silver (I) stock solution was prepared 10-4 mol/mL from silver nitrate as a precursor. In the typical experiment, 5.6 x 10-4 mol (0.0900) was dissolved in 88.8 ml acetic acid (1.5%) and the mixture was stirred overnight to obtain a homogeneous solution. Furthermore, 90 ml of methanol slowly added with vigorous stirring. Afterward, 1.12 ml (1.12 x 10-4 mol) of the silver stock solution was added little by little to the mixture. Then 0.0265 g of sodium borohydride was added slowly while kept stirring vigorously to reduce the silver ions to silver particles to yield brown yellowish colloid. These silver nanoparticles were characterized with TEM.

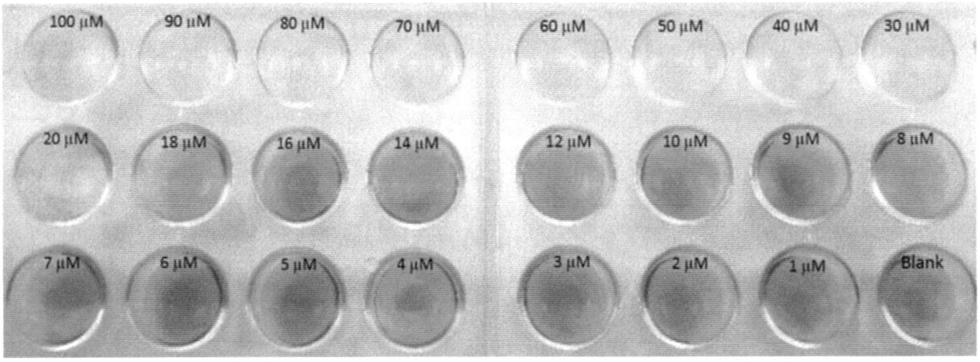

Figure 1. Sensitivity of Chi-AgNPs toward Hg^{2+} ions at different concentration.

2.2 Color Level Identification

The main sensors used was the TCS2300 which is attached on the board. It is an RGB light to frequency sensor which generates an output frequency (1 Hz and 600 kHz) proportional to the measured light intensity of the selected channel (red, green, and blue or clear) [9]. In this study, the color reflectance of sixteen different concentrations of mercury ion for several treatments is shown in Figure 1. Based on the flowchart programming using Arduino-Uno (Figure 2), any number of color mercury can be determined by TCS3200. The data that was detected on the TCS3200 sensor will be sent to the Arduino. Furthermore, the Arduino will transmit the data to the computer via serial communication of Universal Synchronous Bus (USB) in form of the RGB values. After the data entered into the computer the data will be processed for statistical regression analysis using Microsoft Excel.

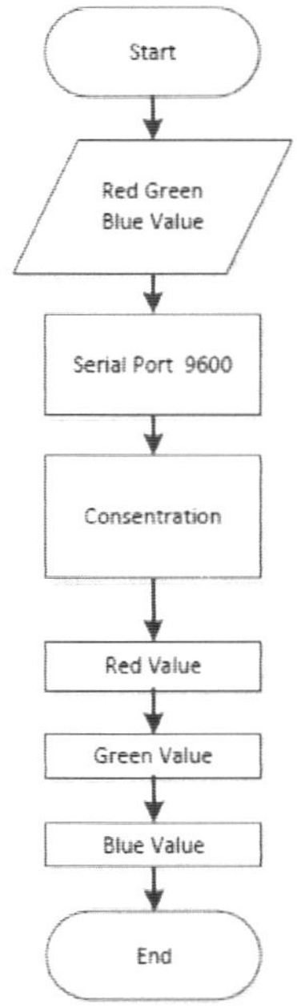

Figure 2. Sensitivity of Chi-AgNPs toward Hg^{2+} ions at different concentration.

The codes for color sensor TCS230 using Arduino-Uno Microcontroller are shown in the Figure 3.

```
FDUEUJ8IVA4KQNO
#include <TCS3200.h>

uint8_t RGBvalue[3];

TCS3200 colSens;

void setup()
{
  Serial.begin(9600);
  Serial.println("BEGIN");
  colSens.begin();

  Serial.begin(9600);
  Serial.println("CLEARDATA");
  Serial.println("LABEL,Red,Green,Blue");
}

void loop()
{
  colSens.loop();
  colSens.getRGB (RGBvalue);
  //colSens.getRGBtoMaxCorrection (RGBvalue);
Serial.print("DATA,");
      Serial.print("\tR:\t"); Serial.print(RGBvalue[0]);
Serial.print(",");
      Serial.print("\tG:\t"); Serial.print(RGBvalue[1]);
Serial.print(",");
```

Figure 3. The codes for an interface between the Arduino-Uno and the TCS230 color sensor.

3. Results and Discussion

The dissolved mercury ions generally do not have the specific color, therefore, biocatalyst as reagents was used to generate color in a solution containing the mercury. In this study, the Silver Nanoparticles was used as a reagent. After reacted with a solution of known concentration of mercury in the value of the solution then carried out on a sample data collection to determine the color reflectance of the solution in RGB values. There were sixteen samples made and tested in this study. Collecting data on the sample are acquired from two conditions that have been arranged, namely at the open and enclosed spaces.

In order to obtain the basic assumption of these components, therefore, it was important to test a relationship between sixteen samples of mercury ion concentrations and the RGB values of the color reflectance of samples. Correlation analysis has been the most commonly used in statistical tests to find the relationship among variables. This test can be performed in order to determine that the relationship between these data is linear or nonlinear. Assessments of the correlation (r^2) between the concentration of mercury ion in micro Molar (µM) and blue values from TCS230 color sensor were correlated to the concentration, except for the green and red channel of TCS230 color sensor. There were no significant correlations between the mercury concentration and R, G values (mercury to R & G). In the samples, the correlation for mercury to R & G was very weak, either from the color reflectance of samples that acquired in open and enclosed spaces.

The regression analysis between mercury concentration and blue values of TCS230 color sensor acquired from open space are presented in Figure 4. The maximum value of the r^2 between the concentration of mercury and the blue channel was 0.66. This means the light intensity of the blue channel had better correlations with concentration than the red and green channel. The open spaces testing showed that there was the ambient light contribution for color reflectance deviation of the TCS230 sensor. In spite of the correlation between them is moderate but significant correlated in this case study.

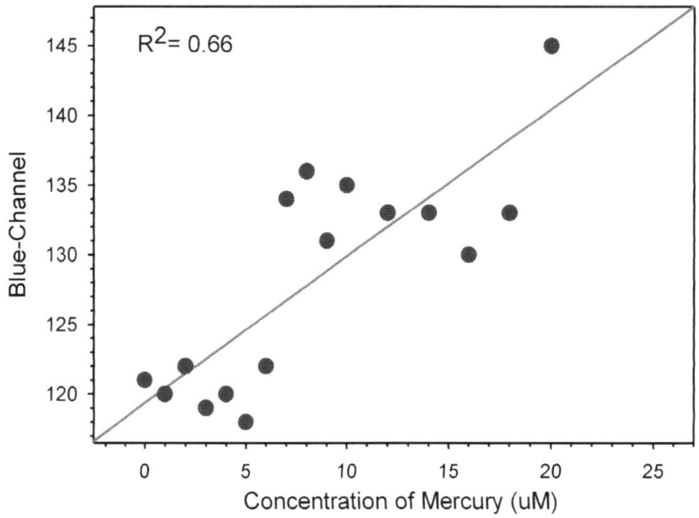

Figure 4. Scatter plot of mercury concentration and Blue Channel in open spaces testing.

Next, mercury concentration was derived using the blue channel for testing at enclosed space. The results show that the accuracy of concentrations was 0.72 (Figure 5). It indicates that the variation of concentrations is also significantly correlated to the blue channel of TCS230. These results also demonstrated that the accuracy of the mercury concentration estimation using enclosed space has improved the retrieval accuracy than using the open spaces testing. This is because the influence of light noise from the outside is negligible in the enclose space so that the sensor only detects the color reflectance possessed by the image. In the state of open space, color possessed by the sample is affected by the light from the outside which causes the color reflectance in the sample experienced a slight discoloration.

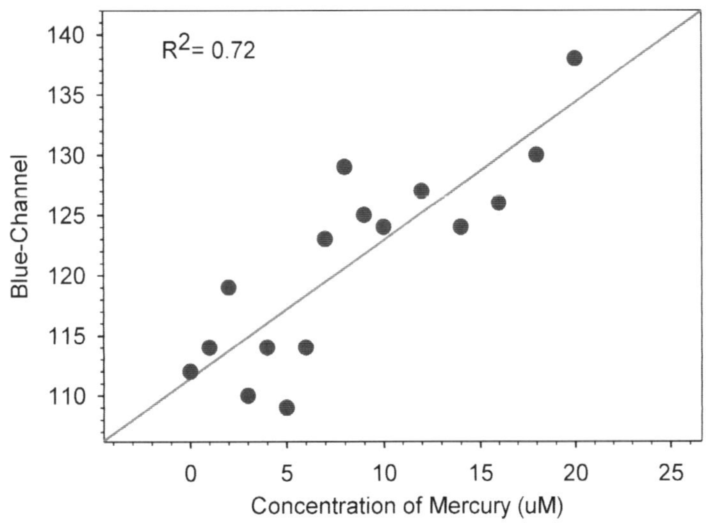

Figure 5. Scatter plot of mercury concentration and Blue Channel in close spaces testing.

4. Conclusion

TCS3200 color sensor can be used to detect the toxic hazards of mercury ions in solution through the imagery data. This tool could be applied to detect mercury ions dissolved in the environment with the support by the biocatalyst as color reagent in a solution containing mercury ions.

References

[1] Kadirvelu K, Kavipriya M, Karthika C, Vennilamani N and Pattabhi S 2004 Mercury (II) adsorption by activated carbon made from sago waste *Carbon* **42** 745-52

[2] Wang X, Gartia M R, Jiang J, Chang T-W, Qian J, Liu Y, Liu X and Liu G L 2015 Audio jack based miniaturized mobile phone electrochemical sensing platform *Sensors and Actuators B: Chemical* **209** 677-85

[3] Gao Y, Shi Z, Long Z, Wu P, Zheng C and Hou X 2012 Determination and speciation of mercury in environmental and biological samples by analytical atomic spectrometry *Microchemical Journal* **103** 1-14

[4] Aruan N M, Andjani D and Yuliora E 2016 Pembuatan album warna dengan menggunakan sensor warna jenis TCS230. In: *PROSIDING SEMINAR NASIONAL FISIKA (E-JOURNAL)*, pp SNF2016-CIP-47-52

[5] Badamasi Y A 2014 The working principle of an Arduino. In: *Electronics, computer and computation (icecco), 2014 11th international conference on*: IEEE) pp 1-4

[6] Comission E 2017 Tackling mercury pollution in the EU and worldwide

[7] Zarlaida F, Adlim M, Surbakti M S and Omar A F 2018 Chitosan-stabilized Silver Nanoparticles for Colorimetric Assay of Mercury (II) Ions in aqueous system. In: *IOP Conference Series: Materials Science and Engineering*: IOP Publishing) p 012049

[8] Adlim A 2012 Preparation of Chitosan-Stabilized Silver (Chi-Ag) Nanoparticles Using Different Reducing Agents And Techniques *Jurnal Sains MIPA Universitas Lampung* **4**

[9] Berns K and Luksch T 2008 *Autonome Mobile Systeme 2007: 20. Fachgespräch Kaiserslautern, 18./19. Oktober 2007*: Springer-Verlag)

Escape Hill as Geotechnical Quick Response Method in Facing Upcoming Tsunami Disaster

R P Munirwan[1,*], **and Munirwansyah**[1]

[1]Department of Civil Engineering, Syiah Kuala University, Banda Aceh, Indonesia

*Corresponding E-mail: r.munirwan@unsyiah.ac.id

Abstract. High number of casualties due to Aceh's Tsunami in 26th December 2004 was because of lack of information regarding to the disaster among Aceh government and people at that time. Escape building as a quick response for tsunami disaster may become an inevitable action in shoreline of Banda Aceh in facing forthcoming disaster. Moreover, as a result of open space need due to environmentally friendly infrastructure campaign nowadays, the construction of green design of escape hill can be considered. Escape hill can also be functionalized as recreation zone like green park. This article study purpose is to design an escape hill with the rule of geotechnical engineering aspect namely bearing capacity of soil below the escape hill and slope soil stability for the designing escape hill. Deah Baro Village of Meuraxa district was chosen as an escape hill design location. Several undisturbed soil samples were taken within this village for obtaining the existing soil parameters. There were eight hills modelling have been conducted by using Plaxis software. Based on the slope stability result from Plaxis software and by using natural soil as subgrade with soil classification of A-1-b as AASHTO method, the settlement value of final model dimension was 9.117 cm with safety factor of 1.507. The safety factor is still acceptable because of still larger that 1.5 as a standard value for slope safety factor. Escape hill geometry dimension is 12 m high which is larger than tsunami run up wave with soil pressure was calculate per layers of 2 m.

Keywords: Earthquake, construction, building, green-design.

1. Introduction

Aceh – Indonesia 26 December 2004 earthquake and tsunami disasters was considered as one of the deadliest disasters of Indonesia in century. This 9.3 Scala Richter magnitude earthquake and follow by huge tsunami waves not only occurred and effected Indonesia but also several neighbour countries like Thailand, Malaysia, Maldives, even the tsunami wave were able to reach distant countries namely Sri Lanka, Somalia and India. In Indonesia, more than 220,000 fatalities occurred because of this natural disaster event. The high number of casualties was because of lack of information regarding to the disaster among Aceh - Indonesia government and people at that time. Learn from the disaster, there have been an increasing of alertness for tsunami risk reduction among to coastal countries such as Indonesia and Japan. Even in 2011 Japan has been considered as the best prepared nation in the world for large tsunami attack survival [1]. Aceh – Indonesia already experienced several large magnitude earthquakes in recent century. As [2] and [3] Aceh also experienced fatal 6.4 Scala Richter earthquake in Pidie Jaya Aceh which also destroy quite a lot important constructions. Moreover [3] plotted based on spectral hazard map that Aceh location is in high risk zone of earthquake. As the result from several natural disaster in the past, best preparation for future disaster must be considered deeply as disaster mitigation in Aceh.

High number of casualties due to Aceh's Tsunami in 26th December 2004 was because of lack of information regarding to the disaster among Aceh government and people at that time. Escape building as a quick response for tsunami disaster may become an inevitable action in shoreline of Banda Aceh in facing forthcoming disaster. Moreover, as a result of open space need due to environmentally friendly infrastructure campaign nowadays, the construction of green design of escape hill can be considered. Escape hill can also be functionalized as recreation zone like green park. Some researches in predicted of upcoming tsunami disaster were also conducted by [4] [5] and [6]. Existing breakwater toughness in Ulee Lheue harbour which is close to Deah Baro was analysis by doing some simulation of earthquake – tsunami [4]. The result shows that the breakwater able to hold up to 8.0 Mw follow by tsunami waves strike.

Several type of tsunami mitigation procedures have been considered nowadays such as constructing sea defence structures, land use management or designing evacuation procedures depending on specific local situations. The local situations which must be considered including society aging, population ratio, land availability, and reconstruction aims chosen by local communities [7]. Resilient construction by the Aceh government were expectantly should be applied for future [4] and [8]. The reinforcing procedure of break water foundation by steel sheet piles and gabions was studied by [8]. This article study purpose is to design an escape hill with the rule of geotechnical engineering aspect namely bearing capacity of soil below the escape hill and slope soil stability for the designing escape hill. Deah Baro Village of Meuraxa district was chosen as an escape hill design location. Several undisturbed soil samples were taken within this village for obtaining the existing soil parameters.

According to [9] the tsunami run up in Deah Baro area was up to 7.0 m. It was recorded from [9] that several Tsunami Height Memorial Poles were constructed along Banda Aceh and Aceh Besar to remember the height of tsunami wave of Aceh tsunami 26 December 2004. The poles constructed for future generations to remember the rare event of tsunami disaster.

Table 1. Deah Baro Existing Soils Parameter Results

No.	Soil Parameter	1.6 m Soil Sample Taken	3.0 m Soil Sample Taken	Unit
1.	Material Model	MC	MC	
2.	Type of Soil Taken	Undrained	Undrained	
3.	Dry Soil Weight (γ_{dry})	10.3	9.7	kN/m^3
4.	Wet Soil Weight (γ_{wet})	13.9	13.9	kN/m^3
5.	Horizontal Permeability (k_x)	8.64E-01	8.64E-01	m/day
6.	Vertical Permeability (k_y)	8.64E-01	8.64E-01	m/day
7.	Young's Modulus (E_{ref})	10000	19610	kN/m^2
8.	Poisson's Ratio (v)	0.3	0.3	-
9.	Cohesion (c)	10.7	0.4	kN/m^2
10.	Friction Angle (ϕ)	32.3	37.6	^{o}
11.	Dilatancy Angle (ψ)	0	0	^{o}

The coastal land use after 2004 tsunami along the research area were already studied by [10] [11] and [12]. Land use if Ulee Lheue bay in 2015 mostly for paddy field, fishponds and coastal forest [10]. It shows an indication that people tend to avoid shore line for residential area in order to avoid upcoming tsunami disaster. Majority of resident tend to change their livelihood activity for adaptation with post disaster condition. Moreover, the designing of escape hill as green structure design in research area are suit with the recent land use. [12] studied about how to lead Banda Aceh to green city especially along tsunami 2004 affected area.

2. Material and Method

There were eight hills modelling have been conducted by using Plaxis software. The finite element and limit equilibrium methods are the most common numerical methods nowadays among the engineers. Material set of escape hill modelling with variating the value of soil cohesion and soil shear angle parameters had been considered. Deah Baro Village of Meuraxa district was chosen as an escape hill design location because this location was one of most destructive area after tsunami 2004. Furthermore, according to the existing Tsunami Height Memorial Poles in this village, the maximum run up of 2004 tsunami wave almost 7.0 m which is highest among Meuraxa district. Several undisturbed soil samples were taken within this village for obtaining the existing soil parameters.

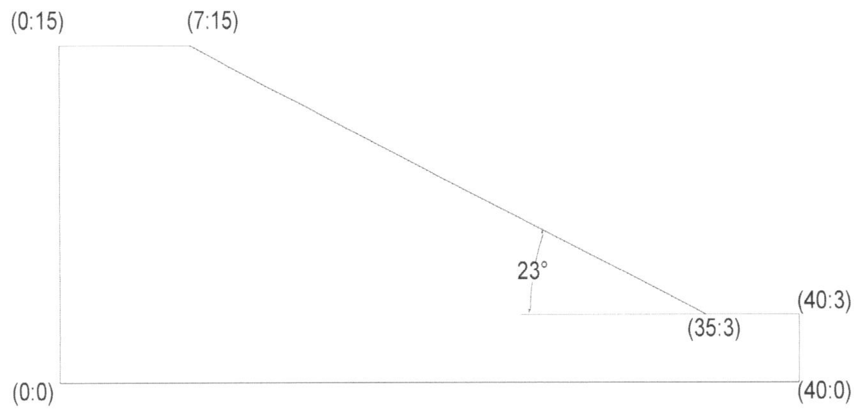

Figure 1. The designing cross section geometry of escape hill

The soil data obtained from this research are from experimental works in Soil Mechanics Laboratory of Syiah Kuala University Indonesia. Existing soil in Deah Baro Village as a subgrade were collected and tested for physical and mechanical soil parameter in order to analysis the base soil stability of escape hill. The slope soil data used was based on natural subgrade as existing soil and natural embankment of sand. The existing soil parameter can be seen in Table 1. The natural soil as subgrade with soil classification of A-1-b as AASHTO method and SW for USCS. The cross-section geometry of escape hill design is shown in Figure 1. The slope angle designing in 23° with the width and height of escape hill are 35.0 m and 12.0 m respectively. The degree of 23° was considered after several trial and error of safest slope angel. As already mention before, height of 12 m chosen because the design escape hill must be higher the maximum run up tsunami 2004 which is 7 m [9].

As the form of escape hill in Figure 1, due to 23o of the design must be analyses in terms of slope stability, underneath soil bearing capacity and also the effect of ground water or pore water pressure to the escape hills. Previous slope stability analysis were conducted by [13], [14] and [15]. In this research embankment installation with 2 m interval was considered as proposed by [13] for spatially variable soils. In spite of the analysis was based on an assumption of all embankment soil and subgrade soils are homogenous and isotropic, in the reality it is very hard to be applied [15]. This article purpose is to design an escape hill with the rule of geotechnical engineering aspect namely bearing capacity of soil below the escape hill and slope soil stability for the designing escape hill. [16] and [17] were also studied a comparison between mechanical soil parameter modelling.

3. Results and Discussion

The analysis was based on natural subgrade and natural embankment parameter. The embankment soil type used is sand. The embankment installation procedure was modelled with an interval of 2 m. embankment up to 12 m. Eight model of soil parameters were tried with the variation of cohesion and

IOP Conf. Series: Earth and Environmental Science **273** (2019) 012053 doi:10.1088/1755-1315/273/1/012053

shear angle parameter. The first model was by using the existing soil conditions and combining the value of cohesion and soil shear angle to established the safest model condition. The result from this model were settlement of 9.117 cm with safety factor of 1.577. After increasing 25% of cohesion parameter for second model, the settlement and safety factor change to 9.112 cm and 1.598 respectively. Moreover, the third model was increasing the cohesion to 50% and get the result of settlement 0f 9.112 cm and safety factor of 1.621 as in Table 2.

Table 2. Result of Cohesion and Shear Angle Modelling for Safety Factor

Slope Model	Cohesion (kg/cm²)		Shear Angle (°)		U_{tot} (m)	Safety Factor
1	C	0.04	φ	30.92	91.17E-03	1.507
2	c + 25%	0.05	φ	30.92	91.12E-03	1.598
3	c + 50%	0.06	φ	30.92	91.12E-03	1.621
4	c + 100%	0.08	φ	30.92	91.12E-03	1.660
5	C	0.04	φ + 10%	34.02	91.12E-03	1.646
6	C	0.04	φ + 20%	38.02	91.12E-03	1.747
7	C	0.04	φ + 30%	40.20	91.12E-03	1.771
8	C + 100%	0.08	φ + 30%	40.20	91.12E-03	1.850

The forth model was added 100% the value of cohesion and obtained settlement for 9.112 cm and safety factor 0f 1.66. The fifth model was used the original cohesion value but increasing the soil shear angle for 10%, the result shows that settlement of soil for 9.112 cm and safety factor of the slope model was 1.646. When adding the shear angle value for 20% in sixth model, the settlement remain 9.112 cm and safety factor was 1.747. 30% of soil shear angle was added for 7th model. The outcome displays that settlement of soil still remain 9.112 cm and safety factor of 1.771. The last model was using the most effective combination within cohesion and soil shear angle which was (c_{nature} + 100%) and (φ_{nature} + 30%). The settlement of hill was 9.112 and safety factor was 1.850. The compilation result of different parameter influence can be seen in Table 2.

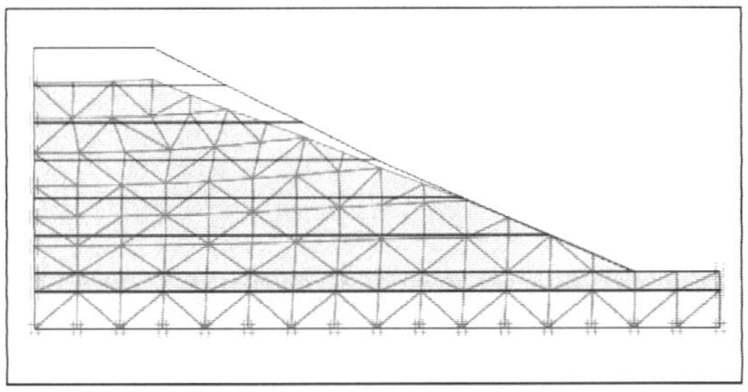

Figure 2. Deformation scenario in mesh element

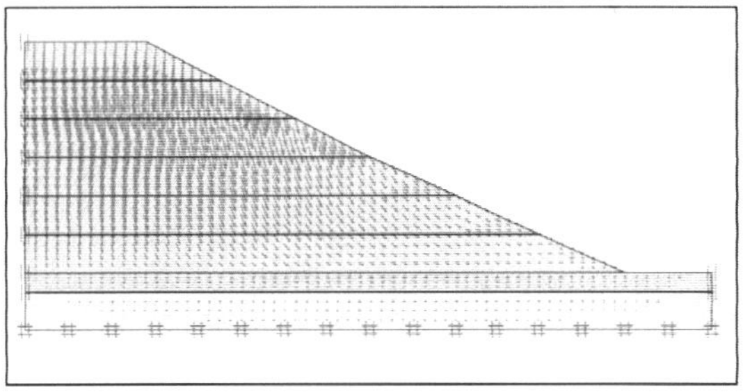

Figure 3. Total Displacements (arrows)

The study of comparison was completed in order to examines the application of finite element method and limit equilibrium method in calculating the effective design of the slope in terms of safety factor and critical failure of surface. From Plaxis analysis of the deformation procedure it had been noted that the extreme scenario of slope movement as deformation mesh element, total displacement of the escape hills is 91.17 x 10^{-3} m as shown in Figure 2 and Figure 3. The safety factor for the slope obtained as 1.507. The displacements effect to natural subgrade is still acceptable for design purposes. Figure 4 explained about the possibility of slope natural embankment to slip. Moreover, it also explained that there is an influence from embankment load distribution above to natural subgrade lied below the escape hill. As a result, the analysis of natural soil must be considered in escape hill design.

The total stresses of escape hill design recorded for 234.7 kN/m as in Figure 5. Homogenous soil height of the slope has a significant effect on safety factor for designing escape hill resting on existing soil layers.

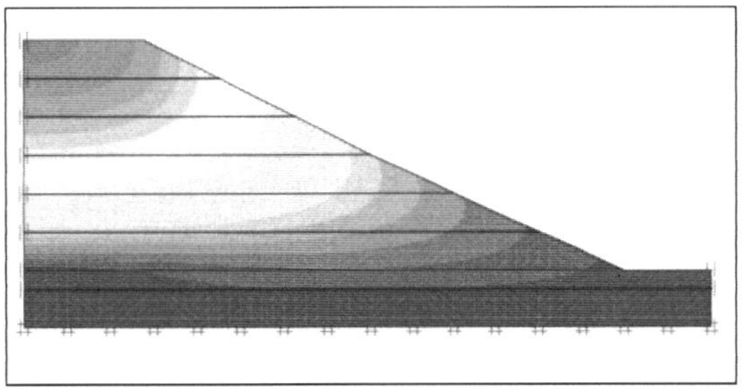

Figure 4. Escape hill total load distribution

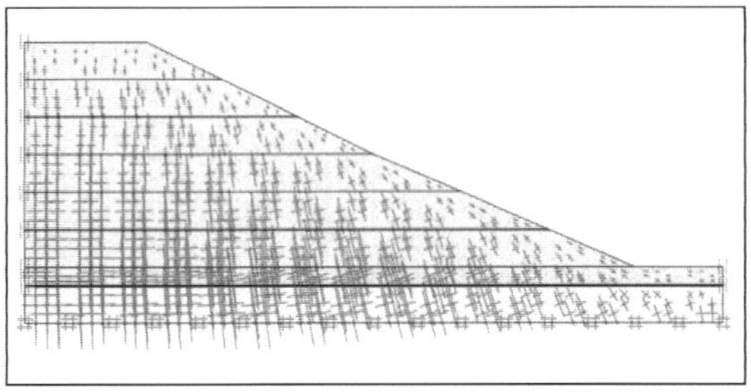

Figure 5. Total Stresses of Escape Hill

The factor of safety from Table 2 were above the required safety factor which is 1.5. Results obtained increases with increasing soil friction angle and soil cohesion. These results are realistic and as predicted.

4. Conclusions

The use of green design of disaster mitigation construction must be considered by local government for future spatial plan. Escape hill can be chosen as one of the solutions for environmentally friendly structures for tsunami evacuation building. However, the design of escape hill must follow several aspects including geotechnical engineering aspects. Based on the slope stability result from Plaxis software and by using natural soil as subgrade with soil classification of A-1-b as AASHTO method and SW for USCS, the settlement value of final model dimension was 9.117 cm with safety factor of 1.507. The safety factor is still acceptable because of still larger that 1.5 as a standard value for slope safety factor. Escape hill geometry dimension is 12 m high which is larger than tsunami run up wave with soil pressure was calculate per layers of 2 m. The deformation of slope stability and safety factor were the output of this article. The safety factor obtained is acceptable since it is larger than the required safety factor larger than 1.5.

5. References

[1] Raby A, Macabuag J, Pomonis A, Wilkinson S and Rossetto T 2015 Implications of the 2011 Great East Japan Tsunami on sea defence design *Int. J. Disaster Risk Reduct.* **14** 332–46
[2] Munirwansyah M, Munirwan R P and Yunita H 2018 Geotechnical Engineering Aspect Related to Pidie Jaya–Aceh Earthquake Disaster and Mitigation *Int. J. Adv. Sci. Eng. Inf. Technol.* **8** 870
[3] Irsyam M, Dangkua D T, Hendriyawan, Hoedajanto D, Hutapea B M, Kertapati E K, Boen T and Petersen M D 2008 Proposed seismic hazard maps of Sumatra and Java islands and microzonation study of Jakarta city, Indonesia *J. Earth Syst. Sci.* **117** 865–78
[4] Fachrurrazi, Syamsidik, Al'ala M and Mahardi W 2017 Numerical simulations of tsunami waves impacts on Ulee Lheue Harbour in Banda Aceh-Indonesia *IOP Conf. Ser. Earth Environ. Sci.* **56** 012015
[5] Schmidtlein M C and Wood N J 2015 Sensitivity of tsunami evacuation modeling to direction and land cover assumptions *Appl. Geogr.* **56** 154–63
[6] Løvholt F, Setiadi N J, Birkmann J, Harbitz C B, Bach C, Fernando N, Kaiser G and Nadim F 2014 Tsunami risk reduction - are we better prepared today than in 2004? *Int. J. Disaster Risk Reduct.* **10** 127–42

[7] Strusińska-Correia A 2017 Tsunami mitigation in Japan after the 2011 Tōhoku Tsunami *Int. J. Disaster Risk Reduct.* **22** 397–411
[8] Hazarika H, Nishimura K and Chaudhary B 2014 Model testing on resilient solution for breakwater protection against tsunami 40–4
[9] Iemura H, Pradono M H, Husen A bin, Jauhari T and Sugimoto4 M 2008 Information Dissemination For Reality-Based Tsunami Disaster Education *14th World Conf. Earthq. Eng.*
[10] Syamsidik, Fahmi M, Fatimah E and Fitrayansyah A 2018 Coastal land use changes around the Ulee Lheue Bay of Aceh during the 10-year 2004 Indian Ocean tsunami recovery process *Int. J. Disaster Risk Reduct.* **29** 24–36
[11] Ismail N, Okazaki K, Ochiai C and Fernandez G 2018 Livelihood changes in Banda Aceh, Indonesia after the 2004 Indian Ocean Tsunami *Int. J. Disaster Risk Reduct.* **28** 439–49
[12] Arif A A 2017 Green city Banda Aceh : city planning approach and environmental aspects Green city Banda Aceh : city planning approach and environmental aspects *Earth Environ. Sci.* **6** 3–10
[13] Jiang S-H and Huang J-S 2016 Efficient slope reliability analysis at low-probability levels in spatially variable soils *Comput. Geotech.* **75** 18–27
[14] Munirwansyah, Irsyam M, Munirwan R P, Yunita H and Zulfan Usrina M 2018 Geotechnical approach for occupational safety risk analysis of critical slope in open pit mining as implication for earthquake hazard *IOP Conf. Ser. Mater. Sci. Eng.* **352**
[15] Xu J shu and Yang X li 2018 Effects of Seismic Force and Pore Water Pressure on Three Dimensional Slope Stability in Nonhomogeneous and Anisotropic Soil *KSCE J. Civ. Eng.* **22** 1720–9
[16] Berisavljević Z, Berisavljević D, Čebašek V and Rakić D 2015 Analize stabilnosti pokosa metodama granične ravnoteže i smanjenja parametara čvrstoće tla *J. Croat. Assoc. Civ. Eng.* **67** 975–83
[17] Maula B H and Zhang L 2011 Assessment of embankment factor safety using two commercially available programs in slope stability analysis *Procedia Eng.* **14** 559–66

Acknowledgments

This research is partly supported by Indonesia Ministry of Research and Technology and Lembaga Penelitian dan Pengabdian Masyarakat (LPPM) Syiah Kuala University Banda Aceh Indonesia for Professor Research Scheme 2018 No: 287/UN11/SP/PNBP/2018 on 29 January 2018. The authors would also like to thank you for any supports and appreciate also to geotechnics laboratory staffs and assistants for all their assistances.

IOP Conf. Series: Earth and Environmental Science **273** (2019) 012054 doi:10.1088/1755-1315/273/1/012054

Coastal Environment and Social Environment Affecting the Vulnerability of Communities to Malaria Events in the Coastal Area of Syiah Kuala Sub-district of Banda Aceh

B Edi[1] Yustina Ida[2] Juanita[3]
Faculty of Public Health, University of Sumatera Utara, Universitas Street No.21 USU Campus Medan 20155, Medan, Sumatera Utara Province, Indonesia
edybakhtiar@ymail.com
Faculty of Public Health, University of Sumatera Utara, Universitas Street No.21 USU Campus Medan 20155, Medan, Sumatera Utara Province, Indonesia
idayust@yahoo.com
Faculty of Public Health, University of Sumatera Utara, Universitas Street No.21 USU Campus Medan 20155, Medan, Sumatera Utara Province, Indonesia
joean_ita@yahoo.com

Abstract. Banda Aceh city is dominated by coastal areas which are vulnerable areas as malaria vector breeding sites, Syiah Kuala Sub-district is one of the Sub-districts located in coastal areas with Annual Malaria Incident of 133 cases in 2017. This study aims to analyze coastal environmental factors and social environment against malaria. This type of research is a survey, using the mixed method model Concurrent Embedded method with a population of all households in 6 Villages of 5,628 households taking a sample of 112 respondents using the proportion formula. Data collection was done through questionnaires, observations, distance measurements and in-depth interviews with 6 informants, the data were analyzed bivariately using the chi-square test. The results showed that there were influences of lagoons, fish ponds, activities at certain times and knowledge of the incidence of malaria. Maintenance of malaria elimination can be optimized with coordinated, integrated planning and implementation efforts in the malaria breakdown forum. Raising and enhancing partnerships with various programs, sectors, NGOs, religious organizations, professional organizations, international organizations, donor institutions, business institutions and all levels of society by conducting migration surveys, investigating at-risk residents and monitoring the mobility of people who come and go to endemic areas to prevent malaria cases from re-entering. Malaria vector control in larva and positive larvae in the location focus by involving the community to support malaria elimination maintenance programs, especially in potential mosquito breeding sites such as lagoons, unproductive fish ponds by maintaining tilapia and blue panchax.

1. Introduction

Malaria is an infectious disease that is a public health problem in the world. High malaria can have a broad impact on quality of life, economy and poverty. According to the 2017 World Health Organization report, malaria is one of the deadliest diseases in the world that can be prevented. The number of malaria causes 105,574 cases and high risk 3.02 billion of the world's 7.6 billion population, the Global Technical Target (GTS) for Malaria 2016-2030 through the Sustainable Development Goals (SDGs) is an eradication effort contained in the third objective of ensuring a healthy life and seeking prosperity for all people until 2030.[1] In 2007 the 60th Conference of World Health Assembly (WHA) held a regional meeting in Geneva to discuss malaria elimination.[2] On December 21, 2011, the Asia-Pacific Malaria Elimination Network (APMEN) took place in the city of Brisbane, Australia discussing malaria as one of the most dangerous deadly diseases in the world.[3] The Government of Indonesia as a member of the APMEN malaria free program issued a policy through the Decree of the Minister of Health of the Republic of Indonesia Number 293 / MENKES / SK / IV / 2009 dated 28 April 2009 concerning Elimination of malaria to realize a healthy living community, gradually free from malaria transmission until 2030.[4]

Five Provinces with Annual Parasite Incident (API) / 1,000 residents, the highest being Papua (31.93), West Papua (31.29), East Nusa Tenggara (7.04), Maluku (5.81) and North Maluku (2 , 77). Provinces with the lowest API are West Java, Banten, DKI Jakarta, Bali and East Java, each with 0.00.[5] Based on the profile of the Ministry of Health of the Republic of Indonesia in 2016, suspected clinical malaria was 1,450,894 cases, blood samples were examined for 1,412,438 (97%) of suspected suspects. The number of positive malaria was 200,378 cases (0.77%) Annual Parasite Incident (API) / 1,000 residents, treated with Artemisinin-based combination therapy (ACT) as many as 172,779 (86%) of the total malaria positive in 412 districts / cities in Indonesia.[6] Historically, malaria in Aceh Province showed a relatively unstable number in the past four years. In 2014 the number of malaria positive 909 cases with API was 0.2 / 1,000 at risk, in 2015 it dropped to 422 cases with API 0.1 / 1,000 at risk, 2016 to 259 cases with API at 0.05 / 1,000 at risk, 2017 there was an increase to 300 cases with API of 0.06 / 1,000 at risk.[7] The form of Aceh Province's support in the malaria program is set out in the Aceh Governor Regulation number 40 of 2010 concerning the guidelines for Malaria Elimination.[8] Banda Aceh city is located in the coastal area and vulnerable to malaria transmission.
Based on the profile of the Banda Aceh City Health Service in the last two years, in 2016 there were 1,135 suspects, in 2017 there were 808 suspects, 2 of them (1 male and 1 female over 15 years) were positive for contracting malaria with the API 0.01 / 1,000 vulnerable population transmitted by malaria.[9] According to residential characteristics, people residing in rural and coastal areas have a higher percentage of 7.1% than people living in urban areas at 5.0%. Job characteristics as a farmer / fisherman / laborer has a higher priority of 7.8% because the type of work has a probability of vulnerability and exposure with a larger malaria vector.[10] According to the United Nations Office for Disaster Risk Reduction, vulnerability is a condition determined by physical, social, economic and environmental factors or processes that can increase the vulnerability of a community to the effects of hazards.[11]

The potential to be affected by damage or loss related to the capacity to anticipate hazards, overcome hazards, prevent hazards and recover from the effects of hazards, both vulnerability and opponents of resilience, is determined by physical, social, political, cultural and institutional factors.[12] Vulnerability determined by physical factors including (risky buildings, insecure infrastructure, insecure important facilities, rapid urbanization), environmental factors including (deforestation, soil, water and air pollution, destruction of natural protection against storms, for example, mangroves) and climate change, social factors include (unsafe area settlements, high-density land settlements, low mobility, low risk perceptions, hazard-prone jobs, vulnerable groups and individuals , corruption, low levels of education, poverty, lack of vulnerability analysis and capacity, inadequate management and leadership, lack of disaster planning and preparedness and economic factors including agriculture with one type of food crop, non-diversified economy, economic subsistence, indebtedness, dependence on assistance / support prevented and cured, thus preventive action is one of the important actions to overcome malaria.[13] Efforts to prevent infectious diseases are a shared responsibility of the government, regional government and society.[14] Indicators of malaria prevention are sleeping using mosquito nets, using mosquito coils / spray / electric / mosquito repellent lotion, installing wire netting and using closed wear. [15]

Syiah Kuala Sub-district is one of the second highest suspected malaria sub-districts out of 9 sub-districts in Banda Aceh, with Annual Malaria Incidence (AMI) of 133 cases suspected of being vulnerable to other coastal communities. The researcher focused on 6 villages from 10 villages in Syiah Kuala Sub-district, 5 of which were directly adjacent to the coast, there were lagoons at several points of settlement with global climate change which had unstable weather variability and resulted in some community fish ponds being unproductive, there was 1 village with several point of swamps, some people work as fishermen and fish pond farmers who are accustomed to activities at night do not use closed clothing. In the breeding cycle, Anopheles mosquitoes need breeding sites to lay eggs. This breeding place becomes an important thing in the life process of mosquitoes from larva then develops into pupa.[16] Then the pupa becomes an adult mosquito in the air. Only mosquito breeding sites have certain criteria that can become a place for Anopheles mosquitoes to breed. Therefore, mosquito breeding sites are one of the keys to analyzing the incidence of malaria. The purpose of this study was to analyze the vulnerability factors of the community including environmental vulnerability (lagoons, fish ponds) and social vulnerability (working at a certain time, knowledge) to the incidence of malaria in the coastal of Syiah Kuala sub-district, Banda Aceh.

2. Methodology

This type of research uses the mixed method Concurrent Embedded model.[17] In this case the author uses quantitative data as the main data and qualitative data as a complement. The population in this study were household in 6 villages in Syiah Kuala sub-district as many as 5,628 households with a sample of 112 respondents and using the proportion formula. Data collection was done through questionnaires, observations, measurements of the distance of houses with malaria mosquito habitat related to the flying ability of anopheles mosquitoes ranging from 0.5 to 2000 meters and in-depth interviews with 6 informants, data were analyzed bivariately using the chi-square test. For coastal environment vulnerability variables including (lagoons and fish ponds) measured distance of respondents 'homes to lagoons, fish farms using digital thrust Meter Rollers as research instruments with not vulnerable categories if the distance of respondents' houses to lagoons and fish ponds> 2000 meters and vulnerable categories if the distance of the house respondents to lagoons and fish ponds <2000 meters, while social vulnerability variables include (work / activity at a certain time and knowledge) to work / move at a certain time when the checklist is categorized as non-vulnerable if the respondent works at <18.00 and the category is vulnerable if the respondent works / moves at> 18:00 a.m.

For knowledge carried out using a structured questionnaire by submitting 15 questions to respondents including knowledge about malaria, ways to transmit malaria, efforts to prevent malaria, the habit of biting of anopheles mosquitoes, people's habits at night, clinical symptoms caused by anopheles mosquito bites, anopheles mosquito habitat , how to control malaria, use vector breeding places, how to treat malaria properly. Observations aim to obtain data on social phenomena related to the control and utilization of lagoons, fish ponds, abandoned fish ponds in focus locations suspected of influencing clinical malaria events. Furthermore, in-depth interviews with 6 informants consisting of 2 main informants namely farm owners who suffer from clinical malaria and 4 supporting informants namely the head of Banda Aceh City Health Office, from malaria program coordinator of Banda Aceh Health Office, head of Kopelma Health Center and head of Jeulingke Health Center in Syiah Kuala Sub-district . The study was conducted from April to August 2018 in 6 villages (Lamgugop, Rukoh, Jeulingke, Tibang, Deah Raya and Aleu Naga) Syiah Kuala Sub-district, Banda Aceh City. The sample is household, the part (subset) of the population chosen in a certain way until it is considered to be able to represent the population and use the Lameshow formula to determine the number of samples with the formula.

$$n = \frac{\left\{z_{1-\alpha}\sqrt{P_0(1-P_0)} + z_{1-\beta}\sqrt{P_a(1-P_a)}\right\}^2}{(P_a - P_0)^2} \quad (1)$$

Meaning:

n = minimum sample size
$Z_{1-\alpha}$ = standard normal distribution value (table Z) at α 5% = 1.96
$Z_{1-\beta}$ = standard normal distribution value (table Z) on β 20% = 0.842
P_0 = proportion of annual malaria incidence (AMI) Syiah Kuala sub-district = 23% (0.23)
P_a = estimated population proportion = 60% (0.6)
P_a-P_0 = estimated difference in the proportion studied in the population proportion = 37% (0.37)

$$n = \frac{\left\{1.96\sqrt{0.23\,(0.77)} + 0.842\sqrt{0.6\,(0.4)}\right\}^2}{0.1367^2} = 112 \quad (2)$$

Based on the sample calculation formula above, the sample size that will be used in the study is 112 heads of household. To determine the number of samples in each village of Syiah Kuala Sub-district, the author used a proportionate sampling because the number of subjects in each region is not equal, it can be determined using the following formula:

$$Sample = \frac{head\ of\ household\ per\ village}{total\ head\ of\ household} \times total\ sample \quad (3)$$

Table 1. Distribution of samples per village in Syiah Kuala coastal area

No	Village	Total HH	Calculation	Sample
1	Lamgugop	1.362	(1362/5628)x112	27
2	Rukoh	1.560	(1560/5628 x112	31
3	Jeulingke	1.552	(1552/5628)x112	31
4	Tibang	451	(451/5628 x112	9
5	Deah Raya	234	(234/5628) x112	5
6	Aleu Naga	468	(468/5628) x112	9
	Total		5628	112

3. Results and Discussion

3.1. Geography

Syiah Kuala Sub-district is one of the sub-districts in Banda Aceh with an area of 14,244 km^2 (1,424.4 Ha) consisting of 3 Settlements and 10 Villages, 5 Villages are the working area of Jeulingke Health Center including (Jeulingke Village, Pineung, Peurada, Tibang , Aleu Naga) and 5 Villages are the working areas of the Kopelma Health Center including (Kopelma Village, Rukoh, Lamgugop, Ie Masen Kayee Adang, Deah Raya). Some villages are directly adjacent to the coast, there are fish pond, lagoons and swamps. The area of Syiah Kuala Sub-district is bordered by: (1) The North is bordered by the Straits of Malaka, (2) The South is bordered by the District of Ulee Kareng, (3) The East is bordered by the District of Aceh Besar (4).

3.2. Demographics

The population of Syiah Kuala Sub-district in Banda Aceh City in July 2018 was 36,477 people, consisting of 18,581 men and 17,896 women with 10,241 households. The largest population is in Jeulingke Village with 6,440 inhabitants and the smallest population is in Deah Raya Village with 1,004 inhabitants.[18]

3.3. Head of Household Characteristics in Coastal Area

The results showed that from 112 respondents, 41 (36.6%) heads of household finished their high school education, the smallest percentage of 3 (2.7%) heads of household did not go to school or finish elementary school and the rest 68 (60.7%) heads of household finished junior high school, assciate degree and undergraduate degree. The largest respondents within 36-45 years of age are 41 (36.6%) heads of household. Related to gender, 93 (83%) respondents are male and for occupation, 46 (41.1%) heads of household work as entrepreneur.

3.4. Distribution of Lagoon

Of the 6 villages, the focus of the research is only Aleu Naga Village, which still has one lagoon with an area of 3 hectares.[19] The number of lagoons and area can be seen in Table 2 below:

Table 2. Distribution of Lagoon Presence in the Coastal sub-district of Syiah Kuala Banda Aceh

No	Village	Total	area (Ha)
1	Lamgugop	0	0
2	Rukoh	0	0
3	Jeulingke	0	0
4	Tibang	0	0
5	Deah Raya	0	0
6	Alue Naga	1	3
	Total	1	3

3.5. Distribution of Fish Ponds

Of 6 Villages, the focus of research in Aleu Naga and Tibang Village having the widest fish ponds of 145 hectares was in Aleu Naga Village and 125 were found in Tibang Village and the smallest fish ponds were in Jeulingke Village of 3 hectares. The existence and area of the fish pond can be seen in Table 3 below:

Table 3. Distribution of Fish Pond Existence in Coastal Syiah Kuala Sub-district, Banda Aceh

No	Village	Area (Ha)
1	Lamgugop	0
2	Rukoh	6
3	Jeulingke	3
4	Tibang	125
5	Deah Raya	42
6	Alue Naga	145
Total		321

3.6. Distribution of Anopheles mosquitoes

Of 6 villages, the focus of research is 5 Villages, there are Anopheles mosquitoes with An.vagus species in fish ponds and An.Subpitus species in the lagoon.[20] The presence of anopheles can be seen in Table 4 below:

Table 4. Distribution of Anopheles Types in the Coastal Sub-district of Syiah Kuala Banda Aceh

No	Village	Kind of Anopheles Mosquito	Place
1	Lamgugop	-	-
2	Rukoh	*An. Vagus*	Fish Pond

3	Jeulingke	*An. Vagus*	Fish Pond
4	Tibang	*An. Vagus*	Fish Pond
5	Deah Raya	*An. Vagus*	Fish Pond
6	Aleu Naga	*An. Subpitus*	Lagoon

3.7. Distribution of clinical malaria

Of the 6 villages, the focus of the study found that the largest clinical malaria in Rukoh Village was 46 cases and the smallest one in Lamgugop Village was 10 cases.[21] The number of clinical malaria can be seen in Table 5 below:

Table 5. Number of Clinical Malaria in the Coastal of Syiah Kuala Sub-district Banda Aceh

No	Village	Total of clinical malaria
1	Lamgugop	10
2	Rukoh	46
3	Jeulingke	14
4	Tibang	18
5	Deah Raya	16
6	Alue Naga	29
	Total	133

3.8. Effect of Lagoon on Malaria Events

Based on the results of the statistical analysis using the Chi Square test, the p value = 0,000 was obtained, meaning that the lagoon factor affected the incidence of malaria (p <0.05) in the coastal area of Syiah Kuala Sub-district can be seen in Table 6 below:

Table 6. Influence of lagoon on malaria cases in Syiah Kuala coastal area

Lagoon	Malaria cases				Number		*P value*
	High		Low				
	n	%	n	%	N	%	
Vulnerable	12	92,3	1	7,7	13	100.0	0,000
Not Vulnerable	24	24,2	75	75,8	99	100.0	

From the results of observations and secondary data in Syiah Kuala Sub-district, Aleu Naga Village still has a large lagoon with a 50-meter sea boundary, which is bordered by sand. The profile data of Syiah Kuala Sub-district in 2018 states that there are around 3 hectares of lagoons on the border of Aleu Naga Village with Bait Kaju Village, Aceh Besar. There are 2 out of 6 coastal villages including Aleu Naga and part of Tibang which their houses / residences are close to the lagoon. Respondents in this study aged 26-65 years were the age in the category of early adults up to the final elderly. The majority of respondents were 9 (8%) houses / dwellings close to the lagoon aged 36-45 years, namely the final adult age group and the smallest respondents of 4 (3.6%) houses / dwellings close to the lagoon aged 45-55 years namely early age groups were dominated by male family heads who were used to doing fishing.

From the results of measurements of the distance of the house / residence of respondents with a lagoon in 112 respondents obtained a description of distance with a vulnerable category as many as 13 families (11.6%). After a bivariate test with Chi-Square statistical test shows the lagoon variable has a relationship with the incidence of malaria (p <0.05), p = 0,000. Details of distance are 9 households in Aleu Naga Village, which are 200-500 meters away from 6 (66.67%) households and 600-1,500 meters in distance (3 (33.33%) households, from 4 families in Tibang Village with 3 (75%) KK is 1,000-1,300 m and 1 (25%) KK is 1,400-2,000 m. Female anopheles mosquitoes usually bite humans at night or from dusk to dawn and the flight distance is no more than 0.5-2 meters from the breeding place.[22] Based on in-depth interviews with the head of the family who served as head of the hamlet with the profession of fish pond farmers in Aleu Naga Village suffering from clinical malaria, it was explained that in the past 6 months, health workers had visited their villages to check / snatch malaria larvae / parasites (plasmodium) in the lagoon. They did not know that the lagoon was a breeding ground for malaria mosquitoes and that for the past 3 months there had been no mutual aid to drain lagoon water into the sea. Some people use lagoons to fish, which are dominated by adult men early to late adulthood and look for oysters in lagoons dominated by mothers from late adulthood to the elderly. None of them collect / use the lagoon to maintain predatory fish larvae such as tin-headed fish, tilapia and blue panchax.

Malaria is an infectious disease that is a public health problem in the world. High malaria can have a broad impact on quality of life, economy and poverty. The WHO's target of the 2016-2030 Global Technical Strategy (GTS) for malaria through the Sustainable Development Goals (SDGs) is eradication efforts contained in the third objective of ensuring a healthy life and seeking prosperity for everyone until 2030. Based on in-depth interviews with the Health Office of Banda Aceh city, which was represented in the field of prevention and infectious diseases (P2P), found that during this time the Health Office Banda Aceh city realized the programs / activities of infectious diseases, especially those related to mosquitoes, especially malaria in the maintenance phase of malaria. Laboratory to read the results of microscopic blood tests. Conduct surveys in the field if there are case reports from the community, conduct examination of thick blood preparations and peripheral blood in patients suspected of clinical malaria. Spraying the IRS in 1000 houses in cases and then running 20 houses in the working area of Kopelma Health Center. Until now there has been no special activity at potential mosquito breeding sites to monitor and pick up plasmodium larvae in the lagoon.

The absence of information dissemination to the good coastal communities of Syiah Kuala Sub-District, especially the control and utilization of lagoons for the coastal communities of Syiah Kuala sub-district, Banda Aceh. This is worsened by the lack of attention and cooperation from other sectors such as the Fisheries Service, Forest Service, Public Works Agency and others in controlling the environment such as lagoons, productive fish ponds and unproductive fish ponds. Whereas by utilizing the lagoon to maintain larvae predator fish such as tin-headed fish, tilapia and blue panchax can add income, reduce vulnerability and prevent good clinical malaria in the coastal areas of Syiah Kuala Sub-district, Banda Aceh.

3.9. The Influence of Fishponds on Malaria Incidence

Based on the measurement results of the distance from the respondent's house to the fishponds, 85 (75.90%) respondents are of vulnerable category and of 27 (20.10%) respondents are of not vulnerable category. The results of statistical analysis using the Chi-squared test showed ($p<0.003$), meaning that there is a significant influence between fishponds and the incidence of malaria ($p <0.05$) in the coastal area of Syiah Kuala. It can be seen in the table 7 below:

Table 7. Influence of fishponds on malaria cases in Syiah Kuala coastal area

Fishponds	Malaria cases				Number		*P value*
	High		Low				
	n	%	n	%	N	%	
Vulnerable	51	60,0	34	40,0	85	100.0	0,003
Not vulnerable	25	92,6	2	7,4	27	100.0	

Based on the results of observations and measurements of the distance from the respondents' house to the fishponds, the distance with a bad category was 85 heads of household (75.90%). After bivariate analysis with Chi-squared test showed that fishpond variables have a significant relationship with the incidence of malaria ($p<0.05$), $p=0.003$. Houses with mosquito breeding vectors around them have a greater proportion of malaria incidence (54.5%), compared to houses that have no mosquito breeding sites around them (49.4%).[23] The malaria case increased because there was a lack of support from related sectors. The community considers malaria is not a serious problem, so if people are affected by malaria, they do not immediately check with health service officials that will lead to higher malaria transmission. This is worsened by the lack of attention from other related sectors such as the Fisheries Agency, Forestry Service, Public Works Agency and other related agencies or services.

From the observation results and secondary data in Syiah Kuala Sub-district, there were 5 out of 6 coastal villages that have both productive and unproductive fishponds. The data from Syiah Kuala Sub-district in 2018, there are around 356 hectares of fishponds scattered in 5 coastal villages in Syiah Kuala. The largest fishponds are located in Aleu Naga Village with 145 hectares and the second ones are in Tibang with 125 hectares while the smallest fishponds are in Jeulingke Village with 3 hectares. Activities in handling of abandoned fishponds in Sidodadi Village, Padang Cermin sub-district, Pesawaran Regency are the removal of moss and the spread of tilapia predatory fish which involves malaria cadres.[24]

Malaria Care Forum (NGO) was in charge of the activity, while the fish spreading activity involving malaria cadres and the community with Pesawaran District Fisheries Service in charge of the activity. The spreading of 10,000 fish in abandoned fishponds in Sidodadi Village is a cross-sector program between the Fisheries Service and the Public Health Office. After the program to remove moss and algae and the spreading fish, cases of clinical malaria and positive plasmodium appear to be decreasing. In addition to handling abandoned fishponds, the other malaria control program was Indoor Residual Spraying (IRS), mass blood survey, larva catching, mangrove planting, and clean Friday movement. Cross-sector support related to malaria treatment is a necessity. The phenomenon of malaria is a continuous sequence. Firtsly, a person falls ill due to contact with the environment, then an agent reacts in the body to fight the disease which may end up in a condition of being sick or healthy. The continuous phenomenon occurs around the world that includes infectious diseases such as malaria incidence on the coast of Syiah Kuala Sub-District, Banda Aceh.

Related to respondents' occupation, 46 heads of household are entrepreneurs (41.1%), others are 24 (21.4%), 15 are civil servants (13.40%), 10 are housewives (8.10%), 9 are fishpond farmers (8%) and the smallest number is 8 fishermen (7.1%). Related to the measurement results of the distance from the house of the respondents to the fishponds, from 9 heads of household in Aleu Naga Village, 7 (77.8%) have a distance of 200-500 meter and 2 (22.2%) have a distance of 600-1,500 meter. From 9 heads of household in Tibang Village, 6 (66.67 %) have a distance of 100-500 meter and 3 (33.3%) have a distance of 600-1,000 meter. From 5 heads of household in Deah Raya Village, 4 (80%) have a distance of 100-500 meter and 1 has a distance of 600-1,000 meter. From 31 heads of household in Rukoh Village, 10 have a distance of 100-500 meter and 21 have a distance of 600-1,000 meter. From 31 heads of household in Jeulingke village, 8 (25.80%) have a distance of 100-500 meter and a 23 (74.20) have a distance of 600-1,000 meter.

Region-based disease management must be carried out in an integrated manner from the planning, implementation, financing and monitoring of its implementation, likewise, the management of abandoned fishponds with an integrated manner must be carried out at all stages of malaria control activities, for example, at the prevention stage, integration can be applied to extension programs both carried out by the Public Health Office and other agencies, from the health aspect, besides explaining the symptoms and treatment of malaria, it is also necessary to explain about the potential mosquito breeding sites and their dangers, whereas from the Forestry Service, it is necessary to explain the effects of mangrove removal and the benefits of mangrove preservation, also from the legal aspect, it is necessary to explain the rules and regulations per se; it also needs to be conveyed about the importance spatial planning of coastal area in the spatial planning of regency/city.[25] Likewise, the Spatial Planning in the coastal area of Syiah Kuala needs to be carried out.

While in the spatial planning of province, the extension program will succeed well if there is an integration from planning, implementation, financing and monitoring. The management of abandoned fishponds so as not to be a place for breeding mosquitoes is a cross-sector program involving various agencies. Besides, it also requires a large amount of funds. Because many parties are involved and require large funds, the government's political will is crucial. Without the support and involvement of all parties, the management activities of abandoned fishponds will not be sustainable or only a temporary activity. The need for community empowerment in handling abandoned fishponds and active participation from the community are also highly expected in mosquito breeding management programs in abandoned fishponds. With the involvement of the community, there is a guarantee for the sustainability of activities because the community can participate in monitoring the activities or at the same time implementing them. Utilizing the fishponds again are not an easy thing, many factors can influence it, in addition to the availability of funds, human resources, and technology.

Constraints on community empowerment programs for the spread of predatory fish such as red tilapia in the fishponds that are not productive are not easy. Although in terms of HR, it is not a problem with the active involvement of the community, but in terms of technology regarding fish farming, both red tilapia and milkfish are not easy, requiring the technology of water flow in the fishponds. The availability of this technology initially received assistance from the Fisheries Service but in its implementation there were obstacles that the community could not immediately resolve. The communities often act passively and no effort to solve the problem on their own.

Malaria is a complex problem so that eradication of malaria must be carried out in an integrated manner by all related components and becomes an integral part of national development for the realization of healthy communities, gradually getting rid of malaria transmission until 2030. Based on in-depth interviews with householders who work as fishpond farmers in Deah Raya Village who suffer from clinical malaria, it has been stated in the last 6 months that health workers have never visited their villages to check malaria and catch parasite larvae (plasmodium) in fishponds. In the past year, moss has never been cleaned on farms, because the fishponds were only partially in use by making barrier on fishponds because their harvest had declined in the last two years, so there was not enough fund to clean the fishponds. While fishpond farmers also experienced the same thing, there was a crop failure due to pest. And until now the village financial assistance has not been disbursed, only few fish seeds are kept and they have much time to go fishing, sometimes also fishing in neglected fishponds for additional income.

In the last three months, they fished or sailed without using long-sleeved clothes and did not always use mosquito repellent lotion, except while going for fishing that they use lotions because there are lots of mosquitoes. They use the neglected fishponds just to fish as a hobby that can give them extra income. Malaria is a serious problem and the handling of abandoned fishponds is necessary to break the chain of transmission of malaria. In addition to reviving abandoned fishponds with shrimp farming, it can also replace it with cultivation of blue panchax, parrot fish and tilapia fish that require funds not as large as shrimp farming. Cross-sector, integrated, and sustainable cooperation are the groundwork in handling abandoned fishponds.

Based on the in-depth interview with a section chief of infectious disease prevention in the Public Health Office, there were funds originating from health operational costs to monitor mosquito breeding vectors, but now the funds are only utilized to monitor the larvae around the house such as bathtubs, used bottles and trenches in rural areas which are more likely to monitor aedes aegipty larvae (DHF).

Whereas in unproductive or neglected fishponds, larvae catching has never been conducted. Until now there is no mutual cooperation program to clean moss on the fishponds, and the community itself either individually or in groups do not clean the moss in fishponds because the cost of cleaning is expensive while the crop are small (crop failure) and there is no socialization to inform coastal communities, both in the Syiah Kuala sub-district and the other sub-districts in Banda Aceh related to the use of abandoned fishponds for coastal communities. The lack of attention from other related sectors such as the Fisheries Service, and others in environmental control has exacerbated the situation. In fact, using fishponds to keep blue panchax, parrot fish and tilapia fish can add up income, reduce vulnerability and prevent clinical malaria and positive malaria in the coast of Syiah Kuala Sub-district of Banda Aceh.

3.10. Working / Activity Factors at a Specific Time with Malaria Events

Based on the results of statistical analysis using the Chi Square test, the value of p = 0,000, meaning that the activity / work at a certain time has an effect on the incidence of malaria (p <0.05) in the coastal area of Syiah can be seen in Table 8 below:

Table 8. Influence of work / activities on malaria cases in the Syiah Kuala coastal area

Work/ activity at specific time	Malaria cases				Number		*P Value*
	High		Low				
	n	%	N	%	n	%	
Vulnerable	34	40,0	51	60,0	85	100.0	0,000
Not vulnerable	2	7,4	25	92,6	27	100.0	

Based on the results of interviews and community observations on 112 households in the coastal area of Syiah Kuala Sub-district, descriptions of work / activities at certain times were obtained with a vulnerable category of 49 (43.75%) households. After bivariate testing using the Chi-Square statistical test shows that the variable work / activity at a certain time has an influence on the incidence of malaria (p <0.05), p = 0,000. Observations found that malaria caused by work / activity outside the home at night was related to the habits of some species of mosquitoes that were exophagic at night. Exophagic mosquitoes are mosquitoes that bite a lot outside the home, but can enter the house if humans are the preferred main host. This is also related to the number of respondents who work / do activities outside the home such as some households of early adulthood and late adulthood who fish in the lagoon, the fish pond is neglected and some mothers look for oysters in the lagoon until late at night on the coastal District Syiah Kuala.

Based on in-depth interviews with the households who served as head of the hamlet with a fish pond farming profession in Aleu Naga Village who suffered from clinical malaria, explained that the farm had not been cleaned for the past 2 years and found many mosquitoes and usually went to the fish pond using short-sleeved clothes and not using mosquito repellent lotion, they only burn dried tree trash to repel mosquitoes. Malaria prevention efforts can be carried out by increasing awareness of the risk of malaria, preventing mosquito bites, controlling vector / suspect vectors and chemoprophylaxis. Prevention of mosquito bites can be done using mosquito nets, mosquito repellent lotion / spray / mosquito repellent / electric mosquito repellent and for those who are accustomed to working / doing activities above 6:00 p.m. you should use closed clothes or long sleeves and trousers to avoid mosquito bites.[26]

3.11. Knowledge with Malaria Events in the Coastal of Syiah Kuala Sub-district Banda Aceh

Based on the results of statistical analysis using the Chi Square test, the value of p = 0.002 is obtained, meaning that there is a significant influence between knowledge and the incidence of malaria (p <0.05) in the coastal area of Syiah Kuala Banda Aceh can be seen in Table 9 below:

Table 9. Influence of knowledge on malaria cases in the Syiah Kuala coastal area

Knowledge	Malaria cases				Number		*P value*
	High		Low				
	n	%	N	%	n	%	
Poor	27	45,8	32	54,2	59	100.0	0,00
Good	9	17,0	44	83,0	53	100.0	

Based on the results of research on 112 respondents obtained a description with a bad category as many as 59 (52.68%) households. After bivariate testing with Chi-Square statistical tests showed that the knowledge variable had a significant effect on the incidence of malaria (p <0.05), p = 0.002 in the Coastal of Syiah Kuala Sub-district, Banda Aceh. Respondents only know partially about understanding, clinical symptoms, modes of transmission, prevention methods, places to breed mosquitoes, how to find the right treatment. Based on the results of the statistical analysis, it can be explained that one's knowledge will influence the incidence of malaria on the coast of Syiah Kuala Sub-district, Banda Aceh. This is supported by Bloom's theory stating that knowledge is knowing what is done and how to do it; knowledge is the result of knowing someone about an object through their senses and is influenced by the intensity of attention and perception of the object.[27]

In this study respondents did not know that malaria was transmitted by parasites to anopheles mosquitoes as many as 41 (36.6%) households, did not know malaria transmission through direct contact with malaria sufferers as many as 63 (56.3%) households, did not know one efforts to prevent malaria are to use mosquito nets during sleep at night as many as 47 (42.0%) households, do not know that malaria mosquitoes actively bite in the morning, afternoon and night as many as 63 (56.3%) households, do not know the habit of doing activities outside the home at night will be at risk of being bitten by malaria mosquitoes as many as 57 (50.9%) households. Not knowing malaria will heal on its own as many as 55 (49.1%) households, not knowing malaria can be affected by all age groups, except toddlers as many as 52 (46.4%) households, do not know the clinical symptoms of malaria, namely the presence of spots in the arms and body as many as 62 respondents (55.4%) households, did not know that malaria mosquitoes could breed in bathtubs and buckets filled with water in the house as many as 56 (50.0%) households, did not know that malaria mosquitoes could not develop multiply in unproductive lagoons / fish ponds and swamps as many as 51 (45.5%) households, do not know the need to hoard as soon as swamps, puddles that can breed malaria mosquitoes as many as 63 respondents (56.3%) households.

Unaware of how to make a lagoon water irrigation channel into the sea is needed as many as 60 respondents (53.6%) households, do not know to maintain tilapia and blue panchax can reduce vulnerability, prevent malaria and increase income by 55 (49.1%) households , not knowing malaria can cause death if untreated immediately as many as 48 (42.9%) households did not know if the fever, shivering (malaria) had to go directly to the public health center, they only had 57 (50 9%) households. This is in line with the results of testing with chi square, the solution is to use fisher exact, the results show that p = 0.02 with continuity correction p value is 0.002, this shows that Ho is rejected with $\alpha = 0.05$, meaning that p-Value is smaller than the level of error that has been determined so that it can be seen that there is a relationship between the level of knowledge with the incidence of malaria in the Public health center area Kasongan Katingan Hilir Sub-district Katingan Regency.[28] This is reinforced by the research results of statistical analysis obtained X2 count (33,885)> X2 table (3,841) and p (0,000) <α (0,05), meaning there is a relationship between knowledge and incidence of malaria in Koeloda Health Center, Golewa Sub-district Ngada Regency.[29]

Knowledge determines a person's behavior, for example preventive measures (health prevention behavior) for malaria, i.e. every action taken by individuals to prevent malaria, among others, sleeping using mosquito nets, using anti-mosquitoes, installing mosquito nets, using long-sleeved clothes when working / doing activities above 18.00 hours, flowing water from the lagoon to lagoon or creating a coastal barrier, cleaning the fish ponds from moss, closing / hoarding marshes and dealing with lagoons, ponds are not productive to maintain predator mosquito larvae like blue panchax , indigo and tilapia. Efforts to increase public knowledge about the dangers of malaria have been carried out by the Banda Aceh City Health Office through the management of eradication and prevention of malaria is to carry out counseling and promotions in the form of solicitation such as morning patrol conducted by Kopelma Public Health Center. However, providing information on lagoon production, productive ponds and unproductive ponds as a place that has the potential and susceptibility to malaria has never been done. Likewise the management / utilization of lagoons, unproductive fish ponds can be utilized to maintain tilapia and blue panchax as predatory fish to eat mosquito larvae which can reduce the malaria vector and can increase additional income.

Based on the results of interviews with the Head of the Health Office represented by the prevention, control of infectious diseases and the Malaria Program Coordinator, the Banda Aceh City Health Office said that counseling was carried out individually when the community was sick with clinical symptoms of malaria coming to the health center. The results of interviews with the Head of the Jeulingke Public Health Center said that counseling was not carried out related to mosquito breeding places such as lagoons, fish ponds, while the explanation of the Kopelma Public Health Center head was now conducted only in the morning to clean the house using ambulances in one neighborhood each Village Kopelma Health Center as a pilot project. Based on the results of the study on 112 respondents 36 (32.14%) Households experienced clinical malaria. The results of in-depth interviews with pond farmers who suffered from clinical malaria in the past 6 months, health workers never came to the village to inform them that potential places for developing malaria mosquitoes such as lagoons, unproductive fish ponds with the reason that they might not have monitoring programs especially on lagoons, fish farms are not productive. Some people do not know that the lagoon is a breeding ground for malaria mosquitoes on the grounds that there is no information that the lagoon is a breeding ground for anopheles mosquitoes. During the last 2 years, some community farms were not operated and did not carry out maintenance, cleaning moss on ponds with the reason of crop failure so that there was not enough funds to clean up the removal of moss because the cost of cleaning fish ponds was expensive.

During the last 3 months the community had never had a mutual aid flowing water from the lagoon to the sea on the grounds that there was no appeal from the Village Head also from Health officials and part of the coast had been made embankments but was cut off in the middle of Aleu Naga Village with the border of Bait Kaju Village, Aceh Besar. When going to unproductive farms /fish ponds at> 18:00 a.m. usually do not use mosquito repellent or do not use clothes on the grounds that when they go to fish farms, they usually burn trash of dried tree branches or use mosquito coils, also accustomed to using open clothing or short sleeve shirt. Non-productive ponds are not utilized to maintain predatory larvae fish such as blue panchax, tilapia for the reason that the place is like a lot of crabs and various other types of fish. They did not know that maintaining blue panchax and tilapia, besides being able to increase income, could also reduce the breeding of plasmodium mosquito vectors so that the incidence of malaria both clinical and malaria positive was reduced. The malaria control program carried out in Syiah Kuala Sub-district based on this research seems to emphasize more on goal rationality, namely health workers are still limited to being part of the "healer" facility, these officers have not been optimal in carrying out preventive and preventive functions on the coast of Syiah Kuala Sub-District, Banda Aceh.

4. Conclusions

Based on the results of the research and discussion described earlier, some conclusions can be drawn as follows:

1. There is an effect of lagoons on the malaria incidence at p=0,000 on the coast of Syiah Kuala
2. There is an effect of fish ponds on malaria incidence at p = 0.003 on the coast of Syiah Kuala Sub-District, Banda Aceh.
3. There is the influence of work / activity at a certain time on the incidence of malaria in the amount of p = 0,000 in the coastal area of Syiah Kuala Sub-District, Banda Aceh.
4. There is an influence of knowledge on the incidence of malaria at p = 0.002 on the coast of Syiah Kuala Sub-District, Banda Aceh.

5. Suggestions

Suggestions that can be given for prevention, control of malaria vectors related to the vulnerability of the community to the incidence of malaria on the coast of Syiah Kuala Sub-District, Banda Aceh are:

1. For the Banda Aceh City Health Office, in planning and implementing efforts to maintain malaria elimination, it should be carried out in a coordinated and integrated manner in the national malaria eradication forum. Raising and enhancing partnerships with various programs, sectors and all levels of society by conducting migration surveys, investigating at-risk residents and monitoring the mobility of people who come and go to endemic areas. Malaria vector control in larva and positive larvae by

involving the community by keeping fish such as blue panchax and tilapia on potential mosquito breeding sites like lagoons, productive fishponds and unproductive fishponds.
2. For the Jeulingke and Kopelma Public Health Center in Syiah Kuala Sub-District, it is expected to increase surveillance that reaches all eliminated areas. Detect cases early and treat appropriately, increase socialization / extension of sustainability. Encourage the community to participate in mutual cooperation in potential breeding places for mosquitoes, support the management of breeding places such as lagoons, productive fishponds and unproductive fishponds so as to produce economic value.
3. For coastal communities in Syiah Kuala Sub-district are expected to avoid mosquito bites by using repellent, wearing closed clothes when they go out above 18.00, using mosquito nets and willing their house be sprayed by the IRS. Using unproductive fishpond and lagoons to maintain blue panchax and tilapia fish as a malaria larvae control. Supporting and getting involved in local malaria snatching in seeking malaria sufferers as early as possible.

References

[1] WHO 2017 *G. S. P. R. B. Mal. 2005-2015* Geneva
[2] WHA 2011 *Prev. Cont. N. Com. Dis. Imp. G. S.* The Sixtieth Conference of World Health Assembly Geneva
[3] Asia-Pacific Malaria Elimination Network 2011 *Disc. Mal. M. Da. Dd. Dis. W.* Brisbane.
[4] Republic of Indonesia 2009 *Min. H. Reg. Con. Mal. Eli.* Jakarta
[5] Ministry of Health 2016 *Min. H. Reg. Ind. H. Pro.* Jakarta
[6] Ministry of Health 2017 *Min. H. Reg. Ind. H. Pro.* Jakarta
[7] Aceh Province Health Office 2017 *Pro. Ac. Prov. H. O.* Banda Aceh
[8] Aceh Government 2010 *Aceh Gov. Reg. No. 40* Con. *Gui. Mal. Eli. Ac.* Banda Aceh
[9] Banda Aceh City Health Office 2017 *Pro. B. Ac. H. O.* Banda Aceh
[10] Republic of Indonesia Sub-directorate of Malaria Control P2B2, Directorate General of PP & PL Ministry of Health of and GF ATM PR of Indonesian Ministry of Health Malaria Component 2015 *Mal. Cont.* Jakarta
[11] UNISDR 2005-2015 *Sum. Hyo. Fra. A. Bui. Resi. N. Comm. Dis. UN/ISDR-07-2007* Geneva
[12] The National Agency for Disaster Countermeasure 2012 *Reg. Nat. Dis. Man. 2012 Con. Gen. Gui. Dis. R. As.* Jakarta
[13] Benson C, John T and Rossetto T 2007 *T. F. Main. D. R. Reduc.* (Geneva: Prevention Consortium Secretariat) p 121
[14] Republic of Indonesia 2009 *Reg. A. H. No.36* Jakarta
[15] Ministry of Health 2014 *Mal. Prev. Indi.* Jakarta
[16] Harijanto P N 2000 *Sym. Cli. Mal. Epi. Pat. Cli. Man. Man.* (Jakarta: EGC) p 151-155
[17] Sugiyono 2014 *Com. Res. Met. (*Bandung: Alfabeta)
[18] Banda Aceh City Statistics Agency 2017 *Tot. Pop. Ave. Pop. Den. Vil. Ave. Pop. Den.*
[19] Government of Syiah Kuala District 2017 *Gov. S. K. Dis. 2017 Con. Syiah Kuala Sub-District Fig.* Banda Aceh
[20] Widya S 2007 *Jou. Stu. T. Ano. Bre. S. Ru. V. S. K.* Banda Aceh
[21] Jeulingke and Kopelma Health Center 2017 *Pro. Jeul. H. C.* Banda Aceh
[22] Harijanto P N 2010 *Sym. Cli. Mal. Epi. Pat. Cli. Man. Man.* (Jakarta: EGC)
[23] Ernawati K, Umar F Achmadi, Tresna P Soemardi, Thayyib H and Sri M R 2012 *A. F. Bre. G. M. M. Ende. (Ca. Hand.)* **10** **(2)**: 8
[24] Ernawati K, Umar F Achmadi, Tresna P Soemardi, Thayyib H and Sri M R 2012 *A. F. Bre. G. M. M. Ende. (Ca. Hand.)* **10** **(2)**: 10-11
[25] Ernawati K, Umar F Achmadi, Herika H 2014 *Rel. B. M. Bre. P. W. M. Pesa.* Public Health Science Paper **5** **(3)**: 8
[26] Republic of Indonesia Ministry of Health Malaria Sub-Directorate of P2PTVZ 2017 *Perf. R. of Gen. Dir. Pre. Cont. Dis.* Surabaya
[27] Notoatmodjo S 2005 *Health Res. Met. Rev. Edi.* (Jakarta: Rineke Cipta)

[28] Darmiah, Baserani, Khair A, Isnawati, Suryatinah Y 2017 *Rel. Lev. Kno. Pat. Beh. Inc. Mal. Kat. Dis. Cen. Kal. Pro.* JHECDs **3** (**2**): 4-5
[29] Idrus M, Getrudis 2014 *Rel. Ind. Fac. Env. Mal. Inc. Koe. H. C. Gol.* Health Journal **7** (**2**)

The Evolution of Localized Housing Recovery in Japan

Elizabeth Maly[1] **and Tsukasa Iwata**[1]

[1]International Research Institute of Disaster Science IRIDeS, Tohoku University, 468-1 Aoba, Aramaki, Aoba-ku, Sendai, 980-8572 Japan.

Abstract. Localized disaster recovery housing refers to housing made using local resources (materials, people, and skills). Using local resources to build temporary and permanent housing for people who lost their homes helps revitalize the disaster-affected area by contributing to the local economy and supporting local businesses. In the context of Japan, localized housing recovery can be understood as using some or all of the following: timber materials; traditional/local wood construction methods; and local companies and craftspeople. The term 'localized housing recovery,' *chiikigata juutaku fukkou* in Japanese, was first used in the recovery of the Yamakoshi area of Nagaoka City after the 2004 Chuetsu Earthquake, where local timber and carpenters were included from the post-disaster planning through rebuilding phases. While wooden temporary housing was rarely used since the 1995 Great Hanshin Earthquake in Kobe, after the 2011 Great East Japan Earthquake a "localized type" of construction used local natural resources for the construction of wooden temporary housing or recovery housing, including over 6,000 units of wooden temporary housing in Fukushima Prefecture alone. A similar approach was used after the 2011 Flood Disaster in the Kii Peninsula. This paper provides an overview of the evolution process of this localized recovery housing in Japan.

1. Introduction

Due to the various benefits for residents and the affected areas, experts often promote the use of local building materials in housing reconstruction after natural disasters [1, 2]. Sourcing materials locally contributes to the economic recovery of the affected area, supports local businesses, and can reduce time needed to start the reconstruction process [3]. Environmental impacts are reduced by shorter transportation distances to get materials to the site initially and in the life cycle of the building as future repairs can also be made using locally available materials. This in turn benefits residents, as repairs will be more affordable (or may be done by residents themselves). Using local materials also supports traditional craftspeople and the cultural heritage of the region, and traditional housing designs can contribute to a comfortable living environment for disaster survivors.

However, especially in using local timber for reconstruction of wooden houses, there may be challenges such as material scarcity or risk of accidentally supporting illegal logging/environmentally damaging impacts, as occurred in Aceh among the donor-driven housing reconstruction projects after the 2004 Indian Ocean Tsunami [4]. Humanitarian shelter providers from outside the disaster area may also make mistaken assumptions that materials will be locally available, or that the use of timber is the best environmental or economic choice even when this is not the case [4, 5]. However, this is not a concern in Japan, where there is a well-established and well-regulated system for all aspects of forestry management and timber production.

The use of local timber materials in post-disaster housing reconstruction in Japan occurs within the specific context of a traditional wood construction culture and local timber industry, paired with specific government-driven approaches to disaster recovery. With houses traditionally built from wood, Japan has a long history of timber production, construction, and carpentry. Even today, wood construction is the standard for building detached housing, although the process has been modernized with the use of pre-cut timbers that can be assembled more quickly than in traditional carpentry. While the construction process has modernised, traditional modules of wooden housing construction are still visible.

Japan also has a distinct post-disaster housing reconstruction context. Although the Japanese government accepts some international support after disasters, recovery projects are almost exclusively funded by the national government. With a long history of disaster experience, Japan has well-established disaster recovery policies that have been modified over time. Japanese post-disaster housing recovery occurs within a clearly delineated process for the provision of temporary housing and permanent housing support. Implementation of localized housing recovery in Japan, therefore, integrates the use of traditional timber construction into the context of housing recovery shaped by government-driven reconstruction policies. Initiatives supporting localized housing recovery include temporary housing and self-funded (by homeowner) and government-funded construction of permanent housing.

Within the context of Japanese post-disaster recovery, this paper considers that 'localized housing recovery' uses a combination of timber materials; traditional/local wood construction methods; and local companies and craftspeople. The term 'localized housing recovery,' *chiikigata juutaku fukkou* in Japanese, was first used in the recovery of the Yamakoshi area of Nagaoka City after the 2004 Chuetsu Earthquake, where local timber and carpenters were included from the post-disaster planning through rebuilding phases. After the 2011 Great East Japan Earthquake, there was a significant increase in a localized type of construction using local natural resources for the construction of wooden temporary housing or recovery housing. Localized housing continued to develop after subsequent disasters including the 2011 Flood Disaster in the Kii Peninsula. Following an overview of Japanese post-disaster housing reconstruction support policies and projects, this paper traces the implementation and development of localized disaster housing recovery in Japan.

2. Purpose and Methodology

Although the use of local materials is commonly supported by international humanitarian agencies and housing recovery experts, there are few studies explaining the Japanese approach and experience with localized housing recovery in international literature on housing reconstruction. The purpose of this paper is to clarify the features of Japanese localized housing recovery and chart the development of this approach through the implementation of key cases.

Through an examination of the factors that contribute to the creation of these projects and their connection to traditional construction methods and local materials, this paper considers the necessary conditions to support and improve the implementation of localized housing recovery, along with lessons from Japan that may be relevant to other countries. This paper relies on a review and synthesis of key literature and data about localized disaster recovery housing in Japan, and also draws on the authors' direct involvement in the promotion of localized housing and implementation of related cases.

3. Japanese System for Post-Disaster Housing Reconstruction Support

With frequent exposure to natural hazards, Japan suffers repeated disaster damage and has established policies and systems to support temporary and permanent housing reconstruction. The national government coordinates and funds housing reconstruction in Japan, with prefectural and municipal governments having clear roles for planning and implementation of recovery projects. Unlike countries where international agencies provide the majority of housing reconstruction support through donor-driven reconstruction projects, government policies dictate each phase of housing recovery and reconstruction in Japan. In the case of a major disaster, after an initial evacuation phase where survivors stay for days, weeks, or even months in evacuation centers such as schools or municipal buildings, housing recovery support focuses on the provision of temporary housing, followed by projects supporting permanent housing reconstruction.

3.1. Temporary Housing

The Act for Support of Disaster Victims stipulates the provision of temporary housing by prefectural governments to disaster survivors at no cost. According to this law, temporary housing is designed to be used for 2 years, and therefore does not require construction to the same building standards as typical housing construction. However, the use of temporary housing can and has been extended — more than 7 years after the 2011 Great East Japan Earthquake there are still tens of thousands of people living in temporary housing [6]. Most temporary housing units are smaller than $30m.^2$

3.2. Permanent Housing: Private Reconstruction, and Public Housing Provision

For the reconstruction of permanent housing, recovery projects mirror the Japanese government' two-tiered approach to housing policy: while the majority of households are expected to secure their own housing in the private market, public housing is provided as welfare housing for those who cannot access this market [7]. After the 1995 Hanshin Awaji Earthquake in Kobe, the government provided very little support for private housing reconstruction, focusing instead on the provision of public housing. In response to the difficulties that this caused for life recovery, subsequent revisions to the Act for Support of Disaster Victims in 1999 allowed families with completely damaged houses to receive up to 3 million yen ($30,000 US) compensation from the national government (local governments can add to this) that can be used for rebuilding [8]. Although allowing government funds to be used for private reconstruction represents a significant change in policy, this amount is not enough to rebuild a new house.

Instead, Japanese government support for post-disaster housing reconstruction continues to focus on the provision of disaster recovery public housing. As all public housing in Japan, rents for qualified residents are subsidized by the government based on residents' income. Public housing includes multi-family apartment-style buildings as well as single family detached houses. In addition to systems supporting the reconstruction of houses or provision of in-kind public housing units, through the use of post-disaster relocation projects, the government provides and prepares new lots for residential reconstruction, including land clearing and preparation of roads and other site infrastructure.

3.3. Localized Recovery Housing in the Context of Contemporary Construction Trends

Although as the common Japanese building material through history wood had been used for temporary housing in the past, after the 1995 Hanshin Awaji Earthquake in Kobe most temporary housing was prefabricated metal construction. Since then, all of Japan's 47 prefectures entered into agreements with the Japan Prefabricated Construction Suppliers and Manufacturers Association (abbreviated as *pre-kyo* in Japanese) which allows member companies to quickly provide temporary housing after a disaster [9].

Along with this shift to the use of prefabricated structures for temporary housing, the Japanese private housing market has also become increasingly dominated by mass-produced and pre-fabricated houses built by Japan's massive "housemaker" corporations [10]; mirroring typical suburban housing development patterns, homeowners rebuilding post-disaster are also likely to choose a house from a housemaker. In contrast to this process of increasing reliance on mass-production and pre-fabrication for both temporary and permanent housing after disaster, localized disaster recovery housing promotes the use of local resources (materials and people). Within Japan's post-disaster housing recovery process, localized housing recovery can therefore include: 1) wooden temporary housing provided by the government; 2) privately rebuilt houses; and 3) single family detached disaster recovery public housing built by the government. The following section introduces cases from these 3 types.

4. Cases of Localized Housing Recovery in Japan

4.1. The Birth of 'Localized Housing Recovery' after the 2004 Chuetsu Earthquake

On October 23, 2004, the Chuetsu Earthquake struck the rural, mountainous areas of Niigata Prefecture. With the goal of supporting as many people as possible to be able to return and rebuild in the heavily damaged Yamakoshi area, Nagaoka City created a committee to develop a "Uplands Localized Recovery House" (hereafter: Uplands House) that was required to be: 1) a "Yamakoshi-like" house; 2) able to co-exist well with snow; 3) have local circulation; 4) built for a low construction cost (under 10 million yen, or about $100,000 US); and 5) a house where people can live safely and comfortably for a long time[11]. Two model units of the Uplands Houses were built with 3 meter and 6 meter floor elevation, respectively, so residents rebuilding their own houses could see and experience these designs.

The creation of this localized model for housing recovery was the result of strong local government leadership and involvement of multiple actors--financers, local builders and designers in cooperation with Nagaoka City and housing developers--to build Yamakoshi's Localized Recovery Housing, shown in Figure 1. Construction of the Uplands House used local materials and companies, which has an ecological benefit and minimizes construction costs. The houses were designed to be comfortable in the local climate, to use low energy and be easy to care for and survive long term exposure to the elements. Using a traditional modular plan, the rooms can be rearranged to respond to potential future changes in household size or lifestyle. The Uplands House design was also used for the construction of disaster recovery public housing in Yamakoshi. In the first 3 years after the earthquake, 19 Uplands Houses were built by residents and 36 as public housing [11].

Figure 1. Yamakoshi style Uplands Localized Recovery House.

4.2. Temporary and Permanent Localized Housing Recovery: the 2011 Great East Japan Earthquake

On March 11, 2011, the Great East Japan Earthquake (GEJE) caused a massive tsunami that devasted communities along the north east Tohoku coast. The GEJE was a wide-area, complex disaster, including earthquake, fires, tsunami, and the nuclear meltdown at the Fukushima Daiichi Nuclear Power Plant. With more than 500 square kilometres inundated by the tsunami, 20,000 people lost their lives and 129,000 houses were completely damaged [12]. 470,000 people evacuated in the few days after the tsunami, and as of August 2018 close to 60,000 people are still living in temporary housing [6]. As of August 2018, in Fukushima where evacuees face longer term displacement from radioactive contamination, more than 10,000 people are living in temporary housing within Fukushima and more than 33,000 continue to live in evacuation outside of the prefecture [6].

Although Fukushima Prefecture, responsible for contracting the construction of temporary housing, had existing agreements with the pre-fabricated housing association, providing the large needed numbers of temporary housing exceeded their capacity. In Fukushima, where nuclear evacuees face long term displacement, the prefecture promoted the use of wooden temporary housing, issuing two public calls for the construction of wooden temporary housing in April of 2011, and again in July of the same year. The first call was for 4000 units, and the 2nd call was for 1000 (later increased to 2000) units. Promoting the use of local materials, the conditions required that applicants be from Fukushima prefecture, and for the 2nd call, the conditions also specified that the houses should be able to be dismantled and relocated. As a result, there were 6,000 wooden temporary housing units built in Fukushima Prefecture, as well as 10,000 prefabricated units.

4.2.1. Wooden Temporary Housing in Miharu Town, Fukushima Prefecture. Miharu Town, located in the inland part of Fukushima Prefecture and not affected by tsunami or nuclear accident, became the location for temporary housing for 770 evacuee households from Tomioka Town and Katsurao Village, areas close to the Fukushima Daiichi Nuclear Power Plant. Through coordination with the Japan Institute of Architects Tohoku branch, the Making Recovery Housing in Miharu Town Committee worked with five local builders to construct 100 units of wooden temporary housing in Miharu Town. Instead of wood, the foundation for these houses was poured concrete, and all construction timbers were square and standard length. In addition to reducing costs, standard-sized timbers don't require a special order but can be easily purchased at local stores. These houses included various improved features compared to usual temporary housing such as: substantial insulation in the floors, walls, exterior; insulation and ventilation in the ceiling; double pane glass; and spaces between each unit which provide noise isolation.

Figure 2. Wooden Temporary Housing in Miharu Town.

4.2.2. Wooden Temporary Housing and Permanent Conversion in Aizu Wakamatsu Town, Fukushima Prefecture. Wooden temporary housing was also provided for nuclear evacuees in Aizu Wakamatsu Town, in the inland area of Fukushima Prefecture. Using the Japanese *itakura* method, in which walls are constructed by setting solid wooden boards within a slotted frame, there was a plan from the beginning to reuse and convert the temporary housing to permanent housing. In the actual conversion, about 66% of the materials were reused. Although this is an ecological approach because materials are reused, the total cost for adjustments and additional materials needed in the conversion process made the cost equivalent to new construction.

4.2.3. Localized Permanent Recovery Housing after the GEJE: Iwanuma City, Miyagi Prefecture. In

Figure 3. Temporary housing converted into public housing in Aizu Wakamatsu, Fukushima.

Miyagi Prefecture, the coastal area of Iwanuma City was devasted by the tsunami, and residents from coastal areas moved to a new residential area inland in a collective relocation project. This new residential area includes both lots for people rebuilding their own houses and disaster recovery public housing. With collaboration between local designers, architects, and builders, designs for localized

Figure 4: "Kuro no Ie" model house in Iwanuma

Figure 5. "Shiro no Ie" model house in Iwanuma

housing recovery were created for Iwanuma by the Japanese Institute of Architects, using a traditional Japanese housing module, and materials from within Miyagi Prefecture. Two housing designs were created to demonstrate this approach for people choosing to rebuild, including the smaller "Kuro no Ie" (Black House, Figure 4) and "Shiro no Ie" (White House, Figure 5) based on the modularity of traditional wooden construction. However, there were no residents who actually chose to use this design.

As the standard building method in Japan, wood construction is not only familiar and versatile, but the modularity of traditional construction can also contribute to flexibility in housing recovery. The traditional module for housing construction is based on an area the size of 8 tatami mats (1 tatami mat is a rectangle slightly smaller than 1 meter by 2 meters). This smallest module of Japanese construction is a 2 by 2 square, where each side of the square is the length of 2 tatami mats. The smallest housing unit using this module is the square shape (4 × 4) plan, where each side of the square is the length of 4 tatami mats. This smallest size housing unit based on the traditional module was the basis for the Kuro no Ie, as shown in Figure 6 (left top, left bottom). As fewer people live in multi-generation households and with an aging society, the number of small families is growing. To reduce costs, simple, small-sized homes can be used as basic housing units, with lofts and a second floor or expanding to a 4 x 6 (4 tatami mats wide and 6 tatami mats long) plan as needed, such as the Shiro no Ie design, Figure 6 (top right, top bottom). In addition, by using modules, materials can be standardized and costs reduced. The Kuro no Ie design was also used for the construction of disaster recovery public housing in Iwanuma, shown in Figure 7. In this case, the localized disaster recovery housing units were also arranged in a configuration to create shared community space and interactions.

Figure 6. Plans showing the design of Kuro no Ie (1st floor, bottom left, loft top left), and Shiro no Ie (1st floor bottom right, loft top right) housing design based on a traditional Japanese module of wooden construction. The Kuro no Ie is composed of 4 square modules, with each module having the dimensions the length of 2 tatami mats on each side. The Shiro no Ie is composed of 6 of the

Figure 7. In Iwanuma City, the localized housing recovery design of the Kuro no Ie was used for disaster recovery public housing, arranged to promote community interaction.

4.3. Totsukawa Village after the 2011 Kii Peninsula Flood

Totsukawa Village in Nara Prefecture suffered severe damage in the 2011 Kii Peninsula Flood, with 18 houses completely destroyed and 30 houses partially damaged. With many small settlements, in some areas entire villages were destroyed. Totsukawa Village's abundant forest resources were used to build localized housing recovery, including 30 wooden temporary houses built with local materials and carpenters, and a model of 'Totsukawa Localized Recovery Housing.' This design was also used for the construction of disaster recovery public housing. In Totsukawa Village, with steep mountainous area and few flatlands, finding land and building housing was difficult. With increasing depopulation, and the percentage of elderly reaching 42.5%, vacant lots and houses were noticeable in each settlement. Based on investigating settlements' history, model houses to promote localized disaster recovery were built on empty land which had not experienced past disasters. These were built of wood according to the disaster recovery public housing specifications. Using localized housing to rebuild on infill sites (Figure 8) is a model for sustainability of depopulated regions and the creation of more compact cities, and is a new approach to address problems of large scale projects such as

Figure 8. Totsukawa Village after reconstruction.

collective relocation.

5. Discussion

5.1. Economic Benefit
The economic benefit of localized housing is not only minimized construction costs but also support for the economy (and recovery) of the affected region. By using local workers (local builders/local carpenters), the majority of money spend on construction will go into the local economy. Secondary expenses which accompany new housing construction, such as construction for electric and water and sewer infrastructure, moving costs, purchase of furniture, and appliances, and paying for housing loans, represent a significant economic impact. According to data from the Japanese government, based on average construction cost of 25 million yen/ house, these secondary costs add up to an addition 51.7 million yen/house [13].

5.2. Design Benefits of Involving Local Designers and Builders
Although Japan is a small country, there are significant climatic variations, and housing built without an understanding of the local area may result in unsuitable designs. For example, after the Great East Japan Earthquake, many prefabricated temporary housing units were built using standard designs, without consideration for the cold winters in the Tohoku area. As a result, additional construction was needed to add exterior insulation (in effect adding an additional external wall), as well as double pane windows. Local builders who have experience with housing construction in the region, will already be familiar with building for the local climate, as well as local housing lifestyles.

5.3. Potential for Reuse of Wooden Temporary Housing
With recent examples such as in Aizu, where wooden temporary housing was converted into permanent housing, there is a growing discussion about the potential for this kind of building and material reuse. While there are still remaining challenges to make this process more effective, the government is also seriously considering measures to make similar processes easier in the future.

5.4. Developments and Challenges for Implementation of Localized Recovery Housing
In the years since 'localized housing recovery' was first introduced after the 2004 Chuetsu Earthquake, a growing number of disaster recovery housing projects have incorporated localized aspects including: the use of local materials/ timber construction; local builders/carpenters; and designs that reflect the local regional culture. After the 2011 Great East Japan Earthquake, there was a significant increase in the share of wooden housing as part of the total number of newly-built temporary housing; this has continued after subsequent disasters. There have also been more concerted efforts to promote the potential of localized housing to residents who are rebuilding post-disaster, as well as to promote localized housing for single family detached disaster recovery public housing. However, there are also challenges to carry out localized disaster recovery housing on a large scale. Key factors for successful implementation include strong leadership supporting of the idea of localized housing recovery from the beginning, timing, careful coordination of multiple stakeholders, and knowledge from those who have past experience with similar projects.

6. Conclusion
Drawing on traditional wood construction techniques, designs, and materials, localized housing recovery in Japan represents a way to not only support the disaster-affected area, but also to quickly provide high quality and comfortable housing to disaster survivors, using familiar materials. Houses designed and built by local builders will be more suitable for the local conditions than houses designed for a different climate/location. Moreover, the economic benefit from using localized housing is significant not only for local carpenters, but for the region's economic recovery and long term sustainability. As ideas of localized housing recovery have advanced both in terms of the use of local human resources and local building materials, the implementation of localized housing has also advanced to be incorporated within the local community, as in the case of Totsukawa Village. The use

of local resources is widely supported within the international disaster recovery community of experts, and Japan's localized housing recovery serves as a useful detailed example of the application of this approach. Although the post-disaster housing recovery context in Japan includes some specific policies regarding provision of temporary and permanent housing, lessons about the coordination of local resources and disaster recovery policies are relevant to other countries, especially where wooden construction is common.

References

[1] Jha AK 2010 *Safer Homes, Stronger Communities: A Handbook for Reconstructing After Natural Disasters* (Washington D.C.: World Bank); Available from: http://dx.doi.org/10.1596/978-0-8213-8045-1.

[2] USAID 2012 Building back housing in post-disaster situations basic engineering principles for development professionals: a primer; Available from: http://www.buildchange.org/pdfs/USAID_Building_Back_Housing_in_Post-Disaster_Situations-A_Primer_FINAL.pdf; 2012

[3] UN Habitat 2008 People's Process in Post-disaster and Post-conflict Recovery and Reconstruction. Available from: https://www.sheltercluster.org/sites/default/files/docs/People%27s%20Process%20in%20Post-disaster%20and%20Post-conflict%20Recovery%20and%20Reconstruction.pdf

[4] Da Silva J 2010 Lessons from Aceh: Key Considerations in Post-Disaster Reconstruction, Arup, Practical Action Publishing Available from: http://dx.doi.org/10.3362/9781780440606

[5] Barakat S 2003 Housing reconstruction after conflict and disaster (London: Humanitarian Practice Network at ODI).

[6]Reconstruction Agency. 2018. http://www.reconstruction.go.jp/topics/main-cat2/sub-cat2-1/20180831_hinansha.pdf. (in Japanese) (Accessed 13 Sept, 2018)

[7] Hirayama Y 2000 Collapse and Reconstruction: Housing Recovery Policy in Kobe after the Hanshin Great Earthquake. *Housing Studies*. (Informa UK Limited) 2000 Jan **15(1)** 111–28. Available from: http://dx.doi.org/10.1080/02673030082504

[8] Iuchi K Maly E and Johnson L 2014 Three Years After a Mega-disaster: Recovery Policies, Programs and Implementation After the Great East Japan Earthquake. *Post Tsunami Hazards; Reconstruction and Restoration* (Cham: Springer) 29–46. Available from: http://dx.doi.org/10.1007/978-3-319-10202-3_3

[9] Maly E and Matsushita T 2014 Innovations in The Temporary Housing Phase of Disaster Recovery: Wooden Temporary Housing in Japan After the 3.11 Great East Japan Earthquake and Tsunami in an International Comparative Context, International Conference of Architecture and Planning in Indonesia, Yogyakarta.

[10] Daniell T 2005 FOB Homes: Brand Recognition. *FOBA*. (Prince Archit.Press)169–71 Available from: http://dx.doi.org/10.1007/1-56898-635-1_14

[11] Takeda K 2004 Reconstruction Housing Models for Earthquake Damaged Areas in Yamakoshi (no1); Available at: http://www.alsed.co.jp/2news/mokuzo_kasetsu/yamakoshi_fukkojutaku.pdf

[12] Reconstruction Agency 2016 Number of disaster related deaths after the Great East Japan earthquake. http://www.reconstruction.go.jp/topics/main-cat2/sub-cat2-6/20160630_kanrenshi.pdf. (Accessed 13 Sept. 2018)

[13] Nihon to Sekai Toukei Data 2018; https://toukeidata.com/

Acknowledgments
This research was supported by JSPS Grants-in-Aid 17H03358.

IOP Conf. Series: Earth and Environmental Science **273** (2019) 012056 doi:10.1088/1755-1315/273/1/012056

What are determining factors of tourist loyalty to tsunami affected tourism destination?

Syafruddin [1], H Fahlevi[2], Jalaluddin[3]

[1,2,3]Universitas Syiah Kuala, Indonesia, 23111.

E-mail: syafruddin.chan@unsyiah.ac.id

Abstract. The rapid development of the tsunami tourism attraction combined by the Islamic tourism of Aceh has contributed to the advancement of local people economy including the victims who live surrounding the attraction sites. However, researches on tourists of tsunami-affected area and its dimension are still scanty. This study is imperative because tsunami tourism offers uniqueness and risks for the tourists. Having understanding the perception of tourists, the quality and attractiveness of tsunami tourism can be optimized. Thus, this study aimed to investigate potential determinants of tourist loyalty to tsunami-affected tourism destination. There are 3 potential determinants tested in this study, namely tsunami attraction image, perceived value, and tourist satisfaction. The purpose of this study is to define the determinant variables (competitive tsunami attraction image, perceived value, and tourist satisfaction) of tourist loyalty to tsunami-affected tourism destination using an SEM approach. A proposed research model in which six direct and four mediations hypothesis was developed and tested. Empirical data collected at major tsunami tourism destinations in Banda Aceh. Data gathered through questioners and interviews with tourists who are visiting Banda Aceh especially the visitors of Museum Tsunami, "Kapal di atas Rumah" (boat on the house) and "Kapal Apung". A total of 328 questionnaires were returned and data were analyzed using Structural Equation Modelling (SEM). The mediation hypothesis was tested by using Bootstrapping with AMOS] and Mediation Test proposed by Barron and Kenny. Results supported the proposed loyalty model: (1) Competitive Tsunami Tourist Attractions has a positive direct influence on Perceived Value, Tourist Satisfaction and Tourist Loyalty to Tsunami Tourism Destination. (2) Perceived Value and Satisfaction Tourists also have a direct influence on Tourist Loyalty to Tsunami Tourism Destination. (3) From 4 mediation hypotheses proposed only 3 of them were supported. This implies that to increase the tourist loyalty to tsunami-affected tourism destination, the government need to focus on increasing the competitiveness of Tsunami Tourist Attractions. This can be done through the improvement of the tourism objects and the supporting business environment (awareness, amenities, accessibility, accommodation, attractiveness).

1. Introduction

The tragedy of the tsunami that occurred in 2004 in Aceh left a lasting impression on all Indonesians, but particularly on the families of the victims. It is estimated by the US Geological Survey that the earthquake that initiated the tsunami has released energy the equivalent of 23,000 atomic bombs. The incident killed more than 200,000 thousand people in 14 countries with Indonesia suffering the majority of the casualties [1].

At the time of the tsunami, the tourism industry in Thailand was in very good condition with tourist arrivals totalling more than 24 million per year, considered the highest in Southeast Asia. In contrast to Thailand, Aceh had a negligible tourism industry in 2004. Any tourism industry that had existed previously had collapsed due to the Aceh Civil war between the Gerakan Aceh Merdeka (GAM or the Free Aceh Movement) and the Indonesian government. This deadly conflict retarded development in the province and simply frightened tourists away [2]. The 2004 Indian Ocean tsunami proved to be a turning point in several ways. When the tsunami hit Phuket, occupancy rates dipped for several years and then recovered, suggestive of a relatively minor setback for the whole Thai tourism industry.

Therefore The relatively low number of deaths and localized devastation meant that the Thai community were able to work, after the tsunami, they hand in hand to quickly recover the industry.

Aceh, like other tsunami-affected areas, has felt a need to retain and interpret prominent remnants of the tsunami ('tsunami objects') for their heritage value and as a solemn reminder to the scale and force of the disaster to be deployed by future generations by maintaining Tsunami heritage [3]. A number of prominent tsunami objects are a direct result of the disaster such as the Floating PLTD ship and the 'ship on a house's while other major buildings and new monuments such as Tsunami Museum have been constructed subsequent to the tragedy. Maintaining tsunami objects can be seen as a smart step in interpreting history related to the disaster [4]. Tsunami objects can also be used as a medium to bridges understanding of disasters in the past for younger generations who may have no direct experience of disaster. The Aceh Government acted quickly to build on the benefits of the new tsunami object through the Aceh Tourism Agency who continued to develop and market the tsunami in Aceh [5]. Tsunami tourism objects such as the Floating PLTD ship and the Tsunami Museum became an icon of marketing campaigns in Aceh shopping as a tsunami tourism destination.

Just like in other products, tourism may also be benefited by 'loyal' tourists [6] , those who make repeat visits to Aceh. Research to determine tourist loyalty factors is increasingly attracting interest. Many researchers have found a relationship between tourist loyalty with perception value, satisfaction and attractiveness of objects [7-13], carried out an assessment of the significant impacts of customer satisfaction on customer loyalty. Other researchers such as Caruana & Ewing [14] ; Hapsari et al.[8] Jin et al.[10]; Kassim & Abdullah [11]; and Yang & Peterson, [15] examined the real influence of perceived value on customer loyalty. While Caruana & Ewing [14]; Hapsari et al.,[8] and Kassim & Abdullah, [11] show that loyalty should become the main objective for promoting tourism destinations.

Tsunami tourism arising from the 2004 tsunami disaster has led to an accelerated the development of new tourist attractions in Banda Aceh. According to the Banda Aceh Department of Tourism, the number of foreign tourists who visited Banda Aceh has grown rapidly from 756 people in 2005 to 706.646 in 2016 (RPID, 2017). This significant increase can be linked to the presence of tsunami-based tourist attractions including the Tsunami museum and "Kapal Apung". The tsunami museum was visited by 28.714 peoples from Indonesian and around in 2017 (Antara Riau, 2017). The rapid development of the tsunami tourist attractions combined with Islamic tourism has contributed to the advancement of the local economy including in particular the survivors of the tsunami who live in the surrounding areas. Despite this important economic and social impact, researches on tourists of a tsunami-affected areas and its various dimensions are This study is imperative because tsunami tourism offers unique opportunities and risks for locals. Through development of an understanding of the perceptions of tourists, the attractiveness of tsunami-based tourism objects can be optimized. Accordingly, this study aimed investigate the potential determinants of tourist loyalty to a tsunami-affected tourism destination. There are 3 potential determinants tested in this study,, namely; the image of the tsunami tourist attraction, perceived value, and tourist satisfaction. These three factors will be assessed and subject to a test of variables to get an idea of their role in influencing the level of tourist.

2. Literature Study

Please follow these instructions as carefully as possible so all articles within a conference have the same style to the title page. This paragraph follows a section title so it should not be indented.

2.1 Tsunami Tourism

Without tourist attractions, there would be no tourism. Without tourism, there would be no tourist attractions. The tsunami that occurred on December 26, 2004, left many stories and historical traces[16]. Adding to this deep background of authentic disaster experience. Abandoned tsunami objects have has been converted into the mainstay memorial / tourist attractions for of the city of Banda Aceh. This includes the Kapal Apung formerly a Power Plant now resting in Punge Blang Cut Banda Aceh [17]. Kapal Apung was a floating electricity generation plan length and 19 m wide. Original moored off the coast of Banda Aceh. It now rests in the city, inland having ploughed through houses and buildings driven by the surge of the tsunami. It is now the central feature of a tsunami education park.

Disaster tourism is a well-documented phenomenon centred on tourist visitation of areas hit by natural disasters [16]. Banda Aceh is a prime example. Reports and information about disasters and their impacts attracts the attention of potential tourists [16]. All areas affected by a natural disaster will inevitably retain some material evidence of the event in the form of objects that can evoke memories and emotions among tourists and locals alike. In particular the retention of these objects can remind younger generations of the tragedy, potentially contributing to ongoing risk awareness and improvements in disaster preparation [18]. At the same time other people who come from outside the area who want to watch and study the impact of disasters in an area can also benefit from the retention of Tsunami disaster.

Individuals who participate in this type of tourism are usually curious to see the impact of disasters, [18]. This open up opportunities for travel agents to make Tsunami tour packages which in turn adds value to both the local economy and the experience of tourists themselves. Through visiting the tsunami tourism objects visitors open a window of understanding to the tsunami disaster.

2.2 Customer Loyalty
"Loyal" literally means faithful. Loyalty can also be interpreted as trustworthiness [20-22] Trustworthiness and loyalty must arises without coercion and must relate to an individual awareness of the past and past actions. Efforts made to create consumer satisfaction were more likely to influence consumer attitude to become loyal. The concept of consumer loyalty is important so far as it late to buying behaviour [23]. Loyalty will develop where a consumer identifies that the supplier anticipates and provides goods or services which are valued by the consumer. Done well, the anticipation of consumer values leads to loyalty and product endorsement to third parties

Customer loyalty is very important for a company to maintain s its business continuity. Faithful customers are those who are very satisfied with certain products and services, leading to an have enthusiasm to introduce the business to anyone they know [24]. At the highest levels customer loyalty will lead customers expand their "loyalty" to other products made by the same manufacturer [22]. In the end, they may become loyal to a certain producer or company for their entire lives. Kotler [25] states that loyal customers are those the one who make frequent purchases from the same company. A company needs to put in greater effort to grow and retain loyal customers compared to the effort devoted to finding new customers [26] but customer loyalty converted from merely satisfied customers is a valuable asset for a company [27].

2.3 Customer Satisfaction
Consumer satisfaction refers to the level of consumer feelings flowing from a comparison of satisfied with the value provided by a product or service, it is very likely that they become long lifetime consumers [20]. According to Yi & La [28] consumer satisfaction is represents a positive or pleasurable feeling which arises after the supply of a product or the performance of a service which exceeds their expectations. Most businesses will have an objective to satisfy their customers [29]. Customers who are satisfied with their products and services will tend toward repeat buying from a business that leaves them with the positive feeling that the expectations have been met or exceeded.

According to Kotler [30], the factors that drive customer satisfaction include: quality (customers will satisfied if they feel the quality of the product meets or exceeds their expectation; price (for sensitive customers, low price is an important source of satisfaction) which represents customers value for money; service quality (which is usually difficult to imitate). Service quality is a driver that has five dimensions, called TERRA. Customers will also find satisfaction through emotional valuation of the brand of the product and finally customers may also be satisfied if obtaining the product or services is relatively easy, comfortable and efficient [31].

2.4 Customer Perceived Value

The customer's perceived value is the customer's relates to their perception of a company's offerings [14]. Customer value is a function of benefit divided by cost. The higher the benefits and the lower the cost, the higher the customer value is [31].

2.5 Relations among Variables.
Kaura, V [32] have examined the factors that influence customer loyalty, and concluded that customer satisfaction was a variable that had a significant influence in establishing customer loyalty. Positive influence on customer loyalty in the Malaysian service sector, in particular the value of the customer. In addition, Pandža Bajs, [33] also saw the attractiveness of the product to the values given to customers after conducting research in the tourist area of Dubrovnik, Croatia. Ramseook et al. [34] found a significant influence caused by an increase in the attractiveness of tourism products to the customer loyalty.

2.6 Hypothesis Development
Ten hypothesis developed for this study:
H1, H2, H3. Tsunami attraction is positively influencing Perceived Value, Customer Satisfaction and Customer Loyalty
H4, H5. Perceived Value is positively influencing Customer Satisfaction and Customer Loyalty
H6. Customer Satisfaction is positively influencing Customer Loyalty
H7. Tsunami attraction is influencing Customer Satisfaction through Perceived Value
H8.H9. Tsunami attraction is positively influencing Customer Loyalty through Perceived Value and Customer SatisfactionH.10 Perceived Value is influencing Customer Loyalty through Customer Satisfaction

3. Methodology
The current research was conducted in the Banda Aceh region which is one of the areas most severely affected by the 2004 tsunami, but has subsequently benefited from the tsunami tragedy that struck. The research shows that, all tourists who come to Banda Aceh find it mandatory to visit the three major tsunami sites; the Tsunami Museum, the Kapal Apung and the "Ship on top of House". The respondents selected by using convenience sampling method. 328 questioners distributed to those who visited the site.

Analysis of data using the Structural Equation Model using AMOS 22 software and backed up by SPSS software (also version 22). In addition, to analysing the validity of each item or indicator, hypotheses were also be tested for direct and indirect relationships (Mediating Effect). The use of SEM for the data analysis number of stages. The first stage is to measure the validity of each indicator against coefficient of loading factor is greater than 0.5 so that it is declared valid and can be included in the next data processing stage.

The next stage is testing the hypothesis to measure the influence of independent variables on dependent variables both directly and indirectly. The indirect hypothesis is the measurement of independent variables against the dependent variable through a mediation variable. Hypothesis testing uses a structural model to produce accurate output. If CR coefficient is >1.96 and P <0.05, the hypothesis is declared acceptable. But if the opposite happens, then the hypothesis is rejected. The last stage is to test the indirect effect. This testing involves mediation variable and uses Bootstrapping analysis to find the CR and p values [35].

4. Result and Discussion

4.1. Validity and Goodness of Fit
In the initial stage, the validity of the indicators for each variable has to be checked to determine which variables are valid and which are invalid. The benchmarks were coefficient loading factors in the measurement model that is run using AMOS 22.0. This was followed by testing the model to examine its compatibility with the existing data (Goodness of Fit). This then followed by hypothesis testing of both direct and indirect effect between variables.

The relationship between items or indicators for each variable can be seen in the following figure. While Table 1 (below) shows the loading factor coefficient that measures the relationship between indicators with their respective variables. The higher the loading factor score, the more closely the indicators relates to the variables.

Table 1. Coefficient Loading Factor for Validity

No	Dependent		Independent	Estimate
1	AttrX11	<---	Tsunami Attractive	,791
2	AttrX12	<---	Tsunami Attractive	,823
3	AttrX13	<---	Tsunami Attractive	,837
4	AttrX14	<---	Tsunami Attractive	,867
5	AttrX15	<---	Tsunami Attractive	,897
6	PercvY11	<---	Perceived Value	,756
7	PercvY12	<---	Perceived Value	,740
8	PercvY13	<---	Perceived Value	,770
9	PercvY14	<---	Perceived Value	,779
10	PercvY15	<---	Perceived Value	,606
11	ToursY21	<---	Satisfaction	,711
12	ToursY22	<---	Satisfaction	,867
13	ToursY23	<---	Satisfaction	,903
14	ToursY24	<---	Satisfaction	,748
15	LoyZ11	<---	Loyalty	,811
16	LoyZ12	<---	Loyalty	,704
17	LoyZ13	<---	Loyalty	,782
18	LoyZ14	<---	Loyalty	,546

One of the requirements in structural equation modelling is to do a Goodness of Fit analysis. The goodness of fit testing can be done using several criteria such as GFI, CMIN/DF, and RMSEA. The model presented to this study has not yet met the criteria of fit yet. Of the 4 (four) criteria used to measure the fit; CMIN/DF, AGFI, RMSEA, and PRATIO, two of them; CMIN / DF and RMSEA, were outside the permitted cut-off limits, therefore it is necessary to modify the model using the Modification Indices technique.

The following table describes the output of the Modification Indices suggested by AMOS. It can be seen that the addition of a covariance line between e10 and e11 results in the highest MI value of 48,229. This was also suggested to connect e14 and e15 that have an MI score of 34.410.

Table 2. Modification Indices Covariance

Modification Indices Connection			M.I.	Par Change
e10	<-->	e11	48,229	,129

e14	<-->	e15	34,410	,074

Referring to table 2, the model needs to be modified by adding covariance lines that connect the indicator errors those two suggested. Because the data has not been fixed, it is necessary to modify the model using a modified indices method, before the test hypothesis are made.

To improve the fitness of the model, a variance line is needed to connect the Perceive Value indicators and Satisfaction indicators as shown in Figure 4. Modifying the model by connecting PerceY15 and ToursY21, can increase the performance of CMIN/DF to below 3 and RMSEA to <0.08, so that it now fits the data.

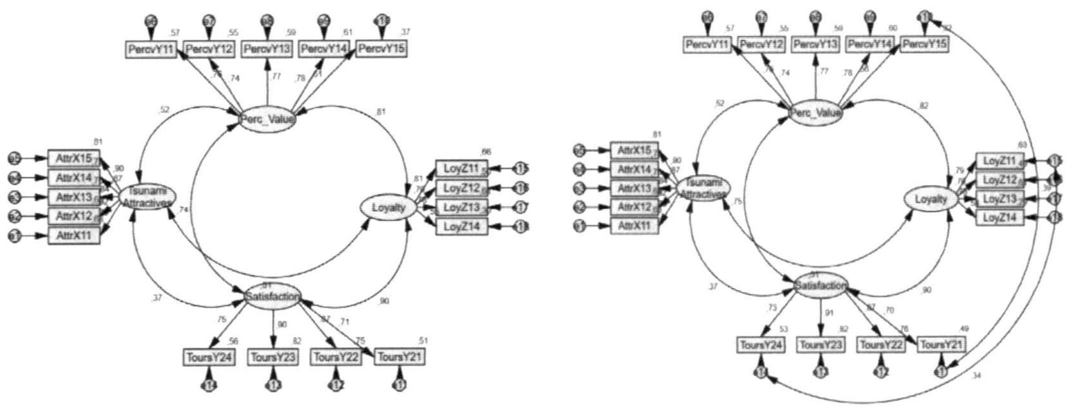

Figure 1. Measurement Model Before and After Modification Indices

Figure 1 consists of two diagrams of measurement models. The left diagram is the original measurement model, while the right one is a modified measurement model created by adding two covariance lines as suggested by AMOS. The results look better, as shown in table 2. CMIN/DF criteria which were previously above 3, now dropped to2.953.While the RMSEA which was still above 0.08 now dropped to 0.071. Both criteria now fall inside of the cut off value. Likewise, the other two indicators, which are AGFI and Pratio have also improved so that they are closer to the ideal number, showing a better fit for s the before.

Table 3. Goodness of Fit

Criteria	Cut-Off Value	Before MI	After MI	Model Evaluation
	<3	3,587	2,935	Good
	Between 0-1	0,849	0,868	Good
	<0,08	0,082	0,071	Good
	Between 0-1	0,843	0,884	Good

Seeing the results, it can be seen that the model built for this study fits with the available data so that the data processing can be proceed to structural modelling to test the H1 - H10 Hypothesis.

4.2 Analysis of Structural Equation Modelling

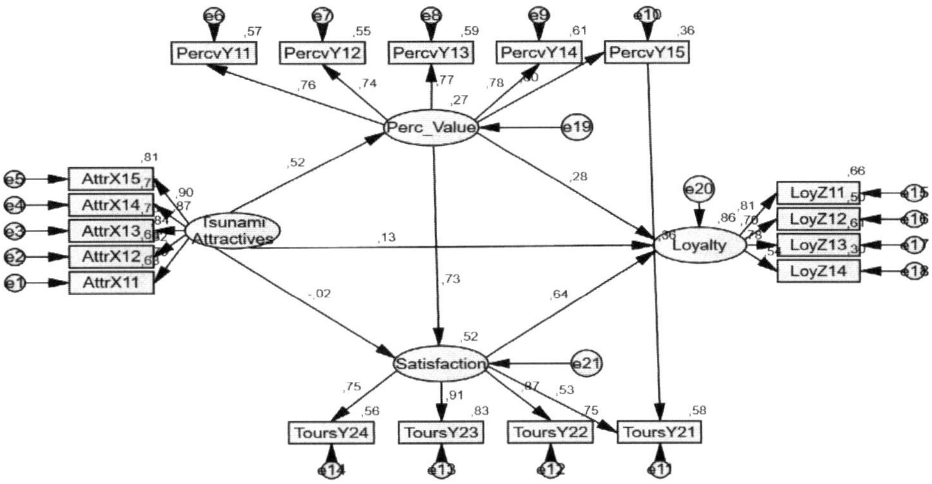

Figure 2. Structural Model

Table 4 illustrates the results of the hypothesis test for direct effect, that is testing Hypothesis H1 - H6 using criteria of CR> 1.96 and a P value <0.05

Table 4. Hypothesis Test

No	Dependent	<---	Independent	Estimate	S.E.	C.R.	P	Beta
1	Perceived Value	<--	Tsunami Attractiveness	,354	,040	8,847	***	0,522
2	Satisfaction	<--	Tsunami Attractiveness	-,012	,030	-,391	,696	0,021
3	Satisfaction	<--	Perceived Value	,599	,069	8,717	***	0,728
4	Loyalty	<--	Perceived Value	,329	,077	4,279	***	0,276
5	Loyalty	<--	Satisfaction	,930	,107	8,672	***	0,642
6	Loyalty	<--	Tsunami Attractiveness	,108	,034	3,211	,001	0,133

IOP Conf. Series: Earth and Environmental Science **273** (2019) 012056 doi:10.1088/1755-1315/273/1/012056

Of the 6 hypotheses tested, 5 of them met the significance requirements. Only 1 did not meet the requirements, namely Hypothesis No2, which tests the effect of Tsunami Attractiveness on Customer Satisfaction.

Meanwhile from the magnitude of the Beta coefficient point of view which illustrates the strength of the relationship among variables, the effect of Customer Satisfaction variable on Customer Loyalty can be seen to be the strongest (0,642) compared to the contribution of other variables such as variable Perceived Value or Tsunami Attractiveness. Thus, it can be concluded that the main variables that have a significant impact on the loyalty of tourists visiting the tsunami objects in Aceh are Customer Satisfaction. The more satisfied the tourists, the higher their chances of becoming loyal tourists.

4.3 Mediation Effect

As suggested by Preacher et al (2008), a bootstrapping procedure is used to test the mediating effects among variables shown in H7 - H10. In this scheme Perceived Value and Customer Satisfaction are the mediating variables.

There are 4 (four) mediation effects that were tested in this study and all of them have a significant effect. The following table presents a summary analysis of indirect impact hypotheses 7-10.

Table 5. Mediation Test Results by using Bootstrapping Approach

No	Variables	Tsunami Attractiveness	Perceived Value
1	Customer Satisfaction	,001	-
2	Customer Loyalty	,001	,001

Table 6. Summary Indirect Effect

Variables	Direct w/o M (Bootstrapping)	Sig. (P Value)	Direct w M (Bootstrapping)	Sig. (P Value)	Indirect	Sig. (P Value)	Result
Tsunami-Perc.Value-Loyalty	0,133	0,01 (**)	0,108	0,014 (**)	0,221	0,001 (***)	Partial Mediation
Tsunami-Perc.Value-Satisfaction	0,133	0,01 (**)	0,108	0,014 (**)	0,221	0,001 (***)	Fully Mediation
Tsunami-Satisfaction-Loyalty	-0,21	0,696 (ns)	-0,12	0,764 (ns)	0,151	0,001 (***)	No Mediation
Perc.Value-Satisfaction-Loyalty	0,276	***	0,329	0,001 (***)	0,397	0,001 (***)	Partial Mediation

Ns: not significant **: significant at 5% ***: significant at 1%

Table 7 shows that of the 4 hypotheses tested as, indirect effects in this study (H7-H10), all of them showed a positive level of significance. In other words, the Ho Hypothesis of H7-H10 were accepted.

5. Discussion and conclusion

The study has produced findings that provide supporting evidence for the development of tsunami tourism in Aceh by focusing on the generation of tourist loyalty. The use of SEM to test 6 hypotheses developed in this study, 5 of it are supported or accepted. This increase in loyalty can be achieved by increasing the attractiveness of products, perceived value, and customer satisfaction.

The analysis has shown that customer satisfaction is the main variable influencing an increase tourist loyalty in Aceh. This is indicated by the largest Beta value among determinant variables such as the Perceived Value and the Tourism object Attractiveness. This is in line with the results of previous studies conducted by Kaura, V., Durga Prasad, CS, & Sharma, S. (2015) which examined the factors that influenced customer loyalty in Chinese Chain restaurants. Customer satisfaction is a variable which was shown to have a significant influence in converting customers to become loyal customers. The same opinion was expressed by Blut et al (2015). The findings of this study also indicate customer satisfaction as a mediating variable between product attractiveness and customer loyalty.

This study also has proven demonstrated a positive influence between perceived values on tourist loyalty. The higher the value of the tourism services they received during a visit to the Aceh tsunami tour, the more likely they were to become loyal and to visit again. This finding is in line with the research results of Meanwhile Rasheed, F. A., & Abadi, M. F. (2014) found other variables that have a positive influence on customer loyalty in the Malaysian services sector. Eid, R. (2015) identified a relationship between customers' perceived value and customer loyalty. Eid specializes in research on Muslim tourism markets and uses variable Muslim Customer Perceived Value (MCPV) as a determinant variable on consumer satisfaction and customer loyalty.

The only direct hypothesis that is not proven relates to the influence of Tsunami object attractiveness on customer satisfaction This means the efforts made by the Aceh government in to increase tourist loyalty must pay attention to perceived value as a key objective.

The influence of indirect effect through variable customer satisfaction is not proven. In other words, there is no indirect relationship between tsunami attractiveness and customer loyalty through customer satisfaction to increase the loyalty of tourist. This finding implies that to increase tourist loyalty for tsunami-affected tourist destinations, the government needs to focus on increasing the competitiveness of Tsunami Tourist Attractions and image (Munhurrun, P. R., Seebaluck, V. N., & Naidoo, P. 2014). This can be done through the improvement of the tourism objects and by supporting the business environment such as awareness, amenities, accessibility, accommodation, attractiveness. (Chen, C. F., & Chen, F. S. 2010).

6. Limitations and Future Study

The findings of this study have produced some empirical evidence that can be used to develop the tsunami tourism industry in Banda Aceh.There are however some limitations that shoud be taken into consideration in future research on the same topic. This includes the method of selecting respondents using convenience sampling techniques that were considered less representative because they did not meet the probability sampling method. The generalization of research results must be done carefully, especially if it is to be applied to different industries or different regions.

References

[1] Régnier P, Neri B, Scuteri S, Miniati S. From emergency relief to livelihood recovery. Gaillard J, editor. Disaster Prev Manag An Int J [Internet]. 2008 Jun 20 [cited 2018 Dec 25];17(3):410–30. Available from: http://www.emeraldinsight.com/doi/10.1108/09653560810887329

[2] Backer E, Morrison AM. VFR Travel: Is It Still Underestimated? Int J Tour Res [Internet]. 2017 Jul [cited 2018 Dec 25];19(4):395–9. Available from: http://doi.wiley.com/10.1002/jtr.2145

[3] Jitpraphai S, ... NA-T in M, 2017 undefined. Tsunami Disaster Risk And Vulnerability In Coastal Tourism Community: The Case Of Khao Lak Area, Thailand. ingentaconnect.com [Internet]. [cited 2018 Dec 25]; Available from: https://www.ingentaconnect.com/contentone/cog/tme/2017/00000012/f0020003/art00002

[4] Lee K, Choo S, Hospitality SH-IJ of, 2016 undefined. Effects of recovery experiences on hotel employees' subjective well-being. Elsevier [Internet]. [cited 2018 Dec 25]; Available from: https://www.sciencedirect.com/science/article/pii/S0278431915000523

[5] Maulida Z, Indah D, Scientific FS-J of A, 2015 undefined. Analysis on the Impact of the

Development of Lamphuuk Beach as the Tourism Location in Aceh against Business Strategy and Business Doers Welfare and. aessweb.com [Internet]. [cited 2018 Dec 25]; Available from: http://www.aessweb.com/pdf-files/jasr-2015-5(7)-349-356.pdf

[6] Zhang H, Fu X, Cai L, Lu L. Destination image and tourist loyalty: A meta-analysis. Tour Manag [Internet]. 2014 [cited 2017 Feb 24]; Available from: http://www.sciencedirect.com/science/article/pii/S0261517713001258

[7] Gronholdt L, Martensen A. The relationship between customer satisfaction and loyalty: cross-industry differences. Total Qual [Internet]. 2000 [cited 2017 Apr 1];

[8] Hapsari R, Clemes M, Dean D. The Mediating Role of Perceived Value on the Relationship between Service Quality and Customer Satisfaction: Evidence from Indonesian Airline. Procedia Econ Financ [Internet]. 2016 [cited 2017 Apr 1]; Available from: http://www.sciencedirect.com/science/article/pii/S2212567116000484

[9] Helander N, Ulkuniemi P. Journal of High Technology Management Research Customer perceived value in the software business. J High Technol Manag Res [Internet]. 2012;23(1):26–35. Available from: http://dx.doi.org/10.1016/j.hitech.2012.03.003

[10] Jin N, Lee S, Lee H. The effect of experience quality on perceived value, satisfaction, image and behavioral intention of water park patrons: new versus repeat visitors. Int J Tour [Internet]. 2015 [cited 2017 Feb 24]; Available from: http://onlinelibrary.wiley.com/doi/10.1002/jtr.1968/full

[11] Kassim N, Abdullah NA. The effect of perceived service quality dimensions on customer satisfaction, trust, and loyalty in e-commerce settings: A cross cultural analysis. Asia Pacific J Mark [Internet]. 2010 [cited 2016 Jun 27]; Available from: http://www.emeraldinsight.com/doi/abs/10.1108/13555851011062269

[12] Vesel P, Zabkar V. Managing customer loyalty through the mediating role of satisfaction in the DIY retail loyalty program. J Retail Consum Serv. 2009;16(5):396–406.

[13] Yang Z, Jun M, Peterson R. Measuring customer perceived online service quality: scale development and managerial implications. Int J [Internet]. 2004 [cited 2016 Jun 24]; Available from: http://www.emeraldinsight.com/doi/pdf/10.1108/01443570410563278

[14] Caruana A, Ewing MT. How corporate reputation, quality, and value influence online loyalty. J Bus Res [Internet]. 2010;63(9–10):1103–10. Available from: http://dx.doi.org/10.1016/j.jbusres.2009.04.030

[15] Yang Z, Peterson R. Customer perceived value, satisfaction, and loyalty: The role of switching costs. Psychol Mark [Internet]. 2004 [cited 2017 Apr 1]; Available from: http://onlinelibrary.wiley.com/doi/10.1002/mar.20030/full

[16] Science DR-J of WA of, Engineering undefined, 2016 undefined. Natural Disaster Tourism as a Type of Dark Tourism. researchgate.net [Internet]. [cited 2018 Dec 25]; Available from: https://www.researchgate.net/profile/Dorota_Rucinska/publication/307577952_Natural_Disaster_Tourism_as_a_Type_of_Dark_Tourism/links/57c9c0d908aedb6d6d97bfee.pdf

[17] Fahmi U, Ginting N, Sitorus R. Local communities and tourists' perception towards to PLTD Apung sites as tsunami disaster tourism in Banda Aceh City. IOP Conf Ser Earth Environ Sci [Internet]. 2018 Mar [cited 2018 Dec 25];126:012177. Available from: http://stacks.iop.org/1755-1315/126/i=1/a=012177?key=crossref.9f7ee831dd0273dc2cfff749c2f2a9a6

[18] Muktaf Z. Disaster Literacy in Communication Perspective. 2017 [cited 2018 Dec 25]; Available from: http://repository.umy.ac.id/handle/123456789/13380

[19] Wright D, Sharpley R. Local community perceptions of disaster tourism: the case of L'Aquila, Italy. Curr Issues Tour [Internet]. 2018 Sep 22 [cited 2018 Dec 25];21(14):1569–85. Available from: https://www.tandfonline.com/doi/full/10.1080/13683500.2016.1157141

[20] Flint DJ, Blocker CP, Boutin PJ. Customer value anticipation, customer satisfaction and loyalty: An empirical examination. Ind Mark Manag. 2011;40(2):219–30.

[21] Moore S, Rodger K, Taplin R. Moving beyond visitor satisfaction to loyalty in nature-based tourism: a review and research agenda. Curr Issues Tour [Internet]. 2015 [cited 2017 Feb 24]; Available from: http://www.tandfonline.com/doi/abs/10.1080/13683500.2013.790346

[22] Srinivasan S, Anderson R, Ponnavolu K. Customer loyalty in e-commerce: an exploration of its antecedents and consequences. J Retail [Internet]. 2002 [cited 2017 Apr 1]; Available from:

http://www.sciencedirect.com/science/article/pii/S0022435901000653
[23] Chang YH, Chen FY. Relational benefits, switching barriers and loyalty: A study of airline customers in Taiwan. J Air Transp Manag. 2007;13(2):104–9.
[24] Libai B, Bolton R, Bügel M, Ruyter K De. Customer-to-customer interactions: broadening the scope of word of mouth research. J Serv [Internet]. 2010 [cited 2017 Feb 11]; Available from: http://jsr.sagepub.com/content/13/3/267.short
[25] Kotler P, Keller KL. Marketing Management. Vol. 22, Organization. 2009. 329-331 p.
[26] Ranaweera C, Prabhu J. The influence of satisfaction, trust and switching barriers on customer retention in a continuous purchasing setting. Int J Serv [Internet]. 2003 [cited 2017 Apr 1]; Available from: http://www.emeraldinsight.com/doi/pdf/10.1108/09564230310489231
[27] Rosenberg L, Czepiel J. A marketing approach for customer retention. J Consum Mark [Internet]. 1984 [cited 2017 Feb 18]; Available from: http://www.emeraldinsight.com/doi/abs/10.1108/eb008094
[28] Yi Y, La S. What influences the relationship between customer satisfaction and repurchase intention? Investigating the effects of adjusted expectations and customer loyalty. Psychol Mark [Internet]. 2004 [cited 2017 Jun 1]; Available from: 1
[29] Faizan, Ali, Kisang, Ryu & Kashif H. Influence of Experiences on Memories, Satisfaction and, Behavioral Intentions: A Study of Creative Tourism. J Travel Tour Mark. 2015;
[30] Kotler P, Keller K, Manceau D, Hémonnet-Goujot A. Marketing management. 2015 [cited 2017 Feb 22]; Available from: http://www.academia.edu/download/32649000/KOTLER-SUMMARY.pdf
[31] Al- Hawari M, Ward T. The effect of automated service quality on Australian banks' financial performance and the mediating role of customer satisfaction. Mark Intell Plan [Internet]. 2006 Feb [cited 2018 Mar 27];24(2):127–47. Available from: https://www.emeraldinsight.com/doi/10.1108/02634500610653991
[32] Kaura V, Durga Prasad CS, Sharma S. Service quality, service convenience, price and fairness, customer loyalty, and the mediating role of customer satisfaction. Int J Bank Mark [Internet]. 2015 Jun [cited 2018 Dec 25];33(4):404–22. Available from: http://www.emeraldinsight.com/doi/10.1108/IJBM-04-2014-0048
[33] Pandža Bajs I. Tourist Perceived Value, Relationship to Satisfaction, and Behavioral Intentions. J Travel Res [Internet]. 2015 Jan 9 [cited 2018 Nov 3];54(1):122–34. Available from: http://journals.sagepub.com/doi/10.1177/0047287513513158
[34] Ramseook-Munhurrun P, … VS-P-S and, 2015 undefined. Examining the structural relationships of destination image, perceived value, tourist satisfaction and loyalty: case of Mauritius. Elsevier [Internet]. [cited 2018 Nov 3]; Available from: https://www.sciencedirect.com/science/article/pii/S1877042815012586
[35] Hair JF, Sarstedt M, Ringle CM, Mena JA. An assessment of the use of partial least squares structural equation modeling in marketing research. J Acad Mark Sci [Internet]. 2012 May 7 [cited 2017 Nov 18];40(3):414–33. Available from: http://link.springer.com/10.1007/s11747-011-0261-

IOP Conf. Series: Earth and Environmental Science **273** (2019) 012057 doi:10.1088/1755-1315/273/1/012057

Scenario based logistic capacity assessment for disaster preparedness

A Nugroho[1,2], H Maulana[3], Fachruddin[4], N Ariyati[3] and D Kurniawan[1]

[1] *Agribusiness Department, Universitas Syiah Kuala, Banda Aceh, 23111, Indonesia; nugroho@unsyiah.ac.id*
[2]*Tsunami and Disaster Mitigation Research Center (TDMRC). Universitas Syiah Kuala. Banda Aceh, 23232, Indonesia*
[3]*Islamic Economics and Business. Ar Raniry Islamic State University, Banda Aceh, 24415, Indonesia.*
[4]*Agriculture Engineering Department, Universitas Syiah Kuala, Banda Aceh, 23111, Indonesia*

Abstract. Disasters that occur in various locations will definitely have impacts on infrastructure damage such as roads, bridges, markets, shopping centers. The perceived disturbances will eventually disrupt logistics distribution (including food) and other basic needs from the point of entering logistics (logistics center) to evacuation sites (victims). In the case of less severe disasters, disturbances can cause logistical delays from producers to consumers. Finally, this can cause price volatility due to the stocks decrease. This study focuses on areas along the two main routes in Aceh: the East and West routes. The object of this study includes all information related to logistic distribution are producers, collectors, traders, wholesalers, consumers, and shipping companies and related local government agencies. This study applied scenario based on the combination of multiple models and geographical information system to identify the optimal location to aid emergency logistics transportation routes. The results of this study indicate that logistics distribution is still very dependent on land transportation modes such as trucks and containers. Ports and airports have not been used optimally while the progress of railway construction is plodding.

1. Introduction

Indonesia faces a higher level of risks triggered by volcanic and tectonic activities such as the earthquake, eruption and tsunami and also experienced more frequent floods, cyclones and landslides due to climate change. Disasters that occur in various locations will have effects on the damage of infrastructures such as roads, bridges, markets, shopping centers. Therefore, the disturbances will ultimately disrupt the distribution of logistics (including foods) and other basic needs from logistic entry points (logistic center) to evacuation sites (victims). In the case of less severity disaster, the disruption may cause delays of logistic from producers to consumers. Eventually, it may lead to a price increase due to decreasing stocks. Previous studies on the relationship between logistics and disaster have been done, [1][2][3][4] however, only a few studies concerned in scenario-based models. Review on various forms and classification of tackling uncertainty in Humanitarian Logistics for Disaster Management and Optimization models utilized in emergency logistics, which are broken down into three parts: facility location, relief distribution and casualty transportation, and other operations, have been done in the past [5][6]. For example, [7] proposed a model to determine the location of relief distribution centers and the allocation of affected area to relief distribution centers using a mixed-integer nonlinear programming to minimize the sum of the expected total cost (which includes charges of location, procurement, transportation, holding, and shortage) and the variance of the total cost. Furthermore, an attempt to improve the previous model by adding the satisfaction level variables using a multi-objective robust stochastic programming approach for disaster relief logistics under uncertainty has been done[8].

The vulnerability of a demand point can also be computed by using fault tree analysis and incorporating the study into the optimization model[9]. The locations produced by the proposed model are compared to those provided by the p-center model concerning risk value, coverage distance, and covered population by using several test problems. The model is also applied to a real problem. Another research [10] introduces a disaster preparedness system based on a combination of multi-objective optimization and geographical information systems to aid multi-organisational decision-making. The results highlighted that the number of government organizations deployed to handle the situation was excessive, leading to high cost without achieving the best possible level of satisfaction. The system proposed showed the potential to deliver better performance regarding the cost and level of service than the approach currently employed by the authorities. A more recent method on the inventory allocation optimization using a case study from the Turkish Red Crescent has been conducted using forecasting methods from the earthquake engineering literature and avoids using probabilistic scenarios to represent the uncertainties related to earthquake occurrences[11].

This research intends to examine the impact of the disaster on logistic distribution emphasizing on food security in Aceh Province. The motivation is to observe the development of logistic capacity after great Sumatra earthquake and tsunami in 2004. It is expected that the disaster impacts on logistic distribution network based on several scenarios can be derived comprehensively. The scenes involve the changing in demand and supply of goods, distance and travel time, as well as cost and benefit as a consequence of changes in distribution channels. Finally, this research will derive the perception of stakeholders along the supply chains regarding the changing in the logistics distribution in Aceh Province.

2. Methods

This study focuses on the areas along the two main roads in Aceh: Eastern and Western routes. The survey was conducted in 11 districts and cities, and the data was collected from 27 business centers and traditional markets (Figure 1). In total, 185 respondents were interviewed during August to September 2017. The object of this study covers information related to the logistics distribution of related stakeholders involved in the production, trading, and distribution of edible commodities. The stakeholders along the supply chains addressed in this study were producers, collector, traders, wholesaler, consumers, and shipping companies as well as related local government agencies.

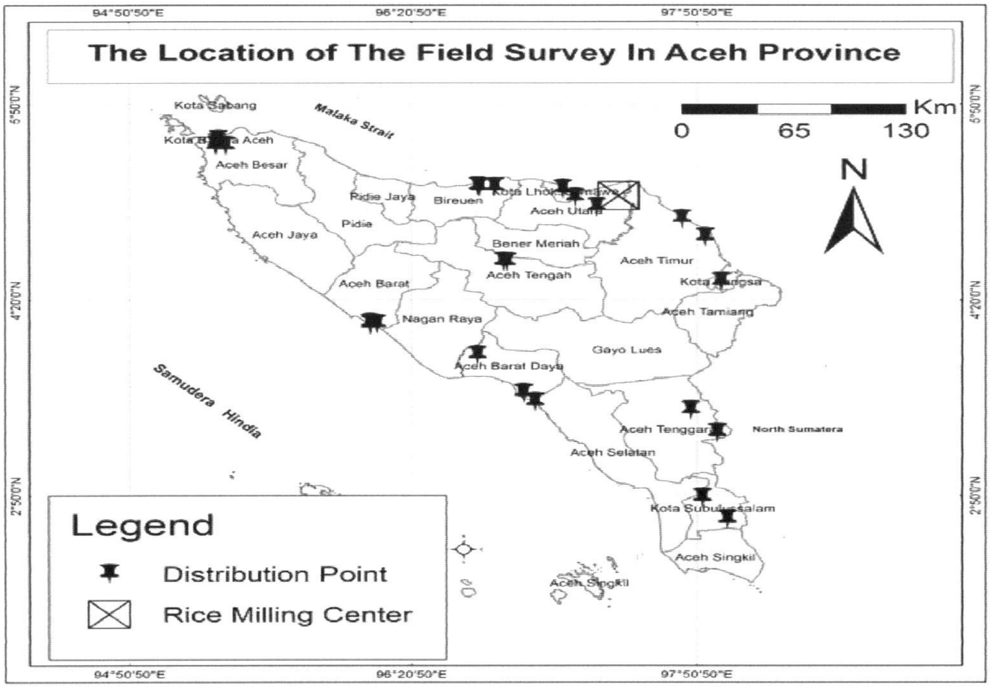

Source: Primary data processed (2017)

Figure 1. Study area

Disaster risk analysis needs to be put in the first step in order to identify hazards, exposures, vulnerability, capacity, risk mitigation that has already been carried out, as well as the impact of the disaster. This risk analysis will be used to develop alternative decision strategies as a preventive measure and mitigation if these risks occur in the future. In this analysis, historical data from governmental bodies will mainly be used to assess the risks. Additionally, respondents' perceptions of the locations for the future development of regional distribution centers in the study areas, during normal conditions and in the event of a disaster, will also be carried. The rationale of this is that the perceptions of respondents are: (1) expected to describe strategic location that are jointly desired from all the stakeholders involved in market activities including producers, collectors, logistics transport of traders and consumers; (2) expected to clarify the missing information that are not recorded in the local statistical data agencies regarding past exposures and impacts of the disasters as well capacities of local communities.

Spatial analysis in this study utilizes a Geographic Information System (GIS) method. Through this method will be approached the location of land that is not employed such as shrubs and open area, slope, in an overlay with an official map and supporting infrastructures such as roads, railroads, ports, and airports. Several alternative locations that are feasible and strategic will be able to be recommended according to the approach of social analysis, disaster risk analysis, and respondents' perceptions

In the base scenario, businesses are treated as usual without counting the impact of the disasters. Logistics, therefore, are delivered based on standard conditions, following the minimum costs and routes preferences. However, during the disaster scenario, logistics are then delivered based on the nearest affected area regardless the difficulty of the routes and costs burden. Therefore, it is important to identify which routes that will give the optimum benefits and to identify which locations that applicable to be developed as Regional Distribution Centers (RDC). These regional logistic centers will eventually minimize the dependency and time of delivery from the surrounding provinces.

IOP Conf. Series: Earth and Environmental Science **273** (2019) 012057 doi:10.1088/1755-1315/273/1/012057

3. Result

The findings show that logistic distributions still rely heavily on land transportation modes such as trucks and containers. Ports and airports have not been utilized optimally whereas the progress of railways construction is stagnant. The geographical features of Aceh and North Sumatra provinces are characterized by mountainous area so that railway connecting the two regions are hard to be developed. In the baseline scenario, the logistics in the eastern areas are transported through the north-east route or known as Medan-Banda Aceh route, while south-west course connecting Medan-Subulussalam- Aceh Barat is used to deliver logistic to western parts of Aceh. In the event of a disaster occurred at these routes, it is certain that there must be adjustments. However, alternative ways are less available. Based on Fig.2, it indicates that the dependency on the two main existing routes put this province in a high level of vulnerability regarding logistics assurance and may trigger food insecurity in the future.

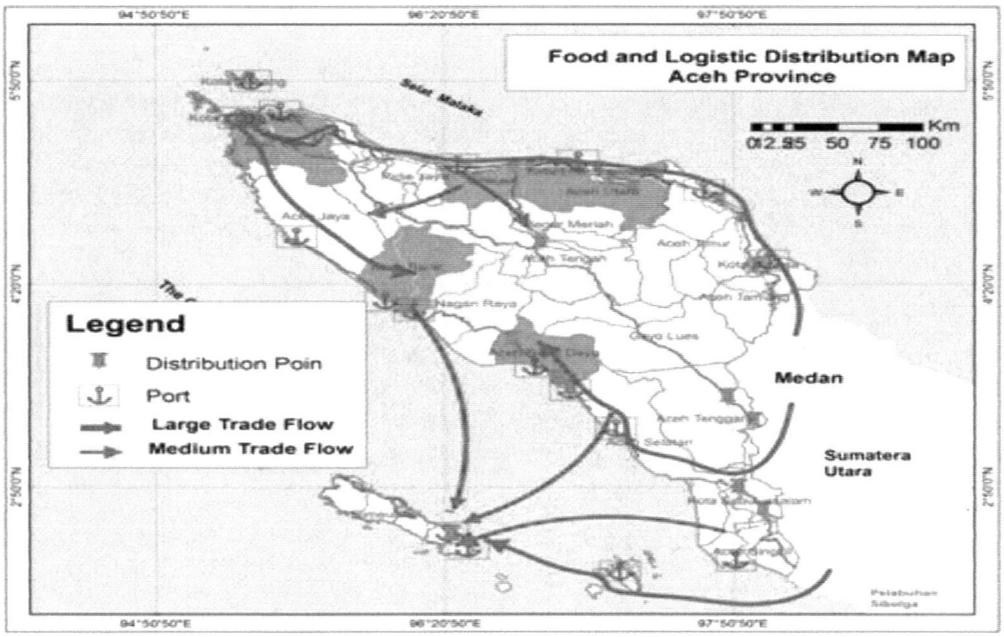

Source: Primary data processed (2017)

Figure 2. Food and logistic distribution in Aceh

3.1. Disaster Risk and Logistics Distribution

On December 26, 2004, Aceh was hit by an earthquake and tsunami which resulted in the loss of hundreds of thousands of lives, damage to physical infrastructure and disturbance in economic activities. Recently, another 6 magnitude earthquake also destroyed Pidie Jaya regency on December 7, 2016. During 2017-2018, approximately 125 events, including floods, cyclones, landslides and earthquakes occurred in several regencies and caused 249,000 people to be evacuated. The disasters that occurred in various regencies had impacts on the damage to infrastructure such as roads and bridges which ultimately disrupted the smooth distribution of goods from producers to consumers in the Aceh region. Aceh also has hundreds of bridges with various specifications as illustrated in Table 7.6. Moreover, access to the distribution of goods in Aceh is only supported by 2 (two) main routes, namely the east and west edges. That is, damage to the infrastructure will lead to smooth access to the distribution of logistics goods in the Aceh region.

This research explores the extent to which disasters disrupt the activities of the respondent's economic activities. One question, for example, is: "To what extent does the following disaster disrupt your economic activities?". If the answer given by the respondent (e.g., producer) is "Never" means that the producer has never been disturbed by the disaster during the production activity. If the answer given is "Rarely-Never" means that some respondents feel annoyed with rare intensity so that they are never disturbed at all. In general, the results of this study conclude that disasters caused by disasters on economic activities of perpetrators are very minor (rarely to never). Spatially, only two regencies which concern the disruption of economic activities due to disasters as shown by yellow signs in Figure 3. Simeulue island is the only regency which may suffer serious food insecurity due to high dependency on neighbour areas. These results are relatively different from expectations because Aceh is one of the areas vulnerable to disaster. Therefore, further analysis and caution are needed in interpreting the results of this study.

3.2. Perceptions of Respondents on RDC locations

The choice of location for an appropriate infrastructure regarding technical, socio-economic and environmental will have a positive impact on the acceleration of development for the region and the surrounding area as well as one of efficient disaster mitigation. Conversely, the selection of locations that are not technically appropriate will increase the risk of incompatibility between the carrying capacity of the land and its use. From an economic aspect, the selection of inappropriate locations will increase the risk of loss and non-use of the infrastructure facilities. Therefore, it is necessary to develop indicators that can represent the range of the right location.

The results of the study on the eastern route of the Aceh region showed that as many as 21 percents of the total respondents stated that the location suitable for the existence of the RDC was a factor of proximity to the site of the primary market in the area. The second factor is that the position of the RDC should be on edge / near the national crossing of Medan-Banda Aceh. Meanwhile, the element of proximity to loading and unloading terminals and near the border of the area is the next consideration. The results related to respondents' perception of the location of the RDC in Aceh province are presented in Figure 3.

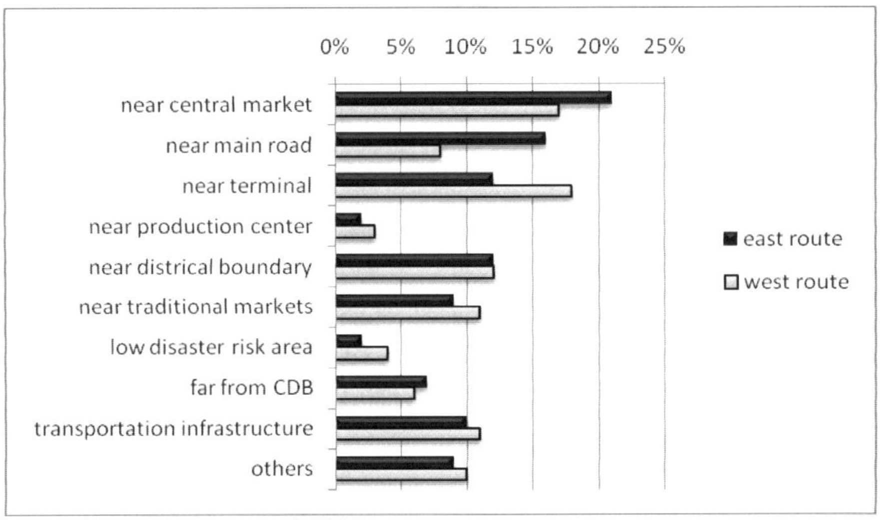

Source: Primary data processed (2017)

Figure 3. Respondents' perception on the RDC location

The results of the perception study on the western route show a slight difference with the east route. On the west line, the central aspect that is important to be considered in the selection of the RDC location is the proximity to the loading and unloading terminal which is 18 percent. Next is the factor of the vicinity of the parent market and is close to the border of the region, which is equal to 17 percent and 12 percent. Whereas the proximity factor to the west-west Medan-Banda Aceh crossing is only around 8 percent.

Based on the two graphs above, it can be concluded that most of the stakeholders related to the distribution network of staple food indicate that the exact and appropriate location criteria for the existence of the RDC are: (1) the location with the primary market in the region; (2) The area is located near / around the freight terminal; (3) the place is at least located in the vicinity of the Medan-Banda Aceh national road, and (4) the situation is at the border of the area. The results of this perception study also show that business and economic motives (costs and benefits) are far more influential than environmental factors such as disaster factors (floods, landslides, earthquake faults). This is relatively surprising because Aceh is a disaster-prone area and several cases show that there is a positive correlation between infrastructure damage due to disasters and delays in supply. Business players' awareness toward risks may influence this phenomenon. After the 2004 tsunami in Aceh, the risk awareness level has not been put into behaviour as well as structural actions.

3.3. Western Route vs Eastern Routes RDC Scenario

The combined results of spatial data analysis with an analysis of the study of logistics distribution channels and the support of adequate facilities and infrastructure to be used as regional distribution centers in western region (see Fig 4) are at point D (city of Meulaboh) and E (city of Blang Pidie). Factors supporting the election of point D (city of Meulaboh) as an alternative location include: (1) Meulaboh is one of the reference cities to monitor the rate of inflation. Even at some time, the rate of increase in the town of Meulaboh had several times exceeded the average inflation rate in the province of Aceh; (2) Total population and readiness of supporting infrastructure such as markets, shops, warehouses, roads, loading and unloading terminals, ports are available.

However, the primary inhibiting factor in the selection of Meulaboh was the supply line. According to the results of interviews with logistic transportation actors, the supply of staple goods (non-rice) is almost 80 percent still using the eastern route. The east route has many advantages such as transportation players are many (availability of helps due to accidence), safe, wide roads and convenience, and less ascend. Another reason is fewer informal levies than using the westbound lane. So, even though the west route has a shortage regarding time (more extended travel), the costs incurred by transportation actors can be more economical (or not much different) compared to the use of westbound lanes. Some of the impacts are: (1) the possibility that the RDC in Meulaboh will depend on the RDC in Regional I at point A (potentially at Banda Aceh city or Aceh Besar regency) so that its existence is less than optimal; (2) the presence of the RDC will be less optimal in the south-west lane in other words the community in the south-west path does not touch the RDC function as the carrying capacity of food security and price protection for consumers. Therefore, if the RDC will later be built in Meulaboh, it is necessary to pay attention to the impact.

The second possible location for the RDC to be built in Region III is at point E (Blang Pidie City). Several supporting factors for the selection of Blang Pidie are: (1) this area becomes the final destination for trucks/modes of transportation from Medan / North Sumatra via Sidikalang-Subulussalam. Therefore, the function of Blang Pidie is similar to the city of Banda Aceh, which is the last stop of the logistic transportation mode via the east route. The impact, transportation, and logistic distribution activities will accelerate economic development along the path that it passes (Aceh Selatan to Subulussalam city); (2) The location of Blang Pidie is midway to regional III. This position is strategic because it has good access and connectivity with the surrounding area (Aceh Selatan-Nagan Raya and Aceh Barat). Some other supporting infrastructures including markets, shops, open space, warehousing, loading terminals, roads with sufficient access and width as well as ports are available.

The combined results of spatial data analysis with an analysis of the study of logistics distribution channels and the support of adequate facilities and infrastructure to be used as regional distribution centers II are at point B (Aceh Utara regency) and at point C (Bireuen regency). These two locations are chosen because they meet several indicators such as the availability of adequate facilities including Krueng Geukuh Port, Malikussaleh airport, industrial development planning areas, as well as low level of historical risk affected regions. The locations are also near the new line of Aceh Utara to Bener Meriah regency intersection which makes travel distance closer and faster, connecting these regencies to Bener Meriah or Aceh Tengah, which are the vegetables producing centers in Aceh, which takes ranges is from 1.5 to 2 hours to travel. The use of the route to the Aceh Utara -Bener Meriah intersection is far more preferable regarding duration and cost than using the Bireuen-Takengon crossing which takes up to 5 hours from Aceh Utara regency. To sum up, the proposed RDC location based on the scenarios is presented in Figure 4.

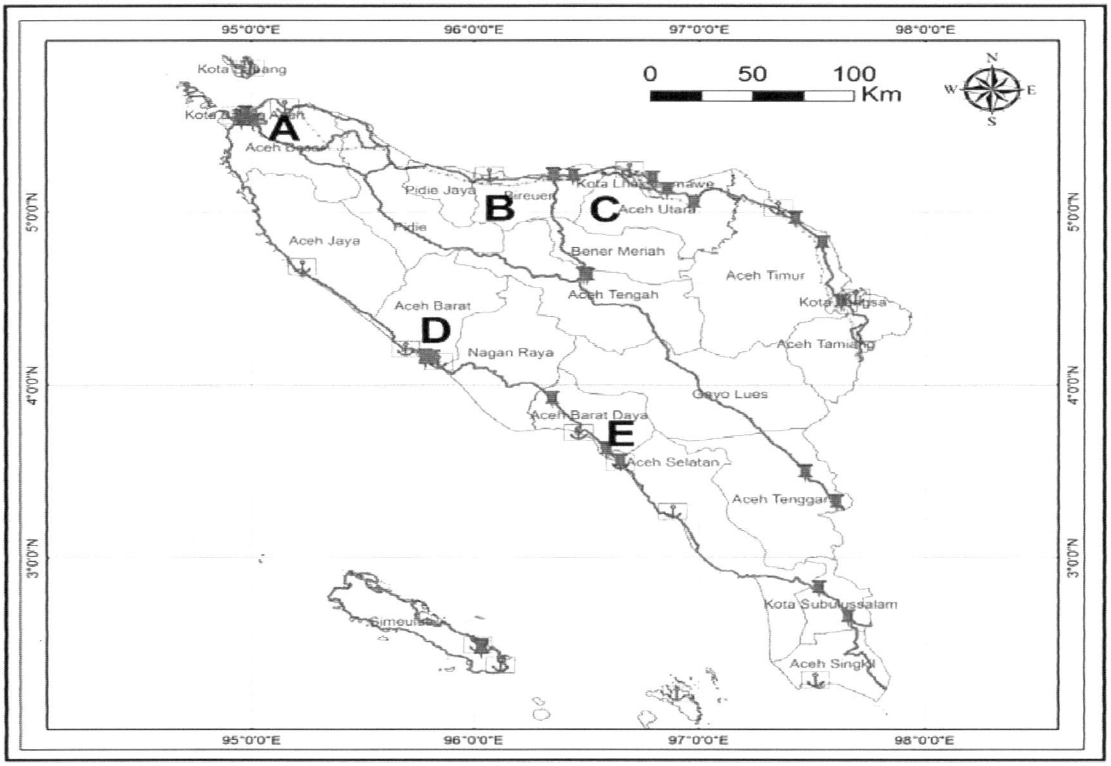

Source: Field assessment (2017)

Figure 4. RDCs proposed location

4. Conclusion

Especially for rice, demand needs can still be met by local rice production in Aceh. However, other commodities much rely on supply from other provinces (North Sumatra). The eastern route is more preferred by the land transportation players, connecting from Medan (North Sumatra) to Meulaboh (Aceh Barat), while the western route is only favored for the logistics from Medan to the Aceh Barat Daya, or vice versa. This study considers the establishment of a regional logistics center to facilitate the flow of supply and distribution of logistics in both eastern and western parts of Aceh Province in the future, as well as to anticipate the negative impacts related to the disruption of logistics distribution due to the natural disaster. Based on disaster risk analysis, respondents' perceptions and spatial analysis using geographic information systems found potential locations that could be used as RDCs. Aceh Besar regency, which is adjacent to the city of Banda Aceh, is one of the locations that have high potential as a Provincial Distribution Center. Whereas for the eastern route, Aceh Utara or Bireuen Regencies are the most suitable location due to the support from the existing facilities and infrastructures, whereas Aceh Barat Daya or Aceh Barat is recommended for the west route.

Keywords: *food security; disaster preparedness; humanitarian logistic; regional distribution center*

References

[1] Balcik, B., & Beamon, B. M. (2008). Facility location in humanitarian relief. *International Journal of Logistics*, *11*(2), 101-121.
[2] Barbarosoğlu, G., & Arda, Y. (2004). A two-stage stochastic programming framework for transportation planning in disaster response. *Journal of the operational research society*, *55*(1), 43-53.
[3] Kovács, G., & Spens, K. M. (2007). Humanitarian logistics in disaster relief operations. *International Journal of Physical Distribution & Logistics Management*, *37*(2), 99-114.
[4] Lin, Y. H., Batta, R., Rogerson, P. A., Blatt, A., & Flanigan, M. (2011). A logistics model for emergency supply of critical items in the aftermath of a disaster. *Socio-Economic Planning Sciences*, *45*(4), 132-145.
[5] Liberatore, F., Pizarro, C., de Blas, C. S., Ortuño, M. T., & Vitoriano, B. (2013). Uncertainty in Humanitarian Logistics for Disaster Management. A Review. In *Atlantis Computational Intelligence Systems*. https://doi.org/10.2991/978-94-91216-74-9_3
[6] Caunhye, A. M., Nie, X., & Pokharel, S. (2012). Optimization models in emergency logistics: A literature review. *Socio-Economic Planning Sciences*. https://doi.org/10.1016/j.seps.2011.04.004
[7] Bozorgi-Amiri, A., Jabalameli, M. S., Alinaghian, M., & Heydari, M. (2012). A modified particle swarm optimization for disaster relief logistics under uncertain environment. International Journal of Advanced Manufacturing Technology. https://doi.org/10.1007/s00170-011-3596-8
[8] Bozorgi-Amiri, A., Jabalameli, M. S., & Mirzapour Al-e-Hashem, S. M. J. (2013). A multi-objective robust stochastic programming model for disaster relief logistics under uncertainty. *OR Spectrum*. https://doi.org/10.1007/s00291-011-0268-x
[9] Akgun, I., Gumusbuga, F., & Tansel, B. (2015). Risk based facility location by using fault tree analysis in disaster management. *Omega-International Journal of Management Science*. https://doi.org/10.1016/j.omega.2014.04.003
[10] Rodriguez-Espindola, O., Albores, P., & Brewster, C. (2018). Disaster preparedness in humanitarian logistics: A collaborative approach for resource management in floods. European Journal of Operational Research. https://doi.org/10.1016/j.ejor.2017.01.021
[11] Battarra, M., Balcik, B., & Xu, H. (2018). Disaster preparedness using risk-assessment methods from earthquake engineering. European Journal of Operational Research. https://doi.org/10.1016/j.ejor.2018.02.014

IOP Conf. Series: Earth and Environmental Science **273** (2019) 012058 doi:10.1088/1755-1315/273/1/012058

Do villages allocate sufficient fund for disaster risk reduction? A survey of 10 villages in Banda Aceh

H Fahlevi[1*], M F Nur[2], M Arifin[2] and A Azril[2]

[1]Accounting Department, Faculty of Economics and Business, Syiah Kuala University, Banda Aceh, Indonesia
[2]Department of Disaster Science, Postgraduate Program, Syiah Kuala University, Banda Aceh, Indonesia

Abstract. This study aims to assess and evaluate the disaster budget in 10 villages in Banda Aceh. By using a qualitative descriptive approach, this study analyzed budgets and other financial information from 10 selected villages in Banda Aceh. The villages were affected by the 2004 tsunami. A survey and series-of interviews were conducted to obtain a more detailed understanding of why the villages do or do not allocate sufficient disaster budget allocations. This study unveils that the villages allocated a very small disaster budgets ranging from 0.00 to 0.88% of their total budget. The disaster budgets were mainly allocated as cash financial assistance to disaster victims, rather than as preventive programs. The reasons behind the small allocation to disaster budgets in the villages are (1) disaster budget is not spent/used, (2) high dependence on the Banda Aceh city in term of disaster management, (3) very strict financial regulations on the use of village funds, (4) insufficient Village Funds, and (5) lack of knowledge of DRR among village government officers. The result of this study can be used to improve the capacity of village apparatus in disaster management as well as to enhance the financing of DRR programs in Indonesia.

*email: hfahlevi@unsyiah.ac.id

1. Introduction

The Indonesian law of Disaster Management Number 24 (2007) states that the implementation of disaster management is a sharing responsibility of all parties, including government, community and business entities [1]. In this context, the central government and local governments are required to allocate funds for covering the disaster management programs and strategies. In parallel, the Indonesian village law urges villages to use village fund for disaster mitigation activities including to revise budget to accommodate special or extraordinary events i.e. disasters [2]. In fact, the central government prioritizes disaster preparedness programs to be financed by the village fund. These regulations indicate the interest and concern of the Indonesian central government on the use of village funds for disaster preparedness and mitigation programs. This funding option is crucial as the local governments have very limited budget [4].

The usage of village funds for disaster-related programs seems to be promising although it has many potential obstacles and challenges. Indonesia as a disaster-prone country needs to find a good formula for financing disaster mitigation program to ensure the effectiveness of DRR nationally, and more importantly to reduce the number of victims and the scale of financial losses. However, research on disaster budgeting is quite limited. There is a research gap relating to how government allocates the disaster funds, especially in local governments in Indonesia [5]. Previous studies on disaster budgeting focus more on how much money has been allocated for disasters and the determinants of disaster budgets at the national and provincial level.

The financing of DRR programs in local government have some obstacles and lack concern from the local governments [4]. In this context, villages can play a more significant role in financing DRR programs through budgeting for disaster preparation. Firstly, each Indonesian village receives a Village Fund ranging from IDR 800 million to more than IDR 1.5 billion every year from the central and local governments. As indicated above, this money may be allocated to DRR program. Secondly, DRR programs could be more relevant and effective when they are planned and decided by the village government as they are the people who know best their needs and problems.

This study aims to investigate whether and how villages in Banda Aceh allocate Village Funds for disaster risk reduction efforts and to understand the reasons, motivations and challenges faced by the villages in allocating these disaster-related expenditures.

2. Data and Methods

This research adopts a mixed research method, namely descriptive quantitative and qualitative methods to analyze the proportion of allocated funds for disaster mitigation efforts in villages within Banda Aceh region. The data were collected from the 2017 village budgets and questionnaires filled by the head of each village or the secretary of the villages. The sample of this research is 10 villages located in the Banda Aceh coastline. The sample criteria is villages that were significantly affected by the 2004 Indian Ocean tsunami. A serial Semi-structured interviews was carried out to uncover the budgeting process, allocation of the DRR program and the motivation behind decisions taken.

The informants/ respondents were all village-leaders actively involved in the village budgeting process. The main purpose of the interviews is to explore why and how the village fund was used for the DRR programs. The data were analyzed by using descriptive analysis and qualitative data analysis approaches. The data were collected between April and August 2018. All interviews were conducted in the village offices or in the village leaders' house. On average, the duration of the interviews was between 30 to 60 minutes and not all interviews were recorded.

3. Results

3.1 Profile of sample and statistic descriptive

This study was carried in 10 selected villages in Banda Aceh. These villages were significantly damaged in the 2004 tsunami and with loss of life ranging from 15% to 50% of the local population. This high casualty rate is due to the villages close proximity to the sea. It is assumed on the basis of this tragic event that the current inhabitants of these 10 villages and their leaders really understand about the tsunami and its potential risk and consequently that they will more motivated to prepare for disasters than the other villages in Banda Aceh (please see Table 1).

Table 1. Profile of samples and amount of village fund in 2017

No.	Village Name	Sub-Districts	Distance to the seashore	Village Budget 2017 in million IDR (in thousand USD)
1.	Gampong Jawa	Kuta Raja	1.21 km	1,644 (113)
2.	Gampong Pande	Kuta Raja	1.03 km	1,509 (104)
3.	Ulee Lheue	Meuraxa	94 m	1,461 (100)
4.	Blang Oi	Meuraxa	1.26 km	1,427 (98)

5.	Lambung	Meuraxa	889 m	1,351 (93)
6.	Deah Baro	Meuraxa	482 m	1,385 (95)
7.	Deah Glumpang	Meuraxa	566 m	1,373 (94)
8.	Alue Deah Teungoh	Meuraxa	568 m	1,409 (97)
9.	Tibang	Syiah Kuala	1.55 km	1,580 (109)
10.	Alue Naga	Syiah Kuala	674 m	788 (54)

From the table 1, it can be seen that the sub-districts located on the seashore such as Kuta Raja, Meuraxa, and Syiah Kuala sub-districts have potentially large risk exposure to tsunami risk. The distance between the villages and the nearest beach is less than 2 km and it would takes only few minutes for tsunami tidal reach the villages. Thus, the 10 villages should anticipate the risk through a well-designed planning and preparedness system.

It can also be clearly seen from the table 1 that the villages have different amount of budget. The reasons for the budget variation is that the central government taking into account the population, poverty level, are, and level of geographical difficulties [6]. Almost half of the villages have more than IDR 1.4 billion or more than USD 100,000 annual budget. This number can be considered as a big budget compared to other villages outside Banda Aceh.

3.2 The allocation of the village budget for DRR related programs

The village budget is proposed and managed by the village officers which include the head of the village, secretary, and treasurer. The budget proposal and proposed programs are discussed with the village supervisory board which comprises the villager representatives and the informal leaders in the villages. This system seeks to ensure that a sufficient level of villager participation in the budgeting process. According to the data analysis shown in the table 2, some villages have used the village funds for DRR programs.

Table 2. Village fund allocation for the DRR program on the 2017 village budget

No	Village Name	Budget for DRR Related Programs in IDR (in USD)	Portion of Total DRR Related Program form the Total Budget
1	Gampong Jawa	10,000,000 (680)	0.61 %
2	Gampong Pande	0	0 %
3	Ulee Lheu	3,000,000 (206)	0.21 %
4	Blang Oi	1,031,817 (71)	0.07 %
5	Lambung	600,257 (41)	0.04 %
6	Deah Baro	1,000,000 (68)	0.07 %
7	Deah Glumpang	0	0 %
8	Alue Deah Teungoh	10,000,000 (680)	0.71 %
9	Tibang	13,995,000 (965)	0.88 %
10	Alue Naga	6,320,000 (430)	0.81 %

This study reveals that the villages, although they are vulnerable to the tsunami and other disasters, allocated a very small amount for disaster related programs. In fact, three out of ten villages did not allocate money for the DRR programs. It can be seen in the table 2 that seven villages have budgeted programs related to disasters but the proportions (and actual amounts) are insignificant. The highest portion is only 0.88 percent or IDR 13,995,000 for a year. Apparently, there is no correlation between village demography or proximity to the coastline and the size of the DRR program allocation.

Table 3 exhibits that some villages have used the village budget for DRR activities. For instance, Tibang village has budgeted money to basic logistics for an emergency response program, while Lambung village has established tsunami evacuation route signs in the village roads. In addition, there are villages that have not spent any of the village budget for the DRR activities, even though they are exposed to the risk of tsunami and other natural disasters.

Table 3. DRR Activities in Realization of Village Budget in 2017

No	Village Name	Name of DRR Related Program	Description of the Programs
1	Gampong Jawa	Grant for disaster victims	Financial assistance for residents whose houses burned (fire).
2	Gampong Pande	-	-
3	Ulee Lheu	Grant for disaster victims	Financial assistance for residents who died while fishing in the sea. The village provided financial aid (cash) of IDR 3,000,000 million for each victim.
4	Blang Oi	Grant for disaster victims	Financial assistance to people affected by the disaster.
5	Lambung	-	-
6	Deah Baro	Preparing medicines	Preparing medicines for residents in the event of a disaster, for example, a fire is injured, medicines will be provided.
7	Deah Glumpang	-	-
8	Alue Deah Teungoh	Grant for disaster victims	Providing assistance to residents whose homes were damaged by storms and fires.
9	Tibang	Procurement of Food and Beverage Supplies, cabinets, cookware, health equipment and equipment for safety procedures	The village prepares basic logistics for disaster responses including medicines and kitchen utensils that distributed to the victim of, for example, fire. The logistics were distributed to residents who were hit by the calamity. The village also paid for a temporary house rental for the victims.
10	Alue Naga	Procurement of house materials	Repairing the victims' homes which were damaged by storm

Based on the tables 2 and table 3, it can be concluded that the amount of budget allocated for disaster management in the villages studied is very small i.e. less than 1% of the total village funds obtained from the upper level of governments. In fact, some villages did not allocate any money for the disaster-related programs. Furthermore, those villages that did allocate money tended to provide for financial aids for victims and logistics for disaster emergency responses rather than preparation. Thus, the disaster-related program budget can be categorized as responsive actions to disasters, rather than preventive. In other words, the DRR budget allocation is small, and of that is focused helping victims after the disaster happen.

This is consistent with the finding of Oktari et al [4] and Dartanto et al [8]. Oktari et al [4] studied the budget of the Regional Disaster Management Agency (or *Badan Penanggulangan Bencana Daerah*/BPBD Banda Aceh. They uncovered that the BPBD receives a relatively small portion of budget from the local government budget i.e. not more than 0.6% of the total budget of Banda Aceh. Meanwhile, Dartanto et al [8] investigated the awareness of 486 Indonesia local governments in promoting disaster mitigation. They found that more than 50% of the local governments did not allocate sufficient budget for disaster-related programs.

3.3 Why did allocated disaster budget difficult?

The village fund provided by the central government is accompanied by an extended financial autonomy for the village government [7]. There are several challenges and difficulties in budgeting and spending village funds for village level disaster risk reduction efforts in those villages studied. Based on the serial interviews, the informants expressed reasons why allocating disaster budget is a difficult task, namely: (1) unspent disaster budget, (2) a high dependence on the BPBD in term of disaster management, (3) very strict financial regulations on the use of Village Funds from the central government, (4) insufficient village fund, and (5) lack of knowledge of the village government in the DRR.

3.3.1 Unspent disaster budget

Most of the budgeted DDR program budget is contingent activities with a high level of uncertainty. In these circumstance the budget allocation will only be spent or realized if the anticipated disasters occur. In other words, if the disaster does not occurs, the money will not be used and will remain as idle fund at the end of the financial. This unrealized budget (surplus) potential discourages village governments to allocate budget for the DRR related programs because a high surplus is considered to be poor financial management by the local government and most probably result in the surplus affecting the next year budget in the form of a budget cut. As a result, some villages such as Ulee Lheue do not budget village funds for DRR anymore because the funds could not be used because there was no disaster in that year, so it would be budget surplus.

"Most of our people are fishermen, and their job is very risky, so we prepare 3 million IDR to provide financial aid or assistance if they get incident in the sea. We can also use the funds if there is a fire, but for 2017 there were no such incidents, so we cannot propose the budget in the current year (2018)".

3.3.2 Highly dependence on the BPBD

This study found a consensus within the villages that disaster management and preparedness should be arranged by the BPBD. The villages do not allocate money for DRR programs because they believe that BPBD has this responsibility. The villages will allocate the money if they receive a direct-order or request from the BPBD. For example, Gampong Pande respondents reported that they do not use their village funds if BPBD does not instruct them to do it. "No one is responsible (for the DRR program), because the BPBD does not direct us to use our own funds". A similar problem is also found in the Alue Deah Teungoh. "It is already budgeted, but so far we had a lot of assistance from outside parties, such as the BPBD for the DRR related activity in this village here and for the establishment of this village as a Disaster Resilient Village. These activities and program were funded by the BPBD".

3.3.3 Strict financial regulations

Some of the interviewees admit that the financial rules of village fund are very strict. The Gampong Jawa village head reported that there are some rules that discourage them to budget for DRR programs. They are reluctant to be more innovative in designing DRR programs because the program my not align with the financial reporting requirement and therefore cannot be fund from the village budget.

3.3.4 Insufficient village funds
The village fund is mostly used for staff expenses (salaries) and capacity building of the village government staffs. The rest of the money tends to be used for building infrastructures i.e. drains and village roads. As a result, the Village Fund is always not sufficient to cover more programs related to DRR. "if I can say, we have a lot of needs, so the budget is still not enough" (Village senior officer of Lambung).

3.3.5 Lack of knowledge of the village government in the DRR
Based on the interviews with the respondents, it is noticed that the villages' staff and key officers have limited knowledge of disaster and DRR. The village government mostly budgeted disaster emergency programs which are sporadic/ reactive. One reason for this could be that they did not know exactly what and when the DRR programs should be proposed and budgeted. For example, the Alue Naga village has a budget to help villagers whose houses are damaged by disasters. This was acknowledged by the village secretary "in the event of a disaster we have funds to help affected victims". Most of the village leaders do not have educational backgrounds and past experiences on disaster management. They might have graduated from university but their major is not related with disaster management. Furthermore, not all villages heads and staff have participated in any DRR training. One of the village leaders, the head of Tibang village, has participated in DRR training in Japan. As a result, he is able and has the motivation to develop DRR related programs in his village. "We make activities that different from what BPBD has, so we make routine activities such as self-training on tsunami day or tsunami reminder when the Hajj month approaching, so our people can reflect their tsunami experience and be more ready [2004 Tsunami was happen in the Hajj month] ".

4. Discussion
In order to be effective and beneficial, the DRR programs require continued financial resources from governments, including from the village governments. The DRR investment and programs can be expensive and it should be financed by each level of government, not only the central government. In the context of Indonesia, each village government has the opportunity to take part in DRR, especially after the village laws enacted recognize the Indonesian village as one of the formal levels of government. In addition, village leaders have more information and understanding on the village and villagers, their characteristics, problems and needs to help them in designing effective DRR programs.

Despite this supportive context, this study has found a very limited role is played by Indonesian villages in DRR programs. This is reflected by their small number and the tiny proportion of disaster budget allocation when compared to their total budget. There are several causes/ reasons for this and it can be classified as internal and external factors. The internal factors are the contingency characteristics of DRR programs, insufficient village Funds and the lack of knowledge of the village government of DRR. Addressing these should a concern for the central governments in its role as the primary regulator. The government, and especially the Indonesian Finance Ministry, should provide specific regulations that capture the uncertainty and contingent characteristic of DRR program budgets. Any 'idle' money or surplus associated with the DRR program should be not seen as an indicator of poor financial performance, but should be allowed to 'roll over' as an ongoing disaster contingency fund for the village.

Moreover, the government should encourage villages to allocate a minimum amount of budget for DRR programs. For example, the government might require each village to allocate at least 2% of its budget for the DRR programs. On top of that, the capacity and knowledge of village officers needs to be improved to enhance the quality of DRR programs they propose and implement [11]. Without adequate education and training, they will be incapable to design the most appropriate DRR programs for their villages, particularly allowing for the limited size of the budget. Octrian et al [9] documented the limited and insignificant impact of Village Fund on the village development. This finding reflects a more general poor quality of planning and monitoring of Village Funds, indicating potential wider benefits associated with improved education and training of village officers [12].

Another external factors are the high dependence on the BPBD for disaster management, and very strict financial regulations on the use of village funds from the central government. These can be solved by designing a map of responsibilities for DRR program and financing. The synchronization and direction among institutions is-also required in order to avoid overlapping disaster programs being delivered to the community [10]. The BPBD should lead the mapping of the resources and capacity of each party. For example, villages with limited capacity in terms of knowledge and financial resources, can propose and manage capacity building program among officers and the villages rather than providing financial aids for the victims after an event.

These research results are consistent with previous studies. Ririn [14] found that there are resilient villages (or *Desa Tanggap Bencana*) that still have not used village funds in disaster budgeting in Yogyakarta, Indonesia, despite the Disaster National Agency's (BNPB) regulation No. 01 [15] stipulating-that designated "resilient villages" can use village funds to carry out DRR activities. This is in line with the results of this study that there are two villages in the Banda Aceh area that have been established as "resilient villages" but they did not allocate a disaster budget and instead rely on assistance from external villages.

The problem of small disaster management budgets is not limited to villages, but also Indonesian local governments. The BPBDs, as the government bodies responsible for DRR in each city/districts, have relatively small budgets compared to other arms of government. This is confirmed by Oktari et al study in BPBD Banda Aceh. A similar case is also found in the national context. According to Darwanto [5] disaster budgets of the central government accounted around 1-2% of the total national budget in 2011- 2012. This number is relatively small when compared to the international standard of disaster budget. One reason for this trend might be a low awareness in local government of the threat of disasters in the region so that the disaster program has not become the development priority [13].

Lastly, the concept of a resilient society is (defined by Twigg [20]) encourages the government to enhance the capacity of the community so that the community can independently carry out DRR programs in their areas. In this context, the village funds are expected to become a source of DRR program financing-to improve the villagers' capacity in dealing with disasters. The disaster programs in villages should be designed to increase the community preparedness and awareness of disasters rather than to give financial aids and logistics for the victims.

5. Conclusion

Although the regulations regarding the allocation and the use of village funds for disaster mitigation program is granted, the studied villages did allocate sufficient allocations for the DRR programs. The allocations for DRR were consistently less than 1 % of the total village budget and in fact, 2 out of 10 villages did-not allocate any budget for DRR programs although they are in the tsunami-affected zone.

The villages with DRR program funds focus primarily on responding to disasters and in particular supporting victims rather than on preparedness and awareness of the community. This study identified several reasons or challenges that contribute to the low level of village-based DRR. These include fear of budget surplus (unspent budget), high dependence on the BPBD, a strict interpretation of financial reporting regulations, the small size of the overall Village Funds and lack of knowledge of DRR concept among village government officers. These findings should provide a compelling narrative and it can be used by the policymakers to map and improve the management of financial resources for DRR programs at the grass-roots level across Indonesia. Further research is imperative to improve the role of villages in DDR strategies and design idea DDR programs for villages.

6. References

[1] Undang-undang Republik Indonesia Nomor 24 Tahun 2007 tentang Penanggulangan Bencana [Indonesian Law No. 24 (2007) on Disaster Response].

[2] Peraturan Pemerintah Republik Indonesia No. 43 Tahun 2014 tentang Peraturan Pelaksanaan Undang-Undang Nomor 6 Tahun 2014 tentang Desa [Indonesian government regulation No.43 (2014) on Regulation of Law Implementation No.6 (2014) on Village].

[3] Peraturan Menteri Desa Pembangunan Daerah Tertinggal No.19 tentang Penetapan Prioritas Penggunaan Dana Desa Tahun 2017 [Regulation of Indonesian Ministry for Village Development and underdevloped region N0.19 (2017 on the Decision on Village Fund Priority year 2017 .

[4] Oktari R S, Fahlevi H and Irawati W 2017 Disaster budgeting of Banda Aceh's local government: Trends and analysis of post-tsunami Aceh 2004 *IOP Conference Series: Earth and Environmental Science* **56**

[5] Darwanto H 2012 Preliminary examination of existing methodologies for allocating and tracking national government budget for disaster risk reduction (DRR) in Indonesia. Study commissioned by UNISDR in collaboration with ADPC under the IAP project—Regional Stocktaking and Mapping of Disaster Risk Reduction Interventions for Asia and the Pacific.

[6] Peraturan Walikota Banda Aceh Nomor 2 Tahun 2016 [Banda Aceh Mayor Regulation No.2 (2016)

[7] Munti F, and Fahlevi H Determinan Kinerja Pengelolaan Keuangan Desa: Studi pada kecamatan Gandapura Kabupaten Bireuen Aceh [Determinants of Village Financial Management: Study at Gandapura sub distict, Aceh] 2017 *Jurnal Akuntansi dan Investasi*, **18** 180281

[8] Dartanto, T, Bastiyan, D. F., and Sofiyandi, Y 2017 Are local governments in Indonesia really aware of disaster risks? AIP Conference Proceedings 1857, 110006 (2017)

[9] Octrian TSL, Mappamiring, and Mappigau Samma 2013 Implementation of Village Fund Allocation (ADD) to Community Empowerment in Punagaya Village Bangkala District Jeneponto Regency *Otoritas Jurnal Ilmu Pemerintahan* **3**(1) p 67-74

[10] Sari R N, Heru R, and Mohammad S 2014 Pengelolaan Alokasi Dana Desa dalam Perspektif Pemberdayaan Masyarakat: Studi Pada Kantor Pemerintahan Desa Ngasem, Kecamatan Ngasem, Kabupaten Kediri [Village Fund Allocation Management in Community Empowerment Perspective: Studi at Village of Ngasem, Kediri] *Jurnal Administrasi Publik* **3**(11) p 1880-1885

[11] Putra C K, Ratih N P, and Suwondo 2013 Pengelolaan Alokasi Dana Desa dalam Pemberdayaan Masyarakat Desa: Studi pada Desa Wonorejo Kecamatan Singosari Kabupaten Malang [Village Fund Allocation Management in Community in Empowerment Perspective: Studi at Village of Singosari], Kediri *Jurnal Administrasi Publik* **1**(6) p 1203-1212

[12] Kholmi M 2016 Akuntabilitas Pengelolaan Alokasi Dana Desa: Studi di Desa Kedungbetik Kecamatan Kesambern Kabupaten Jombang [Accountability of Village Fund Management: Study at Kedingbetik Village] *Ekonomika Bisnis* **7**(2) p 143-152

[13] Meutia I 2017 Pengelolaan Keuangan Dana Desa [Village Fund Financial Management] *Jurnal Akuntansi Multiparadigma* **8**(2) p 227-429

[14] Ririn T N 2013 *Disaster resilient village as a strong modality for Building Disaster Resilient Community in Indonesia Case Study: Disaster resilient village in Yogyakarta* p 1-21

[15] Indonesia Disaster Management Agency (BNPB) 2012 *Pedoman Umum Desa/Kelurahan Tangguh Bencana* [General Guideline of village/sub village Disaster Resilience] (BNPB: Jakarta, Indonesia)

[16] Sofiyanto M 2016 Pengelolaan Dana Desa dalam Upaya Meningkatkan Pembangunan di Desa Banyuates Kecamatan Banyuates Kabupaten Sampang [Village Fund Management to improve development of Banyuates village] *Jurnal Riset Manajemen* p 28-38
[17] Syamsidik, Rusydy I, Arief S, Munadi K, and Melianda E 2017 Disaster risk reduction policies and regulations in Aceh after the 2004 Indian Ocean *Tsunami IOP Conf. Series: Earth and Environmental Science* **56** p 1-11
[18] Sunarti E 2009 Evaluasi Penanggulangan Bencana di Indonesia (Lesson Learned 2006-2007) [Evaluation of Disaster Management in Indonesia] *Lembaga Penelitian dan Pengabdian Masyarakat* (Institut Pertanian Bogor: Indonesia)
[19] Ishinabe K 2017 *Resilience in Disaster: Japan Offers the World Lessons From Experience* (Japan Forwards)
[20] Twigg J 2009 *Character of a Disaster-Resilient Community* vol 2, University College London (London: Lattitude) p 84

IOP Conf. Series: Earth and Environmental Science **273** (2019) 012059 doi:10.1088/1755-1315/273/1/012059

What is the role of accounting in disaster recovery and relief? A literature review

H Fahlevi[1*], M Indriani[1], R Mulyany[1], Nadirsyah[1]

[1]Accounting Department, Syiah Kuala University, Jl. Tgk. Syech Abdul Rauf, Darussalam, Banda Aceh, 23111, Indonesia

*E-mail: hfahlevi@unsyiah.ac.id

Abstract. This study explores the role of accounting system and its output on government's and donors' responses to disasters across the world. It discusses potential contribution and limitation of accounting in the disaster relief and recovery process. To do so, a literature review of 22 papers on accounting and disaster is performed. The review papers were search and gathered from reputable journals indexed in Scopus database. This study discovers that accounting can play a more extensive role, rather than dealing merely with financial loss calculation and budget preparation for disaster recovery and relief program. Accounting has a broader role and this extended role has not been sufficiently covered in several previous studies. The role of accounting in responding to disasters is beyond the calculative function and passive recording of events. Accounting can support the establishment of exceptional governance, facilitate dialogue and create mutual understanding between stakeholders through financial information. In addition, accounting can assist the projects monitoring and evaluation and its outputs (budget, unit cost calculation, and storage bookkeeping) can be used by researchers to understand what really happens and how society responses to disasters.

1. Introduction

Natural disasters have been occurring more frequently during the last decades and become common topic and discussion in major media channels. The disasters have also attracted attention of researchers from different fields and encouraged multidisciplinary research collaborations. Many efforts have been made to mitigate the impact of the disasters on the society through research, seminars, workshops and even simulations. Relevant experts and government entities have been working together to formulate the best disaster preparedness and responses procedures. Both parties are concerned with several dimensions of disaster and sectors including health, shelters, city mapping, food reserve and many more.

One of important aspects in disaster preparedness and responses is accounting. Accounting might be interpreted as reporting and auditing and most probably link to the financial responsibility to the donors and other organisations which provide aids. Accounting has been greatly misunderstood and misinterpreted as bookkeeping and financial responsibility, rather than a system, tool or technology. The function of accounting seems to be narrowed into reporting and responsibility although it has potential to contribute beyond those functions. Due to this misunderstanding and the complicated procedures within accounting, thus there has been a limited attention in the role of accounting and its development in disaster discourse.

In responding to disasters, accounting procedures perhaps become the last thing that government takes into consideration. In most cases, accounting plays a role in reporting the use of resources to the donors and upper-level government entities. However, recent studies have revealed a more active function of accounting in disaster setting and discourse [1]. There have been several studies on the role of accounting and its products in disaster. For instance, Sargiacomo [2] investigates the function of accounting in 2009 earthquake in Abruzzo, Italy and Walker [3] explores how the victim of drought in the 1930s in the US assist the government effort in establishing a successful recovery program. To fill the gap in the existing literature, this study aims to explore the role of accounting and its evidence in the response to disasters around the globe. The results of this study can be used to formulate policy and establish a more capable accounting system that helps the government and other key actors in implementing their recovery and relief efforts. Besides, this type of study can potentially build a basis for future theory building and for future comparative research related to the response to more recent natural disasters [1].

As this study attempts to reveal the broadened role of accounting in disaster responses, previous studies on the role of accounting in response to disaster is reviewed and discussed. The results of this study are expected to create a stepping-stone in raising governments and other stakeholders' attention towards the broadened role of accounting in responding to natural disasters.

2. Data and Methods

This study uses literature review approach. A literature review is a systematic review of previous relevant studies to identify, evaluate and interpret the results and finding a research gap [5]. The aims of a literature review are to summarize the result of relevant former studies on the specific topic and to build conceptual content for theory development [6]. To ensure the quality of reviewed papers and ease the process of searching the papers, this study reviews only scientific papers that have been published in international reputable journals indexed in Scopus.

The reviewed papers were gathered from the application of two searching methods, namely database search and snowballing techniques. Firstly, the papers were collected from Scopus database by using keywords and filter features. The keywords used are 'accounting' and 'disaster' and the period of paper search was set up between 1990 and 2018. Based on this method, 63 papers were obtained. However, there were only 14 papers relevant to the research question. Later on, from the 14 papers, the second step was carried through snowballing techniques. The snowballing technique is a literature searching technique that uses the reference list or the citations of the paper that previously has been documented in the database search method [7]. In this second step, 8 relevant papers have been found and thus, the final number of the reviewed papers is 22 papers. Following are the list of the reviewed papers:

Table 1. List of reviewed papers

No	Authors	Title	Source title (year)
1	Paik G.H., Lee B., & Krumwiede K.R.	Corporate social responsibility performance and outsourcing: The case of the Bangladesh tragedy	Journal of International Accounting Research (2017)
2	Vosslamber R.	After the earth moved: Accounting and accountability for earthquake relief and recovery in early twentieth-century New Zealand	Accounting History (20150
3	Sargiacomo M.	Earthquakes, exceptional government and extraordinary accounting	Accounting, Organizations and Society (2015)
4	Atkins J., Atkins B.C., Thomson I., & Maroun W.	"Good" news from nowhere: Imagining utopian sustainable accounting	Accounting, Auditing and Accountability Journal (2015)
5	Sargiacomo M., Ianni L., & Everett J.	Accounting for suffering: Calculative practices in the field of disaster relief	Critical Perspectives on Accounting (20140
6	Lai A., Leoni G., Stacchezzini R.	The socializing effects of accounting in flood recovery	Critical Perspectives on Accounting (2014)
7	Cooper C., & Coulson A.B.	Accounting activism and Bourdieu's 'collective intellectual' - Reflections on the ICL Case	Critical Perspectives on Accounting (2014)

8	Walker S.P.	Drought, resettlement and accounting	Critical Perspectives on Accounting (2014)
9	Miley F., & Read A.	After the quake: The complex dance of local government, national government and accounting	Accounting History (2013)
10	Hsu Y.-S., Liu C.Z., Yang Y.-J., & Chou Y.-Y.	Implications of the British petroleum oil spill disaster for its industry peers - evidence from the market reaction and earnings quality	Asia-Pacific Journal of Accounting and Economics (2013)
11	Anderson S.E., Jones L.C., & Reed T.N.	Insurance fraud: Losses, liabilities, and September 11	Issues in Accounting Education (2012)
12	Cooper C., Coulson A., & Taylor P.	Accounting for human rights: Doxic health and safety practices - The accounting lesson from ICL	Critical Perspectives on Accounting (2011)
13	Matilal S., & Höpfl H.	Accounting for the Bhopal disaster: Footnotes and photographs	Accounting, Auditing and Accountability Journal (2009)
14	Philip O'Regan	'A dense mass of petty accountability': Accounting in the service of cultural imperialism during the Irish Famine, 1846–1847	Accounting, Organizations and Society (2010)
15	Bisman	Budgeting for famine in Tudor England, 1527–1528: social and policy perspectives	Accounting History Review (2012)
16	HÓgartaigh, C Ó, HÓgartaigh M Ó, & Tyson, T	'Irish property should pay for Irish poverty': accounting for the poor in pre-famine Ireland	Accounting History Review (2012)
17	Baker, C R	Break downs of accountability in the face of natural disasters: The case of Hurricane Katrina	Critical Perspectives on Accounting (2014)
18	Everett, J, & Friesen C	Humanitarian accountability and performance in the Théâtre de l'Absurde	Critical Perspectives on Accounting (2014)
19	Sciulli, N	Weathering the storm: Accountability implications for flood relief and recovery from a local government perspective	Financial Acc & Man. 2017;1–15. (2018)
20	Perkiss, S, & Moerman L	Hurricane Katrina: Exploring justice and fairness as a sociology of common good(s)	Critical Perspectives on Accounting (2017)
21	Taylor, D, Tharapos, M & Sidaway, S	Downward accountability for a natural disaster recovery effort: evidence and issues from Australia's Black Saturday	Critical Perspectives on Accounting (2014)
22	Bebbington, J & Harrison, J	Global Climate Change Responsiveness in the USA: An Estimation of Population Coverage and Implications for Environmental Accountants	Social and Environmental Accountability Journal (2017)

3. Results and Discussions

Based on prior research that explores the role of accounting in disaster response, relief and recovery program, this study finds that the role accounting can be expanded and is not merely dealt with financial loss calculation and budget for disaster recovery and relief program. Accounting has a broader role and this extended role has been uncovered in several previous studies. The results of literature review can be classified into two main findings, namely the role of accounting in the response to disasters and the limitation of accounting in the context of disaster.

3.1. The role of accounting in response to disasters

Accounting is not a merely a technique and a system of calculating, reporting, and controlling the usage of financial resources, but also a social activity that mirrors and constructs the social organizations that shape the response of a society to a situation [1]. In fact, researchers and experts have acknowledged since 1980s that accounting practice is shaped by the situation and the context of setting where it is operated [10]. Thus, the review of papers in this study attempted to capture the role of accounting in the emergency situation and its contribution to the society responses. The broadened role of accounting has been documented by several papers and thus, it can be classified into three functions.

Firstly, it is discovered that the role of accounting in responding to disasters is more than calculative function and inactive recording of events [1]. For example, Sargiacomo [2] investigated the contribution of accounting in aftermath of the earthquake in Abruzzo, Italy that occurred in 2009. This study unveiled how accounting classification, incorporation with scientific classification (i.e. the Mercalli scale), become the crucial part of the establishment of exceptional governance. In this case, accounting classification technique is used by the local health departments to classify the affected areas based on the resulted damages and to determine how much money should be given for the disaster relief funding by the federal government. This involves distinguishing areas and people from those significantly affected and to those insignificantly affected at all and finally attaching financial and costing measures based on the level of damage severity [2]. In an emergency response situation where an extraordinary government is established, accounting provides a tool for calculating the cost of earthquake as well for justification of financial provision for the victims.

Secondly, accounting has the ability to facilitate dialogue and create mutual understanding between stakeholders. This can be achieved through financial report and calculation for damage reimbursement program. The study of Lai, Leoni, & Stracchezzini [8] has demonstrated this situation. They studied accounting procedures in determining and calculating financial loss and reimbursement for the victims of Italian flood in November 2010. In this case, accounting was utilized in the recovery phase for calculating the consequential damages bore by the individual and organizations as well as for the collection of funds from the Italian central government and other donators [8]. This mission required all the key actors i.e. the central government, municipal officers, victims and sponsors to do their role in their recovery process. The dialogue between these key actors was fostered by accounting data; create mutual understanding, corporation and trust that support the accountability process [Lai, Leoni, & Stracchezzini]. The victims and other actors who are responsible for the recovery process have involved in intensive discussion and coordination to accelerate the reimbursement for the incurred damages. Accounting is able to unveil the people, business and local government financial needs and condition after the disaster, and thus, it is welcomed. Indeed, they need to collaborate and support each other to ease the reimbursement process.

Thirdly, accounting can become a key facilitating technology in monitoring and evaluating recovery projects. The output of accounting system provides a rich data that can be used to trace and evaluate the need of victims [3]. Accounting procedures make visible the recovery process, the resulted damage, the resources' need and the outcomes [8]. Moreover, the accounting data and documents can provide important information to understand beneficiaries' condition and monitor the progress as well as the effectiveness of a recovery project. Walker [3] studied a resettlement project namely Red River Valley Farms Project which aimed to relocate families affected by the severe drought disaster occurred in North Dakota back in 1930s. In this study, Walker explored how accounting records maintained by the beneficiaries used by the government and expert to monitor the successfulness of the project, to measure the progress of its beneficiaries and identify the most needed assistance. The accounting records were called a Farm and Home Plan and a Farm Family Record Book. The former is a financial assessment of the farmstead status and a budget of business and domestic operations, whereas the latter is a record of all transaction, production, inventory, assets, liabilities, and cash flows relating to both farm and home activities. These records were maintained by the farmers and has become the basis for the resettlement project establishment and evaluation [3].

Finally, the products of accounting system such as budget, unit cost calculation, and storage bookkeeping can be used by researchers to understand what really happens and how society responses to disasters. It gives an insight on historical events that can be a lesson learnt for the next generation especially in responding to disasters. Most of the reviewed papers collect the data by using documentation analysis approach where accounting records are the main ingredient. For instance, Walker [3] analysed the accounting records produced by farmers and other supporting project reports to uncover the role of accounting as a key facilitating technology in the drought recovery project. Meanwhile, Bisman [9] investigated the England government policy on the East Anglian Famine of 1527-1528 by using a specific budget document prepared by the Kingdom in the 16th century to assist decision making and resource management in the period of severe scarcity of food. Taken together, the available accounting records so far can be used, accompanied with other documents, to obtain a complex picture of disaster recovery process and situation faced by the government.

3.2. The limitation of accounting

Although accounting has an ability to facilitate our understanding on disaster recovery phase, it has some limitations. Firstly, accounting may potentially be used for political purposes. Accounting standards and regulations as the basis for accounting practices and reports are not free from political interest. The standards can be changed substantially to align with the politicians' and government interest. These situations can be found in the New Zealand case where the government change accounting practice at all levels of government [11]. These changes have influenced the scope of financial disclosures of financial realities and thus, change the quantity of information of the accounting records. For example is a change from cash basis to accrual basis accounting. The accrual basis might provide more detailed information about the expenditures but it is prone to manipulation.

Secondly, accounting artifacts cannot be used independently without any support from other information and documents. Budget and financial reports mostly consist of numbers and researchers need other source of information to gain the whole story [12]. In fact, the use of accounting information as the single source of information to understand disaster responses and recovery can be problematic and misleading. In a study conducted by Matilal & Höpfl [13], it is found that the use of both statements of account and the photographic image of Bhopal gas exposition tragedy can capture the tragic situation and consequences of the disaster more comprehensively.

4. Conclusions

This study concludes that accounting has a role beyond a mere passive recording of events. In the context of disasters, accounting is able to provide not only a calculative mechanism for reporting, but also facilitates communication among the stakeholders, monitors and evaluates disaster recovery projects. Apart from that, accounting within the context of disaster also contributes into the development of research in accounting and disaster management. However, some drawbacks are noted namely it can be politically constructed and it needs other supporting information to obtain the complex picture of the situation. Finally, given these important roles, accounting is supposed to receive more attention and concern by government and donators. This includes proper accounting information storage and security especially for emergency situation, accounting standards and procedures for disaster recovery phase and projects and the involvement of accountants or accounting experts in disaster recovery activities.

References

[1] Vosslamber R 2015 After the earth moved : Accounting and accountability for earthquake relief and recovery in early twentieth-century New Zealand *Accounting History* **20** 518-535
[2] Sargiacomo M 2015 Earthquakes, exceptional government and extraordinary accounting *Accounting, Organizations and Society* **42** 67–89.

[3] Walker S. P. 2014 Drought, resettlement and accounting *Critical Perspectives on Accounting* ***25*** 604–619.
[4] Jalali, S., and Wohlin, C. 2012 Systematic Literature Studies: Database Searches vs. Backward Snowballing *Proceedings of the ACM-IEEE International Symposium on Empirical Software Engineering and Measurement* 29–38
[5] Fink A. 1998 Conducting research literature reviews: from paper to the internet. Thousand Oaks: Sage
[6] Seuring, S and Müller M From a literature review to a conceptual framework for sustainable supply chain management 2008 *Journal of Cleaner Production* **16** 1699–1710
[7] Wohlin, C 2014. Guidelines for snowballing in systematic literature studies and a replication in software engineering EASE'14 Proc. 18th Int. Conf. Eval. Assess. Softw. Eng
[8] Lai, A., Leoni, G., and Stracchezzini, R. 2014 The socializing effects of accounting in flood recovery *Critical Perspectives on Accounting*, **25** 579–603
[9] Bisman, J. E. 2012 Budgeting for famine in Tudor England, 1527 – 1528 : social and policy perspectives *Accounting History Review* **22** 105–126
[10] Mennicken, A. 2002 Bringing calculation back in: Sociological studies in accounting *Sociological Studies in Accounting, Economic Sociology: European Electronic Newsletter* 17–27
[11] Miley, F., and Read, A. 2013 The complex dance of local government, national government and accounting *Accounting History* **18** 447–471
[12] Oktari R S, Fahlevi H and Irawati W 2017 Disaster budgeting of Banda Aceh's local government: Trends and analysis of post-tsunami Aceh 2004 IOP Conference Series: Earth and Environmental Science 56
[13] Matilal, S., and Höpfl, H 2009 Accounting for the Bhopal disaster : footnotes and photographs *Accounting, Auditing & Accountability Journal* **22** 953–972

Why tsunami survivors remain living in the same location? Evidences from *Lampulo Village, Banda Aceh*

Muhammad Ilhamsyah Siregar[1,*], Taufiq Carnegie Dawood[1], Fakhruddin[1], Cut Zakia Rizki[1], Fitriyani[1], Muhammad Reza[1]

Faculty of Economics and Business, Syiah Kuala University, Banda Aceh, Indonesia

*Corresponding author: ilham@unsyiah.ac.id

Abstract. This research aims to analyse factors influencing the Indian Ocean tsunami survivors to remain in the same location after the tsunami. Logistic model is utilized to measure factors influencing the decision to choose the tsunami affected village after the tsunami. Those factors are land ownership, income level, and profession. The concerned location is *Lampulo Village* in Banda Aceh, which was severely affected by the tsunami in 26 December 2004. Data is primary data generated from field interview for 200 respondents, divided evenly for respondents who used to live in the village but decided to move away and live in different locations, and respondents who remain living in the village. The results suggest that the main reason to choose resettling in the affected village is merely the land ownership. In other words, people who have lands in the village will have higher likelihood to remain living in the village 1.5 times higher than those who do not have lands. Other factors are not statistically important in decision making to remain in the village. The implication of this research show that disaster survivors will tend to choose the same location after disaster. It implies the stronger need to reduce the risks of potential disaster in those locations, making the human settlement safer against future disaster. This paper recommend policy makers to avoid relocating survivors to different locations, and to invest public funds to strengthen disaster preparedness in the disaster prone areas.

Keywords: Housing Choice, Land Ownership, Income, Profession, Logistic Model, Banda Aceh.

1. Introduction

The decision to choose a place to live, especially in urban areas, is a decision that involves many aspects and considerations. The most important consideration is the price of land parcels [1]. In addition to land prices, consideration of location amenities, proximity of dwellings to various important locations of economic and social activities, as well as socio-cultural proximity are also important. Choices will be available to humans and the decision to settle will be largely determined by the location that provides the most optimal locational satisfaction from the available budget [2].

However, in disaster settings, the availability of options becomes curbed. Disaster conditions present difficulties in making rational decisions. Material and non-material damage and loss, loss of productive assets and sources of livelihood provide difficulties in making decision, especially decision to place a new dwelling. These difficulties occur from narrowed options available in the post-disaster

IOP Conf. Series: Earth and Environmental Science **273** (2019) 012060 doi:10.1088/1755-1315/273/1/012060

conditions. The available options are usually to return to the location affected by the disaster, or move to a new location.

Based on the results of a study conducted by [3], relocation decisions were made with consideration to reduce vulnerability to disasters faced by the community. So, by moving people in disaster-prone locations to safer locations, vulnerability is reduced. Based on the relocation experience in Indonesia and India reviewed by Boen and Jigyasu, relocation decisions tend to not last long. So it can be concluded that although the purpose of relocation is good enough, namely to reduce the vulnerability of disasters, the relocation areas tend not to last long.

In addition, [4] and [5] explained that the limited choice of residential locations would be exacerbated by several factors, including: first, loss of home and land and deterioration of sanitation, which caused deterioration in health conditions; Second, damage to educational facilities and employment opportunities which caused poor access to education and decreased income; and third, the destruction of social support networks which causes various social difficulties. Thus, the decision to relocate in a new residential area becomes very difficult to choose as a voluntary choice.

The financial limitations of disaster survivors often encourage the survivors to depend their choice of housing on the availability of assistance, both from the government and other parties, such as non-governmental organizations or NGOs, donor agencies and other social institutions. In such difficult conditions, post-disaster reconstruction was also carried out without considering important aspects of the survivors' preferences. Aid providers tend to use a top-down planning approach that has often been proven to ignore the importance of socio-economic, development and cultural aspects. In fact, these aspects are also essential aspects in the post-disaster reconstruction process. Thus, the choice to settle in the original place is a very difficult choice [6]. However, in the context of disasters such as tsunamis, the location of houses in a landscape that is prone to tsunamis always requires consideration of human safety. Therefore, the choice of relocation is always an option that is often part of policy choices.

In post-disaster event, [7] explains that housing remains essential for almost all disaster survivors. Furthermore, [7] and [8] further explained that houses are important assets for families, determinants of welfare, health, education, family and social status. In developing countries, houses are considered to be the most important asset, so that if a disaster occurs and destroys the house, the impact of the loss will be enormous. [9] asserts that nearly 98 percent of house losses due to disasters are common in developing countries, and usually become the focus of post-disaster recovery.

The earthquake and tsunami disaster in Aceh on December 26, 2004 was one of the worst disasters during modern human civilization. According to [10], the total losses due to the disaster were equivalent to 2.2 percent of Indonesia's 2004 total GDP. The greatest losses were caused by damage and loss of houses. The estimated loss is equivalent to more than US $ 1.5 billion, three times greater than the loss in the transportation sector. From the data released by (2005), more than 160 thousand people died and disappeared, and nearly 180 thousand homes were destroyed or heavily damaged. The loss was also added to damage to health, religious and education facilities (see Table 1.)

Table 1. Estimated Losses from the Indian Ocean Tsunami.

No	Type of losses	Unit	Losses
1	Died or missing	person	165,945
2	Injured	person	4,662
3	Displaced	Person	522,462
4	Damaged houses	Unit	179,312
5	Damaged health facilities	Unit	240
6	Damaged religious facilities	Unit	2,742
7	Damaged education facilities	Unit	1.226

Sources: BNPB (2005)

The authors calculated the degree of losses from damaged houses and social economic facilities and found that the losses are enormous. From the estimation of tsunami losses assessment conducted by World Bank, the value of losses from damaged houses in Aceh is equivalent of 30.06 percent of 2005

GRDP of Aceh. This loss is considered massive, as the tsunami has created housing loss equivalent to almost one third of total economy in Aceh. Compared to other value of losses from damaged health, education, social and religious facilities, the value in housing losses is beyond comparison (see Table 2).

Table 2. Estimated Value of Losses from the Indian Ocean Tsunami.

No	Type of losses	Value (billion IDR)	% of GRDP [1)]
1	Damaged houses	13,365	30.06
2	Damaged health facilities	854	1.92
3	Damaged religious facilities	776	1.75
4	Damaged education facilities	1,194	2.69

Sources: World Bank (2005)

Note: [1)] the 2005 GRDP of Aceh was IDR 44,467.16 billion; data was from BPS Aceh

According to [11], relocation of disaster survivors to safer locations is part of the post-tsunami reconstruction policy in Aceh. However, difficulties in implementation caused the relocation process did not proceed smoothly. Most survivors decided to stay at their original location, and reject the option of relocating to a safer place.

One of the locations that suffered total damage in the tsunami event was *Lampulo* Village which is located in Syiah Kuala District, Banda Aceh. *Lampulo* is inhabited by the majority of fishermen and aquaculture, in addition to traders and other professions. Lampulo is one of the areas directly affected by the tsunami and indirectly affected from water pollution and deteriorating environment. However, it is observed that despite the massive losses and waning environment, people decided to resettle in the location. This paper aims to investigate factors affecting the decision making of staying in the same location after disaster. This paper, to our best knowledge, is the first paper to analyse this phenomenon. Thus, this paper serves to fill the knowledge gap and help decision makers to design more efficient housing provision aftermath devastating disasters.

2. Literature Review

2.1. Housing demand theories and Land Ownership

Some researchers have documented the theories on housing demand, for example [2] describe the developments of housing demand thoughts. Quoting [12], [2] concluded that house is the most important commodity purchased by a consumer, because the house provides shelter, a place for various economic and life activities, and a form of investment that is beneficial for the owner. The basic theory of housing demand was developed from the basic assumption that the number of houses was limited but homogeneous, so that the known demand theory could be used as an analytical tool (see [13] and [14]). [15] argue that the demand for a house, besides being influenced by the price of the house, the demand for the house is also determined by the physical characteristics and location of the house. Location factors play an important role because each house offers different utilities due to location.

In addition to price, [2] further explained that based on the findings of various empirical studies, in addition to price and income factors, it is also important to add spatial variables as stated by [14], [13] [16]. [17] used location variables (ie distance to medical services) as variable that influenced the willingness to pay for housing for retirees in Spain. [18] used the variable of the proximity to the location of a landfill as one of the determinants of the willingness to pay for the urban area of Calabar, Nigeria. [19] argue that distance to the workplace is an important variable and determinant of housing demand by low-income people in the State of Benue, Nigeria, because transportation costs are an important factor.

One of the philosophical foundations underlying the development of the most influential home demand theory is the theory developed by Rosen [20]. Rosen developed a market theory for goods that have various characteristics, such as houses. Rosen argues that housing demand is derived from optimizing consumer satisfaction derived from home purchases and consumption, as well as composite goods at a unit price level. Rosen developed the bid-rent theory developed by British classical economist David Ricardo. The magnitude of the desire to pay a buyer indicates the optimum bid-rent that buyers can offer to a variety of house characteristics.

The theory developed by Rosen was refined by Quigley [16] who argued that the characteristics of a house would vary due to spatial factors. Quigley criticizes the view of empirical studies that developed the explanation of variations in house prices from the factor of house prices. Quigley argues that spatial factors provide differences in home demand. Polycentric urban characteristics will encourage prospective home buyers to consider distance and transportation costs into the decision to buy a house. Quigley concluded that the desire to buy a home actually shows the effective price, namely the level of house prices adjusted for the cost of transportation that must be paid to travel to the workplace.

Further empirical findings supported [13], such as [21], [14] and [16] and argues that in theory, settlement locations in urban areas are derivations from optimizing satisfaction from consumption of houses and other goods and services. Consumers who optimize the satisfaction of choosing a house with spatial characteristics, or house amenities. Therefore, Straszheim believes that the location of the house is a determining aspect of the desire to pay and demand for a house.

Another factor affecting housing demand is socio-economic and demographic aspects. [22] explain that there is a close and significant relationship between housing demand and socio-economic and demographic aspects, such as income level, household size, education level and profession. In Sweden, house demand is influenced by age distribution. The older a consumer is, the consumer has a higher demand for a home, both to be occupied by himself, and as a form of investment. Lindh and Malmberg's research is consistent with research conducted by [23] in Austria, [24] in the Netherlands and [25] in Japan.

2.2. Location Proximity and profession of the (future) owner of the house

Determination of the selection of land parcels for the city is largely determined by the distance between the land parcels and the main activity locations. The activities in question are: the location of the workplace, the location of the market for shopping and public facilities such as schools and hospitals, in addition, the price of land, income is a determining factor, see [26] for details.

Lamb [27] states that choosing a good place or location is an important decision, because: first, the place is a long-term resource commitment that can reduce the flexibility of the future of the business. Second, the location will affect future growth. The area chosen must be able to grow economically so that it can maintain business viability and third, the local environment may change at any time, if the location value deteriorates, the business location must be moved or closed.

According to [28] study in Sri Lanka rebuilding tsunami-damaged homes in Sri Lanka was carried out in the original place because of the focus of NGOs and international donor agencies providing housing assistance in their original locations. NGOs and international donor agencies did not relocate tsunami victims to safer locations. The quality of the houses being built is also substandard because of the rushed development. This observation was not only observed in in Sri Lanka but also in India. Such substandard housing resulted from poor toilets and kitchens situation.

3. Methods

Population size is not well documented. Authors found no reliable records of total survivors in *Lampulo,* making the sample size determination rather difficult. It was also difficult to find the records of survivors who moved away from the village and where they have moved to. However, authors selected 200 sample sizes to allow logistic modelling to work, considering the model nature of non-linearity, which tend to work well with large sample size. Authors selected 100 samples for each category: survivors who moved away from *Lampulo* and survivors who remain in the village.

Logistic model is employed and modelled as the following:

$$E\,(Y|x) = \frac{e^{\left(\beta_0 + \sum_{k=1}^{p} \beta_k x_k\right)}}{1+e^{\left(\beta_0 + \sum_{k=1}^{p} \beta_k x_k\right)}} \quad (1)$$

The value of E(Y|x) is the odd of success, namely the odd of respondents to choose to remain in *Lampulo*, so the expression can be rewritten as:

$$p(x) = \frac{e^{\left(\beta_0 + \sum_{k=1}^{p} \beta_k x_k\right)}}{1+e^{\left(\beta_0 + \sum_{k=1}^{p} \beta_k x_k\right)}} \quad (2)$$

Eq.(2) is transformed into:

$$y = \ln\left[\frac{p(x)}{1-p(x)}\right] = \beta_0 + \sum_{k=1}^{p} \beta_k\, x_k \quad (3)$$

And operationalized into the following regression:

$$y = \ln\left[\frac{p(x)}{1-p(x)}\right] = \beta_0 + \beta_1 X_1 + \beta_2 X_2 + \beta_3 X_3 \quad (4)$$

$$y = \ln\left[\frac{p(x)}{1-p(x)}\right] = \beta_0 + \beta_1 LO + \beta_2 INC + \beta_3 PROFF \quad (5)$$

Where y is the binary value of moving away respondents (zero) and remaining in *Lampulo* respondents (one); LO is binary value of people who own land (one) and without land (zero), INC is income and PROFF is binary value of people who work in fishery, aquaculture and the relevant (one) otherwise zero.

4. Results and Discussions

4.1. Selected Respondents Characteristics

According to sex, most of the respondents who were willing to provide information for this study were men. A total of 141 of the 200 respondents (or equivalent to 70.5 percent) are men. The remaining 59 out of 200 respondents (or equivalent to 29.5 percent) are women.

Based on the age of the respondents, the majority of respondents aged between 30-38 years. As many as 80 people or 40 percent of the total respondents were in that age group. A small proportion of respondents are relatively elderly. There were four respondents or two percent of the total respondents older than 56 years. The rest, evenly distributed between adults, with the proportion of 44 respondents or 22 percent aged 21-29 years and 42 respondents or 21 percent aged between 39-47 years.

Furthermore, the majority of respondents completed education up to high school. A total of 112 respondents or 56 percent had such an education level. A balanced proportion is also seen in respondents who have a junior high school; 40 respondents or 20 percent, and those who earned higher education degree which is 39 people. Only nine respondents were equivalent to primary school education, or 4.5 percent of the total respondents. In term of income, the majority of respondents have an income level of between IDR 1,600,000 and IDR 2,600,000. A total of 125 people or 62.5 percent of the total respondents are in the income group. A small percentage of respondents have a high level of income, which is greater than IDR 2,600,000, which is 23 respondents or equivalent to 11.5 percent. However, the other majority, as many as 52 respondents or 26 percent have the lowest income level of between IDR 600,000 to IDR 1,600,000.

4.2. Model Estimation

Table 3 show the estimated coefficient from the logistic regression model.

Table 3. Regression Results

Variables	Coefficient	Odd Ratio	Std. Error	t-value	Critical t-value
Land Ownership	**0,43262**	**1,54129041905**	**0,19840**	**2,1806**	**1,96**
Income	-0,00000020441	0,99999979559	0,00000016156	-1,2652	**1,96**
Profession	-0,087331	0,91637372572	0,18128	-0,48174	**1,96**

Source: Authors' calculation (2018)

The coefficient of land ownership variables is positive and estimated at 0.433, which means that land ownership is related to the decision to settle in *Lampulo* after the tsunami. The estimated odds ratio is 1.541 which means that respondents who own land in *Lampulo* before the tsunami have a 1.541 times greater chance of settling in the village after the tsunami compared to respondents who do not own land. The calculated t-ratio of this variable is 2.1806 and the critical value for the 95 percent confidence level is 1.96. Thus, this variable is statistically significant at 95 percent confidence level in explaining the decision of respondents who settled in *Lampulo* after the tsunami.

While income and profession variables have negative values and are estimated to be -0.00000020441 and -0.087331, respectively, which means that the two variables have an opposite direction to the decision to settle in *Lampulo* after the tsunami. The opportunity value of the income variable is 0.999 which means that respondents who have a higher income level tend to leave *Lampulo* 0.999 times greater than lower income respondents. The opportunity value of the profession variable is 0.916 indicating that respondents with the fishing profession tend to leave this location 0.916 times greater than the respondents who are non-fishermen profession. However, both of these variables are not statistically significant.

The empirical results show that the major determinant for dwellers to remain in *Lampulo* was the land ownership. The results suggest that those who own lands in *Lampulo* will be more likely to remain in *Lampulo* than those who does not own lands. The majority of profession is linked with fishery and aquaculture and it suggests that people who work in these fields also have likelihood to remain in *Lampulo*, however the variable is not statistically significant. These findings provide the empirical evidence that the decision to remain in tsunami affected village is mainly determined by the factor of land ownership. The fact that after more than a decade aftermath the event of disaster people have settled in tsunami prone and tsunami devastated village suggest more challenging policy for reducing the hazard and improving the quality of disaster preparedness. Further researches are suggested to evaluate the degree of hazard and vulnerability, to ensure that policy makers have better understanding to plan for disaster risk reduction and mitigation.

5. Conclusions and Recommendation

This study analyses why the survivors return to the original village after the tsunami. This paper found that first, the main factor that determines the decision of disaster survivors to settle at the disaster affected location or move to another location is the ownership or non-ownership of land at the tsunami affected location. The results of logistic regression showed that disaster survivors who owned land in *Lampulo* had a tendency or opt to resettle in *Lampulo* 1.541 times greater than those who did not own land in the village. Disaster survivors in *Lampulo* who do not own land are usually tenants of rented property, or live with relatives and family.

IOP Conf. Series: Earth and Environmental Science **273** (2019) 012060 doi:10.1088/1755-1315/273/1/012060

Based on this trend, and also reinforced by the availability of assistance for housing development by both the government and non-government at that time, the tsunami-affected areas, including the *Lampulo*, were resettled after the tsunami. Empirical results also show that occupation factors are not statistically dominant in influencing the decision to resettle in the tsunami affected areas.

Second, the type of occupation factors that are closely related to the location of the affected area have no direct relationship with the decision to resettle in the tsunami affected area. This phenomenon is different from the empirical findings in Sri Lanka and several other tsunami affected locations. This shows that in the period before the tsunami, fishermen were not able to buy land in *Lampulo*. Most of them only rent houses in these locations because the village facilities are proximate to the coastal areas where fishermen run their economic activities. So, after the disaster, the tenant fisherman cannot decide to settle in *Lampulo*, because of the absence of the land.

Reference

[1] Siregar, M I 2015 Permintaan Rumah Sewa di Kota Banda Aceh. Syiah Kuala University Press. Banda Aceh
[2] Siregar, M I and Rizki, C Z 2014 Permintaan Rumah Bagi Kelompok Masyarakat Berpendapatan Rendah di Kota Banda Aceh. Syiah Kuala University Press, Banda Aceh
[3] Boen, T and Jigyasu, R 2005 Cultural considerations for post disaster reconstruction post-tsunami challenges. *In UNDP Conference* (pp. 1-10)
[4] Zaman, M Q 1996 Development and Displacement in Bangladesh toward a Resettlement Policy. *Asian Survey*, 36(7)
[5] Badri, S A, Asgary A, Eftekhari, A R, and Levy, J 2006 Post-disaster resettlement, development and change: a case study of the 1990 Manjil earthquake in Iran. *Disasters*, 30(4), 451-468
[6] El-Masri, S, and Kellett, P 2001 Post-war reconstruction. Participatory approaches to rebuilding the damaged villages of Lebanon: a case study of al-Burjain. *Habitat International*, 25(4), 535-557
[7] Barakat, S 2000 Housing reconstruction after conflict and disaster. Humanitarian Policy Group Network Papers 43: 1-40
[8] Peacock, W G, Dash, N, Zhang, Y 2007 Sheltering and housing recovery following disaster in Rodriguez, H., Quarantelli, E.L., Dynes, R.R., (eds.) Handbook of disaster research, Springer, New York
[9] Gilbert, R 2001 Doing More for Those Made Homeless by Natural Disasters, Disaster Risk Management Series No. 1, Washington, DC, World Bank
[10] World Bank 2005 Indonesia: Notes on Reconstruction, The December 26, 2004 Natural Disaster, A Technical Report Prepared by BAPPENAS and the International Donor Community, The Consultative Group on Indonesia, January 19-20, 2005
[11] Bappenas 2005 Rancangan Mekanisme Pendanaan Rehabilitasi dan Rekonstruksi Pasca Tsunami, Bappenas, Jakarta
[12] Brueckner, J 2011 Lectures on Urban Economics, The MIT Press
[13] Muth, R F 1969 Cities and housing: the spatial pattern of urban residential land use
[14] Mills, E S (1967). An aggregative model of resource allocation in a metropolitan area. *The American Economic Review*, 197-210
[15] Follain JR, Jimenez E 1985 The demand for housing characteristics in developing countries. Urban Studies. Oct; 22(5):421-32
[16] Quigley, J M 1976 Housing demand in the short run: An analysis of polytomous choice. *Explorations in Economic Research*, Volume 3, number 1. NBER , 76-102
[17] Gibler KM, Manuel Casado-Díaz J, Angeles Casado-Díaz M, Rodríguez V, Taltavull P. Late life second move housing choices of international retiree migrants. International Journal of Housing Markets and Analysis. 2009 May 29;2(2):190-210
[18] Andrew-Essien EE, Akintoye OA. Urban poverty and residential environment degradation in Calabar Area of Cross River State, Nigeria. Global Journal of Human-Social Science Research. 2012 Mar 27;12(6)

[19] Onu V, Onu AJ. Urban Residential Housing and Low-Income Earners: A Study of Makurdi Metropolis, Benue State, Nigeria. European Scientific Journal, ESJ. 2012 Dec 28;8(28)
[20] Rosen, S. (1974). Hedonic prices and implicit markets: product differentiation in pure competition. *The journal of political economy*, 34-55
[21] Straszheim M. Straszheim. Urban Economics. Elsevier, North Holland. 1987;2:717-57
[22] Lindh T, Malmberg B. Demography and housing demand—what can we learn from residential construction data?. Journal of Population Economics. 2008 Jul 1;21(3):521-39
[23] Lee GS, Schmidt-Dengler P, Felderer B, Helmenstein C. Austrian demography and housing demand: is there a connection. Empirica. 2001 Sep 1;28(3):259-76.
[24] Rouwendal J. Housing wealth and household portfolios in an ageing society. De Economist. 2009 Mar 1;157(1):1-48
[25] Moriizumi Y. Current wealth, housing purchase, and private housing loan demand in Japan. The Journal of Real Estate Finance and Economics. 2000 Jul 1;21(1):65-86
[26] O'sullivan A. Urban economics. McGraw-Hill/Irwin; 2007
[27] Lamb CM. Housing segregation in suburban America since 1960: Presidential and judicial politics. Cambridge University Press; 2005 Jan 24
[28] Shaw R. Post disaster recovery: issues and challenges. In Disaster Recovery 2014 (pp. 1-13). Springer, Tokyo

Post-Disaster Recovery and A New Economic Growth Path: A Comparative Study of Three Provinces in Indonesia

B S Nazamuddin[1] and Agus Nugroho[2,3]

[1]Department of Development Economics, Universitas Syiah Kuala, Jl. Syech Abdurrauf, Banda Aceh, 23111, Indonesia
[2]Department of Agribusiness, Universitas Syiah Kuala, Jl. Syech Abdurrauf, Banda Aceh, 23111, Indonesia
[3]Tsunami and Disaster Mitigation Research Center (TDMRC), Universitas Syiah Kuala, Jl. Prof. Dr. Ibrahim Hasan, Gampong Pie, Banda Aceh, 23233, Indonesia

*E-mail: nazamuddin@unsyiah.ac.id

Abstract. The purpose of this study is to examine whether there exists a new regional output growth path in post-disaster situation in three provinces of Indonesia affected by relatively large-scale natural disasters, i.e. Aceh, West Sumatera (Sumbar), and Yogyakarta. By using real Gross Regional Domestic Product time series for the period 2001-2017 obtained from the Central Agency of Statistics, a comparative analysis was made from the differences in the effects across those provinces between pre-disaster and post-disaster periods. This study used a simple model with deterministic trend, post-disaster recovery dummy and autoregressive component to estimate changes in the coefficients of trends for each province's output growth path. The results show that there is no significant difference in the overall effects of trend changes across provinces, while the effects of trend changes between pre- and post-disaster periods vary across provinces and in general post-disaster interventions do not move upward the long-term effects on regional output trends.

Keywords: post-disaster recovery; new growth path; regional economic growth

1. Introduction

It has been a common case that a post-disaster recovery leads to an increased per capita income and higher employment. A disaster may serve not only as an opportunity to build back the economic foundations better and faster [1], but also for positive community change [2]. However, do interventions to revitalize the economy and disaster risk reduction related policies result in a long-term new trend of output. [3] show that "while the short-term growth consequences of natural disasters are comparatively well studied, little is known about the long-run perspective". [4] find that even large natural disasters would have no long-term effect on economic growth. Only when they are followed by radical political revolution do they have long-lasting negative effects on the economic growth. In the similar line of thought, post-disaster recovery per se, regardless of its scale, may not lead to large and long-lasting positive effect on output growth.

Many recent studies have focused on disaster loss, disaster resilience, and changes in the government's behaviour in response to natural disaster shocks, which can be classified [5-10] as studies on the negative effects of disasters. Meanwhile, little has been discussed on the positive effects of post-disaster recovery, such as the shift in the long-term economic growth. This study is aimed at investigating whether post-disaster recovery leads to a new growth path.

Specifically, this study examines the changes in the Gross Regional Domestic Product (GRDP) growth of three provinces in Indonesia, i.e. Aceh, West Sumatera (Sumbar), and Yogyakarta, that were stricken by natural disasters at different points and of different scales. In other words, whether a post-disaster recovery shock shifts the regional economies to a new long-term growth path. A comparative analysis is made from the differences in the effects of for the three provincial cases. The specific research questions to be addressed are the following: 1) Whether post-disaster recovery intervention, which is assumed to have been a norm following any disaster in a country, has effects of on the paths of regional output (measured by Gross Regional Domestic Product); and 2) Are there differences in the effects between pre-disaster and post-disaster as well as across regions.

2. Method

Estimation of trend changes

The procedures taken were as follows. First, the data for different provinces were assumed to exhibit random variation and were independent from one another. To compare three provincial GRDP time series trends in terms of magnitude of increases in each province's GRDP, tests were conducted to check whether trend parameters were significantly different between one province and another by applying a model with deterministic trend and autoregressive component as follows:

$$GRDP_t^i = \beta_1^i + \beta_2^i TREND + \beta_3^i GRDP_{t-1}^i + \mu_t^i \quad (1)$$

$$i = Province\ i$$
$$t = 2001, \ldots ,2017$$

where $GRDP$ is Gross Regional Domestic Product, TREND is time trend, and μ_t is the white noise error term.

The deterministic trend is represented by $\beta_2^i TREND$ and a stochastic component by μ_t^i which both affect the long-run course of GRDP of province i. Under the assumption of independence of the samples from three different provinces, the difference between coefficients (slopes) of TREND between any two regressions were calculated and divided by the square root of the sum of the squared standard errors. Under the normal assumption, then a Z statistic with n-2 degrees of freedom could be obtained (Clogg, 1995) and (Paternoster, 1998), as follows:

Trend slope difference between Aceh and Sumbar: $(\beta^A - \beta^S)/\sqrt{(SSE_A + (SSE_S)}$ (2)

Trend slope difference between Aceh and Yogya: $(\beta^A - \beta^Y)/\sqrt{(SSE_A + (SSE_Y)}$ (3)

Trend slope difference between Sumbar and Yogya: $(\beta^S - \beta^Y)/\sqrt{(SSE_S + (SSE_Y)}$ (4)

The standard errors of the difference between slopes were then computed, for example between Aceh and Sumbar as

$$SE = s\sqrt{SE(\beta^A)/s_A)^2 + SE(\beta^S)/s_S)^2} \quad (5)$$

where β^A and β^S are estimated coefficients of Aceh and Sumbar regressions respectively. s_A and s_S are the estimated standard errors of the two regressions, and s is computed as follows

$$s = \sqrt{\frac{(n_A - k)s_A^2 + (n_S - k)s_S^2}{n_A + n_S - 2k}} \quad (6)$$

where n_A, n_S are sample sizes of Aceh and Sumbar, and k the number of parameters, including constant. The above SE is used as the denominator for the t-test on n-2k degrees of freedom.

IOP Conf. Series: Earth and Environmental Science **273** (2019) 012061 doi:10.1088/1755-1315/273/1/012061

The comparison is made only to the regressions that give slopes that are significantly different from zero. For the cases being examined, this procedure is used to determine if the effects (coefficients) of the trends are the same for the following cases:

a) If the coefficients of trends for Aceh and Sumbar are statistically equal.
b) If the coefficients of trends for Aceh and Yogya are statistically equal.
c) If the coefficients of trends of Sumbar and Yogya are statistically equal.

For each province, tests were also conducted to check if the slopes were statistically equal between two time points (before and after the disaster). If, for example, βA >βB of the trend then we could say that the trend of GRDP growth for one province is greater than another. If for a particular province, for example, the trend coefficient for post-disaster is greater than the post-disaster period, then the impact of recovery in terms of growth path change can be compared between before and after the disaster.

Recovery Dummy Variable
The second way of testing the effect of post-disaster recovery is to construct a dummy variable, RECOVERY, such that the following two equations were estimated:

$$GRDP_t^i = \beta_1^i + \beta_2^i RECOVERY + \beta_3^i TREND + \beta_4^i GRDP_{t-1}^i + \mu_t^i \quad (7)$$

$$GRDP_t^i = \beta_1^i + +\beta_2^i TREND + \beta_3^i RECOVERY * GRDP_t^i + \beta_4^i GRDP_{t-1}^i + \mu_t^i \quad (8)$$

The RECOVERY variable takes the value of 1 for the period one year after the occurrence of the disaster to year 2017 (i.e. 2005-2017 for Aceh, 2010-2017 for West Sumatera, and 2007-2017 for Yogyakarta) and 0 elsewhere.

3. Brief overview of the disasters

The December 26, 2004 tsunami greatly devastated the economy of Aceh province. All economic sectors were affected. The total material loss was estimated at around 4.5 billion US dollars, i.e. 2.2% of Indonesia's Gross Domestic Product or 97% of Aceh's Gross Regional Domestic Product [11]. Around 78% of the losses were suffered by the private sector, while the rest was endured by the public sector. Material losses were due to destruction or damage caused by the unprecedented disaster to the private and public infrastructure and facilities. The sectors that suffered the most damage were the productive sectors such as agriculture, fisheries and the trade industry with the scale of the damage of 37%. Before the disaster, the agricultural sector was an important sector because around 62.05% of the Aceh's population depended on agriculture and 32.17% of Aceh's GRDP originated from the sector. As the result of the destruction, the number of the poor increased drastically due to losses of jobs and shelters. Fishermen lost their jobs due to the destruction of facilities and infrastructure to earn a living (Directorate General of Fisheries and Maritime Affairs). Many corporate businesses, supermarkets, and small-scale businesses were destroyed and even the businesses that survived the disaster became out of business because their places and equipment was either destroyed or disappeared. Substantial support by the Indonesian government through the Agency for Aceh and Nias Reconstruction (BRR) as well as direct assistance by foreign governments and international communities or nongovernmental organizations has built back Aceh better and put Aceh on a new economic growth path.

The 6.3 magnitude earthquake in Yogyakarta province that took place on May 27, 2006, caused a total damage and losses of 3,134 million US dollars [12]. Housing, productive sectors, and social facilities were the worst affected. 88.6% of the damage and losses was sustained by the private sector. Therefore, similar to Aceh's recovery, in Yogyakarta relatively large efforts were devoted to rebuilding houses. As many as 66,000 people added up to the already high poverty rate, around 130,000 jobs were lost, many of them in small scale manufacturing and services. In total districts' Gross Regional Domestic Products dropped by 5 to 18% varying according to their scales of destruction. Post-disaster recovery took place quicker because of the lesser scale of devastation and lessons-learned from Aceh's recovery. Quicker loss assessment also made it possible for the government to make a quicker decision with regards to Yogyakarta recovery interventions.

The 7.6 magnitude earthquake that struct West Sumatera on September 30, 2009 caused 2.3 billion US dollars of losses. Almost 80% of the damage and losses affected infrastructure and housing, while the productive sectors suffered 11% loss. The worst affected area was the provincial capital Padang, the center of businesses. Over 88% of the total damage and losses were suffered by the private sector, including trade, tourism, and financial services. Around 31% of houses were damaged at various scales, 115,000 houses destroyed and 135,000 damaged [13].

To compare the effects on regional output of the three disasters, the impact of them on the regional economy's output varies across provinces. Figure 1 shows that the GRDP of Aceh dropped drastically as the impact of 2004 tsunami. The regional real output fell by 19.4 % in 2005. Only in 2006 after a huge influx of post-disaster funding, real GRDP rose by a very high rate of growth of 66.5% (see Table 1). Very different phenomena were experienced by Sumbar and Yogyakarta provinces where the two provinces recorded a high positive growth in the same year of disaster occurrence. Yogyakarta's real GRDP grew by 20.5% in 2006 and Sumbar's grew by 19.4% in 2009. However, on the average annual the post-disaster real output grew slower in all three provinces (see Table 2). Although the long-term trend of output seems to be positively upward, the long-term effect of post-disaster interventions seems to be very small. Real output only returns to its normal rate as before.

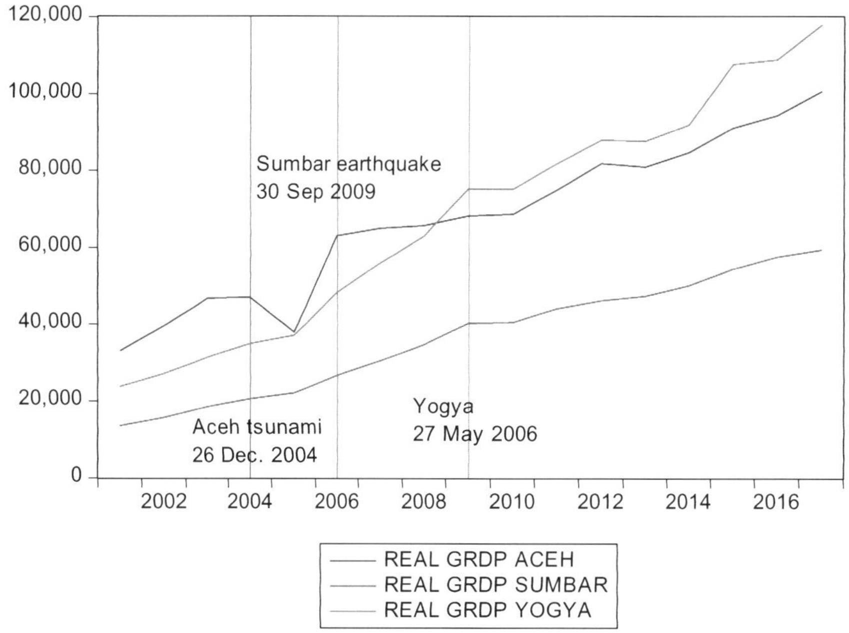

Figure 1. Real GRDPs of three provinces

IOP Conf. Series: Earth and Environmental Science **273** (2019) 012061 doi:10.1088/1755-1315/273/1/012061

Table 1. GRDP Growth Rates of Aceh, Sumbar, and Yogyakarta

Year	Growth of real GRDP Aceh (%)	Growth of real GRDP Sumbar (%)	Growth of real GRDP Yogya (%)
2002	19.7%	13.9%	15.6%
2003	18.2%	15.7%	18.6%
2004	0.6%	11.3%	11.0%
2005	**-19.4%**	6.2%	7.4%
2006	66.4%	29.9%	**20.5%**
2007	3.0%	16.2%	14.4%
2008	1.1%	12.6%	13.8%
2009	3.8%	**19.4%**	16.1%
2010	0.7%	-0.1%	0.5%
2011	9.1%	8.8%	8.7%
2012	9.2%	7.5%	4.9%
2013	-1.1%	-0.3%	2.5%
2014	4.6%	4.9%	5.9%
2015	7.6%	17.1%	8.5%
2016	3.6%	1.1%	5.9%
2017	6.7%	8.4%	3.3%

Source: Author's calculation

Table 2. GRDP Average Growth Rates of Aceh, Sumbar, and Yogyakarta

Year	Pre-disaster average growth (%) *	Post-disaster recovery average growth (%) *
Aceh (2001-2004)	12.9%	-
Yogya (2001-2006)	14.6%	-
Sumbar (2001-2009)	15.6%	-
Aceh (2005-2017)	-	7.3%
Yogya (2007-2017)	-	5.9%
Sumbar (2010-2017)	-	7.7%

Source: Author's calculation

IOP Conf. Series: Earth and Environmental Science **273** (2019) 012061 doi:10.1088/1755-1315/273/1/012061

4. Estimation results and discussion

As can be seen from Table 3, the effects of Trend variable are significantly positive for all three provinces in 4 different models. In model 1, the effect of recovery dummy is significant and has positive sign only for Aceh and Yogyakarta. In addition to the trend effects, the shift in intercepts mean that post-disaster recovery has positive upward effects. When recovery dummy is multiplied by GRDP to examine the effect in terms of slope changes, only the coefficient for Yogyakarta regression is significantly positive. Hence, for Yogyakarta in addition to fast post-disaster recovery effect as indicated in Table 1, both trend and slope affect the long-term real output growth. It is not the case for Aceh and Sumbar, although the latter also experiences a fast recovery of real output growth. However, an examination of slope differences between provinces (see Table 4) suggests that the differences between any two provinces in terms of the effects of trend are not statistically significant. Surprisingly, when two periods (pre- and post-disaster) are examined (Models 3 and 4), the regressions for Sumbar and Yogyakarta show significantly positive trend coefficient differences, and not for Aceh (Table 5). However, as can be shown in Table 3 the coefficients in fact fall, which implies that the post-disaster recovery does have impacts, but the impacts are even making the trend slopes lower.

Post-disaster recovery interventions move real regional output trend upward, but for Yogyakarta the slope of the trend has also moved upward. Nevertheless, for all the three regions the average growth of output during the period of recovery is lower than before (see Table 2). The comparison of slope of Trend between Model 3 and Model 4 in Table 3 suggests that the effects of trend changes between two time points are significantly positive for all three provinces. Nevertheless, the effects become smaller in the post-disaster recovery period for Aceh and Sumbar, while for Yogyakarta the effect becomes a little greater. All of these results imply that the post-disaster interventions only put the regional economy's output back to normal path without changing long term directions. This result is inconsistent with [1] in the sense that long term output growth path is attributable to factors that determine productivity growth as suggested in the endogenous growth theory. The magnitude of negative effect of natural disasters on economic growth differs across disasters [10], it may also differ across regions. Likewise, the magnitude of positive effect of post-disaster recovery may differ across regions even within the same country and across different time periods. In addition, countries and regions may be quite different in their pre- and post-disaster characteristics, and therefore the post-disaster recovery also differs across places. The nonexistence of permanent effect on output growth of post-disaster recovery is similar to nonexistence of the pre-disaster effect on the economic growth found by [4] unless the recovery efforts are accompanied by other reforms that in the long-term determine changes in output growth path.

IOP Conf. Series: Earth and Environmental Science **273** (2019) 012061 doi:10.1088/1755-1315/273/1/012061

Table 3. Estimation results

Variable	Model 1 (Dep. Var. Real GRDP, full sample 2002-2017)			Model 2 (Dep. Var. Real GRDP with Recovery_dummy*Real_GRDP and lag Real GRDP, full sample 2002-2017)			Model 3 (Dep. Var. Real GRDP, split samples)			Model 4 (Dep. Var. Real GRDP, split samples)		
	Aceh	Sumbar	Yogya	Aceh	Sumbar	Yogya	Aceh (2001-2004)	Sumbar (2001-2009)	Yogya (2001-2006)	Aceh (2005-2017)	Sumbar (2010-2017	Yogya (2007-2017
Constant	38270.75*	11537.75**	9986.202*	38982.68*	11744.99**	12288.86*	34197.12*	19034.7*	13294.98*	34843.88*	20696.23*	15768.08*
Trend	3782.036*	4198.277**	1975.83**	3906.948**	4568.173**	2035.67*	4921.896**	6258.002*	2486.778*	4033.843*	5935.318*	2737.699*
Lag Real GRDP				-0.112887	0.353787	0.125515						
Recovery dummy	8963.174**	-5167.337	2788.854**									
Recovery_dummy*Real_GRDP				0.075036	-0.058526	0.1272**						
Real GRDP	-0.118487	0.402483	0.265117									

* significant at 1%
** significant at 5%

Source: Author's calculation

Table 4. Differences between slopes of two provinces' regressions

	Difference between slopes	SE	t statistics	t-table at 5%
Aceh-Sumbar	-416.241	19882.1856	-0.0209354	2.306
Aceh-Yogya	1975.83	26792.43324	0.06741478	2.306
Sumbar-Yogya	2222.447	183079497	0.00001214	2.306

Source: Author's calculation

Table 5. Differences between slopes of two periods – pre and post disaster

	Difference between slopes	SE	t statistics	t-table at 5%
Aceh pre- and post-disaster	888.053	29764.5883	0.02983589	2.179
Sumbar pre- and post-disaster	322.684	10.901232	223.380074	2.179
Yogya pre- and post-disaster	-250.921	183079497	479.00	2.179

Source: Author's calculation

5. Conclusions

There is no significant difference in the overall effects of trend changes of output growth paths between any two disaster-affected regions. Post-disaster recovery interventions move real regional output trend upward, but the magnitude of the effects differs across regions and across time periods. This implies that the post-disaster interventions in general only put the economies of disaster-affected regions back to normal growth path without changing long term directions. In order for post-disaster recovery to have positive permanent effects, it should always be followed by concomitant reforms as an integral part of recovery.

References

[1] Hosoya K. Recovery from natural disaster: A numerical investigation based on the convergence approach. Econ Model [Internet]. 2016;55:410–20. Available from: http://dx.doi.org/10.1016/j.econmod.2016.02.029
[2] Mochizuki J, Chang SE. Disasters as opportunity for change: Tsunami recovery and energy transition in Japan. Int J Disaster Risk Reduct. 2017;21:331–9.
[3] Berlemann M, Wenzel D. Hurricanes, Economic Growth and Transmission Channels - Empirical Evidence for Developed and Underdeveloped Countries. Work Pap [Internet]. 2016;105(October 2016):231–47. Available from: https://www.cesifo-group.de/de/ifoHome/publications/working-papers/CESifoWP/CESifoWPdetails?wp_id=19241745
[4] Cavallo E, Galiani S, Noy I, Pantano J. Catastrophic Natural Disasters and Economic Growth. 2013;95(December):1549–61. Available from: http://www.mitpress
[5] Songwathana K. The Relationship between Natural Disaster and Economic Development: A Panel Data Analysis. Procedia Eng [Internet]. 2018;212(2017):1068–74. Available from: https://doi.org/10.1016/j.proeng.2018.01.138
[6] Marin G, Modica M. Socio-economic exposure to natural disasters. Environ Impact Assess Rev [Internet]. 2017;64:57–66. Available from: http://dx.doi.org/10.1016/j.eiar.2017.03.002
[7] Keerthiratne S, Tol RSJ. Impact of natural disasters on income inequality in Sri Lanka. World Dev [Internet]. 2018;105:217–30. Available from: https://doi.org/10.1016/j.worlddev.2018.01.001
[8] Fang Y ping, Zhu F biao, Qiu X ping, Zhao S. Effects of natural disasters on livelihood resilience of rural residents in Sichuan. Habitat Int [Internet]. 2018;76(May):19–28. Available from: https://doi.org/10.1016/j.habitatint.2018.05.004
[9] Benali N, Abdelkafi I, Feki R. Natural-disaster shocks and government's behavior: Evidence from middle-income countries. Int J Disaster Risk Reduct. 2018;27(August):1–6.
[10] Klomp J, Valckx K. Natural disasters and economic growth: A meta-analysis. Glob Environ Chang [Internet]. 2014;26(1):183–95. Available from: http://dx.doi.org/10.1016/j.gloenvcha.2014.02.006
[11] The Consultative Group on Indonesia. Indonesia: Preliminary Damage and Loss Assessment The December 26, 2004 Natural Disaster. Consult Gr. 2005;
[12] Bappenas. Preliminary Damage and Loss Assessment Yogyakarta and Central Java Natural Disaster. 2006;(June):140. Available from: http://siteresources.worldbank.org/INTINDONESIA/Resources/226271-1150196584718/PackageJune13_HIRES_FINAL.pdf
[13] WHO. West Sumatra earthquake: humanitarian response plan. 2009; Available from: http://www.who.int/hac/crises/idn/indonesia_response_plan_2009.pdf

Acknowledgments

I wish to thank the participants of the 11th Aceh International Workshop and Expo on Sustainable Tsunami Disaster Recovery - AIWEST-DR 2018, held at Banda Aceh, Indonesia, 10-12 October 2018 for useful comments and feedback.

IOP Conf. Series: Earth and Environmental Science **273** (2019) 012062 doi:10.1088/1755-1315/273/1/012062

Farmers' knowledge, perceptions, and participation on the implementation of crop insurance program in Aceh Besar

Z Fadhliani[1,2], E Yustika[1], A Nugroho[1,2] and A H Hamid[1]

[1] Departement of Agribusiness, Syiah Kuala University, Kopelma Darussalam, Banda Aceh, Indonesia
[2] Tsunami Disaster Mitigation and Research Center (TDMRC), Syiah Kuala University, Jl. Prof. Dr. Ibrahim Hasan, Gampong Pie, Banda Aceh, Indonesia

Abstract. The evaluation toward the implementation of crop insurance in Indonesia have been received relatively little attention. This study analyzes knowledge, perception and participation of farmers toward the implementation of the crop insurance program (AUTP) by using data from 80 subjects in Suka Makmur Subdistrict in the Regency of Aceh Besar. Utilizing a questionnaire, we display farmers' knowledge, perception, and participation reported based on frequencies. One-way ANOVA tests with the associated post-hoc tests, is performed to examine significant differences among means of knowledge, perception and participation levels with the independent variables being age, gender, education, household income, rice field area, and land ownership. The results indicate that only less than half of farmers have overall good knowledge, despite their participation in AUTP. Most of subjects respond relatively low level of participation toward AUTP. Knowledge, perception and participation levels are varied significantly based on the independent variables, but gender. In general, this study has shown farmers' low level of knowledge, perception and participation toward the implementation of AUTP crop insurance which can significantly affect the sustainability of the crop insurance program. Therefore, emphasis on the program to boost the level of knowledge, perception and participation of the farmers is needed.

1. Introduction

Crop insurance has been widely use in high-income country for decades. In Indonesia, crop insurance is just recently introduced covering limited regions. The Ministry of Agriculture has developed the pilot Asuransi Usaha Tani Padi (AUTP) insurance program as part of mitigation efforts in response to global climate change [1]. AUTP is an indemnity based crop insurance given to a group of farmers (kelompok tani/Poktan), where 80 % of the premium cost is subsidized by the government (farmers required to pay 20% of total premium cost). The first pilot Asuransi Usaha Tani Padi (AUTP) or rice insurance program was implemented in East Java Province and South Sumatra Province in the planting season of 2012-2013 [2]. In Aceh Province, the pilot project is administered since 2015 covering five regencies including the regency of Aceh Besar.

Although years of pilot programs and billions of expenditures in government subsidies, the evaluation toward the implementation of crop insurance in Indonesia have been received relatively little attention. Measuring farmers' behavior toward the program is necessary to observe the progress. Therefore, assessing farmers' knowledge, perception and participation of the implementation of the rice insurance program in Indonesia will be valuable to improve its effectiveness and benefits to farmers.

Past studies on crop insurance in Indonesia narrowly focused on the possibility of implementing a weather index insurance [2,3], while very few studies have analyzed knowledge, perception and participation of farmers on crop insurance program (AUTP). We also examine significant differences among means of knowledge, perception and participation levels with age, gender, education, household income, rice field area, and land ownership since voluminous literature on crop insurance [4-6] indicates that such variables is believed to be related to crop insurance knowledge, perception and participation.

2. Methodology

2.1. Population and sample

The present study is performed in Suka Makmur Subdistrict in the Regency of Aceh Besar. Out of twelve subdistricts in the Aceh Besar Regency, the subdistrict of Suka Makmur has the greatest number of farmers who are joining the government-subsidized crop insurance program that is introduced since 2015 in Aceh Province. The target population in this study is rice farmers joining Rice Insurance program in 2017 (400 farmers). A sample of 80 farmers was interviewed for this study during August 2017 using a questionnaire.

2.2. Instrument

The questionnaire used in the survey consisted of 4 sections. Section one covers the questions investigating the socioeconomic and demographic information of the household such as age, education, and farming experience of the household. Section two places the questions about farmers' knowledge of AUTP crop insurance program in the study area. Section three covers the questions exploring farmers' perception of AUTP crop insurance program. In sections four, respondents are asked about crop insurance experiences regarding their participation on AUTP crop insurance program in the study area.

2.3. Data Analysis

The data obtained from the questionnaires are numerically scored and statistically analyzed using the Statistical Package for Social Sciences (SPSS v. 23). To determine the knowledge, perception and participation of the respondents, total scores are obtained by summating the scores of all questions within each of the sections. The total scores for each section varied for the section on knowledge, responses to statements (n=8) are scored as follows: yes = 1 and no = 0, and the scores are combined to give a score range of 8-16. For the section on perception, responses to statements (n=8) are scored as follows: agree = 3, neutral = 2, disagree = 1, and the scores are combined to give a score range of 8-24. For the section on participation, responses to statements (n = 9) are scored variably depends on the statements ordered from highest (3) to lowest (1), and the scores are combined to give a score range of 9 to 27.

Results regarding knowledge, perception, and participation are reported based on frequencies. One-way ANOVA tests with the associated post-hoc tests, assuming unequal variances, are also performed to examine significant differences among means of knowledge, perception and participation levels with the independent variables being age, gender, education, household income, rice field area, and land ownership.

3. Results and discussion

3.1. Characteristics of respondents

Descriptive statistics of the sample are presented in Table 1. The majority of farmers are female and most farmers reported to be in the age group of 40-59 years. Only few farmers earned a degree and most of them attain primary or secondary education. This situation may affect farmers' knowledge, perception, and participation regarding crop insurance since education has been marked as one expedient indicator to measure farmers' ability in using sophisticated financial insurance tools [5,7].

Farmers in the study area engage in agriculture as their main livelihood. Around 56% of farmers earn between Rp. 1.000.000 – Rp 2.000.000 monthly. These households also work in very small farms with some of them cultivate in the area of 0.5 hectare and above. Moreover, most of the respondents are tenant farmers.

Table 1. Characteristics of respondents

Variable	Scale	Percentage (%)
Age	less than 40	29%
	40 to 59	65%
	60 or above	6%
Gender	Male	24%
	Female	76%
Education	Elementary	28%
	Middle	25%
	High	34%
	College or university	14%
Income	less than 1 million rupiah	39%
	1 million to 2 million rupiah	56%
	more than 2 million rupiah	5%
Area of rice field	<0.5	48%
	≥ 0.5	53%
Land ownership	Self-owned	40%
	Rent/yield share	60%

3.2. Farmers' knowledge about AUTP program

Overall level of knowledge mean frequencies (Table 2) suggests that only less than half of farmers (49.5%) have good knowledge, despite their participation in AUTP rice insurance program. Farmers' best knowledge is regarding to, "the advantage of joining AUTP crop insurance program", "the amount of premium paid by farmers", "the amount of sum insured per Ha, "the term of insurance given", and "the aim of AUTP that is to compensate crop lost because of flood, drought, pest and diseases. However, when the question is changed to whether they know or not that the aim of AUTP is to protect farmers from the risk of flood, drought, and pest and diseases, the majority of farmers show little knowledge (6.3% opt for yes), indicating that farmers still confused and unfamiliar with the program. Lowest knowledge is also shown when farmers are asked conceptually about AUTP crop insurance program. Farmers low level of knowledge is due to the limited information they obtain. Therefore, it is imperative to implement awareness and extension program to assist the farmers getting information about the crop insurance. Extension programs educate farmers about rules and regulations of the insurance plan and provide a direct linkage between farmers and the insurance company [8].

3.3. Farmers' perceptions about AUTP program

Our findings (Table 2) report that farmers still perceive a negative perception toward AUTP rice insurance program because only 43.7% (less than 50%) of the farmers agreed on some level with the positively worded statements. This situation is attributed to the poor administration of AUTP crop insurance program. Farmers argue about the delayed and unsuitable expected indemnity payment. Some even express they gain no benefit from the program since they never receive the compensation payments, although suffering from crop loss. Consequently, farmers need to find another funding (banks, cooperative enterprise, and middleman) to compensate the cost they suffer from agricultural activities.

3.4. Farmers' participation in AUTP program

Response toward questions which determine participation (Table 2) informs that the majority of farmers (more than 60%) opt for negatively worded statements, indicating low participation on the implementation of AUTP crop insurance program. This condition may related to farmer's perception of no benefit received from the crop insurance, causing the farmers unwilling to convey the benefits of the program to other communities. The study also shows that most farmers are never invited to get involve in premium payment process, making farmers tend to be indifferent to all activities related to the crop Insurance program. This is in line with the statement put forward by Azwar [9] that farmers will trust a project or development program more if they feel they need to be involved both in the preparation process and its implementation so that there will be a sense of ownership of a project or development program.

Table 2. Knowledge, perception and participation of the respondents

	Statement	Yes		No			
		N	%	N	%		
Knowledge	Do you know about AUTP crop insurance program?	8	10.0	72	90.0		
	Crop insurance is implemented to protect farmers from the risk of flood, drought, and pest and diseases.	5	6.3	75	93.8		
	The aim of AUTP crop insurance is to compensate crop lost because of flood, drought, pest and diseases.	43	53.8	37	46.3		
	The advantage of joining AUTP program is that farmers will get a payment for crop loss.	48	60.0	32	40.0		
	Farmers need to pay Rp 36.000 of premium rate	48	60.0	32	40.0		
	Farmers could make an insurance claim if they pay the Rp 36.000 premium rate and suffer the total crop damage greater than or equal to 75% of the total planted area	55	68.8	25	31.3		
	The indemnity payment received by farmers if such damage occur is Rp. 6.000.000/hectare	55	68.8	25	31.3		
	The insurance covers only for one planting season (4 months)	55	68.8	25	31.3		
		Disagree		Unsure		Agree	
		N	%	N	%	N	%
Perception	The insurance company uses a proper technique to measure the damage occurs in the field	3	3.8	51	63.8	26	32.5
	AUTP crop insurance provides coverage for the risks (flood, drought, pest and diseases) attributed to the prevalence of the problem faced by farmers	3	3.8	52	65	25	31.3
	The period of coverage is appropriate	5	6.3	48	60	27	33.8
	AUTP registration procedure is uncomplicated	5	6.3	8	10	67	83.8
	The premium rate is affordable	4	5.0	10	12.5	66	82.5
	The payment of insurance claims is made on time	48	60.0	7	8.8	25	31.3
	The amount of indemnity payment is as expected	48	60.0	7	8.8	25	31.3
	The indemnity payment sufficiently covers the loss (no need to find additional funding from banks, cooperative enterprise, and middleman)	55	68.8	6	7.5	19	23.8
		Frequent		Rare		Never	
		N	%	N	%	N	%
Participation	How often do you read information about AUTP in magazines	6	7.5	26	32.5	48	60.0
	How often have you heard information about AUTP on TV/Radio	4	5	28	35	48	60.0
		more than 2 times		2 times		less than 2 times	
		N	%	N	%	N	%
	How many times have you attended meetings with farmer groups to discuss AUTP?	25	31.3	7	8.8	48	60.0
		more than 5 times		5 times		less than 5 times	
		N	%	N	%	N	%
	How many times have you attended socialization / counseling activities to discuss AUTP?	16	20	8	10	56	70.0
		Yes		Rare		No	
		N	%	N	%	N	%
	Do you take decisions and give opinions during socialization / counseling activities?	25	31.3	7	8.8	48	60.0
	Have you ever invited by the head of the farmer group to pay the premium fee to the insurance company?	25	31.3	7	8.8	48	60.0
	Did you join the mentoring program to register as the member of AUTP?	25	31.3	7	8.8	48	60.0
		Yes		Unsure		No	
		N	%	N	%	N	%
	Are you willing to apply good farming practice to meet insurance policy	25	31.3	7	8.8	48	60.0
	Are you willing to inform other farmers about the benefit of AUTP	25	31.3	7	8.8	48	60.0

3.5. Socioeconomic factors associated with farmers' knowledge, perception, and participation on AUTP crop insurance

Agricultural insurance is a fairly new tool that can be used by farmers for an effective risk management in agriculture. In this section, the factors differing farmers' knowledge, perception, and participation of crop insurance in Suka Makmur Subdistrict of Aceh Besar are presented.

ANOVA tests (Table 3) indicate that variable age, education, income, area of rice field, and land ownership ($p < 0.05$) significantly differ farmers' mean knowledge scores and farmers' mean perception scores. Meanwhile, farmers' mean participation scores are significantly different ($p < 0.05$) based on education, income, area of rice field, and land ownership. Although a study revealed that gender influence farmers' preference for crop insurance [10], none of the scores (knowledge, perception, and participation) are significantly different based on gender. This might be related to the respondents' characteristic of this study, where most of them are better educated female. Women fail in understanding insurance contracts is attributed to their lack of confidence and low level of literacy [10].

Our findings in Table 3 show that younger farmers have more knowledge and more favorable perception of crop insurance. Previous studies have found mixed results in terms of the effect of age on the adoption of risk management tools [11, 12]. Studies also shown that there is a linkage between education with the adoption of crop insurance [13, 14]. Better educated farmers are more likely to adopt crop insurance given their more ability to understand the scheme [6]. Not only education, income is also believed as a factor that drives insurance demand through risk aversion. Farmers with high level of education and more income would be more risk averse compared to older farmers with low level of literacy and income [11]. When farmers experience more risks, they are more likely to adopt risk management strategies such as crop insurance. As expected, our results in Table 3 indicate that farmers graduating from university, and farmers earning more income, have significantly high level of knowledge, perception and participation regarding AUTP crop insurance programs.

The results also suggest that the amount of agricultural land yields a significant association with knowledge, perception and participation in crop insurance. This result is in accordance with Sadati [11]. Larger farm will normally obtain higher indemnity payment due to the higher area insured, inducing farmers' interest in AUTP crop insurance. In addition, farmers working on their own land are also significantly more knowledgeable, having more positive perception, and having more participation in crop insurance program. This is in parallel with previous study indicating that farmers' endowments are the mainspring to opt for crop insurance [6].

Table 3. ANOVA model of several independent variables on farmers' knowledge levels, attitude levels and perception levels

Variables		Knowledge					Perception					Participation				
		MKL	SD	F	*p*	Post hoc	MPeL	SD	F	*p*	Post hoc	MPaL	SD	F	*p*	Post hoc
Age	Less than 40	13.0	2.7	4.62	0.013	a	18.5	4.3	3.63	0.031	a	17.0	7.1	2.98	0.057	
	40 to 59	11.9	2.9			a	17.3	4.4			a	14.2	7.4			
	60 or above	9.0	0.7				12.8	1.3				9.0	0.0			
Gender	Male	11.3	2.4	1.81	0.182		16.2	3.5	1.71	0.195		12.5	6.4	2.25	0.137	
	Female	12.3	3.0				17.7	4.6				15.4	7.5			
Education	Elementary	9.9	1.6	39.09	0.000	c,d	14.2	2.6	31.98	0.000	c,d	9.7	3.4	35.9	0.000	c,d
	Middle	10.3	1.1			c,d	14.8	1.3			c,d	9.9	2.8			c,d
	High	13.5	2.6			d,e	19.3	4.4			d,e	18.2	7.1			d,e
	College or university	16.0	0.0			c,e	23.5	0.9			c,e	24.8	1.7			c,e
Income	Less than 1 million rupiah	9.9	0.5	23.24	0.000	b	14.2	1.3	22.96	0.000	b	9.0	0.0	27.2	0.000	b
	1 million to 2 million rupiah	13.6	3.0				19.7	4.5				18.7	7.3			
	More than 2 million rupiah	11.0	2.4				15.8	2.2				13.5	5.2			
Area of rice field	<0.5	10.8	2.1	15.30	0.000		15.3	3.3	18.48	0.000		11.0	5.2	24.2	0.000	
	≥0.5	13.1	3.1				19.2	4.5				18.0	7.3			
Land ownership	Self-owned	14.6	2.2	87.58	0.000		21.0	3.6	67.90	0.000		21.3	5.6	93.1	0.000	
	Rent/yield share	10.3	1.8				14.9	3.0				10.3	4.5			

MKL: mean knowledge level; MPeL: mean perception level, MPaL: mean participation level;
[a] p value <0.05 analysed using Post Hoc test (95% confidence interval) for comparison with group of 60 or above;
[b] p value <0.05 analysed using Post Hoc test (95% confidence interval) for comparison with group of 1 million to 2 million rupiah;
[c] p value <0.05 analysed using Post Hoc test (95% confidence interval) for comparison with group of high school;
[d] p value <0.05 analysed using Post Hoc test (95% confidence interval) for comparison with group of college or university;
[e] p value <0.05 analysed using Post Hoc test (95% confidence interval) for comparison with group of elementary and middle school

4. Conclusion

Our study demonstrates a low level of knowledge, perception and participation among farmers in Suka Makmur Subdistrict toward the implementation of AUTP crop insurance. This may affect the long-term sustainability of the crop insurance program in village level, despite its role as an effective tool for risk management in agriculture. Emphasizing on program to boost the level of knowledge, perception and participation of the farmers is imperative. This can be achieved through improvement on implementation awareness and extension program of AUTP, policies and practices by government and its related agencies. Moreover, the insurers will also need to improve their services, thereby gaining and retaining the trust and loyalty from farmers. This study also reveals that knowledge, perception and participation levels are varied significantly based on age, education, household income, rice field area, land ownership. Therefore, an individualized specific education and assistantship program tailored to farmers' level of education and understanding is preferable.

5. Reference

[1] Pasaribu S M and Sudijanto A 2013 Rice crop insurance pilot project: an implementation review. *Technical report of JICA project of capacity development for climate change strategies in Indonesia* (Jakarta, Indonesia: Ministry of Agriculture - JICA)
[2] Kawanishi M and Mimura N 2015 Assessment of insurance for paddy production: a case study in Indonesia *Clim. Dev.* **7** 257–66.
[3] Putri I Dharmawan K and Tastrawati N 2017 Perhitungan harga premi asuransi pertanian yang berbasis indeks curah hujan menggunakan metode black scholes *E-Jurnal Mat.* **6** 161-7
[4] Mishra A K and Goodwin B K 2006 Revenue insurance purchase decisions of farmers *Appl. Econ.* **38** 149–59
[5] Sherrick B J, Barry P J, Ellinger P N, and Schnitkey G D 2004 Factors influencing farmers' crop insurance decisions *Am. J. Agric. Econ.* **86** 103–14
[6] Ye T, Liu Y, Wang J, Wang M and Shi P 2016 Farmers' crop insurance perception and participation decisions: empirical evidence from Hunan, China *J. Risk Res.* 1-14
[7] Enjolras G and Sentis P 2011 Crop insurance policies and purchases in France *Agric. Econ.* **42** 475–86.
[8] Jayathilaka D P D M and Abeynayake N R 2013 Assessing farmers' perception, awareness and influential factors to purchase a crop insurance as a tool for risk management *Proc. 12th Agric. Res. Symp.* 394-398
[9] Azwar A, Muljono P and Herawati T 2016 Persepsi dan partisipasi petani dalam pelaksanaan kegiatan rehabilitasi tanaman kakao di Kabupaten Sigi Provinsi Sulawesi Tengah *J. Penyul* **2** 157-67
[10] Akter S, Krupnik J T, Rossi F, and Khanam F 2016 The influence of gender and product design on farmers' preferences for weather-indexed crop insurance *Glob. Environ. Chang.* **38** 217-29.
[11] Sadati A S, Ghobadi R F, Mohamadi Y, Sharif O, and Asakereh A 2010 Survey of effective factors on adoption of crop insurance among farmers: A case study of Behbahan County *African J. Agric. Res.* **5** 2237-42
[12] Torkani 2002 Final reports of plan efficiency of wheat insurance *Planning Research and Agricultural Economic Institution of Agricultural Department* (Tehran, Iran)
[13] Goodwin B K and Kastens T L 1993 Adverse selection, disaster relief, and the demand for multiple peril crop insurance *Contract report for the Federal Crop Insurance Corporation*
[14] Smith V H and Bacquet A E 1996 The demand for multiple peril crop insurance: Evidence from Montana wheat farms' *Am. J. Agric. Econ.* **78** 428– 438

Acknowledgments
We would like to thank to Suka Makmur subdistrict government officials, colleagues from Agribusiness Department, Syiah Kuala University and all parties who have contributed to this study. Special thanks to Yusuf Nawawi for his meticulous review onto this manuscript development.

AUTHOR INDEX

AUTHOR INDEX

AUTHOR INDEX